The Distance Scale

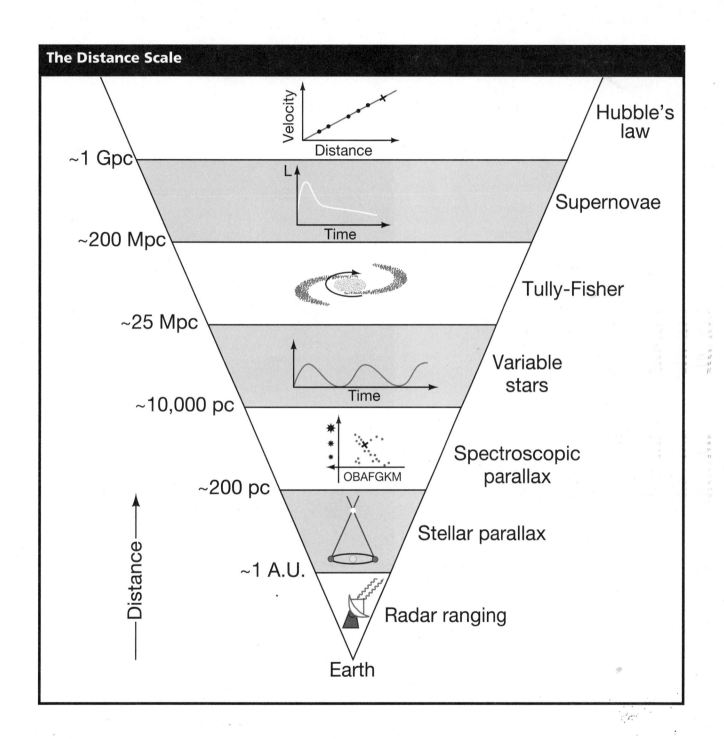

ASTRONOMY TODAY

SEVENTH EDITION

Stars and Galaxies
VOLUME II

ERIC CHAISSON
TUFTS UNIVERSITY

STEVE MCMILLAN
DREXEL UNIVERSITY

Addison-Wesley

Boston   Columbus   Indianapolis   New York   San Francisco   Upper Saddle River
Amsterdam   Cape Town   Dubai   London   Madrid   Milan   Munich   Paris   Montréal   Toronto
Delhi   Mexico City   São Paulo   Sydney   Hong Kong   Seoul   Singapore   Taipei   Tokyo

Publisher: Jim Smith
Executive Editor: Nancy Whilton
Director of Development: Michael Gillespie
Editorial Manager: Laura Kenney
Project Editor: Tema Goodwin
Media Producer: Kelly Reed
Director of Marketing: Christy Lawrence
Senior Marketing Manager: Kerry Chapman
Associate Director of Publishing: Erin Gregg
Managing Editor: Corinne Benson
Production Supervisor: Mary O'Connell
Compositor: Progressive Information Technologies
Production Service: Progressive Publishing Alternatives
Illustrations: Dartmouth Publishing, Inc.
Interior and Cover Design: Derek Bacchus

Manufacturing Buyer: Jeffrey Sargent
Manager, Rights and Permissions: Zina Arabia
Image Permissions Coordinator: Elaine Soares
Manager, Cover Visual Research & Permissions: Karen Sanatar
Photo Research: Kristin Piljay
Cover Printer: Phoenix Color
Printer and Binder: Quad/Graphics, Dubuque
Cover Images:
Main Edition: Carina Nebula, NASA, ESA, and M. Livio and the Hubble 20th Anniversary Team (STScI)
The Solar System: Solar flares, SOHO (ESA & NASA)
Stars and Galaxies: NGC 6302, NASA, ESA, and the Hubble SM4 ERO Team

Library of Congress Cataloging-in-Publication Data

Chaisson, Eric.
  Astronomy today / Eric Chaisson, Steve McMillan.—7th ed.
    p. cm.
  Includes bibliographical references and index.
  ISBN 978-0-13-240085-5 (hardcover : alk. paper) 1. Astronomy. I. McMillan, S. (Stephen), 1955–II. Title.
  QB43.3.C48 2011
  520—dc22   2010017125

ISBN 10-digit 0-321-69143-1; 13-digit 978-0-321-69143-9 (Student edition)
ISBN 10-digit 0-321-70471-1; 13-digit 978-0-321-70471-9 (Exam copy)
ISBN 10-digit 0-321-71862-3; 13-digit 978-0-321-71862-4 (Volume 1)
ISBN 10-digit 0-321-71863-1; 13-digit 978-0-321-71863-1 (Volume 2)
ISBN 10-digit 0-13-212006-2; 13-digit 978-0-13-212006-7 (NASTA)

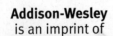

Addison-Wesley
is an imprint of

BRIEF CONTENTS VOLUME II

CONTENTS

GRAY TYPE INDICATES
NOT IN VOLUME TWO *

PART 2
Our Planetary System 130

PART 3
Stars and Stellar Evolution 382

16 THE SUN
OUR PARENT STAR 384

17 THE STARS
GIANTS, DWARFS, AND THE MAIN SEQUENCE 416

18 THE INTERSTELLAR MEDIUM
GAS AND DUST AMONG THE STARS 444

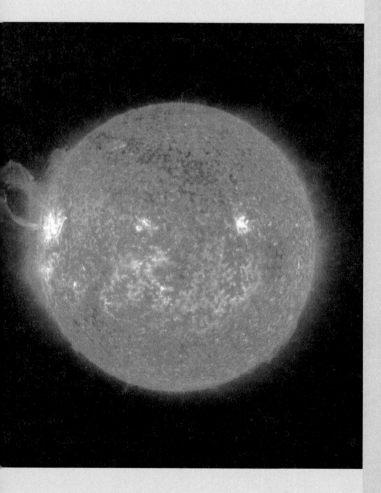

PART 4
Galaxies and Cosmology 570

23 THE MILKY WAY GALAXY
A SPIRAL IN SPACE 572

24 GALAXIES
BUILDING BLOCKS OF THE UNIVERSE 602

22 NEUTRON STARS AND BLACK HOLES
STRANGE STATES OF MATTER 538

This volume (Vol. II: Stars and Galaxies) includes Chapters 1–5 and 16–28 of the *Astronomy Today* text. The gray type in the Table of Contents for Chapters 6–15 is for informational purposes only; those chapters appear in Vol I: The Solar System.

ABOUT THE AUTHORS

Eric Chaisson

Eric holds a doctorate in astrophysics from Harvard University, where he spent ten years on the faculty of Arts and Sciences. For several years thereafter, he was a senior scientist and director of educational programs at the Space Telescope Science Institute and adjunct professor of physics at Johns Hopkins University. For the past decade, Eric has been at Tufts University, where he is a research professor in the Department of Physics and in the School of Education, and director of the Wright Center for Innovative Science Education. He has written 12 books on astronomy, which have received such literary awards as the Phi Beta Kappa Prize, two American Institute of Physics Awards, and Harvard's Smith-Weld Prize for Literary Merit. He has published more than 100 scientific papers in professional journals and has also received Harvard's Bok Prize for original contributions to astrophysics.

Steve McMillan

Steve holds a bachelor's and master's degree in mathematics from Cambridge University and a doctorate in astronomy from Harvard University. He held postdoctoral positions at the University of Illinois and Northwestern University, where he continued his research in theoretical astrophysics, star clusters, and high-performance computing. Steve is currently Distinguished Professor of Physics at Drexel University and a frequent visiting researcher at Princeton's Institute for Advanced Study and Leiden University. He has published more than 70 scientific papers in professional journals.

PREFACE

Astronomy is a science that thrives on new discoveries. Fueled by new technologies and novel theoretical insights, the study of the cosmos continues to change our understanding of the universe. We are pleased to have the opportunity to present in this book a representative sample of the known facts, evolving ideas, and frontier discoveries in astronomy today.

Astronomy Today has been written for students who have taken no previous college science courses and who will likely not major in physics or astronomy. It is intended for use in a one- or two-semester, nontechnical astronomy course. We present a broad view of astronomy, straightforwardly descriptive and without complex mathematics. The absence of sophisticated mathematics, however, in no way prevents discussion of important concepts. Rather, we rely on qualitative reasoning as well as analogies with objects and phenomena familiar to the student to explain the complexities of the subject without oversimplification. We have tried to communicate the excitement we feel about astronomy and to awaken students to the marvelous universe around us.

We are very gratified that the first six editions of this text have been so well received by many in the astronomy education community. In using those earlier texts, many teachers and students have given us helpful feedback and constructive criticisms. From these, we have learned to communicate better both the fundamentals and the excitement of astronomy. Many improvements inspired by these comments have been incorporated into this new edition.

Focus of the Seventh Edition

From the first edition, we have tried to meet the challenge of writing a book that is both accurate and approachable. To the student, astronomy sometimes seems like a long list of unfamiliar terms to be memorized and repeated. Many new terms and concepts will be introduced in this course, but we hope students will also learn and remember how science is done, how the universe works, and how things are connected. In the seventh edition, we have taken particular care to show how astronomers know what they know, and to highlight both the scientific principles underlying their work and the process used in discovery.

New and Revised Material

Astronomy is a rapidly evolving field and, in the three years since the publication of the sixth edition of *Astronomy Today*, has seen many new discoveries covering the entire spectrum of astronomical research. Almost every chapter in the seventh edition has been substantially updated with new information. Several chapters have also seen significant reorganization in order to streamline the overall presentation, strengthen our focus on the process of science, and reflect new understanding and emphases in contemporary astronomy. Among the many changes are:

- The Big Picture feature on each chapter opening spread explains how chapter content fits in with an overall perspective on introductory astronomy, helping students see how each chapter is connected to a broad understanding of the universe. The theme is then called out again in selected figures throughout the chapter.

- Streamlined coverage—a reduction of over 50 pages—keeps students focused on the essentials they need to learn.

- Further emphasis throughout on the process of science and how astronomers "know what they know."

- Updates on the *Hubble Space Telescope, James Webb Space Telescope, Spitzer Space Telescope,* and *Fermi Gamma-Ray Space Telescope* in Chapter 5.

- A more complete account of the condensation theory of solar system formation in Chapter 6.

- Coverage of the *Lunar Reconnaissance Orbiter (LRO)* and *Lunar Crater Observation and Sensing Satellite (LCROSS)* missions in Chapter 8.

- New data and imagery from the *Messenger* mission on Mercury's atmosphere, plains, and magnetosphere in Chapter 8.

- New material on Mars's Borealis Basin and NASA's *Phoenix* lander in Chapter 10.

- Updates on the 2009 comet collision with Jupiter and new "red spot" storms on the planet in Chapter 11.

- Updates throughout Chapter 12 reflecting new *Cassini* data and imagery, including discussion of storms on Saturn, surface conditions on Titan and the other moons, and new discoveries in Saturn's ring system.

- Updates on new discoveries in the Kuiper belt in Chapter 14.

- Rewritten and expanded coverage in Chapter 15 of extrasolar planets, with new discussion of hot Jupiters, super-Earths, exoplanetary properties, and the search for Earth-like planets orbiting in the habitable zones of their parent stars.

- Updated discussion of brown dwarfs in Chapter 19.

- New coverage in Chapter 20 of the evolution of low-mass stars and the discovery of multiple stellar populations in globular star clusters.
- Updates on gamma-ray burst observations and models in Chapter 22.
- Updated discussion of Hubble's constant in Chapter 24 reflecting the recent convergence of measurement techniques.
- Further emphasis in Chapter 24 on active galaxies as part of a continuum with normal galaxies, and quasars as a type of active galaxy.
- Updated description of the standard model of active galactic nuclei in Chapter 24.
- Update on the Sloan Digital Sky Survey and successor surveys in Chapter 25.

The Illustration Program

Visualization plays an important role in both the teaching and the practice of astronomy, and we continue to place strong emphasis on this aspect of our book. We have tried to combine aesthetic beauty with scientific accuracy in the artist's conceptions that adorn the text, and we have sought to present the best and latest imagery of a wide range of cosmic objects. Each illustration has been carefully crafted to enhance student learning; each is pedagogically sound and tied tightly to the nearby discussion of important scientific facts and ideas. This edition contains more than 100 revised figures that show the latest imagery and the results learned from them

Compound Art It is rare that a single image, be it a photograph or an artist's conception, can capture all aspects of a complex subject. Wherever possible, multiple-part figures are used in an attempt to convey the greatest amount of information in the most vivid way:

- Visible images are often presented along with their counterparts captured at other wavelengths.
- Interpretive line drawings are often superimposed on or juxtaposed with real astronomical photographs, helping students to really "see" what the photographs reveal.
- Breakouts—often multiple ones—are used to zoom in from wide-field shots to close-ups so that detailed images can be understood in their larger context.

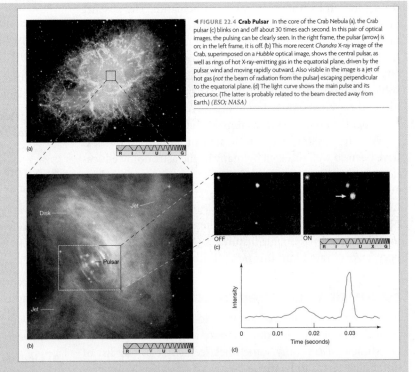

◄ FIGURE 22.4 **Crab Pulsar** In the core of the Crab Nebula (a), the Crab pulsar (c) blinks on and off about 30 times each second. In this pair of optical images, the pulsing can be clearly seen. In the right frame, the pulsar (arrow) is on; in the left frame, it is off. (b) This more recent *Chandra* X-ray image of the Crab, superimposed on a *Hubble* optical image, shows the central pulsar, as well as rings of hot X-ray-emitting gas in the equatorial plane, driven by the pulsar wind and moving rapidly outward. Also visible in the image is a jet of hot gas (*not* the beam of radiation from the pulsar) escaping perpendicular to the equatorial plane. (d) The light curve shows the main pulse and its precursor. (The latter is probably related to the beam directed away from Earth.) *(ESO; NASA)*

Interactive Figures and Photos Icons throughout the text direct students to dynamic versions of art and photos on Mastering Astronomy®. Using online applets, students can manipulate factors such as time, wavelength, scale, and perspective to increase their understanding of these figures.

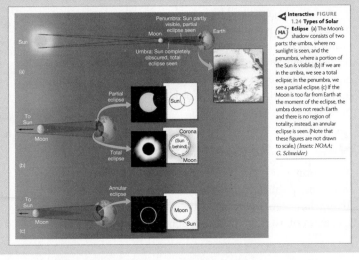

◄ Interactive FIGURE 1.24 **Types of Solar Eclipse** (a) The Moon's shadow consists of two parts: the umbra, where no sunlight is seen, and the penumbra, where a portion of the Sun is visible. (b) If we are in the umbra, we see a total eclipse; in the penumbra, we see a partial eclipse. (c) If the Moon is too far from Earth at the moment of the eclipse, the umbra does not reach Earth and there is no region of totality; instead, an annular eclipse is seen. (Note that these figures are not drawn to scale.) *(Insets: NOAA; G. Schneider)*

Full Spectrum Coverage and Spectrum Icons.
Increasingly, astronomers exploit the full range of the electromagnetic spectrum to gather information about the cosmos. Throughout this book, images taken at radio, infrared, ultraviolet, X-ray, or gamma-ray wavelengths are used to supplement visible-light images. As it is sometimes difficult (even for a professional) to tell at a glance which images are visible-light photographs and which are false-color images created with other wavelengths, each photo in the text is accompanied by an icon that identifies the wavelength of electromagnetic radiation used to capture the image.

Other Pedagogical Features

As with many other parts of our text, instructors have helped guide us toward what is most helpful for effective student learning. With their assistance, we have revised both our in-chapter and end-of-chapter pedagogical apparatus to increase its utility to students.

Learning Goals. Studies indicate that beginning students have trouble prioritizing textual material. For this reason, a few (typically five or six) well-defined Learning Goals are provided at the start of each chapter. These help students structure their reading of the chapter and then test their mastery of key concepts. The Learning Goals are numbered and keyed to the items in the Chapter Summary, which in turn refer back to passages in the text. This highlighting of the most important aspects of the chapter helps students prioritize information and also aids in their review. The Learning Goals are organized and phrased in such a way as to make them objectively testable, affording students a means of gauging their own progress.

The Big Picture. The Big Picture feature on every chapter opening spread encapsulates the overarching message that each chapter imparts, helping students see how chapter content is connected to a broad understanding of the universe. The feature is then called out again in selected chapter figures that explore the big picture.

Concept Checks. We incorporate into each chapter a number of "Concept Checks"—key questions that require the reader to reconsider some of the material just presented or attempt to place it into a broader context. Answers to these in-chapter questions are provided at the back of the book.

CONCEPT CHECK
✔ In what ways might observations of extrasolar planets help us understand our own solar system?

Process of Science Checks. Each chapter now also includes one or two "Process of Science Checks," similar to the Concept Checks but aimed specifically at clarifying the questions of how science is done and how scientists reach the conclusions they do. Answers to these in-chapter questions are also provided at the back of the book.

PROCESS OF SCIENCE CHECK
✔ How do Newton's and Einstein's theories differ in their descriptions of gravity?

Concept Links. In astronomy, as in many scientific disciplines, almost every topic seems to have some bearing on almost every other. In particular, the connection between the astronomical material and the physical principles set forth early in the text is crucial. Practically everything in Chapters 6–28 of this text rests on the foundation laid in the first five chapters. For example, it is important that students, when they encounter the discussion of high-redshift objects in Chapter 25, recall not only what they just learned about Hubble's law in Chapter 24 but also refresh their memories, if necessary, about the inverse-square law (Chapter 17), stellar spectra (Chapter 4), and the Doppler shift (Chapter 3). Similarly, the discussions of the mass of binary-star components (Chapter 17) and of galactic rotation (Chapter 23) both depend on the discussion of Kepler's and Newton's laws in Chapter 2. Throughout, discussions of new astronomical objects and concepts rely heavily on comparison with topics introduced earlier in the text.

It is important to remind students of these links so that they recall the principles on which later discussions rest and, if necessary, review them. To this end, we have inserted "concept links" throughout the text-symbols that mark key intellectual bridges between material in different chapters. The links, denoted by the symbol ∞ together with a section reference, signal that the topic under discussion is related in some significant way to ideas developed earlier, and provide direction to material to review before proceeding.

Key Terms. Like all subjects, astronomy has its own specialized vocabulary. To aid student learning, the most important astronomical terms are boldfaced at their first appearance in the text. Boldfaced Key Terms in the Chapter Summary are linked with the page number where the term was defined. In addition, an expanded alphabetical glossary, defining each Key Term and locating its first use in the text, appears at the end of the book.

H-R Diagrams and Acetate Overlays. All of the book's H-R diagrams are drawn in a uniform format, using real data. In addition, a unique set of transparent acetate overlays dramatically demonstrates to students how the H-R diagram helps us to organize our information about the stars and track their evolutionary histories.

More Precisely Boxes. These boxes provide more quantitative treatments of subjects discussed qualitatively in the text. Removing these more challenging topics from the main flow of the narrative and placing them within a separate modular element of the chapter design (so that they can be covered in class, assigned as supplementary material, or simply left as optional reading for those students who find them of interest) will allow instructors greater flexibility in setting the level of their coverage.

Discovery Boxes. Exploring a wide variety of interesting supplementary topics, Discovery boxes provide the reader with insight into how scientific knowledge evolves and emphasizes the process of science.

End-of-Chapter Questions and Problems. Many elements of the end-of-chapter material have seen substantial reorganization:

- Each chapter incorporates **Review and Discussion Questions,** which may be used for in-class review or for assignment. As with the Self-Test Questions, the material needed to answer Review Questions may be found within the chapter. The Discussion Questions explore particular topics more deeply, often asking for opinions, not just facts. As with all discussions, these questions usually have no single "correct" answer. Questions identified with a **POS** icon encourage students to explore the Process of Science.

- Each chapter also contains **Conceptual Self-Test Questions** in a multiple choice format, including select questions that are tied directly to a specific figure or diagram in the text, allowing students to assess their understanding of the chapter material. These questions are identified with a **VIS** icon. Answers to all these questions appear at the end of the book.

Chapter Review Summaries. The Chapter Review Summaries, a primary review tool, are linked to the Learning Goals at the beginning of each chapter. All Key Terms introduced in each chapter are listed again, in context and in boldface, along with key figures and page references to the text discussion.

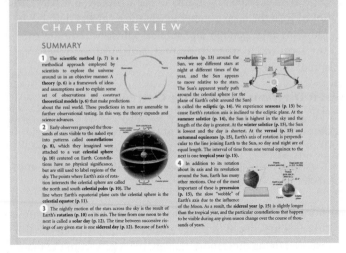

- The end-of-chapter material includes **Problems,** based on the chapter contents and requiring some numerical calculation. In many cases the problems are tied directly to quantitative statements made (but not worked out in detail) in the text. The solutions to the problems are not contained verbatim within the chapter, but the information necessary to solve them has been presented in the text. Answers to odd-numbered Problems appear at the end of the book.

Instructor Resources

 **MasteringAstronomy®**
www.masteringastronomy.com

MasteringAstronomy is the most widely used and most advanced astronomy tutorial and assessment system in the world. By capturing the step-by-step work of students nationally, MasteringAstronomy has established an unparalleled database of learning challenges and patterns. Using this student data, a team of renowned astronomy education researchers has refined every activity and problem. The result is a library of activities of unique educational effectiveness and assessment accuracy. MasteringAstronomy provides students with two learning systems in one: a dynamic self-study area and the ability to participate in online assignments.

MasteringAstronomy provides instructors with a fast and effective way to assign uncompromising, wide-ranging online homework assignments of just the right difficulty and duration. The tutorials coach 90 percent of students to the correct answer with specific wrong-answer feedback. Powerful post-diagnostics allow instructors to assess the progress of their class as a whole or to quickly identify an individual student's areas of difficulty. Tutorials built around text content and all the end-of-chapter problems from the text are available in MasteringAstronomy. A media-rich self-study area is included that students can use whether the instructor assigns homework or not.

Instructor Guide. Revised by James Heath (Austin Community College), this online guide provides: sample syllabi and course schedules; an overview of each chapter; pedagogical tips; useful analogies; suggestions for classroom demonstrations; writing questions, selected readings, and answers/solutions to the end-of-chapter Review and Discussion Questions and Problems; and additional references and resources.
ISBN 0-321-72514-X

Test Bank. An extensive file of approximately 2500 test questions, newly compiled and revised for the seventh edition by Lauren Jones (Ohio Department of Education, Office of Curriculum and Assessment). The questions are organized and referenced by chapter section and by question type. The sixth edition Test Bank has been thoroughly revised and includes many new Multiple Choice and Essay questions for added conceptual emphasis. This Test Bank is available in both Microsoft® Word and TestGen® formats (see description of Instructor Resource DVD). ISBN 0-321-71866-6

Instructor Resource DVD. This DVD provides virtually every electronic asset professors will need in and out of the classroom. The disc contain all text figures in jpeg and PowerPoint formats, as well as the animations and videos from the Mastering Astronomy® Study Area. The IR-DVD also contains TestGen®, an easy-to-use, fully networkable program for creating tests ranging from short quizzes to long exams. Questions from the Test Bank are supplied, and professors can use the Question Editor to modify existing questions or create new questions. This disc set also contains chapter-by-chapter lecture outlines in PowerPoint and conceptual "clicker" questions in PowerPoint. ISBN 0-321-68746-9

Learner-Centered Astronomy Teaching: Strategies for ASTRO 101
Timothy F. Slater, *University of Arizona*
Jeffrey P. Adams, *Montana State University*
Strategies for ASTRO 101 is a guide for instructors of the introductory astronomy course for nonscience majors. Written by two leaders in astronomy education research, this book details various techniques instructors can use to increase students' understanding and retention of astronomy topics, with an emphasis on making the lecture a forum for active student participation. Drawing from the large body of recent research to discover how students learn, this guide describes the application of multiple classroom-tested techniques to the task of teaching astronomy to predominantly nonscience students. ISBN 0-13-046630-1

Peer Instruction for Astronomy
Paul Green, *Harvard Smithsonian Center for Astrophysics*
Peer Instruction is a simple yet effective method for teaching science. Techniques of peer instruction for introductory physics were developed primarily at Harvard and have aroused interest and excitement in the physics education community. This approach involves students in the teaching process, making science more accessible to them. This book is an important vehicle for providing a large number of thought-provoking, conceptual short-answer questions aimed at a variety of class levels. While significant numbers of such questions have been published for use in physics, *Peer Instruction for Astronomy* provides the first such compilation for astronomy. ISBN 0-13-026310-9

Student Resources

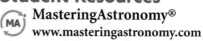

MasteringAstronomy®
www.masteringastronomy.com
This homework, tutorial, and assessment system is uniquely able to tutor each student individually by providing students with instantaneous feedback specific to their wrong answers, simpler subproblems upon request when they get stuck, and partial credit for their method(s) used. Students also have access to a self-study area that contains practice quizzes, self-guided tutorials, animations, videos, and more.

Pearson eText is available through MasteringAstronomy, either automatically when MasteringAstronomy is packaged with new books, or available as a purchased upgrade online. Allowing the students to access the text wherever they have access to the Internet, Pearson eText comprises the full text, including figures that can be enlarged for better viewing. Within Pearson eText students are also able to pop up definitions and terms to help with vocabulary and the reading of the material. Students also can take notes in Pearson eText using the annotation feature.

Starry Night College® Student Access Code Card This best-selling planetarium software lets you escape the Milky Way and travel within 700 million light-years of space. View more than 16 million stars in stunningly realistic star fields. Zoom in on thousands of galaxies, nebulae, and star clusters. Move through 200,000 years of time to see key celestial events in a dynamic and ever-changing universe. Blast off from Earth and see the motions of the planets from a new perspective. Hailed for its breathtaking realism, powerful suite of features, and intuitive ease of use, Starry Night College lives up to its reputation as astronomy software's brightest . . . night after night. ISBN 0-321-71295-1

Starry Night College® Activities & Observation and Research Projects This supplement contains activities for Starry Night College planetarium software by Erin O'Connor (Santa Barbara City College), as well as observation and research projects authored by Steve McMillan. ISBN 0-321-75307-0

Sky and Telescope This supplement contains nine articles that originally appeared in the popular amateur astronomy magazine, plus a summary and four question sets focusing on the issues professors most want to address: general review, process of science, scale of the universe, and our place in the universe. Edited by Evan Skillman. ISBN 0-321-70620-X

Edmund Scientific Star and Planet Locator The famous rotating roadmap of the heavens shows the location of the stars, constellations, and planets relative to the horizon for the

exact hour and date you determine. This 8 square star chart was plotted by the late astronomer and cartographer George Lovi. The reverse side of the locator is packed with additional data on the planets, meteor showers, and bright stars. Included with each star chart is a 16-page, fully illustrated, pocket-size instruction booklet.
ISBN 0-13-140235-8

Lecture Tutorials for Introductory Astronomy, 2nd Edition
Edward E. Prather, *University of Arizona*
Timothy F. Slater, *University of Arizona*
Jeffrey P. Adams, *Montana State University*
Gina Brissenden, *NASA Center for Astronomy Education (CAE) at the University of Arizona*
CAPER, *Conceptual Astronomy and Physics Education Research Team*
Funded by the National Science Foundation, *Lecture Tutorials for Introductory Astronomy* is designed to help make large lecture-format courses more interactive. The second edition features new Tutorials on the Doppler Shift, Extrasolar Planets, Kepler's 2nd Law, and Seasons. Each of the 38 Lecture Tutorials is presented in a classroom-ready format, asks students to work in groups of 2 to 3, takes from 10 to 15 minutes,

challenges students with a series of carefully designed questions that spark classroom discussion, engages students in critical reasoning, and requires no equipment.
ISBN 0-13-239226-7

Observation Exercises in Astronomy This workbook by Lauren Jones contains a series of astronomy exercises that integrate technology from planetarium software such as Stellarium, Starry Night College, WorldWide Telescope, and SkyGazer. Using these online products adds an interactive dimension to students' learning.
ISBN: 0-321-63812-3

Acknowledgments

Throughout the many drafts that have led to this book, we have relied on the critical analysis of many colleagues. Their suggestions ranged from the macroscopic issue of the book's overall organization to the minutiae of the technical accuracy of each and every sentence. We have also benefited from much good advice and feedback from users of the first six editions of the text. To these many helpful colleagues, we offer our sincerest thanks.

Reviewers of the Seventh Edition

Nadine G. Barlow
Northern Arizona University

Thomasanna Hail
Parkland College

Lynn Higgs
University of Utah

Kristine Larsen
Central Connecticut State University

Steve Mellema
Gustavus Adolphus College

John Scalo
University of Texas at Austin

Jie Zhang
George Mason University

Reviewers of Previous Editions

Stephen G. Alexander
Miami University of Ohio

William Alexander
James Madison University

Robert H. Allen
University of Wisconsin, La Crosse

Cecilia Barnbaum
Valdosta State University

Peter A. Becker
George Mason University

Timothy C. Beers
University of Evansville

William J. Boardman
Birmingham Southern College

Donald J. Bord
University of Michigan, Dearborn

Elizabeth P. Bozyan
University of Rhode Island

Malcolm Cleaveland
University of Arkansas

Anne Cowley
Arizona State University

Bruce Cragin
Richland College

Ed Coppola
Community College of Southern Nevada

David Curott
University of North Alabama

Norman Derby
Bennington College

John Dykla
Loyola University, Chicago

Kimberly Engle
Drexel University

Michael N. Fanelli
University of North Texas

Richard Gelderman
Western Kentucky University

Harold A. Geller
George Mason University

David Goldberg
Drexel University

Martin Goodson
Delta College

David G. Griffiths
Oregon State University

Donald Gudehus
Georgia State University

Clint D. Harper
Moorpark College

Marilynn Harper
Delaware County Community College

Susan Hartley
University of Minnesota, Duluth

Joseph Heafner
Catawaba Valley Community College

James Heath
Austin Community College

Fred Hickok
Catonsville Community College

Darren L. Hitt
Loyola College, Maryland

F. Duane Ingram
Rock Valley College

Steven D. Kawaler
Iowa State University

William Keel
University of Alabama

Marvin Kemple
Indiana University-Purdue University, Indianapolis

Mario Klairc
Midlands Technical College

Kristine Larsen
Central Connecticut State University

Andrew R. Lazarewicz
Boston College

Robert J. Leacock
University of Florida

Larry A. Lebofsky
University of Arizona

Matthew Lister
Purdue University

M. A. Lohdi
Texas Tech University

Michael C. LoPresto
Henry Ford Community College

Phillip Lu
Western Connecticut State University

Fred Marschak
Santa Barbara College

Matthew Malkan
University of California, Los Angeles

Chris Mihos
Case Western Reserve University

Milan Mijic
California State University, Los Angeles

Scott Miller
Pennsylvania State University

Mark Moldwin
University of California, Los Angeles

Richard Nolthenius
Cabrillo College

Edward Oberhofer
University of North Carolina, Charlotte

Andrew P. Odell
Northern Arizona University

Gregory W. Ojakangas
University of Minnesota, Duluth

Ronald Olowin
Saint Mary's College of California

Robert S. Patterson
Southwest Missouri State University

Cynthia W. Peterson
University of Connecticut

Lawrence Pinsky
University of Houston

Andreas Quirrenback
University of California, San Diego

Richard Rand
University of New Mexico

James A. Roberts
University of North Texas

Gerald Royce
Mary Washington College

Dwight Russell
Baylor University

Vicki Sarajedini
University of Florida

Malcolm P. Savedoff
University of Rochester

John C. Schneider
Catonsville Community College

Larry Sessions
Metropolitan State College of Denver

Harry L. Shipman
University of Delaware

C. G. Pete Shugart
Memphis State University

Stephen J. Shulik
Clarion University

Tim Slater
University of Arizona

Don Sparks
Los Angeles Pierce College

George Stanley, Jr.
San Antonio College

Maurice Stewart
Williamette University

Jack W. Sulentic
University of Alabama

Andrew Sustich
Arkansas State University

Donald Terndrup
Ohio State University

Craig Tyler
Fort Lewis College

Stephen R. Walton
California State University, Northridge

Peter A. Wehinger
University of Arizona

Louis Winkler
Pennsylvania State University

Robert Zimmerman
University of Oregon

The publishing team at Addison-Wesley/Benjamin Cummings has assisted us at every step along the way in creating this text. Many thanks to Tema Goodwin, who managed the many variables that go into a multifaceted publication such as this. Production managers Crystal Clifton and Mary O'Connell have done an excellent job of tying together the threads of this very complex project, made all the more complex by the necessity of combining text, art, and electronic media into a coherent whole. Special thanks are also in order to cover and interior designer Derek Bacchus for making the seventh edition look spectacular.

We would also like to express our gratitude to Kelly Reed for updating and maintaining the media resources in the MasteringAstronomy® Study Area.

Finally, we would like to express our gratitude to renowned space artist Dana Berry for allowing us to use many of his beautiful renditions of astronomical scenes, and to Lola Judith Chaisson for assembling and drawing all the H–R diagrams (including the acetate overlays) for this edition.

Eric Chaisson
Steve McMillan

PART 1

ASTRONOMY AND THE UNIVERSE

Galileo's sketch of Saturn

Galileo's sketch of Orion

Galileo Galilei

It is often said that we live in a golden age of astronomy. Yet the dawn of the 21st century is actually the second such period of rich discovery and rapid exploration. The late Renaissance began the first era of stunning scientific advancement, when modern astronomy was born. Most notable among those spearheading the rebirth of astronomy at the time was the Italian scientist Galileo Galilei (1564–1642). By turning his telescope to the heavens, he changed radically and forever humankind's view of the universe in which we live.

Although he did not invent the telescope, in 1610 Galileo was the first to record what he saw when he aimed a small (5-centimeter-diameter) lens at the sky. His findings created nothing less than a revolution in astronomy. Viewing for the first time dark blemishes on the Sun, rugged mountains on the Moon, and whole new worlds orbiting Jupiter, he demolished the Aristotelian view of cosmic immutability—the notion that the heavens were perfect and unchanging. To be sure, it was with the philosophers of the day, as much as with the theologians, that Galileo had trouble. In championing the scientific method, he used a tool to test his ideas, and what he found disagreed greatly with the leading thoughts and beliefs of the time.

Galileo's advance was simple yet profound: He used a telescope to focus, magnify, and study radiation reaching Earth from the heavens—in particular, light from the Sun, the Moon, and the planets. Light is the most familiar kind of radiation to humans on Earth, since it enables us to get around on the surface of our planet. But light also enables telescopes to see objects deep in space, allowing us to probe farther than the eye can alone. With his simple optical telescope, Galileo changed completely the way that the oldest science—astronomy—is pursued.

Among other "wondrous things" he found were crowded star clusters along the Milky Way, moons and rings around the big planets,

Galileo's sketch of the Pleiades

Saturn in the ultraviolet *(STScI)*

and colorful nebulae aglow unlike anything anyone had seen before. Some of Galileo's sketches are reproduced here at the left and compared with modern views at the right.

Today, we are again in the midst of another period of unsurpassed scientific achievement—a revolution in which modern astronomers are revealing the invisible universe as Galileo once spied the visible universe. We have learned how to detect, measure, and analyze invisible radiation streaming to us from dark objects in space. And once again our perceptions are changing.

Astronomy no longer evokes visions of plodding intellectuals peering through long telescope tubes. Nor does the cosmos any longer refer to that seemingly inactive, immutable regime seen visually when we gaze at the nighttime sky. Modern astronomers now decipher a more vibrant, changing universe—one in which stars emerge and perish much like living things, galaxies spew forth vast quantities of energy, and life itself is thought to be a natural consequence of the evolution of matter.

New discoveries are rapidly advancing our understanding of the universe, but they also raise new questions. Astronomers will encounter many problems in the decades ahead, but this should neither dismay nor frustrate us, for it is precisely how science operates. Each discovery adds to our storehouse of information, generating a host of questions that lead in turn to more discoveries, and so on, causing an acceleration of basic knowledge.

Most notably, we are beginning to perceive the universe in all its multivaried ways. A single generation—not the generation of our parents and not that of our children, but our generation—has opened up the whole electromagnetic spectrum beyond visible light. And what we, too, have found are "wondrous things."

Orion in the infrared *(Caltech)*

Emerging largely from studies of the invisible universe, our view of the cosmos in its full splendor is one of many new scientific insights that we have recently been privileged to attain. Historians of the future may well regard our generation as the one that took a great leap forward, providing a whole new glimpse of our richly endowed universe. In all of history, there have been only two periods in which our perception of the universe has been so revolutionized within a single human lifetime. The first occurred four centuries ago at the time of Galileo; the second is now under way.

Pleiades in the optical *(AURA)*

1

CHARTING THE HEAVENS

THE FOUNDATIONS OF ASTRONOMY

LEARNING GOALS

Studying this chapter will enable you to

1 Describe how scientists combine observation, theory, and testing in their study of the universe.

2 Explain the concept of the celestial sphere and how we use angular measurement to locate objects in the sky.

3 Describe how and why the Sun and the stars appear to change their positions from night to night and from month to month.

4 Explain why Earth's rotation axis shifts slowly with time, and say how this affects Earth's seasons.

5 Show how the relative motions of Earth, the Sun, and the Moon lead to eclipses.

6 Explain the simple geometric reasoning that allows astronomers to measure the distances and sizes of otherwise inaccessible objects.

Nature offers no greater splendor than the starry sky on a clear, dark night. Silent and jeweled with the constellations of ancient myth and legend, the night sky has inspired wonder throughout the ages—a wonder that leads our imaginations far from the confines of Earth and the pace of the present day and out into the distant reaches of space and cosmic time itself.

Astronomy, born in response to that wonder, is built on two of the most basic traits of human nature: the *need to explore* and the *need to understand*. Through the interplay of curiosity, discovery, and analysis—the keys to exploration and understanding—people have sought answers to questions about the universe since the earliest times. Astronomy is the oldest of all the sciences, yet never has it been more exciting than it is today.

THE BIG PICTURE Stars are the most fundamental visible component of the universe. Roughly as many stars reside in the observable universe as there are grains of sand in all the beaches of the world—around a hundred sextillion, or about 10^{23}.

LEFT: *High overhead on a clear, dark night, we can see a rich band of stars known as the Milky Way—so-called for its resemblance to a milky band of countless stars. All these stars (and more) are part of a much larger system called the Milky Way Galaxy, of which our star, the Sun, is one member. This single exposure, dubbed "Going to the Stars Road," was made at night with only the Moon's light illuminating the terrain on the continental divide at Logan Pass in Glacier National Park, near the Montana/Alberta border. (© Tyler Nordgren)*

Mastering**ASTRONOMY**

Visit the Study Area in www.masteringastronomy.com for quizzes, animations, videos, interactive figures, and self-guided tutorials.

1.1 Our Place in Space

Of all the scientific insights attained to date, one stands out boldly: Earth is neither central nor special. We inhabit no unique place in the universe. Astronomical research, especially within the past few decades, strongly suggests that we live on what seems to be an ordinary rocky *planet* called Earth, one of eight known planets orbiting an average *star* called the Sun, a star near the edge of a huge collection of stars called the Milky Way Galaxy, which is one *galaxy* among billions of others spread throughout the observable universe. To begin to get a feel for the relationships among these very different objects, consult Figures 1.1 through 1.4; put them in perspective by studying Figure 1.5.

We are connected to the most distant realms of space and time not only by our imaginations but also through a common cosmic heritage. Most of the chemical elements that make up our bodies (hydrogen, oxygen, carbon, and many more) were created billions of years ago in the hot centers of long-vanished stars. Their fuel supply spent, these giant stars died in huge explosions, scattering the elements created deep within their cores far and wide. Eventually, this matter collected into clouds of gas that slowly collapsed to give birth to new generations of stars. In this way, the Sun and its family of planets formed nearly five billion years ago. Everything on Earth embodies atoms from other parts of the universe and from a past far more remote than the beginning of human evolution. Elsewhere, other beings—perhaps with intelligence much greater than our own—may at this very moment be gazing in wonder at their own night sky. Our own Sun may be nothing more than an insignificant point of light to them—if it is visible at all. Yet if such beings exist, they must share our cosmic origin.

Simply put, the **universe** is the totality of all space, time, matter, and energy. **Astronomy** is the study of the universe. It is a subject unlike any other, for it requires us to profoundly change our view of the cosmos and to consider matter on scales totally unfamiliar from everyday experience. Look again at the galaxy in Figure 1.3. It is a swarm of about a hundred billion stars—more stars than the number of people who have ever lived on Earth. The entire assemblage is spread across a vast expanse of space 100,000 **light-years** in diameter. Although it sounds like a unit of time, a light-year is in fact the *distance* traveled by light in a year, at a speed of about 300,000 kilometers per second. Multiplying out, it follows that a light-year is equal to 300,000 kilometers/second × 86,400 seconds/day × 365 days or about 10 trillion kilometers, or roughly 6 trillion miles. Typical

15,000 kilometers

R I V U X G

▲ **FIGURE 1.1 Earth** Earth is a planet, a mostly solid object, although it has some liquid in its oceans and core and gas in its atmosphere. In this view, the North and South American continents are clearly visible, though most of the scene shows Pacific waters. *(NASA)*

1,500,000 kilometers

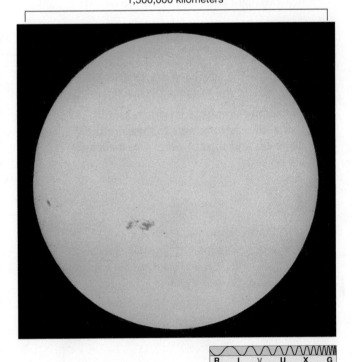

R I V U X G

▲ **FIGURE 1.2 The Sun** The Sun is a star, a very hot ball of gas composed mainly of hydrogen and helium. Much bigger than Earth—more than 100 times larger in diameter—the Sun is held together by its own gravity. The dark blemishes are sunspots (see Chapter 16). *(AURA)*

galactic systems are truly "astronomical" in size. For comparison, Earth's roughly 13,000-km diameter is less than one-twentieth of a light-*second*.

The light-year is a unit introduced by astronomers to help them describe immense distances. We will encounter many such custom units in our studies. As discussed in more detail in Appendix 2, astronomers frequently augment the standard SI (Système Internationale) metric system with additional units tailored to the particular problem at hand.

A thousand (1000), a million (1,000,000), a billion (1,000,000,000), and even a trillion (1,000,000,000,000)—these words occur regularly in everyday speech. But let's take a moment to understand the magnitude of the numbers and appreciate the differences among them. One thousand is easy enough to understand: At the rate of one number per second, you could count to a thousand in 1000 seconds—about 16 minutes. However, if you wanted to count to a million, you would need more than two weeks of counting at the rate of one number per second, 16 hours per day (allowing 8 hours per day for sleep). To count from one to a billion at the same rate of one number per second and 16 hours per day would take nearly 50 years—the better part of an entire human lifetime.

In this book, we consider *distances* in space spanning not just billions of kilometers, but billions of light-years; *objects* containing not just trillions of atoms, but trillions of stars; and *time intervals* of not just billions of seconds or hours, but billions of years. You will need to become familiar—and comfortable—with such enormous numbers. A good way to begin is learning to recognize just how much larger than a thousand is a million, and how much larger still is a billion. Appendix 1 explains the convenient method used by scientists for writing and manipulating very large and very small numbers. If you are unfamiliar with this method, please read that appendix carefully—the *scientific notation* described there will be used consistently throughout our text, beginning in Chapter 2.

Lacking any understanding of the astronomical objects they observed, early skywatchers made up stories to explain them: The Sun was pulled across the heavens by a chariot drawn by winged horses, and patterns of stars traced heroes and animals placed in the sky by the gods. Today, of course, we have a radically different conception of the universe. The stars we see are distant, glowing orbs hundreds of times larger than our entire planet, and the patterns they form span hundreds of light-years. In this first chapter we present some basic methods used by astronomers to chart the space around us. We describe the slow progress of scientific knowledge, from chariots and gods to today's well-tested theories and physical laws, and explain why we now rely on science rather than on myth to help us explain the universe.

About 1000 quadrillion kilometers, or 100,000 light-years

R I V U X G

▲ **FIGURE 1.3 Galaxy** A typical galaxy is a collection of a hundred billion stars, each separated by vast regions of nearly empty space. Our Sun is a rather undistinguished star near the edge of another such galaxy, called the Milky Way. (*NASA*)

About 1,000,000 light-years

R I V U X G

▲ **FIGURE 1.4 Galaxy Cluster** This photograph shows a typical cluster of galaxies, spread across roughly a million light-years of space. Each galaxy contains hundreds of billions of stars, probably planets, and possibly living creatures. (*NASA*)

◄ **FIGURE 1.5 Size and Scale** This artist's conception puts each of the previous four figures in perspective. The bottom of this figure shows spacecraft (and astronauts) in Earth orbit, a view that widens progressively in each of the next five cubes drawn from bottom to top—Earth, our planetary system, the local neighborhood of stars, the Milky Way Galaxy, and the closest cluster of galaxies. The image at the top right depicts the spread of galaxies (white dots) in the universe on extremely large scales—the field of view in this final frame is hundreds of millions of light-years across. The numbers indicate approximately the increase in scale between successive images: Earth is 500,000 times larger than the spacecraft in the foreground, the solar system in turn is some 1,000,000 times larger than Earth, and so on. Modern astronomy is the story of how humans have come to understand the immense scale of the universe, and our place in it. *(D. Berry)*

1.2 Scientific Theory and the Scientific Method

How have we come to know the universe around us? How do we know the proper perspective sketched in Figure 1.5? The earliest known descriptions of the universe were based largely on imagination and mythology and made little attempt to explain the workings of the heavens in terms of known earthly experience. However, history shows that some early scientists did come to realize the importance of careful observation and testing to the formulation of their ideas. The success of their approach changed, slowly but surely, the way science was done and opened the door to a fuller understanding of nature. As knowledge from all sources was sought and embraced for its own sake, the influence of logic and reasoned argument grew and the

power of myth diminished. People began to inquire more critically about themselves and the universe. They realized that *thinking* about nature was no longer sufficient—*looking* at it was also necessary. Experiments and observations became a central part of the process of inquiry.

To be effective, a **theory**—the framework of ideas and assumptions used to explain some set of observations and make predictions about the real world—must be continually tested. Scientists accomplish this by using a theory to construct a **theoretical model** of a physical object (such as a planet or a star) or phenomenon (such as gravity or light) that accounts for its known properties. The model then makes further predictions about the object's properties, or perhaps how it might behave or change under new circumstances. If experiments and observations favor those predictions, the theory can be further developed and refined. If they do not, the theory must be reformulated or rejected, no matter how appealing it originally seemed. This approach to investigation, combining thinking and doing—that is, theory and experiment—is known as the **scientific method.** The process, combining theoretical reasoning with experimental testing, is illustrated schematically in Figure 1.6. It lies at the heart of modern science, separating science from pseudoscience, fact from fiction.

The notion that theories must be tested and may be proven wrong sometimes leads people to dismiss their importance. We have all heard the expression, "Of course, it's only a theory," used to deride or dismiss an idea that someone finds unacceptable. Don't be fooled! Gravity (see Section 2.7) is "only" a theory, but calculations based on it have guided human spacecraft throughout the solar system. Electromagnetism (Chapter 3) and quantum mechanics (Chapter 4) are theories, too, yet they form the foundation for most of 20th- (and 21st-) century technology. Facts about the universe are a dime a dozen. Theories are the intellectual "glue" that combine seemingly unrelated facts into a coherent and interconnected whole.

Notice that there is no end point to the process depicted in Figure 1.6. A theory can be invalidated by a single wrong prediction, but no amount of observation or experimentation can ever prove it "correct." Theories simply become more and more widely accepted as their predictions are repeatedly confirmed. As a class, modern scientific theories share several important defining characteristics:

- They must be *testable*—that is, they must admit the possibility that their underlying assumptions and their predictions can, in principle, be exposed to experimental verification. This feature separates science from, for example, religion, since, ultimately, divine revelations or scriptures cannot be challenged within a religious framework—we can't design an experiment to "verify the mind of God." Testability also distinguishes science from a pseudoscience such as astrology, whose underlying assumptions and predictions have been repeatedly tested and never verified, with no apparent impact on the views of those who continue to believe in it!

- They must continually be *tested,* and their consequences tested, too. This is the basic circle of scientific progress depicted in Figure 1.6.

- They should be *simple.* Simplicity is less a requirement than a practical outcome of centuries of scientific experience—the most successful theories tend to be the simplest ones that fit the facts. This viewpoint is often encapsulated in a principle known as *Occam's razor:* If two competing theories both explain the facts and make the same predictions, then the simpler one is better. Put another way—"Keep it simple!" A good theory should contain no more complexity than is absolutely necessary.

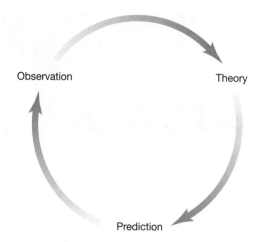

▲ **FIGURE 1.6 Scientific Method** Scientific theories evolve through a combination of observation, theoretical reasoning, and prediction, which in turn suggests new observations. The process can begin at any point in the cycle (although it usually starts with observations), and it continues forever—or until the theory fails to explain an observation or makes a demonstrably false prediction.

- Finally, most scientists have the additional bias that a theory should in some sense be *elegant.* When a clearly stated simple principle naturally ties together and explains several phenomena previously thought to be completely distinct, this is widely regarded as a strong point in favor of the new theory.

You may find it instructive to apply these criteria to the many physical theories—some old and well established, others much more recent and still developing—we will encounter throughout the text.

The birth of modern science is usually associated with the Renaissance, the historical period from the late 14th to the mid-17th century that saw a rebirth (*renaissance* in French) of artistic, literary, and scientific inquiry in European culture following the chaos of the Dark Ages. However, one of the first documented uses of the scientific method in an astronomical context was made by Aristotle (384–322 B.C.) some 17 centuries earlier. Aristotle is not normally remembered as a strong proponent of this approach—many of his best known ideas were based on pure thought, with no attempt at experimental test or verification. Nevertheless, his brilliance extended into many areas now thought of as modern science. He noted that, during a lunar eclipse (Section 1.6), Earth casts a curved shadow onto the surface of the Moon. Figure 1.7 shows a series of photographs taken during a recent lunar eclipse. Earth's shadow, projected onto the Moon's surface, is indeed slightly curved. This is what Aristotle must have seen and recorded so long ago.

Because the observed shadow seemed always to be an arc of the same circle, Aristotle theorized that Earth, the cause of the shadow, must be round. Don't underestimate the scope of this apparently simple statement. Aristotle also had to reason that the dark region was indeed a shadow and

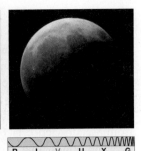

▲ **FIGURE 1.7 A Lunar Eclipse** This series of photographs shows Earth's shadow sweeping across the Moon during a lunar eclipse. By observing this behavior, Aristotle reasoned that Earth was the cause of the shadow and concluded that Earth must be round. His theory has yet to be disproved. *(G. Schneider)*

that Earth was its cause—facts we regard as obvious today, but far from clear 25 centuries ago. On the basis of this *hypothesis*—one possible explanation of the observed facts—he then predicted that any and all future lunar eclipses would show Earth's shadow to be curved, regardless of our planet's orientation. That prediction has been tested every time a lunar eclipse has occurred. It has yet to be proved wrong. Aristotle was not the first person to argue that Earth is round, but he was apparently the first to offer observational proof using the lunar eclipse method.

This basic reasoning forms the basis of all modern scientific inquiry. Armed only with naked-eye observations of the sky (the telescope would not be invented for almost another 2000 years), Aristotle first made an observation. Next, he formulated a hypothesis to explain that observation. Then he tested the validity of his hypothesis by making predictions that could be confirmed or refuted by further observations. *Observation, theory, and testing*—these are the cornerstones of the scientific method, a technique whose power will be demonstrated again and again throughout our text.

Today, scientists throughout the world use an approach that relies heavily on testing ideas. They gather data, form a working hypothesis that explains the data, and then proceed to test the implications of the hypothesis using experiment and observation. Eventually, one or more "well-tested" hypotheses may be elevated to the stature of a physical law and come to form the basis of a theory of even broader applicability. The new predictions of the theory will in turn be tested, as scientific knowledge continues to grow. Experiment and observation are integral parts of the process of scientific inquiry. Untestable theories, or theories unsupported by experimental evidence, rarely gain any measure of acceptance in scientific circles. Used properly over a period of time, this rational, methodical approach enables us to arrive at conclusions that are mostly free of the personal bias and human values of any one scientist. The scientific method is designed to yield an objective view of the universe we inhabit.

PROCESS OF SCIENCE CHECK

✔ Can a theory ever become a "fact," scientifically speaking?

1.3   The "Obvious" View

To see how astronomers have applied the scientific method to understand the universe around us, let's start with some very basic observations. Our study of the cosmos, the modern science of astronomy, begins simply, with looking at the night sky. The overall appearance of the night sky is not so different now from what our ancestors would have seen hundreds or even thousands of years ago, but our *interpretation* of what we see has changed immeasurably as the science of astronomy has evolved and grown.

Constellations in the Sky

Between sunset and sunrise on a clear night, we can see about 3000 points of light. If we include the view from the opposite side of Earth, nearly 6000 stars are visible to the unaided eye. A natural human tendency is to see patterns and relationships among objects even when no true connection exists, and people long ago connected the brightest stars into configurations called **constellations,** which ancient astronomers named after mythological beings, heroes, and animals—whatever was important to them. Figure 1.8 shows a constellation especially prominent in the nighttime sky from October through March: the hunter named Orion. Orion was a mythical Greek hero famed, among other things, for his amorous pursuit of the Pleiades, the seven daughters of the giant Atlas. According to Greek mythology, to protect the Pleiades from Orion, the gods placed them among the stars, where Orion nightly stalks them across the sky. Many constellations have similarly fabulous connections with ancient lore.

Perhaps not surprisingly, the patterns have a strong cultural bias—the astronomers of ancient China saw mythical figures different from those seen by the ancient Greeks, the Babylonians, and the people of other cultures, even though they were all looking at the same stars in the night sky. Interestingly, different cultures often made the same basic *groupings* of stars, despite widely varying interpretations of what they saw. For example, the group of seven stars usually known in North America as "the Dipper" is known as "the Wagon" or "the Plough" in western Europe. The ancient

(a)

R I V U X G

(b)

Greeks regarded these same stars as the tail of "the Great Bear," the Egyptians saw them as the leg of an ox, the Siberians as a stag, and some Native Americans as a funeral procession.

Early astronomers had very practical reasons for studying the sky. Some constellations served as navigational guides. The star Polaris (part of the Little Dipper) indicates north, and the near constancy of its location in the sky, from hour to hour and night to night, has aided travelers for centuries. Other constellations served as primitive calendars to predict planting and harvesting seasons. For example, many cultures knew that the appearance of certain stars on the horizon just before daybreak signaled the beginning of spring and the end of winter.

In many societies, people came to believe that there were other benefits in being able to trace the regularly changing positions of heavenly bodies. The relative positions of stars and planets at a person's birth were carefully studied by *astrologers*, who used the data to make predictions about that person's destiny. Thus, in a sense, astronomy and astrology arose from the same basic impulse—the desire to "see" into the future—and, indeed, for a long time they were indistinguishable from one another. Today, most people recognize that astrology is nothing more than an amusing diversion (although millions still study their horoscope in the newspaper every morning!). Nevertheless, the ancient astrological terminology—the names of the constellations and many terms used to describe the locations and motions of the planets—is still used throughout the astronomical world.

Generally speaking, as illustrated in Figure 1.9 for the case of Orion, the stars that make up any particular constellation are not actually close to one another in space, even by astronomical standards. They merely are bright enough to observe with the naked eye and happen to lie in roughly the same direction in the sky as seen from Earth.

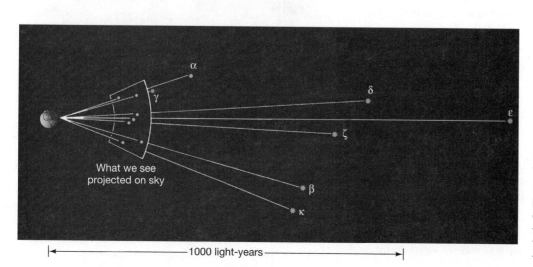

What we see
projected on sky

|← 1000 light-years →|

◄ **FIGURE 1.9 Orion in 3-D**
The true three-dimensional
relationships among the most
prominent stars in Orion. The
distances were determined by
the *Hipparcos* satellite in the
1990s. (See Chapter 17.)

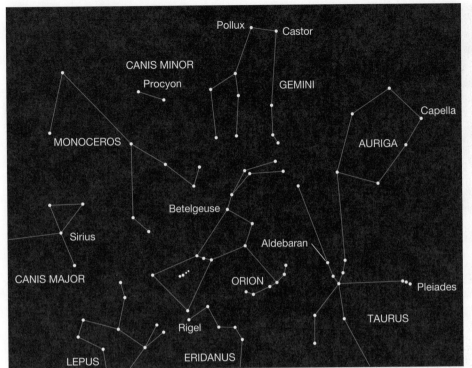

◄ **FIGURE 1.10 Constellations Near Orion** The region of the sky conventionally associated with the constellation Orion, together with some neighboring constellations (labeled in all capital letters). Some prominent stars are also labeled in lowercase letters. The 88 constellations span the entire sky, so that every astronomical object lies in precisely one of them.

but of Earth. Polaris indicates the direction—due north—in which Earth's rotation axis points. Even though we now know that the celestial sphere is an incorrect description of the heavens, we still use the idea as a convenient fiction that helps us visualize the positions of stars in the sky. The points where Earth's axis intersects the celestial sphere are called the **celestial poles.** In the Northern Hemisphere, the north celestial pole lies directly above Earth's North Pole. The extension of Earth's axis in the opposite direction defines the south celestial pole, directly above Earth's South Pole. Midway between the north

Still, the constellations provide a convenient means for astronomers to specify large areas of the sky, much as geologists use continents or politicians use voting precincts to identify certain localities on planet Earth. In all, there are 88 constellations, most of them visible from North America at some time during the year. Figure 1.10 shows how the conventionally defined constellations cover a portion of the sky in the vicinity of Orion.

The Celestial Sphere

Over the course of a night, the constellations seem to move smoothly across the sky from east to west, but ancient skywatchers were well aware that the *relative* locations of stars remained unchanged as this nightly march took place.* It was natural for those observers to conclude that the stars must be firmly attached to a **celestial sphere** surrounding Earth—a canopy of stars resembling an astronomical painting on a heavenly ceiling. Figure 1.11 shows how early astronomers pictured the stars as moving with this celestial sphere as it turned around a fixed, unmoving Earth. Figure 1.12 shows how all stars appear to move in circles around a point very close to the star Polaris (better known as the Pole Star or North Star). To the ancients, this point represented the axis around which the entire celestial sphere turned.

Today we recognize that the apparent motion of the stars is the result of the spin, or **rotation,** not of the celestial sphere,

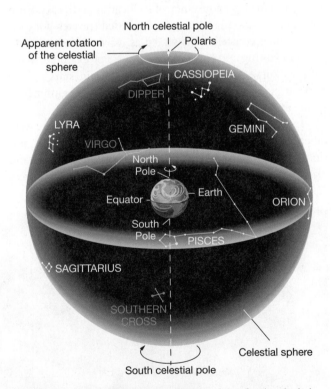

▲ **FIGURE 1.11 Celestial Sphere** Planet Earth sits fixed at the hub of the celestial sphere, which contains all the stars. This is one of the simplest possible models of the universe, but it doesn't agree with all the facts that astronomers now know about the universe.

*We now know that stars do in fact move relative to one another, but this proper motion *across the sky* is too slow to be discerned with the naked eye (see Section 17.1).

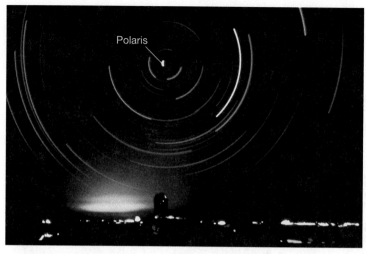

Interactive FIGURE 1.12 Northern Sky A time-lapse photograph of the northern sky. Each curved trail is the path of a single star across the night sky. The duration of the exposure is about 5 hours, and each star traces out approximately 20 percent of a circle. The concentric circles are centered near the North Star, Polaris, whose short, bright arc is labeled. *(AURA)*

and south celestial poles lies the **celestial equator,** representing the intersection of Earth's equatorial plane with the celestial sphere. These parts of the celestial sphere are marked on Figure 1.11.

When discussing the locations of stars "on the sky," astronomers naturally talk in terms of *angular* positions and separations. *More Precisely 1-1* presents some basic information on angular measure.

MORE PRECISELY 1-1

Angular Measure

Size and scale are often specified by measuring lengths and angles. The concept of length measurement is fairly intuitive to most of us. The concept of *angular measurement* may be less familiar, but it, too, can become second nature if you remember a few simple facts:

- A full circle contains 360 *degrees* (360°). Thus, the half-circle that stretches from horizon to horizon, passing directly overhead and spanning the portion of the sky visible to one person at any one time, contains 180°.

- Each 1° increment can be further subdivided into fractions of a degree, called *arc minutes.* There are 60 arc minutes (written 60′) in 1°. (The term "arc" is used to distinguish this angular unit from the unit of time.) Both the Sun and the Moon project an angular size of 30 arc minutes (half a degree) on the sky. Your little finger, held at arm's length, has a similar angular size, covering about a 40′ slice of the 180° horizon-to-horizon arc.

- An arc minute can be divided into 60 *arc seconds* (60″). Put another way, an arc minute is $\frac{1}{60}$ of a degree, and an arc second is $\frac{1}{60} \times \frac{1}{60} = \frac{1}{3600}$ of a degree. An arc second is an extremely small unit of angular measure—the angular size of a centimeter-sized object (a dime, say) at a distance of about 2 kilometers (a little over a mile).

The accompanying figure illustrates this subdivision of the circle into progressively smaller units.

Don't be confused by the units used to measure angles. Arc minutes and arc seconds have nothing to do with the measurement of time, and degrees have nothing to do with temperature. Degrees, arc minutes, and arc seconds are simply ways to measure the size and position of objects in the universe.

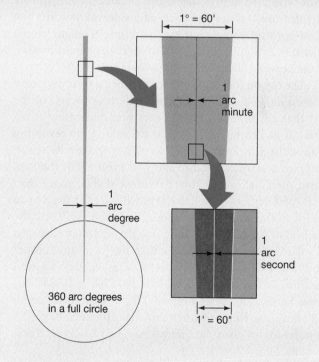

The angular size of an object depends both on its actual size and on its distance from us. For example, the Moon at its present distance from Earth has an angular diameter of 0.5°, or 30′. If the Moon were twice as far away, it would appear half as big—15′ across—even though its actual size would be the same. Thus, *angular size by itself is not enough to determine the actual diameter of an object—the distance to the object must also be known.* We return to this topic in more detail in *More Precisely 1-2.*

1.4 Earth's Orbital Motion

Day-to-Day Changes

We measure time by the Sun. Because the rhythm of day and night is central to our lives, it is not surprising that the period from one noon to the next, the 24-hour **solar day,** is our basic social time unit. The daily progress of the Sun and the other stars across the sky is known as *diurnal motion.* As we have just seen, it is a consequence of Earth's rotation. But the stars' positions in the sky do *not* repeat themselves exactly from one night to the next. Each night, the whole celestial sphere appears to be shifted a little relative to the horizon compared with the night before. The easiest way to confirm this difference is by noticing the stars that are visible just after sunset or just before dawn. You will find that they are in slightly different locations from those of the previous night. Because of this shift, a day measured by the stars—called a **sidereal day** after the Latin word *sidus,* meaning "star"—differs in length from a solar day. Evidently, there is more to the apparent motion of the heavens than simple rotation.

The reason for the difference between a solar day and a sidereal day is sketched in Figure 1.13. It is a result of the fact that Earth moves in two ways simultaneously: It rotates on its central axis while at the same time **revolving** around the Sun. Each time Earth rotates once on its axis, it also moves a small distance along its orbit about the Sun. Earth therefore has to rotate through slightly more than 360° (360 degrees—see *More Precisely 1-1*) for the Sun to return to the same apparent location in the sky. Thus, the interval of time between noon one day and noon the next (a solar day) is slightly greater than one true rotation period (one sidereal day). Our planet takes 365 days to orbit the Sun, so the additional angle is 360°/365 = 0.986°. Because Earth, rotating at a rate of 15° per hour, takes about 3.9 minutes to rotate through this angle, the solar day is 3.9 minutes longer than the sidereal day (i.e., 1 sidereal day is roughly 23^h56^m long).

Seasonal Changes

Figure 1.14(a) illustrates the major stars visible from most locations in the United States on clear summer evenings. The brightest stars—Vega, Deneb, and Altair—form a conspicuous triangle high above the constellations Sagittarius and Capricornus, which are low on the southern

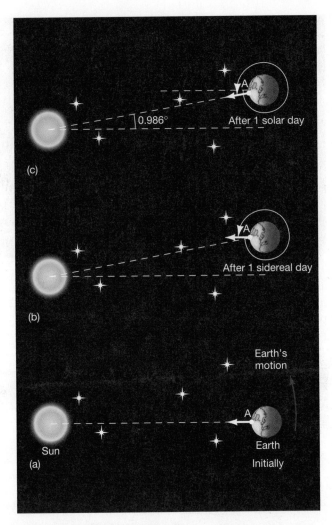

▲ **FIGURE 1.13 Solar and Sidereal Days** A sidereal day is Earth's true rotation period—the time taken for our planet to return to the same orientation in space relative to the distant stars. A solar day is the time from one noon to the next. The difference in length between the two is easily explained once we understand that Earth revolves around the Sun at the same time as it rotates on its axis. Frames (a) and (b) are one sidereal day apart. During that time, Earth rotates exactly once on its axis and also moves a little in its solar orbit—approximately 1°. Consequently, between noon at point A on one day and noon at the same point the next day, Earth actually rotates through about 361° (frame c), and the solar day exceeds the sidereal day by about 4 minutes. Note that the diagrams are not drawn to scale; the true 1° angle is in reality much smaller than shown here.

horizon. In the winter sky, however, these stars are replaced as shown in Figure 1.14(b) by several other, well-known constellations, including Orion, Leo, and Gemini. In the constellation Canis Major lies Sirius (the Dog Star), the brightest star in the sky. Year after year, the same stars and constellations return, each in its proper season. Every winter evening, Orion is high overhead; every summer, it is gone. (For more detailed maps of the

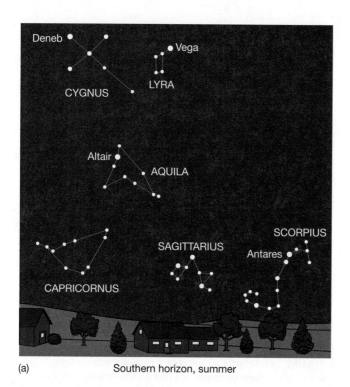

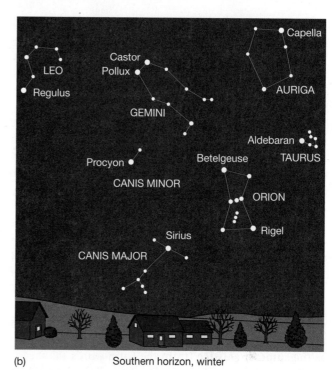

(a)   Southern horizon, summer

(b)   Southern horizon, winter

▲ **FIGURE 1.14 Typical Night Sky** (a) A typical summer sky above the United States. Some prominent stars (labeled in lowercase letters) and constellations (labeled in all capital letters) are shown. (b) A typical winter sky above the United States.

sky at different seasons, consult the star charts at the end of the book.)

These regular seasonal changes occur because of Earth's **revolution** around the Sun: Earth's darkened hemisphere faces in a slightly different direction in space each evening. The change in direction is only about 1° per night (Figure 1.13)—too small to be easily noticed with the naked eye from one evening to the next, but clearly noticeable over the course of weeks and months, as illustrated in Figure 1.15. After 6 months, Earth has reached the opposite side of

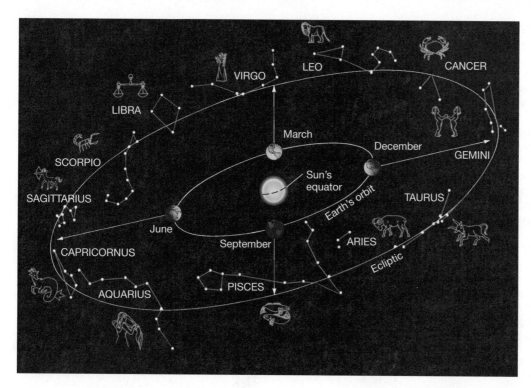

◀ **Interactive FIGURE 1.15 The Zodiac** The view of the night sky changes as Earth moves in its orbit about the Sun. As drawn here, the night side of Earth faces a different set of constellations at different times of the year. The 12 constellations named here make up the astrological zodiac. The arrows indicate the most prominent zodiacal constellations in the night sky at various times of the year. For example, in June, when the Sun is "in" Gemini, Sagittarius and Capricornus are visible at night.

its orbit, and we face an entirely different group of stars and constellations at night. Because of this motion, the Sun appears (to an observer on Earth) to move relative to the background stars over the course of a year. This apparent motion of the Sun on the sky traces out a path on the celestial sphere known as the **ecliptic.**

The 12 constellations through which the Sun passes as it moves along the ecliptic—that is, the constellations we would see looking in the direction of the Sun if they weren't overwhelmed by the Sun's light—had special significance for astrologers of old. These constellations are collectively known as the **zodiac.**

As illustrated in Figure 1.16, the ecliptic forms a great circle on the celestial sphere, inclined at an angle of 23.5° to the celestial equator. In reality, as illustrated in Figure 1.17, the plane of the ecliptic is *the plane of Earth's orbit around the Sun.* Its tilt is a consequence of the *inclination* of our planet's rotation axis to the plane of its orbit.

The point on the ecliptic where the Sun is at its northernmost point above the celestial equator is known as the **summer solstice** (from the Latin words *sol,* meaning "sun," and *stare,* "to stand"). As indicated in Figure 1.17, it represents the location in Earth's orbit where our planet's North Pole comes closest to pointing in the direction of the Sun. This occurs on or near June 21—the exact date varies slightly from year to year because the actual length of a year is not a whole number of days. As Earth rotates, points north of the equator spend the greatest fraction of their time in sunlight on that date, so the summer solstice corresponds to the longest day of the year in

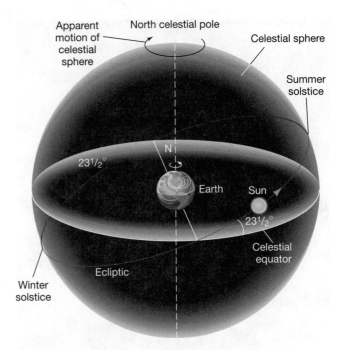

▲ **FIGURE 1.16 Ecliptic** The apparent path of the Sun on the celestial sphere over the course of a year is called the ecliptic. As indicated on the diagram, the ecliptic is inclined to the celestial equator at an angle of 23.5°. In this picture of the heavens, the seasons result from the changing height of the Sun above the celestial equator. At the summer solstice, the Sun is at its northernmost point on its path around the ecliptic; it is therefore highest in the sky, as seen from Earth's Northern Hemisphere, and the days are longest. The reverse is true at the winter solstice. At the vernal and autumnal equinoxes, when the Sun crosses the celestial equator, day and night are of equal length.

▶ **Interactive FIGURE 1.17 Seasons**
In reality, the Sun's apparent motion along the ecliptic is a consequence of Earth's orbital motion around the Sun. The seasons result from the inclination of our planet's rotation axis with respect to its orbit plane. The summer solstice corresponds to the point on Earth's orbit where our planet's North Pole points most nearly toward the Sun. The opposite is true of the winter solstice. The vernal and autumnal equinoxes correspond to the points in Earth's orbit where our planet's axis is perpendicular to the line joining Earth and the Sun. The insets show how rays of sunlight striking the ground at an angle (e.g., during northern winter) are spread over a larger area than rays coming nearly straight down (e.g., during northern summer). As a result, the amount of solar heat delivered to a given area of Earth's surface is greatest when the Sun is high in the sky.

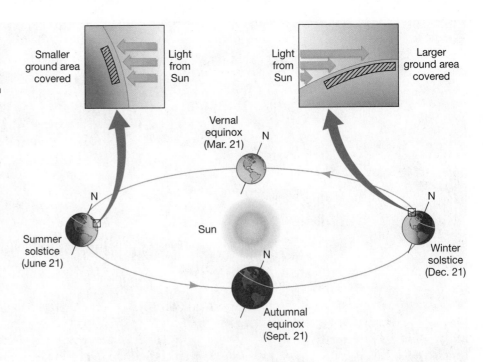

the Northern Hemisphere and the shortest day in the Southern Hemisphere.

Six months later, the Sun is at its southernmost point below the celestial equator (Figure 1.16)—or, equivalently, the North Pole points farthest from the Sun (Figure 1.17). We have reached the **winter solstice** (December 21), the shortest day in Earth's Northern Hemisphere and the longest in the Southern Hemisphere.

The tilt of Earth's rotation axis relative to the ecliptic is responsible for the **seasons** we experience—the marked difference in temperature between the hot summer and cold winter months. As illustrated in Figure 1.17, two factors combine to cause this variation. First, there are more hours of daylight during the summer than in winter. To see why this is, look at the yellow lines on the surfaces of the drawings of Earth in the figure. (For definiteness, they correspond to a latitude of 45°—roughly that of the Great Lakes or the south of France.) A much larger fraction of the line is sunlit in the summertime, and more daylight means more solar heating. Second, as illustrated in the insets in Figure 1.17, when the Sun is high in the sky in summer, rays of sunlight striking Earth's surface are more concentrated—spread out over a smaller area—than in winter. As a result, the Sun feels hotter. Therefore summer, when the Sun is highest above the horizon and the days are longest, is generally much warmer than winter, when the Sun is low and the days are short.

A popular misconception is that the seasons have something to do with Earth's distance from the Sun. Figure 1.18 illustrates why this is *not* the case. It shows Earth's orbit "face on," instead of almost edge-on, as in Figure 1.17. Notice that the orbit is almost perfectly circular, so the distance from Earth to the Sun varies very little (in fact, by only about 3 percent) over the course of a year—not nearly enough to explain the seasonal changes in temperature. What's more, Earth is actually *closest* to the Sun in early January, the dead of winter in the Northern Hemisphere, so distance from the Sun cannot be the main factor controlling our climate.

The two points where the ecliptic intersects the celestial equator (Figure 1.16)—that is, where Earth's rotation axis is perpendicular to the line joining Earth to the Sun (Figure 1.17)—are known as **equinoxes**. On those dates, day and night are of equal duration. (The word *equinox* derives from the Latin for "equal night.") In the fall (in the Northern Hemisphere), as the Sun crosses from the Northern into the Southern Hemisphere, we have the **autumnal equinox** (on September 21). The **vernal equinox** occurs in northern spring, on or near March 21, as the Sun crosses the celestial equator moving north. Because of its association with the end of winter and the start of a new growing season, the vernal equinox was particularly important to early astronomers and astrologers. It also plays an important role

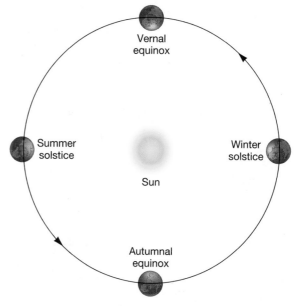

▲ **FIGURE 1.18 Earth's Orbit** Seen face on, Earth's orbit around the Sun is almost indistinguishable from a perfect circle. The distance from Earth to the Sun varies only slightly over the course of a year and is *not* the cause of the seasonal temperature variations we experience.

in human timekeeping: The interval of time from one vernal equinox to the next—365.2422 mean solar days—is known as one **tropical year.**

Long-Term Changes

Earth has many motions—it spins on its axis, it travels around the Sun, and it moves with the Sun through our Galaxy. We have just seen how some of these motions can account for the changing nighttime sky and the changing seasons. In fact, the situation is even more complicated. Like a spinning top that rotates rapidly on its own axis while that axis slowly revolves about the vertical, Earth's axis changes its *direction* over the course of time (although the angle between the axis and a line perpendicular to the plane of the ecliptic always remains close to 23.5°). Illustrated in Figure 1.19, this change is called **precession.** It is caused by torques (twisting forces) on Earth due to the gravitational pulls of the Moon and the Sun, which affect our planet in much the same way as the torque due to Earth's own gravity affects a top. During a complete cycle of precession—about 26,000 years—Earth's axis traces out a cone.

The time required for Earth to complete exactly one orbit around the Sun, relative to the stars, is called a **sidereal year.** One sidereal year is 365.256 mean solar days long—about 20 minutes longer than a tropical year. The

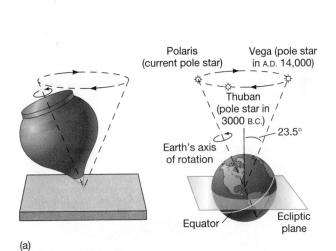

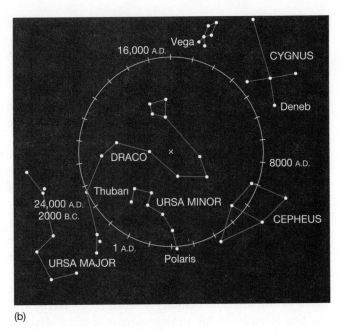

(a)

(b)

Interactive FIGURE 1.19 Precession (a) Earth's axis currently points nearly toward the star Polaris. About 12,000 years from now—almost halfway through one cycle of precession—Earth's axis will point toward a star called Vega, which will then be the "North Star." Five thousand years ago, the North Star was a star named Thuban in the constellation Draco. (b) The circle shows the precessional path of the north celestial pole among some prominent northern stars. Tick marks indicate intervals of a thousand years.

reason for this slight difference is Earth's precession. Recall that the vernal equinox occurs when Earth's rotation axis is perpendicular to the line joining Earth and the Sun, and the Sun is crossing the celestial equator moving from south to north. In the absence of precession, this combination of events would occur exactly once per sidereal orbit, and the tropical and sidereal years would be identical. However, because of the slow precessional shift in the orientation of Earth's rotation axis, the instant when the axis is next perpendicular to the line from Earth to the Sun occurs slightly *sooner* than we would otherwise expect. Consequently, the vernal equinox drifts slowly around the zodiac over the course of the precession cycle.

The tropical year is the year that our calendars measure. If our timekeeping were tied to the sidereal year, the seasons would slowly march around the calendar as Earth precessed—13,000 years from now, summer in the Northern Hemisphere would be at its height in late February! By using the tropical year, we ensure that July and August will always be (northern) summer months. However, in 13,000 years' time, Orion will be a summer constellation.

CONCEPT CHECK

✔ In astronomical terms, what are *summer* and *winter,* and why do we see different constellations during those seasons?

1.5 The Motion of the Moon

The Moon is our nearest neighbor in space. Apart from the Sun, it is the brightest object in the sky. Like the Sun, the Moon appears to move relative to the background stars. Unlike the Sun, however, the Moon really does revolve around Earth. It crosses the sky at a rate of about 12° per day, which means that it moves an angular distance equal to its own diameter—30 arc minutes—in about an hour.

Lunar Phases

The Moon's appearance undergoes a regular cycle of changes, or **phases,** taking roughly 29.5 days to complete. Figure 1.20 illustrates the appearance of the Moon at different times in this monthly cycle. Starting from the *new Moon,* which is all but invisible in the sky, the Moon appears to *wax* (or grow) a little each night and is visible as a growing *crescent* (photo 1 of Figure 1.20). One week after new Moon, half of the lunar disk can be seen (photo 2). This phase is known as a *quarter Moon.* During the next week, the Moon continues to wax, passing through the *gibbous* phase (photo 3) until, 2 weeks after new Moon, the *full Moon* (photo 4) is visible. During the next 2 weeks, the Moon *wanes* (or shrinks), passing in turn through the gibbous, quarter, crescent phases (photos 5–7) and eventually becoming new again.

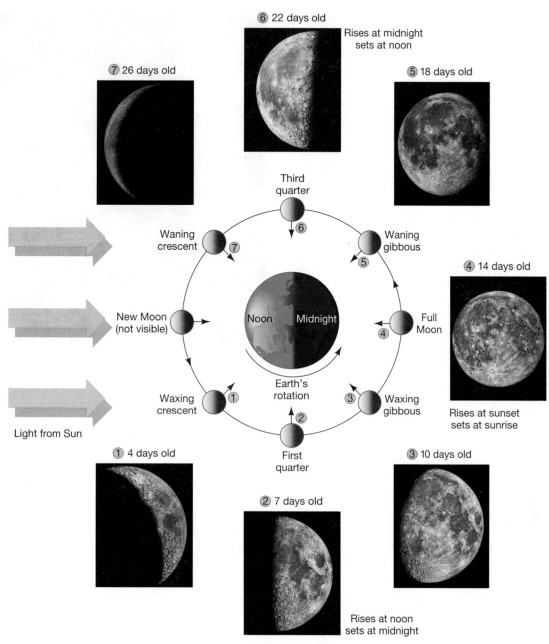

▲ **FIGURE 1.20 Lunar Phases** Because the Moon orbits Earth, the visible fraction of the lunar sunlit face varies from night to night, although the Moon always keeps the same face toward our planet. (Note the location of the small, straight arrows, which mark the same point on the lunar surface at each phase shown.) The complete cycle of lunar phases, shown here starting at the waxing crescent phase and following the Moon's orbit counterclockwise, takes 29.5 days to complete. Rising and setting times for some phases are also indicated. *(UC/Lick Observatory)*

The position of the Moon in the sky relative to the Sun, as seen from Earth, varies with lunar phase. For example, the full Moon rises in the east as the Sun sets in the west, while the first quarter Moon actually rises at noon, but sometimes becomes visible only late in the day as the Sun's light fades. By this time the Moon is already high in the sky. Some connections between the lunar phase and the rising and setting times of the Moon are indicated in Figure 1.20.

The Moon doesn't actually change its size and shape from night to night, of course. Its full circular disk is present at all times. Why, then, don't we always see a full Moon? The answer to this question lies in the fact that, unlike the Sun and the other stars, the Moon emits no light of its own. Instead, it shines by reflected sunlight. As illustrated in Figure 1.20, half of the Moon's surface is illuminated by the Sun at any instant. However, not all of the Moon's sunlit face can be seen because of the Moon's position with respect to Earth and the Sun. When the Moon is full, we see the entire "daylit" face because the Sun and the Moon are in opposite directions from Earth in the sky. In the case of a new Moon, the Moon and the Sun are in almost the same part of the sky, and the sunlit side of the Moon is oriented away from us. At new Moon, the Sun must be almost behind the Moon, from our perspective.

As the Moon revolves around Earth, our satellite's position in the sky changes with respect to the stars. In 1 **sidereal month** (27.3 days), the Moon completes one revolution and returns to its starting point on the celestial sphere, having traced out a great circle in the sky. The time required for the Moon to complete a full cycle of phases, one **synodic month,** is a little longer—about 29.5 days. The synodic month is a little longer than the sidereal month for the same reason that a solar day is slightly longer than a sidereal day: Because of Earth's motion around the Sun, the Moon must complete slightly more than one full revolution to return to the same phase in its orbit (Figure 1.21).

Eclipses

From time to time—but only at new or full Moon—the Sun and the Moon line up precisely as seen from Earth, and we observe the spectacular phenomenon known as an **eclipse.** When the Sun and the Moon are in exactly *opposite* directions, as seen from Earth, Earth's shadow sweeps across the Moon, temporarily blocking the Sun's light and darkening the Moon in a **lunar eclipse,** as illustrated in Figure 1.22. From Earth, we see the curved edge of Earth's shadow begin to cut across the face of the full Moon and slowly eat its way into the lunar disk. Usually, the alignment of the Sun, Earth, and Moon is imperfect, so the shadow

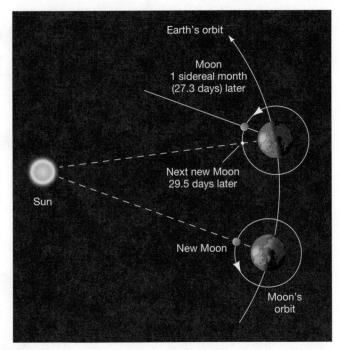

Interactive FIGURE 1.21 Sidereal Month The difference between a *synodic* and a *sidereal* month stems from the motion of Earth relative to the Sun. Because Earth orbits the Sun in 365 days, in the 29.5 days from one new Moon to the next (1 synodic month), Earth moves through an angle of approximately 29°. Thus, the Moon must revolve more than 360° between new Moons. The sidereal month, which is the time taken for the Moon to revolve through exactly 360°, relative to the stars, is about 2 days shorter.

never completely covers the Moon. Such an occurrence is known as a **partial lunar eclipse.** Occasionally, however, the entire lunar surface is obscured in a **total lunar eclipse,** such as that shown in the inset of Figure 1.22. Total lunar

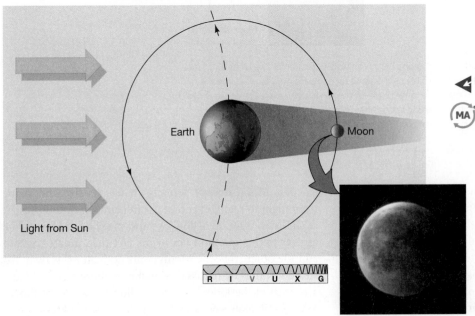

Interactive FIGURE 1.22 Lunar Eclipse A lunar eclipse occurs when the Moon passes through Earth's shadow. At these times, we see a darkened, copper-colored Moon, as shown by the partial eclipse in the inset photograph. The red coloration is caused by sunlight deflected by Earth's atmosphere onto the Moon's surface. An observer on the Moon would see Earth surrounded by a bright, but narrow, ring of orange sunlight. Note that this figure is not drawn to scale, and only Earth's umbra (see text and Figure 1.24) is shown. (*Inset: G. Schneider*)

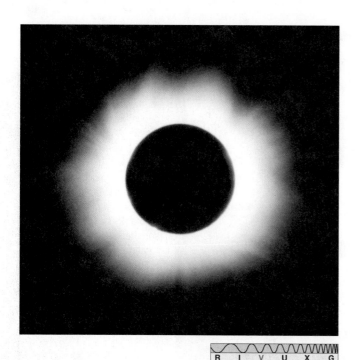

◄ **FIGURE 1.23 Total Solar Eclipse** During a total solar eclipse, the Sun's corona becomes visible as an irregularly shaped halo surrounding the blotted-out disk of the Sun. This was the August 1999 eclipse, as seen from the banks of the Danube River near Sofia, Bulgaria. (*M. Tsavkova/B. Angelov*)

eclipses last only as long as is needed for the Moon to pass through Earth's shadow—no more than about 100 minutes. During that time, the Moon often acquires an eerie, deep red coloration—the result of a small amount of sunlight reddened by Earth's atmosphere (for the same reason that sunsets appeared—see *More Precisely 7-1*) and refracted (bent) onto the lunar surface, preventing the shadow from being completely black.

When the Moon and the Sun are in exactly the *same* direction, as seen from Earth, an even more awe-inspiring event occurs. The Moon passes directly in front of the Sun, briefly turning day into night in a **solar eclipse.** In a *total solar eclipse,* when the alignment is perfect, planets and some stars become visible in the daytime as the Sun's light is reduced to nearly nothing. We can also see the Sun's ghostly outer atmosphere, or *corona* (Figure 1.23).* In a *partial solar eclipse,* the Moon's path is slightly "off center," and only a portion of the Sun's face is covered. In either case, the sight of the Sun apparently being swallowed up by the black disk of the Moon is disconcerting even today. It must surely have inspired fear in early observers. Small wonder that the ability to predict such events was a highly prized skill.

Unlike a lunar eclipse, which is simultaneously visible from all locations on Earth's night side, a total solar eclipse can be seen from only a small portion of Earth's daytime side. The Moon's shadow on Earth's surface is about 7000 kilometers wide—roughly twice the diameter

Actually, although a total solar eclipse is undeniably a spectacular occurrence, the visibility of the corona is probably the most important astronomical aspect of such an event today. It enables us to study this otherwise hard-to-see part of our Sun (see Chapter 16).

of the Moon. Outside of that shadow, no eclipse is seen. However, within the central region of the shadow, called the **umbra,** the eclipse is total. Within the shadow, but outside the umbra, in the **penumbra,** the eclipse is partial, with less and less of the Sun obscured the farther one travels from the shadow's center.

The connections among the umbra, the penumbra, and the relative locations of Earth, Sun, and Moon are illustrated in Figure 1.24. The umbra is always very small. Even under the most favorable circumstances, its diameter never exceeds 270 kilometers. Because the shadow sweeps across Earth's surface at over 1700 kilometers per hour, the duration of a total eclipse at any given point on our planet can never exceed 7.5 minutes.

The Moon's orbit around Earth is not exactly circular. Thus, the Moon may be far enough from Earth at the moment of an eclipse that its disk fails to cover the disk of the Sun completely, even though their centers coincide. In that case, there is no region of totality—the umbra never reaches Earth at all, and a thin ring of sunlight can still be seen surrounding the Moon. Such an occurrence, called an **annular eclipse,** is illustrated in Figure 1.24(c) and shown more clearly in Figure 1.25. Roughly half of all solar eclipses are annular.

Eclipse Seasons

Why isn't there a solar eclipse at every new Moon and a lunar eclipse at every full Moon? That is, why doesn't the Moon pass directly between Earth and the Sun once per orbit and directly through Earth's shadow 2 weeks later?

The answer is that the Moon's orbit is slightly inclined to the ecliptic (at an angle of 5.2°), so the chance that a new (or full) Moon will occur just as the Moon happens to cross the plane of the ecliptic (with Earth, Moon, and Sun perfectly aligned) is quite low. Figure 1.26 illustrates some possible configurations of the three bodies. If the Moon happens to lie above or below the plane of the ecliptic when new (or full), a solar (or lunar) eclipse cannot occur. Such a configuration is termed *unfavorable* for producing an eclipse. In a *favorable* configuration, the Moon is new or full just as it crosses the plane of the ecliptic, and eclipses are seen. Unfavorable configurations are much more common than favorable ones, so eclipses are relatively rare events.

As indicated on Figure 1.26(b), the two points on the Moon's orbit where it crosses the plane of the ecliptic are known as the *nodes* of the orbit. The line joining the nodes, which is also the line of intersection of Earth's and the Moon's orbital planes, is known as the *line of nodes*. When the line of

Penumbra: Sun partly visible, partial eclipse seen

Sun

Moon

Earth

Umbra: Sun completely obscured, total eclipse seen

(a)

Partial eclipse

Sun

To Sun

Moon

Corona

(Sun behind)

Moon

Total eclipse

(b)

Annular eclipse

To Sun

Moon

Moon

Sun

(c)

◀ Interactive FIGURE 1.24 Types of Solar Eclipse (a) The Moon's shadow consists of two parts: the umbra, where no sunlight is seen, and the penumbra, where a portion of the Sun is visible. (b) If we are in the umbra, we see a total eclipse; in the penumbra, we see a partial eclipse. (c) If the Moon is too far from Earth at the moment of the eclipse, the umbra does not reach Earth and there is no region of totality; instead, an annular eclipse is seen. (Note that these figures are not drawn to scale.) *(Insets: NOAA; G. Schneider)*

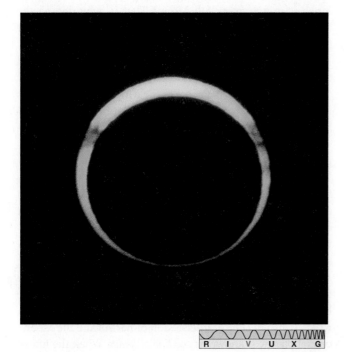

R I V U X G

nodes is not directed toward the Sun, conditions are unfavorable for eclipses. However, when the line of nodes briefly lies along the Earth–Sun line, eclipses are possible. These two periods, known as **eclipse seasons,** are the only times at which an eclipse can occur. Notice that there is no guarantee that an eclipse *will* occur. For a solar eclipse, we must have a new Moon during an eclipse season. Similarly, a lunar eclipse can occur only at full Moon during an eclipse season.

Because we know the orbits of Earth and the Moon to great accuracy, we can predict eclipses far into the future. Figure 1.27 shows the location and duration of all total eclipses of the Sun between 2010 and 2030. It is interesting to note that the eclipse tracks run from west to east—just the opposite of more familiar phenomena such as sunrise and sunset, which are seen earlier by observers located farther east. The reason is that the Moon's shadow sweeps

◀ **FIGURE 1.25 Annular Solar Eclipse** During an annular solar eclipse, the Moon fails to completely hide the Sun, so a thin ring of light remains. No corona is seen in this case because even the small amount of the Sun still visible completely overwhelms the corona's faint glow. This was the December 1973 eclipse, as seen from Algiers. (The gray fuzzy areas at the top left and right are clouds in Earth's atmosphere.) *(G. Schneider)*

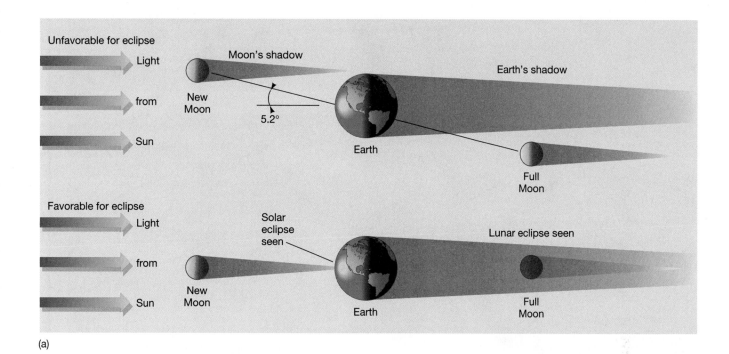

(a)

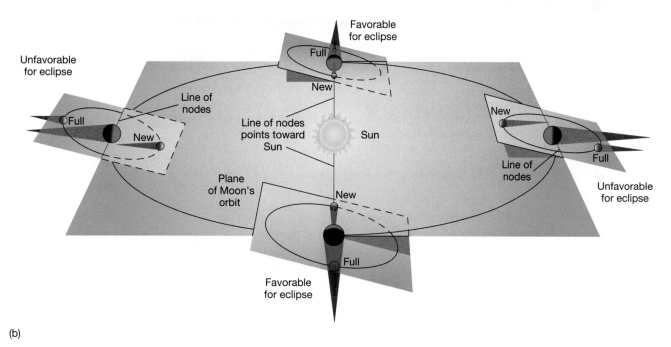

(b)

▲ **FIGURE 1.26 Eclipse Geometry** (a) An eclipse occurs when Earth, Moon, and Sun are precisely aligned. If the Moon's orbital plane lay in exactly the plane of the ecliptic, this alignment would occur once a month. However, the Moon's orbit is inclined at about 5° to the ecliptic, so not all configurations are favorable for producing an eclipse. (b) For an eclipse to occur, the line of intersection of the two planes must lie along the Earth–Sun line. Thus, eclipses can occur just at specific times of the year. Only the umbra of each shadow is shown, for clarity (see Figure 1.24).

across Earth's surface faster than our planet rotates, so the eclipse actually *overtakes* observers on the ground.

The solar eclipses that we do see highlight a remarkable cosmic coincidence. Although the Sun is many times farther away from Earth than is the Moon, it is also much larger. In fact, the ratio of distances is almost exactly the same as the ratio of sizes, so the Sun and the Moon both have roughly the *same* angular diameter—about half a degree, seen from Earth. Thus, the Moon covers the face of the Sun almost exactly. If the Moon were larger, we would never see annular

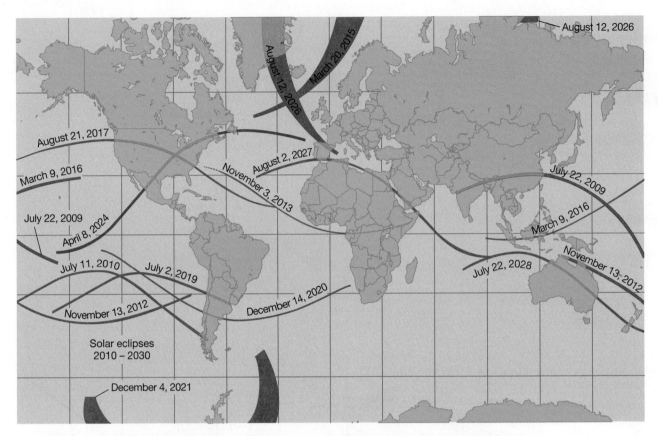

▲ FIGURE 1.27 **Eclipse Tracks** Regions of Earth that saw or will see total solar eclipses between the years 2010 and 2030. Each track represents the path of the Moon's umbra across Earth's surface during an eclipse. The width of the track depends upon the latitude on Earth and the distance from Earth to the Moon during the eclipse. High-latitude tracks are broader because sunlight strikes Earth's surface at an oblique angle near the poles (and also because of the projection of the map). The closer the Moon is to Earth during a total eclipse, the wider is the umbra (see Figure 1.24).

eclipses, and total eclipses would be much more common. If the Moon were a little smaller, we would see only annular eclipses.

The gravitational tug of the Sun causes the Moon's orbital orientation, and hence the direction of the line of nodes, to change slowly with time. As a result, the time between one orbital configuration with the line of nodes pointing at the Sun and the next (with the Moon crossing the ecliptic in the same sense in each case) is not exactly 1 year, but instead is 346.6 days—sometimes called one *eclipse year*. Thus, the eclipse seasons gradually progress backward through the calendar, occurring about 19 days earlier each year. For example, in 1999 the eclipse seasons were in February and August, and on August 11 much of Europe and southern Asia was treated to the last total eclipse of the millennium (Figure 1.23). By 2002, those seasons had drifted into December and June, and eclipses actually occurred on June 10 and December 4 of that year. By studying Figure 1.27, you can follow the progression of the eclipse seasons through the calendar. (Note that two partial eclipses in 2004 and two in 2007 are not shown in the figure.)

The combination of the eclipse year and the Moon's synodic period leads to an interesting long-term cycle in solar (and lunar) eclipses. A simple calculation shows that 19 eclipse years is almost exactly 223 lunar months. Thus, every 6585 solar days (actually 18 years, 11.3 days) the "same" eclipse recurs, with Earth, the Moon, and the Sun in the same relative configuration. Several such repetitions are evident in Figure 1.27—see, for example, the similarly shaped December 4, 2002, and December 14, 2020, tracks. (Note that we must take leap years properly into account to get the dates right!) The roughly 120° offset in longitude corresponds to Earth's rotation in 0.3 day. This recurrence is called the *Saros cycle*. Well known to ancient astronomers, it undoubtedly was the key to their "mystical" ability to predict eclipses!

CONCEPT CHECK

✔ What types of solar eclipses would you expect to see if Earth's distance from the Sun were to double? What if the distance became half its present value?

1.6 The Measurement of Distance

We have seen a little of how astronomers track and record the positions of the stars in the sky. But knowing the direction to an object is only part of the information needed to locate it in space. Before we can make a systematic study of the heavens, we must find a way of measuring *distances,* too. One distance-measurement method, called **triangulation,** is based on the principles of Euclidean geometry and finds widespread application today in both terrestrial and astronomical settings. Surveyors use these age-old geometric ideas to measure the distance to faraway objects indirectly. Triangulation forms the foundation of the family of distance-measurement techniques making up the **cosmic distance scale.**

Triangulation and Parallax

Imagine trying to measure the distance to a tree on the other side of a river. The most direct method is to lay a tape across the river, but that's not the simplest way (nor, because of the current, may it even be possible). A smart surveyor would make the measurement by visualizing an *imaginary* triangle (hence *triangulation*), sighting the tree on the far side of the river from two positions on the near side, as illustrated in Figure 1.28. The simplest possible triangle is a right triangle, in which one of the angles is exactly 90°, so it is usually convenient to set up one observation position directly opposite the object, as at point A. The surveyor then moves to another observation position at point B, noting the distance covered between points A and B. This distance is called the **baseline** of the imaginary triangle. Finally, the surveyor, standing at point B, sights toward the tree and notes the angle at point B between this line of sight and the baseline. Knowing the value of one side (AB) and two angles (the right angle at point A and the angle at point B) of the right triangle, the surveyor geometrically constructs the remaining sides and angles and establishes the distance from A to the tree.

To use triangulation to measure distances, a surveyor must be familiar with *trigonometry,* the mathematics of geometrical angles and distances. However, even if we knew no trigonometry at all, we could still solve the problem by graphical means, as shown in Figure 1.29. Suppose that we pace off the baseline AB, measuring it to be 450 meters, and measure the angle between the baseline and the line from B to the tree to be 52°, as illustrated in the figure. We can transfer the problem to paper by letting one box on our graph represent 25 meters on the ground. Drawing the line AB on paper and completing the other two sides of the triangle, at angles of 90° (at A) and 52° (at B), we measure the distance on paper from A to the tree to be 23 boxes—that is, 575 meters. We have solved the real problem by *modeling* it on paper. The point to remember here is this: Nothing more complex than basic geometry is needed to infer the distance,

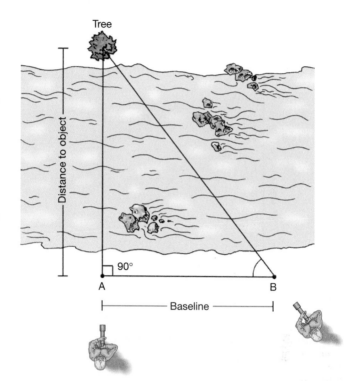

▲ **FIGURE 1.28 Triangulation** Surveyors often use simple geometry and trigonometry to estimate the distance to a faraway object by triangulation. By measuring the angles at A and B and the length of the baseline, the distance can be calculated without the need for direct measurement.

the size, and even the shape of an object that is too far away or inaccessible for direct measurement.

Obviously, for a fixed baseline the triangle becomes longer and narrower as the tree's distance from A increases. Narrow triangles cause problems, because it becomes hard to measure the angles at A and B with sufficient accuracy. The measurements can be made easier by "fattening" the triangle—that is, by lengthening the baseline—but there are limits on how long a baseline we can choose in astronomy. For example, consider an imaginary triangle extending from Earth to a nearby object in space, perhaps a neighboring planet. The triangle is now extremely long and narrow, even for a relatively nearby object (by cosmic standards). Figure 1.30(a) illustrates a case in which the longest baseline possible on Earth—Earth's diameter, measured from point A to point B—is used.

In principle, two observers could sight the planet from opposite sides of Earth, measuring the triangle's angles at A and B. However, in practice it is easier to measure the third angle of the imaginary triangle. Here's how. The observers sight toward the planet, taking note of its position *relative to some distant stars* seen on the plane of the sky. The observer at point A sees the planet at apparent location A′ relative to those stars, as indicated in Figure 1.30(a). The observer at B sees the planet at point B′. If each observer takes a photograph of the

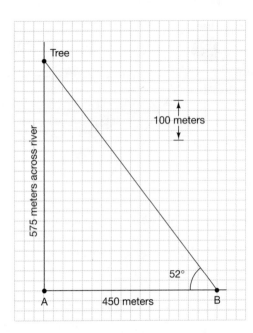

▲ **FIGURE 1.29 Geometric Scaling** Not even trigonometry is needed to estimate distances indirectly. Scaled estimates, like this one on a piece of graph paper, often suffice.

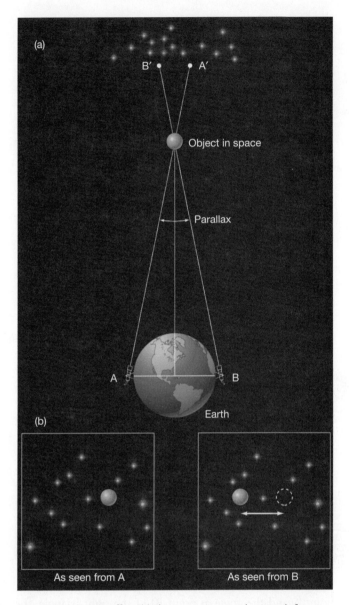

▲ **FIGURE 1.30 Parallax** (a) This imaginary triangle extends from Earth to a nearby object in space (such as a planet). The group of stars at the top represents a background field of very distant stars. (b) Hypothetical photographs of the same star field showing the nearby object's apparent displacement, or shift, relative to the distant undisplaced stars. Elementary geometry is one of the basic distance-measurement techniques that enable astronomers to study the immense universe in which we live.

appropriate region of the sky, the planet will appear at slightly different places in the two images. The planet's photographic image is slightly displaced, or shifted, relative to the field of distant background stars, as shown in Figure 1.30(b). The background stars themselves appear undisplaced because of their much greater distance from the observer.

This apparent displacement of a foreground object relative to the background as the observer's location changes is known as **parallax.** The size of the shift in Figure 1.30(b), measured as an angle on the celestial sphere, is the third, small angle in Figure 1.30(a). In astronomical contexts, the parallax is usually very small. For example, the parallax of a point on the Moon, viewed using a baseline equal to Earth's diameter, is about 2°; the parallax of the planet Venus at closest approach (45 million km), is just 1′ (see *More Precisely 1-2*).

The closer an object is to the observer, the larger is the parallax. Figure 1.31 illustrates how you can see this for yourself. Hold a pencil vertically in front of your nose and concentrate on some far-off object—a distant wall, perhaps. Close one eye, and then open it while closing the other. You should see a large shift in the apparent position of the pencil projected onto the distant wall—a large parallax. In this example, one eye corresponds to point A, the other eye to point B, the distance between your eyeballs to the baseline, the pencil to the planet, and the distant wall to a remote field of stars. Now hold the pencil at arm's length, corresponding to a more distant object (but still not as far away as the even more distant stars). The apparent shift of the pencil will be less. You might even be able to verify that the apparent shift is in-

versely proportional to the distance to the pencil. By moving the pencil farther away, we are narrowing the triangle and decreasing the parallax (and also making accurate measurement more difficult). If you were to paste the pencil to the wall, corresponding to the case where the object of interest is as far away as the background star field, blinking would produce no apparent shift of the pencil at all.

The amount of parallax is thus inversely proportional to an object's distance. Small parallax implies large distance, and

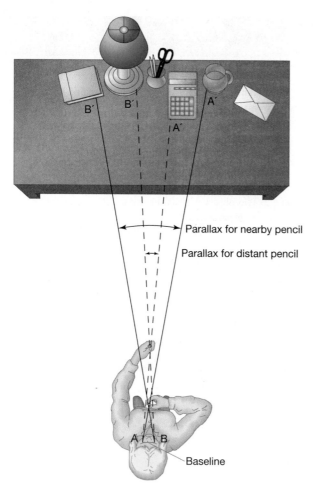

▲ **FIGURE 1.31 Parallax Geometry** Parallax is inversely proportional to an object's distance. An object near your nose has a much larger parallax than an object held at arm's length.

large parallax implies small distance. Knowing the amount of parallax (as an angle) and the length of the baseline, we can easily derive the distance through triangulation. *More Precisely 1-2* explores the connection between angular measure and distance in more detail, showing how we can use elementary geometry to determine both the distances and the dimensions of far away objects.

Surveyors of the land routinely use such simple geometric techniques to map out planet Earth. As surveyors of the sky, astronomers use the same basic principles to chart the universe.

▶ **FIGURE 1.32 Measuring Earth's Radius** The Sun's rays strike different parts of Earth's surface at different angles. The Greek philosopher Eratosthenes realized that the difference was due to Earth's curvature, enabling him to determine Earth's radius by using simple geometry.

Sizing Up Planet Earth

Now that we have studied some of the tools available to astronomers, let's end the chapter with a classic example of how the scientific method, combined with the basic geometric techniques just described, enabled an early scientist to perform a calculation of truly "global" proportions.

In about 200 B.C., a Greek philosopher named Eratosthenes (276–194 B.C.) used simple geometric reasoning to calculate the size of our planet. He knew that at noon on the first day of summer observers in the city of Syene (now called Aswan) in Egypt saw the Sun pass directly overhead. This was evident from the fact that vertical objects cast no shadows and sunlight reached to the very bottoms of deep wells, as shown in the insets in Figure 1.32. However, at noon of the same day in Alexandria, a city 5000 *stadia* to the north, the Sun was seen to be displaced slightly from the vertical. (The *stadium* was a Greek unit of length, roughly equal to 0.16 km—the modern town of Aswan lies about 780 km, or 490 miles, south of Alexandria.) By measuring the length of the shadow of a vertical stick and applying elementary trigonometry, Eratosthenes determined the angular displacement of the Sun from the vertical at Alexandria to be 7.2°.

What could have caused this discrepancy between the two measurements? It was not the result of measurement error—the same results were obtained every time the observations were repeated. Instead, as illustrated in Figure 1.32, the explanation is simply that Earth's surface is not flat, but *curved*. Our planet is a sphere. Eratosthenes was not the first person to

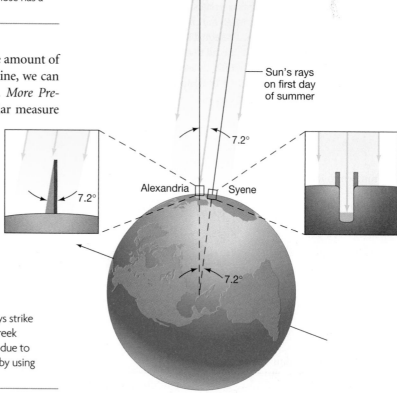

MORE PRECISELY 1-2

Measuring Distances with Geometry

Simple geometrical reasoning forms the basis for almost every statement made in this book about size and scale in the universe. In a very real sense, our modern knowledge of the cosmos depends on the elementary mathematics of ancient Greece. Let's take a moment to look in a little more detail at how astronomers use geometry to measure the distances to, and sizes of, objects near and far.

We can convert baselines and parallaxes into distances, and vice versa, by using arguments made by the Greek geometer Euclid. The first figure represents Figure 1.30(a), but we have changed the scale and added the circle centered on the target planet and passing through our baseline on Earth:

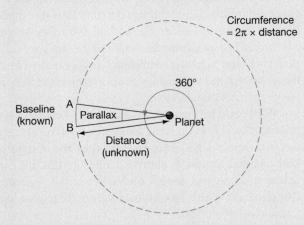

To see how the planet's parallax relates to its distance, we note that the ratio of the baseline AB to the circumference of the large circle shown in the figure must be equal to the ratio of the parallax to one full revolution, 360°. Recall that the circumference of a circle is always 2π times its radius (where π—the Greek letter "pi"—is approximately equal to 3.142). Applying this relation to the large circle in the figure, we find that

$$\frac{\text{baseline}}{2\pi \times \text{distance}} = \frac{\text{parallax}}{360°},$$

from which it follows that

$$\text{parallax} = (360°/2\pi) \times \frac{\text{baseline}}{\text{distance}}.$$

The angle $360°/2\pi \approx 57.3°$ in the preceding equation is usually called 1 *radian*.

EXAMPLE 1 The planet Venus lies roughly 45,000,000 km from Earth at closest approach. Two observers 13,000 km apart (i.e., at opposite ends of Earth's diameter) looking at the planet would measure a parallax of $57.3° \times (13{,}000 \text{ km}/45{,}000{,}000 \text{ km}) = 0.017° = 1.0$ arc minutes, as stated in the text.

Alternatively, if we know the parallax (from direct measurement, such as the photographic technique described in Section 1.6), we can rearrange the above equation to tell us the distance to the planet:

$$\text{distance} = \text{baseline} \times \frac{57.3°}{\text{parallax}}.$$

EXAMPLE 2 Two observers 1000 km apart looking at the Moon might measure a parallax of 9.0 arc minutes—that is, 0.15°. It then follows that the distance to the Moon is 1000 km $\times$ (57.3/0.15) $\approx$ 380,000 km. (More accurate measurements, based on laser ranging using equipment left on the lunar surface by *Apollo* astronauts, yield a mean distance of 384,000 km.)

Knowing the distance to an object, we can determine many other properties. For example, by measuring the object's *angular diameter*—the angle from one side of the object to the other as we view it in the sky—we can compute its size. The second figure illustrates the geometry involved:

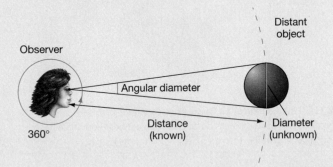

Notice that this is basically the same diagram as the previous one, except that now the angle (the angular diameter) and distance are known, instead of the angle (the parallax) and baseline. Exactly the same reasoning as before then allows us to calculate the diameter. We have

$$\frac{\text{diameter}}{2\pi \times \text{distance}} = \frac{\text{angular diameter}}{360},$$

so

$$\text{diameter} = \text{distance} \times \frac{\text{angular diameter}}{57.3°}.$$

EXAMPLE 3 The Moon's angular diameter is measured to be about 31 arc minutes—a little over half a degree. From the preceding discussion, it follows that the Moon's actual diameter is 380,000 km $\times$ (0.52°/57.3°) $\approx$ 3450 km. A more precise measurement gives 3476 km.

Study the foregoing reasoning carefully. We will use these simple arguments, in various forms, many times throughout this text.

realize that Earth is spherical—the philosopher Aristotle had done that over 100 years earlier (see Section 1.2), but he was apparently the first to build on this knowledge, combining geometry with direct measurement to infer the size of our planet. Here's how he did it.

Rays of light reaching Earth from a very distant object, such as the Sun, travel almost parallel to one another. Consequently, as shown in the figure, the angle measured at Alexandria between the Sun's rays and the vertical (i.e., the line joining Alexandria to the center of Earth) is equal to the angle between Syene and Alexandria, as seen from Earth's center. (For the sake of clarity, the angle has been exaggerated in the figure.) As discussed in *More Precisely 1-2*, the size of this angle in turn is proportional to the fraction of Earth's circumference that lies between Syene and Alexandria:

$$\frac{7.2° \text{ (angle between Syene and Alexandria)}}{360° \text{ (circumference of a circle)}}$$

$$= \frac{5000 \text{ stadia}}{\text{Earth's circumference}}.$$

Earth's circumference is therefore 50 × 5000, or 250,000 stadia, or about 40,000 km, so Earth's radius is 250,000/2π stadia, or 6366 km. The correct values for Earth's circumference and radius, now measured accurately by orbiting spacecraft, are 40,070 km and 6378 km, respectively.

Eratosthenes' reasoning was a remarkable accomplishment. More than 20 centuries ago, he estimated the circumference of Earth to within 1 percent accuracy, using only simple geometry and basic scientific reasoning. A person making measurements on only a small portion of Earth's surface was able to compute the size of the entire planet on the basis of observation and pure logic—an early triumph of the scientific method.

CONCEPT CHECK

✔ Why is elementary geometry essential for measuring distances in astronomy?

CHAPTER REVIEW

SUMMARY

1 The **scientific method (p. 7)** is a methodical approach employed by scientists to explore the universe around us in an objective manner. A **theory (p. 6)** is a framework of ideas and assumptions used to explain some set of observations and construct **theoretical models (p. 6)** that make predictions about the real world. These predictions in turn are amenable to further observational testing. In this way, the theory expands and science advances.

2 Early observers grouped the thousands of stars visible to the naked eye into patterns called **constellations (p. 8)**, which they imagined were attached to a vast **celestial sphere (p. 10)** centered on Earth. Constellations have no physical significance, but are still used to label regions of the sky. The points where Earth's axis of rotation intersects the celestial sphere are called the north and south **celestial poles (p. 10)**. The line where Earth's equatorial plane cuts the celestial sphere is the **celestial equator (p. 11)**.

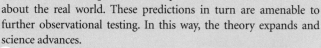

3 The nightly motion of the stars across the sky is the result of Earth's **rotation (p. 10)** on its axis. The time from one noon to the next is called a **solar day (p. 12)**. The time between successive risings of any given star is one **sidereal day (p. 12)**. Because of Earth's

revolution (p. 13) around the Sun, we see different stars at night at different times of the year, and the Sun appears to move relative to the stars. The Sun's apparent yearly path around the celestial sphere (or the plane of Earth's orbit around the Sun) is called the **ecliptic (p. 14)**. We experience **seasons (p. 15)** because Earth's rotation axis is inclined to the ecliptic plane. At the **summer solstice (p. 14)**, the Sun is highest in the sky and the length of the day is greatest. At the **winter solstice (p. 15)**, the Sun is lowest and the day is shortest. At the **vernal (p. 15)** and **autumnal equinoxes (p. 15)**, Earth's axis of rotation is perpendicular to the line joining Earth to the Sun, so day and night are of equal length. The interval of time from one vernal equinox to the next is one **tropical year (p. 15)**.

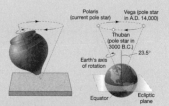

4 In addition to its rotation about its axis and its revolution around the Sun, Earth has many other motions. One of the most important of these is **precession (p. 15)**, the slow "wobble" of Earth's axis due to the influence of the Moon. As a result, the **sidereal year (p. 15)** is slightly longer than the tropical year, and the particular constellations that happen to be visible during any given season change over the course of thousands of years.

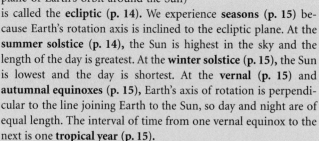

5 The Moon emits no light of its own, but instead shines by reflected sunlight. As the Moon orbits Earth, we see **lunar phases (p. 16)** as the amount 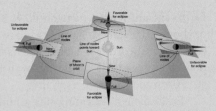 of the Moon's sunlit face visible to us varies. A **lunar eclipse (p. 18)** occurs when the Moon enters Earth's shadow. A **solar eclipse (p. 19)** occurs when the Moon passes between Earth and the Sun. An eclipse may be **total (p. 18)** if the body in question (Moon or Sun) is completely obscured, or **partial (p. 18)** if only a portion of the surface is affected. If the Moon happens to be too far from Earth for its disk to completely hide the Sun, an **annular eclipse (p. 19)** occurs. Because the Moon's orbit around Earth is slightly inclined with respect to the ecliptic, solar and lunar eclipses do not occur every month, but only during **eclipse seasons (p. 20)** (twice per year).

6 Surveyors on Earth use **triangulation (p. 23)** to determine the distances to far-away objects. Astronomers use the same technique to measure the distances to planets and stars. The **cosmic distance scale (p. 23)** is the family of distance-measurement techniques by which astronomers chart the universe. **Parallax (p. 24)** is the 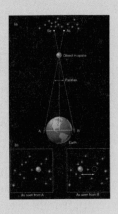 apparent motion of a foreground object relative to a distant background as the observer's position changes. The larger the **baseline (p. 23)**—the distance between the two observation points—the greater is the parallax. The same basic geometric reasoning is used to determine the sizes of objects whose distances are known. The Greek philosopher Eratosthenes used elementary geometry to determine Earth's radius.

MasteringASTRONOMY   *For instructor-assigned homework go to* **www.masteringastronomy.com**

Problems labeled **POS** explore the process of science | **VIS** problems focus on reading and interpreting visual information

REVIEW AND DISCUSSION

1. Compare the size of Earth with that of the Sun, the Milky Way Galaxy, and the entire universe.

2. What does an astronomer mean by "the universe"?

3. How big is a light-year?

4. **POS**   What is the scientific method, and how does science differ from religion?

5. What is a constellation? Why are constellations useful for mapping the sky?

6. Why does the Sun rise in the east and set in the west each day? Does the Moon also rise in the east and set in the west? Why? Do stars do the same? Why?

7. How and why does a day measured with respect to the Sun differ from a day measured with respect to the stars?

8. How many times in your life have you orbited the Sun?

9. Why do we see different stars at different times of the year?

10. Why are there seasons on Earth?

11. What is precession, and what causes it?

12. Why don't the seasons move around the calendar as Earth precesses?

13. If one complete hemisphere of the Moon is always lit by the sun, why do we see different phases of the Moon?

14. What causes a lunar eclipse? A solar eclipse?

15. Why aren't there lunar and solar eclipses every month?

16. **POS**   Do you think an observer on another planet might see eclipses? Why or why not?

17. What is parallax? Give an everyday example.

18. Why is it necessary to have a long baseline when using triangulation to measure the distances to objects in space?

19. What two pieces of information are needed to determine the diameter of a faraway object?

20. **POS**   If you traveled to the outermost planet in our solar system, do you think the constellations would appear to change their shapes? What would happen if you traveled to the next-nearest star? If you traveled to the center of our Galaxy, could you still see the familiar constellations found in Earth's night sky?

CONCEPTUAL SELF-TEST: MULTIPLE CHOICE

1. If Earth rotated twice as fast as it currently does, but its motion around the Sun stayed the same, then **(a)** the night would be twice as long; **(b)** the night would be half as long; **(c)** the year would be half as long; **(d)** the length of the day would be unchanged.

2. A long, thin cloud that stretched from directly overhead to the western horizon would have an angular size of **(a)** 45°; **(b)** 90°; **(c)** 180°; **(d)** 360°.

3. **VIS**   According to Figure 1.15 ("The Zodiac"), in January the Sun is in the constellation **(a)** Cancer; **(b)** Gemini; **(c)** Leo; **(d)** Aquarius.

4. If Earth orbited the Sun in 9 months instead of 12, then, compared with a sidereal day, a solar day would be **(a)** longer; **(b)** shorter; **(c)** unchanged.

5. When a thin crescent of the Moon is visible just before sunrise, the Moon is in its (a) waxing phase; (b) new phase; (c) waning phase; (d) quarter phase.

6. If the Moon's orbit were a little larger, solar eclipses would be (a) more likely to be annular; (b) more likely to be total; (c) more frequent; (d) unchanged in appearance.

7. If the Moon orbited Earth twice as fast but in the same orbit the frequency of solar eclipses would (a) double; (b) be cut in half; (c) stay the same.

8. **VIS** In Figure 1.28 ("Triangulation"), using a longer baseline would result in (a) a less accurate distance to the tree; (b) a more accurate distance to the tree; (c) a smaller angle at point B; (d) a greater distance across the river.

9. **VIS** In Figure 1.30 ("Parallax"), a smaller Earth would result in (a) a smaller parallax angle; (b) a shorter distance measured to the object; (c) a larger apparent displacement; (d) stars appearing closer together.

10. Today, distances to stars are measured by (a) bouncing radar signals; (b) reflected laser beams; (c) travel time by spacecraft; (d) geometry.

PROBLEMS

The number of dots preceding each Problem indicates its approximate level of difficulty.

1. • In 1 second, light leaving Los Angeles reaches approximately as far as (a) San Francisco, about 500 km; (b) London, roughly 10,000 km; (c) the Moon, 384,000 km; (d) Venus, 45,000,000 km from Earth at closest approach; or (e) the nearest star, about 4 light-years from Earth. Which is correct?

2. • (a) Write the following numbers in scientific notation (see Appendix 1 if you are unfamiliar with this notation): 1000; 0.000001; 1001; 1,000,000,000,000,000; 123,000; 0.000456. (b) Write the following numbers in "normal" numerical form: 3.16×10^7; 2.998×10^5; 6.67×10^{-11}; 2×10^0. (c) Calculate: $(2 \times 10^3) + 10^{-2}$; (1.99×10^{30}) (5.98×10^{24}); $(3.16 \times 10^7) \times (2.998 \times 10^5)$.

3. •• How, and by roughly how much, would the length of the solar day change if Earth's rotation were suddenly to reverse direction?

4. • The vernal equinox is now just entering the constellation Aquarius. In what constellation will it lie in the year A.D. 10,000?

5. • Relative to the stars, through how many degrees, arc minutes, or arc seconds does the Moon move in (a) 1 hour of time; (b) 1 minute; (c) 1 second? How long does it take for the Moon to move a distance equal to its own diameter?

6. • A surveyor wishes to measure the distance between two points on either side of a river, as illustrated in Figure 1.28. She measures the distance AB to be 250 m and the angle at B to be 30°. What is the distance between the two points?

7. • At what distance is an object if its parallax, as measured from either end of a 1000-km baseline, is (a) 1°; (b) 1′; (c) 1″?

8. • Given that the angular size of Venus is 55″ when the planet is 45,000,000 km from Earth, calculate Venus's diameter (in kilometers).

9. • The Moon lies roughly 384,000 km from Earth and the Sun lies 150,000,000 km away. If both have the same angular size as seen from Earth, how many times larger than the Moon is the Sun?

10. • Estimate the angular diameter of your thumb, held at arm's length.

THE COPERNICAN REVOLUTION

THE BIRTH OF MODERN SCIENCE

LEARNING GOALS

Studying this chapter will enable you to

1. Describe how some ancient civilizations attempted to explain the heavens in terms of Earth-centered models of the universe.

2. Explain how the observed motions of the planets led to our modern view of a Sun-centered solar system.

3. Describe the major contributions of Galileo and Kepler to our understanding of the solar system.

4. State Kepler's laws of planetary motion.

5. Explain how astronomers have measured the true size of the solar system.

6. State Newton's laws of motion and universal gravitation and explain how they account for Kepler's laws.

7. Explain how the law of gravitation enables us to measure the masses of astronomical bodies.

THE BIG PICTURE Astronomers today unquestionably know a great deal more about the universe than did their predecessors. We have better equipment to aid our eyes and other senses, and better theories to help us interpret what we see, enabling us to understand the cosmos in ways the ancients could have only imagined.

Living in the Space Age, we have become accustomed to the modern view of our place in the universe. Images of our planet taken from space leave little doubt that Earth is round, and no one seriously questions the idea that we orbit the Sun. Yet there was a time, not so long ago, when some of our ancestors maintained that Earth was flat and lay at the center of all things.

Our view of the universe—and of ourselves—has undergone a radical transformation since those early days. Earth has become a planet like many others, and humankind has been torn from its throne at the center of the cosmos and relegated to a rather unremarkable position on the periphery of the Milky Way Galaxy. But we have been amply compensated for our loss of prominence: We have gained a wealth of scientific knowledge in the process. The story of how all this came about is the story of the rise of the scientific method and the genesis of modern astronomy.

LEFT: *Astronomy came into its own as a viable and important academic subject during the 20th century. This collage, clockwise from upper left, shows four outstanding astronomers who helped lead the way: Harlow Shapley (1885–1972) discovered our place in the "suburbs" of the Milky Way. Annie Cannon (1863–1941) classified nearly a million stars over the course of a 50-year career. Karl Jansky (1905–1950) first detected radiation at radio wavelengths coming from the Milky Way. And Edwin Hubble (1889–1953) principally discovered the expansion of the universe. (Harvard Observatory; NRAO; Caltech)*

2.1   Ancient Astronomy

Many ancient cultures took a keen interest in the changing nighttime sky. The records and artifacts that have survived until the present make that abundantly clear. But unlike today, the major driving force behind the development of astronomy in those early societies was probably neither scientific nor religious. Instead, it was decidedly practical and down to earth. Seafarers needed to navigate their vessels, and farmers had to know when to plant their crops. In a real sense, then, human survival depended on knowledge of the heavens. The ability to predict accurately the arrival of the seasons, as well as other astronomical events, was undoubtedly a highly prized, perhaps jealously guarded, skill.

In Chapter 1, we saw that the human mind's ability to perceive patterns in the stars led to the "invention" of constellations as a convenient means of labeling regions of the celestial sphere. ∞ (Sec. 1.3) The realization that these patterns returned to the night sky at the same time each year met the need for a practical means of tracking the seasons. Widely separated cultures all over the world built elaborate structures to serve, at least in part, as primitive calendars. Often the keepers of the secrets of the sky enshrined their knowledge in myth and ritual, and these astronomical sites were also used for religious ceremonies.

Perhaps the best-known such site is *Stonehenge*, located on Salisbury Plain in England, and shown in Figure 2.1. This ancient stone circle, which today is one of the most popular tourist attractions in Britain, dates from the Stone Age.

Researchers think it was an early astronomical observatory of sorts—not in the modern sense of the term (a place for making new observations and discoveries pertaining to the heavens), but rather a kind of three-dimensional calendar or almanac, enabling its builders and their descendants to identify important dates by means of specific celestial events. Its construction apparently spanned a period of about 17 centuries, beginning around 2800 B.C. Additions and modifications continued to about 1100 B.C., indicating its ongoing importance to the Stone Age and, later, Bronze Age people who built, maintained, and used Stonehenge. The largest stones shown in Figure 2.1 weigh up to 50 tons and were transported from quarries many miles away.

Many of the stones are aligned so that they point toward important astronomical events. For example, the line joining the center of the inner circle to the so-called heel stone, set off some distance from the rest of the structure, points in the direction of the rising Sun on the summer solstice. Other alignments are related to the rising and setting of the Sun and the Moon at other times of the year. The accurate alignments (within a degree or so) of the stones of Stonehenge were first noted in the 18th century, but it was only relatively recently—in the second half of the 20th century, in fact—that the scientific community began to credit Stone Age technology with the ability to carry out such a precise feat of engineering. Although some of Stonehenge's purposes remain uncertain and controversial, the site's function as an astronomical almanac seems well established. Although Stonehenge is the most impressive and the best preserved,

◀ FIGURE 2.1 **Stonehenge** This remarkable site in the south of England was probably constructed as a primitive calendar or almanac. The inset shows sunrise at Stonehenge at the summer solstice. As seen from the center of the stone circle, the Sun rose directly over the "heel stone" on the longest day of the year. (*English Heritage*)

other stone circles, found all over Europe, are thought to have performed similar functions.

Many North American cultures were interested in the heavens. The Big Horn Medicine Wheel in Wyoming (Figure 2.2a) is similar to Stonehenge in design—and, perhaps, intent—although it is somewhat simpler in execution. Some researchers have identified alignments between the Medicine Wheel's spokes and the rising and setting Sun at solstices and equinoxes, and with some bright stars, suggesting that its builders—the Plains Indians—had much more than a passing familiarity with the changing nighttime sky. Other experts disagree, however, arguing that the alignments are quite inaccurate and consistent with pure chance and that the Medicine Wheel's purpose was more likely symbolic, rather than practical. A similar controversy swirls around the Caracol temple (Figure 2.2b) in the famous Mayan city of Chitzen Itza, built around A.D. 1000 on Mexico's Yucatán peninsula. Was it an observatory, as some suggest, perhaps tied to human sacrifices when Venus appeared in the morning or evening sky? Or are the claimed alignments of its windows

just wishful thinking and the temple's purpose simply religious, rather than astronomical?

Experts do seem to agree—for now, at least—that the Sun Dagger (Figure 2.2c), in Chaco Canyon, New Mexico, is a genuine astronomical calendar. It is constructed so that the sliver of light passes precisely through the center of the carved stone spiral at noon on the summer solstice. Numerous similar sites have been found throughout the American Southwest.

The ancient Chinese also observed the heavens. Their astrology attached particular importance to "omens" such as comets and "guest stars"—stars that appeared suddenly in the sky and then slowly faded away—and they kept careful and extensive records of such events. Twentieth-century astronomers still turn to the Chinese records to obtain observational data recorded during the Dark Ages (roughly from the 5th to the 10th century A.D.), when turmoil in Europe largely halted the progress of Western science. Perhaps the best-known guest star was one that appeared in A.D. 1054 and was visible in the daytime sky for many months. We now know that the event was actually a *supernova:* the explosion of

(a)

(c)

(b)

▲ FIGURE 2.2 **Observatories in the Americas** (a) The Big Horn Medicine Wheel in Wyoming, built by the Plains Indians, has spokes and other features that roughly align with risings and settings of the Sun and other stars. (b) The Caracol temple in Mexico, built by the Mayan civilization, has some windows that seem to align with astronomical events, suggesting that at least part of Caracol's function may have kept track of the seasons and the heavens. (c) This thin streak of light and shadow, created by the Sun's rays playing off the cliffs in Chaco Canyon of America's Southwest, aligns exactly with a carved rock pattern at noon on the summer solstice—almost certainly an intentional sign for astronomical or agricultural purposes. (*G. Gerster/Comstock*)

▲ **FIGURE 2.3 Turkish Astronomers at Work**   During the Dark Ages, much scientific information was preserved and new discoveries were made by astronomers in the Islamic world, as depicted in this illustration from a 16th-century manuscript. (*The Granger Collection*)

a giant star, which scattered most of its mass into space (see Chapter 21). It left behind a remnant that is still detectable today, nine centuries later. The Chinese data are a prime source of historical information for supernova research.

A vital link between the astronomy of ancient Greece and that of medieval Europe was provided by astronomers in the Muslim world (see Figure 2.3). For six centuries, from the depths of the Dark Ages to the beginning of the Renaissance, Islamic astronomy flourished and grew, preserving and augmenting the knowledge of the Greeks. Its influence on modern astronomy is subtle, but quite pervasive. Many of the mathematical techniques involved in trigonometry were developed by Islamic astronomers in response to practical problems, such as determining the precise dates of holy days or the direction of Mecca from any given location on Earth. Astronomical terms such as *zenith* and *azimuth* and the names of many stars—for example,

Rigel, Betelgeuse, and Vega—all bear witness to this extended period of Muslim scholarship.

Astronomy is not the property of any one culture, civilization, or era. The same ideas, the same tools, and even the same misconceptions have been invented and reinvented by human societies all over the world in response to the same basic driving forces. Astronomy came into being because people knew that there was a practical benefit in being able to predict the positions of the stars, but its roots go much deeper than that. The need to understand where we came from and how we fit into the cosmos is an integral part of human nature.

2.2   The Geocentric Universe

The Greeks of antiquity, and undoubtedly civilizations before them, built models of the universe. The study of the workings of the universe on the largest scales is called *cosmology*. Today, cosmology entails looking at the universe on scales so large that even entire galaxies can be regarded as mere points of light scattered throughout space. To the Greeks, however, the universe was basically the *solar system*—the Sun, Earth, and Moon, and the planets known at that time. The stars beyond were surely part of the universe, but they were considered to be fixed, unchanging beacons on the celestial sphere. The Greeks did not consider the Sun, the Moon, and the planets to be part of this mammoth celestial dome, however. Those objects had patterns of behavior that set them apart.

Observations of the Planets

Greek astronomers observed that over the course of a night, the stars slid smoothly across the sky. Over the course of a month, the Moon moved smoothly and steadily along its path on the sky relative to the stars, passing through its familiar cycle of phases. Over the course of a year, the Sun progressed along the ecliptic at an almost constant rate, varying little in brightness from day to day. In short, the behavior of both Sun and Moon seemed fairly simple and orderly. But ancient astronomers were also aware of five other bodies in the sky—the planets Mercury, Venus, Mars, Jupiter, and Saturn—whose behavior was not so easy to grasp. Their motions ultimately led to the downfall of an entire theory of the solar system and to a fundamental change in humankind's view of the universe.

To the naked eye (or even through a telescope), planets do not behave in as regular and predictable a fashion as the Sun, Moon, and stars. They vary in brightness, and they don't maintain a fixed position in the sky. Unlike the Sun and Moon, the planets seem to wander around the celestial sphere—indeed, the word *planet* derives from the Greek word *planetes*, meaning "wanderer." Planets never stray far from the ecliptic and generally traverse the celestial sphere from west to east, like the Sun. However, they seem to speed up and slow down during their journeys, and at times

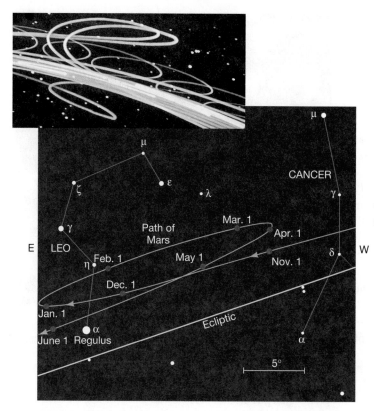

Most of the time, planets move from west to east relative to the background stars. Occasionally—roughly once per year—however, they change direction and temporarily undergo retrograde motion (east to west) before looping back. The main illustration shows an actual retrograde loop in the motion of the planet Mars. The inset depicts the movements of several planets over the course of several years, as reproduced on the inside dome of a planetarium. The motion of the planets relative to the stars (represented as unmoving points) produces continuous streaks on the planetarium "sky." *(Boston Museum of Science)*

early astronomers, the key observations of planetary orbits were the following:

- An inferior planet never strays too far from the Sun, as seen from Earth. As illustrated in the inset to Figure 2.5, because its path on the celestial sphere is close to the ecliptic, an inferior planet makes two *conjunctions* (or close approaches) with the Sun during each orbit. (It doesn't actually come close to the Sun, of course. Conjunction is simply the occasion when the planet and the Sun are in the same direction in the sky.) At *inferior conjunction*, the planet is closest to Earth and moves past the Sun from east to west—that is, in the retrograde sense. At *superior conjunction*, the planet is farthest from Earth and passes the Sun in the opposite (prograde) direction.

- Seen from Earth, the superior planets are not "tied" to the Sun as the inferior planets are. The superior planets make one prograde conjunction with the Sun during each trip around the celestial sphere.

they even appear to loop back and forth relative to the stars, as shown in Figure 2.4. In other words, there are periods when a planet's eastward motion (relative to the stars) stops, and the planet appears to move westward in the sky for a month or two before reversing direction again and continuing on its eastward journey. Motion in the eastward sense is usually referred to as *direct*, or *prograde*, motion; the backward (westward) loops are known as **retrograde motion.**

Ancient astronomers knew well that the periods of retrograde motion were closely correlated with other planetary properties, such as apparent brightness and position in the sky. Figure 2.5 (a modern view of the solar system, note!) shows three schematic planetary orbits and defines some time-honored astronomical terminology describing a planet's location relative to Earth and the Sun. Mercury and Venus are referred to as *inferior* ("lower") *planets* because their orbits lie between Earth and the Sun. Mars, Jupiter, and Saturn, whose orbits lie outside Earth's, are known as *superior* ("higher") *planets*. For

▶ **FIGURE 2.5 Inferior and Superior Orbits** Diagram of Earth's orbit and two other possible planetary orbits. An "inferior" orbit lies between Earth's orbit and the Sun. Mercury and Venus move in such orbits. A "superior" orbit (such as the orbit of Mars, Jupiter, or Saturn) lies outside that of Earth. The points noted on the orbits indicate times when a planet appears to come close to the Sun (conjunction) or is diametrically opposite the Sun on the celestial sphere (opposition); they are discussed further in the text. The inset is a schematic representation of the orbit of an inferior planet relative to the Sun, as seen from Earth.

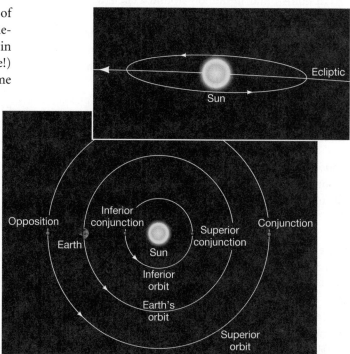

However, they exhibit retrograde motion (Figure 2.4) when they are at *opposition*, diametrically opposite the Sun on the celestial sphere.

- The superior planets are brightest at opposition, during retrograde motion. By contrast, the inferior planets are brightest a few weeks before and after inferior conjunction.

The challenge facing astronomers—then as now—was to find a solar system model that could explain all the existing observations and that could also make testable and reliable predictions of future planetary motions. ∞ (Sec. 1.2)

Ancient astronomers correctly reasoned that the changing brightness of a planet in the night sky is related to the planet's distance from Earth. Like the Moon, the planets produce no light of their own. Instead, they shine by reflected sunlight and, generally speaking, appear brightest when closest to us. Looking at Figure 2.5, you may already be able to discern the basic reasons for some of the planetary properties just listed; we'll return to the "modern" explanation in the next section. However, as we now discuss, the ancients took a very different path in their attempts to explain planetary motion.

A Theoretical Model

The earliest models of the solar system followed the teachings of the Greek philosopher Aristotle (384–322 B.C.) and were **geocentric,** meaning that Earth lay at the center of the universe and all other bodies moved around it. ∞ (Sec. 1.3) The celestial sphere, shown in Figures 1.11 and 1.16, illustrates the basic geocentric view. These models employed what Aristotle, and Plato before him, had taught was the perfect form: the circle. The simplest possible description—uniform motion around a circle with Earth at its center—provided a fairly good approximation to the orbits of the Sun and the Moon, but it could not account for the observed variations in planetary brightness or the retrograde motion of the planets. A more complex model was needed to describe these heavenly "wanderers."

In the first step toward this new model, each planet was taken to move uniformly around a small circle, called an **epicycle,** whose *center* moved uniformly around Earth on a second and larger circle, known as a **deferent** (Figure 2.6). The motion was now composed of two separate circular orbits, creating the possibility that, at some times, the planet's apparent motion could be retrograde. Also, the distance from the planet to Earth would vary, accounting for changes in brightness. By tinkering with the relative sizes of the epicycle and deferent, with the planet's speed on the epicycle, and with the epicycle's speed along the deferent, early astronomers were able to bring this "epicyclic" motion into fairly good agreement with the observed paths of the planets in the sky. Moreover, the model had good predictive power, at least to the accuracy of observations at the time.

However, as the number and the quality of observations increased, it became clear that the simple epicyclic model was

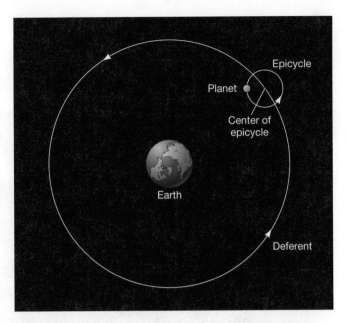

▲ **FIGURE 2.6 Geocentric Model** In the geocentric model of the solar system, the observed motions of the planets made it impossible to assume that they moved on simple circular paths around Earth. Instead, each planet was thought to follow a small circular orbit (the epicycle) about an imaginary point that itself traveled in a large, circular orbit (the deferent) about Earth.

not perfect. Small corrections had to be introduced to bring it into line with new observations. The center of the deferents had to be shifted slightly from Earth's center, and the motion of the epicycles had to be imagined uniform with respect to yet another point in space, not Earth. Furthermore, in order to explain the motions of the inferior planets, the model simply had to assume that the deferents of Mercury and Venus were, for some (unknown) reason, tied to that of the Sun. Similar assumptions also applied to the superior planets, to ensure that their retrograde motion occurred at opposition.

Around A.D. 140, a Greek astronomer named Ptolemy constructed perhaps the most complete geocentric model of all time. Illustrated in simplified form in Figure 2.7, it explained remarkably well the observed paths of the five planets then known, as well as the paths of the Sun and the Moon. However, to achieve its explanatory and predictive power, the full **Ptolemaic model** required a series of no fewer than 80 distinct circles. To account for the paths of the Sun, Moon, and all eight planets (and their moons) that we know today would require a vastly more complicated set. Nevertheless, Ptolemy's comprehensive text on the topic, *Syntaxis* (better known today by its Arabic name, *Almagest*, "the greatest"), provided the intellectual framework for all discussion of the universe for well over a thousand years.

Evaluating the Geocentric Model

Today, our scientific training leads us to seek simplicity, because, in the physical sciences, simplicity has so often proved to be an indicator of truth. We would regard the intricacy of a

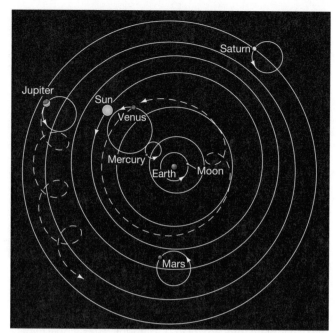

△ Interactive FIGURE 2.7 Ptolemaic Model The basic features, drawn roughly to scale, of Ptolemy's geocentric model of the inner solar system, a model that enjoyed widespread popularity prior to the Renaissance. Only the five planets visible to the naked eye and hence known to the ancients—Mercury, Venus, Mars, Jupiter, and Saturn—are shown. The planets' deferents were considered to move on spheres lying within the celestial sphere that held the stars. The celestial sphere carried all interior spheres around with it, but the planetary (and solar) spheres had additional motions of their own, causing the Sun and planets to move relative to the stars. To avoid confusion, partial paths (dashed) of only two planets—Venus and Jupiter—are drawn here.

model as complicated as the Ptolemaic system as a clear sign of a fundamentally flawed theory. ∞ (Sec. 1.2) Why was the Ptolemaic model so complex? With the benefit of hindsight, we now recognize that its major error lay in its assumption of a geocentric universe. This misconception was compounded by the insistence on uniform circular motion, whose basis was largely philosophical, rather than scientific, in nature.

Actually, history records that some ancient Greek astronomers reasoned differently about the motions of heavenly bodies. Foremost among them was Aristarchus of Samos (310–230 B.C.), who proposed that all the planets, including Earth, revolve around the Sun and, furthermore, that Earth rotates on its axis once each day. This combined revolution and rotation, he argued, would create an *apparent* motion of the sky—a simple idea that is familiar to anyone who has ridden on a merry-go-round and watched the landscape appear to move past in the opposite direction. However, Aristarchus's description of the heavens, though essentially correct, did not gain widespread acceptance during his lifetime. Aristotle's influence was too strong, his followers too numerous, and his writings too comprehensive. The geocentric model went largely unchallenged until the 16th century A.D.

The Aristotelian school did present some simple and (at the time) compelling arguments in favor of their views. First, of course, Earth doesn't *feel* as if it's moving—and if it were moving, wouldn't there be a strong wind as the planet revolves at high speed around the Sun? Also, considering that the vantage point from which we view the stars changes over the course of a year, why don't we see stellar parallax? ∞ (Sec. 1.4)

Nowadays we might dismiss the first points as merely naive, but the last is a valid argument and the reasoning essentially sound. Indeed, we now know that there *is* stellar parallax as Earth orbits the Sun. However, because the stars are so distant, it amounts to less than 1 arc second (1'), even for the closest stars. Early astronomers simply would not have noticed it. (In fact, stellar parallax was conclusively measured only in the middle of the 19th century.)

We will encounter many other instances in astronomy wherein correct reasoning led to the wrong conclusions because it relied on inadequate data. Even when the scientific method is properly applied and theoretical predictions are tested against reality, a theory can be only as good as the observations on which it is based. ∞ (Sec. 1.2)

2.3 The Heliocentric Model of the Solar System

The Ptolemaic picture of the universe survived, more or less intact, for almost 14 centuries, until a 16th-century Polish cleric, Nicolaus Copernicus (Figure 2.8), rediscovered Aristarchus's **heliocentric** (Sun-centered) model and showed how, in its harmony and organization, it provided a more

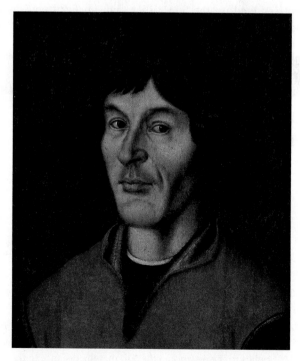

▲ FIGURE 2.8 Nicolaus Copernicus (1473–1543). (*E. Lessing/Art Resource, NY*)

natural explanation of the observed facts than did the tangled geocentric cosmology. Copernicus asserted that Earth spins on its axis and, like the other planets, orbits the Sun. Only the Moon, he said, orbits Earth. As we will see, not only does this model explain the observed daily and seasonal changes in the heavens, but it also naturally accounts for retrograde motion and variations in brightness of the planets. ∞ (Sec. 1.4)

The critical realization that Earth is not at the center of the universe is now known as the **Copernican revolution.** The seven crucial statements that form its foundation are summarized in *Discovery 2-1*.

Figure 2.9 shows how the Copernican view explains the varying brightness of a planet (in this case, Mars), its observed looping motions, and the fact that the retrograde motion of a superior planet occurs at opposition. If we suppose that Earth moves faster than Mars, then every so often Earth "overtakes" that planet. Mars will then appear to move backward in the sky, in much the same way as a car we overtake on the highway seems to slip backward relative to us. Replace Mars by Earth and Earth by Venus, and you should also be able to extend the explanation to the inferior planets. (To complete the story with a full explanation of their apparent brightnesses, however, you'll have to wait until Section 9.1!) Notice that, in the Copernican picture, the planet's looping motions are only apparent. In the Ptolemaic view, they are real.

Copernicus's major motivation for introducing the heliocentric model was simplicity. Even so, he was still influenced by Greek thinking and clung to the idea of circles to model the planets' motions. To bring his theory into agreement with observations of the night sky, he was forced to retain the idea of epicyclic motion, although with the deferent centered on the Sun rather than on Earth and with smaller epicycles than in the Ptolemaic picture. Thus, he

retained unnecessary complexity and actually gained little in predictive power over the geocentric model. The heliocentric model did rectify some small discrepancies and inconsistencies in the Ptolemaic system, but for Copernicus, the primary attraction of heliocentricity was its simplicity—its being "more pleasing to the mind." His theory was more something he *felt* than he could *prove*. To the present day, scientists still are guided by simplicity, symmetry, and beauty in modeling all aspects of the universe.

Despite the support of some observational data, neither his fellow scholars nor the general public easily accepted Copernicus's model. For the learned, heliocentricity went against the grain of much previous thinking and violated many of the religious teachings of the time, largely because it relegated Earth to a noncentral and undistinguished place within the solar system and the universe. And Copernicus's work had little impact on the general populace of his time, at least in part because it was published in Latin (the standard language of academic discourse at the time), which most people could not read. Only long after Copernicus's death, when others—notably Galileo Galilei—popularized his ideas, did the Roman Catholic Church take them seriously enough to bother banning them. Copernicus's writings on the heliocentric universe were placed on the Church's *Index of Prohibited Books* in 1616, 73 years after they were first published. They remained there until the end of the 18th century.

CONCEPT CHECK
✔ How do the geocentric and heliocentric models of the solar system differ in their explanations of planetary retrograde motion?

DISCOVERY 2-1

Foundations of the Copernican Revolution

The following seven points are essentially Copernicus's own words, with the italicized material providing additional explanation:

1. The celestial spheres do not have just one common center. *Specifically, Earth is not at the center of everything.*

2. The center of Earth is not the center of the universe, but is instead only the center of gravity and of the lunar orbit.

3. All the spheres revolve around the Sun. *By spheres, Copernicus meant the planets.*

4. The ratio of Earth's distance from the Sun to the height of the firmament is so much smaller than the ratio of Earth's radius to the distance to the Sun that the distance to the Sun is imperceptible compared with the height of the firmament.

By firmament, Copernicus meant the distant stars. The point he was making is that the stars are very much farther away than the Sun.

5. The motions appearing in the firmament are not its motions, but those of Earth. Earth performs a daily rotation around its fixed poles, while the firmament remains immobile as the highest heaven. *Because the stars are so far away, any apparent motion we see in them is the result of Earth's rotation.*

6. The motions of the Sun are not its motions, but the motion of Earth. *Similarly, the Sun's apparent daily and yearly motion are actually due to the various motions of Earth.*

7. What appears to us as retrograde and forward motion of the planets is not their own, but that of Earth. *The heliocentric picture provides a natural explanation for retrograde planetary motion, again as a consequence of Earth's motion.*

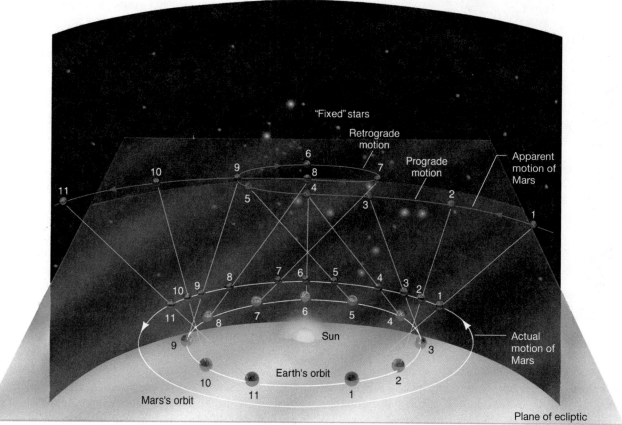

Interactive **FIGURE 2.9 Retrograde Motion** The Copernican model of the solar system explains both the varying brightnesses of the planets and the phenomenon of retrograde motion. Here, for example, when Earth and Mars are relatively close to one another in their respective orbits (as at position 6), Mars seems brighter. When they are farther apart (as at position 1), Mars seems dimmer. Also, because the (light blue) line of sight from Earth to Mars changes as the two planets orbit the Sun, Mars appears to loop back and forth in retrograde motion. Follow the lines in numerical order, and note how the line of sight moves backward relative to the stars between locations 5 and 7. The line of sight changes because Earth, on the inside track, moves faster in its orbit than does Mars. The actual planetary orbits are shown as white curves. The apparent motion of Mars, as seen from Earth, is indicated by the red curve. The heliocentric explanation of retrograde motion was pivotal in changing humanity's conception of the universe and our place in it.

2.4 The Birth of Modern Astronomy

In the century following the death of Copernicus and the publication of his theory of the solar system, two scientists— Galileo Galilei and Johannes Kepler—made indelible imprints on the study of astronomy. Contemporaries, they were aware of each other's work and corresponded from time to time about their theories. Each achieved fame for his discoveries and made great strides in popularizing the Copernican viewpoint, yet in their approaches to astronomy they were as different as night and day.

Galileo's Historic Observations

Galileo Galilei (Figure 2.10) was an Italian mathematician and philosopher. By his willingness to perform experiments to test his ideas—a rather radical approach in those days—and by

embracing the brand-new technology of the telescope, he revolutionized the way in which science was done, so much so that he is now widely regarded as the father of experimental science.

The telescope was invented in Holland in the early 17th century. Hearing of the invention (but without having seen one), Galileo built a telescope for himself in 1609 and aimed it at the sky. What he saw conflicted greatly with the philosophy of Aristotle and provided much new data to support the ideas of Copernicus.*

Using his telescope, Galileo discovered that the Moon had mountains, valleys, and craters—terrain in many ways reminiscent of that on Earth. Looking at the Sun (something that should *never* be done directly and that may have eventually blinded Galileo), he found imperfections—dark blemishes

---

In fact, Galileo had already abandoned Aristotle in favor of Copernicus, although he had not published his opinions at the time he began his telescopic observations.

▲ FIGURE 2.10 **Galileo Galilei (1564–1642).** *(Art Resource, NY)*

now known as *sunspots*. These observations ran directly counter to the orthodox wisdom of the day. By noting the changing appearance of sunspots from day to day, Galileo inferred that the Sun *rotates*, approximately once per month, around an axis roughly perpendicular to the ecliptic plane.

Galileo also saw four small points of light, invisible to the naked eye, orbiting the planet Jupiter and realized that they were moons. Figure 2.11 shows some sketches of these

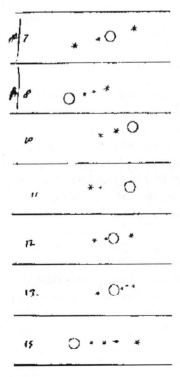

◄ FIGURE 2.11 **Galilean Moons** The four Galilean moons of Jupiter, as sketched by Galileo in his notebook. The sketches show what Galileo saw on seven nights between January 7 and 15, 1610. The orbits of the moons (sketched here as asterisks, and now called Io, Europa, Ganymede, and Callisto) around the planet (open circle) can clearly be seen. More of Galileo's remarkable sketches of Saturn, star clusters, and the Orion constellation can be seen on the first page of Part 1. (*From* Sidereus Nuncius)

moons, taken from Galileo's notes. To Galileo, the fact that another planet had moons provided the strongest support for the Copernican model. Clearly, Earth was not the center of all things. He also found that Venus varied in apparent size and showed a complete cycle of phases, like those of our Moon (Figure 2.12), findings that could be explained only by the planet's motion around the Sun. These observations were more strong evidence that Earth is not the center of all things and that at least one planet orbited the Sun. For more of Galileo's sketches, with comparisons to modern photographs, see the Part 1 Opener on p. 1.

In 1610, Galileo published a book called *Sidereus Nuncius (The Starry Messenger)*, detailing his observational findings and his controversial conclusions supporting the Copernican theory. In reporting and interpreting the wondrous observations made with his new telescope, Galileo was directly challenging both the scientific orthodoxy and the religious dogma of his day. He was (literally) playing with fire—he must certainly have been aware that only a few years earlier, in 1600, the astronomer Giordano Bruno had been burned at the stake in Rome, in part for his heretical teaching that Earth orbited the Sun. However, by all accounts, Galileo delighted in publicly ridiculing and irritating his Aristotelian colleagues. In 1616 his ideas were judged heretical, Copernicus's works were banned by the Roman Catholic Church, and Galileo was instructed to abandon his astronomical pursuits.

But Galileo would not desist. In 1632 he raised the stakes by publishing *Dialogue Concerning the Two Chief World Systems*, which compared the Ptolemaic and Copernican models. The book presented a discussion among three people, one of them a dull-witted Aristotelian whose views (which were in fact the stated opinions of the then Pope, Urban VIII) time and again were roundly defeated by the arguments of one of his two companions, an articulate proponent of the heliocentric system. To make the book accessible to a wide popular audience, Galileo wrote it in Italian rather than Latin. These actions brought Galileo into direct conflict with the authority of the Church. Eventually, the Inquisition forced him, under threat of torture, to retract his claim that Earth orbits the Sun, and he was placed under house arrest in 1633. He remained imprisoned for the rest of his life. Not until 1992 did the Church publicly forgive Galileo's "crimes." But the damage to the orthodox view of the universe was done, and the Copernican genie was out of the bottle once and for all.

Ascendancy of the Copernican System

Although Renaissance scholars were correct, they could not *prove* that our planetary system is centered on the Sun or even that Earth moves through space. The observational consequences of Earth's orbital motion were just too small for the technology of the day to detect. Direct evidence of Earth's motion was obtained only in 1728, when English

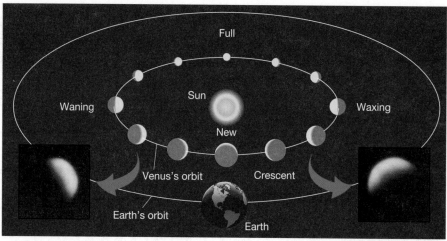

(a) Sun-centered model

| R | I | V | U | X | G |

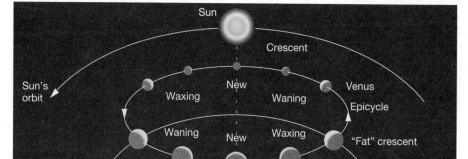

(b) Ptolemy's model

◄ **Interactive FIGURE 2.12 Venus Phases** Both the Ptolemaic and the Copernican models of the solar system predict that Venus should show phases as it moves in its orbit. (a) In the Copernican picture, when Venus is directly between Earth and the Sun, its unlit side faces us and the planet is invisible to us. As Venus moves in its orbit (at a faster speed than Earth moves in its orbit), progressively more of its illuminated face is visible from Earth. Note the connection between the orbital phase and the apparent size of the planet: Venus seems much larger in its crescent phase than when it is full because it is much closer to us during its crescent phase. This is the behavior actually observed. (The insets at bottom left and right are actual photographs of Venus taken at two of its crescent phases.) (b) The Ptolemaic model (see also Figure 2.7) is unable to account for these observations. In particular, the full phase of the planet cannot be explained. Seen from Earth, Venus reaches only a "fat crescent" phase, then begins to wane as it nears the Sun. (Both these views are from a sideways perspective; from overhead, both orbits are very nearly circular, as shown in Figure 2.18) *(Images from R. Beebe/Courtesy of New Mexico State University)*

astronomer James Bradley discovered the **aberration of starlight**—a slight (roughly 20″) shift in the observed direction to a star, caused by Earth's motion perpendicular to the line of sight, much as rain drops falling vertically leave slanted tracks on the passenger window of a moving car or train. Bradley's observation was the first proof that Earth revolves around the Sun; subsequent observations of many stars in many different directions have repeatedly confirmed the effect. Additional proof of Earth's orbital motion came in 1838, with the first unambiguous determination of stellar parallax (see Figure 1.30) by German astronomer Friedrich Bessel. ∞ (Sec. 1.6)

Following those early measurements, support for the heliocentric solar system has grown steadily, as astronomers have subjected the theory to more and more sophisticated observational tests, culminating in the interplanetary expeditions of our unmanned space probes of the 1960s, 1970s, and 1980s. Today, the evidence is overwhelming. The development and eventual acceptance of the heliocentric model were milestones in human thinking. This removal of Earth

from any position of great cosmic significance is generally known, even today, as the *Copernican principle*. It has become a cornerstone of modern astrophysics.

The Copernican revolution is a prime example of how the scientific method, though affected at any given time by the subjective whims, human biases, and even sheer luck of researchers, does ultimately lead to a definite degree of objectivity. ∞ (Sec. 1.2) Over time, many groups of scientists checking, confirming, and refining experimental tests can neutralize the subjective attitudes of individuals. Usually, one generation of scientists can bring sufficient objectivity to bear on a problem, although some especially revolutionary concepts are so swamped by tradition, religion, and politics that more time is necessary. In the case of heliocentricity, objective confirmation was not obtained until about three centuries after Copernicus published his work and more than 2000 years after Aristarchus had proposed the concept. Nonetheless, objectivity *did in fact* eventually prevail, and our knowledge of the universe has expanded immeasurably as a result.

2.5 The Laws of Planetary Motion

At about the same time as Galileo was becoming famous—or notorious—for his pioneering telescopic observations and outspoken promotion of the Copernican system, Johannes Kepler (Figure 2.13), a German mathematician and astronomer, was developing the laws of planetary motion that now bear his name. Galileo was in many ways the first "modern" observer. He used emerging technology, in the form of the telescope, to achieve new insights into the universe. In contrast, Kepler was a pure theorist. His groundbreaking work that so clarified our knowledge of planetary motion was based almost entirely on the observations of others, principally an extensive collection of data compiled by Tycho Brahe (1546–1601), Kepler's employer and arguably one of the greatest observational astronomers that has ever lived.

Brahe's Complex Data

Tycho, as he is often called, was both an eccentric aristocrat and a skillful observer. Born in Denmark, he was educated at some of the best universities in Europe, where he studied astrology, alchemy, and medicine. Most of his observations, which predated the invention of the telescope by several decades, were made at his own observatory, named *Uraniborg*, in Denmark (Figure 2.14). There, using instruments of his own design, Tycho maintained meticulous and accurate records of the stars, planets, and other noteworthy celestial events, including a comet and a supernova (see Chapter 21), the appearance of which helped convince him that the Aristotelean view of the universe could not be correct.

In 1597, having fallen out of favor with the Danish court, Tycho moved to Prague as Imperial Mathematician of the Holy Roman Empire. Prague happens to be fairly close to Graz, in Austria, where Kepler lived and worked. Kepler joined Tycho in Prague in 1600 and was put to work trying to find a theory that could explain Brahe's planetary data. When Tycho died a year later, Kepler inherited not

▲ FIGURE 2.13 **Johannes Kepler (1571–1630).** *(E. Lessing/Art Resource, NY)*

▲ FIGURE 2.14 **Tycho Brahe** The astronomer in his observatory Uraniborg, on the island of Hveen in Denmark. Brahe's observations of the positions of stars and planets on the sky were the most accurate and complete set of naked-eye measurements ever made. *(Royal Ontario Museum)*

only his position, but also his most priceless possession: the accumulated observations of the planets, spanning several decades. These observations, though made with the naked eye, were nevertheless of very high quality. In most cases, his measured positions of stars and planets were accurate to within about 1′. Kepler set to work seeking a unifying principle to explain in detail the motions of the planets, without the need for epicycles. The effort was to occupy much of the remaining 29 years of his life.

Kepler had already accepted the heliocentric picture of the solar system. His goal was to find a simple and elegant description of planetary motion within the Copernican framework that fit Brahe's complex mass of detailed observations. In the end, he found it necessary to abandon Copernicus's simple idea of circular planetary orbits. However, even greater simplicity emerged as a result. After long years of studying Brahe's planetary data, and after many false starts and blind alleys, Kepler developed the laws that now bear his name.

Kepler determined the shape of each planet's orbit by triangulation—not from different points on Earth, but from different points on Earth's orbit, using observations made at many different times of the year. ∞ (Sec. 1.6) By using a portion of Earth's orbit as a baseline for his triangle, Kepler was able to measure the relative sizes of the other planetary orbits. Noting where the planets were on successive nights, he found the speeds at which the planets move. We do not know how many geometric shapes Kepler tried for the orbits before he hit upon the correct one. His difficult task was made even more complicated because he had to determine Earth's own orbit, too. Nevertheless, he eventually succeeded in summarizing the motions of all the known planets, including Earth, in just three laws: the **laws of planetary motion.**

Kepler's Simple Laws

Kepler's first law of planetary motion has to do with the *shapes* of the planetary orbits:

The orbital paths of the planets are elliptical (*not* circular), with the Sun at one focus.

An **ellipse** is simply a flattened circle. Figure 2.15 illustrates a means of constructing an ellipse with a piece of string and two thumbtacks. Each point at which the string is pinned is called a **focus** (plural: *foci*) of the ellipse. The long axis of the ellipse, containing the two foci, is known as the *major axis.* Half the length of this long axis is referred to as the **semimajor axis,** a measure of the ellipse's size. A circle is a special case in which the two foci happen to coincide; its semimajor axis is simply its radius.

The **eccentricity** of an ellipse is simply a measure of how flattened it is. Technically, eccentricity is defined as the

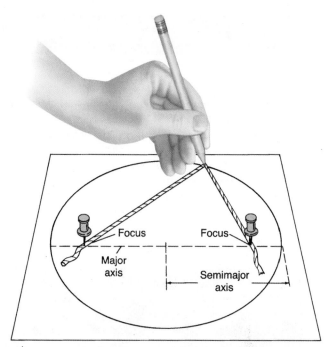

▲ **Interactive FIGURE 2.15 Ellipse** An ellipse can be drawn with the aid of a string, a pencil, and two thumbtacks. The wider the separation of the foci, the more elongated, or eccentric, is the ellipse. In the special case where the two foci are at the same place, the curve drawn is a circle.

ratio of the distance between the foci to the length of the major axis, but the most important thing to remember here is that an eccentricity of zero corresponds to no flattening—a perfect circle—whereas an eccentricity of one means that the circle has been squashed all the way down to a straight line. Note that, while the Sun resides at one focus of the elliptical orbit, the other focus is empty and has no particular physical significance. (However, we can still figure out where it is, because the two foci are symmetrically placed about the center, along the major axis.)

The length of the semimajor axis and the eccentricity are all we need to describe the size and shape of a planet's orbital path (see *More Precisely 2-1*). In fact, no planet's elliptical orbit is nearly as elongated as the one shown in Figure 2.15. With one exception (the orbit of Mercury), planetary orbits in our solar system have such small eccentricities that our eyes would have trouble distinguishing them from true circles. Only because the orbits are so nearly circular were the Ptolemaic and Copernican models able to come as close as they did to describing reality.

Kepler's substitution of elliptical for circular orbits was no small advance. It amounted to abandoning an aesthetic bias—the Aristotelian belief in the perfection of the circle—that had governed astronomy since Greek antiquity. Even Galileo Galilei, not known for his conservatism in scholarly matters, clung to the idea of circular motion and never accepted the notion that the planets move in elliptical paths.

MORE PRECISELY 2-1

Some Properties of Planetary Orbits

Two numbers—*semimajor axis* and *eccentricity*—are all that are needed to describe the size and shape of a planet's orbital path. From them, we can derive many other useful quantities. Two of the most important are the planet's *perihelion* (its point of closest approach to the Sun) and its *aphelion* (point of greatest distance from the Sun). From the definitions presented in the text, it follows that if the planet's orbit has semimajor axis a and eccentricity e, the planet's perihelion is at a distance $a(1 - e)$ from the Sun, while its aphelion is at $a(1 + e)$. These points and distances are illustrated in the following figure:

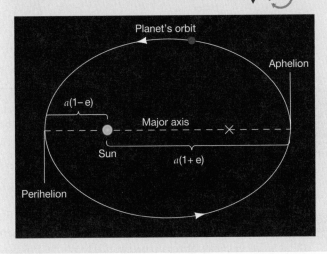

EXAMPLE 1 We can locate the other focus of the ellipse in the diagram and hence determine the eccentricity from the definition in the text quite simply. The second focus is placed symmetrically along the major axis, at the point marked with an "X." With a ruler, measure (1) the length of the major axis and (2) the distance between the two foci. Dividing the second distance by the first, you should find an eccentricity of 3.4 cm/6.8 cm = 0.5. Alternatively, we could use the formula given for the perihelion. Measure the perihelion distance to be $a(1 - e) = 1.7$ cm. Dividing this by $a = 3.4$ cm, we obtain $1 - e = 0.5$, so once again, $e = 0.5$.

EXAMPLE 2 A (hypothetical) planet with a semimajor axis of 400 million km and an eccentricity of 0.5 (i.e., with an orbit as shown in the diagram) would range between $400 \times (1 - 0.5) = 200$ million km and $400 \times (1 + 0.5) = 600$ million km from the Sun over the course of one complete orbit. With $e = 0.9$, the range in distances would be 40 to 760 million km, and so on.

No planet has an orbital eccentricity as large as 0.5—the planet with the most eccentric orbit is Mercury, with $e = 0.206$ (see Table 2.1). However, many meteoroids and all comets (see Chapter 14) have eccentricities considerably greater than that. In fact, most comets visible from Earth have eccentricities very close to $e = 1$. Their highly elongated orbits approach within a few astronomical units of the Sun at perihelion, yet these tiny frozen worlds spend most of their time far beyond the orbit of Pluto.

The second law, illustrated in Figure 2.16, addresses the *speed* at which a planet traverses different parts of its orbit:

> An imaginary line connecting the Sun to any planet sweeps out equal areas of the ellipse in equal intervals of time.

While orbiting the Sun, a planet traces the arcs labeled A, B, and C in the figure in equal times. Notice, however, that the distance traveled by the planet along arc C is greater than the distance traveled along arc A or arc B. Because the time is the same and the distance is different, the speed must vary. When a planet is close to the Sun, as in sector C, it moves much faster than when farther away, as in sector A.

By taking into account the relative speeds and positions of the planets in their elliptical orbits about the Sun, Kepler's first two laws explained the variations in planetary brightness and some observed peculiar nonuniform motions that could not be accommodated within the assumption of circular motion, even with the inclusion of epicycles. Gone at last were the circles within circles that rolled across the sky. Kepler's modification of the Copernican theory to allow the possibility of elliptical orbits both greatly simplified the

model of the solar system and at the same time provided much greater predictive accuracy than had previously been possible. Note, too, that these laws are not restricted to planets. They apply to *any* orbiting object. Spy satellites, for example, move very rapidly as they swoop close to Earth's surface, not because they are propelled with powerful onboard rockets, but because their highly eccentric orbits are governed by Kepler's laws.

Kepler published his first two laws in 1609, stating that he had proved them only for the orbit of Mars. Ten years later, he extended them to all the then-known planets (Mercury, Venus, Earth, Mars, Jupiter, and Saturn) and added a third law relating the size of a planet's orbit to its sidereal orbital **period**—the time needed for the planet to complete one circuit around the Sun:

> The square of a planet's orbital period is proportional to the cube of its semimajor axis.

This law becomes particularly simple when we choose the (Earth sidereal) year as our unit of time and the *astronomical unit* as our unit of length. One **astronomical unit** (AU) is the semimajor axis of Earth's orbit around the

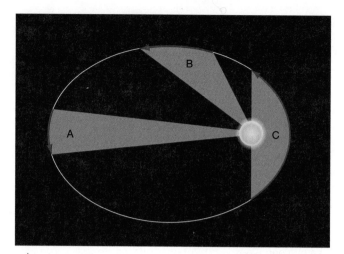

⚡ Interactive FIGURE 2.16 Kepler's Second Law A line joining a planet to the Sun sweeps out equal areas in equal intervals of time. The three shaded areas A, B, and C are equal. Any object traveling along the elliptical path would take the same amount of time to cover the distance indicated by the three red arrows. Therefore, planets move faster when closer to the Sun.

TABLE 2.1 Some Solar System Dimensions

Planet	Orbital Semimajor Axis, a (AU)	Orbital Period, P (years)	Orbital Eccentricity, e	P^2/a^3
Mercury	0.387	0.241	0.206	1.002
Venus	0.723	0.615	0.007	1.001
Earth	1.000	1.000	0.017	1.000
Mars	1.524	1.881	0.093	1.000
Jupiter	5.203	11.86	0.048	0.999
Saturn	9.537	29.42	0.054	0.998
Uranus	19.19	83.75	0.047	0.993
Neptune	30.07	163.7	0.009	0.986

Sun—essentially the average distance between Earth and the Sun. Like the light-year, the astronomical unit is custom made for the vast distances encountered in astronomy. Using these units for time and distance, we can rewrite Kepler's third law for any planet as

$$P^2 \text{ (in Earth years)} = a^3 \text{ (in astronomical units)},$$

where P is the planet's sidereal orbital period and a is the length of its semimajor axis. The law implies that a planet's "year" P increases more rapidly than does the size of its orbit, a. For example, Earth, with an orbital semimajor axis of 1 AU, has an orbital period of 1 Earth year. The planet Venus, orbiting at a distance of roughly 0.7 AU, takes only 0.6 Earth year—about 225 days—to complete one circuit. By contrast, Saturn, almost 10 AU from the Sun, takes considerably more than 10 Earth years—in fact, nearly 30 years—to orbit the Sun just once.

Table 2.1 presents basic data describing the orbits of the eight planets now known. Renaissance astronomers knew these properties for the innermost six planets and used them to construct the currently accepted heliocentric model of the solar system. The second column presents each planet's orbital semimajor axis, measured in astronomical units; the third column gives the orbital period, in Earth years. The fourth column lists the planets' orbital eccentricities. For purposes of verifying Kepler's third law, the fifth column lists the ratio P^2/a^3. As we have just seen, the third law implies that this number should always be unity in the units used in the table.

The main points to be grasped from Table 2.1 are these: (1) With the exception of Mercury, the planets' orbits are nearly circular (i.e., their eccentricities are close to zero)

and (2) the farther a planet is from the Sun, the greater is its orbital period, in agreement with Kepler's third law to within the accuracy of the numbers in the table. (The small, but significant, deviations of P^2/a^3 from unity in the cases of Uranus and Neptune are caused by the gravitational attraction between those two planets; see Chapter 13.) Most important, note that Kepler's laws are obeyed by *all* the known planets, *not just by the six on which he based his conclusions.*

The laws developed by Kepler were far more than mere fits to existing data. They also made definite, testable predictions about the future locations of the planets. Those predictions have been borne out to high accuracy every time they have been tested by observation—the hallmark of any credible scientific theory. ∞ (Sec. 1.2)

CONCEPT CHECK

✔ Why is it significant that Kepler's laws also apply to Uranus and Neptune?

2.6 The Dimensions of the Solar System

Kepler's laws allow us to construct a scale model of the solar system, with the correct shapes and *relative* sizes of all the planetary orbits, but they do not tell us the *actual* size of any orbit. We can express the distance to each planet only in terms of the distance from Earth to the Sun. Why is this? Because Kepler's triangulation measurements all used a portion of Earth's orbit as a baseline, distances could be expressed only relative to the size of that orbit, which was itself not determined by the method. Thus, our model of the solar system would be like a road map of the United States showing the *relative* positions of cities and towns, but lacking the all-important scale marker indicating distances in kilometers or miles. For example, we would know that Kansas City is about three times more distant from New York than it is from Chicago, but we would not know the actual mileage between any two points on the map.

If we could somehow determine the value of the astronomical unit—in kilometers, say—we would be able to add the vital scale marker to our map of the solar system and compute the precise distances between the Sun and each of the planets. We might propose using triangulation to measure the distance from Earth to the Sun directly. However, we would find it impossible to measure the Sun's parallax using Earth's diameter as a baseline. The Sun is too bright, too big, and too fuzzy for us to distinguish any apparent displacement relative to a field of distant stars. To measure the Sun's distance from Earth, we must resort to some other method.

Before the middle of the 20th century, the most accurate measurements of the astronomical unit were made by using triangulation on the planets Mercury and Venus during their rare *transits* of the Sun—that is, during the brief periods when those planets passed directly between the Sun and Earth (as shown for the case of Mercury in Figure 2.17). Because the time at which a transit occurs can be measured with great precision, astronomers can use this information to make accurate measurements of a planet's position in the sky. They can then employ simple geometry to compute the distance to the planet by combining observations made from different locations on Earth, as discussed earlier in Chapter 1. ∞ (Sec. 1.6) For example, the parallax of Venus at closest approach to Earth, as seen from two diametrically opposite points on Earth (separated by about 13,000 km), is about 1′ (1/60°)—at the limit of naked-eye capabilities, but easily measurable

telescopically. Using the second formula presented in *More Precisely 1-2*, we find that this parallax represents a distance of 13,000 km × 57.3°/(1/60°), or approximately 45,000,000 km.

Knowing the distance to Venus, we can compute the magnitude of the astronomical unit. Figure 2.18 is an idealized diagram of the Sun–Earth–Venus orbital geometry. The planetary orbits are drawn as circles here, but in reality they are slight ellipses. This is a subtle difference, and we can correct for it using detailed knowledge of orbital motions. Assuming for the sake of simplicity that the orbits are perfect circles, we see from the figure that the distance from Earth to Venus at closest approach is approximately 0.3 AU. Knowing that 0.3 AU is 45,000,000 km makes determining 1 AU straightforward—the answer is 45,000,000/0.3, or 150,000,000 km.

The modern method for deriving the absolute scale (that is, the scale expressed in kilometers, rather than just relative to Earth's orbit) of the solar system uses radar rather than triangulation. The word **radar** is an acronym for **ra**dio **d**etection **a**nd **r**anging. In this technique, radio waves are transmitted toward an astronomical body, such as a planet. (We cannot use radar ranging to measure the distance to the Sun directly, because radio signals are absorbed at the solar

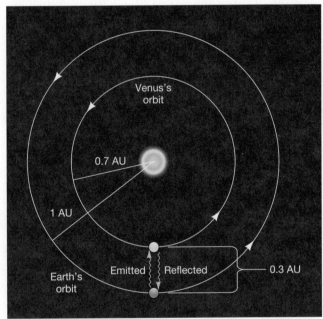

▲ FIGURE 2.18 **Astronomical Unit** Simplified geometry of the orbits of Earth and Venus as they move around the Sun. The wavy blue lines represent the paths along which radar signals are transmitted toward Venus and received back at Earth at the particular moment (chosen for simplicity) when Venus is at its minimum distance from Earth. Because the radius of Earth's orbit is 1 AU and that of Venus is about 0.7 AU, we know that this distance is 0.3 AU. Thus, radar measurements allow us to determine the astronomical unit in kilometers. This extraterrestrial measurement sets the scale of the solar system—and, as we will see, the entire universe.

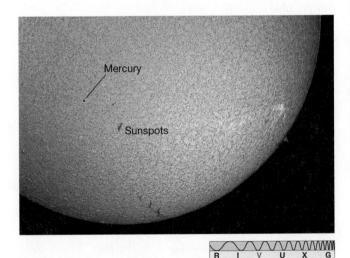

▲ FIGURE 2.17 **Solar Transit** The transit of Mercury across the face of the Sun. Such transits happen only about once per decade because Mercury's orbit does not quite coincide with the plane of the ecliptic. Transits of Venus are even rarer, occurring only about twice per century. The most recent took place in 2006. (*P. Jones*)

surface and are not reflected to Earth.) The returning echo indicates the body's direction and range, or distance, in absolute terms—that is, in kilometers rather than in astronomical units. Multiplying the 300-second round-trip travel time of the radar signal (the time elapsed between transmission of the signal and reception of the echo) by the speed of light (300,000 km/s, which is also the speed of radio waves), we obtain twice the distance to the target planet.

Venus, whose orbit periodically brings it closest to Earth, is the most common target for radar ranging. The round-trip travel time (for example, at closest approach, as indicated by the wavy lines in Figure 2.18) can be measured with high precision—in fact, well enough to determine the planet's distance to an accuracy of about 1 km. In this way, the astronomical unit is now known to be 149,597,870 km. We will use the rounded-off value of 1.5×10^8 km in this text.

Having determined the value of the astronomical unit, we can reexpress the sizes of the other planetary orbits in terms of more familiar units, such as miles or kilometers. The entire scale of the solar system can then be calibrated to high precision.

CONCEPT CHECK
✔ Why don't Kepler's laws tell us the value of the astronomical unit?

2.7 Newton's Laws

Kepler's three laws, which so simplified the solar system, were discovered *empirically*. In other words, they resulted solely from the analysis of observational data and were not derived from any theory or mathematical model. Indeed, Kepler did not have any appreciation of the physics underlying his laws. Nor did Copernicus understand *why* his heliocentric model of the solar system worked. Even Galileo, often called the father of modern physics, failed to understand why the planets orbit the Sun (although Galileo's work laid vital groundwork for Newton's theories).

What prevents the planets from flying off into space or from falling into the Sun? What causes them to revolve about the Sun, apparently endlessly? To be sure, the motions of the planets obey Kepler's three laws, but only by considering something more fundamental than those laws can we really understand planetary motion. The heliocentric system was secured when, in the 17th century, the British mathematician Isaac Newton (Figure 2.19) developed a deeper understanding of the way *all* objects move and interact with one another.

The Laws of Motion

Isaac Newton was born in Lincolnshire, England, on Christmas Day in 1642, the year Galileo died. Newton studied at Trinity College of Cambridge University, but when the

▲ **FIGURE 2.19 Isaac Newton (1642–1727).** (*The Granger Collection*)

bubonic plague reached Cambridge in 1665, he returned to the relative safety of his home for 2 years. During that time he made probably the most famous of his discoveries, the law of gravity (although it is but one of the many major scientific advances for which Newton was responsible). However, either because he regarded the theory as incomplete or possibly because he was afraid that he would be attacked or plagiarized by his colleagues, he did not tell anyone of his monumental achievement for almost 20 years. It was not until 1684, when Newton was discussing the leading astronomical problem of the day—*Why* do the planets move according to Kepler's laws?—with Edmund Halley (of Halley's comet fame) that he astounded his companion by revealing that he had solved the problem in its entirety nearly two decades before!

Prompted by Halley, Newton published his theories in perhaps the most influential physics book ever written: *Philosophiae Naturalis Principia Mathematica* (*The Mathematical Principles of Natural Philosophy*—what we would today call "science"), usually known simply as Newton's *Principia*. The ideas expressed in that work form the basis for what is now known as **Newtonian mechanics.** Three basic laws of motion, the law of gravity, and a little calculus (which Newton also developed) are sufficient to explain and quantify virtually all the complex dynamic behavior we see on Earth and throughout the universe.

Figure 2.20 illustrates Newton's first law of motion:

Every body continues in a state of rest or in a state of uniform motion in a straight line, unless it is compelled to change that state by a force acting on it.

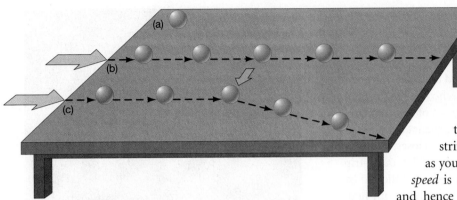

▲ FIGURE 2.20 **Newton's First Law** An object at rest will remain at rest (a) until some force acts on it. When a force (represented by the red arrow) does act (b), the object will remain in that state of uniform motion until another force acts on it. When a second force (green arrow) acts in a direction different from the first force (c), the object changes its direction of motion.

The first law simply states that a moving object will move forever in a straight line, unless some external **force**—a push or a pull—changes its speed or direction of motion. For example, the object might glance off a brick wall or be hit with a baseball bat; in either case, a force changes the original motion of the object. Another example of a force, well known to most of us, is **weight**—the force (commonly measured in pounds in the United States) with which gravity pulls you toward Earth's center.

The tendency of an object to keep moving at the same speed and in the same direction unless acted upon by a force is known as **inertia.** Newton's first law implies that it requires no force to maintain motion in a straight line with constant speed. This contrasts sharply with the view of Aristotle, who maintained (incorrectly) that the natural state of an object was to be *at rest*—most probably an opinion based on Aristotle's observations of the effect of friction. In our discussion, we will neglect friction—the force that slows balls rolling along the ground, blocks sliding across tabletops, and baseballs moving through the air. In any case, it is not an issue for the planets because there is no appreciable friction in outer space—there is no air or any other matter to impede a planet's motion. The fallacy in Aristotle's argument was first realized and exposed by Galileo, who conceived of the notion of inertia long before Newton formalized it into a law.

A familiar measure of an object's inertia is its **mass**—loosely speaking, the total amount of matter the object contains. The greater an object's mass, the more inertia it has, and the greater is the force needed to change its state of motion.

Newton's first law describes motion in a straight line with constant speed—that is, motion with constant **velocity.** An object's velocity includes both its speed (in miles per hour or meters per second, say) *and* its direction in space (up, down, northwest, and so on). In everyday speech, we tend to use the terms "speed" and "velocity" more or less interchangeably, but we must realize that they are actually different quantities and that Newton's laws are always stated in terms of the latter. As a specific illustration of the difference, consider a rock tied to a string, moving at a constant rate in a circle as you whirl it around your head. The rock's *speed* is constant, but its *direction* of motion, and hence its velocity, is continually changing. Thus, according to Newton's first law, a force must be acting. That force is the tension you feel in the string. In a moment, we'll see reasoning similar to this applied to the problem of planetary motion.

The rate of change of the velocity of an object—speeding up, slowing down, or simply changing direction—is called the object's **acceleration** and is the subject of Newton's second law, which states that the acceleration of an object is directly proportional to the applied force and inversely proportional to its mass:

When a force F acts on a body of mass m, it produces in it an acceleration a equal to the force divided by the mass. Thus, $a = F/m$, or $F = ma$.

Hence, the greater the force acting on the object or the smaller the mass of the object, the greater is the acceleration of the object. If two objects are pulled with the same force, the more massive one will accelerate less; if two identical objects are pulled with different forces, the one acted on by the greater force will accelerate more.

Acceleration is the rate of change of velocity, so its units are velocity units per unit of time, such as meters per second *per second* (usually written as m/s^2). In honor of Newton, the SI unit of force is named after him. By definition, 1 newton (N) is the force required to cause a mass of 1 kilogram to accelerate at a rate of 1 meter per second every second ($1 \ m/s^2$). One newton is approximately 0.22 pound.

At Earth's surface, the force of gravity produces a downward acceleration of approximately 9.8 m/s^2 on *all* bodies, regardless of mass. According to Newton's second law, this means that your weight (in newtons) is directly proportional to your mass (in kilograms). We will return to this very important point later.

Finally, Newton's third law simply tells us that forces cannot occur in isolation:

To every action, there is an equal and opposite reaction.

In other words, if body *A* exerts a force on body *B*, then body *B* necessarily exerts a force on body *A* that is equal in magnitude, but oppositely directed. For example, when a baseball player hits a home run, Newton's third law says that the bat and the ball exert equal and opposite forces on one another during the instant they are in contact. According to the second law, the ball subsequently moves away much faster than the bat because the *mass* of the ball is much less than the combined mass of the bat plus batter (whose body absorbs much of the reaction force), so the ball's *acceleration* is much greater.

Only in extreme circumstances—when speeds approach the speed of light—do Newton's laws break down, and this fact was not realized until the 20th century, when Albert Einstein's theories of relativity once again revolutionized our view of the universe (see Chapter 22). Most of the time, however, Newtonian mechanics provides an excellent description of the motions of planets, stars, and galaxies through the cosmos.

Gravity

Forces may act *instantaneously* or *continuously*. The force from the baseball bat that hits the home run can reasonably be thought of as being instantaneous. A good example of a continuous force is the one that prevents the baseball from zooming off into space—**gravity,** the phenomenon that started Newton on the path to the discovery of his laws. Newton hypothesized that any object having mass always exerts an attractive **gravitational force** on all other massive objects. The more massive an object, the stronger is its gravitational pull.

Consider a baseball thrown upward from Earth's surface, as illustrated in Figure 2.21. In accordance with Newton's first law, the downward force of Earth's gravity continuously modifies the baseball's velocity, slowing the initial upward motion and eventually causing the ball to fall back to the ground. Of course, the baseball, having some mass of its own, also exerts a gravitational pull on Earth. By Newton's third law, this force is equal and opposite to the weight of the ball (the force with which Earth attracts it). But, by Newton's second law, Earth has a much greater effect on the light baseball than the baseball has on the much more massive Earth. The ball and Earth act upon each other with the same gravitational force, but Earth's acceleration is much smaller.

Now consider the trajectory of the same baseball batted from the surface of the Moon. The pull of gravity is about one-sixth as great on the Moon as on Earth, so the baseball's velocity changes more slowly—a typical home run in a ballpark on Earth would travel nearly half a mile on the Moon. Less massive than Earth, the Moon has less gravitational influence on the baseball. The magnitude of the gravitational force, then, depends on the *masses* of the attracting bodies. In fact, the force is *directly proportional* to the product of the two masses.

Studying the motions of the planets uncovers a second aspect of the gravitational force. At locations equidistant from the Sun's center, the gravitational force has the same strength

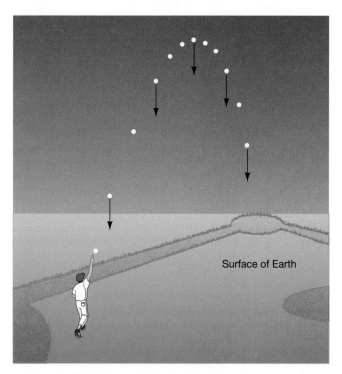

▲ **FIGURE 2.21 Gravity** A ball thrown up from the surface of a massive object, such as a planet, is pulled continuously downward (arrows) by the gravity of that planet (and, conversely, the gravity of the ball continuously pulls the planet).

and is always directed toward the Sun. Furthermore, detailed calculation of the planets' accelerations as they orbit the Sun reveals that the strength of the Sun's gravitational pull decreases in proportion to the *square* of the distance from the Sun. The force of gravity is said to obey an **inverse-square law.** As shown in Figure 2.22, inverse-square forces decrease rapidly with distance from their source. For example, tripling the distance makes the force $3^2 = 9$ times weaker, whereas multiplying the distance by five results in a force that is $5^2 = 25$ times weaker. Despite this rapid decrease, the force never quite reaches zero. The gravitational pull of an object having some mass can never be completely extinguished.

We can combine the preceding statements about mass and distance to form a law of gravity that dictates the way in which *all* massive objects (i.e., objects having some mass) attract one another:

> Every particle of matter in the universe attracts every other particle with a force that is directly proportional to the product of the masses of the particles and inversely proportional to the square of the distance between their centers.

As a proportionality, the law of gravity may be written as

$$\text{gravitational force} \propto \frac{\text{mass of object 1} \times \text{mass of object 2}}{\text{distance}^2}$$

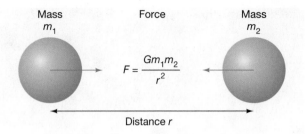

Mass m_1 Force   Mass m_2

$$F = \frac{Gm_1 m_2}{r^2}$$

Distance r

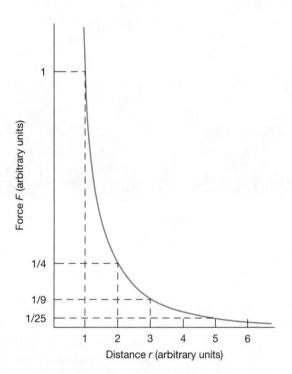

Force F (arbitrary units)

Distance r (arbitrary units)

▲ FIGURE 2.22 **Gravitational Force** (a) The gravitational force between two bodies is proportional to the mass of each and is inversely proportional to the square of the distance between them (b). Inverse-square forces rapidly weaken with distance from their source. The strength of the Sun's gravitational force decreases with the square of the distance from the Sun, but never quite reaches zero, no matter how far away from the Sun.

(The symbol $\propto$ here means "is proportional to.") The rule for computing the force F between two bodies of masses m_1 and m_2, separated by a distance r, is usually written more compactly as

$$F = \frac{Gm_1 m_2}{r^2}.$$

The quantity G is known as the *gravitational constant*, or, often, simply as Newton's constant. It is one of the fundamental constants of the universe. The value of G has been measured in extremely delicate laboratory experiments as 6.67×10^{-11} newton meter^2/kilogram^2 ($\text{N} \cdot \text{m}^2/\text{kg}^2$).

2.8 Newtonian Mechanics

Newton's three laws of motion and the law of gravitation provide a solid theoretical foundation upon which we can base a deeper understanding of planetary orbits, the laws of planetary motion, and many other important aspects of orbital motion. With the development of Newtonian mechanics, the transition from geocentric lore to heliocentric fact was complete.

Planetary Motion

The mutual gravitational attraction between the Sun and the planets, as expressed by Newton's law of gravity, is responsible for the observed planetary orbits. As depicted in Figure 2.23, this gravitational force continuously pulls each planet toward the Sun, deflecting its forward motion into a curved orbital path. Because the Sun is much more massive than any of the planets, it dominates the interaction. We might say that the Sun "controls" the planets, not the other way around.

The Sun–planet interaction sketched here is analogous to our earlier example of the rock whirling on a string. The Sun's gravitational pull is your hand and the string, and the planet is the rock at the end of that string. The tension in the string provides the force necessary for the rock to move in a circular path. If you were suddenly to release the string—which would be like eliminating the Sun's gravity—the rock would fly away along a tangent to the circle, in accordance with Newton's first law.

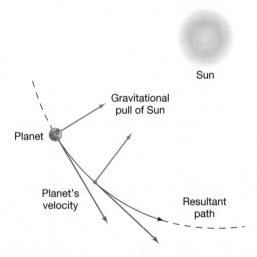

Sun

Gravitational pull of Sun

Planet

Planet's velocity

Resultant path

▲ FIGURE 2.23 **Solar Gravity** The Sun's inward pull of gravity on a planet competes with the planet's tendency to continue moving in a straight line. These two effects combine, causing the planet to move smoothly along an intermediate path, which continuously "falls around" the Sun. This unending tug-of-war between the Sun's gravity and the planet's inertia results in a stable orbit. Newtonian mechanics and the theory of gravity are fundamental to our understanding of the cosmos on every scale.

In the solar system, at this very moment, Earth is moving under the combined influence of gravity and inertia. The net result is a stable orbit, despite our continuous rapid motion through space. (In fact, Earth orbits the Sun at a speed of about 30 km/s, or approximately 70,000 mph. You can verify this for yourself by calculating how fast Earth must move to complete a circle of radius 1 AU—and hence of circumference 2π AU, or 940 million km—in 1 year, or 3.2×10^7 seconds. The answer is 9.4×10^8 km / 3.2×10^7 s, or 29.4 km/s.) *More Precisely 2-2* describes how astronomers can use Newtonian mechanics and the law of gravity to quantify planetary motion and measure the masses of Earth, the Sun, and many other astronomical objects by studying the orbits of objects near them.

Kepler's Laws Reconsidered

Newton's laws of motion and law of universal gravitation provide a theoretical explanation for Kepler's empirical laws of planetary motion. Kepler's three laws follow directly from Newtonian mechanics, as solutions of the equations describing the motion of a body moving in response to an inverse-square force. However, just as Kepler modified the Copernican model by introducing ellipses rather than circles, so, too, did Newton make corrections to Kepler's first and third laws. It turns out that a planet does not orbit the exact center of the Sun. Instead, both the planet and the Sun orbit their common **center of mass**—the "average" position of all the matter making up the two bodies (Figure 2.24). Because the Sun and the planet are acted upon by equal and

opposite gravitational forces (by Newton's third law), the Sun must also move (by Newton's first law), driven by the gravitational influence of the planet. The Sun, however, is so much more massive than any planet that the center of mass of the planet–Sun system is very close to the center of the Sun, which is why Kepler's laws are so accurate. Thus, Kepler's first law becomes

> The orbit of a planet around the Sun is an ellipse, with the *center of mass of the planet–Sun system* at one focus.

As shown in Figure 2.24, the center of mass of two objects of comparable mass does not lie within either object. For identical masses orbiting one another (Figure 2.25a), the orbits are identical ellipses, with a common focus located midway between the two objects. For unequal masses (as in Figure 2.25b), the elliptical orbits still share a focus, and both have the same eccentricity, but the more massive object moves more slowly and on a tighter orbit. (Note that Kepler's second law, as stated earlier, continues to apply without modification to each orbit separately, but the *rates* at which the two orbits sweep out areas are different.)

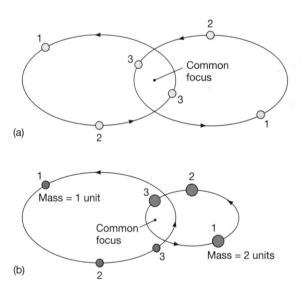

(a)

(b)

▲ **FIGURE 2.25 Orbits** (a) The orbits of two bodies (stars, for example) with equal masses, under the influence of their mutual gravity, are identical ellipses with a common focus. That focus is not at the center of either star, but instead at their center of mass, located midway between them. The pairs of numbers indicate the positions of the two bodies at three different times. (Note that a line joining the bodies at any give time always passes through the common focus.) (b) The orbits of two bodies, one of which is twice as massive as the other. Again, the elliptical orbits have a common focus (at the center of mass of the two-body system), and the two ellipses have the same eccentricity. However, in accordance with Newton's laws of motion, the more massive body moves more slowly and in a smaller orbit. In this particular case, the larger ellipse is twice the size of the smaller one.

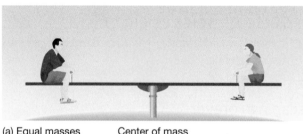

(a) Equal masses   Center of mass

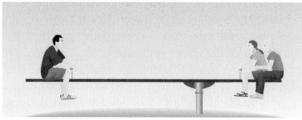

(b) Unequal masses   Center of mass

▲ **Interactive FIGURE 2.24 Center of Mass** (a) The center of mass of two bodies of equal mass lies midway between them. (b) As the mass of one body increases, the center of mass moves toward it. The situation is analogous to riding a seesaw—when both sides are balanced, the center of mass is at the pivot point.

MORE PRECISELY 2-2

Weighing the Sun

We can use Newtonian mechanics to calculate some useful formulae relating the properties of planetary orbits to the mass of the Sun. Again for simplicity, let's assume that the orbits are circular (not a bad approximation in most cases, and Newton's laws easily extend to cover the more general case of eccentric orbits). Consider a planet of mass m moving at speed v in an orbit of radius r around the Sun, of mass M. Even though the planet's *speed* is constant, the *direction* of its motion is not, so the planet's velocity is changing—it is accelerating. In fact, the planet's acceleration is

$$a = \frac{v^2}{r},$$

so, by Newton's second law, the force required to keep the planet in orbit is

$$F = ma = \frac{mv^2}{r}.$$

Setting this equation equal to the gravitational force due to the Sun, we obtain

$$\frac{mv^2}{r} = \frac{GmM}{r^2},$$

so the speed of the planet in the circular orbit is

$$v = \sqrt{\frac{GM}{r}}.$$

Now let's turn the problem around. Because we have measured G in the laboratory on Earth, and because we know the length of a year and the size of the astronomical unit, we can use Newtonian mechanics to *weigh* the Sun—that is, find its mass by measuring its gravitational influence on another body (in this case, Earth). Rearranging the last equation to read

$$M = \frac{rv^2}{G}$$

and substituting the known values of $v = 30$ km/s, $r = 1$ AU $= 1.5 \times 10^{11}$ m, and $G = 6.7 \times 10^{-11}$ Nm²/kg², we calculate the mass of the Sun to be 2.0×10^{30} kg—an enormous mass by terrestrial standards.

EXAMPLE   Similarly, knowing the distance from Earth to the Moon ($r = 384{,}000$ km) and the length of the (sidereal) month ($P = 27.3$ days), we can calculate the Moon's orbital speed to be $v = 2\pi r/P = 1.02$ km/s, and hence, using the preceding formula, measure Earth's mass to be 6.0×10^{24} kg.

In fact, this is basically how *all* masses are measured in astronomy. Because we can't just go out and attach a scale to an astronomical object when we need to know its mass, we must look for its gravitational influence on something else. This principle applies to planets, stars, galaxies, and even clusters of galaxies—very different objects, but all subject to the same physical laws.

The change to Kepler's third law is also small in the case of a planet orbiting the Sun, but very important in other circumstances, such as the orbital motion of two stars that are gravitationally bound to each other. Following through the mathematics of Newton's theory, we find that the true relationship between the semimajor axis a (measured in astronomical units) of the planet's orbit relative to the Sun and its orbital period P (in Earth years) is

$$P^2 \text{ (in Earth years)} = \frac{a^3 \text{ (in astronomical units)}}{M_{total} \text{ (in solar units)}},$$

where M_{total} is the *combined* mass of the two objects. Notice that Newton's restatement of Kepler's third law preserves the proportionality between P^2 and a^3, but now the proportionality includes M_{total}, so it is *not* quite the same for all the planets. The Sun's mass is so great, however, that the differences in M_{total} among the various combinations of the Sun and the other planets are almost unnoticeable, so Kepler's third law, as originally stated, is a very good approximation. This modified form of Kepler's third law is true in all circumstances, inside or outside the solar system.

PROCESS OF SCIENCE CHECK

✔ In what ways did Newtonian mechanics supersede Kepler's laws as a model of the solar system?

Escaping Forever

The law of gravity that describes the orbits of planets around the Sun applies equally well to natural moons and artificial satellites orbiting any planet. All our Earth-orbiting, human-made satellites move along paths governed by a combination of the inward pull of Earth's gravity and the forward motion gained during the rocket launch. If the rocket initially imparts enough speed to the satellite, it can go into orbit. Satellites not given enough speed at launch, by accident or design (e.g., intercontinental ballistic missiles) fail to achieve orbit and fall back to Earth (see Figure 2.26).

Some space vehicles, such as the robot probes that visit other planets, attain enough speed to escape our planet's gravity and move away from Earth forever. This speed, known as the **escape speed,** is about 41% greater (actually,

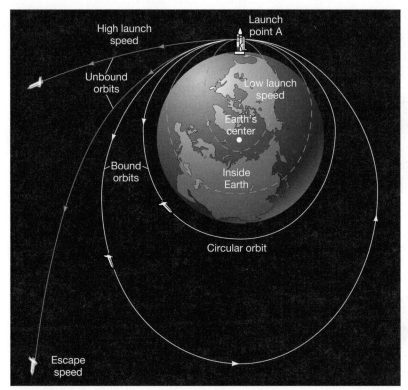

$\sqrt{2}$ = 1.414... times greater) than the speed of an object traveling in a circular orbit at any given radius.* At less than the escape speed, the old adage "What goes up must come down" (or at least stay in orbit) still applies. At more than the escape speed, however, a spacecraft will leave Earth for good. Planets, stars, galaxies—all gravitating bodies—have escape speeds. No matter how massive the body, gravity decreases with distance. As a result, the escape speed diminishes with increasing separation. The farther we go from Earth (or any gravitating body), the easier it becomes to escape.

The speed of a satellite in a circular orbit just above Earth's atmosphere is 7.9 km/s (roughly 18,000 mph). The satellite would have to travel at 11.2 km/s (about 25,000 mph) to escape from Earth altogether. If an object exceeds the escape speed, its motion is said to be **unbound,** and the orbit is no longer an ellipse. In fact, the path of the spacecraft relative to Earth is a related geometric figure called a *hyperbola*. If we simply change the word *ellipse* to

In terms of the formula presented in More Precisely 2-2, the escape speed is given by $v_{escape} = \sqrt{2\,GM/r}$.

hyperbola, the modified version of Kepler's first law still applies, as does Kepler's second law. (Kepler's third law does not extend to unbound orbits because it doesn't make sense to talk about a period in those cases.)

The Circle of Scientific Progress

The progression from the complex Ptolemaic model of the universe to the elegant simplicity of Newton's laws is a case study in the scientific method. ∞ (Sec. 1.2) Copernicus made a radical conceptual leap away from the Ptolemaic view, gaining much in insight but little in predictive power. Kepler made critical changes to the Copernican picture and gained both accuracy and predictive power but still fell short of a true physical explanation of planetary motion within the solar system, or of orbital motion in general. Eventually, Newton showed how all known planetary motion could be explained in detail by the application of four, simple, fundamental laws—the three laws of motion and the law of gravity. The process was slow, with many starts and stops and a few wrong turns, but it worked!

In a sense, then, the development of Newton's laws and their application to planetary motion represented the end of the first "loop" around the schematic diagram presented in Figure 1.6. The long-standing practical and conceptual questions raised by ancient observations of retrograde motion were finally resolved, and new predictions, themselves amenable to observational testing, became possible. And the laws are still being tested today. Every time a comet appears in the night sky right on schedule, or a spacecraft reaches the end of a billion-kilometer journey within meters of its target and seconds of the predicted arrival time, our confidence in Newtonian mechanics is further strengthened. But unlike the essentially descriptive models of Ptolemy, Copernicus, and Kepler, Newtonian mechanics is not limited to the motions of planets, or even to events occurring within our own solar system. They apply to moons, comets, spacecraft, stars, and even the most distant galaxies, extending the range of our scientific inquiries across the observable universe—as well as apples falling to the ground.

CONCEPT CHECK

✔ Explain, in terms of Newton's laws of motion and gravity, why planets orbit the Sun.

CHAPTER REVIEW

SUMMARY

1 **Geocentric (p. 36)** models of the universe were based on the view that the Sun, the Moon, and the planets all orbit Earth. The most successful and long lived of these was the **Ptolemaic model (p. 36)**. Unlike the Sun and the Moon, planets sometimes appear to temporarily reverse their direction of motion (from night to night) relative to the stars and then resume their normal "forward" course. This phenomenon is called **retrograde motion (p. 35)**. To account for retrograde motion within the geocentric picture, it was necessary to suppose that planets moved on small circles called **epicycles (p. 36)**, whose centers orbited Earth on larger circles called **deferents (p. 36)**.

2 The **heliocentric (p. 37)** view of the solar system holds that Earth, like all the planets, orbits the Sun. This model accounts for retrograde motion and the observed sizes and variations in brightness of the planets in a much more straightforward way than the geocentric model. The widespread realization during the Renaissance that the solar system is Sun centered, and not Earth centered, is known as the **Copernican revolution (p. 38)**, in honor of Nicolaus Copernicus, who laid the foundations of the modern heliocentric model.

3 Galileo Galilei is often regarded as the father of experimental science. His telescopic observations of the Moon, the Sun, Venus, and Jupiter played a crucial role in supporting and strengthening the Copernican picture of the solar system. Johannes Kepler improved on Copernicus's model by condensing the observational data of Tycho Brahe into three **laws of planetary motion (p. 43)**.

4 Kepler's Laws state: (1) Planetary orbits are **ellipses (p. 43)** with the Sun at one **focus (p. 43)**; (2) a planet moves faster as its orbit takes it closer to the Sun; (3) the **semimajor axis (p. 43)** of the orbit is related in a simple

way to the planet's orbital **period (p. 44)**. Most planets move on orbits whose **eccentricities (p. 43)** are quite small, so their paths differ only slightly from perfect circles.

5 Kepler's laws tell us only the relative sizes of the planetary orbits. To determine the true scale of the solar system, astronomers must find an independent measure of the distance between two bodies. The distance from Earth to the Sun is called the **astronomical unit (p. 44)**. Nowadays, the astronomical unit is determined by bouncing **radar (p. 46)** signals off the planet Venus and measuring the time the signal takes to return.

6 Isaac Newton succeeded in explaining planetary motion in terms of a few general physical principles, now known as **Newtonian mechanics (p. 47)**. Newton's laws of motion are: (1) To change the body's velocity, a **force (p. 48)** must be applied; (2) the rate of change of velocity, called **acceleration (p. 48)**, is equal to the applied force divided by the body's **mass (p. 48)**; (3) when bodies interact, the forces between them are always equal and opposite to one another. To explain Kepler's laws, Newton postulated that **gravity (p. 49)** attracts the planets to the Sun. Every object with any mass exerts a **gravitational force (p. 49)** on all other objects in the universe. The strength of this force decreases with increasing distance according to an **inverse-square law (p. 49)**.

7 For one object to escape from the gravitational pull of another, its speed must exceed the **escape speed (p. 52)** of that other object. By determining the gravitational force needed to keep one body orbiting another, Newton's laws allow astronomers to measure the masses of distant objects.

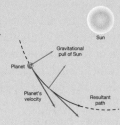

MasteringASTRONOMY *For instructor-assigned homework go to www.masteringastronomy.com*

Problems labeled **POS** explore the process of science | **VIS** problems focus on reading and interpreting visual information

REVIEW AND DISCUSSION

1. What contributions to modern astronomy were made by Chinese and Islamic astronomers during the Dark Ages of medieval Europe?

2. Briefly describe the geocentric model of the universe.

3. **POS** The benefit of our current knowledge lets us see flaws in the Ptolemaic model of the universe. What is its basic flaw?

4. What was the great contribution of Copernicus to our knowledge of the solar system? What was still a flaw in the Copernican model?

5. What is a theory? Can a theory ever be proved to be true?

6. When were Copernicus's ideas finally accepted?

7. What is the Copernican principle?

8. **POS** What discoveries of Galileo helped confirm the views of Copernicus, and how did they do so?

9. Briefly describe Kepler's three laws of planetary motion.

10. How did Tycho Brahe contribute to Kepler's laws?

11. If radio waves cannot be reflected from the Sun, how can radar be used to find the distance from Earth to the Sun?

12. How did astronomers determine the scale of the solar system prior to the invention of radar?

13. **POS** What does it mean to say that Kepler's laws are empirical?

14. What are Newton's laws of motion and gravity?

15. List the two modifications made by Newton to Kepler's laws.

16. Why do we say that a baseball falls toward Earth, and not Earth toward the baseball?

17. Why would a baseball go higher if it were thrown up from the surface of the Moon than if it were thrown with the same velocity from the surface of Earth?

18. In what sense is the Moon falling toward Earth?

19. What is the meaning of the term *escape speed*?

20. What would happen to Earth if the Sun's gravity were suddenly "turned off"?

CONCEPTUAL SELF-TEST: MULTIPLE CHOICE

1. Planets near opposition (**a**) rise in the east; (**b**) rise in the west; (**c**) do not rise or set; (**d**) have larger deferents.

2. A major flaw in Copernicus's model was that it still had (**a**) the Sun at the center; (**b**) Earth at the center; (**c**) retrograde loops; (**d**) circular orbits.

3. **VIS** As shown in Figure 2.12 ("Venus Phases"), Galileo's observations of Venus demonstrated that Venus must be (**a**) orbiting Earth; (**b**) orbiting the Sun; (**c**) about the same diameter as Earth; (**d**) similar to the Moon.

4. An accurate sketch of Jupiter's orbit around the Sun would show (**a**) the Sun far off center; (**b**) an oval twice as long as it is wide; (**c**) a nearly perfect circle; (**d**) phases.

5. A calculation of how long it takes a planet to orbit the Sun would be most closely related to Kepler's (**a**) first law of orbital shapes; (**b**) second law of orbital speeds; (**c**) third law of planetary distances; (**d**) first law of inertia.

6. An asteroid with an orbit lying entirely inside Earth's (**a**) has an orbital semimajor axis of less than 1 AU; (**b**) has a longer orbital period than Earth's; (**c**) moves more slowly than Earth; (**d**) must have a highly eccentric orbit.

7. If Earth's orbit around the Sun were twice as large as it is now, the orbit would take (**a**) less than twice as long; (**b**) two times longer; (**c**) more than two times longer to traverse.

8. **VIS** Figure 2.21 ("Gravity"), showing the motion of a ball near Earth's surface, depicts how gravity (**a**) increases with altitude; (**b**) causes the ball to accelerate downward; (**c**) causes the ball to accelerate upward; (**d**) has no effect on the ball.

9. If the Sun *and its mass* were suddenly to disappear, Earth would (**a**) continue in its current orbit; (**b**) suddenly change its orbital speed; (**c**) fly off into space; (**d**) stop spinning.

10. **VIS** Figure 2.25(b) ("Orbits") shows the orbits of two stars of unequal masses. If one star has twice the mass of the other, then the more massive star (**a**) moves more slowly than; (**b**) moves more rapidly than; (**c**) has half the gravity of; (**d**) has twice the eccentricity of the less massive star.

PROBLEMS

The number of dots preceding each Problem indicates its approximate level of difficulty.

1. • Tycho Brahe's observations of the stars and planets were accurate to about 1 arc minute (1'). To what distance does this angle correspond at the distance of (a) the Moon; (b) the Sun; and (c) Saturn (at closest approach)?

2. •• To an observer on Earth, through what angle will Mars appear to move relative to the stars over the course of 24 hours when the two planets are at closest approach? Assume for simplicity that Earth and Mars move on circular orbits of radii 1.0 AU and 1.5 AU, respectively, in exactly the same plane. Will the apparent motion be prograde or retrograde?

3. •• A spacecraft has an orbit that just grazes Earth's orbit at aphelion and just grazes Venus's orbit at perihelion. Assuming that Earth and Venus are in the right places at the right times, how long will the spacecraft take to travel from Earth to Venus?

4. •• Halley's comet has a perihelion distance of 0.6 AU and an orbital period of 76 years. What is its greatest distance from the Sun?

5. •• What is the maximum possible parallax of Mercury during a solar transit, as seen from either end of a 3000-km baseline on Earth?

6. • How long would a radar signal take to complete a round-trip between Earth and Mars when the two planets are 0.7 AU apart?

7. •• Jupiter's moon Callisto orbits the planet at a distance of 1.88 million kilometers. Callisto's orbital period about Jupiter is 16.7 days. What is the mass of Jupiter? [Assume that Callisto's mass is negligible compared with that of Jupiter, and use the modified version of Kepler's third law (Section 2.7).]

8. • The acceleration due to gravity at Earth's surface is 9.80 m/s². What is the gravitational acceleration at altitudes of (a) 100 km; (b) 1000 km; (c) 10,000 km? Take Earth's radius to be 6400 km.

9. • The Moon's mass is 7.4×10^{22} kg and its radius is 1700 km. What is the speed of a spacecraft moving in a circular orbit just above the lunar surface? What is the escape speed from the Moon?

10. • Use Newton's law of gravity to calculate the force of gravity between you and Earth. Convert your answer, which will be in newtons, to pounds, using the conversion 4.45 N equals 1 pound (1 lb). What do you normally call this force?

RADIATION

INFORMATION FROM THE COSMOS

LEARNING GOALS

Studying this chapter will enable you to

1. Describe the basic properties of wave motion.

2. Tell how electromagnetic radiation transfers energy and information through interstellar space.

3. Describe the major regions of the electromagnetic spectrum and explain how Earth's atmosphere affects our ability to make astronomical observations at different wavelengths.

4. Explain what is meant by the term "blackbody radiation" and describe the basic properties of such radiation.

5. Tell how we can determine the temperature of an object by observing the radiation it emits.

6. Show how the relative motion between a source of radiation and its observer can change the perceived wavelength of the radiation, and explain the importance of this phenomenon to astronomy.

Astronomical objects are more than just things of beauty in the night sky. Planets, stars, and galaxies are of vital significance if we are to fully understand our place in the big picture—the "grand design" of the universe. Each object is a source of information about the material aspects of our universe—its state of motion, its temperature, its chemical composition, and even its past history.

This information comes to us in the form of light. When we look at the stars, the light we see actually began its journey to Earth decades, centuries—even millennia—ago. The faint rays from the most distant galaxies have taken billions of years to reach us. The stars and galaxies in the night sky show us the far away and the long ago. In this chapter, we begin our study of how astronomers extract information from the light emitted by astronomical objects. These basic concepts of radiation are central to modern astronomy.

THE BIG PICTURE Astronomical objects emit a wealth of visible and invisible information that travels through the vast expanse of space. Detailed study of this radiation is the principal way in which astronomers study stars and other distant astronomical objects, from birth to death.

Visit the Study Area in www.masteringastronomy.com for quizzes, animations, videos, interactive figures, and self-guided tutorials.

LEFT: *About 5 billion years ago, the Sun emerged from a stellar nursery much like the one shown here. The* Spitzer Space Telescope, *now orbiting Earth, took this infrared image in order to gain insight about the origins of our solar system. By capturing radiation longer in wavelength than visible light, astronomers can peer inside this young nebula, named RCW 49, where its gas and dust glow in the warmth of starlight from more than 2200 stars. Red color depicts hydrogen gas near the center, and cyan maps the presence of organic molecules within the dusty outer regions. (SSC/JPL)*

3.1   Information from the Skies

Figure 3.1 shows a galaxy in the constellation Andromeda. On a dark, clear night, far from cities or other sources of light, the Andromeda galaxy, as it is generally called, can be seen with the naked eye as a faint, fuzzy patch on the sky, comparable in diameter to the full Moon. Yet the fact that it is visible from Earth belies this galaxy's enormous distance from us: It lies roughly 2.5 million *light-years* away.

An object at such a distance is truly inaccessible in any realistic human sense. Even if a space probe could miraculously travel at the speed of light, it would need 2.5 million years to reach this galaxy and 2.5 million more to return with its findings. Considering that civilization has existed on Earth for less than 10,000 years, and its prospects for the next 10,000 are far from certain, even this unattainable technological feat would not provide us with a practical means of exploring other galaxies. Even the farthest reaches of our own galaxy, "only" a few tens of thousands of light-years distant, are effectively off limits to visitors from Earth, at least for the foreseeable future.

Given the practical impossibility of traveling to such remote parts of the universe, how do astronomers know anything about objects far from Earth? How do we obtain detailed information about planets, stars, or galaxies too distant for a personal visit or any kind of controlled experiment? The answer is that we use the laws of physics, as we know them here on Earth, to interpret the **electromagnetic radiation** emitted by those objects.

R   I   V   U   X   G

▲ **FIGURE 3.1 Andromeda Galaxy** The pancake-shaped Andromeda Galaxy lies about 2.5 million light-years away and contains a few hundred billion stars. All of our knowledge of this galaxy derives from detailed study of the light it emits. (*R. Gendler*)

Light and Radiation

Radiation is any way in which energy is transmitted through space from one point to another without the need for any physical connection between the two locations. The term *electromagnetic* just means that the energy is carried in the form of rapidly fluctuating *electric* and *magnetic* fields (to be discussed in more detail later in Section 3.2). Virtually all we know about the universe beyond Earth's atmosphere has been gleaned from painstaking analysis of electromagnetic radiation received from afar. Our understanding depends completely on our ability to decipher this steady stream of data reaching us from space.

How bright are the stars (or galaxies, or planets), and how hot? What are their masses? How rapidly do they spin, and what is their motion through space? What are they made of, and in what proportion? The list of questions is long, but one fact is clear: Electromagnetic theory is vital to providing the answers—without it, we would have no way of testing our models of the cosmos, and the modern science of astronomy simply would not exist. ∞ (Sec. 1.2)

Visible light is the particular type of electromagnetic radiation to which our human eyes happen to be sensitive. As light enters our eye, small chemical reactions triggered by the incoming energy send electrical impulses to the brain, producing the sensation of sight. But modern instruments (see Chapter 5) can also detect many forms of *invisible* electromagnetic radiation, which goes completely unnoticed by our eyes. **Radio, infrared,** and **ultraviolet** waves, as well as **X rays** and **gamma rays,** all fall into this category.

Note that, despite the different names, the words *light, rays, radiation,* and *waves* all really refer to the same thing. The names are just historical accidents, reflecting the fact that it took many years for scientists to realize that these apparently very different types of radiation are in reality one and the same physical phenomenon. Throughout this text, we will use the general terms *light* and *electromagnetic radiation* more or less interchangeably.

Wave Motion

Despite the early confusion still reflected in current terminology, scientists now know that all types of electromagnetic radiation travel through space in the form of **waves.** To understand the behavior of light, then, we must know a little about wave motion.

Simply stated, a wave is a way in which energy is transferred from place to place without the physical movement of material from one location to another. In wave motion, the energy is carried by a *disturbance* of some sort. This disturbance, whatever its nature, occurs in a distinctive repeating pattern. Ripples on the surface of a pond, sound waves in air, and electromagnetic waves in space, despite their many obvious differences, all share this basic defining property.

Imagine a twig floating in a pond (Figure 3.2). A pebble thrown into the pond at some distance from the twig disturbs the surface of the water, setting it into up-and-down motion. This disturbance will move outward from the point of impact in the form of waves. When the waves reach the twig, some of the pebble's energy will be imparted to it, causing the twig to bob up and down. In this way, both energy and *information*—the fact that the pebble entered the water—are transferred from the place where the pebble landed to the location of the twig. We could tell that a pebble (or, at least, some object) had entered the water just by observing the twig. With a little additional physics, we could even estimate the pebble's energy.

A wave is not a physical object. No water traveled from the point of impact of the pebble to the twig—at any location on the surface, the water surface simply moved up and down as the wave passed. What, then, *did* move across the surface of the pond? As illustrated in the figure, the answer is that the wave was the *pattern* of up-and-down motion. This pattern was transmitted from one point to the next as the disturbance moved across the water.

Figure 3.3 shows how wave properties are quantified and illustrates some standard terminology. The wave's **period** is the number of seconds needed for the wave to repeat itself at any given point in space. The **wavelength** is the number of meters needed for the wave to repeat itself at a given moment in time. It can be measured as the distance between two adjacent wave *crests,* two adjacent wave *troughs,* or any other two similar points on adjacent wave cycles (e.g., the points marked × in the figure). A wave moves

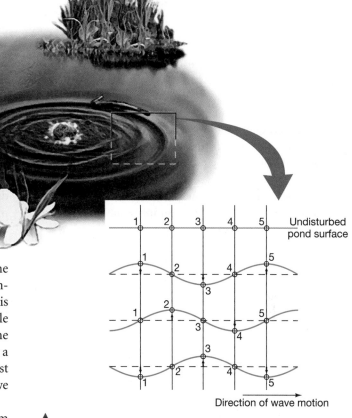

Interactive FIGURE 3.2 Water Wave The passage of a wave across a pond causes the surface of the water to bob up and down, but there is no movement of water from one part of the pond to another. Here waves ripple out from the point where a pebble has hit the water to the point where a twig is floating. The inset shows a simplified series of "snapshots" of the pond surface as the wave passes by. The points numbered 1 through 5 represent surface locations that move up and down with the passage of the wave.

a distance equal to one wavelength in one period. The maximum departure of the wave from the undisturbed state—still air, say, or a flat pond surface—is called the wave's **amplitude.**

The number of wave crests passing any given point per unit time is called the wave's **frequency.** If a wave of a given wavelength moves at high speed, then many crests pass per second and the frequency is high. Conversely, if the same wave

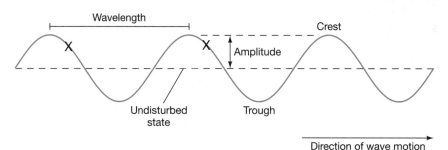

Interactive FIGURE 3.3 Wave Properties Representation of a typical wave, showing its direction of motion, wavelength, and amplitude. In one wave period, the entire pattern shown here moves one wavelength to the right.

moves slowly, then its frequency will be low. The frequency of a wave is just the reciprocal of the wave's period; that is,

$$\text{frequency} = \frac{1}{\text{period}}.$$

Frequency is expressed in units of inverse time (that is, 1/second, or cycles per second), called hertz (Hz) in honor of the 19th-century German scientist Heinrich Hertz, who studied the properties of radio waves. Thus, a wave with a period of 5 seconds (5 s) has a frequency of (1/5) cycles/s = 0.2 Hz, meaning that one wave crest passes a given point in space every 5 seconds.

Because a wave travels one wavelength in one period, it follows that the *wave velocity* is simply equal to the wavelength divided by the period:

$$\text{velocity} = \frac{\text{wavelength}}{\text{period}}.$$

Since the period is the reciprocal of the frequency, we can equivalently (and more commonly) write this relationship as

$$\text{velocity} = \text{wavelength} \times \text{frequency}.$$

Thus, if the wave in our earlier example had a wavelength of 0.5 m, its velocity would be (0.5 m)/(5 s), or (0.5 m) × (0.2 Hz) = 0.1 m/s. In the case of electromagnetic radiation, the velocity is the speed of light. Notice that wavelength and wave frequency are *inversely* related—doubling one halves the other.

The Components of Visible Light

White light is a mixture of colors, which we conventionally divide into six major hues: red, orange, yellow, green, blue, and violet. As shown in Figure 3.4, we can separate a beam of white light into a rainbow of these basic colors—called a *spectrum* (plural, *spectra*)—by passing it through a prism. This experiment was first reported by Isaac Newton over 300 years ago. In principle, the original beam of white light could be recovered by passing the spectrum through a second prism to recombine the colored beams.

What determines the color of a beam of light? The answer is its frequency (or alternatively, its wavelength). We see different colors because our eyes react differently to electromagnetic waves of different frequencies. A prism splits a beam of light up into separate colors because light rays of different frequencies are bent, or *refracted*, slightly differently as they pass through the prism—red light the least, violet light the most. Red light has a frequency of roughly 4.3×10^{14} Hz, corresponding to a wavelength of about 7.0×10^{-7} m. Violet light, at the other end of the visible range, has nearly double the frequency—7.5×10^{14} Hz—and (since the speed of light is always the same) just over half the wavelength—4.0×10^{-7} m. The other colors we see have frequencies and wavelengths intermediate between these two extremes, spanning the entire *visible spectrum* shown in Figure 3.4. Radiation outside this range is invisible to human eyes.

Scientists often use a unit called the *nanometer* (nm) in describing the wavelength of light (see Appendix 2). There are 10^9 nanometers in 1 meter. An older unit called the

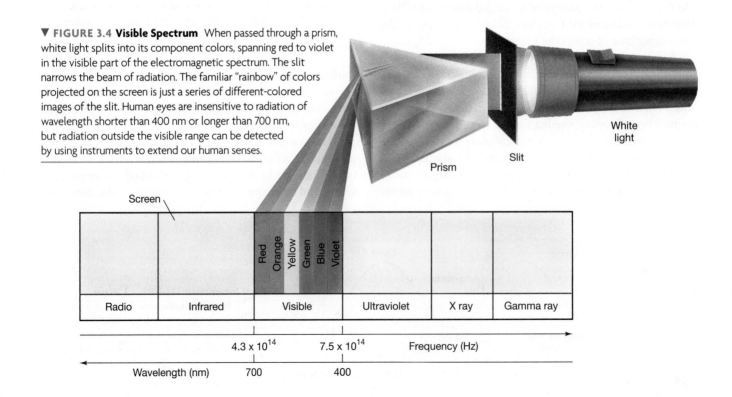

▼ **FIGURE 3.4 Visible Spectrum** When passed through a prism, white light splits into its component colors, spanning red to violet in the visible part of the electromagnetic spectrum. The slit narrows the beam of radiation. The familiar "rainbow" of colors projected on the screen is just a series of different-colored images of the slit. Human eyes are insensitive to radiation of wavelength shorter than 400 nm or longer than 700 nm, but radiation outside the visible range can be detected by using instruments to extend our human senses.

angstrom (1 Å = 10^{-10} m = 0.1 nm) is also widely used. (The unit is named after the 19th-century Swedish physicist Anders Ångstrom—pronounced "ong·strem.") However, in SI units, the nanometer is preferred. Thus, the visible spectrum covers the range of wavelengths from 400 nm to 700 nm (4000 Å to 7000 Å). The radiation to which our eyes are most sensitive has a wavelength near the middle of this range, at about 550 nm (5500 Å), in the yellow-green region of the spectrum. It is no coincidence that this wavelength falls within the range of wavelengths at which the Sun emits most of its electromagnetic energy—our eyes have evolved to take greatest advantage of the available light.

3.2 Waves in What?

Waves of radiation differ fundamentally from water waves, sound waves, or any other waves that travel through a material medium. Radiation needs *no* such medium. When light travels from a distant galaxy, or from any other cosmic object, it moves through the virtual vacuum of space. Sound waves, by contrast, cannot do this, despite what you have probably heard in almost every sci-fi movie ever made! If we were to remove all the air from a room, conversation would be impossible (even with suitable breathing apparatus to keep our test subjects alive!) because sound waves cannot exist without air or some other physical medium to support them. Communication by flashlight or radio, however, would be entirely feasible.

The ability of light to travel through empty space was once a great mystery. The idea that light, or any other kind of radiation, could move as a wave through nothing at all seemed to violate common sense, yet it is now a cornerstone of modern physics.

Interactions Between Charged Particles

To understand more about the nature of light, consider for a moment an *electrically charged* particle, such as an **electron** or a **proton.** Like mass, electrical charge is a fundamental property of matter. Electrons and protons are elementary particles—"building blocks" of atoms and all matter—that carry the basic unit of charge. Electrons are said to carry a *negative* charge, whereas protons carry an equal and opposite *positive* charge.

Just as a massive object exerts a gravitational force on every other massive body, an electrically charged particle exerts an *electrical* force on every other charged particle in the universe. ∞ (Sec. 2.7) The buildup of electrical charge (a net excess of positive over negative, or vice versa) is what causes "static cling" on your clothes when you take them out of a hot clothes dryer; it also causes the shock you sometimes feel when you touch a metal door frame on a particularly dry day.

Unlike the gravitational force, which is always attractive, electrical forces can be either attractive or repulsive. As illustrated in Figure 3.5(a), particles with *like* charges (i.e., both negative or both positive—for example, two electrons or two protons) repel one another. Particles with *unlike* charges (i.e., having opposite signs—an electron and a proton, say) attract.

How is the electrical force transmitted through space? Extending outward in all directions from any charged particle is an **electric field,** which determines the electrical force exerted by the particle on all other charged particles in the universe (Figure 3.5b). The strength of the electric field, like the strength of the gravitational field, decreases with increasing distance from the charge according to an inverse-square law. By means of the electric field, the particle's presence is "felt" by all other charged particles, near and far.

Now, suppose our particle begins to vibrate, perhaps because it becomes heated or collides with some other particle.

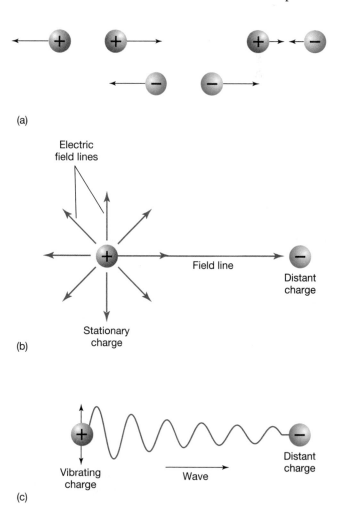

▲ **FIGURE 3.5 Charged Particles** (a) Particles carrying like electrical charges repel one another, whereas particles with unlike charges attract. (b) A charged particle is surrounded by an electric field, which determines the particle's influence on other charged particles. We represent the field by a series of field lines. (c) If a charged particle begins to vibrate, its electric field changes. The resulting disturbance travels through space as a wave, eventually interacting with distant charged particles.

Its changing position causes its associated electric field to change, and this changing field in turn causes the electrical force exerted on other charges to vary (Figure 3.5c). If we measure the change in the force on these other charges, we learn about our original particle. Thus, *information about the particle's state of motion is transmitted through space via a changing electric field.* This *disturbance* in the particle's electric field travels through space as a wave.

Electromagnetic Waves

The laws of physics tell us that a **magnetic field** must accompany every changing electric field. Magnetic fields govern the influence of *magnetized* objects on one another, much as electric fields govern interactions among charged particles. The fact that a compass needle always points to magnetic north is the result of the interaction between the magnetized needle and Earth's magnetic field (Figure 3.6). Magnetic fields also exert forces on *moving* electric charges (i.e., electric currents)—electric meters and motors rely on this basic fact. Conversely, moving charges *create* magnetic fields (electromagnets are a familiar example). In short, electric and magnetic fields are inextricably linked to one another: A change in either one necessarily creates the other.

Thus, as illustrated in Figure 3.7, the disturbance produced by the moving charge in Figure 3.5(c) actually consists of vibrating electric *and* magnetic fields, moving together through space. Furthermore, as shown in the diagram, these fields are always oriented *perpendicular* to one another and to the direction in which the wave is traveling. The fields do not exist as independent entities; rather, they are different aspects of a single physical phenomenon: **electromagnetism.** Together, they constitute an *electromagnetic wave* that carries energy and information from one part of the universe to another.

Now consider a real cosmic object—a star, say. When some of its charged contents move around, their electric fields change, and we can detect that change. The resulting electromagnetic ripples propagate (travel) outward as waves through space, requiring no material medium in which to move. Small charged particles, either in our eyes or in our experimental equipment, eventually respond to the electromagnetic field changes by vibrating in tune with the radiation that is received. This response is how we detect the radiation—how we see.

How *quickly* is one charge influenced by the change in the electromagnetic field when another charge begins to move? This is an important question because it is equivalent to asking how fast an electromagnetic wave travels. Does it propagate at some measurable speed, or is it instantaneous? Both theory and experiment tell us that all electromagnetic

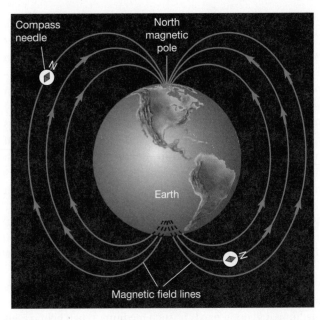

▲ **FIGURE 3.6 Magnetism** Earth's magnetic field interacts with a magnetic compass needle, causing the needle to become aligned with the field—that is, to point toward Earth's north (magnetic) pole. The north magnetic pole actually lies at latitude 80° N, longitude 107° W, some 1140 km from the geographic North Pole.

waves move at a very specific speed—the **speed of light** (always denoted by the letter *c*). Its exact value is 299,792.458 km/s in a vacuum (and somewhat less in material substances such as air or water). We will round this value off to $c = 3.00 \times 10^5$ km/s, an extremely high speed. In the time needed to snap your fingers (about a tenth of a second), light

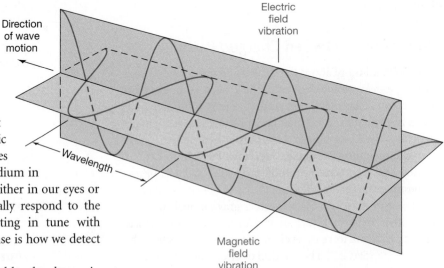

▲ **FIGURE 3.7 Electromagnetic Wave** Electric and magnetic fields vibrate perpendicularly to each other. Together they form an electromagnetic wave that moves through space at the speed of light in the direction perpendicular to both the electric and the magnetic fields comprising it.

can travel three-quarters of the way around our planet! If the currently known laws of physics are correct, then the speed of light is the fastest speed possible (see *More Precisely 22-1*).

The speed of light is very large, but it is still *finite*. That is, light does not travel instantaneously from place to place. This fact has some interesting consequences for our study of distant objects. It takes time—often lots of time—for light to travel through space. The light we see from the nearest large galaxy—the Andromeda Galaxy, shown in Figure 3.1— left that object about 2.5 million years ago, around the time our first human ancestors appeared on planet Earth. We can know nothing about this galaxy as it exists today. For all we know, it may no longer even exist! Only our descendants, 2.5 million years into the future, will know whether it exists now. So as we study objects in the cosmos, remember that the light we see left those objects long ago. We can never observe the universe as it is—only as it was.

The Wave Theory of Radiation

The description presented in this chapter of light and other forms of radiation as electromagnetic waves traveling through space is known as the **wave theory of radiation.** It is a spectacularly successful scientific theory, full of explanatory and predictive power and deep insight into the complex interplay between light and matter—a cornerstone of modern physics.

Two centuries ago, however, the wave theory stood on much less solid scientific ground. Before about 1800, scientists were divided in their opinions about the nature of light. Some thought that light was a wave phenomenon (although at the time, electromagnetism was unknown), whereas others maintained that light was in reality a stream of particles that moved in straight lines. Given the experimental apparatus available at the time, neither camp could find conclusive evidence to disprove the other's theory. *Discovery 3-1* discusses some wave properties that are of particular importance to modern astronomers and describes how their detection in experiments using visible light early in the 19th century tilted the balance of scientific opinion in favor of the wave theory.

But that's not the end of the story. The wave theory, like all good scientific theories, can and must continually be tested by experiment and observation. ∞ (Sec. 1.2) Around the turn of the 20th century, physicists made a series of discoveries about the behavior of radiation and matter on very small (atomic) scales that simply could not be explained by the "classical" wave theory just described. Changes had to be made. As we will see in Chapter 4, the modern theory of radiation is actually a hybrid of the once-rival wave and particle views, combining key elements of each in a unified and—for now—undisputed whole.

PROCESS OF SCIENCE CHECK

✔ Describe the scientific reasoning leading to the conclusion that light is an electromagnetic wave.

3.3 The Electromagnetic Spectrum

Figure 3.8 plots the entire range of electromagnetic radiation, illustrating the relationships among the different types of electromagnetic radiation listed earlier. Notice that the only characteristic distinguishing one from another is wavelength, or frequency. To the low-frequency, long-wavelength side of visible light lie *radio* and *infrared* radiation. Radio frequencies include radar, microwave radiation, and the familiar AM, FM, and TV bands. We perceive infrared radiation as heat. At higher frequencies (shorter wavelengths) are the domains of *ultraviolet, X-ray,* and *gamma-ray* radiation. Ultraviolet radiation, lying just beyond the violet end of the visible spectrum, is responsible for suntans and sunburns. The shorter-wavelength X rays are perhaps best known for their ability to penetrate human tissue and reveal the state of our insides without resorting to surgery. Gamma rays are the shortest-wavelength radiation. They are often associated with radioactivity and are invariably damaging to living cells they encounter.

The Spectrum of Radiation

All these spectral regions, including the visible spectrum, collectively make up the **electromagnetic spectrum.** Remember that, despite their greatly differing wavelengths and the different roles they play in everyday life on Earth, all are basically the same phenomenon, and all move at the same speed—the speed of light, *c.*

Figure 3.8 is worth studying carefully, as it contains a great deal of information. Note that wave frequency (in hertz) increases from left to right, and wavelength (in meters) increases from right to left. Scientists often disagree on the "correct" way to display wavelengths and frequencies in diagrams of this type. In picturing wavelengths and frequencies, this book consistently adheres to the convention that frequency increases toward the *right.*

Notice also that the wavelength and frequency scales in Figure 3.8 do not increase by equal increments of 10. Instead, successive values marked on the horizontal axis differ by *factors of 10*—each is 10 times greater than its neighbor. This type of scale, called a *logarithmic* scale, is often used in science to condense a large range of some quantity into a manageable size. Had we used a linear scale for the wavelength range shown in the figure, it would have been many light-years long! Throughout the text, we will often find it convenient to use a logarithmic scale to compress a wide range of some quantity onto a single, easy-to-view plot.

Figure 3.8 shows that wavelengths extend from the size of mountains (radio radiation) to the size of an atomic nucleus (gamma-ray radiation). The box at the upper right emphasizes how small the visible portion of the electromagnetic spectrum is. Most objects in the universe emit large amounts of invisible radiation. Indeed, many of them emit

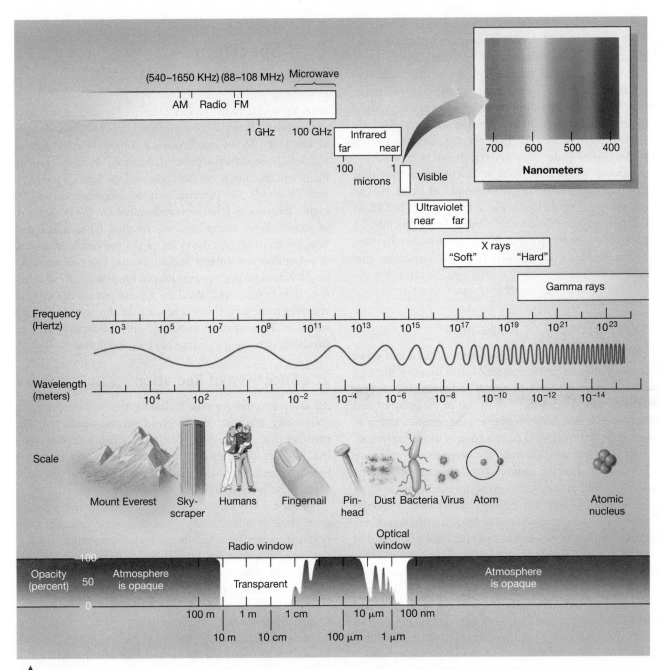

Interactive FIGURE 3.8 Electromagnetic Spectrum The entire electromagnetic spectrum, running from long-wavelength, low-frequency radio waves to short-wavelength, high-frequency gamma rays.

only a tiny fraction of their total energy in the visible range. A wealth of extra knowledge can be gained by studying the invisible regions of the electromagnetic spectrum.

To remind you of this important fact and to identify the region of the electromagnetic spectrum in which a particular observation was made, we have attached the following spectrum icon—an idealized version of the wavelength scale in Figure 3.8—to every astronomical image presented in this text [R I V U X G]. Hence, we can tell at a glance from the highlighted "V" that, for example, Figure 3.1 (p. 58) is an image made with the use of visible

light, whereas the first image in Figure 3.11 (p. 69) was captured in the radio ("R") part of the spectrum. Chapter 5 discusses in more detail how astronomers actually make such observations, using telescopes and sensitive detectors tailored to different electromagnetic waves.

Atmospheric Opacity

Only a small fraction of the radiation produced by astronomical objects actually reaches Earth's surface, because of the **opacity** of our planet's atmosphere. Opacity is the extent

DISCOVERY 3-1

The Wave Nature of Radiation

Until the early 19th century, debate raged in scientific circles regarding the true nature of light. On the one hand, the particle, or *corpuscular,* theory, proposed by Isaac Newton, held that light consisted of tiny particles moving in straight lines. Different colors were presumed to correspond to different particles. On the other hand, the *wave* theory, championed by the 17th-century Dutch astronomer Christian Huygens, viewed light as a wave phenomenon, in which color was determined by frequency, or wavelength. During the first few decades of the 19th century, growing experimental evidence that light displayed two key wave properties—*diffraction* and *interference*—argued strongly in favor of the wave theory.

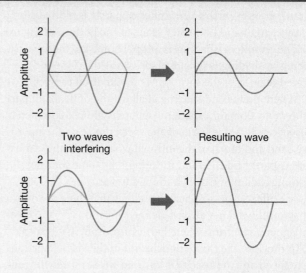

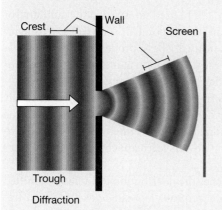

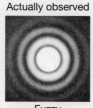

Diffraction is the deflection, or "bending," of a wave as it passes a corner or moves through a narrow gap. We might expect that light passing through a sharp-edged hole in a barrier would produce a sharp shadow, especially if radiation were composed of rays or particles moving in perfectly straight lines. As depicted in the first figure, however, closer inspection reveals that the shadow actually has a "fuzzy" edge, as shown in the photograph at the right—the diffraction pattern produced by a small circular opening.

We are not normally aware of such effects in everyday life, because diffraction is generally very small for visible light. For any wave, the amount of diffraction is proportional to the ratio of the wavelength to the width of the gap. The longer the wavelength or the smaller the gap, the greater is the angle through which the wave is diffracted. Thus, visible light, with its extremely short wavelengths, shows perceptible diffraction only when passing through very narrow openings. (The effect is much more noticeable for sound waves. No one thinks twice about our ability to hear people, even when they are around a corner and out of our line of sight.)

Interference is the ability of two or more waves to reinforce or diminish each other. The second figure shows two sets of waves moving through the same region of space. The waves are positioned so that their crests and troughs exactly coincide.

In the upper frames, the waves have the same wavelength, but the green one has twice the amplitude in the opposite direction to the orange one. The net effect is that the two wave motions interfere with each other, resulting in the wave at the right. This phenomenon is known as *destructive interference*. When the waves reinforce each other instead, as in the lower frames, the effect is called *constructive interference*.

As with diffraction, interference between waves of visible light is not noticeable in everyday experience, but it is readily measured in the laboratory. The final photograph shows the characteristic *interference pattern* that results when two identical light sources are placed side by side. The light and dark bands are formed by constructive and destructive interference of the beams from the two sources. This classic experiment, first performed by English physicist Thomas Young around 1805, was instrumental in establishing the wave nature of radiation.

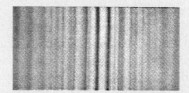

Both diffraction and interference are predicted by the wave theory of light. The particle theory did not predict them; in fact, it predicted that they should *not* occur. By the 1830s, experimenters had reported clear measurements of both phenomena, convincing most scientists that the wave theory was the proper description of electromagnetic radiation. It would be almost a century before the particle description of radiation would resurface, but in a radically different form, as we will see in Chapter 4.

ANIMATION/VIDEO Fresnel Diffraction

MA

to which radiation is blocked by the material through which it is passing—in this case, air. The more opaque an object is, the less radiation gets through it: Opacity is just the opposite of transparency. Earth's atmospheric opacity is plotted along the wavelength and frequency scales at the bottom of Figure 3.8. The extent of shading is proportional to the opacity. Where the shading is greatest (such as at the X-ray or "far" infrared regions of the spectrum), no radiation can get in or out. Where there is no shading at all (in the optical and part of the radio domain), the atmosphere is almost completely transparent. In some parts of the spectrum (e.g., the microwave band and much of the infrared portion), Earth's atmosphere is partly transparent, meaning that some, but not all, incoming radiation makes it to the surface.

The effect of atmospheric opacity is that there are only a few *spectral windows* at well-defined locations in the electromagnetic spectrum where Earth's atmosphere is transparent. In much of the radio domain and in the visible portions of the spectrum, the opacity is low, and we can study the universe at those wavelengths from ground level. In parts of the infrared range, the atmosphere is partially transparent, so we can make certain infrared observations from the ground. Moving to the tops of mountains, above as much of the atmosphere as possible, improves observations. In the rest of the spectrum, however, the atmosphere is opaque: Ultraviolet, X-ray, and gamma-ray observations can be made only from above the atmosphere, from orbiting satellites.

What causes opacity to vary along the spectrum? Certain atmospheric gases absorb radiation very efficiently at some wavelengths. For example, water vapor (H_2O) and oxygen (O_2) absorb radio waves having wavelengths less than about a centimeter, whereas water vapor and carbon dioxide (CO_2) are strong absorbers of infrared radiation. Ultraviolet, X-ray, and gamma-ray radiation are completely blocked by the *ozone* (O_3) *layer* high in Earth's atmosphere (see Section 7.2). A passing, but unpredictable, source of atmospheric opacity in the visible part of the spectrum is the blockage of light by atmospheric clouds.

In addition, the interaction between the Sun's ultraviolet radiation and the upper atmosphere produces a thin, electrically conducting layer at an altitude of about 100 km. The *ionosphere*, as this layer is known, reflects long-wavelength radio waves (wavelengths greater than about 10 m) as well as a mirror reflects visible light. In this way, extraterrestrial waves are kept out, and terrestrial waves—such as those produced by AM radio stations—are kept in. (That's why it is possible to transmit some radio frequencies beyond the horizon—the broadcast waves bounce off the ionosphere.)

CONCEPT CHECK

✔ In what sense are radio waves, visible light, and X rays one and the same phenomenon?

3.4 Thermal Radiation

All macroscopic objects—fires, ice cubes, people, stars—emit radiation at all times, regardless of their size, shape, or chemical composition. They radiate mainly because the microscopic charged particles they are made up of are in constantly varying random motion, and whenever charges interact ("collide") and change their state of motion, electromagnetic radiation is emitted. The **temperature** of an object is a direct measure of the amount of microscopic motion within it (see *More Precisely 3-1*). The hotter the object—that is, the higher its temperature—the faster its component particles move, the more violent are their collisions, and the more energy they radiate.

The Blackbody Spectrum

Intensity is a term often used to specify the amount or strength of radiation at any point in space. Like frequency and wavelength, intensity is a basic property of radiation. No natural object emits all its radiation at just one frequency. Instead, because particles collide at many different speeds—some gently, others more violently—the energy is generally spread out over a range of frequencies. By studying how the intensity of this radiation is distributed across the electromagnetic spectrum, we can learn much about the object's properties.

Figure 3.9 sketches the distribution of radiation emitted by an object. The curve peaks at a single, well-defined frequency and falls off to lesser values above and below that frequency. Note that the curve is not shaped like a symmetrical bell that declines evenly on either side of the peak. Instead, the intensity falls off more slowly from the peak to lower frequencies than it does on the high-frequency side. This overall shape is characteristic of the thermal radiation emitted by *any* object, regardless of its size, shape, composition, or temperature.

The curve drawn in Figure 3.9(a) is the radiation-distribution curve for a mathematical idealization known as a *blackbody*—an object that absorbs all radiation falling on it. In a steady state, a blackbody must reemit the same amount of energy it absorbs. The **blackbody curve** shown in the figure describes the distribution of that reemitted radiation. (The curve is also known as the *Planck curve*, after Max Planck, the German physicist whose mathematical analysis of such thermal emission in 1900 played a key role in the development of modern physics.)

No real object absorbs and radiates as a perfect blackbody. For example, the Sun's actual curve of emission is shown in Figure 3.9(b). However, in many cases, the blackbody curve is a good approximation to reality, and the properties of blackbodies provide important insights into the behavior of real objects.

MORE PRECISELY 3-1

The Kelvin Temperature Scale

The atoms and molecules that make up any piece of matter are in constant random motion. This motion represents a form of energy known as *thermal energy*, or, more commonly, *heat*. The quantity we call *temperature* is a direct measure of an object's internal motion: The higher the object's temperature, the faster, on average, is the random motion of its constituent particles. Note that the two concepts, though obviously related, are different. The temperature of a piece of matter specifies the *average* thermal energy of the particles it contains.

Our familiar Fahrenheit temperature scale, like the archaic English system in which length is measured in feet and weight in pounds, is of somewhat dubious value. In fact, the "degree Fahrenheit" is now a peculiarity of American society. Most of the world uses the Celsius scale of temperature measurement (also called the centigrade scale). In the Celsius system, water freezes at 0 degrees (0°C) and boils at 100 degrees (100°C), as illustrated in the accompanying figure.

There are, of course, temperatures below the freezing point of water. In principle, temperatures can reach as low as −273.15°C (although we know of no matter anywhere in the universe that is actually that cold). Known as *absolute zero*, this is the temperature at which, theoretically, all thermal atomic and molecular motion ceases. Since no object can have a temperature below that value, scientists find it convenient to use a temperature scale that takes absolute zero as its starting point. This scale is called the *Kelvin scale*, in honor of the 19th-century British physicist Lord Kelvin. Since it starts at absolute zero, the Kelvin scale differs from the Celsius scale by 273.15°. In this book, we round off the decimal places and simply use

$$\text{kelvins} = \text{degrees Celsius} + 273.$$

Thus,

- all thermal motion ceases at 0 kelvins (0 K).
- water freezes at 273 kelvins (273 K).
- water boils at 373 kelvins (373 K).

Note that the unit is *kelvins*, or *K*, but not *degrees kelvin* or *°K*. (Occasionally, the term *degrees absolute* is used instead.)

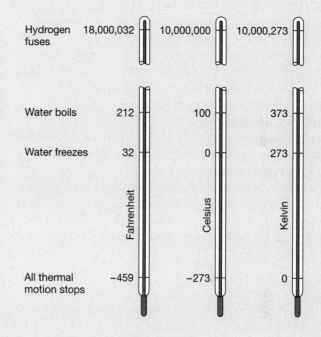

	Fahrenheit	Celsius	Kelvin
Hydrogen fuses	18,000,032	10,000,000	10,000,273
Water boils	212	100	373
Water freezes	32	0	273
All thermal motion stops	−459	−273	0

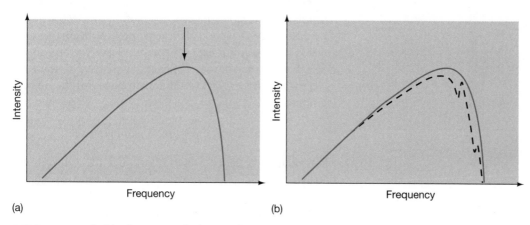

(a)   (b)

▲ FIGURE 3.9 **Blackbody Curves, Ideal vs. Reality** The blackbody, or Planck, curve represents the spread of the intensity of radiation emitted by any object over all possible frequencies. The arrow indicates the frequency of peak emission. Note the contrast between the clean, "textbook" case (a) with a real graph (dashed) of the Sun's emission (b). Absorption in the atmospheres of the Sun and Earth causes the difference.

The Radiation Laws

The blackbody curve shifts toward higher frequencies (shorter wavelengths) and greater intensities as an object's temperature increases. Even so, the *shape* of the curve remains the same. This shifting of radiation's peak frequency with temperature is familiar to us all: Very hot glowing objects, such as toaster filaments or stars, emit visible light. Cooler objects, such as warm rocks, household radiators, or people, produce invisible radiation—warm to the touch, but not glowing hot to the eye. These latter objects emit most of their radiation in the lower frequency infrared part of the electromagnetic spectrum (Figure 3.8).

Imagine a piece of metal placed in a hot furnace. At first, the metal becomes warm, although its visual appearance doesn't change. As it heats up, the metal begins to glow dull red, then orange, brilliant yellow, and finally white. How do we explain this phenomenon? As illustrated in Figure 3.10, when the metal is at room temperature (300 K—see *More Precisely 3-1* for a discussion of the Kelvin temperature scale), it emits only invisible infrared radiation. As the metal becomes hotter, the peak of its blackbody curve shifts toward higher frequencies. At 1000 K, for instance, most of the emitted radiation is still infrared, but now there is also a small amount of visible (dull red) radiation being emitted. (Note that the high-frequency portion of the 1000 K curve just overlaps the visible region of the graph.)

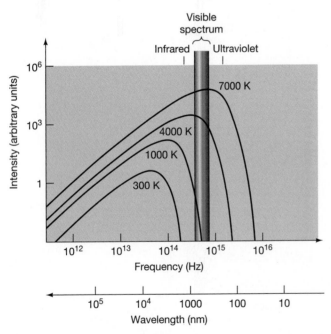

▲ FIGURE 3.10 **Multiple Blackbody Curves** As an object is heated, the radiation it emits peaks at higher and higher frequencies. Shown here are curves corresponding to temperatures of 300 K (room temperature), 1000 K (beginning to glow dull red), 4000 K (red hot), and 7000 K (white hot).

As the temperature continues to rise, the peak of the metal's blackbody curve moves through the visible spectrum, from red (the 4000 K curve) through yellow. Eventually, the metal becomes white hot because, when its blackbody curve peaks in the blue or violet part of the spectrum (the 7000 K curve), the low-frequency tail of the curve extends through the entire visible spectrum (to the left in the figure), meaning that substantial amounts of green, yellow, orange, and red light are also emitted. Together, all these colors combine to produce white.

From studies of the precise form of the blackbody curve, we obtain a very simple connection between the wavelength at which most radiation is emitted and the absolute temperature (i.e., the temperature measured in kelvins) of the emitting object:

$$\text{wavelength of peak emission} \propto \frac{1}{\text{temperature}}.$$

(Recall that the symbol "$\propto$" here just means "is proportional to.") This relationship is called **Wien's law,** after Wilhelm Wien, the German scientist who formulated it in 1897.

Simply put, Wien's law tells us that the hotter the object, the bluer is its radiation. For example, an object with a temperature of 6000 K emits most of its energy in the visible part of the spectrum, with a peak wavelength of 480 nm. At 600 K, the object's emission would peak at a wavelength of 4800 nm, well into the infrared portion of the spectrum. At a temperature of 60,000 K, the peak would move all the way through the visible spectrum to a wavelength of 48 nm, in the ultraviolet range (see Figure 3.11).

It is also a matter of everyday experience that, as the temperature of an object increases, the *total* amount of energy it radiates (summed over all frequencies) increases rapidly. For example, the heat given off by an electric heater increases very sharply as it warms up and begins to emit visible light. Careful experimentation leads to the conclusion that the total amount of energy radiated per unit time is actually proportional to the fourth power of the object's temperature:

$$\text{total energy emission} \propto \text{temperature}^4.$$

This relation is called **Stefan's law,** after the 19th-century Austrian physicist Josef Stefan. From the form of Stefan's law, we can see that the energy emitted by a body rises dramatically as its temperature increases. Doubling the temperature causes the total energy radiated to increase by a factor of $2^4 = 16$; tripling the temperature increases the emission by $3^4 = 81$, and so on.

The radiation laws are presented in more detail in *More Precisely 3-2.*

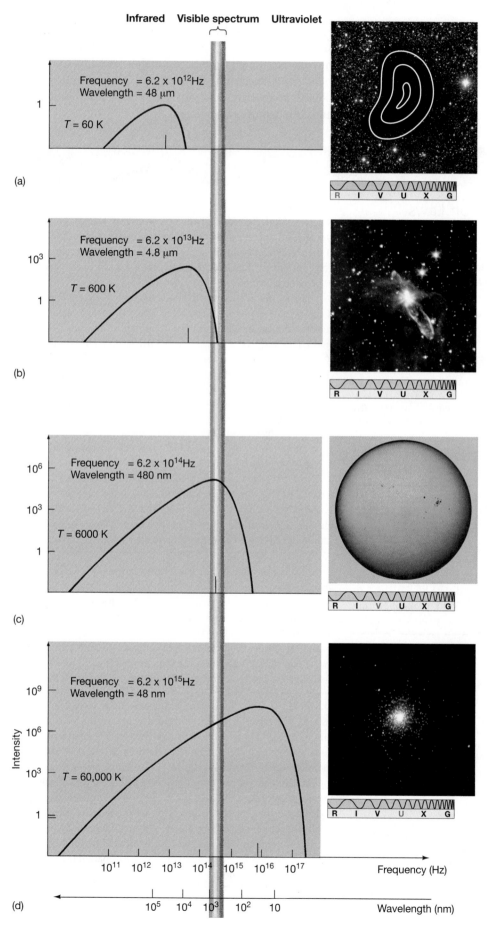

Infrared   Visible spectrum   Ultraviolet

(a)

Frequency = 6.2 x 10¹²Hz
Wavelength = 48 μm

T = 60 K

(b)

Frequency = 6.2 x 10¹³Hz
Wavelength = 4.8 μm

T = 600 K

(c)

Frequency = 6.2 x 10¹⁴Hz
Wavelength = 480 nm

T = 6000 K

(d)

Frequency = 6.2 x 10¹⁵Hz
Wavelength = 48 nm

T = 60,000 K

Intensity

Frequency (Hz)

Wavelength (nm)

Interactive **FIGURE 3.11**
Astronomical Thermometer
Comparison of blackbody curves of four cosmic objects. The frequencies and wavelengths corresponding to peak emission are marked. (a) A cool, dark galactic gas cloud called Barnard 68. At a temperature of 60 K, it emits mostly radio radiation, shown here as overlaid contours. (b) A dim, young star (shown white in the inset photograph) called Herbig-Haro 46. The star's atmosphere, at 600 K, radiates mainly in the infrared. (c) The Sun's surface, at approximately 6000 K, is brightest in the visible region of the electromagnetic spectrum. (d) Some very hot, bright stars in a cluster called Messier 2, as observed by an orbiting space telescope above Earth's atmosphere. At a temperature of 60,000 K, these stars radiate strongly in the ultraviolet. In all cases, application of the radiation laws allows astronomers to probe the properties of otherwise inaccessible objects. *(ESO; AURA; SST; GALEX)*

Astronomical Applications

No known natural terrestrial objects reach temperatures high enough to emit very high frequency radiation. Only human-made thermonuclear explosions are hot enough for their spectra to peak in the X-ray or gamma-ray range. (Most human inventions that produce short-wavelength, high-frequency radiation, such as X-ray machines, are designed to emit only a specific range of wavelengths and do not operate at high temperatures. They are said to produce a *nonthermal* spectrum of radiation.) Many extraterrestrial objects, however, do emit large amounts of ultraviolet, X-ray, and even gamma-ray radiation.

MORE PRECISELY 3-2

More About the Radiation Laws

As mentioned in Section 3.4, *Wien's law* relates the temperature T of an object to the wavelength λ_{max} at which the object emits the most radiation. (The Greek letter λ—lambda—is conventionally used to denote wavelength.) Mathematically, if we measure T in kelvins and λ_{max} in millimeters, we can determine the constant of proportionality in the relation presented in the text, to find that

$$\lambda_{max} = \frac{2.9 \text{ mm}}{T}.$$

We could also convert Wien's law into an equivalent statement about frequency f, using the relation $f = c/\lambda$ (see Section 3.1), where c is the speed of light, but the law is most commonly stated in terms of wavelength and is probably easier to remember that way.

EXAMPLE 1 For a blackbody with the same temperature T as the surface of the Sun (≈ 6000 K), the wavelength of maximum intensity is $\lambda_{max} = (2.9/6000)$ mm, or 480 nm, corresponding to the yellow-green part of the visible spectrum. A cooler star with a temperature of $T = 3000$ K has a peak wavelength of $\lambda_{max} = (2.9/3000)$ mm ≈ 970 nm, just beyond the red end of the visible spectrum, in the near infrared. The blackbody curve of a hotter star with a temperature of 12,000 K peaks at 242 nm, in the near ultraviolet, and so on.

In fact, this application—simply looking at the spectrum and determining where it peaks—is an important way of estimating the temperature of planets, stars, and other objects throughout the universe and will be used extensively throughout the text.

We can also give *Stefan's law* a more precise mathematical formulation. With T measured in kelvins, the total amount of energy emitted per square meter of the body's surface per second (a quantity known as the *energy flux F*) is given by

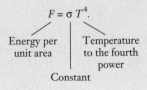

$$F = \sigma T^4.$$

Energy per unit area | Temperature to the fourth power

Constant

This equation is usually referred to as the *Stefan-Boltzmann* equation. Stefan's student, Ludwig Boltzmann, was an Austrian physicist who played a central role in the development of the laws of thermodynamics during the late 19th and early 20th centuries. The constant σ (the Greek letter sigma) is known as the Stefan-Boltzmann constant.

The SI unit of energy is the *joule* (J). Probably more familiar is the closely related unit called the *watt* (W), which measures power—the *rate* at which energy is emitted or expended by an object. One watt is the emission of 1 J per second. For example, a 100-W lightbulb emits energy (mostly in the form of infrared and visible light) at a rate of 100 J/s. In SI units, the Stefan-Boltzmann constant has the value $\sigma = 5.67 \times 10^{-8}$ W/m$^2 \cdot$ K^4.

EXAMPLE 2 Notice just how rapidly the energy flux increases with increasing temperature. A piece of metal in a furnace, when at a temperature of $T = 3000$ K, radiates energy at a rate of $\sigma T^4 \times (1 \text{ cm})^2 = 5.67 \times 10^{-8}$ W/m$^2 \cdot$ K$^4 \times (3000 \text{ K})^4 \times (0.01 \text{ m})^2 = 460$ W for every square centimeter of its surface area. Doubling this temperature to 6000 K (so that the metal becomes yellow hot, by Wien's law), the surface temperature of the Sun, increases the energy emitted by a factor of 16 (four "doublings"), to 7.3 *kilo*watts (7300 W) per square centimeter.

Finally, note that Stefan's law relates to energy emitted *per unit area*. The flame of a blowtorch is considerably hotter than a bonfire, but the bonfire emits far more energy *in total* because it is much larger. Thus, in computing the total energy emitted from a hot object, both the object's temperature *and* its surface area must be taken into account. This fact is of great importance in determining the "energy budget" of planets and stars, as we will see in later chapters. The next example illustrates the point in the case of the Sun.

EXAMPLE 3 The Sun's temperature is approximately $T = 5800$ K. (The earlier example used a rounded-off version of this number.) Thus, by Stefan's law, each square meter of the Sun's surface radiates energy at a rate of $\sigma T^4 = 6.4 \times 10^7$ W (64 megawatts). By measuring the Sun's angular size and knowing its distance, we can employ simple geometry to determine the solar radius. ∞ (Secs. 1.6, 2.6) The answer is $R = 700,000$ km, or 7×10^8 m, allowing us to calculate the Sun's total surface area as $4\pi R^2 = 6.2 \times 10^{18}$ m^2. Multiplying by the energy emitted per unit area, we find that the Sun's total energy emission (or *luminosity*) is 4×10^{26} W—a remarkable number that we obtained without ever leaving Earth!

Astronomers often use blackbody curves as thermometers to determine the temperatures of distant objects. For example, an examination of the solar spectrum indicates the temperature of the Sun's surface. Observations of the radiation from the Sun at many frequencies yield a curve shaped somewhat like that shown in Figure 3.9. The Sun's curve peaks in the visible part of the electromagnetic spectrum; the Sun also emits a lot of infrared and a little ultraviolet radiation. Using Wien's law, we find that the temperature of the Sun's surface is approximately 6000 K. (A more precise measurement, applying Wien's law to the blackbody curve that best fits the solar spectrum, yields a temperature of 5800 K.)

Other cosmic objects have surfaces very much cooler or hotter than the Sun's, emitting most of their radiation in invisible parts of the spectrum. For example, the relatively cool surface of a very young star may measure 600 K and emit mostly infrared radiation. Cooler still is the interstellar gas cloud from which the star formed; at a temperature of 60 K, such a cloud emits mainly long-wavelength radiation in the radio and infrared parts of the spectrum. The brightest stars, by contrast, have surface temperatures as high as 60,000 K and hence emit mostly ultraviolet radiation (see Figure 3.11).

CONCEPT CHECK

✔ Describe, in terms of the radiation laws, how and why the appearance of an incandescent lightbulb changes as you turn a dimmer switch to increase its brightness from "off" to "maximum."

3.5 The Doppler Effect

Imagine a rocket ship launched from Earth with enough fuel to allow it to accelerate to speeds approaching that of light. As the ship's speed increased, a remarkable thing would happen (Figure 3.12). Passengers would notice that the light from the star system toward which they were traveling seemed to be getting *bluer*. In fact, *all* stars in front of the ship would appear bluer than normal, and the greater the ship's speed, the greater the color change would be. Furthermore, stars behind the vessel would seem *redder* than normal, but stars to either side would be unchanged in appearance. As the spacecraft slowed down and came to rest relative to Earth, all stars would resume their usual appearance.

The travelers would have to conclude that the stars had changed their colors, not because of any real change in their physical properties, but because of the spacecraft's own *motion*. This motion-induced change in the observed frequency of a wave is known as the **Doppler effect,** in honor of Christian Doppler, the 19th-century Austrian physicist who first explained it in 1842. This phenomenon is not restricted to electromagnetic radiation and fast-moving

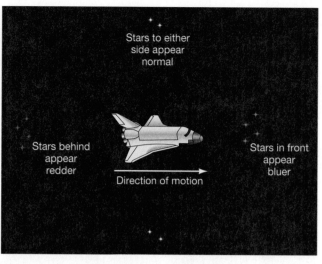

▲ FIGURE 3.12 **High-Speed Observers** Observers in a fast-moving spacecraft see the stars ahead of them bluer than normal, while those behind are reddened. The stars have not actually changed their properties—the color changes result from the observers' motion relative to the stars.

spacecraft. Waiting at a railroad crossing for an express train to pass, most of us have had the experience of hearing the pitch of a train whistle change from high shrill (high frequency, short wavelength) to low blare (low frequency, long wavelength) as the train approaches and then recedes. The explanation is basically the same. Applied to cosmic sources of electromagnetic radiation, the Doppler effect has become one of the most important measurement techniques in all of modern astronomy. Here's how it works.

Imagine a wave moving from the place where it is created toward an observer who is not moving with respect to the source of the wave, as shown in Figure 3.13(a). By noting the distances between successive crests, the observer can determine the wavelength of the emitted wave. Now suppose that not just the wave, but the *source* of the wave, also is moving. As illustrated in Figure 3.13(b), because the source moves between the times of emission of one crest and the next, successive crests in the direction of motion of the source will be seen to be *closer together* than normal, whereas crests behind the source will be more widely spaced. An observer in front of the source will therefore measure a *shorter* wavelength than normal, whereas one behind will see a *longer* wavelength. The numbers indicate (a) successive crests emitted by the source and (b) the location of the source at the instant each crest was emitted.

The greater the relative speed between source and observer, the greater is the observed shift. If the other velocities involved are not too large compared with the wave speed—less than a few percent, say—we can write down a particularly simple formula for what the observer sees. In

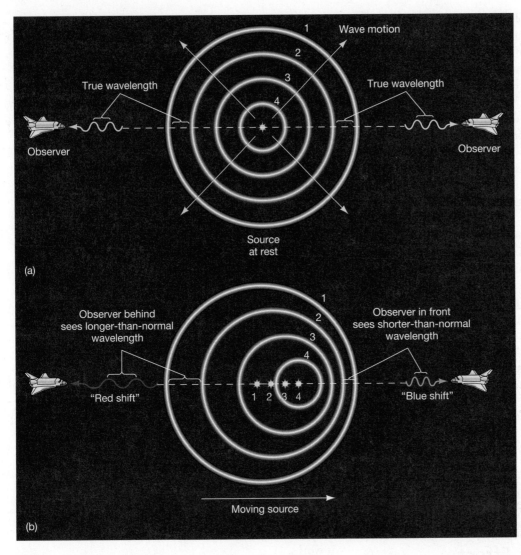

(a)

(b)

Interactive FIGURE 3.13 Doppler Effect

(a) Wave motion from a source toward an observer at rest with respect to the source. The four numbered circles represent successive wave crests emitted by the source. At the instant shown, the fifth crest is just about to be emitted. As seen by the observer, the source is not moving, so the wave crests are just concentric spheres (shown here as circles). (b) Waves from a moving source tend to "pile up" in the direction of motion and be "stretched out" on the other side. (The numbered points indicate the location of the source at the instant each wave crest was emitted.) As a result, an observer situated in front of the source measures a shorter-than-normal wavelength—a *blueshift*—while an observer behind the source sees a *redshift*. In this diagram, the source is shown in motion. However, the same general statements hold whenever there is any relative motion between source and observer, allowing astronomers to probe the motions of distant objects.

terms of the net velocity of *recession* between source and observer, the apparent wavelength and frequency (measured by the observer) are related to the true quantities (emitted by the source) as follows:

$$\frac{\text{apparent wavelength}}{\text{true wavelength}} = \frac{\text{true frequency}}{\text{apparent frequency}}$$

$$= 1 + \frac{\text{recession velocity}}{\text{wave speed}}.$$

The recession velocity measures the rate at which the distance between the source and the observer is changing. A positive recession velocity means that the two are moving apart; a negative velocity means that they are approaching.

The wave speed is the speed of light, *c*, in the case of electromagnetic radiation. For most of this text, the assumption that the recession velocity is small compared to the speed of light will be a good one. Only when we discuss the properties of black holes (Chapter 22) and the structure of

the universe on the largest scales (Chapters 25 and 26) will we have to reconsider this formula.

Note that in Figure 3.13 the *source* is shown in motion (as in our train analogy), whereas in our earlier spaceship example (Figure 3.12) the *observers* were in motion. For electromagnetic radiation, the result is the same in either case—only the *relative* motion between source and observer matters. Note also that only motion along the line joining source and observer—known as *radial* motion—appears in the foregoing equation. Motion that is *transverse* (perpendicular) to the line of sight has no significant effect.* Notice, incidentally, that the Doppler effect depends only on the *relative motion* between source and

*In fact, Einstein's theory of relativity (see Chapter 22) implies that when the transverse velocity is comparable to the speed of light, a change in wavelength, called the transverse Doppler shift, does occur. For most terrestrial and astronomical applications, however, this shift is negligibly small, and we will ignore it here.

MORE PRECISELY 3-3

Measuring Velocities with the Doppler Effect

Because the speed of light, c, is so large—300,000 km/s—the Doppler effect is extremely small for everyday terrestrial velocities. For example, consider a source receding from the observer at Earth's orbital speed of 30 km/s, a velocity much greater than any encountered in day-to-day life. Using the formula in the text, we find that the shift in wavelength of a beam of blue light would be just

$$\frac{\text{change in wavelength}}{\text{true wavelength}} = \frac{\text{recession velocity}}{\text{wave speed}}$$

$$= \frac{30 \text{ km/s}}{300,000 \text{ km/s}} = 0.01 \text{ percent.}$$

That is, the wavelength would lengthen from 400 nm to 400.04 nm—a very small change indeed, and one that the human eye cannot distinguish. However, it *is* easily detectable with modern instruments.

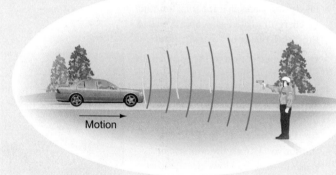

Astronomers can use the Doppler effect to find the line-of-sight speed of any cosmic object simply by measuring the extent to which its light is redshifted or blueshifted. To see how, let's use a simple example.

EXAMPLE Suppose that the beam of blue light just mentioned is observed to have a wavelength of 401 nm, instead of the 400 nm with which it was emitted. (Let's defer until the next chapter the question of *how* an observer might know the wavelength of the emitted light.) Using the earlier equation, rewritten as

$$\frac{\text{recession velocity}}{\text{wave speed, c}} = \frac{\text{change in wavelength}}{\text{true wavelength}}$$

$$= \frac{1 \text{ nm}}{400 \text{ nm}} = 0.0025,$$

the observer could calculate the source's recession velocity to be 0.0025 times 300,000 km/s, or 750 km/s.

The basic reasoning is simple, but very powerful. The motions of nearby stars and distant galaxies—even the expansion of the universe itself—have all been measured in this way.

Motorists stopped for speeding on the highway have experienced another, much more down-to-earth application: As illustrated in the accompanying figure, police radar measures speed by means of the Doppler effect, as do the radar guns used to clock the velocity of a pitcher's fastball or a tennis player's serve. As shown in the illustration, the reflected radiation (blue crests) from the oncoming car is shifted to shorter wavelengths by an amount proportional to the car's speed. Gotcha!

---

observer; it does not depend on the *distance* between them in any way.

A wave measured by an observer situated in front of a moving source is said to be *blueshifted*, because blue light has a shorter wavelength than red light. Similarly, an observer situated behind the source will measure a longer-than-normal wavelength—the radiation is said to be *redshifted*. This terminology is used even for invisible radiation, for which "red" and "blue" have no meaning. Any shift toward shorter wavelengths is called a blueshift, and any shift toward longer wavelengths is called a redshift. For example, ultraviolet radiation might be blueshifted into the X-ray part of the spectrum or redshifted into the visible; infrared radiation could be redshifted into the microwave range, and so on. *More Precisely 3-3* describes how the Doppler effect is used in practice to measure velocities in astronomy.

In practice, it is hard to measure the Doppler shift of an entire blackbody curve, simply because it is spread over many wavelengths, making small shifts hard to determine with any accuracy. However, if the radiation is more narrowly defined and takes up just a narrow "sliver" of the spectrum, then precise measurements of Doppler effect *can* be made. We will see in the next chapter that in many circumstances this is precisely what does happen, making the Doppler effect one of the observational astronomer's most powerful tools.

CONCEPT CHECK

✔ Astronomers observe two stars orbiting one another. How might the Doppler effect be useful in determining the masses of the stars?

CHAPTER REVIEW

SUMMARY

1 **Visible light (p. 58)** is a particular type of **electromagnetic radiation (p. 58)** and travels through space in the form of a **wave (p. 58)**. A wave is char-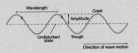
acterized by its **period (p. 59)**, the length of time taken for one complete cycle; its **wavelength (p. 59)**, the distance between successive wave crests; and its **amplitude (p. 59)**, which measures the size of the disturbance associated with the wave. A wave's **frequency (p. 59)** is the reciprocal of the period—it counts the number of wave crests that pass a given point in one second.

2 **Electrons (p. 61)** and **protons (p. 61)** are elementary particles that carry equal and opposite electrical charges. Any elec-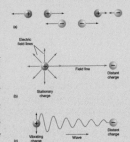
trically charged object is surrounded by an **electric field (p. 61)** that determines the force the object exerts on other charged objects. When a charged particle moves, information about its motion is transmitted throughout the universe by the particle's changing electric and magnetic fields. According to the **wave theory of radiation (p. 63)**, the information travels through space in the form of a wave at the **speed of light (p. 62)**. Because both electric and **magnetic fields (p. 62)** are involved, the phenomenon is known as **electromagnetism (p. 62)**. **Diffraction (p. 65)** and **interference (p. 65)** are properties of radiation that mark it as a wave phenomenon.

3 A beam of white light is bent, or re-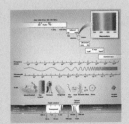
fracted, as it passes through a prism. Different frequencies of light within the beam are refracted by different amounts, so the beam is split up into its component colors—the visible spectrum. The color of visible light is simply a measure of its wavelength—red light has a longer wavelength than blue light. The entire **electromagnetic spectrum**

(p. 63) consists of (in order of increasing frequency) **radio waves, infrared radiation, visible light, ultraviolet radiation, X rays,** and **gamma rays (p. 58)**. Only radio waves, some infrared wavelengths, and visible light can penetrate the atmosphere and reach the ground from space.

4 The **temperature (p. 66)** of an object is a measure of the speed with which its constituent particles move. The intensity of radiation of different frequencies emitted by a hot object has a characteristic distribution,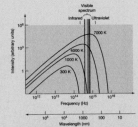
called a **blackbody curve (p. 66)**, that depends only on the object's temperature.

5 **Wien's law (p. 68)** tells us that the wavelength at which the object radiates most of its energy is inversely propor-
tional to the temperature of the object. Measuring that peak wavelength tells us the object's temperature. **Stefan's law (p. 68)** states that the total amount of energy radiated is proportional to the fourth power of the temperature.

6 Our perception of the wavelength of a beam of light can be altered by our velocity relative to the source. This motion-induced change in the observed frequency of a wave is called the **Doppler effect (p. 71)**. Any net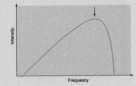
motion away from the source causes a redshift—a shift to lower frequencies—in the received beam. Motion toward the source causes a blueshift. The extent of the shift is directly proportional to the observer's recession velocity relative to the source, providing a way to measure the velocity of a distant object by observing the radiation it emits.

Mastering ASTRONOMY *For instructor-assigned homework go to* **www.masteringastronomy.com**

Problems labeled **POS** explore the process of science | **VIS** problems focus on reading and interpreting visual information

REVIEW AND DISCUSSION

1. What is a wave?

2. Define the following wave properties: period, wavelength, amplitude, and frequency.

3. What is the relationship between wavelength, wave frequency, and wave velocity?

4. What is diffraction, and how does it relate to the behavior of light as a wave?

5. **POS** What's so special about c?

6. Name the colors that combine to make white light. What is it about the various colors that causes us to perceive them differently?

7. What effect does a positive charge have on a nearby negatively charged particle?

8. Compare and contrast the gravitational force with the electric force.

9. Describe the way in which light leaves a star, travels through the vacuum of space, and finally is seen by someone on Earth.

10. Why is light referred to as an electromagnetic wave?

11. What do radio waves, infrared radiation, visible light, ultraviolet radiation, X rays, and gamma rays have in common? How do they differ?

12. In what regions of the electromagnetic spectrum is the atmosphere transparent enough to allow observations from the ground?

13. What is a blackbody? What are the main characteristics of the radiation it emits?

14. **POS** What does Wien's law reveal about stars in the sky?

15. What does Stefan's law tell us about the radiation emitted by a blackbody?

16. In terms of its blackbody curve, describe what happens as a red-hot glowing coal cools.

17. What is the Doppler effect, and how does it alter the way in which we perceive radiation?

18. How do astronomers use the Doppler effect to determine the velocities of astronomical objects?

19. A source of radiation and an observer are traveling through space at precisely the same velocity, as seen by a second observer. Would you expect the first observer to measure a Doppler shift in the light received from the source?

20. **POS** If Earth were completely blanketed with clouds and we couldn't see the sky, could we learn about the realm beyond the clouds? What forms of radiation might be received?

CONCEPTUAL SELF-TEST: MULTIPLE CHOICE

1. Compared with ultraviolet radiation, infrared radiation has a greater (a) wavelength; (b) amplitude; (c) frequency; (d) energy.

2. Compared with red light, blue wavelengths of visible light travel (a) faster; (b) slower; (c) at the same speed.

3. An electron that collides with an atom will (a) cease to have an electric field; (b) produce an electromagnetic wave; (c) change its electric charge; (d) become magnetized.

4. **VIS** According to Figure 3.8 ("Electromagnetic Spectrum"), the wavelength of green light is about the size of (a) an atom; (b) a bacterium; (c) a fingernail; (d) a skyscraper.

5. An X-ray telescope located in Antarctica would not work well because of (a) the extreme cold; (b) the ozone hole; (c) continuous daylight; (d) Earth's atmosphere.

6. **VIS** In Figure 3.10 ("Multiple Blackbody Curves"), an object at 1000 K emits mostly (a) infrared light; (b) red light; (c) multiple green light; (d) blue light.

7. According to Wien's law, the hottest stars also have (a) the longest peak wavelength; (b) the shortest peak wavelength; (c) maximum emission in the infrared region of the spectrum; (d) the largest diameters.

8. Stefan's law says that if the Sun's temperature were to double, its energy emission would (a) become half its present value; (b) double; (c) increase four times; (d) increase 16 times.

9. A star much cooler than the Sun would appear (a) red; (b) blue; (c) smaller; (d) larger.

10. The blackbody curve of a star moving toward Earth would have its peak shifted (a) to a higher intensity; (b) toward higher energies; (c) toward longer wavelengths; (d) to a lower intensity.

PROBLEMS

The number of dots preceding each Problem indicates its approximate level of difficulty.

1. • A sound wave moving through water has a frequency of 256 Hz and a wavelength of 5.77 m. What is the speed of sound in water?

2. • What is the wavelength of a 100-MHz ("FM 100") radio signal?

3. • Estimate the frequency of an electromagnetic wave having a wavelength equal to the size of the period at the end of this sentence. In what part of the electromagnetic spectrum would such a wave lie?

4. • Normal human body temperature is about 37°C. What is this temperature in kelvins? What is the peak wavelength emitted by a person with this temperature? In what part of the spectrum does this lie?

5. •• Estimate the total amount of energy you radiate to your surroundings.

6. • Two otherwise identical bodies have temperatures of 300 K and 1500 K, respectively. Which one radiates more energy, and by what factor does its emission exceed the emission of the other?

7. • The Sun has a temperature of 5800 K, and its blackbody emission peaks at a wavelength of approximately 500 nm. At what wavelength does a protostar with a temperature of 1000 K radiate most strongly?

8. • Radiation from the nearby star Alpha Centauri is observed to be reduced in wavelength (after correction for Earth's orbital motion) by a factor of 0.999933. What is the recession velocity of Alpha Centauri relative to the Sun?

9. • At what velocity and in what direction would a spacecraft have to be moving for a radio station on Earth transmitting at 100 MHz to be picked up by a radio tuned to 99.9 MHz?

10. ••• Imagine that you are observing a spacecraft moving in a circular orbit of radius 100,000 km around a distant planet. You happen to be located in the plane of the spacecraft's orbit. You find that the spacecraft's radio signal varies periodically in wavelength between 2.99964 m and 3.00036 m. Assuming that the radio is broadcasting normally, at a constant wavelength, what is the mass of the planet?

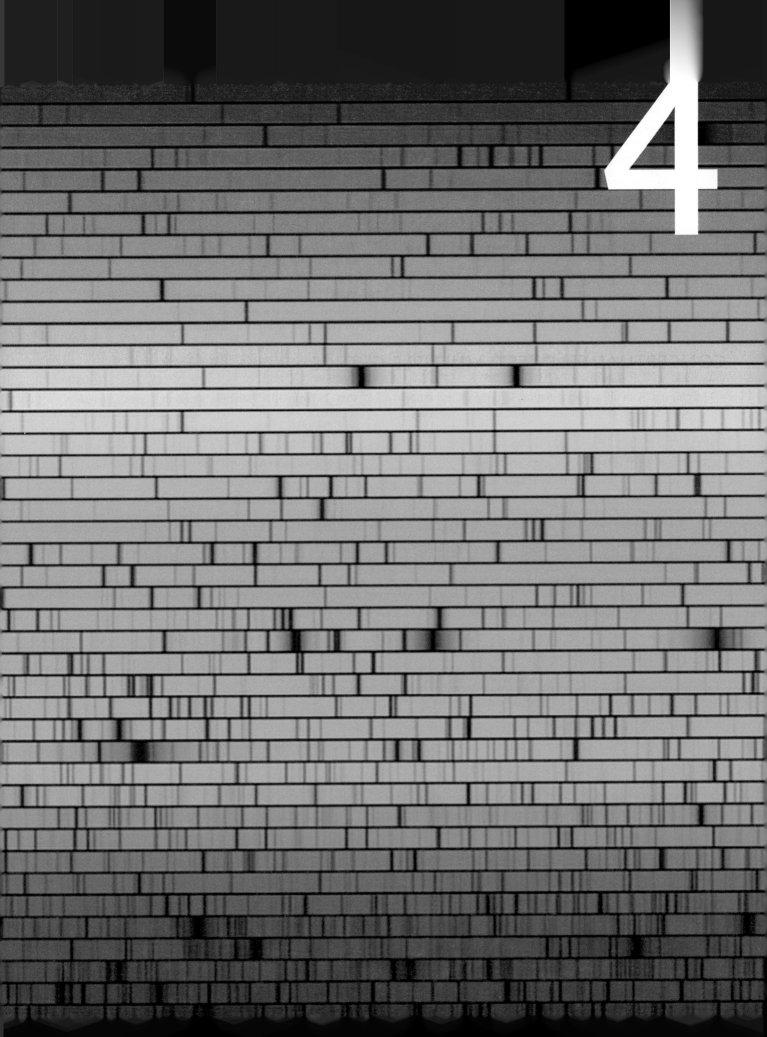

SPECTROSCOPY

THE INNER WORKINGS OF ATOMS

LEARNING GOALS

Studying this chapter will enable you to

1 Describe the characteristics of continuous, emission, and absorption spectra and the conditions under which each is produced.

2 Explain the relation between emission and absorption lines and what we can learn from those lines.

3 Specify the basic components of the atom and describe our modern conception of its structure.

4 Discuss the observations that led scientists to conclude that light has particle as well as wave properties.

5 Explain how electron transitions within atoms produce unique emission and absorption features in the spectra of those atoms.

6 Describe the general features of spectra produced by molecules.

7 List and explain the kinds of information that can be obtained by analyzing the spectra of astronomical objects.

THE BIG PICTURE Spectroscopy is a powerful observational technique enabling scientists to infer the nature of matter by the way it emits or absorbs radiation. Not only can spectroscopy reveal the chemical composition of distant stars, but it can also provide a wealth of information about the origin, evolution, and destiny of stars in the universe.

Visit the Study Area in www.masteringastronomy.com for quizzes, animations, videos, interactive figures, and self-guided tutorials.

The wave description of radiation allowed 19th-century astronomers to begin to decipher the information reaching Earth from the cosmos in the form of visible and invisible light. However, early in the 20th century, it became clear that the wave theory of electromagnetic phenomena was incomplete—some aspects of light simply could not be explained in purely wave terms. When radiation interacts with matter on atomic scales, it does so not as a continuous wave, but in a jerky, discontinuous way—in fact, as a particle. With this discovery, scientists quickly realized that atoms, too, must behave in a discontinuous way, and the stage was set for a scientific revolution—*quantum mechanics*—that has affected virtually every area of modern life.

The collection of observational and theoretical techniques that enable researchers to determine the nature of distant atoms by the way they emit and absorb radiation is called *spectroscopy*. It is the indispensable foundation of modern astrophysics.

LEFT: *The beautiful visible spectrum of the star Procyon is shown here from red to blue, interrupted by hundreds of dark lines caused by the absorption of light in the hot star's cooler atmosphere. The whole spectrum is normally 6 meters (20 feet) across, but to display it on a single page the full spectrum is cut into dozens of horizontal segments and stacked vertically. (NOAO/AURA)*

4.1 Spectral Lines

In Chapter 3, we saw something of how astronomers can analyze electromagnetic radiation received from space to obtain information about distant objects. A vital step in this process is the formation of a *spectrum*—a splitting of the incoming radiation into its component wavelengths. But in reality, *no* cosmic object emits a perfect blackbody spectrum like those discussed earlier. ∞ (Sec 3.4) All spectra deviate from this idealized form—some by only a little, others by a lot. Far from invalidating our earlier studies, however, these deviations contain a wealth of detailed information about physical conditions within the source of the radiation. Because spectra are so important, let's examine how astronomers obtain and interpret them.

Radiation can be analyzed with an instrument known as a **spectroscope.** In its most basic form, this device consists of an opaque barrier with a slit in it (to define a beam of light), a prism (to split the beam into its component colors), and an eyepiece or screen (to allow the user to view the resulting spectrum). Figure 4.1 shows such an arrangement. The research instruments called *spectrographs,* or *spectrometers,* used by professional astronomers are rather more complex, consisting of a telescope (to capture the radiation), a dispersing device (to spread the radiation out into a spectrum), and a detector (to record the result). Despite their greater sophistication, their basic operation is conceptually similar to the simple spectroscope shown in the figure.

In many large instruments, the prism is replaced by a device called a *diffraction grating,* consisting of a sheet of transparent material with numerous closely spaced parallel lines ruled on it. The spacing between the lines is typically a few microns (10^{-6} m), comparable to the wavelength of visible light. The spaces act as many tiny openings, and light is diffracted as it passes through the grating (or is reflected from it, depending on the design of the device). ∞ (Discovery 3-1) Because different wavelengths of electromagnetic radiation are diffracted by different amounts on encountering the grating, the effect is

to split a beam of light into its component colors. You are probably more familiar with diffraction gratings than you think—the "rainbow" of colors seen in light reflected from a compact disk is the result of precisely this process.

Emission Lines

The spectra we encountered in Chapter 3 are examples of **continuous spectra.** A lightbulb, for example, emits radiation of all wavelengths (mostly in the visible range), with an intensity distribution that is well described by the blackbody curve corresponding to the bulb's temperature. ∞ (Sec. 3.4) Viewed through a spectroscope, the spectrum of the light from the bulb would show the familiar rainbow of colors, from red to violet, without interruption, as presented in Figure 4.2(a).

Not all spectra are continuous, however. For instance, if we took a glass jar containing pure hydrogen gas and passed an electrical discharge through it (a little like a lightning bolt arcing through Earth's atmosphere), the gas would begin to glow—that is, it would emit radiation. If we were to examine that radiation with our spectroscope, we would find that its spectrum consists of only a few bright lines on an otherwise dark background, quite unlike the continuous spectrum described for the incandescent lightbulb. Figure 4.2(b) shows the experimental arrangement and its result schematically. (A more detailed rendering of the spectrum of hydrogen appears in the top panel of Figure 4.3.) Note that the light produced by the hydrogen in this experiment does *not* consist of all possible colors, but instead includes only a few narrow, well-defined **emission lines**—thin "slices" of the continuous spectrum. The black background represents all the wavelengths *not* emitted by hydrogen.

After some experimentation, we would also find that, although we could alter the *intensity* of the lines—for example, by changing the amount of hydrogen in the jar or the strength of the electrical discharge—we could not alter their *color* (in other words, their frequency or wavelength). The pattern of spectral emission lines shown is a property of the element hydrogen. Whenever we perform this experiment, the same characteristic colors result.

By the early 19th century, scientists had carried out similar experiments on many different gases.

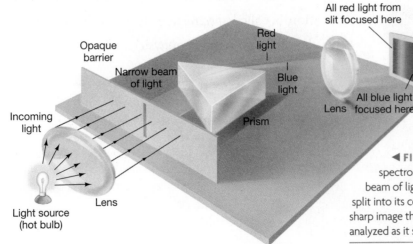

◀ **FIGURE 4.1 Spectroscope** Diagram of a simple spectroscope. A thin slit in the barrier at the left allows a narrow beam of light to pass. The light continues through a prism and is split into its component colors. A lens then focuses the light into a sharp image that is either projected onto a screen, as shown here, or analyzed as it strikes a detector.

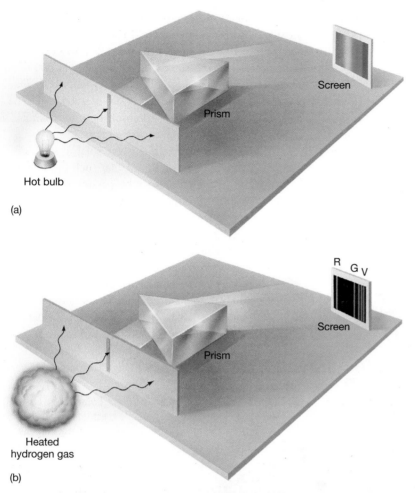

(a)

(b)

Interactive FIGURE 4.2 Continuous and Emission Spectra When passed through a slit and split up by a prism, light from a source of continuous radiation (a) gives rise to the familiar rainbow of colors. By contrast, the light from excited hydrogen gas (b) consists of a series of distinct bright spectral lines called emission lines. (The focusing lenses have been omitted for clarity—see Section 5.1.)

By vaporizing solids and liquids in a flame, they extended their inquiries to include materials that are not normally found in the gaseous state. Sometimes the pattern of lines was fairly simple, and sometimes it was complex, but it was always *unique* to that element. Even though the origin of the lines was not understood, researchers quickly realized that the lines provided a one-of-a-kind "fingerprint" of the substance under investigation. They could detect the presence of a particular atom or molecule (a group of atoms held together by chemical bonds—see Section 4.4) solely through the study of the light it emitted. Scientists have accumulated extensive catalogs of the specific wavelengths at which many different hot gases emit radiation. The particular pattern of light emitted by

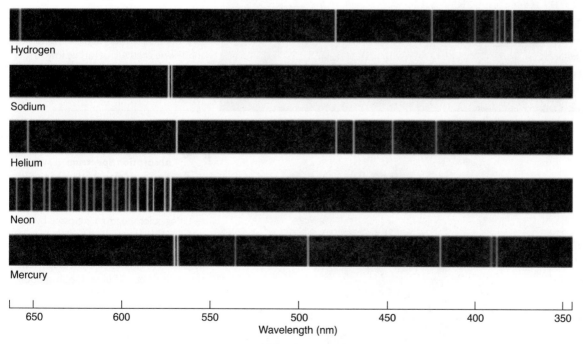

Hydrogen

Sodium

Helium

Neon

Mercury

▲ **FIGURE 4.3 Elemental Emission** The emission spectra of some well-known elements. In accordance with the convention adopted throughout this text, frequency increases to the right. Note that wavelengths shorter than approximately 400 nm, shown here in shades of purple, are actually in the ultraviolet part of the spectrum and are not visible to the human eye. (*Wabash Instrument Corp.*)

a gas of a given chemical composition is known as the **emission spectrum** of the gas. The emission spectra of some common substances are shown in Figure 4.3.

Absorption Lines

When sunlight is split by a prism, at first glance it appears to produce a continuous spectrum. However, closer scrutiny with a spectroscope shows that the solar spectrum is interrupted vertically by a large number of narrow dark lines, as shown in Figure 4.4. We now know that many of these lines represent wavelengths of light that have been removed (absorbed) by gases present either in the outer layers of the Sun or in Earth's atmosphere. These gaps in the spectrum are called **absorption lines.**

The English astronomer William Wollaston first noticed the solar absorption lines in 1802. They were studied in greater detail about 10 years later by the German physicist Joseph von Fraunhofer, who measured and cataloged over 600 of them. They are now referred to collectively as *Fraunhofer lines*. Although the Sun is by far the easiest star to study, and so has the most extensive set of observed absorption lines, similar lines are known to exist in the spectra of all stars.

At around the same time as the solar absorption lines were discovered, scientists found that such lines could also be produced in the laboratory by passing a beam of light from a source that produces a continuous spectrum through a cool gas, as shown in Figure 4.5. The scientists quickly observed an intriguing connection between emission and absorption lines: The absorption lines associated with a given gas occur at precisely the *same* wavelengths as the emission lines produced when the gas is heated.

As an example, consider the element sodium, whose emission spectrum appears in Figure 4.6. When heated to high temperatures, a sample

◀ **FIGURE 4.4 Solar Spectrum** This *THE BIG PICTURE* visible spectrum of the Sun shows hundreds of vertical dark absorption lines superimposed on a bright continuous spectrum. The high-resolution spectrum is displayed in a series of 48 horizontal strips stacked vertically; each strip covers a small portion of the entire spectrum from left to right. The scale extends from long wavelengths (red) at the upper left to short wavelengths (blue) at the lower right. Analysis of these absorption lines reveals the detailed properties of the Sun. *(AURA)*

◀ **Interactive FIGURE 4.5 Absorption Spectrum** (a) When cool gas is placed between a source of continuous radiation (such as a hot lightbulb) and a detector/screen, the resulting color spectrum is crossed by a series of dark absorption lines. These lines are formed when the intervening cool gas absorbs certain wavelengths (colors) from the original beam of light. The absorption lines appear at precisely the same wavelengths as the emission lines that would be produced if the gas were heated to high temperatures (see Figure 4.2). (b) An everyday analogy for any of these line spectra is a supermarket bar code that uniquely determines the cost of some product.

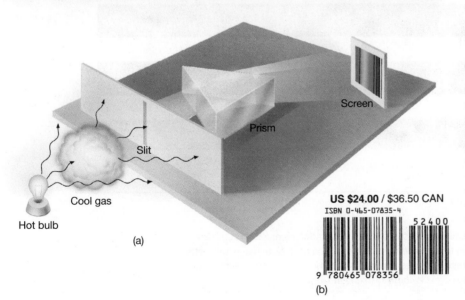

Screen

Prism

Slit

Cool gas

Hot bulb

(a)

US $24.00 / $36.50 CAN
ISBN 0-465-07835-4

9 780465 078356

52400

(b)

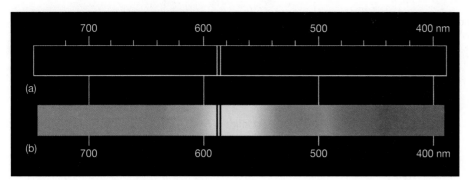

Interactive **FIGURE 4.6 Sodium Spectrum** (a) The characteristic emission lines of sodium. The two bright lines in the center appear in the yellow part of the spectrum. (b) The absorption spectrum of sodium. The two dark lines appear at exactly the same wavelengths as the bright lines in the sodium emission spectrum.

of sodium vapor emits visible light strongly at just two wavelengths—589.6 nm and 589.0 nm—lying in the yellow part of the spectrum. When a continuous spectrum is passed through some relatively cool sodium vapor, two sharp, dark absorption lines appear at precisely the same wavelengths. The emission and absorption spectra of sodium are compared in Figure 4.6, clearly showing the relation between emission and absorption features.

Kirchhoff's Laws

The analysis of the ways in which matter emits and absorbs radiation is called **spectroscopy.** One early spectroscopist, the German physicist Gustav Kirchhoff, summarized the observed relationships among the three types of spectra—continuous, emission line, and absorption line—in 1859. He formulated three spectroscopic rules, now known as **Kirchhoff's laws,** governing the formation of spectra:

1. A luminous solid or liquid, or a sufficiently dense gas, emits light of all wavelengths and so produces a *continuous spectrum* of radiation.

2. A low-density, hot gas emits light whose spectrum consists of a series of bright *emission lines* that are characteristic of the chemical composition of the gas.

3. A cool, thin gas absorbs certain wavelengths from a continuous spectrum, leaving dark *absorption lines* in their place, superimposed on the continuous spectrum. Once again, these lines are characteristic of the composition of the intervening gas—they occur at precisely the same wavelengths as the emission lines produced by that gas at higher temperatures.

Figure 4.7 illustrates Kirchhoff's laws and the relationship between absorption and emission lines. Viewed directly, the light source, a hot solid (the filament of the bulb), has a continuous (blackbody) spectrum. When the light source is viewed through a cloud of cool hydrogen gas, a series of dark absorption lines appear, superimposed on the spectrum at wavelengths characteristic of hydrogen. The lines appear because the light at those wavelengths is absorbed by the hydrogen. As we will see later in this chapter, the absorbed energy is

subsequently reradiated into space—but in all directions, not just the original direction of the beam. Consequently, when the cloud is viewed from the side against an otherwise dark background, a series of faint emission lines is seen. These lines contain the energy lost by the forward beam. If the gas was heated to incandescence, it would produce stronger emission lines at precisely the same wavelengths.

Identifying Starlight

By the late 19th century, spectroscopists had developed a formidable arsenal of techniques for interpreting the radiation received from space. Once astronomers knew that spectral lines were indicators of chemical composition, they set about identifying the observed lines in the solar spectrum. Almost all the lines in light from extraterrestrial sources could be attributed to known elements. For example, many of the Fraunhofer lines in sunlight are associated with the element iron, a fact first recognized by Kirchhoff and coworker Robert Bunsen (of Bunsen burner fame) in 1859. However, some unfamiliar lines also appeared in the solar spectrum. In 1868, astronomers realized that those lines must correspond to a previously unknown element. It was given the name helium, after the Greek word *helios,* meaning "Sun." Not until 1895, almost three decades after its detection in sunlight, was helium discovered on Earth! (A laboratory spectrum of helium is included in Figure 4.3.)

Yet, for all the information that 19th-century astronomers could extract from observations of stellar spectra, they still lacked a theory explaining how the spectra themselves arose. Despite their sophisticated spectroscopic equipment, they knew scarcely any more about the physics of stars than did Galileo or Newton. To understand how spectroscopy can be used to extract detailed information about astronomical objects from the light they emit, we must delve more deeply into the processes that produce line spectra.

CONCEPT CHECK

✔ What are absorption and emission lines, and what do they tell us about the composition of the gas producing them?

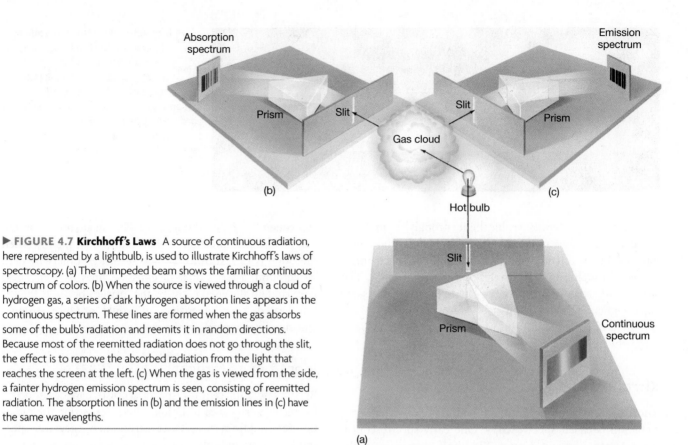

▶ FIGURE 4.7 **Kirchhoff's Laws** A source of continuous radiation, here represented by a lightbulb, is used to illustrate Kirchhoff's laws of spectroscopy. (a) The unimpeded beam shows the familiar continuous spectrum of colors. (b) When the source is viewed through a cloud of hydrogen gas, a series of dark hydrogen absorption lines appears in the continuous spectrum. These lines are formed when the gas absorbs some of the bulb's radiation and reemits it in random directions. Because most of the reemitted radiation does not go through the slit, the effect is to remove the absorbed radiation from the light that reaches the screen at the left. (c) When the gas is viewed from the side, a fainter hydrogen emission spectrum is seen, consisting of reemitted radiation. The absorption lines in (b) and the emission lines in (c) have the same wavelengths.

4.2   Atoms and Radiation

By the start of the 20th century, physicists had accumulated substantial evidence that light sometimes behaves in a manner that cannot be explained by the wave theory. As we have just seen, the production of absorption and emission lines involves only certain very specific frequencies or wavelengths of light. This would not be expected if light behaved like a continuous wave and matter always obeyed the laws of Newtonian mechanics. Other experiments conducted around the same time strengthened the conclusion that the notion of radiation as a wave was incomplete. It became clear that when light interacts with matter on very small scales, it does so not in a continuous way, but in a discontinuous, "stepwise" manner. The challenge was to find an explanation for this unexpected behavior. The eventual solution revolutionized our view of nature and now forms the foundation for all of physics and astronomy—indeed, for virtually all modern science.

Atomic Structure

To explain the formation of emission and absorption lines, we must understand not just the nature of light, but also the structure of **atoms**—the microscopic building blocks from which all matter is constructed. Let's start with the simplest atom of all: hydrogen. A hydrogen atom consists of an electron with a negative electrical charge orbiting a proton carrying a

positive charge. The proton forms the central **nucleus** (plural: nuclei) of the atom. The hydrogen atom as a whole is electrically neutral. The equal and opposite charges of the proton and the orbiting electron produce an electrical attraction that binds them together within the atom.

How does this picture of the hydrogen atom relate to the characteristic emission and absorption lines associated with hydrogen gas? If an atom absorbs some energy in the form of radiation, that energy must cause some internal change. Similarly, if the atom emits energy, that energy must come from somewhere within the atom. It is reasonable (and correct) to suppose that the energy absorbed or emitted by the atom is associated with changes in the motion of the orbiting electron.

The first theory of the atom to provide an explanation of hydrogen's observed spectral lines was set forth by the Danish physicist Niels Bohr in 1912. Now known simply as the *Bohr model* of the atom, its essential features are as follows:

1.   There is a state of lowest energy—the **ground state**— which represents the "normal" condition of the electron as it orbits the nucleus.

2.   There is a maximum energy that the electron can have and still be part of the atom. Once the electron acquires more than that maximum energy, it is no longer bound to the nucleus, and the atom is said to be **ionized;** an atom missing one or more of its electrons is called an *ion.*

3. Most important (and also least intuitive), between those two energy levels, the electron can exist only in certain sharply defined energy states, often referred to as **orbitals.**

This description of the atom contrasts sharply with the predictions of Newtonian mechanics, which would permit orbits with *any* energy, not just at certain specific values. ∞ (Sec. 2.8) In the atomic realm, such discontinuous behavior is the norm. In the jargon of the field, the orbital energies are said to be **quantized.** The rules of **quantum mechanics,** the branch of physics governing the behavior of atoms and subatomic particles, are far removed from everyday experience.

In Bohr's original model, each electron orbital was pictured as having a specific radius, much like a planetary orbit in the solar system, as shown in Figure 4.8. However, the modern view is not so simple. Although each orbital *does* have a precise energy, the orbits are not sharply defined, as indicated in the figure. Rather, the electron is now envisioned as being smeared out in an "electron cloud" surrounding the nucleus, as illustrated in Figure 4.9. We cannot tell "where" the electron is—we can only speak of the *probability* of finding it in a certain location within the cloud. It is common to speak of the average distance from the cloud to the nucleus as the "radius" of the electron's orbit. When a hydrogen atom is in its ground state, the radius of the orbit is about 0.05 nm (0.5 Å). As the orbital energy increases, the radius increases, too.

For the sake of clarity in the diagrams that follow, we will represent electron orbitals in this chapter as solid lines. (See *More Precisely 4-1* on p. 86 for a more detailed rendition of hydrogen's energy levels.) However, you should always bear in mind that Figure 4.9 is a more accurate depiction of reality.

Atoms do not always remain in their ground state. An atom is said to be in an **excited state** when an electron occupies an orbital at a greater-than-normal distance from its parent nucleus. An atom in such an excited state has a greater-than-normal amount of energy. The excited state

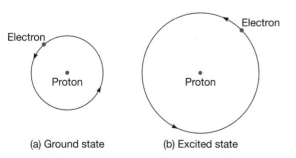

(a) Ground state   (b) Excited state

▲ **FIGURE 4.8 Classical Atom** An early-20th-century conception of the hydrogen atom—the Bohr model—pictured its electron orbiting the central proton in a well-defined orbit, rather like a planet orbiting the Sun. Two electron orbitals of different energies are shown: (a) the ground state and (b) an excited state.

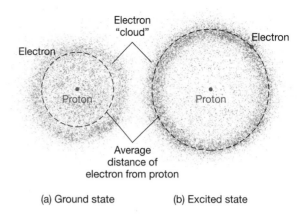

(a) Ground state   (b) Excited state

▲ **FIGURE 4.9 Modern Atom** The modern view of the hydrogen atom sees the electron as a "cloud" surrounding the nucleus. The same two energy states are shown as in Figure 4.8.

with the lowest energy (that is, the state closest in energy to the ground state) is called the *first excited state,* that with the second-lowest energy is the *second excited state,* and so on. An atom can become excited in one of two ways: by absorbing some energy from a source of electromagnetic radiation or by colliding with some other particle—another atom, for example. However, the electron cannot stay in a higher orbital forever; the ground state is the only level where it can remain indefinitely. After about 10^{-8} s, an excited atom returns to its ground state.

Radiation as Particles

Because electrons can exist only in orbitals having specific energies, atoms can absorb only specific amounts of energy as their electrons are boosted into excited states. Likewise, atoms can emit only specific amounts of energy as their electrons fall back to lower energy states. Thus, the amount of light energy absorbed or emitted in these processes *must correspond precisely to the energy difference between two orbitals.* The atom's quantized energy levels require that light be absorbed and emitted in the form of distinct "packets" of electromagnetic radiation, each carrying a specific amount of energy. We call these packets **photons.** A photon is, in effect, a "particle" of electromagnetic radiation.

The idea that light sometimes behaves not as a continuous wave, but as a stream of particles, was proposed by Albert Einstein in 1905 to explain a number of experimental results (especially the *photoelectric effect*—see *Discovery 4-1*) then puzzling physicists. Furthermore, Einstein was able to quantify the relationship between the two aspects of light's double nature. He found that the energy carried by a photon had to be proportional to the *frequency* of the radiation:

photon energy ∝ radiation frequency.

For example, a "deep red" photon having a frequency of 4 × 10^{14} Hz (or a wavelength of approximately 750 nm) has half the energy of a violet photon of frequency of 8×10^{14} Hz (wavelength = 375 nm) and 500 times the energy of an 8×10^{11} Hz (wavelength = 375 μm) microwave photon.

The constant of proportionality in the preceding relation is now known as *Planck's constant,* in honor of the German physicist Max Planck, who determined its numerical value. It is always denoted by the symbol *h*, and the equation relating the photon energy *E* to the radiation frequency *f* is usually written

$$E = hf.$$

Like the gravitational constant *G* and the speed of light, *c*, Planck's constant is one of the fundamental physical constants of the universe.

In SI units, the value of Planck's constant is a very small number: $h = 6.63 \times 10^{-34}$ joule seconds (J · s). Consequently, the energy of a single photon is tiny. Even a very high-frequency gamma ray (the most energetic type of electromagnetic radiation) with a frequency of 10^{22} Hz has an energy of just $(6.63 \times 10^{-34}) \times 10^{22} \approx 7 \times 10^{-12}$ J—about the same energy carried by a flying gnat. Nevertheless, this energy is more than enough to damage a living cell. The basic reason that gamma rays are so much more dangerous to life than visible light is that each gamma-ray photon typically carries millions, if not billions, of times more energy than a photon of visible radiation.

The equivalence between the energy and frequency (or inverse wavelength) of a photon completes the connection between atomic structure and atomic spectra. Atoms absorb and emit radiation at characteristic wavelengths determined by their own particular internal structure. Because this structure is *unique* to each element, the colors of the absorbed and emitted photons—that is, the spectral lines we observe—are characteristic of that element *and only that element.* The spectrum we see is thus a unique identifier of the atom involved.

Many people find it confusing that light can behave in two such different ways. To be truthful, modern physicists don't yet fully understand *why* nature displays this wave–particle duality. Nevertheless, there is irrefutable experimental evidence for both of these aspects of radiation. Environmental conditions ultimately determine which description—wave or stream of particles—better fits the behavior of electromagnetic radiation in a particular instance. As a general rule of thumb, in the macroscopic realm of everyday experience, radiation is more usefully described as a wave, whereas in the microscopic domain of atoms, it is best characterized as a series of particles.

PROCESS OF SCIENCE CHECK

✔ Describe the scientific reasoning leading to the conclusion that light behaves both as a wave and a particle.

4.3 The Formation of Spectral Lines

With quantum mechanics as our guide to the internal structure of atoms, we can now explain quantitatively the spectral lines we see. Let's start with hydrogen, the simplest element, then move on to more complex systems.

The Spectrum of Hydrogen

The full spectrum of hydrogen consists of many lines, spread across much of the electromagnetic spectrum, from ultraviolet to radio; we focus here on just a few of those lines. The energy levels and spectrum of hydrogen are discussed in more detail in *More Precisely 4-1.*

Figure 4.10 illustrates schematically the absorption and emission of photons by a hydrogen atom. Figure 4.10(a) shows the atom absorbing a photon and making a transition from the ground state to the first excited state. It then emits a photon of precisely the same energy and drops back to the ground state. The energy difference between the two states corresponds to an ultraviolet photon of wavelength 121.6 nm (1216 Å).

Absorption may also boost an electron into an excited state higher than the first excited state. Figure 4.10(b) depicts the absorption of a more energetic (higher frequency, shorter wavelength) ultraviolet photon, one with a wavelength of 102.6 nm (1026 Å). The absorption of this photon causes the atom to jump to the *second* excited state. As before, the atom returns rapidly to the ground state, but this time, because there are two states lying below the excited state, the atom can do so in one of two possible ways:

1. It can proceed directly back to the ground state, in the process emitting an ultraviolet photon identical to the one that excited the atom in the first place.

2. Alternatively, the electron can *cascade* down, one orbital at a time. If this occurs, the atom will emit *two* photons: one with an energy equal to the difference between the second and first excited states and the other with an energy equal to the difference between the first excited state and the ground state.

Either possibility can occur, with roughly equal probability. The second step of the cascade process produces a 121.6-nm ultraviolet photon, just as in Figure 4.10(a). However, the first transition—the one from the second to the first excited state—produces a photon with a wavelength of 656.3 nm (6563 Å), which is in the visible part of the spectrum. This photon is seen as red light. An individual atom—if one could be isolated—would emit a momentary red flash. This is the origin of the red line in the hydrogen spectrum shown in Figure 4.3.

The absorption of additional energy can boost the electron to even higher orbitals within the atom. As the excited

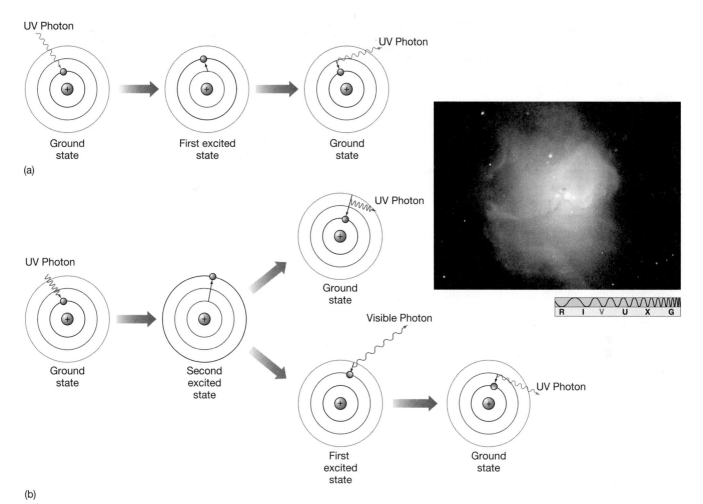

Interactive FIGURE 4.10 Atomic Excitation (a) Absorption of an ultraviolet photon (left) by a hydrogen atom causes the momentary excitation of the atom into its first excited state (center). After about 10^{-8} s, the atom returns to its ground state (right), in the process emitting a photon having exactly the same energy as the original photon. (b) Absorption of a higher-energy ultraviolet (UV) photon may boost the atom into a higher excited state, from which there are several possible paths back to the ground state. (Remember, the sharp lines used for the orbitals here and in similar figures that follow are intended merely as a schematic representation of the electron energy levels and are not meant to be taken literally. In actuality, electron orbitals are "clouds," as shown in Figure 4.9.) At the top, the electron falls immediately back to the ground state, emitting a photon identical to the one it absorbed. At the bottom, the electron initially falls into the first excited state, producing visible radiation of wavelength 656.3 nm—the characteristic (Hα) red glow of excited hydrogen. Subsequently, the atom emits another photon (having the same energy as in part (a) as it falls back to the ground state. The object shown in the inset, designated N81, is an emission nebula—an interstellar cloud made mostly of hydrogen gas excited by absorbing radiation emitted by some extremely hot stars (the white areas) near the center. Its red coloration is the direct result of the bottom sequence sketched in part (b). *(Inset: NASA)*

electron cascades back down to the ground state, the atom may emit many photons, each with a different energy and hence a different wavelength, and the resulting spectrum shows many spectral lines. In a sample of heated hydrogen gas, at any instant atomic collisions ensure that atoms are found in many different excited states. The complete emission spectrum therefore consists of wavelengths corresponding to all possible transitions between those states and states of lower energy.

In the case of hydrogen, all transitions ending at the ground state produce ultraviolet photons. However, downward transitions ending at the *first* excited state give rise to spectral lines in or near the visible portion of the electromagnetic spectrum (Figure 4.3; *More Precisely 4-1*). Other transitions ending in higher states generally give rise to infrared and radio spectral lines.

The inset in Figure 4.10 shows an astronomical object whose red coloration is the result of precisely the process mentioned in step 2 on p. 84. As ultraviolet photons from a young, hot star pass through the surrounding cool hydrogen gas out of which the star recently formed, some photons are

MORE PRECISELY 4-1

The Hydrogen Atom

By observing the emission spectrum of hydrogen and using the connection between photon energy and color first suggested by Einstein (Section 4.2), Niels Bohr determined early in the 20th century what the energy differences between the various energy levels must be. Using that information, he was then able to infer the actual energies of the excited states of hydrogen.

A unit of energy often used in atomic physics is the *electron volt* (eV). (The name actually has a rather technical definition: the amount of energy gained by an electron when it accelerates through an electric potential of 1 volt. For our purposes, however, it is just a convenient quantity of energy.) One electron volt (1 eV) is equal to 1.60×10^{-19} J (joule)—roughly half the energy carried by a single photon of red light. The minimum amount of energy needed to ionize hydrogen from its ground state is 13.6 eV. Bohr numbered the energy levels of hydrogen, with level 1 the ground state, level 2 the first excited state, and so on. He found that, by assigning zero energy to the ground state, the energy of any state (the *n*-th, say) could then be written as follows:

$$E_n = 13.6 \left(1 - \frac{1}{n^2} \right) \text{eV}.$$

Thus, the ground state ($n = 1$) has energy $E_1 = 0$ eV, the first excited state ($n = 2$) has energy $E_2 = 13.6 \times (1 - 1/4)$ eV = 10.2 eV, the second excited state has energy $E_3 = 13.6 \times (1 - 1/9)$ eV = 12.1 eV, and so on. There are infinitely many excited states between the ground state and the energy at which the atom is ionized, crowding closer and closer together as *n* increases and E_n approaches 13.6 eV.

EXAMPLE Using Bohr's formula for the energy of each electron orbital, we can reverse his reasoning and calculate the energy associated with a transition between any two given states. To boost an electron from the first excited state to the second, an atom must be supplied with $E_3 - E_2 = 12.1$ eV $- 10.2$ eV $= 1.9$ eV of energy, or 3.0×10^{-19} J. Now, from the formula $E = hf$ presented in the text, we find that this energy corresponds to a photon with a frequency of 4.6×10^{14} Hz, having a wavelength of 656 nm, and lying in the red portion of the spectrum. (A more precise calculation gives the value 656.3 nm reported in the text.)

The accompanying diagram summarizes the structure of the hydrogen atom. The increasing energy levels are depicted as a series of circles of increasing radius. The electronic transitions between these levels (indicated by arrows) are conventionally grouped into families, named after their discoverers, that define the terminology used to identify specific spectral lines. (Note that the spacings of the energy levels are not drawn to scale here to provide room for all labels on the diagram. In reality, the circles should become more and more closely spaced as we move outward.)

Transitions starting from or ending at the ground state (level 1) form the *Lyman series*, named after American spectroscopist Theodore Lyman, who discovered these lines in 1914. The first is *Lyman alpha* (Lyα), corresponding to the transition between the first excited state (level 2) and the ground state. The energy difference is 10.2 eV, and the Lyα photon has a wavelength of 121.6 nm (1216 Å). The Lyβ (beta) transition, between level 3 (the second excited state) and the ground state, corresponds to an energy change of 12.10 eV and a photon of wavelength 102.6 nm (1026 Å). Lyγ (gamma) corresponds to a jump from level 4 to level 1, and so on. All Lyman-series energies lie in the ultraviolet region of the spectrum.

The next series of lines, the *Balmer series*, involves transitions down to (or up from) level 2, the first excited state. The series is named after the Swiss mathematician Johann Balmer, who didn't discover these lines (they were well known to spectroscopists early in the 19th century), but who published a mathematical formula for their wavelengths in 1885. All the Balmer series lines lie in or close to the visible portion of the electromagnetic spectrum.

Because they form the most easily observable part of the hydrogen spectrum and were the first to be discovered, the Balmer lines are often referred to simply as the *Hydrogen series*, denoted by the letter H. As with the Lyman series, the individual transitions are labeled with Greek letters. An Hα photon (level 3 to level 2) has a wavelength of 656.3 nm, in the red part of the visible spectrum, Hβ (level 4 to level 2) has a wavelength of 486.1 nm (green), Hγ (level 5 to level 2) has a wavelength of 434.1 nm (blue), and so on. We will use these designations (especially Hα and Hβ) frequently in later chapters. The most energetic Balmer series photons have energies that place them just beyond the blue end of the visible spectrum, in the near ultraviolet.

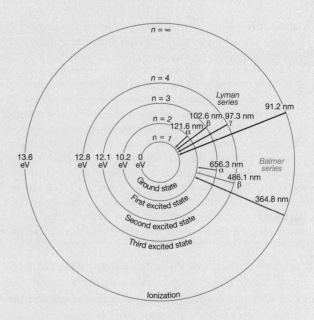

A few of the transitions making up the Lyman and Balmer (Hydrogen) series are marked on the figure. There are infinitely many other families of lines, all lying in the infrared and radio regions of the spectrum, but astronomically, the Lyman and Balmer sequences are the most important.

absorbed by the gas, boosting its atoms into excited states or ionizing them completely. The 656.3-nm red glow characteristic of excited hydrogen gas results as the atoms cascade back to their ground states. The phenomenon is called *fluorescence*.

Kirchhoff's Laws Explained

Let's reconsider our earlier discussion of emission and absorption lines in terms of the model just presented. In Figure 4.7, a beam of continuous radiation shines through a cloud of hydrogen gas. The beam contains photons of all energies, but most of them cannot interact with the gas—the gas can absorb only those photons having just the right energy to cause a change in an electron's orbit from one state to another. All other photons in the beam—with energies that cannot produce a transition—do not interact with the gas at all, but pass through it unhindered. Photons having the right energies are absorbed, excite the gas, and are removed from the beam. This sequence is the cause of the dark absorption lines in the spectrum of Figure 4.7(b). The lines are direct indicators of the energy differences between orbitals in the atoms making up the gas.

The excited atoms rapidly return to their original states, each emitting one or more photons in the process. Most of these reemitted photons leave at angles that do *not* take them through the slit and onto our detector. A second detector looking at the cloud from the side would record the reemitted energy as an emission spectrum, as illustrated in Figure 4.7(c). An astronomical example is the *emission nebula* shown in the inset to Figure 4.10. Like the absorption spectrum, the emission spectrum is characteristic of the gas, not of the original beam. The type of spectrum we see depends on our chance location with respect to both the source and the intervening cloud.

Figure 4.7(a) shows a *continuous* spectrum, in which emitted photons escape from the bulb without further interaction with matter. Actually, the situation in a dense source of radiation (a thick gas cloud or in a liquid or solid body) is more complex. There, a photon is likely to interact with atoms, free electrons, and ions in the body many times before finally escaping, exchanging some energy with the matter at each encounter. The net result is that the emitted radiation displays a continuous spectrum, in accordance with Kirchhoff's first law. The spectrum is approximately that of a blackbody with the same temperature as the source.

▶ **FIGURE 4.11 Helium and Carbon** (a) A helium atom in its ground state. Two electrons occupy the lowest-energy orbital around a nucleus containing two protons and two neutrons. (b) A carbon atom in its ground state. Six electrons orbit a six-proton, six-neutron nucleus, two of the electrons in an inner orbital, the other four at a greater distance from the center.

More Complex Spectra

All hydrogen atoms have basically the same structure—a single electron orbiting a single proton—but, of course, there are many other kinds of atoms, each kind having a unique internal structure. The number of protons in the nucleus of an atom determines the **element** that it represents. Just as all hydrogen atoms have a single proton, all oxygen atoms have 8 protons, all iron atoms have 26 protons, and so on.

The next simplest element after hydrogen is helium. The central nucleus of the most common form of helium is made up of two protons and two **neutrons** (another kind of elementary particle having a mass slightly larger than that of a proton, but having no electrical charge). Two electrons orbit this nucleus. As with hydrogen and all other atoms, the "normal" condition for helium is to be electrically neutral, with the negative charge of the orbiting electrons exactly canceling the positive charge of the nucleus (Figure 4.11a).

More complex atoms contain more protons (and neutrons) in the nucleus and have correspondingly more orbiting electrons. For example, an atom of carbon, shown in Figure 4.11(b), consists of six electrons orbiting a nucleus containing six protons and six neutrons. As we progress to heavier and heavier elements, the number of orbiting electrons increases, and the number of possible electron transitions rises rapidly. The result is that very complicated spectra can be produced. The complexity of atomic spectra generally reflects the complexity of the atoms themselves. A good example is the element iron, which contributes nearly 800 of the Fraunhofer absorption lines seen in the solar spectrum (Figure 4.4).

Atoms of a single element such as iron can yield many lines for two main reasons. First, the 26 electrons of a normal iron atom can make an enormous number of different transitions among available energy levels. Second, many iron atoms are *ionized*, with some of their 26 electrons stripped away. The removal of electrons alters an atom's electromagnetic structure, and the energy levels of ionized iron are quite different from those of neutral iron. Each new level of ionization introduces a whole new set of

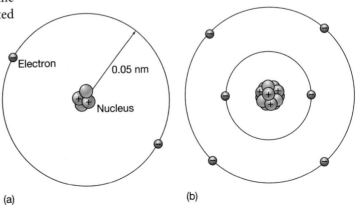

(a)　　　　　　　　　　(b)

DISCOVERY 4-1

The Photoelectric Effect

Einstein developed his breakthrough insight into the nature of radiation partly as a means of explaining a puzzling experimental result known as the **photoelectric effect.** This effect can be demonstrated by shining a beam of light on a metal surface (as shown in the accompanying figure). When high-frequency ultraviolet light is used, bursts of electrons are dislodged from the surface by the beam, much as when one billiard ball hits another, knocking it off the table. However, the speed with which the particles are ejected from the metal is found to depend only on the *color* of the light, and not on its intensity. For lower-frequency light—blue, say—an electron detector still records bursts of electrons, but now their speeds, and hence their *energies,* are less. For even lower frequencies—red or infrared light—*no* electrons are kicked out of the metal surface at all.

These results are difficult to reconcile with a wave model of light, which would predict that the energies of the ejected electrons should increase steadily with increasing intensity at any frequency. Instead, the detector shows an abrupt cutoff in ejected electrons as the frequency of the incoming radiation drops below a certain level. Einstein realized that the only way to explain the cutoff, and the increase in electron speed with frequency above the cutoff, was to envision radiation as traveling as "bullets," or particles, or *photons.* Furthermore, to account for the experimental findings, the energy of any photon had to be proportional to the *frequency* of the radiation. Low-frequency, long-wavelength photons carry less energy than high-frequency, short-wavelength ones.

If we also suppose that some minimum amount of energy is needed just to "unglue" the electrons from the metal, then we can see why no electrons are emitted below some critical frequency: The photons associated with red light in the diagram just don't carry enough energy. Above the critical frequency, photons do have enough energy to dislodge the electrons. Moreover, any energy they possess above the necessary minimum is imparted to the electrons as *kinetic energy,* the energy of motion. Thus, as the frequency of the radiation increases, so, too, does the photon's energy and hence the speed of the electrons that they liberate from the metal.

The realization and acceptance of the fact that light can behave both as a wave and as a particle is another example of the scientific method at work. Despite the enormous success of the wave theory of radiation in the 19th century, the experimental evidence led 20th-century scientists to the inevitable conclusion that the theory was incomplete—it had to be modified to allow for the fact that light sometimes acts like a particle. Although Einstein is perhaps best known today for his theories of relativity, in fact his 1919 Nobel prize was for his work on the photoelectric effect. In addition to bringing about the birth of a whole new branch of physics—the field of quantum mechanics—Einstein's explanation of the photoelectric effect radically changed the way physicists view light and all other forms of radiation.

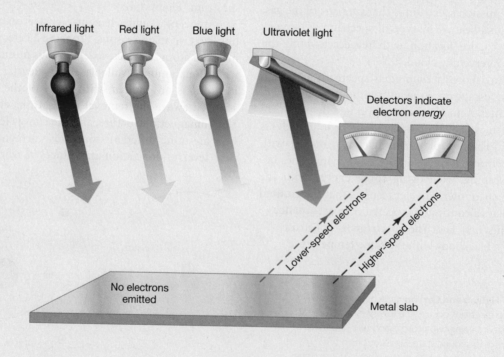

Infrared light   Red light   Blue light   Ultraviolet light

Detectors indicate electron *energy*

Lower-speed electrons

Higher-speed electrons

No electrons emitted

Metal slab

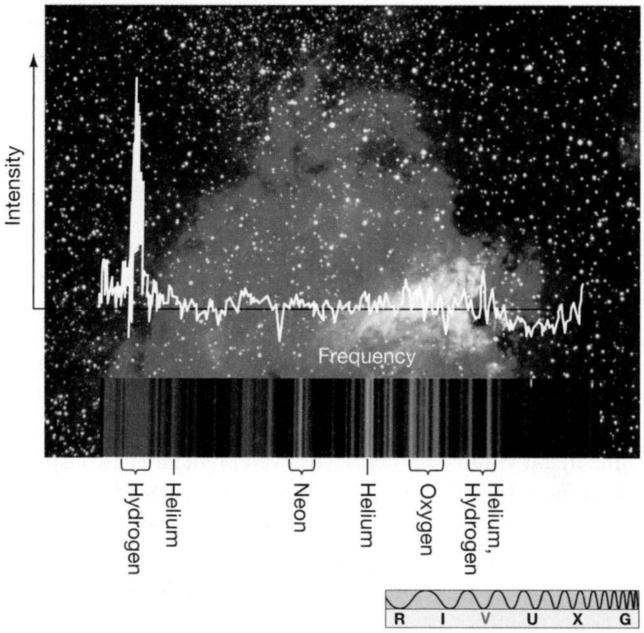

▲ FIGURE 4.12 Emission Nebula The visible spectrum of the hot gases in a nearby gas cloud known as the Omega Nebula (M17). (The word *nebula* means "gas cloud"—one of many sites in our Galaxy where new stars are forming today.) Shining by the light of several very hot stars, the gas in the nebula produces a complex spectrum of bright and dark lines (bottom). That same spectrum can also be displayed, as shown here, as a white graph of intensity versus frequency, spanning the spectrum from red to blue. *(Adapted from ESO)*

spectral lines. Besides iron, many other elements, also in different stages of excitation and ionization, absorb photons at visible wavelengths. When we observe the entire Sun, all these atoms and ions absorb simultaneously, yielding the rich spectrum we see.

The power of spectroscopy is most apparent when a cloud contains many different gases mixed together, because it enables us to study one kind of atom or ion to the exclusion of all others simply by focusing on specific wavelengths of radiation. By identifying the superimposed absorption and emission spectra of many different atoms, we can determine the cloud's composition (and much more—see Section 4.4). Figure 4.12 shows an actual spectrum observed from a real cosmic object. As in Figure 4.10, the characteristic red glow of this emission nebula comes from the Hα transition in hydrogen, the nebula's main constituent.

Spectral lines occur throughout the entire electromagnetic spectrum. Usually, electron transitions among the lowest orbitals of the lightest elements, such as hydrogen and helium, produce visible and ultraviolet spectral lines. Transitions among very highly excited states of hydrogen and other elements can produce spectral lines in the infrared and radio parts of the electromagnetic spectrum. Conditions on Earth make it all but impossible to detect these radio and

infrared features in the laboratory, but they are routinely observed by radio and infrared telescopes (see Chapter 5) in radiation coming from space. Electron transitions among lower energy levels in heavier, more complex elements produce X-ray spectral lines, which have been observed in the laboratory. Some have also been observed in stars and other cosmic objects.

CONCEPT CHECK

✔ How does the structure of an atom determine the atom's emission and absorption spectra?

4.4   Molecules

A **molecule** is a tightly bound group of atoms held together by interactions among their orbiting electrons—interactions that we call *chemical bonds*. Much like atoms, molecules can exist only in certain well-defined energy states, and again like atoms, molecules produce characteristic emission or absorption spectral lines when they make a transition from one state to another. Because molecules are more complex than individual atoms, the rules of molecular physics are also more complex. Nevertheless, as with atomic spectral lines, painstaking experimental work over many decades has determined the precise frequencies (or wavelengths) at which millions of molecules emit and absorb radiation.

In addition to the lines resulting from electron transitions, molecular lines result from two other kinds of change not possible in atoms: Molecules can *rotate,* and they can *vibrate.* Figure 4.13 illustrates these basic molecular motions. Molecules rotate and vibrate in specific ways. Just as with atomic states, only certain spins and vibrations are allowed by the rules of molecular physics. When a molecule *changes* its rotational or vibrational state, a photon is emitted or absorbed. Spectral lines characteristic of the specific kind of molecule result. Like their atomic counterparts, these lines are unique molecular fingerprints, enabling researchers to identify and study one kind of molecule to the exclusion of all others. As a rule of thumb,

- *electron transitions* within molecules produce visible and ultraviolet spectral lines (the largest energy changes).

- changes in molecular *vibration* produce infrared spectral lines.

- changes in molecular *rotation* produce spectral lines in the radio part of the electromagnetic spectrum (the smallest energy changes).

Molecular lines usually bear little resemblance to the spectral lines associated with their component atoms. For example, Figure 4.14(a) shows the emission spectrum of the simplest

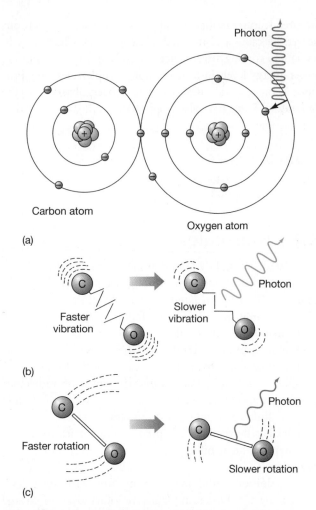

Carbon atom

Oxygen atom

(a)

Faster
vibration

Slower
vibration

Photon

(b)

Faster rotation

Photon

Slower rotation

(c)

▲ FIGURE 4.13 **Molecular Emission** Molecules can change in three ways while emitting or absorbing electromagnetic radiation. The colors and wavelengths of the emitted photons represent the relative energies involved. Sketched here is the molecule carbon monoxide (CO) undergoing (a) a change in which an electron in the outermost orbital of the oxygen atom drops to a lower energy state (emitting a photon of shortest wavelength, in the visible or ultraviolet range), (b) a change in vibrational state (of intermediate wavelength, in the infrared), and (c) a change in rotational state (of longest wavelength, in the radio range).

molecule known: molecular hydrogen. Notice how different it is from the spectrum of atomic hydrogen shown in part (b) of the figure.

CONCEPT CHECK

✔ What kinds of internal changes within a molecule can cause radiation to be emitted or absorbed?

4.5   Spectral-Line Analysis

Astronomers apply the laws of spectroscopy in analyzing radiation from beyond Earth. A nearby star or a distant galaxy takes the place of the lightbulb in our previous examples. An interstellar cloud or a stellar (or even planetary) atmosphere plays the role of the intervening cool gas, and a spectrograph attached to a telescope replaces our simple prism and detector. We began our study of electromagnetic radiation by stating that virtually all we know about planets, stars, and galaxies is gleaned from studies of the light we receive from them, and we have presented some of the ways in which that knowledge is obtained. Here, we describe a few of the ways in which the properties of emitters and absorbers can be determined by careful analysis of radiation received on (or near) Earth. We will encounter other important examples as our study of the cosmos unfolds.

A Spectroscopic Thermometer

In the hot interior of a star, atoms are fully ionized. Electrons travel freely through the gas, unbound to any nucleus, and the spectrum of radiation is continuous. However, near the relatively cool stellar surface, some atoms retain a few, or even most, of their orbital electrons. As noted earlier, astronomers can determine the star's chemical composition by matching the spectral lines they see with the laboratory spectra of known atoms, ions, and molecules.

The strength of a spectral line (brightness or darkness, depending on whether the line is seen in emission or absorption) depends on the number of atoms giving rise to it.

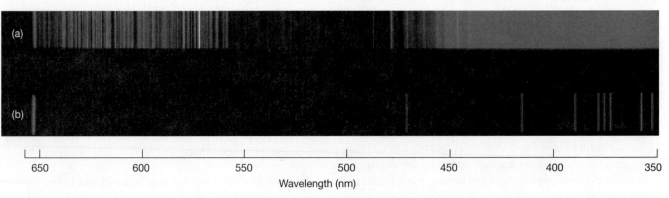

(a)

(b)

650   600   550   500   450   400   350

Wavelength (nm)

GURE 4.14 **Hydrogen Spectra** (a) The emission spectrum of molecular hydrogen. Notice how it s from the spectrum of the simpler atomic hydrogen (b). (© *Bausch & Lomb, Inc.*)

The more atoms there are to emit or absorb photons of the appropriate frequency, the stronger the line. But the strength of line also depends critically on the *temperature* of the gas containing the atoms, because temperature determines how many atoms at any instant are in the right orbital to undergo any particular transition. Simply put, at low temperatures, only low-lying energy states tend to be populated, and transitions into and out of those states dominate the spectrum. At higher temperatures, more atoms are in excited states, and some may be ionized, radically changing the character of the possible transitions and hence the spectrum we see.

Spectroscopists have developed mathematical formulas that relate the number of emitted or absorbed photons to the energy levels of the atoms involved and the temperature of the gas. Once an object's spectrum is measured, astronomers can interpret it by matching the observed intensities of the spectral lines with those predicted by the formulas. In this way, astronomers can refine their measurements of both the composition *and* the temperature of the gas producing the lines. These temperature measurements are generally much more accurate than crude estimates based on the radiation laws and the assumption of blackbody emission. ∞ (Sec. 3.4) In Chapter 17, we will see how these ideas are put to use in the classification and interpretation of stellar spectra.

Measurement of Radial Velocity

The Doppler effect—the apparent shift in the frequency of a wave due to the motion of the source relative to the observer—is a classical phenomenon common to all waves. ∞ (Sec. 3.5) However, by far its most important astronomical application comes when it is combined with observations of atomic and molecular spectral lines.

The spectra of many atoms, ions, and molecules are well known from laboratory measurements. Often, however, a familiar pattern of lines appears, but the lines are displaced from their usual locations. In other words, as illustrated in Figure 4.15, a set of spectral lines may be recognized as belonging to a particular element, except that the lines are all offset— blueshifted or redshifted—by the same fractional amount from their normal wavelengths. These shifts are due to the Doppler effect, and they allow astronomers to measure how fast the source of the radiation is moving along the line of sight from the observer (the *radial velocity* of the source).

For example, in Figure 4.15, the 486.1-nm Hβ line of hydrogen in the spectrum of a distant galaxy is received on Earth at a wavelength of 485.1 nm—blueshifted to a slightly shorter wavelength. (Remember, we know that it is the Hβ line because *all* the hydrogen lines are observed to have the same fractional shift—the characteristic line pattern identifies the spectrum as that of hydrogen.) We can compute the galaxy's line-of-sight velocity relative to Earth by using the Doppler equation presented in Section 3.5. The calculation is essentially the same as that presented in *More Precisely 3-3:* The change in wavelength (apparent minus true) is 485.1 nm − 486.1 nm = −1.0 nm (the negative sign simply indicating that the wavelength has decreased). It then follows that the recession velocity is

$$\frac{-1.0 \text{ nm}}{486.1 \text{ nm}} \times c = -620 \text{ km/s}.$$

In other words, the galaxy is *approaching* us (this is the meaning of the negative sign) at a speed of 620 km/s.

This book will have a lot to say about the motions of planets, stars, and galaxies throughout the universe. Just bear

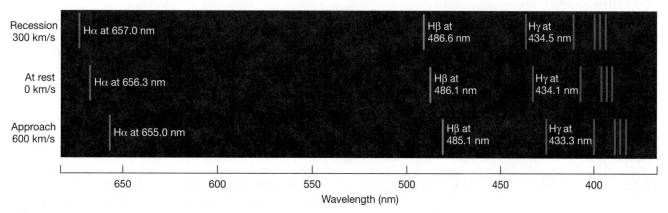

⚠ Interactive FIGURE 4.15 Doppler Shift Because of the Doppler effect, the entire spectrum of a moving object is shifted to higher or lower frequencies. The spectrum at the center is the unshifted emission spectrum of pure hydrogen, corresponding to the object at rest. The top spectrum shows the slight redshift of the hydrogen lines from an object moving at a speed of 300 km/s away from the observer. Once we recognize the spectrum as that of hydrogen, we can identify specific lines and measure their shifts. The amount of the shift (0.1 percent here) tells us the object's recession velocity—0.001 c. The spectrum at the bottom shows the blueshift of the same set of lines from an object approaching us at 600 km/s. The shift is twice as large (0.2 percent), because the speed has doubled, and in the opposite sense because the direction has reversed.

in mind that almost all of that information is derived from telescopic observations of Doppler-shifted spectral lines in many different parts of the electromagnetic spectrum.

Line Broadening

The structure of the lines themselves reveals still more information. At first glance the emission lines shown earlier may seem uniformly bright, but more careful study shows that this is in fact not the case. As illustrated in Figure 4.16, the brightness of a line is greatest at the center and falls off toward either side. Earlier, we stressed that photons are emitted and absorbed at very precise energies, or frequencies. Why, then, aren't spectral lines extremely narrow, occurring only at specific wavelengths? This *line broadening* is not the result of some inadequacy of our experimental apparatus; rather, it is caused by the *environment* in which the emission or absorption occurs—the physical state of the gas or star in which the line is formed. For definiteness, we have drawn Figure 4.16 and subsequent figures to refer to emission lines, but realize that the ideas apply equally well to absorption features.

Several processes can broaden spectral lines. The most important involve the Doppler effect. Imagine a hot cloud of gas containing individual atoms in random thermal motion in every possible direction, as illustrated in Figure 4.17(a). If an atom happens to be moving away from us as it emits a photon, that photon is redshifted by the Doppler effect—we do not record it at the precise wavelength predicted by atomic physics, but rather at a slightly longer wavelength. The extent of this redshift is proportional to the atom's instantaneous velocity away from the detector. Similarly, if the atom is moving toward us at the instant of emission, its light is blueshifted. In short, because of thermal motion within

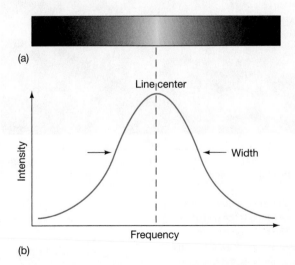

▲ FIGURE 4.16 **Line Profile** By tracing the changing brightness across a typical emission line (a) and expanding the scale, we obtain a graph of the line's intensity versus its frequency (b).

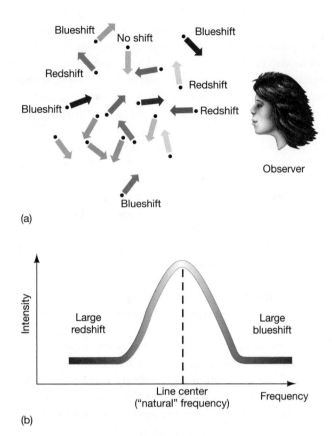

▲ FIGURE 4.17 **Thermal Broadening** Atoms moving randomly (a) produce broadened spectral lines (b) as their individual redshifted and blueshifted emission lines merge in our detector. The hotter the gas, the greater is the degree of thermal broadening.

the gas, emission and absorption lines are observed at frequencies slightly different from those we would expect if all atoms in the cloud were motionless.

Most atoms in a typical cloud have small thermal velocities, so in most cases the line is Doppler shifted just a little. Only a few atoms have large shifts. As a result, the center of a spectral line is much more pronounced than its "wings," producing a bell-shaped spectral feature like that shown in Figure 4.17(b). Thus, even if all atoms emitted and absorbed photons at only one precise wavelength, the effect of their thermal motion would be to smear the line out over a range of wavelengths. The hotter the gas, the larger the spread of Doppler motions and the greater the width of the line. ∞ *(More Precisely 3-1)* By measuring a line's width, astronomers can estimate the average speed of the particles and hence the temperature of the gas producing it.

Other processes, such as *rotation* and *turbulence*, can produce similar effects. Consider an astronomical object (a star or a gas cloud) that is spinning about some axis as sketched in Figure 4.18 or that has some other internal motion, such as turbulent eddies or vortices on many scales. Photons emitted from regions that happen to be moving toward us are blueshifted by the Doppler effect; photons emitted

▶ FIGURE 4.18 **Rotational Broadening** The rotation of a star can cause spectral line broadening. Since most stars are unresolved—that is, they are so distant that we cannot distinguish one part of the star from another—light rays from all parts of the star merge to produce broadened lines. The more rapid the rotation, the greater is the broadening.

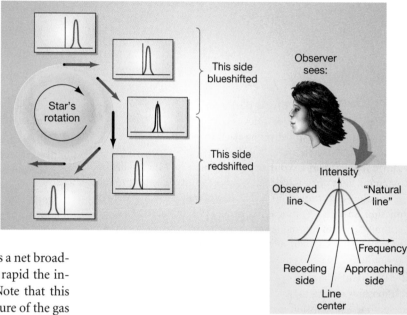

from regions moving away from us are red-shifted. Often the object under study is so small or far away that our equipment cannot distinguish, or *resolve*, different parts from one another—all the emitted light is blended together in our detector. In that case, the result is a net broadening of the observed spectral lines. The more rapid the internal motion, the more broadening we see. Note that this broadening has *nothing* to do with the temperature of the gas producing the lines and is generally superimposed on the thermal broadening just discussed.

Still other broadening mechanisms do not depend on the Doppler effect at all. For example, if electrons are moving between orbitals while their parent atom is colliding with another atom, the energy of the emitted or absorbed photons changes slightly, blurring the spectral lines. This mechanism, which occurs most often in dense gases where collisions are most frequent, is usually referred to as *collisional broadening*. The amount of broadening increases as the density of the emitting or absorbing gas rises.

Finally, *magnetic fields* can also broaden spectral lines by a process called the *Zeeman effect*. The electrons and nuclei within atoms behave as tiny spinning magnets, and the basic emission and absorption rules of atomic physics change slightly whenever atoms are immersed in a magnetic field, as is the case in many stars to greater or lesser extents. The result is a slight splitting of a spectral line, which then blurs into an overall line broadening. Generally, the stronger the magnetic field, the more pronounced is the broadening.

The Message of Starlight

Given sufficiently sensitive equipment, there is almost no end to the wealth of data that can be obtained from starlight. Table 4.1 lists some basic measurable properties of an incoming beam of radiation and indicates what sort of information can be obtained from them.

It is important to realize, however, that deciphering the extent to which each of the factors just described influences a spectrum can be a very difficult task. Typically, the spectra of many elements are superimposed on one another, and often several competing physical effects are occurring simultaneously, each modifying the spectrum in its own way. Further analysis is generally required to disentangle them. For example,

if we know the temperature of the emitting gas (perhaps by comparing intensities of different spectral lines, as discussed earlier), then we can calculate how much of the broadening is due to thermal motion and therefore how much is due to the other mechanisms just described. In addition, it is often possible to distinguish between the various broadening mechanisms by studying the detailed *shapes* of the lines.

The challenge facing astronomers is to decode spectral-line profiles to obtain meaningful information about the sources of the lines. In the next chapter, we will discuss some of the means by which astronomers obtain the raw data they need in their quest to understand the cosmos.

CONCEPT CHECK

✔ Why is it so important for astronomers to analyze spectral lines in detail?

TABLE 4.1 Spectral Information Derived from Starlight	
Observed Spectral Characteristic	**Information Provided**
Peak frequency or wavelength (continuous spectra only)	Temperature (Wien's law)
Lines present	Composition, temperature
Line intensities	Composition, temperature
Line width	Temperature, turbulence, rotation speed, density, magnetic field
Doppler shift	Line-of-sight velocity

CHAPTER REVIEW

SUMMARY

1 A **spectroscope (p. 78)** is a device for splitting a beam of radiation into its component frequencies and delivering them to a screen or detector for detailed study. 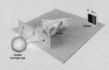 Many hot objects emit a **continuous spectrum (p. 78)** of radiation, containing light of all wavelengths. A hot gas may instead produce an **emission spectrum (p. 80)**, consisting of only a few well-defined **emission lines (p. 78)** of specific frequencies, or colors. Passing a continuous beam of radiation through cool gas will produce **absorption lines (p. 80)** at precisely the same frequencies as are present in the gas's emission spectrum.

2 **Kirchhoff's laws (p. 81)** describe the relationships among these different types of spectra. The emission and absorption lines produced by each element are unique—they provide a "fingerprint" of that element. The study of the spectral 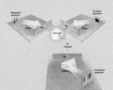 lines produced by different substances is called **spectroscopy (p. 81)**. Spectroscopic studies of the Fraunhofer lines in the solar spectrum yield detailed information about the Sun's composition.

3 **Atoms (p. 82)** are made up of negatively charged electrons orbiting a positively charged heavy **nucleus (p. 82)** consisting of positively charged protons and electrically neutral **neutrons (p. 87)**. Usually, the num- 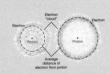 ber of orbiting electrons equals the number of protons in the nucleus, and the atom as a whole is electrically neutral. The number of protons in the nucleus determines the particular **element (p. 87)** of which the atom is a constituent. An atom has a minimum-energy **ground state (p. 82)**, representing its "normal" condition. If an orbiting electron is given enough energy, it can escape from the atom, which is then **ionized (p. 82)**. Between these two states, the electron can exist only in certain well-defined **excited states (p. 83)**, each with a specific energy—the electron's energy is **quantized (p. 83)**. In the modern

view, the electron is envisaged as being spread out in a "cloud" around the nucleus, but still with a sharply defined energy.

4 Electromagnetic radiation exhibits both wave and particle properties. Particles of radiation are called **photons (p. 83)**. In 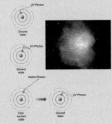 order to explain the **photoelectric effect (p. 88)**, Einstein found that the energy of a photon must be directly proportional to the photon's frequency.

5 As electrons move between energy levels within an atom, the difference in energy between the states is emitted or absorbed in the form of photons. Because the energy levels have definite energies, the photons 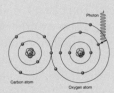 also have definite energies, and hence colors, that are characteristic of the type of atom involved. More complex atoms generally produce more complex spectra.

6 **Molecules (p. 89)** are groups of two or more atoms bound together by electromagnetic forces. Like atoms, molecules exist in energy states that obey rules similar to those governing the internal structure of atoms. 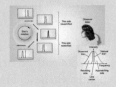 Again like atoms, when molecules make transitions between energy states, they emit or absorb a characteristic spectrum of radiation that identifies them uniquely.

7 Astronomers apply the laws of spectroscopy in analyzing radiation from beyond Earth. Several physical mechanisms can broaden spectral lines. The most im-  portant is the Doppler effect, which occurs because stars are hot and their atoms are in motion or because the object being studied is rotating or in turbulent motion.

Mastering ASTRONOMY *For instructor-assigned homework go to* **www.masteringastronomy.com**

Problems labeled **POS** explore the process of science | **VIS** problems focus on reading and interpreting visual information

REVIEW AND DISCUSSION

1. What is spectroscopy? Explain how astronomers might use spectroscopy to determine the composition and temperature of a star.

2. Describe the basic components of a simple spectroscope.

3. What is a continuous spectrum? An absorption spectrum?

4. Why are gamma rays generally harmful to life-forms, but radio waves generally harmless?

5. What is a photon?

6. In the particle description of light, what is color?

7. In what ways does the Bohr model of atomic structure differ from the modern view?

8. Give a brief description of a hydrogen atom.

9. What does it mean to say that a physical quantity is quantized?

10. What is the normal condition for atoms? What is an excited atom? What are orbitals?

11. Why do excited atoms absorb and reemit radiation at characteristic frequencies?

12. **POS** How are absorption and emission lines produced in a stellar spectrum? What information might absorption lines in the spectrum of a star reveal about a cloud of cool gas lying between us and the star?

13. According to Kirchhoff's laws, what are the necessary conditions for a continuous spectrum to be produced?

14. Explain how a beam of light passing through a diffuse cloud may give rise to both absorption and emission spectra.

15. Why might the spectral lines of an element in a star's spectrum be weak, even though that element is relatively abundant in the star?

16. How do molecules produce spectral lines unrelated to the movement of electrons between energy levels?

17. **POS** How does the intensity of a spectral line yield information about the source of the line?

18. How can the Doppler effect cause broadening of a spectral line?

19. Describe what happens to a spectral line from a star as the star's rotation rate increases.

20. **POS** List three properties of a star that can be determined from observations of its spectrum.

CONCEPTUAL SELF-TEST: MULTIPLE CHOICE

1. Compared with a spectrum from a ground-based observation, the spectrum of a star observed from above Earth's atmosphere would show (**a**) no absorption lines; (**b**) fewer emission lines; (**c**) slightly fewer absorption lines; (**d**) many more absorption lines.

2. The visible spectrum of sunlight reflected from Saturn's cold moon Titan would be expected to be (**a**) continuous; (**b**) an emission spectrum; (**c**) an absorption spectrum.

3. **VIS** Figure 4.3 ("Elemental Emission") shows the emission spectrum of neon gas. If the temperature of the gas were increased, we would observe (**a**) fewer red lines and more blue lines; (**b**) even more red lines; (**c**) some faint absorption features; (**d**) no significant change.

4. Compared with a star having many blue absorption lines, a star with many red and blue absorption lines must be (**a**) moving away from the observer; (**b**) cooler than the other star; (**c**) of different composition than that of the other star; (**d**) moving away from the other star.

5. An atom that has been ionized (**a**) has equal numbers of protons and electrons; (**b**) has more protons than electrons; (**c**) is radioactive; (**d**) is electrically neutral.

6. **VIS** In Figure 4.10 ("Atomic Excitation"), compared with an electron transition from the first excited state to the ground state, a transition from the third excited state to the second excited state emits a photon of (**a**) greater energy; (**b**) lower energy; (**c**) identical energy.

7. Compared with a complex atom like neon, a simple atom such as hydrogen has (**a**) more excited states; (**b**) fewer excited states; (**c**) the same number of excited states.

8. Compared with cooler stars, the hottest stars have absorption lines that are (**a**) thin and distinct; (**b**) broad and fuzzy; (**c**) identical to the lines in the cooler stars.

9. Compared with slowly rotating stars, the fastest spinning stars have absorption lines that are (**a**) thin and distinct; (**b**) broad and fuzzy; (**c**) identical to the lines in the slowly rotating stars.

10. Astronomers analyze starlight to determine a star's (**a**) temperature; (**b**) composition; (**c**) motion; (**d**) all of the above.

PROBLEMS

The number of dots preceding each Problem indicates its approximate level of difficulty.

1. • What is the energy (in electron volts—see *More Precisely 4-1*) of a 450-nm blue photon? A 200-nm ultraviolet photon?

2. • What is the energy (in electron volts) of a 100-GHz (1 gigahertz $= 10^9$ Hz) microwave photon?

3. • What is the wavelength of a 2-eV red photon? Repeat your calculation for an 0.1-eV infrared photon and a 5000-eV (5-keV) X ray.

4. • How many times more energy has a 1-nm gamma ray than a 10-MHz radio photon?

5. •• Calculate the wavelength and frequency of the radiation emitted by the electronic transition from the 10th to the 9th excited state of hydrogen. In what part of the electromagnetic spectrum does this radiation lie? Repeat the question for transitions from the 100th to the 99th excited state.

6. •• How many different photons (i.e., photons of different frequencies) can be emitted as a hydrogen atom in the third excited state falls back, directly or indirectly, to the ground state? What are the wavelengths of those photons?

7. •• List all the spectral lines of hydrogen that lie in the visible range (taken to run from 400 to 700 nm in wavelength).

8. • A distant galaxy is receding from Earth with a radial velocity of 3000 km/s. At what wavelength would its Lyα line be received by a detector above Earth's atmosphere?

9. •• At a temperature of 5800 K, hydrogen atoms in the solar atmosphere have typical random speeds of about 12 km/s. Assuming that spectral-line broadening is sim ply the result of atoms moving toward us or away from us at this random speed, estimate the thermal width (in nanometers) of the 656.3-nm solar Hα line.

10. •• In a demonstration of the photoelectric effect, suppose that a minimum energy of 5×10^{-19} J (3.1 eV) is required to dislodge an electron from a metal surface. What is the minimum frequency (and longest wavelength) of radiation for which the detector registers a response?

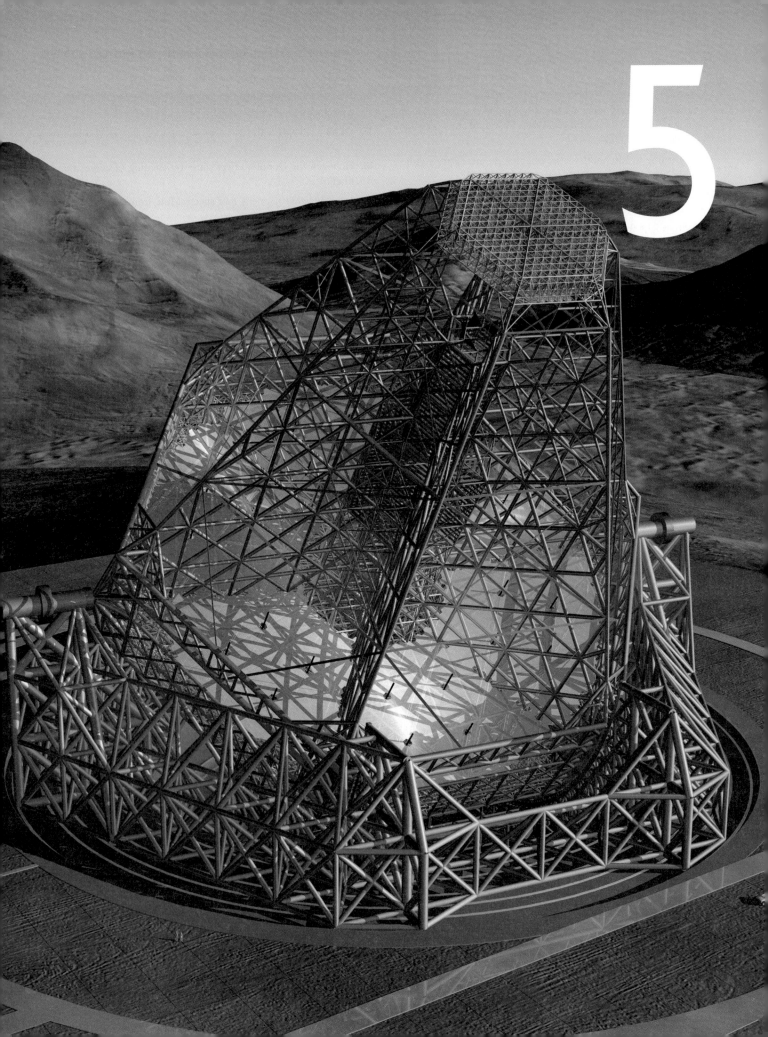

5

TELESCOPES

THE TOOLS OF ASTRONOMY

LEARNING GOALS

Studying this chapter will enable you to

1 Sketch and describe the basic designs of optical telescopes used by astronomers.

2 Explain the particular advantages of reflecting telescopes for astronomical use, and specify why very large telescopes are needed for most astronomical studies.

3 Explain the purpose of some of the detectors used in astronomical telescopes.

4 Describe how Earth's atmosphere affects astronomical observations, and discuss some of the current efforts to improve ground-based astronomy.

5 Discuss the relative advantages and disadvantages of radio and optical astronomy.

6 Explain how interferometry can enhance the usefulness of astronomical observations.

7 Explain why some astronomical observations are best done from space, and discuss the advantages and limitations of space-based astronomy.

8 Say why it is important to make astronomical observations in different regions of the electromagnetic spectrum.

THE BIG PICTURE Telescopes are time machines and astronomers, in a sense, are historians. Telescopes enhance our senses, enabling us to look out in space—and hence back in time—much farther than would be possible with our unaided eyes, and to perceive radiation at wavelengths far beyond human vision.

At its heart, astronomy is an observational science. More often than not, observations of cosmic phenomena precede any clear theoretical understanding of their nature. As a result, our detecting instruments—our telescopes—have evolved to observe as broad a range of wavelengths as possible.

Until the middle of the 20th century, telescopes were limited to collecting visible light. Since then, technological advances have expanded our view of the universe to all regions of the electromagnetic spectrum. Some telescopes are sited on Earth, whereas others must be placed in space, and design considerations vary widely from one part of the spectrum to another. Whatever the details of their construction, however, telescopes are devices whose basic purpose is to collect electromagnetic radiation and deliver it to a detector for detailed study.

LEFT: *Astronomers like to think big, and really big telescopes are now on the drawing board. This artist's conception for the European Southern Observatory shows OWL, the OverWhelmingly Large telescope. With a mirror diameter of 100 meters, OWL would combine unrivaled light-gathering power with the ability to examine cosmic objects with unprecedented detail. This gargantuan project has an estimated cost of 1 billion Euros and would use building techniques pioneered in 1899 by Gustave Eiffel for his famous tower in Paris. (ESO)*

Mastering**ASTRONOMY**

Visit the Study Area in www.masteringastronomy.com for quizzes, animations, videos, interactive figures, and self-guided tutorials.

97

5.1 Optical Telescopes

In essence, a **telescope** is a "light bucket" whose primary function is to capture as many photons as possible from a given region of the sky and concentrate them into a focused beam for analysis. Much like a water bucket that collects only the rain falling into it, a telescope intercepts only that radiation falling onto it.

Optical telescopes are designed specifically to collect the wavelengths that are visible to the human eye. These telescopes have a long history, stretching back to the days of Galileo in the early 17th century, and for most of the past four centuries astronomers have built their instruments primarily for use in the narrow, visible, portion of the electromagnetic spectrum. ∞ (Sec. 3.3) Optical telescopes are probably also the best-known type of astronomical hardware, so it is fitting that we begin our study with them.

Although the various telescope designs presented in this section all come to us from optical astronomy, the discussion applies equally well to many instruments designed to capture invisible radiation, particularly in the infrared and ultraviolet regimes. Many large ground-based optical facilities are also used extensively for infrared work.* Indeed, many ground-based observatories have recently been constructed with infrared observing as their principal function.

Refracting and Reflecting Telescopes

Optical telescopes fall into two basic categories: *refractors* and *reflectors*. **Refraction** is the bending of a beam of light as it passes from one transparent medium (e.g., air) into

Recall from Chapter 3 that, while Earth's atmosphere effectively blocks all ultraviolet, and most infrared, radiation, there remain several fairly broad spectral windows through which ground-based infrared observations can be made. ∞ (Sec. 3.3)

another (e.g., glass). Consider for example how a straw that is half immersed in a glass of water looks bent (Figure 5.1). The straw is straight, of course, but the light by which we see it is bent—refracted—as that light leaves the water and enters the air. When the light then enters our eyes, we perceive the straw as being bent.

A **refracting telescope** uses a *lens* to gather and concentrate a beam of light. Figure 5.2(a) shows how refraction at two faces of a prism can be used to change the direction of a beam of light. As illustrated in Figure 5.2(b), we can think of a lens as a series of prisms combined in such a way that all light rays arriving parallel to its axis (the imaginary line through the center of the lens), regardless of their distance from that axis, are refracted to pass through a single point, called the *focus*. The distance between the primary mirror and the focus is the *focal length*.

Figure 5.3 shows how a **reflecting telescope** uses a curved *mirror* instead of a lens to focus the incoming light. As shown in Figure 5.3(a), light striking a polished surface is reflected back, leaving the mirror at the same angle at which it arrived. The mirror in a reflecting telescope is constructed so that all light rays arriving parallel to its axis are reflected to pass through the focus (Figure 5.3b). In astronomical contexts, the mirror that collects the incoming light is usually called the *primary mirror*, because telescopes often contain more than one mirror. The focus of the primary mirror is referred to as the **prime focus**.

Astronomical telescopes are often used to make **images** of their field of view (simply, the portion of the sky that the telescope "sees"). Figure 5.4 illustrates how that is accomplished, in this case by the mirror in a reflecting telescope. Light from a distant object (here, a comet) reaches us as parallel, or very nearly parallel, rays. Any ray of light entering the instrument parallel to the telescope's axis strikes the mirror and is reflected through the prime focus.

(a)

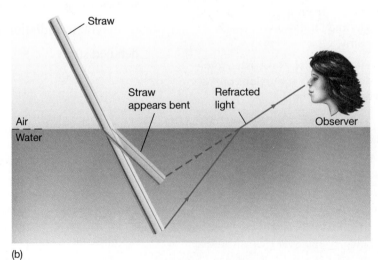

(b)

▲ **FIGURE 5.1 Refraction** A straw placed in a bowl of water appears bent (a) because the light from the part of the straw under the surface is refracted as it leaves the water and enters the air. (b) Consequently, the image formed in our eyes is displaced relative to the true position of the straw. (*R. Megna/Fundamental Photographs, NYC*)

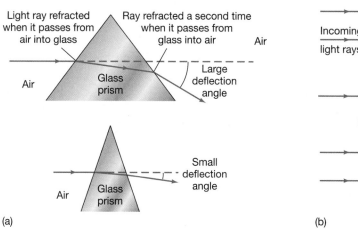

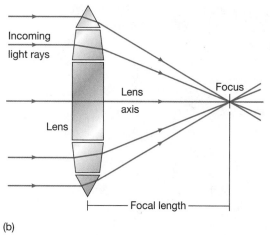

▲ **FIGURE 5.2 Refracting Lens** (a) Refraction by a prism changes the direction of a light ray by an amount that depends on the angle between the prism's faces. When the angle between the faces is large, the deflection is large; when the angle is small, so is the deflection. (b) A lens can be thought of as a series of prisms. A light ray traveling along the axis of a lens is undeflected as it passes through the lens. Parallel rays arriving at progressively greater distances from the axis are refracted by increasing amounts in such a way that they all pass through a single point—the focus.

Light coming from a slightly different direction—inclined slightly to the axis—is focused to a slightly different point. In this way, an image is formed near the prime focus. Each point on the image corresponds to a different point in the field of view.

The prime-focus images produced by large telescopes are actually quite small—the image of the entire field of view may be as little as 1 cm across. Often, the image is magnified with a lens known as an *eyepiece* before being observed by eye or, more likely, recorded as a photograph or digital image. The angular diameter of the magnified image is much greater than the telescope's field of view, allowing much more detail to be discerned. Figure 5.5(a) shows the basic design of a simple refracting telescope, illustrating how a small eyepiece is used to view the image focused by the lens. Figure 5.5(b) shows how a reflecting telescope accomplishes the same function.

Comparing Refractors and Reflectors

The two telescope designs shown in Figure 5.5 achieve the same result: Light from a distant object is captured and focused to form an image. On the face of it, then, it might appear that there is little to choose between the two in deciding which type to buy or build. However, as the sizes of telescopes have increased steadily over the years (for reasons to be discussed in Section 5.3), a number of important factors have tended to favor reflecting instruments over refractors:

1. The fact that light must pass through the lens is a major disadvantage of refracting telescopes. Just as a prism disperses white light into its component colors, the lens in a refracting telescope tends to focus red and blue light differently. This deficiency is known as *chromatic aberration*. Careful design and choice of materials can

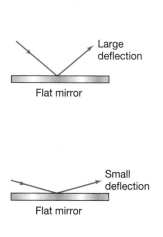

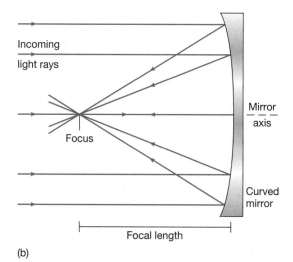

◀ **FIGURE 5.3 Reflecting Mirror** (a) Reflection of light from a flat mirror occurs when light is deflected, depending on its angle of incidence. (b) A curved mirror can be used to focus to a single point all rays of light arriving parallel to the mirror axis. Light rays traveling along the axis are reflected back along the axis, as indicated by the arrowheads pointing in both directions. Off-axis rays are reflected through greater and greater angles the farther they are from the axis, so that they all pass through the focus.

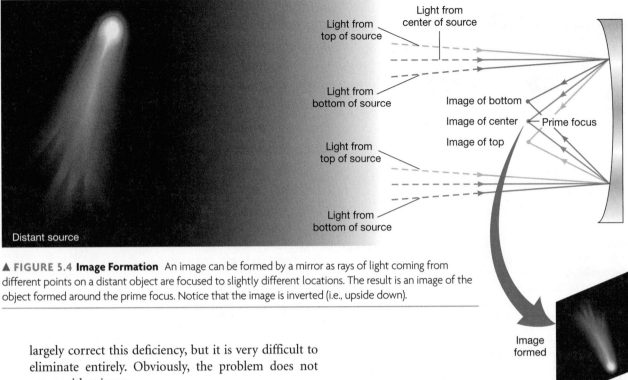

Light from
center of source

Light from
top of source

Light from
bottom of source

Light from
top of source

Light from
bottom of source

Distant source

Image of bottom
Image of center — Prime focus
Image of top

Image
formed

▲ **FIGURE 5.4 Image Formation** An image can be formed by a mirror as rays of light coming from different points on a distant object are focused to slightly different locations. The result is an image of the object formed around the prime focus. Notice that the image is inverted (i.e., upside down).

largely correct this deficiency, but it is very difficult to eliminate entirely. Obviously, the problem does not occur with mirrors.

2. As light passes through the lens, some of it is absorbed by the glass. This absorption is a relatively minor problem for visible radiation, but it can be severe for infrared and ultraviolet observations because glass blocks most of the radiation in those regions of the electromagnetic spectrum. Again, the problem does not affect mirrors.

3. A large lens can be quite heavy. Because it can be supported only around its edge (so as not to block

the incoming radiation), the lens tends to deform under its own weight. A mirror does not have this drawback because it can be supported over its entire back surface.

4. A lens has two surfaces that must be accurately machined and polished—a task that can be very difficult indeed—but a mirror has only one.

For these reasons, *all* large modern telescopes use mirrors as their primary light gatherers. The largest refractor ever built, installed in 1897 at the Yerkes Observatory in Wisconsin and still in use today, has a lens diameter of just over 1 m (40 inches). By contrast, many

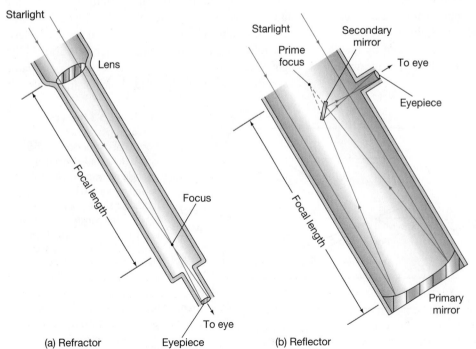

Starlight

Lens

Focal length

Focus

To eye

Eyepiece

(a) Refractor

Starlight

Prime focus

Secondary mirror

To eye

Eyepiece

Focal length

Primary mirror

(b) Reflector

◀ **FIGURE 5.5 Refractors and Reflectors** Comparison of (a) refracting and (b) reflecting telescopes. Both types are used to gather and focus electromagnetic radiation—to be observed by human eyes or recorded on photographs or in computers. In both cases, the image formed at the focus is viewed with a small magnifying lens called an eyepiece.

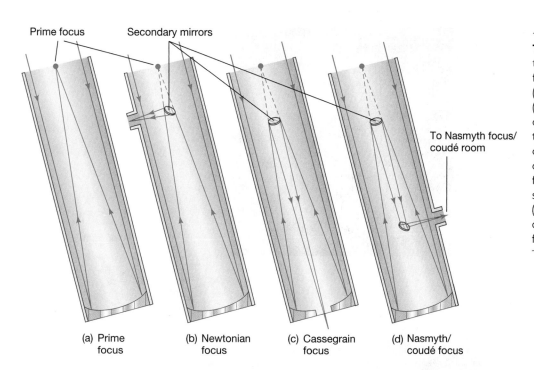

Prime focus   Secondary mirrors

To Nasmyth focus/ coudé room

(a) Prime focus

(b) Newtonian focus

(c) Cassegrain focus

(d) Nasmyth/ coudé focus

◀ FIGURE 5.6 **Reflecting Telescopes** Four reflecting telescope designs: (a) prime focus, (b) Newtonian focus, (c) Cassegrain focus, and (d) Nasmyth/coudé focus. Each design uses a primary mirror at the bottom of the telescope to capture radiation, which is then directed along different paths for analysis. Notice that the secondary mirrors shown in (c) and (d) are actually slightly diverging, so that they move the focus outside the telescope.

TUTORIAL Reflecting Telescopes

MA

recently constructed reflecting telescopes have mirror diameters in the 10-m range, and still larger instruments are on the way.

Types of Reflecting Telescope

Figure 5.6 shows some basic reflecting telescope designs. Radiation from a star enters the instrument, passes down the main tube, strikes the primary mirror, and is reflected back toward the prime focus, near the top of the tube. Sometimes astronomers place their recording instruments at the prime focus; however, it can be inconvenient, or even impossible, to suspend bulky pieces of equipment there. More often, the light is intercepted on its path to the focus by a *secondary mirror* and redirected to a more convenient location, as in Figure 5.6(b) through 5.6(d).

In a **Newtonian telescope** (named after Sir Isaac Newton, who invented this particular design), the light is intercepted before it reaches the prime focus and then is deflected by 90°, usually to an eyepiece at the side of the instrument. This is a popular design for smaller reflecting telescopes, such as those used by amateur astronomers, but it is relatively uncommon in large instruments. On a large telescope, the Newtonian focus may be many meters above the ground, making it an inconvenient place to attach equipment (or place an observer).

Alternatively, astronomers may choose to work on a rear platform where they can use equipment, such as a spectroscope, that is too heavy to hoist to the prime focus. In this case, light reflected by the primary mirror toward the prime focus is intercepted by a smaller secondary mirror, which reflects it back down through a small hole at the center of the primary mirror. This arrangement is known as a **Cassegrain telescope** (after Guillaume Cassegrain, a French lensmaker). The point behind the primary mirror where the light from the star finally converges is called the *Cassegrain focus.*

A more complex observational configuration requires starlight to be reflected by several mirrors. As in the Cassegrain design, light is first reflected by the primary mirror toward the prime focus and is then reflected back down the tube by a secondary mirror. Next, a third, much smaller, mirror reflects the light out of the telescope, where (depending on the details of the telescope's construction) the beam may be analyzed by a detector mounted alongside, at the *Nasmyth focus,* or it may be directed via a further series of mirrors into an environmentally controlled laboratory known as the *coudé* room (from the French word for "bent"). This laboratory is separate from the telescope itself, enabling astronomers to use very heavy and finely tuned equipment that cannot be placed at any of the other foci (all of which necessarily move with the telescope). The arrangement of mirrors is such that the light path to the coudé room does not change as the telescope tracks objects across the sky.

To illustrate some of these points, Figure 5.7(a) shows the twin 10-m-diameter optical/infrared telescopes of the Keck Observatory on Mauna Kea in Hawaii, operated jointly by the California Institute of Technology and the University of California. The diagram in part (b) illustrates the light paths and some of the foci. Observations may be made at the Cassegrain, Nasmyth, or coudé focus, depending on the needs of the user. As the size of the person in part (c) indicates, this is indeed a very large telescope—in fact, the two mirrors are currently the largest on Earth. We will see numerous examples throughout this text of Keck's many important discoveries.

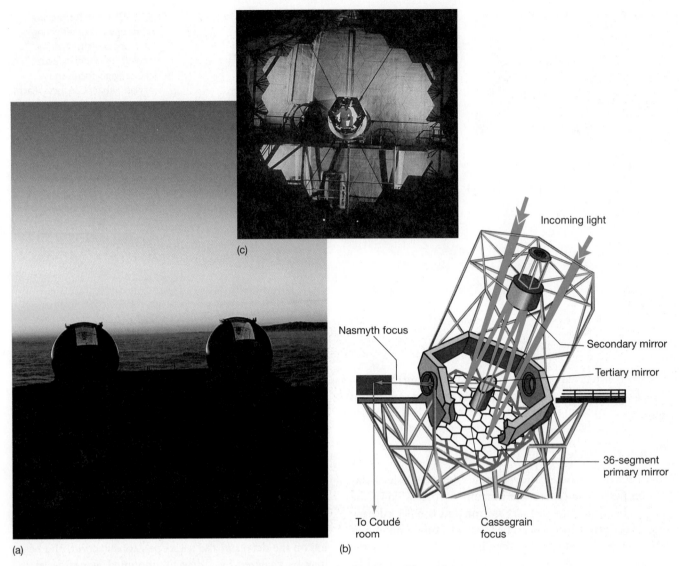

▲ FIGURE 5.7 Keck Telescope (a) The two 10-m telescopes of the Keck Observatory. (b) Artist's illustration of the telescope, the path taken by an incoming beam of starlight, and some of the locations where instruments may be placed. (c) One of the 10-m mirrors. (The odd shape is explained in Section 5.3.) Note the technician in orange coveralls at center. *(W. M. Keck Observatory)*

Perhaps the best-known telescope on (or, rather, near) Earth is the *Hubble Space Telescope (HST)*, named for one of America's most notable astronomers, Edwin Hubble. The device was placed in Earth orbit by NASA's space shuttle *Discovery* in 1990 and is still in operation (as of 2010; see *Discovery 5-1*). *HST* is a Cassegrain telescope in which all the instruments are located directly behind the primary mirror. The telescope's detectors are capable of making measurements in the optical, infrared, and ultraviolet parts of the spectrum.

CONCEPT CHECK

✔ Why do all modern telescopes use mirrors to gather and focus light?

5.2 Telescope Size

Modern astronomical telescopes have come a long way from Galileo's simple apparatus (see the Part 1 opening text). Their development over the years has seen a steady increase in *size*, for two main reasons. The first has to do with the amount of light a telescope can collect—its *light-gathering power*. The second is related to the amount of detail that can be seen—the telescope's *resolving power*. Simply put, large telescopes can gather and focus more radiation than can their smaller counterparts, allowing astronomers to study fainter objects and to obtain more detailed information about bright ones. This fact has played a central role in determining the design of contemporary instruments.

Light-Gathering Power

One important reason for using a larger telescope is simply that it has a greater **collecting area,** which is the total area capable of gathering radiation. The larger the telescope's reflecting mirror (or refracting lens), the more light it collects, and the easier it is to measure and study an object's radiative properties. Astronomers spend much of their time observing very distant—and hence very *faint*—cosmic sources. In order to make detailed observations of such objects, very large telescopes are essential. Figure 5.8 illustrates the effect of increasing the size of a telescope by comparing images of the Andromeda Galaxy taken with two different instruments. A large collecting area is particularly important for spectroscopic work, as the radiation received in that case must be split into its component wavelengths for further analysis.

The observed brightness of an astronomical object is directly proportional to the area of our telescope's mirror and therefore to the *square* of the mirror diameter. Thus, a 5-m telescope will produce an image 25 times as bright as a 1-m instrument, because a 5-m mirror has $5^2 = 25$ times the collecting area of a 1-m mirror. We can also think of this relationship in terms of the length of *time* required for a telescope to collect enough energy to create a recognizable image on a photographic plate. Our 5-m telescope will produce an image 25 times faster than the 1-m device because it gathers energy at a rate 25 times greater. Put another way, a 1-hour exposure with a 1-m telescope is roughly equivalent to a 2.4-minute exposure with a 5-m instrument.

Until the 1980s, the conventional wisdom was that telescopes with mirrors larger than 5 or 6 m in diameter were simply too expensive and impractical to build. The problems involved in casting, cooling, and polishing a huge block of quartz or glass to very high precision (typically less than the width of a human hair) were just too great. However, new, high-tech manufacturing techniques, coupled with radically new mirror designs, make the construction of telescopes in the 8- to 12-m range almost a routine matter. Experts can now make large mirrors much lighter for their size than had previously been thought feasible and can combine many smaller mirrors into the equivalent of a much larger single-mirror telescope. Several large-diameter instruments now exist, and many more are planned.

The Keck telescopes, shown in detail in Figure 5.7 and in a larger view in Figure 5.9, are a case in point. Each telescope combines 36 hexagonal 1.8-m mirrors into the equivalent collecting area of a single 10-m reflector. The first Keck telescope became fully operational in 1992; the second was completed in 1996. The large size of these devices and the high altitude at which they operate make them particularly well suited to detailed spectroscopic studies of very faint objects, in both the optical and infrared parts of the spectrum. Mauna Kea's 4.2-km (13,800 feet) altitude minimizes

(a)

(b)

R I V U X G

▲ **FIGURE 5.8 Sensitivity** Effect of increasing telescope size on an image of the Andromeda Galaxy. Both photographs had the same exposure time, but image (b) was taken with a telescope twice the size of that used to make image (a). Fainter detail can be seen as the diameter of the telescope mirror increases because larger telescopes are able to collect more photons per unit time, greatly extending our view of the universe. *(Adapted from AURA)*

atmospheric absorption of infrared radiation, making this site one of the finest locations on Earth for infrared astronomy.

Numerous other large telescopes can be seen in the figure. Some are designed exclusively for infrared work; others, like Keck, operate in both the optical and the infrared. To the right of the Keck domes is the 8.3-m Subaru (the Japanese name for the Pleiades) telescope, part of the National Astronomical Observatory of Japan. Its mirror, shown in Figure 5.9(b), is the largest single mirror (as opposed to the seg-

DISCOVERY 5-1

The Hubble Space Telescope

The *Hubble Space Telescope (HST)* is the largest, most complex, most sensitive observatory ever deployed in space. At over $9 billion (including the cost of several missions to service and refurbish the system), it is also one of the most expensive scientific instruments ever built and operated. A joint project of NASA and the European Space Agency, *HST* was designed to allow astronomers to probe the universe with at least 10 times finer resolution and some 30 times greater sensitivity to light than existing Earth-based devices. *HST* is operated remotely from the ground; there are no astronauts aboard the telescope, which orbits Earth about once every 95 minutes at an altitude of some 600 km (380 miles).

The telescope's overall dimensions approximate those of a city bus or railroad tank car: 13 m (43 feet) long, 12 m (39 feet) across with solar arrays extended, and 11,000 kg (12.5 tons when weighed on Earth). At the heart of *HST* is a 2.4-m-diameter mirror designed to capture optical, ultraviolet, and infrared radiation before it reaches Earth's murky atmosphere. The first figure shows the telescope being lifted out of the cargo bay of the space shuttle *Discovery* in the spring of 1990.

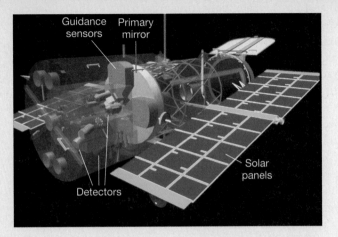

(NASA)

The second "see-through" diagram displays *HST*'s main features. The telescope's Cassegrain design (Section 5.1) reflects light from its large mirror back to a smaller, 0.3-m secondary mirror, which sends the light through a small hole in the main mirror and into the rear portion of the spacecraft. There, any of five major scientific instruments wait to analyze the incoming radiation. Most of these instruments are about the size of a telephone booth. They are designed to be maintained by NASA astronauts, and indeed, most of the telescope's instruments have been upgraded or replaced since *HST* was launched. The current detectors on the telescope span the

visible, near-infrared, and near-ultraviolet regions of the electromagnetic spectrum, from about 100 nm (UV) to 2200 nm (IR).

Soon after the launch of *HST,* astronomers discovered that the telescope's primary mirror had been polished to the wrong shape. The mirror was too flat by 2 μm, about 1/50 the width of a human hair, making it impossible to focus light as well as expected. The optical flaw (known as *spherical aberration*) produced by this design error meant that *HST* was not as sensitive as designed, although it could still see many objects in the universe with unprecedented resolution. In 1993, astronauts aboard the space shuttle *Endeavour* visited *HST* and succeeded in repairing some of its ailing equipment. They replaced *Hubble*'s gyroscopes to help the telescope point more accurately, installed sturdier versions of the solar panels that power the telescope's electronics, and—most importantly—inserted an intricate set of small mirrors (each about the size of a coin) to compensate for the faulty primary mirror. *Hubble*'s resolution is now close to the original design specifications, and the telescope has regained much of its lost sensitivity. Additional service missions were performed in 1997, 1999, 2002, and 2009 to replace instruments and repair faulty systems.

A good example of *Hubble*'s scientific capabilities today can be seen by comparing the two images of the spiral galaxy M100, shown in the accompanying figure. On the left is perhaps the best ground-based photograph of this beautiful galaxy, showing rich detail and color in its spiral arms. On the right, to the same scale and orientation, is an *HST* image showing improvement in both resolution and sensitivity. (The chevron-shaped field of view is caused by the corrective optics inserted into the telescope; an additional trade-off is that *Hubble*'s field of view is smaller than those of ground-based telescopes.) The inset shows *Hubble*'s exquisite resolution of small fields of view.

mented design used in Keck) yet built. Subaru saw "first light" in 1999. In the distance is another large single-mirror instrument: the 8.1-m Gemini North telescope, completed in 1999 by a consortium of seven nations, including the United States. Its twin, Gemini South, in the Chilean Andes, went into service in 2002.

In terms of total available collecting area, the largest telescope currently available is the European Southern Observatory's optical-infrared Very Large Telescope (VLT), located at Cerro Paranal, in Chile (Figure 5.10). The VLT consists of four separate 8.2-m mirrors that can function as a single instrument. The last of its four mirrors was completed in 2001.

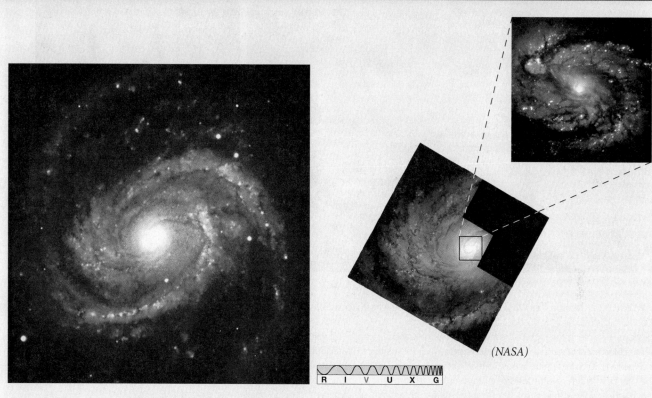

(NASA)

(D. Malin/Angio-Australian Telescope)

During its decade and a half of operation, *Hubble* has revolutionized our view of the sky, helping to rewrite some theories of the universe along the way. The space-based instrument has studied newborn galaxies almost at the limit of the observable universe with unprecedented clarity, allowing astronomers to see the interactions and collisions that may have shaped the evolution of our Milky Way. Turning its gaze to the hearts of galaxies closer to home, *Hubble* has provided strong evidence for supermassive black holes in their cores. Within our own Galaxy, it has given astronomers stunning new insights into the physics of star formation and the evolution of stellar systems and stars of all sizes, from superluminous giants to objects barely more massive than planets. Finally, in our solar system, *Hubble* has afforded scientists new views of the planets, their moons, and the tiny fragments from which they formed long ago. Many spectacular examples of the telescope's remarkable capabilities appear throughout this book.

With *Hubble* now ending its second decade of highly productive service, NASA has formally approved plans for the telescope's successor. The Next Generation Space Telescope (now known as the *James Webb Space Telescope*, or *JWST*, after the administrator who led NASA's *Apollo* program during the 1960s and 1970s) will dwarf *Hubble* in both scale and capability. Sporting a 6.5-m segmented mirror with seven times the collecting area of *Hubble*'s mirror, and containing a formidable array of detectors optimized for use at visible and infrared wavelengths, *JWST* will orbit at a distance of 1.5 million km from Earth, far beyond the Moon. NASA hopes to launch the instrument in 2014. The telescope's primary mission aims to study the formation of the first stars and galaxies, measuring the large-scale structure of the universe and investigating the evolution of planets, stars, and galaxies.

The *JWST* launch schedule will likely mean an interruption of a year or more in space-based astronomical observations at near-optical wavelengths. *HST* mission scientists expect that the 2009 servicing mission has extended *Hubble*'s operating lifetime well beyond 2011, but few expect that the telescope will survive until *JWST* becomes fully operational in 2015.

Resolving Power

A second advantage of large telescopes is their finer **angular resolution.** In general, *resolution* refers to the ability of any device, such as a camera or telescope, to form distinct, separate images of objects lying close together in the field of view. The finer the resolution, the better we can distinguish the objects and the more detail we can see. In astronomy, where we are always concerned with angular measurement, "close together" means "separated by a small angle in the sky," so angular resolution is the factor that determines our ability to see fine structure. Figure 5.11 illustrates how the

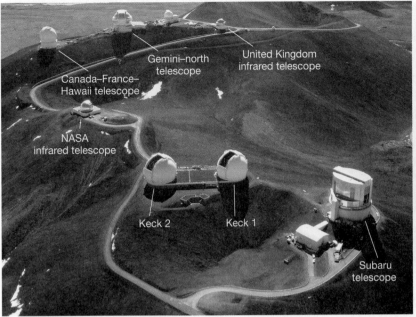

(a)

(b)

▲ FIGURE 5.9 **Mauna Kea Observatory** (a) The world's highest ground-based observatory, at Mauna Kea, Hawaii, is perched atop a dormant volcano more than 4 km (nearly 14,000 feet) above sea level. Among the domes visible in the picture are those housing the Canada–France–Hawaii 3.6-m telescope, the 8.1-m Gemini North instrument, the 2.2-m telescope of the University of Hawaii, Britain's 3.8-m infrared facility, and the twin 10-m Keck telescopes. To the right of the twin Kecks is the Japanese 8.3-m Subaru telescope. The thin air at this high-altitude site guarantees less atmospheric absorption of incoming radiation and hence a clearer view than at sea level, but the air is so thin that astronomers must occasionally wear oxygen masks while working. (b) The mirror in the Subaru telescope, currently the largest one-piece mirror in any astronomical instrument. *(R. Wainscoat; R. Underwood/Keck Observatory; NAOJ)*

▲ FIGURE 5.10 **VLT Observatory** Located at the Paranal Observatory in Atacama, Chile, the European Southern Observatory's Very Large Telescope (VLT) is the world's largest optical telescope. It comprises four 8.2-m reflecting telescopes, which can be used in tandem to create the effective area of a single 16-m mirror. *(ESO)*

appearance of two objects—stars, say—might change as the angular resolution of our telescope varies. Figure 5.12 shows the result of increasing resolving power with views of the Andromeda Galaxy at several different resolutions.

What limits a telescope's resolution? One important factor is *diffraction,* the tendency of light—and all other waves, for that matter—to bend around corners. ∞ (*Discovery 3-1*) Because of diffraction, when a parallel beam of light enters a telescope, the rays spread out slightly, making it impossible to focus the beam to a sharp point, even with a perfectly constructed mirror. Diffraction introduces a certain "fuzziness," or loss of resolution, into any optical system. The degree of fuzziness—the minimum angular separation that can be distinguished—determines the angular resolution of the telescope. The amount of diffraction is proportional to the wavelength of the radiation and inversely proportional to the diameter of the telescope mirror. For a circular mirror and otherwise perfect optics, we can write (in convenient units):

$$\text{angular resolution (arcsec)} = 0.25 \frac{\text{wavelength (μm)}}{\text{diameter (m)}},$$

where 1 μm (1 micron) $= 10^{-6}$ m (see Appendix 2).

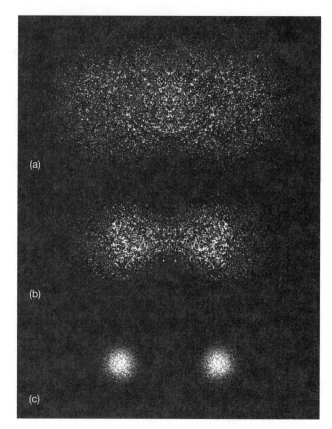

▲ **FIGURE 5.11 Resolving Power** Two comparably bright light sources become progressively clearer when viewed at finer and finer angular resolution. When the angular resolution is much poorer than the separation of the objects, as in (a), the objects appear as a single fuzzy "blob." As the resolution improves, through (b) and (c), the two sources become discernible as separate objects.

For a given telescope size, the amount of diffraction increases in proportion to the wavelength used, and observations in the infrared or radio range are often limited by its effects. For example, the best possible angular resolution of blue light (with a wavelength of 400 nm) that can be obtained using a 1-m telescope is about 0.1″. This quantity is called the **diffraction-limited resolution** of the telescope. But if we were to use our 1-m telescope to make observations in the near infrared, at a wavelength of 10 μm (10,000 nm), the best resolution we could obtain would be only 2.5″. A 1-m radio telescope operating at a wavelength of 1 cm would have an angular resolution of just under 1°.

For light of any given wavelength, large telescopes produce less diffraction than small ones. A 5-m telescope

▶ **FIGURE 5.12 Resolution** Detail becomes clearer in the Andromeda Galaxy as the angular resolution is improved some 600 times, from (a) 10′, to (b) 1′, (c) 5″, and (d) 1″. The resolution of the human eye is approximately that of part (b)—if only our eyes were sensitive enough to see this view. *(Adapted from AURA)*

R I V U X G

observing in blue light would have a diffraction-limited resolution five times finer than the 1-m telescope just discussed—about 0.02″. A 0.1-m (10-cm) telescope would have a diffraction limit of 1″, and so on. For comparison, the angular resolution of the human eye in the middle of the visual range is about 0.5′.

PROCESS OF SCIENCE CHECK

✔ Give two reasons why astronomers need to build very large telescopes.

5.3   Images and Detectors

In the previous section, we saw how telescopes gather and focus light to form an image of their field of view. In fact, most large observatories use many different instruments to analyze the radiation received from space—including detectors sensitive to many different wavelengths of light, spectroscopes to study emission and absorption lines, and other custom-made equipment designed for specialized studies. These devices may be placed at various points along the light path outside the telescope—see, for example, the multiple foci and light paths in Figure 5.7(b), or the more compact arrangement of detectors within the *Hubble Space Telescope* in *Discovery 5-1*. In this section, we look in a little more detail at how telescopic images are actually produced and at some other types of detectors that are in widespread use.

Image Acquisition

Computers play a vital role in observational astronomy. Most large telescopes today are controlled either by computers or by operators who rely heavily on computer assistance, and images and data are recorded in a form that can be easily read and manipulated by computer programs.

It is becoming rare for photographic equipment to be used as the primary means of data acquisition at large observatories. Instead, electronic detectors known as **charge-coupled devices,** or **CCDs,** are in widespread use. Their output goes directly to a computer. A CCD (Figure 5.13a and b) consists of a wafer of silicon divided into a two-dimensional array of many tiny picture elements, known as **pixels.** When light strikes a pixel, an electric charge builds up on it. The amount of charge is directly proportional to the number of photons striking each pixel—in other words, to the intensity of the light at that point. The buildup of charge is monitored electronically, and a two-dimensional image is obtained (Figure 5.13c and d).

A CCD is typically a few square centimeters in area and may contain several million pixels, generally arranged on a square grid. As the technology improves, both the areas of CCDs and the number of pixels they contain continue to increase. Incidentally, the technology is not limited to astronomy: Many home video cameras contain CCD chips similar in basic design to those in use at the great astronomical observatories of the world.

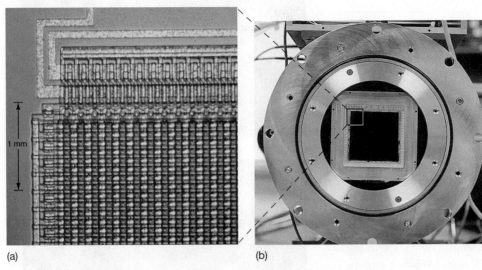

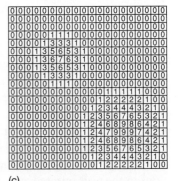

(c)

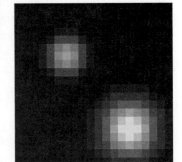

(d)

(a)   (b)

▲ **FIGURE 5.13 CCD Chip** A charge-coupled device (CCD) consists of hundreds of thousands, or even millions, of tiny light-sensitive cells called pixels, usually arranged in a square array. Light striking a pixel causes an electrical charge to build up on it. By electronically reading out the charge on each pixel, a computer can reconstruct the pattern of light—the image—falling on the chip. (a) Detail of a CCD array. (b) A CCD chip mounted for use at the focus of a telescope. (c) Typical data from the chip consist of an array of numbers, running from 0 to 9 in this simplified example. Each number represents the intensity of the radiation striking that particular pixel. (d) When interpreted as intensity levels on a computer screen, an image of the field of view results. *(MIT Lincoln Lab)*

CCDs have two important advantages over photographic plates, which were the staple of astronomers for over a century. First, CCDs are much more *efficient* than photographic plates, recording as many as 90 percent of the photons striking them, compared with less than 5 percent for photographic methods. This difference means that a CCD image can show objects 10 to 20 times fainter than a photograph made with the same telescope and the same exposure time. Alternatively, a CCD can record the same level of detail in less than a tenth of the time required by photographic techniques, or it can record that detail with a much smaller telescope. Second, CCDs produce a faithful representation of an image in a digital format that can be placed directly on magnetic tape or disk or, more commonly, sent across a computer network to an observer's home institution.

Image Processing

Computers are also widely used to reduce *background noise* in astronomical images. Noise is anything that corrupts the integrity of a message, such as static on an AM radio or "snow" on a television screen. The noise corrupting telescopic images has many causes. In part, it results from faint, unresolved sources in the telescope's field of view and from light scattered into the line of sight by Earth's atmosphere. It can also be caused by imperfections within the detector itself, which may result in an electronic "hiss" similar to the faint background hiss you might hear when you listen to a particularly quiet piece of music on your stereo.

Even though astronomers often cannot determine the origin of the noise in their observations, they can at least measure its characteristics. For example, if we observe a part of the sky where there are no known sources of radiation, then whatever signal we do receive is (almost by definition) noise. Once the properties of the signal have been measured, the effects of noise can be partially removed with the aid of high-speed computers, allowing astronomers to see features in their data that would otherwise remain hidden.

Using computer processing, astronomers can also compensate for known instrumental defects. In addition, the computer can often carry out many of the relatively simple, but tedious and time-consuming, chores that must be performed before an image (or spectrum) reaches its final "clean" form. Figure 5.14 illustrates how computerized image-processing techniques were used to correct for known instrumental problems in *HST,* allowing much of the planned resolution of the telescope to be recovered even before its repair in 1993 (see *Discovery 5-1*).

Wide-Angle Views

Large reflectors are very good at forming images of narrow fields of view, wherein all the light that strikes the mirror surface moves almost parallel to the axis of the instrument. However, as the angle at which the light enters increases, the accuracy of the focus decreases, degrading the overall quality of the image. The effect (called *coma*) worsens as we move farther from the center of the field of view. Eventually, the quality of the image is reduced to the point where the image is no longer usable. The distance from the center to where the image becomes unacceptable defines the useful field of view of the telescope—typically, only a few arc minutes for large instruments.

Photometry

When a CCD is placed at the focus of a telescope to record an image of the instrument's field of view, the telescope is acting, in effect, as a high-powered camera. However,

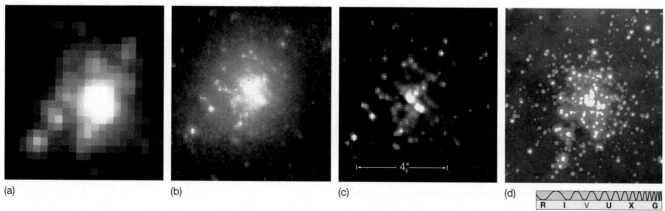

(a)   (b)   (c)   (d)

R   I   V   U   X   G

▲ **FIGURE 5.14 Image Processing** (a) Ground-based view of the star cluster R136, a group of stars in the Large Magellanic Cloud (a nearby galaxy). (b) The "raw" image of this same region as seen by the *Hubble Space Telescope* in 1990, before its first repair mission. (c) The same image after computer processing that partly compensated for imperfections in the mirror. (d) The same region as seen by the repaired *HST* in 1994, here observed at a somewhat bluer wavelength. *(AURA/NASA)*

astronomers often want to carry out more specific measurements of the radiation received from space.

One very fundamental property of a star (or any other astronomical object) is its *brightness*—the amount of light energy from the star striking our detector every second. The measurement of brightness is called **photometry** (literally, "light measurement"). In principle, determining a star's brightness is just a matter of adding up the values in all the CCD pixels corresponding to the star (see Figure 5.13c). However, in practice, the process is more complicated, as stellar images may overlap, and computer assistance is generally need to disentangle them.

Astronomers often combine photometric measurements with the use of colored *filters* in order to limit the wavelengths they measure. (A filter simply blocks out all incoming radiation, except in some specific range of wavelengths; see Section 17.3 for a more detailed discussion.) Many standard filters exist, covering various "slices" of the spectrum, from near-infrared through visible to near-ultraviolet wavelengths. By confining their attention to these relatively narrow ranges, astronomers can often estimate the shape of an object's blackbody curve and hence determine, at least approximately, the object's temperature. ∞ (Sec. 3.4) Filters are also used with CCD images in order to simulate natural color. For example, most of the visible-light *HST* images in this text are actually composites of *three* raw images, taken through red, green, and blue filters, respectively, and combined afterwards to reconstruct a single color frame.

Astronomical objects are generally faint, and most astronomical images entail long exposures—minutes to hours—in order to see fine detail (Section 5.2). Thus, the brightness we measure from an image is really an average over the entire exposure. Short-term fluctuations (if any) cannot be seen. When highly accurate and rapid measurements of light intensity are required, a specialized device known as a **photometer** is used. The photometer measures the total amount of light received in all or part of the field of view. When only a portion of the field is of interest, that region is selected simply by masking (blocking) out the rest of the field of view. Using a photometer often means "throwing away" spatial detail—usually no image is produced—but in return, more information is obtained about the intensity and time variability of a source, such as a pulsating star or a supernova explosion.

Spectroscopy

Often, astronomers want to study the *spectrum* of the incoming light. Large **spectrometers** work in tandem with optical telescopes. Light collected by the primary mirror may be redirected to the coudé room, defined by a narrow slit, dispersed (split into its component colors) by means of a prism or a diffraction grating, and then sent on to a detector—a process not so different in concept from the operation of the simple spectroscope described in Chapter 4. ∞ (Sec. 4.1) The spectrum can be studied in real time (i.e., as it is being

received at the telescope) or recorded using a CCD (or, less commonly nowadays, a photographic plate) for later analysis. Astronomers can then apply the analysis techniques discussed in Chapter 4 to extract detailed information from the spectral lines they record. ∞ (Sec. 4.5)

CONCEPT CHECK

✔ Why aren't astronomers satisfied with just taking photographs of the sky?

5.4 High-Resolution Astronomy

Even large telescopes have limitations. For example, according to the discussion in the preceding section, the 10-m Keck telescope should have an angular resolution of around 0.01″ in blue light. In practice, however, without further technological advances, it could not do better than about 1″. In fact, apart from instruments using special techniques developed to examine some particularly bright stars, *no* ground-based optical telescope built before 1990 can resolve astronomical objects to much better than 1″. The reason is *turbulence* in Earth's atmosphere—small-scale eddies of swirling air all along the line of sight, which blur the image of a star even before the light reaches our instruments.

Atmospheric Blurring

As we observe a star, atmospheric turbulence produces continual small changes in the optical properties of the air between the star and our telescope (or eye). As a result, the light from the star is refracted slightly as it travels toward us, so the stellar image dances around on our detector (or on our retina). This continual deflection is the cause of the well-known "twinkling" of stars. It occurs for the same reason that objects appear to shimmer when viewed across a hot roadway on a summer day: The constantly shifting rays of light reaching our eyes produce the illusion of motion.

On a good night at the best observing sites, the maximum amount of deflection produced by the atmosphere is slightly less than 1″. Consider taking a photograph of a star. After a few minutes of exposure (long enough for the intervening atmosphere to have undergone many small random changes), the image of the star has been smeared out over a roughly circular region an arc second or so in diameter. Astronomers use the term **seeing** to describe the effects of atmospheric turbulence. The circle over which a star's light (or the light from any other astronomical source) is spread is called the **seeing disk.** Figure 5.15 illustrates the formation of the seeing disk for a small telescope.*

---

*In fact, for a large instrument—more than about 1 m in diameter—the situation is more complicated, because rays striking different parts of the mirror have actually passed through different turbulent regions of the atmosphere. The end result is still a seeing disk, however.

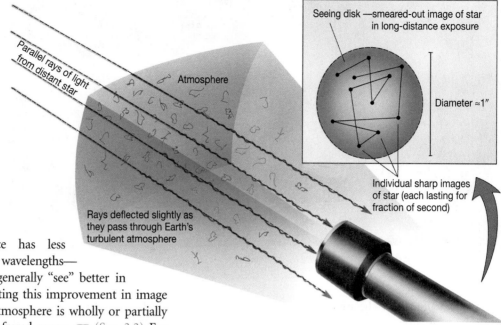

► **FIGURE 5.15 Atmospheric Turbulence** Light rays from a distant star strike a telescope detector at slightly different locations because of turbulence in Earth's atmosphere. Over time, the light covers a roughly circular region on the detector, and even the pointlike image of a star is recorded as a small disk, called the *seeing disk*.

Atmospheric turbulence has less effect on light of longer wavelengths—ground-based astronomers generally "see" better in the infrared. However, offsetting this improvement in image quality is the fact that the atmosphere is wholly or partially opaque over much of the infrared range. ∞ (Sec. 3.3) For these reasons, to achieve the best possible observing conditions, telescopes are sited on mountaintops (to get above as much of the atmosphere as possible) in regions of the world where the atmosphere is known to be fairly stable and relatively free of dust and moisture. Another reason for the choice of remote locations is the growing problem of **light pollution** in populated areas—unwanted upward-directed light from streets, parking lots, homes, and businesses, that scatters back from dust in the air into our telescopes, literally drowning out the faint signals from distant stars and galaxies that astronomers want to observe.

In the continental United States, these sites tend to be in the desert Southwest. The U.S. National Observatory for optical astronomy in the Northern Hemisphere, completed in

1973, is located high on Kitt Peak near Tucson, Arizona. The site was chosen because of its many dry, clear nights. Seeing less than 1″ from such a location is regarded as good, and seeing a few arc seconds is tolerable for many purposes. Even better conditions are found on Mauna Kea, Hawaii (Figure 5.9), and at numerous sites in the Andes Mountains of Chile (Figures 5.10 and 5.16), which is why many large telescopes have recently been constructed at these exceptionally clear locations.

An optical telescope placed in orbit about Earth or on the Moon could obviously overcome the limitations imposed by the atmosphere on ground-based instruments. Without atmospheric blurring, extremely fine resolution—close to

▲ **FIGURE 5.16 European Southern Observatory** Located in the Andes Mountains of Chile, the European Southern Observatory at La Silla is run by a consortium of European nations. Numerous domes house optical telescopes of different sizes, each with varied support equipment, making this one of the most versatile observatories south of the equator. The largest telescope at La Silla—the square building to the right of center—is the New Technology Telescope, a 3.5-m state-of-the-art active optics device. *(ESO)*

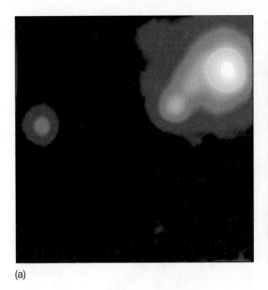

(a)

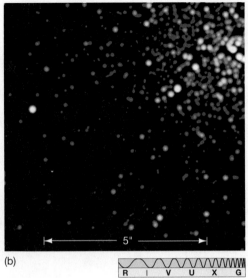

▶ **FIGURE 5.17 Active Optics** These false-color infrared photographs of part of the star cluster R136—the same object shown in Figure 5.14—contrast the resolution obtained (a) without and (b) with an active optics system. Both images were taken with the New Technology Telescope shown in Figure 5.16. *(ESO)*

(b)

5″

R I V U X G

the diffraction limit—can be achieved, subject only to the engineering restrictions associated with building or placing large structures in space. The 2.4-m mirror in the *Hubble Space Telescope* has a (blue-light) diffraction limit of only 0.050, giving astronomers a view of the universe as much as 20 times sharper than that normally available from even much larger ground-based instruments.

Active Optics

The latest techniques for producing ultrasharp images take the ideas of computer control and image processing (see Section 5.2) several stages further. By analyzing the image formed by a telescope *while the light is still being collected,* it is now possible to adjust the telescope from moment to moment to avoid or compensate for the effects of mirror distortion, temperature changes in the dome, and even atmospheric turbulence. By these means, some recently constructed telescopes have achieved resolutions very close to their theoretical (diffraction) limits.

Even under conditions of perfect seeing, most telescopes would not achieve diffraction-limited resolution. The temperature of the mirror or in the dome may fluctuate slightly during the many minutes or even hours required for the image to be exposed, and the precise shape of the mirror may change slightly as the telescope tracks a source across the sky. The effect of these changes is that the mirror's focus may shift from minute to minute, blurring the eventual image in much the same way as atmospheric turbulence creates a seeing disk (Figure 5.15). At the best observing sites, the seeing is often so good that these tiny effects may be the main cause of image blurring. The collection of techniques aimed at controlling such environmental and mechanical fluctuations is known as **active optics.**

The first telescope designed to incorporate active optics was the New Technology Telescope (NTT), constructed in 1989 at the European Southern Observatory in Chile and upgraded in 1997. (NTT is the most prominent instrument visible in Figure 5.16.) This 3.5-m instrument, employing the latest in real-time telescope controls, can achieve a resolution as sharp as 0.2″ by making minute modifications to the tilt of its mirror as its temperature and orientation change, thus maintaining the best possible focus at all times. Figure 5.17 shows how active optics can dramatically improve the resolution of an image. Active-optics techniques now include improved dome design to control airflow, precise control of the mirror temperature, and the use of pistons behind the mirror to maintain its precise shape. All of the large telescopes described earlier include active optics systems, improving their resolution to a few tenths of an arc second.

Real-Time Control

With active optics systems in place, Earth's atmosphere once again becomes the main agent limiting a telescope's resolution. Remarkably, even this problem can now be addressed, using an approach known as **adaptive optics,** a technique that actually deforms the shape of a mirror's surface under computer control, while the image is being exposed, in order to undo the effects of atmospheric turbulence. The mirror in question is generally not the large primary mirror of the telescope. Rather, for both economic and technical reasons, a much smaller (typically 20- to 50-cm-diameter) mirror is inserted into the light path and manipulated to achieve the desired effect. Adaptive optics presents formidable theoretical and practical problems, but the rewards are so great that it has been the subject of intense research since the 1970s.

The effort received an enormous boost in the 1990s from declassified military technology from the Strategic Defensive Initiative, a Reagan-era missile defense program (dubbed "Star Wars" by its detractors) intended to target and

ANIMATION/VIDEO Adaptive Optics

◀ **FIGURE 5.18 Adaptive Optics System** In this daytime photo, a test is being conducted at the Lick Observatory 3-m Shane telescope in California. A laser is used to create an "artificial star" (light reflected from the atmosphere back into the telescope) to improve guiding. The laser beam probes the atmosphere above the telescope, allowing tiny computer-controlled changes to be made in the shape of the mirror surface thousands of times each second. *(Lick Observatory)*

shoot down incoming ballistic missiles. In the system shown in Figure 5.18, a laser probes the atmosphere above the telescope, creating an "artificial star" that allows astronomers to gauge atmospheric conditions and pass that information to a computer that modifies the telescope mirror thousands of times per second to compensate for poor seeing.

Adaptive corrections are somewhat easier to apply in the infrared than in the optical, because atmospheric distortions are smaller (stars "twinkle" less in the infrared) and because the longer infrared wavelengths impose less stringent

requirements on the precise shape of the mirror. Infrared adaptive optics systems already exist in many large telescopes. For example, Gemini and Subaru have reported adaptive optics resolutions of around 0.06″ in the near infrared—not quite at the diffraction limit (0.03″ for an 8-m telescope at 1 μm, according to the earlier equation), but already better than the resolution of *HST* at the same wavelengths (see Figure 5.19a). Both Keck and the VLT incorporate adaptive optics instrumentation that will ultimately be capable of producing diffraction-limited images at near-infrared wavelengths.

Visible-light adaptive optics has been demonstrated experimentally, and some astronomical telescopes may incorporate the technology within the next few years. Figure 5.19(b) compares a pair of visible-light observations of a nearby double star called Castor. The observations were made with a relatively modest 1.5-m telescope. The adaptive optics system clearly distinguishes the two stars. Remarkably, adaptive optics techniques are giving astronomers the "best of both worlds," achieving with large ground-based optical telescopes the kind of resolution once attainable only from space.

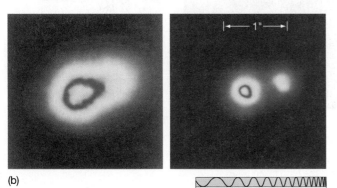

(a)

(b)

◀ **FIGURE 5.19 Adaptive Optics in Action** (a) The improvement in image quality produced by adaptive optics systems can be seen in these images acquired by the 8-m Gemini telescope atop Mauna Kea in Hawaii. The uncorrected visible-light image (left) of the star cluster NGC 6934 is resolved to a little less than 1″. With adaptive optics applied (right), the resolution in the infrared is improved by nearly a factor of 10, allowing more stars to be seen more clearly. (b) These visible-light images were acquired at a military observatory atop Mount Haleakala in Maui, Hawaii. The uncorrected image (left) of the double star Castor is a blur spread over several arc seconds, giving only a hint of its binary nature. With adaptive compensation applied (right), the resolution is improved to a mere 0.1″, and the two stars are clearly separated. *(NOAO; MIT Lincoln Laboratory)*

CONCEPT CHECK

✔ What steps do optical astronomers take to overcome the obscuring and blurring effects of Earth's atmosphere?

5.5   Radio Astronomy

In addition to the visible radiation that penetrates Earth's atmosphere on a clear day, radio radiation also reaches the ground. Indeed, as indicated in Figure 3.8, the radio window in the electromagnetic spectrum is much wider than the optical window. ∞ (Sec. 3.3) Because the atmosphere is no hindrance to long-wavelength radiation, radio astronomers have built many ground-based **radio telescopes** capable of detecting radio waves reaching us from space. These devices have all been constructed since the 1950s—radio astronomy is a much younger subject than optical astronomy.

Early Observations

The field of radio astronomy originated with the work of Karl Jansky at Bell Labs in 1931. Jansky was studying the causes of shortwave-radio interference when he discovered a faint static "hiss" that had no apparent terrestrial (Earthly) source. He noticed that the strength of the hiss varied in time and that its peak occurred about 4 minutes earlier each day. He soon realized that the peaks were coming exactly one *sidereal day* apart, and he concluded that the hiss was indeed not of terrestrial origin, but came from a definite direction in space. ∞ (Sec. 1.4) That direction is now known to correspond to the center of our Galaxy.

Some astronomers were intrigued by Jansky's discovery, but with the limited technology of the day—and even more limited Depression-era budgets—progress was slow. Jansky himself was moved to another project at Bell Labs and never returned to astronomical studies. However, by 1940, the first systematic surveys of the radio sky were under way. After a series of technological breakthroughs made during World War II, these studies rapidly grew into a distinct branch of astronomy.

During the 1930s, astronomers became aware that the space between the stars in our Galaxy is not empty, but instead is filled with extremely diffuse (low-density) gas (see Chapter 18). The growing realization in the 1940s that this otherwise completely invisible part of the Galaxy could be observed and mapped in detail at radio wavelengths established the true importance of Jansky's pioneering work. Today he is regarded as the father of radio astronomy.

Essentials of Radio Telescopes

Figure 5.20(a) shows the world's largest steerable radio telescope: the large 105-m-diameter (340-foot-diameter) telescope located at the National Radio Astronomy Observa-

(a)

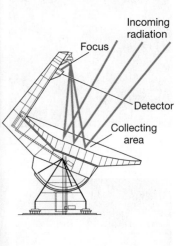

(b)

▲ **FIGURE 5.20 Radio Telescope** (a) The 105-m-diameter device at the National Radio Astronomy Observatory in Green Bank, West Virginia, is 150 m tall—taller than the Statue of Liberty and nearly as tall as the Washington Monument. (b) Schematic diagram of the telescope, showing the path taken by an incoming beam of radio radiation. *(NRAO)*

tory in West Virginia. Although much larger than reflecting optical telescopes, most radio telescopes are built in basically the same way. They have a large, horseshoe-shaped mount supporting a huge curved metal dish that serves as the collecting area. As illustrated in Figure 5.20(b), the dish captures incoming radio waves and reflects them to the focus, where a receiver detects the signals and channels them to a computer for storage and analysis.

Conceptually, the operation of a radio telescope is similar to the operation of an optical reflector with the detecting instruments placed at the prime focus (Figure 5.6a). However, unlike optical instruments, which can detect all visible wavelengths simultaneously, radio detectors normally register only a narrow band of wavelengths at any one time. To observe radiation with a different frequency, we must retune the equipment, much as we tune a radio or television set to a different channel.

Radio telescopes must be built large partly because cosmic radio sources are extremely faint. In fact, the total amount of radio energy received by Earth's entire surface is less than a trillionth of a watt. Compare this with the roughly 10 *million* watts our planet's surface receives in the form of infrared and visible light from any of the bright stars seen in the night sky. To capture enough radio energy to allow detailed measurements to be made, a large collecting area is essential. Figure 5.21 shows an even larger, but unmovable, radio telescope strung among the hills of Arecibo, Puerto Rico. Constructed in 1963 in a natural depression in the hillside, the Arecibo telescope is approximately 300 m (1000 feet) in diameter. Its reflecting surface spans nearly 20 acres.

The angular resolution of radio telescopes is generally quite poor compared with that of their optical counterparts because of the effects of diffraction. Typical wavelengths of

▲ FIGURE 5.21 **Arecibo Observatory** An aerial photograph of the 300-m-diameter dish at the National Astronomy and Ionospheric Center near Arecibo, Puerto Rico. The receivers that detect the focused radiation are suspended nearly 150 m (about 45 stories) above the center of the dish. The left inset shows a close-up of the radio receivers hanging high above the dish. The right inset shows technicians adjusting the dish surface to make it smoother. *(D. Parker/T. Acevedo/NAIC; Cornell)*

radio waves are about a million times longer than those of visible light, and these longer wavelengths impose a corresponding crudeness in angular resolution. (Recall from Section 5.2 that the longer the wavelength, the greater the amount of diffraction.) Even the enormous sizes of radio dishes only partly offset this effect. The radio telescope shown in Figure 5.20 can achieve a resolution of about 1′ when receiving radio waves having wavelengths of around 3 cm. However, it was designed to operate most efficiently (i.e., it is most sensitive to radio signals) at wavelengths closer to 1 cm, where the resolution is approximately 20″. The best angular resolution obtainable with a single radio telescope is about 10″ (for the largest instruments operating at millimeter wavelengths)—at least 100 times coarser than the capabilities of some large optical systems.

Radio telescopes can be built so much larger than their optical counterparts because their reflecting surfaces need not be as smooth as is necessary for light waves of shorter wavelength. Provided that surface irregularities (dents, bumps, and the like) are much smaller than the wavelength of the waves to be detected, the surface will reflect them without distortion. Because the wavelength of visible radiation is short (less than 10^{-6} m), extremely smooth mirrors are needed to reflect the waves properly, and it is difficult to construct very large mirrors to such exacting tolerances. However, even rough metal surfaces can focus 1-cm waves accurately, and radio waves of wavelength a meter or more can be reflected and focused perfectly well by surfaces having irregularities even as large as your fist. The Arecibo instrument was originally surfaced with chicken wire, which was lightweight and cheap. Although fairly rough, the chicken wire was adequate for proper reflection because the openings between adjacent strands of wire were much smaller than the long-wavelength radio waves that were to be detected.

The entire Arecibo dish was resurfaced with thin metal panels in 1974 and was further upgraded in 1997 so that it can now be used to study radio radiation of shorter wavelength. Since the 1997 upgrade, the panels can be adjusted to maintain a precise spherical shape to an accuracy of about 3 mm over the entire surface. At a frequency of 5 GHz (corresponding to a wavelength of 6 cm—the shortest wavelength that can be studied, given the properties of the dish surface), the telescope's angular resolution is about 1′. The huge size of the dish creates one distinct disadvantage, however: The Arecibo telescope cannot be pointed very well to follow cosmic objects across the sky. The detectors can move roughly 10° on either side of the focus, restricting the telescope's observations to those objects that happen to pass within about 20° of overhead as Earth rotates.

Arecibo is an example of a rough-surfaced telescope capable of detecting long-wavelength radio radiation. At the other extreme, Figure 5.22 shows the 36-m-diameter Haystack dish in northeastern Massachusetts. Constructed

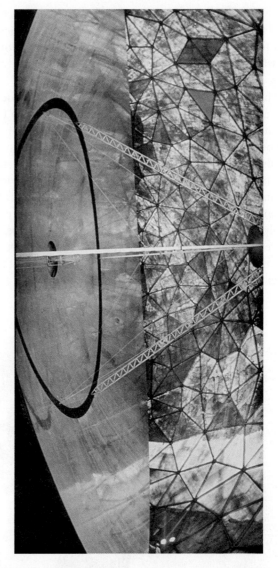

▲ FIGURE 5.22 **Haystack Observatory** Photograph of the Haystack dish, inside its protective radome. For scale, note the engineer standing at the bottom. Note also the dull shine on the telescope surface, indicating its smooth construction. Haystack is a poor optical mirror, but a superb radio telescope. Accordingly, it can be used to reflect and accurately focus radiation having short radio wavelengths, even as small as a fraction of a centimeter. (*MIT*)

of polished aluminum, this telescope maintains a parabolic curve to an accuracy of about a millimeter all the way across its solid surface. It can reflect and accurately focus radio radiation with wavelengths as short as a few millimeters. The telescope is contained within a protective shell, or radome, that protects the surface from the harsh New England weather. The radome acts much like the protective dome of an optical telescope, except that there is no slit through which the telescope "sees." Incoming cosmic radio signals pass virtually unimpeded through the radome's fiberglass construction.

The Value of Radio Astronomy

Despite the inherent disadvantage of relatively poor angular resolution, radio astronomy enjoys many advantages. Radio telescopes can observe 24 hours a day. Darkness is not needed for receiving radio signals because the Sun is a relatively weak source of radio energy, so its emission does not swamp radio signals arriving at Earth from elsewhere in the sky. In addition, radio observations can often be made through cloudy skies, and radio telescopes can detect the longest-wavelength radio waves even during rain or snowstorms. Poor weather causes few problems because the wavelength of most radio waves is much larger than the typical size of atmospheric raindrops or snowflakes. Optical astronomy cannot be performed under these conditions because the wavelength of visible light is smaller than a raindrop, a snowflake, or even a minute water droplet in a cloud.

However, perhaps the greatest value of radio astronomy (and, in fact, of all astronomies concerned with nonvisible regions of the electromagnetic spectrum) is that it opens up a whole new window on the universe. There are three main reasons for this. First, just as objects that are bright in the visible part of the spectrum (e.g., the Sun) are not necessarily strong radio emitters, many of the strongest radio sources in the universe emit little or no visible light. Second, visible light may be strongly absorbed by interstellar dust along the line of sight to a source. Radio waves, by contrast, are generally unaffected by intervening matter. Third, as mentioned earlier, many parts of the universe cannot be seen at all by optical means, but are easily detectable at longer wavelengths. The center of the Milky Way Galaxy is a prime example of such a totally invisible region—our knowledge of the galactic center is based almost entirely on radio and infrared observations. Thus, these observations not only afford us the opportunity to study the same objects at different wavelengths, but also allow us to see whole new classes of objects that would otherwise be completely unknown.

Figure 5.23 shows an optical photograph of the Orion Nebula (a huge cloud of interstellar gas) taken with the 4-m telescope on Kitt Peak. Superimposed on the optical image is a radio map of the same region, obtained by scanning the Haystack radio telescope (Figure 5.22) back and forth across the nebula and taking many measurements of radio intensity. The map is drawn as a series of contour lines connecting locations of equal radio brightness, similar to pressure contours drawn by meteorologists on weather maps or height contours drawn by cartographers on topographic maps. The inner contours represent stronger radio signals, the outside contours weaker signals. Note, however, how the radio map is much less detailed than its optical counterpart; that's because the acquired radio radiation has such a long wavelength compared to light.

The radio map in Figure 5.23 has many similarities to the visible-light image of the nebula. For instance, the radio

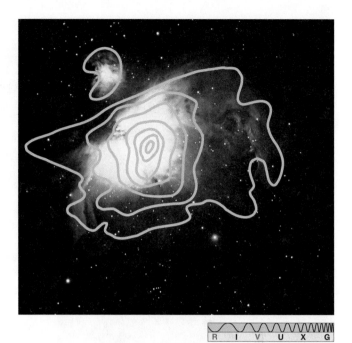

▲ **FIGURE 5.23 Orion Nebula in Radio and Visible** The Orion Nebula is a star-forming region about 1500 light-years from Earth. (The nebula is located in the constellation Orion and can be seen as a small smudge in Figure 1.8.) The bright regions in this photograph are stars and clouds of glowing gas. The dark regions are not empty, but their visible emission is obscured by interstellar matter. Superimposed on the optical image is a radio contour map (blue lines) of the same region. Each curve of the contour map represents a different intensity of radio emission. Note how the radio contours in some places enclose regions that are dark in visible light, allowing us to "see through" the obscuring material. The resolution of the optical image is about 1″; that of the radio map is 1′. (*Background photo: AURA*)

emission is strongest near the center of the optical image and declines toward the nebular edge. But there are also subtle differences between the radio and optical images. The two differ chiefly toward the upper left of the main cloud, where visible light seems to be absent, despite the existence of radio waves. How can radio waves be detected from locations not showing any emission of light? The answer is that this particular nebular region is known to be especially dusty in its top left quadrant. The dust obscures the short-wavelength visible light, but not the long-wavelength radio radiation. Thus, we have a trade-off typical of many in astronomy: Although the long-wave radio signals provide a less resolved map of the region, those same radio signals can pass relatively unhindered through dusty regions, in this case allowing us to see the true extent of the Orion Nebula.

CONCEPT CHECK

✔ In what ways does radio astronomy complement optical observations?

(a)

(b)

◀ FIGURE 5.24 VLA Interferometer
(a) This large interferometer, located on the Plain of San Augustin in New Mexico, comprises 27 separate dishes spread along a Y-shaped pattern about 30 km across. The most sensitive radio device in the world, it is called the Very Large Array, or VLA for short. (b) A close-up view from ground level shows how some of the VLA dishes are mounted on railroad tracks so that they can be repositioned easily. (NRAO)

In interferometry, two or more radio telescopes are used in tandem to observe the *same* object at the *same* wavelength and at the *same* time. The combined instruments together make up an **interferometer.** Figure 5.24 shows a large interferometer—many separate radio telescopes working together as a team. By means of electronic cables or radio links, the signals received by each antenna in the array making up the interferometer are sent to a central computer that combines and stores the data.

A Radio Interferometer

Interferometry works by analyzing how the signals interfere with each other when added together. ∞ *(Discovery 3-1)* Consider an incoming wave striking two detectors (Figure 5.25). Because the detectors lie at different distances from the source, the signals they record will, in general, be out of step with one another. In that case, when

5.6 Interferometry

The main disadvantage of radio astronomy compared with optical work is its relatively poor angular resolution. However, in some circumstances, radio astronomers can overcome this limitation with a technique known as **interferometry.** This technique makes it possible to produce radio images of angular resolution higher than can be achieved with even the best optical telescopes, on Earth or in space.

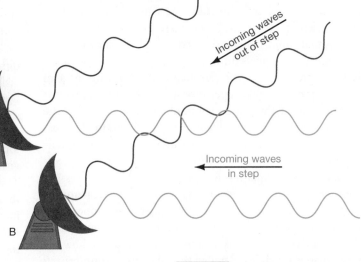

▶ FIGURE 5.25 **Interferometry** Two detectors, A and B, record different signals from the same incoming wave because of the time it takes the radiation to traverse the distance between them. When the signals are combined, the amount of interference depends on the wave's direction of motion, providing a means of measuring the position of the source in the sky. Here, the dark-blue waves come from a source high in the sky and interfere destructively when captured by antennas A and B. But when the same source has moved because of Earth's rotation (light-blue waves), the interference can be constructive.

the signals are combined, they will interfere destructively, partly canceling each other out. Only if the detected radio waves happen to be exactly in step will the signals combine constructively to produce a strong signal. Notice that the amount of interference depends on the *direction* in which the wave is traveling relative to the line joining the detectors. Thus—in principle, at least—careful analysis of the strength of the combined signal can provide an accurate measurement of the source's position in the sky.

As Earth rotates and the antennae track their target, the interferometer's orientation relative to the source changes, and a pattern of peaks and troughs emerges. In practice, extracting positional information from the data is a complex task, as multiple antennae and several sources are usually involved. Suffice it to say that, after extensive computer processing, the interference pattern translates into a high-resolution image of the target object.

An interferometer is, in essence, a substitute for a single huge antenna. As far as resolving power is concerned, the effective diameter of an interferometer is the distance between its outermost dishes. In other words, two small dishes can act as opposite ends of an imaginary, but huge, single radio telescope, dramatically improving angular resolution. For example, a resolution of a few arc seconds can be achieved at typical radio wavelengths (such as 10 cm), either by using a single radio telescope 5 km in diameter (which is impossible to build) or by using two or more much smaller dishes separated by the same 5 km, but connected electronically. The larger the distance separating the telescopes—that is, the longer the *baseline* of the interferometer—the better is the resolution attainable.

Large interferometers like the instrument shown in Figure 5.24 now routinely attain radio resolution comparable to that of optical images. Figure 5.26 shows an interferometric radio map of a nearby galaxy and a photograph of that same galaxy, made with a large optical telescope. The radio clarity is much better than in the contour map of Figure 5.23—in fact, the radio resolution in part (a) is comparable to that of the optical image in part (b).

Astronomers have created radio interferometers spanning great distances, first across North America and later between continents. A typical very-long-baseline interferometry experiment (usually known by its abbreviation, VLBI) might use radio telescopes in North America, Europe, Australia, and Russia to achieve an angular resolution on the order of 0.001″. It seems that even Earth's diameter is no limit: Radio astronomers have successfully used an antenna in orbit, together with several antennae on the ground, to construct an even longer baseline and achieve still better resolution. Proposals exist to place interferometers entirely in Earth orbit and even on the Moon.

Interferometry at Other Wavelengths

Although the technique was originally developed by radio astronomers, interferometry is no longer restricted to the radio domain. Radio interferometry became feasible when

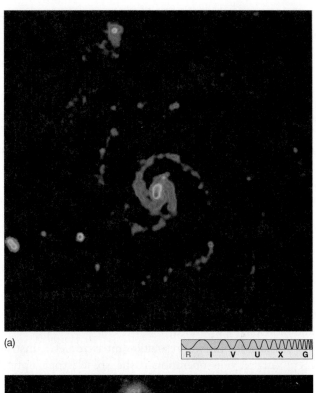

(a)

(b)

▲ FIGURE 5.26 **Radio–Optical Comparison** (a) VLA radio "image" (or radiograph) of the spiral galaxy M51 observed at radio frequencies with an angular resolution of a few arc seconds. (b) Visible-light photograph of that same galaxy, made with the 4-m Kitt Peak optical telescope and displayed on the same scale as (a). *(NRAO; AURA)*

electronic equipment and computers achieved speeds great enough to combine and analyze radio signals from separate radio detectors without loss of data. As the technology has improved, it has become possible to apply the same methods to radiation of higher frequency. Millimeter-wavelength interferometry has already become an established and important observational technique, and both the Keck telescopes and the VLT are expected to be used for near-infrared interferometry within the next few years.

Optical interferometry is currently the subject of intensive research. In 1997, a group of astronomers in Cambridge, England, succeeded in combining the light from three small optical telescopes to produce a single, remarkably clear, image. Each telescope of the Cambridge Optical Aperture Synthesis Telescope (COAST) was only 0.4 m in diameter, but with the mirrors positioned 6 m apart, the resulting resolution was a stunning 0.01″— better than the resolution of the best adaptive-optics systems on the ground or of *HST* operating above Earth's atmosphere. The three telescopes working together enabled astronomers to "split" the binary star Capella, whose two member stars are separated by only 0.05″. Normally Capella is seen from the ground only as a slightly oblong blur. With the COAST array, the two stars that make up the system were cleanly and individually separated.

Perhaps the highest resolution interferometric instrument currently in operation is the six-telescope optical array operated by the Center for High Angular Resolution Astronomy (CHARA) on Mount Wilson in California (Figure 5.27). Although each telescope is only 1 m in diameter, the placement of the array over the mountain results in a combined light beam having resolution equivalent to a single telescope 300 m across. CHARA is not designed to produce images of the stars it studies; however, it can resolve details as small as 0.0002″ across, allowing the positions, orbits, and even radii of some stars to be measured with exquisite accuracy.

▲ **FIGURE 5.27 Optical Interferometry** This aerial photo shows the facilities of the CHARA array intermingled with the existing equipment of the historic Mount Wilson Observatory, in California. The small 1-m telescopes in the array are numbered. *(E. Simison/Sea West Enterprises)*

CONCEPT CHECK
✔ What is the main reason for the poor angular resolution of radio telescopes? How do radio astronomers overcome this problem?

5.7 Space-Based Astronomy

Optical and radio astronomy are the oldest branches of astronomy, but since the 1970s there has been a virtual explosion of observational techniques spanning the rest of the electromagnetic spectrum. Today, all portions of the spectrum are studied, from radio waves to gamma rays, to maximize the amount of information available about astronomical objects.

As noted earlier, the types of astronomical objects that can be observed differ quite markedly from one wavelength range to another. Full-spectrum coverage is essential not only to see things more clearly, but even to see some things at all. Because of the transmission characteristics of Earth's atmosphere, astronomers must study practically all regions of the electromagnetic spectrum, from gamma rays through X rays to visible light, and on down to infrared and radio waves, from space. The rise of these "other astronomies" has therefore been closely tied to the development of the space program.

Infrared Astronomy

Infrared studies are a vital component of modern observational astronomy. Infrared astronomy spans a broad range of phenomena in the universe, from planets and their parent stars to the vast regions of interstellar space where new stars are forming, to explosive events occurring in faraway galaxies. Generally, **infrared telescopes** resemble optical telescopes, but their detectors are designed to be sensitive to radiation of longer wavelengths. Indeed, as we have seen, many ground-based "optical" telescopes are also used for infrared work, and some of the most useful infrared observing is done from the ground (e.g., from Mauna Kea—see Figure 5.9), even though the radiation is somewhat diminished in intensity by our atmosphere.

As with radio observations, the longer wavelength of infrared radiation often enables us to perceive objects that are partially hidden from optical view. As a terrestrial example of the penetrating properties of infrared radiation, Figure 5.28(a) shows a dusty and hazy region in California, hardly viewable optically, but easily seen in the infrared (Figure 5.28b). Figures 5.28(c) and (d) show a similar comparison for an astronomical object—the dusty regions of the Orion Nebula, where visible light is blocked by interstellar clouds, but which is clearly distinguishable in the infrared.

▲ FIGURE 5.28 **Smog Revealed** An optical photograph (a) taken near San Jose, California, and an infrared photo (b) of the same area taken at the same time. Infrared radiation of long wavelength can penetrate smog much better than short-wavelength visible light. The same advantage pertains to astronomical observations: An optical view (c) of an especially dusty part of the central region of the Orion Nebula is more clearly revealed in this infrared image (d) showing a cluster of stars behind the obscuring dust. (*Lick Observatory; NASA*)

Astronomers can make better infrared observations if they can place their instruments above most or all of Earth's atmosphere, using balloon-, aircraft-, rocket-, and satellite-based telescopes. As might be expected, the infrared telescopes that can be carried above the atmosphere are considerably smaller than the massive instruments found in ground-based observatories. Figure 5.29 shows an infrared image of the Orion region as seen by the 0.6-m British–Dutch–U.S. *Infrared Astronomy Satellite (IRAS)* in 1983 and compares it with the same region made in visible light. Figure 5.29(a) is a composite of IRAS images taken at three different infrared wavelengths (12 μm, 60 μm, and 100 μm). It is represented in *false color*, a technique commonly used for displaying images taken in non-visible light. The colors do not represent the actual wavelength of the radiation emitted, but instead some other property of the source, in this case temperature, descending from white to orange to black. The whiter regions thus denote higher temperatures and hence greater strength of infrared radiation. At about 1′ angular resolution, the fine details of the Orion

nebula evident in the visible portion of Figure 5.23 cannot be perceived. Nonetheless, the infrared image shows clouds of warm dust and gas thought to play a critical role in the formation of stars and also shows groups of bright young stars that are completely obscured at visible wavelengths.

In August 2003, NASA launched the 0.85-m *Space Infrared Telescope Facility (SIRTF)*. In December of that year, when the first images from the instrument were made public, NASA renamed the spacecraft the *Spitzer Space Telescope (SST)*, in honor of Lyman Spitzer, Jr., renowned astrophysicist and the first person to propose (in 1946) that a large telescope be located in space. The facility's detectors are designed to operate at wavelengths between 3.6 and 160 μm. Unlike previous space-based observatories, the *SST* does not orbit Earth, but instead follows our planet in its orbit around the Sun, trailing millions of kilometers behind in order to minimize Earth's heating effect on the detectors. The spacecraft is currently drifting away from Earth at the rate of 0.1 AU per year. The detectors were cooled to near absolute zero in order to observe infrared signals

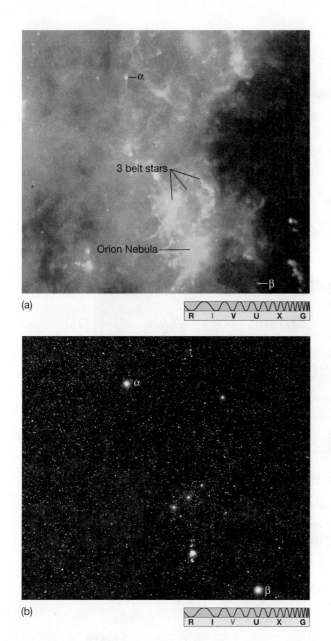

(a)

R I V U X G

(b)

R I V U X G

▲ FIGURE 5.29 **Infrared Image** (a) This infrared image of the Orion region was made by the *Infrared Astronomy Satellite*. In this false-color image, colors denote different temperatures, descending from white to orange to black. The resolution is about 1′. (b) The same region photographed in visible light with 1″ resolution. The labels α and β refer, respectively, to Betelgeuse and Rigel, the two brightest stars in the constellation. Note how the red star Betelgeuse is easily seen in the infrared (part a), whereas the blue star Rigel is very faint. *(NASA; J. Sanford)*

(a)

(b)

R I V U X G

▲ FIGURE 5.30 *Spitzer* **Images** These images from the *Spitzer Space Telescope*, now in orbit around the Sun, clearly show its camera's capabilities. (a) The magnificent spiral galaxy, M81, about 12 million light-years away. (b) Its companion, M82, is not so serene, rather resembling a "smoking hot cigar." *(JPL)*

from space without interference from the telescope's own heat. Figure 5.30 shows some examples of the spectacular imagery obtained from NASA's latest eye on the universe.

Unfortunately, the liquid helium keeping *SST*'s detectors cool could not be confined indefinitely, and (as expected) it slowly leaked away into space. In early 2009, *Spitzer* entered a new "warm" phase of operation as its temperature increased to roughly 30 K—still very cool by Earth standards, but warm enough for the telescope's own thermal emission to overwhelm the long-wavelength detectors on board. (Recall how Wien's law tells us that the thermal emission of a 30-K object peaks at a wavelength of roughly 100 μm.) ∞ (*Sec. 3.4*) However, the craft's shorter-wavelength detectors (at around 3.6 and 4.5 μm) are expected to remain operational for a further 3–4 years, and the telescope will continue to be an important astronomical resource during that time.

The present instrument package aboard the *Hubble Space Telescope (see Discovery 5-1)* also includes a high-resolution (0.1″) near-infrared camera and spectroscope.

Ultraviolet Astronomy

On the short-wavelength side of the visible spectrum lies the ultraviolet domain. Extending in wavelength from 400 nm (blue light) down to a few nanometers ("soft" X rays), this spectral range has only recently begun to be explored. Because Earth's atmosphere is partially opaque to radiation below 400 nm and is totally opaque below about 300 nm (due in part to the ozone layer), astronomers cannot conduct any useful ultraviolet observations from the ground, even from the highest mountaintop. Rockets, balloons, or satellites are therefore essential to any **ultraviolet telescope**—a device designed to capture and analyze that high-frequency radiation.

One of the most successful ultraviolet space missions was the *International Ultraviolet Explorer (IUE)*, placed in Earth orbit in 1978 and shut down for budgetary reasons in late 1996. Like all ultraviolet telescopes, its basic appearance and construction were quite similar to those of optical and infrared devices. Several hundred astronomers from all over the world used *IUE*'s near-ultraviolet spectroscopes to explore a variety of phenomena in planets, stars, and galaxies. In subsequent chapters, we will learn what this relatively new window on the universe has shown us about the activity and even the violence that seems to pervade the cosmos.

Figure 5.31 shows images captured by two more recent ultraviolet satellites. Figure 5.31(a) shows an image of a supernova remnant—the remains of a violent stellar explosion that occurred some 12,000 years ago—obtained by the *Extreme Ultraviolet Explorer (EUVE)* satellite, launched in 1992. Since its launch, *EUVE* has mapped out our local cosmic neighborhood as it presents itself in the far ultraviolet and has radically changed astronomers' conception of interstellar space in the vicinity of the Sun. Figure 5.31(b) shows an image of two relatively nearby galaxies, called M81 and M82, captured by the *Galaxy Evolution Explorer (GALEX)* satellite, launched in 2003. *HST*, best known as an optical telescope, is also a superb imaging and spectroscopic ultra violet instrument.

PROCESS OF SCIENCE CHECK

✔ Why are observations made at many different electromagnetic wavelengths useful to astronomers?

High-Energy Astronomy

High-energy astronomy studies the universe as it presents itself to us in X rays and gamma rays—the types of radiation whose photons have the highest frequencies and hence the greatest energies. How do we detect radiation of such short

wavelengths? First, it must be captured high above Earth's atmosphere, because none of it reaches the ground. Second, its detection requires the use of equipment fundamentally different in design from that used to capture the relatively low-energy radiation discussed up to this point.

The difference in the design of **high-energy telescopes** comes about because X rays and gamma rays cannot be reflected easily by any kind of surface. Rather, these rays tend to pass straight through, or be absorbed by, any material they strike. When X rays barely graze a surface, however, they can be reflected from it in a way that yields an image, although the mirror design is fairly complex. As illustrated in Figure 5.32, to ensure that all incoming rays are reflected at grazing angles, the telescope is constructed as a series of nested cylindrical mirrors,

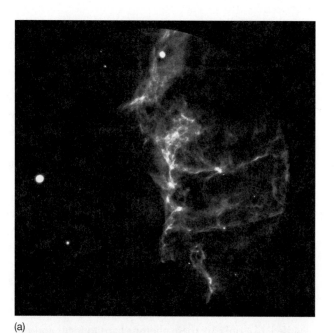

(a)

(b)

▶ FIGURE 5.31 **Ultraviolet Images** (a) A camera on board the *Extreme Ultraviolet Explorer* satellite captured this image of the Cygnus loop supernova remnant, the result of a massive star that blew itself virtually to smithereens. The release of energy was enormous, and the afterglow has lingered for centuries. The glowing field of debris shown here within the telescope's circular field of view lies some 1500 light-years from Earth. Based on the velocity of the outflowing debris, astronomers estimate that the explosion itself must have occurred about 12,000 years ago. (b) This false-color image of the spiral galaxy M81 and its companion M82, (see also Figure 5.30) made by the *Galaxy Evolution Explorer* satellite, reveals stars forming in the blue arms well away from the galaxy's center. *(NASA; GALEX)*

R I V U X G

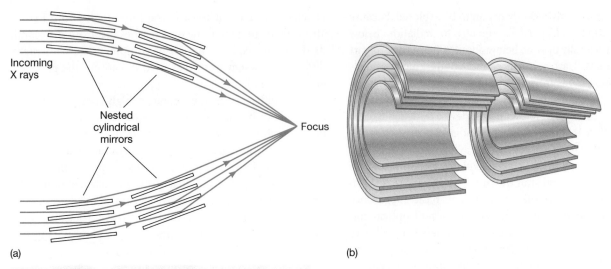

(a)

(b)

▲ FIGURE 5.32 **X-Ray Telescope** (a) The arrangement of nested mirrors in an X-ray telescope allows the rays to be reflected at grazing angles and focused to form an image. (b) A cutaway 3-D rendition of the mirrors, showing their shape more clearly.

carefully shaped to bring the X rays to a sharp focus. For gamma rays, no such method of producing an image has yet been devised; present-day gamma-ray telescopes simply point in a specified direction and count the photons they collect.

In addition, detection methods using photographic plates or CCD devices do not work well for hard (high-frequency) X rays and gamma rays. Instead, individual photons are counted by electronic detectors on board an orbiting device, and the results are then transmitted to the ground for further processing and analysis. Furthermore, the number of photons in the universe seems to be inversely related to their frequency. Trillions of visible (starlight) photons reach the detector of an optical telescope on Earth each second, but hours or even days are sometimes needed for a single gamma-ray photon to be recorded. Not only are these photons hard to focus and measure, but they are also very scarce.

The *Einstein Observatory*, launched by NASA in 1978, was the first X-ray telescope capable of forming an image of its field of view. During its 2-year lifetime, this spacecraft drove major advances in our understanding of high-energy phenomena throughout the universe. The German *ROSAT* (short for "Röntgen Satellite," after Wilhelm Röntgen, the discoverer of X rays) was launched in 1991. During its 7-year lifetime, this instrument generated a wealth of high-quality observational data. It was turned off in 1999, a few months after its electronics were irreversibly damaged when the telescope was accidentally pointed too close to the Sun.

In July 1999, NASA launched the *Chandra X-Ray Observatory* (named in honor of the Indian astrophysicist Subramanyan Chandrasekhar and shown in Figure 5.33). With greater sensitivity, a wider field of view, and better resolution than either *Einstein* or *ROSAT, Chandra* is providing high-energy astronomers with new levels of observational detail. Figure 5.34 shows a typical image returned by *Chandra:* a supernova remnant in the constellation Cassiopeia. Known

▲ FIGURE 5.33 *Chandra* **Observatory** The *Chandra* X-ray telescope is shown here during the final stages of its construction in 1998. The left end of the mirror arrangement, depicted in Figure 5.32, is at the bottom of the satellite as oriented here. *Chandra's* effective angular resolution is 1″, allowing this spacecraft to produce images of quality comparable to that of optical photographs. *Chandra* now occupies an elliptical orbit high above Earth; its farthest point from our planet, 140,000 km out, reaches almost one-third of the way to the Moon. Such an orbit takes *Chandra* above most of the interfering radiation belts girdling Earth (see Section 7.5) and allows more efficient observations of the cosmos. *(NASA)*

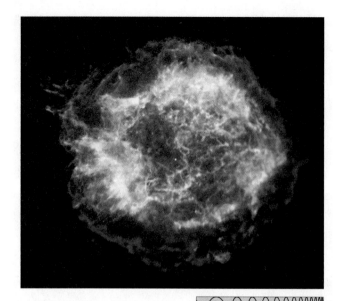

◀ FIGURE 5.34 **X-Ray Image** A false-color *Chandra* X-ray image of the supernova remnant Cassiopeia A, a debris field of scattered, hot gases that were once part of a massive star. Here, color represents the intensity of the X rays observed, from white (brightest) through red (faintest). Roughly 10,000 light-years from Earth and barely visible in the optical part of the spectrum, Cas A is now awash in brilliantly glowing X rays spread across some 10 light-years. (*NASA*)

as Cas A, the ejected gas is all that now remains of a star that was observed to explode about 320 years ago. The false-color image shows 50-million-K gas in wisps of ejected stellar material; the bright white point at the very center of the debris may be a black hole. The *XMM-Newton* satellite is more sensitive to X rays than is *Chandra* (that is, it can detect fainter X-ray sources), but it has significantly poorer angular resolution (5″, compared to 0.5″ for *Chandra*), making the two missions complementary to one another.

Gamma-ray astronomy is the youngest entrant into the observational arena. As just mentioned, imaging gamma-ray telescopes do not exist, so only fairly coarse (1° resolution) observations can be made. Nevertheless, even at this resolution, there is much to be learned. Cosmic gamma rays were originally detected in the 1960s by the U.S. *Vela* series of satellites, whose primary mission was to monitor illegal nuclear detonations on Earth. Since then, several X-ray telescopes have also been equipped with gamma-ray detectors.

NASA's *Compton Gamma-Ray Observatory* (*CGRO*) was placed in orbit by the space shuttle in 1991. At 17 tons, it was at the time the largest astronomical instrument ever launched. *CGRO* scanned the sky and studied individual objects in much greater detail than had previously been attempted. Many examples of *CGRO's* imagery appear throughout this book. The mission ended on June 4, 2000, when, following a failure of one of the satellite's three gyroscopes, NASA opted for a controlled reentry and dropped *CGRO* into the Pacific Ocean. In August 2008, NASA launched the *Fermi Gamma-Ray Space Telescope* (Figure 5.35a). With greater sensitivity to a broader range of

(a)

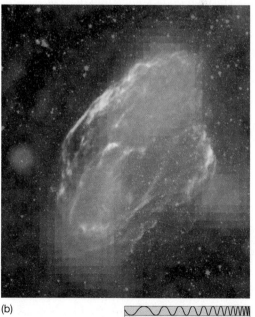

(b)

▲ FIGURE 5.35 **Gamma-Ray Astronomy** (a) The *Fermi Gamma-Ray Space Telescope,* shown here in an artist's conception, was named after Enrico Fermi, an Italian-American scientist who did pioneering work in high-energy physics. The wide arrays are solar panels to power the spacecraft; the box amidship contains layers of tungsten that detect the gamma rays. (b) A typical false-color gamma-ray image—this one showing the remains of a violent event (a supernova) in a region named W44. The gamma rays are shown mainly in magenta. (*NASA*)

gamma-ray energies than CGRO, Fermi's capabilities are greatly expanding astronomers' view of the high-energy universe. Figure 5.35(b) shows a false-color gamma-ray image of a violent stellar explosion in a distant galaxy. An early all-sky image obtained by Fermi appears in Figure 5.36(e).

CONCEPT CHECK

✔ List some scientific benefits of placing telescopes in space. What are the drawbacks of space-based astronomy?

5.8 Full-Spectrum Coverage

Many astronomical objects are now routinely observed at many different electromagnetic wavelengths. As we proceed through the text, we will discuss more fully the wealth of information that high-precision astronomical instruments can provide us.

It is reasonable to suppose that the future holds many further improvements in both the quality and the availability of astronomical data and that many new discoveries will be made. The current and proposed pace of technological progress presents us with the following exciting prospect: Within the next decade, if all goes according to plan, it will be possible, for the first time ever, to make *simultaneous* high-quality measurements of any astronomical object at *all* wavelengths, from radio ray to gamma ray. The consequences of this development for our understanding of the workings of the universe may be little short of revolutionary.

As a preview of the comparison that full-spectrum coverage allows, Figure 5.36 shows a series of images of our own Milky Way Galaxy. The images were made by several different instruments, at wavelengths ranging from radio to gamma rays, over a period of about 5 years. By comparing the features visible in each, we immediately see how multiwavelength observations can complement one another, greatly extending our perception of the dynamic universe around us.

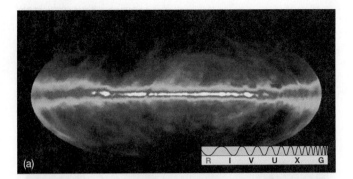

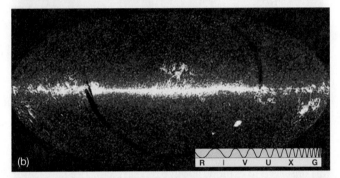

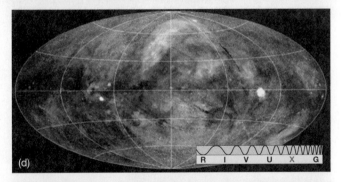

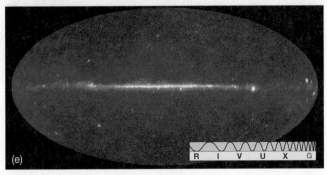

▶ FIGURE 5.36 **Multiple Wavelengths** The Milky Way Galaxy as it appears at (a) radio, (b) infrared, (c) visible, (d) X-ray, and (e) gamma-ray wavelengths. Each frame is a panoramic view covering the entire sky. The differences between these views clearly illustrate the value of multiwavelength astronomy. The center of our Galaxy, which lies in the direction of the constellation Sagittarius, is at the center of each map. *(NRAO; NASA; Lund Observatory; MPI; NASA)*

CHAPTER REVIEW

SUMMARY

1 A **telescope (p. 98)** is a device designed to collect as much light as possible from some distant source and deliver it to a detector for detailed study. A **refracting telescope (p. 98)** uses a lens to concentrate and focus the light; **reflecting telescopes (p. 98)** use mirrors. The **Newtonian (p. 101)** and **Cassegrain (p. 101)** reflecting telescope designs employ secondary mirrors to avoid placing detectors at the prime focus. More complex light paths are also used to allow the use of large or heavy equipment that cannot be placed near the telescope.

2 All large astronomical telescopes are reflectors, because large mirrors are lighter and much easier to construct than large lenses, and they also suffer from fewer optical defects. The light-gathering power of a telescope depends on its **collecting area (p. 103)**, which is proportional to the square of the mirror diameter. To study the faintest sources of radiation, astronomers must use large telescopes. Large telescopes also suffer least from the effects of diffraction and hence can achieve better **angular resolution (p. 105)** once the blurring effects of Earth's atmosphere are overcome. The amount of diffraction is proportional to the wavelength of the radiation under study and inversely proportional to the size of the mirror.

3 Most modern telescopes use **charge-coupled devices,** or **CCDs (p. 108),** instead of photographic plates to collect their data. The field of view is divided into an array of millions of **pixels (p. 108)** that accumulate an electric charge when light strikes them. CCDs are many times more sensitive than photographic plates, and the resultant data are easily saved directly in digital form image processing. The light collected by a telescope may be processed in a number of ways. It can be made to form an **image (p. 98)**. **Photometry (p. 110)** may be performed either on a stored image or during the observation itself, using a specialized detector, or a **spectrometer (p. 110)** may be used to analyze the spectrum of the radiation received.

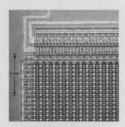

4 The resolution of most ground-based optical telescopes is limited by **seeing (p. 110)**—the blurring effect of Earth's turbulent atmosphere, which smears the pointlike images of stars out into **seeing disks (p. 110)** a few arc seconds in diameter. Radio and space-based telescopes do not suffer from this effect, so their resolution is determined by the effects of diffraction. Astronomers can greatly improve a telescope's resolution by using **active optics (p. 112),** in which a telescope's environment and focus are carefully monitored

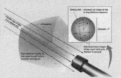

and controlled, and **adaptive optics (p. 112),** in which the blurring effects of atmospheric turbulence are corrected for in real time.

5 **Radio telescopes (p. 114)** are conceptually similar in construction to optical reflectors. However, they are generally much larger than optical instruments, in part because so little radio radiation reaches Earth from space, so a large collecting area is essential. Their main disadvantage is that diffraction of long-wavelength radio waves limits their resolution, even for very large radio telescopes. Their principal advantage is that they allow astronomers to explore a whole new part of the electromagnetic spectrum and of the universe—many radio emitters are completely undetectable in visible light. In addition, radio observations are largely unaffected by Earth's atmosphere and weather or by the position of the Sun.

6 In order to increase the effective area of a telescope, and hence improve its resolution, several separate instruments may be combined into a device called an **interferometer (p. 118),** in which the interference pattern of radiation received by two or more detectors is used to reconstruct a high-precision map of the source. Using **interferometry (p. 118),** radio telescopes can produce images sharper than those from the best optical telescopes. Infrared and optical interferometric systems are now in use at many observatories.

7 **Infrared telescopes (p. 120)** and **ultraviolet telescopes (p. 123)** are generally similar in design to optical systems. Studies undertaken in some parts of the infrared range can be carried out using large ground-based systems. Ultraviolet astronomy *must* be carried out from space. **High-energy telescopes (p. 123)** study the X-ray and gamma-ray regions of the electromagnetic spectrum. X-ray telescopes can form images of their field of view, although the mirror design is more complex than that of optical instruments. Gamma-ray telescopes simply point in a certain direction and count the photons they collect. Because the atmosphere is opaque at these short wavelengths, both types of telescopes must be placed in space.

8 Different physical processes can produce very different types of electromagnetic radiation, and the appearance of an object at one wavelength may bear little resemblance to its appearance at another. Observations at wavelengths spanning the spectrum are essential to a complete understanding of astronomical events.

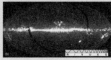

Mastering ASTRONOMY *For instructor-assigned homework go to* **www.masteringastronomy.com**

Problems labeled **POS** explore the process of science | **VIS** problems focus on reading and interpreting visual information

REVIEW AND DISCUSSION

1. **POS**   Cite two reasons that astronomers are continually building larger and larger telescopes.

2. List three advantages of reflecting telescopes over reflectors.

3. What and where are the largest optical telescopes in use today?

4. How does Earth's atmosphere affect what is seen through an optical telescope?

5. What advantages does the *Hubble Space Telescope (HST)* have over ground-based telescopes? List some disadvantages.

6. What are the advantages of a CCD over a photograph?

7. What is image processing?

8. What determines the resolution of a ground-based telescope?

9. How do astronomers use active optics to improve the resolution of telescopes?

10. How do astronomers use adaptive optics to improve the resolution of telescopes?

11. Why do radio telescopes have to be very large?

12. Which astronomical objects are best studied with radio techniques?

13. What is interferometry, and what problem in radio astronomy does it address?

14. Is interferometry limited to radio astronomy?

15. Compare the highest resolution attainable with optical telescopes with the highest resolution attainable with radio telescopes (including interferometers).

16. Why do infrared satellites have to be cooled?

17. Are there any ground-based ultraviolet observatories?

18. In what ways do the mirrors in X-ray telescopes differ from those found in optical instruments?

19. **POS**   What are the main advantages of studying objects at many different wavelengths of radiation?

20. **POS**   Our eyes can see light with an angular resolution of 1′. Suppose our eyes detected only infrared radiation, with 1° angular resolution. Would we be able to make our way around on Earth's surface? To read? To sculpt? To create technology?

CONCEPTUAL SELF-TEST: MULTIPLE CHOICE

1. **VIS**   According to Figure 5.2 ("Refracting Lens") the thickest lenses deflect and bend light (**a**) the fastest; (**b**) the slowest; (**c**) the most; (**d**) the least.

2. The main reason that most professional research telescopes are reflectors is that (**a**) mirrors produce sharper images than lenses do; (**b**) their images are inverted; (**c**) they do not suffer from the effects of seeing; (**d**) large mirrors are easier to build than large lenses.

3. If telescope mirrors could be made of odd sizes, the one with the *most* light-gathering power would be (**a**) a triangle with 1-m sides; (**b**) a square with 1-m sides; (**c**) a circle 1 m in diameter; (**d**) a rectangle with two 1-m sides and two 2-m sides.

4. **VIS**   The image shown in Figure 5.12 ("Resolution") is sharpest when the ratio of wavelength to telescope size is (**a**) large; (**b**) small; (**c**) close to unity; (**d**) none of these.

5. The primary reason professional observatories are built on the highest mountaintops is to (**a**) get away from city lights; (**b**) be above the rain clouds; (**c**) reduce atmospheric blurring; (**d**) improve chromatic aberration.

6. Compared with radio telescopes, optical telescopes can (**a**) see through clouds; (**b**) be used during the daytime; (**c**) resolve finer detail; (**d**) penetrate interstellar dust.

7. When multiple radio telescopes are used for interferometry, resolving power is most improved by increasing (**a**) the distance between telescopes; (**b**) the number of telescopes in a given area; (**c**) the diameter of each telescope; (**d**) the electrical power supplied to each telescope.

8. The *Spitzer Space Telescope (SST)* is stationed far from Earth because (**a**) this increases the telescope's field of view; (**b**) the telescope is sensitive to electromagnetic interference from terrestrial radio stations; (**c**) doing so avoids the obscuring effects of Earth's atmosphere; (**d**) Earth is a heat source and the telescope must be kept very cool.

9. The best way to study warm (1000 K) young stars forming behind an interstellar dust cloud would be to use (**a**) X rays; (**b**) infrared light; (**c**) ultraviolet light; (**d**) blue light.

10. The best frequency range in which to study the hot (million-kelvin) gas found among the galaxies in the Virgo galaxy cluster would be in the following region of the electromagnetic spectrum: (**a**) radio; (**b**) infrared; (**c**) X-ray; (**d**) gamma-ray.

PROBLEMS

The number of dots preceding each Problem indicates its approximate level of difficulty.

1. • A certain telescope has a 10′ × 10′ field of view that is recorded using a CCD chip having 2048 × 2048 pixels. What angle on the sky corresponds to 1 pixel? What would be the diameter, in pixels, of a typical seeing disk (1″ radius)?

2. • The *SST*'s planned operating temperature is 5.5 K. At what wavelength (in micrometers) does the telescope's own blackbody emission peak? How does this wavelength compare with the wavelength range in which the telescope is designed to operate? ∞ (*More Precisely 3-2*)

3. • A 2-m telescope can collect a given amount of light in 1 hour. Under the same observing conditions, how much time would be required for a 6-m telescope to perform the same task? A 12-m telescope?

4. • A space-based telescope can achieve a diffraction-limited angular resolution of 0.05″ for red light (wavelength 700 nm). What would the resolution of the instrument be (a) in the infrared, at 3.5 μm, and (b) in the ultraviolet, at 140 nm?

5. Based on the numbers given in the text, estimate the angular resolution of the Gemini North telescope (a) in red light (700 nm) and (b) in the near infrared (2 μm).

6. •• Two identical stars are moving in a circular orbit around one another, with an orbital separation of 2 AU. ∞ (Sec. 2.6) The system lies 200 light-years from Earth. If we happen to view the orbit head-on, how large a telescope would we need to resolve the stars, assuming diffraction-limited optics at a wavelength of 2 μm?

7. •• What is the greatest distance at which *HST*, in blue light (400 nm), could resolve the stars in the previous question?

8. • What is the equivalent single-mirror diameter of a telescope constructed from two separate 10-m mirrors? Four separate 8-m mirrors?

9. • The Moon lies about 380,000 km away. To what distances do the angular resolutions of *SST* (3″), *HST* (0.05″), and a radio interferometer (0.001″) correspond at that distance?

10. • Estimate the angular resolutions of (a) a radio interferometer with a 5000-km baseline, operating at a frequency of 5 GHz, and (b) an infrared interferometer with a baseline of 50 m, operating at a wavelength of 1 μm.

Life can be hard for graduate students in astronomy. They take some tough courses, many in physics, and they assist in teaching undergraduate courses, but mostly they strive to do original research. Ideally, on the (typically 5- or 6-year) road to their Ph.D. degree, they make a discovery or gain some unique insight that they then write up as part of their doctoral dissertation. The process is exhausting, and some leave the field after the grad school grind, never again to publish in a scientific journal. Others, though, also find it exhilarating, and go on to highly productive careers in astronomy.

Arguably one of the most brilliant doctoral theses in astronomy was written in 1925 by a student at Harvard—and she did it in 2 years. Cecilia Payne-Gaposchkin (1900–1979) was an English student who crossed the ocean to pursue graduate studies at Radcliffe College, and she quickly gravitated to the nearby Harvard Observatory, then perhaps the leading center for research on stars. It was also a place where women, though they often did not get the credit at the time, were making some of the most fundamental advances in stellar astronomy. It was Nirvana for her, and she never left.

Cecilia Payne knew far more physics than most astronomers of the time. She was one of the first to apply the then revolutionary quantum theory of atoms to the spectra of stars, thereby ascertaining stellar temperatures and chemical abundances. Of fundamental importance, her work proved that hydrogen and helium—not the heavy elements, as was then supposed—are the most abundant elements in stars and, therefore, in the universe. Her findings were so revolutionary that the leading theorist of the time, Henry Norris Russell of Princeton, declared her work to be "clearly impossible." It took years to convince the astronomical community that hydrogen is about a million times more abundant in stars than are most of the common elements found on Earth. Yet we take these findings for granted today.

In collaboration with her husband, the exiled Russian astronomer Sergei Gaposchkin, Cecilia Payne-Gaposchkin spent decades making literally millions of observations of thousands of star clusters, variable stars, and galactic novae. Her analysis provided a firm theoretical basis for many properties of stars and their use as distance indicators in the universe. Much of her work has stood the test of time. Despite a flood of new data and new theoretical ideas, it remains the bedrock of modern astronomy.

Portrait of Cecilia Payne-Gaposchkin *(Harvard)*

Women "computers" at work; Payne at the inclined desk *(Harvard)*

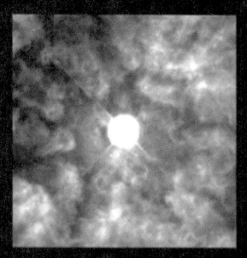

Variable Star Wr 124 *(STScI)*

Pistol Star *(STScI)*

Today, the landscape of stellar research is even richer than Cecilia Payne-Gaposchkin knew. We now see stars much more clearly, their spectra in much finer detail, and all of it to much greater distances. Not only do we know what stars are made of, in most cases we also know *why* their composition is as it is and how many of those stars are born, live, and die. Yet the picture is still very much unfinished, as 21st-century astronomy continues to uncover new and exciting features of stars and stellar systems.

In 1976, well past her retirement, Payne-Gaposchkin was chosen (perhaps ironically, given Russell's initial reaction to her work half a century earlier) as the Henry Norris Russell Lecturer, the highest accolade of the American Astronomical Society. Her talk was an enthusiastic summary of a lifetime of astronomical research—an encyclopedic talk with no notes and no prompts, given in perfectly punctuated English. It was very clear that she knew some individual giant stars as well as she knew her best friends. Her talk ended with the following advice to astronomers young and old:

> "The reward of the young scientist is the emotional thrill of being the first person in the history of the world to see something or to understand something. Nothing can compare with that experience. The reward of the old scientist is the sense of having seen a vague sketch grow into a masterly landscape. Not a finished picture, of course; a picture that is still growing in scope and detail with the application of new techniques and new skills. The old scientist cannot claim that the masterpiece is his own work. He may have roughed out part of the design, laid on a few strokes, but he has learned to accept the discoveries of others with the same delight that he experienced on his own when he was young."

Illustrated on this page are some recent findings in stellar research—work that undoubtedly would have caused Cecilia Payne-Gaposchkin to express more of her trademark enthusiasm. Today's research also would have made her justly proud, for so much of it relies on the insights gained by her and her colleagues during the first half of the 20th century.

Rosebud Nebula *(JPL)*

Henize Nebula *(JPL)*

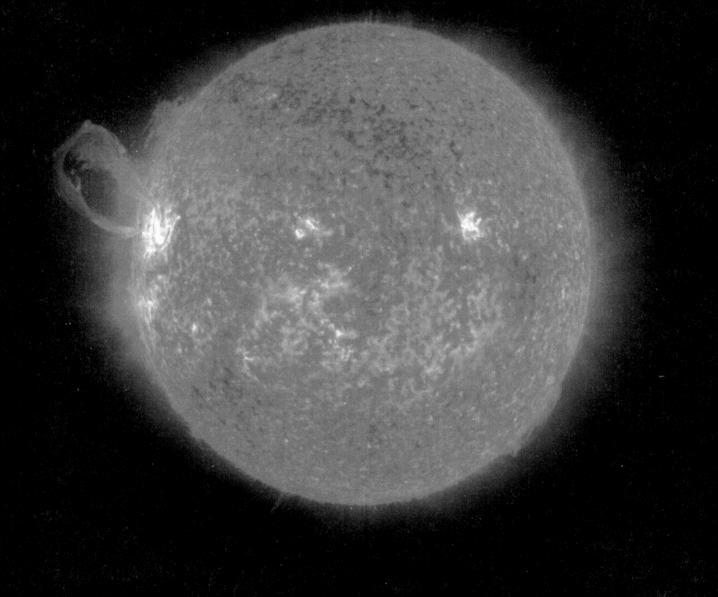

16

THE SUN

OUR PARENT STAR

LEARNING GOALS

Studying this chapter will enable you to

1 Summarize the overall properties and internal structure of the Sun.

2 Describe the concept of luminosity, and explain how it is measured.

3 Explain how studies of the solar surface tell us about the Sun's interior.

4 List and describe the outer layers of the Sun.

5 Discuss the nature and variability of the Sun's magnetic field.

6 Describe the various types of solar activity and their relation to solar magnetism.

7 Outline the process by which energy is produced in the Sun's interior.

8 Explain how observations of the Sun's core changed our understanding of fundamental physics.

THE BIG PICTURE The Sun is our star—the main source of energy that powers weather, climate, and life on Earth. Although we take it for granted each and every day, the Sun is vitally important to us in the cosmic scheme of things. Simply put, without it, we would not exist.

Visit the Study Area in www.masteringastronomy.com for quizzes, animations, videos, interactive figures, and self-guided tutorials.

Living in the solar system, we have the chance to study at close range perhaps the most common type of cosmic object—a star. Our Sun is a star, and a fairly average one at that, but with a unique feature: It is very close to us—some 300,000 times closer than our next nearest neighbor, Alpha Centauri. Whereas Alpha Centauri is 4.3 light-years distant, the Sun is only 8 light-minutes away from us. Consequently, astronomers know far more about the properties of the Sun than about any of the other distant points of light in the universe.

A good fraction of all our astronomical knowledge is based on modern studies of the Sun—from the production of seemingly boundless energy in its core to the surprisingly complex activity in its atmosphere. Just as we studied our parent planet, Earth, to set the stage for our exploration of the solar system, we now examine our parent star, the Sun, as the next step in our exploration of the universe.

LEFT: *The Sun, much like the planets, experiences a kind of weather, including storms. This spectacular image shows a particularly large solar prominence on the limb of the Sun, arising from a powerful active region in 2005. The image was acquired by the Solar and Heliospheric Observatory—a 2-ton robot "parked" in space between Earth and the Sun. Its job is to stare at the Sun unblinkingly 24 hours a day, eavesdropping on its atmosphere, surface, and interior. (ESA/NASA)*

16.1 Physical Properties of the Sun

The Sun is the sole source of light and heat for the maintenance of life on Earth. The Sun is a **star**—a glowing ball of gas held together by its own gravity and powered by nuclear fusion at its center. In its physical and chemical properties, the Sun is similar to most other stars, regardless of when and where they formed. Indeed, our Sun appears to be a rather typical star, lying right in the middle of the observed ranges of stellar mass, radius, brightness, and composition. Far from detracting from our interest in the Sun, this very mediocrity is one of the main reasons that astronomers study it—they can apply knowledge of solar phenomena to many other stars in the universe.

Overall Properties

The Sun's radius, roughly 700,000 km, is determined most directly by measuring the angular size (0.5°) of the Sun and then employing elementary geometry. ∞ (Sec. 1.6) The Sun's mass, 2.0×10^{30} kg, follows from Newton's laws of motion and gravity, applied to the observed orbits of the planets. ∞ (More Precisely 2-2) The average solar density derived from its mass and volume, approximately 1400 kg/m³, is quite similar to that of the jovian planets and about one-quarter the average density of Earth.

Solar rotation can be measured by timing sunspots and other surface features as they traverse the solar disk. ∞ (Sec. 2.4) These observations indicate that the Sun rotates in about a month, but it does not do so as a solid body. Instead, it spins *differentially,* like Jupiter and Saturn—faster at the equator and slower at the poles. ∞ (Sec. 11.1) The equatorial rotation period at the equator is about 25 days. Sunspots are never seen above latitude 60° (north or south), but at that latitude they indicate a 31-day period. Other measurement techniques, such as those discussed in Section 16.2, reveal that the Sun's rotation period continues to increase as we approach the poles. The polar rotation period is not known with certainty, but it may be as long as 36 days.

The Sun's surface temperature is measured by applying the radiation laws to the observed solar spectrum. ∞ (Sec. 3.4) The distribution of solar radiation has the approximate shape of a blackbody curve for an object at about 5800 K. The average solar temperature obtained in this way is known as the Sun's *effective temperature.*

Having a radius of more than 100 Earth radii, a mass of more than 300,000 Earth masses, and a surface temperature well above the melting point of any known material, the Sun is clearly a body that is very different from any other we have encountered so far.

Solar Structure

The Sun has a surface of sorts—not a solid surface (the Sun contains no solid material), but rather that part of the brilliant gas ball we perceive with our eyes or view through a heavily filtered telescope. This "surface"—the part of the Sun that emits the radiation we see—is called the **photosphere.** Its radius is about 700,000 km. However, the thickness of the photosphere is probably no more than 500 km, less than 0.1 percent of the radius, which is why we perceive the Sun as having a well-defined, sharp edge (Figure 16.1).

The main regions of the Sun are illustrated in Figure 16.2 and summarized in Table 16.1. We will discuss them all in more detail later in the chapter. Just above the photosphere is the Sun's lower atmosphere, called the **chromosphere,** about 1500 km thick. From 1500 km to 10,000 km above the top of the photosphere lies a region called the **transition zone,** in which the temperature rises dramatically. Above 10,000 km, and stretching far beyond, is a tenuous (thin), hot upper atmosphere: the solar **corona.** At still greater distances, the corona turns into the **solar wind,** which flows away from the Sun and permeates the entire solar system. ∞ (Sec. 6.5) Extending down some 200,000 km below the photosphere is the **convection zone,** a region where the material of the Sun is in constant convective motion. Below the convection zone lies the **radiation zone,** in which solar energy is transported toward the surface by radiation rather than by convection. The term *solar interior* is often used to mean both the radiation and convection zones. The central **core,** roughly 200,000 km in radius, is the site of powerful nuclear reactions that generate the Sun's enormous energy output.

Luminosity

The properties of size, mass, density, rotation rate, and temperature are familiar from our study of the planets. But the Sun has an additional property, perhaps the most important of all from the point of view of life on Earth: The Sun *radiates* a great deal of energy into space, uniformly (we assume) in all directions. By holding a light-sensitive device—a photoelectric cell, perhaps—perpendicular to the Sun's rays, we can measure how much solar energy is received per square meter of surface area every second. Imagine our detector as having a surface area of 1 square meter (1 m²) and as being placed at the top of Earth's atmosphere. The amount of solar energy reaching this surface each second is a quantity known as the **solar constant,** whose value is approximately 1400 watts per square meter (W/m²).

About 50 to 70 percent of the incoming energy from the Sun reaches Earth's surface; the rest is intercepted by the atmosphere (30 percent) or reflected away by clouds (0 to 20 percent). Thus, on a clear day, a sunbather's body having a total surface area of about 0.5 m² receives solar energy at a rate of roughly 1400 W/m² × 0.70 (70 percent) × 0.5 m² ≈ 500 W, equivalent to the output of a small electric room heater or five 100-watt lightbulbs.

Let us now ask about the *total* amount of energy radiated in all directions from the Sun, not just the small fraction

▲ **FIGURE 16.1 The Sun** The inner part of this composite, filtered image of the Sun shows a sharp solar edge, although our star, like all stars, is made of a gradually thinning gas. The edge appears sharp because the solar photosphere is so thin. The outer portion of the image is the solar corona, normally too faint to be seen, but visible during an eclipse, when the light from the solar disk is blotted out. Note the blemishes; they are sunspots. ∞ (Sec. 2.4) *(NOAO)*

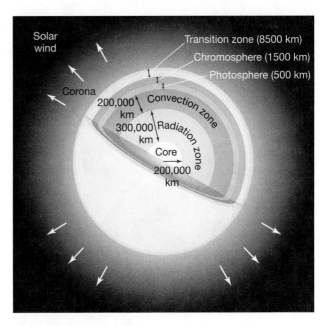

▲ **FIGURE 16.2 Solar Structure** The main regions of the Sun, not drawn to scale, with some physical dimensions labeled. The photosphere is the visible "surface" of the Sun. Below it lie the convection zone, the radiation zone, and the core. Above the photosphere, the solar atmosphere consists of the chromosphere, the transition zone, and the corona.

intercepted by our detector or by Earth. Imagine a three-dimensional sphere is centered on the Sun and just large enough that its surface intersects Earth's center (Figure 16.3). The sphere's radius is 1 AU, and its surface area is therefore $4\pi \times (1 \text{ AU})^2$, or approximately 2.8×10^{23} m². Multiplying the rate at which solar energy falls on each square meter of

the sphere (i.e., the solar constant) by the total surface area of our imaginary sphere, we can determine the total rate at which energy leaves the Sun's surface. This quantity is known as the **luminosity** of the Sun. It turns out to be just under 4×10^{26} W.

The Sun is an enormously powerful source of energy. *Every second*, it produces an amount of energy equivalent to the detonation of about 10 billion 1-megaton nuclear

Region	Inner Radius (km)	Temperature (K)	Density (kg/m³)	Defining Properties
Core	0	15,000,000	150,000	Energy generated by nuclear fusion
Radiation zone	200,000	7,000,000	15,000	Energy transported by electromagnetic radiation
Convection zone	496,000*	2,000,000	150	Energy carried by convection
Photosphere	696,000*	5800	2×10^{-4}	Electromagnetic radiation can escape—the part of the Sun we see
Chromosphere	696,500*	4500	5×10^{-6}	Cool lower atmosphere
Transition zone	698,000*	8000	2×10^{-10}	Rapid increase in temperature
Corona	706,000*	3,000,000	10^{-12}	Hot, low-density upper atmosphere
Solar wind	10,000,000	>1,000,000	10^{-23}	Solar material escapes into space and flows outward through the solar system

TABLE 16.1 The Standard Solar Model

* These radii are based on the accurately determined radius of the photosphere. The other radii quoted are approximate, round numbers.

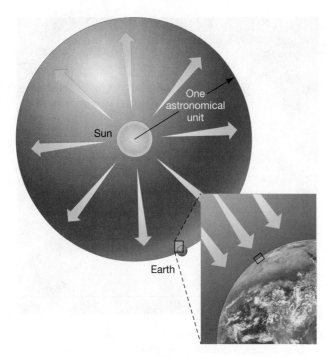

▲ FIGURE 16.3 **Solar Luminosity** We can draw an imaginary sphere around the Sun so that the sphere's surface passes through Earth's center. The radius of this imaginary sphere equals 1 AU. Overall, Earth receives energy from the Sun at a rate of roughly 200 million gigawatts—ten thousand times the current energy consumption of our entire planet. The "solar constant" is the amount of power striking a 1-m² detector at Earth's distance, as suggested by the inset. By multiplying the sphere's surface area by the solar constant, we can measure the Sun's luminosity—the amount of energy it emits each second.

bombs. Six seconds worth of solar energy output, suitably focused, would evaporate all of Earth's oceans. Three minutes would melt our planet's crust. The scale on which the Sun operates simply defies earthly comparison. Let's begin our more detailed study with a look at where all this energy comes from.

PROCESS OF SCIENCE CHECK

✔ Why must we assume that the Sun radiates equally in all directions when we compute the solar luminosity from the solar constant?

16.2 The Solar Interior

How do astronomers know about conditions in the interior of the Sun? As we have just seen, the fact that the Sun shines tells us that its center must be very hot, but our direct knowledge of the solar interior is actually quite limited. (See Section 16.7 for a discussion of one important "window" we do have into the solar core.) Lacking direct measurements, researchers must use other means to probe the inner workings of our parent star. To this end, they construct

mathematical models of the Sun, combining all available data with theoretical insight into solar physics to find the model that agrees most closely with observations. ∞ (Sec. 1.2) Recall from Chapter 11 how similar techniques are used to infer the structures of the jovian planets. ∞ (Sec. 11.3) The result in the case of the Sun is the **standard solar model,** which has gained widespread acceptance among astronomers.

Modeling the Structure of the Sun

The Sun's bulk properties—its mass, radius, temperature, and luminosity—do not vary much from day to day or from year to year. Although we will see in Chapter 20 that stars like the Sun do change significantly over periods of *billions* of years, for our purposes here this slow evolution may be ignored. On "human" time scales, the Sun may reasonably be thought of as unchanging.

Based on this simple observation, as illustrated in Figure 16.4, theoretical models generally begin by assuming that the Sun is in a state of **hydrostatic equilibrium,** in which pressure's outward push exactly counteracts gravity's inward pull. This stable balance between opposing forces is the basic reason that the Sun neither collapses under its own weight nor explodes into interstellar space. ∞ *(More Precisely 8-1)* The assumption of hydrostatic equilibrium, coupled with our knowledge of some basic physics, then lets us predict the density and temperature in the solar interior. This information, in turn, allows the model to make predictions about other observable solar properties—luminosity, radius, spectrum, and so on—and the internal details of the model are fine-tuned until the predictions agree with

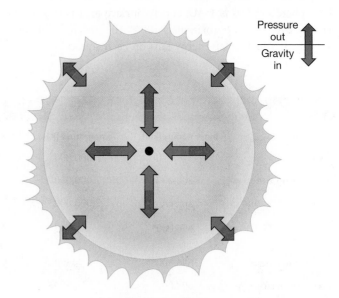

▲ FIGURE 16.4 **Hydrostatic Equilibrium** In the interior of a star such as the Sun, the outward pressure of hot gas exactly balances the inward pull of gravity. This is true at every point within the star, guaranteeing its stability.

observations. This is the scientific method at work; the standard solar model is the result. ∞ (Sec. 1.2)

Hydrostatic equilibrium has an important consequence for the solar interior. Because the Sun is very massive, its gravitational pull is very strong, so very high internal pressure is needed to maintain the balance. This high pressure in turn requires a very high central temperature, a fact crucial to our understanding of solar energy generation (Section 16.6). Indeed, calculations of this sort carried out by British astrophysicist Sir Arthur Eddington around 1920 provided astronomers with the first inkling that fusion might be the process that powers the Sun.

To test and refine the standard solar model, astronomers are eager to obtain information about the solar interior. However, with so little direct information about conditions below the photosphere, we must rely on more indirect techniques. In the 1960s, measurements of the Doppler shifts of solar spectral lines revealed that the surface of the Sun oscillates, or vibrates, like a complex set of bells. ∞ (Secs. 3.5, 4.5) These vibrations, illustrated in Figure 16.5(a), are the result of internal pressure waves (somewhat like sound waves in air) that reflect off the photosphere and repeatedly cross the solar interior (Figure 16.5b). Because the waves can penetrate deep inside the Sun, analysis of their surface patterns allows scientists to study conditions far below the Sun's surface. The process is similar to the way in which seismologists learn about the interior of Earth by observing the P- and S-waves produced by earthquakes. ∞ (Sec. 7.3) For this reason, the study of solar surface patterns is usually called **helioseismology,** even though solar pressure waves have

nothing whatever to do with solar seismic activity—there is no such thing.

The most extensive study of solar oscillations is the ongoing Global Oscillations Network Group (GONG) project. By making continuous observations of the Sun from many clear sites around Earth, solar astronomers can obtain uninterrupted high-quality solar data spanning many days and even weeks—almost as though Earth were not rotating and the Sun never set. The *Solar and Heliospheric Observatory (SOHO),* launched by the European Space Agency in 1995 and now permanently stationed between Earth and the Sun some 1.5 million km from our planet (see *Discovery 16-1),* also provides continuous monitoring of the Sun's surface and atmosphere. Analysis of ground- and space-based data provides important additional information about the temperature, density, rotation, and convective state of the solar interior, allowing detailed comparisons between theory and reality to be made. Direct comparison is possible throughout a large portion of the Sun, and the agreement between model and observations is spectacular: The frequencies and wavelengths of observed solar oscillations are within 0.1 percent of the predictions of the standard solar model.

Figure 16.6 shows the solar density and temperature, plotted as functions of distance from the Sun's center, according to the standard solar model. Notice how the density drops rather sharply at first and then decreases more slowly near the solar photosphere, some 700,000 km from the center. The variation in density is large, ranging from a core value of about 150,000 kg/m^3, 20 times the density of iron, to an intermediate value (at 350,000 km) of about 1000 kg/m^3, the density of water, to an extremely small photospheric value of 2×10^{-4} kg/m^3, 10,000 times less dense than air at the surface of Earth. Because the density is so high in the core, roughly 90 percent of the Sun's mass is contained within the inner half of its radius. The solar density continues to decrease beyond the photosphere, reaching values as low as 10^{-23} kg/m^3 in the far corona—about as thin as the best vacuum physicists can create in laboratories on Earth.

The solar temperature also decreases with increasing radius in the solar interior, but not as rapidly as the density. Computer models indicate a temperature of about 15 million K at the core, consistent with the minimum 10 million K needed to initiate the nuclear reactions known to power most stars, decreasing to the observed value of about 5800 K at the photosphere.

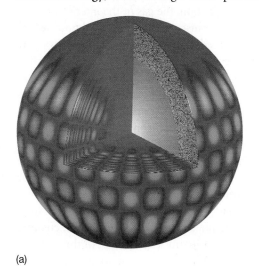

(a)

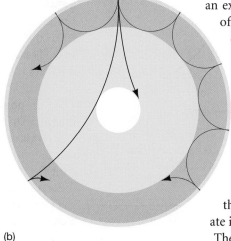

(b)

▲ **FIGURE 16.5 Solar Oscillations** (a) The Sun has been found to vibrate in a very complex way. By observing the motion of the solar surface, scientists can determine the wavelengths and the frequencies of the individual waves and deduce information about the Sun not obtainable by other means. The alternating patches represent gas moving down (red) and up (blue). (See also *Discovery 16-1.*) (b) Depending on their initial directions, the waves contributing to the observed oscillations may travel deep inside the Sun, providing vital information about the solar interior. The wave shown closest to the surface here corresponds approximately to the vibration pattern depicted in part (a). *(National Solar Observatory)*

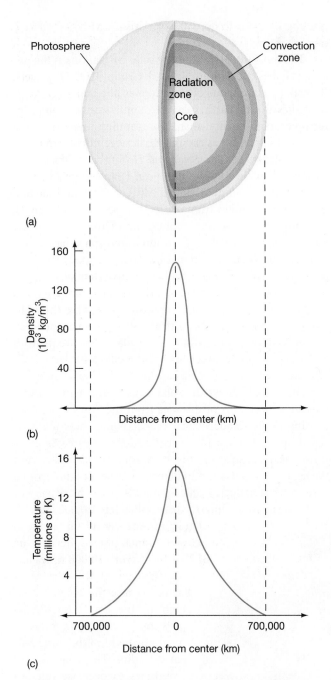

(a)

(b)

(c)

▲ **FIGURE 16.6 Solar Interior** (a) A cross-sectional cut through the middle of the Sun, with corresponding graphs of (b) density and (c) temperature across the cut, according to the standard solar theoretical model.

As the data improve and old mysteries are resolved, new ones often emerge. For example, helioseismology indicates that the Sun's rotation speed varies with depth—perhaps not too surprising, given the surface differential rotation mentioned earlier and the fact that similar behavior has been noted in the outer planets. What is puzzling, though, is the *complexity* of the differential motion. The surface layers show a "zonal flow" of sorts, with alternating bands of higher- and lower-than-average rotation

rates. Just below the surface are wide "rivers" of lower speed (at the equator) and higher speed (polar) rotation. The material at the base of the convection zone appears to oscillate in rotation speed, sometimes moving faster (by about 10 percent) than the surface layers, sometimes slower, with a period of about 1.3 years. Deeper still, the radiative interior rotates more or less as a solid body, once every 26.9 days. A full explanation of the Sun's rotation currently eludes theorists.

Energy Transport

The very hot solar interior ensures violent and frequent collisions among gas particles. Particles move in all directions at high speeds, bumping into one another unceasingly. In and near the core, the extremely high temperatures guarantee that the gas is completely ionized. Recall from Chapter 4 that, under less extreme conditions, atoms absorb photons that can boost their electrons to more excited states. ∞ (Sec. 4.2) With no electrons left on atoms to capture the photons, however, the deep solar interior is relatively transparent to radiation. Only occasionally does a photon encounter and scatter off of a free electron or proton. The energy produced by nuclear reactions in the core travels outward toward the surface in the form of radiation with relative ease.

As we move outward from the core, the temperature falls, atoms collide less frequently and less violently, and more and more electrons manage to remain bound to their parent nuclei. With more and more atoms retaining electrons that can absorb the outgoing radiation, the gas in the interior changes from being relatively transparent to being almost totally opaque. By the outer edge of the radiation zone, roughly 500,000 km from the center (actually, 496,000 km, according to the best available *SOHO* data), *all* the photons produced in the Sun's core have been absorbed. Not one of them reaches the surface. But what happens to the energy they carry?

The photons' energy must travel beyond the Sun's interior: That we see sunlight—visible energy—proves that energy escapes. The escaping energy reaches the surface by *convection*—the same basic physical process we saw in our study of Earth's atmosphere, although it operates in a very different environment in the Sun. ∞ (Sec. 7.2) Hot solar gas moves outward while cooler gas above it sinks, creating a characteristic pattern of convection cells. All through the convection zone, energy is transported to the surface by physical motion of the solar gas. (Note that this actually represents a departure from hydrostatic equilibrium, as defined above, but it can still be handled within the standard solar model.) Remember that there is no physical movement of material when radiation is the energy-transport mechanism; convection and radiation are *fundamentally different* ways in which energy can be transported from one place to another.

DISCOVERY 16-1

SOHO: Eavesdropping on the Sun

Throughout the few decades of the Space Age, various nations, led by the United States, have sent spacecraft to all but one of the major bodies in the solar system. That unexplored body is the Sun. Currently, the next best thing to a dedicated reconnoitering spacecraft is the *Solar and Heliospheric Observatory (SOHO),* which has radioed back to Earth volumes of new data—and more than a few new puzzles—about our parent star since the spacecraft's launch in 1995.

SOHO is a billion-dollar mission operated primarily by the European Space Agency. The 2-ton robot is now on station about 1.5 million km sunward of Earth—about 1 percent of the distance from Earth to the Sun. This is the so-called L_1 Lagrangian point, where the gravitational pull of the Sun and Earth are

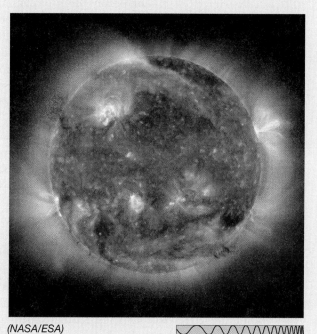

(NASA/ESA)

R I V U X G

precisely equal—a good place to park a monitoring platform. ∞ (Sec. 14.1) From this vantage point, *SOHO* continuously monitors our parent star, studying its surface, atmosphere, and interior. The automated vehicle carries a dozen instruments capable of measuring almost everything from the Sun's corona and magnetic field to its solar wind and internal vibrations. The accompanying figure shows a false-color image of the Sun's lower corona, obtained by combining *SOHO* data captured at three different ultraviolet wavelengths.

SOHO is positioned just beyond Earth's magnetosphere, so its instruments can cleanly study the charged particles of the solar wind—high-speed matter escaping from the Sun, flowing outward from the corona. Coordinating these on-site measurements with *SOHO* images of the Sun itself, astronomers now think they can follow solar magnetic field loops expanding and breaking as the Sun prepares itself for mass ejections several days before they actually occur (see Section 16.5). Given that such coronal storms can wreak havoc on communications, power grids, satellite electronics, and other human activities, the prospect of having accurate forecasts of disruptive solar events is a welcome development.

Section 16.2 discusses how astronomers can "take the pulse" of the Sun by measuring its complex rhythmic motions. *SOHO* also has the ability to study the weak sound waves that echo and resonate inside the Sun and can map these vibrations with much higher resolution than was previously possible. It does so not by sensing sound itself, but by watching the Sun's surface move up and down ever so slightly. The Sun's "loudest" vibrations are extremely low pitched (0.003 Hz, or one oscillation every 5 minutes), more like rolling rumbles, just as we might expect from such a huge and massive object.

As of early 2010, *SOHO* is still operating, 12 years beyond its planned lifetime, having survived assaults from both the Sun and Earth. Twice so far, skillful engineers have brought the spacecraft back from apparent death after mission controllers mistakenly sent incorrect commands from the ground. It is a good thing that they did, for this remarkable spacecraft has radioed back to Earth a wealth of new scientific insight into our parent star.

Figure 16.7 is a schematic diagram of the solar convection zone. There is a hierarchy of convection cells, organized in tiers of many different sizes at different depths. The deepest tier, lying approximately 200,000 km below the photosphere, is thought to contain large cells some tens of thousands of kilometers in diameter. Heat is then successively carried upward through a series of progressively smaller cells, stacked one on another, until, at a depth of about 1000 km, the individual cells are about 1000 km across. The top of this uppermost tier of convection is the visible surface of the Sun, where astronomers can directly observe the cell sizes. Information about convection below that level is inferred mostly from computer models of the solar interior.   .

At some distance from the core, the solar gas becomes too thin to sustain further upwelling by convection. Theory suggests that this distance roughly coincides with the photospheric surface we see. Convection does not proceed into the solar atmosphere; there is simply not enough gas there—the density is so low that there are too few atoms or ions to intercept much sunlight, so the gas becomes transparent again and radiation once more becomes the mechanism of energy transport. Photons reaching the photosphere escape more or less freely into space, and the photosphere emits thermal radiation, like any other hot object. The photosphere is narrow, and the "edge" of the Sun sharp, because this transition from opacity to complete transparency is very rapid. Just

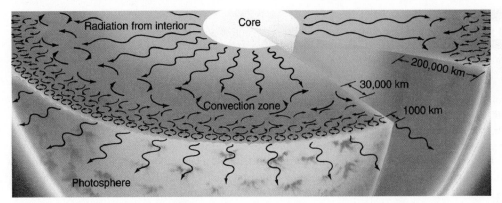

▲ **FIGURE 16.7 Solar Convection** Physical transport of energy in the Sun's convection zone. The upper-interior region is visualized as a boiling, seething sea of gas. Near the surface, each convective cell is about 1000 km across. The sizes of the convective cells become progressively larger at greater depths, reaching some 30,000 km in diameter at the base of the convection zone, 200,000 km below the photosphere. (This is a highly simplified diagram; there are many different cell sizes, and they are not so neatly arranged.)

below the bottom of the photosphere the gas is still convective, and radiation does not reach us directly. A few hundred kilometers higher, the gas is too thin to emit or absorb any significant amount of radiation.

Granulation

Figure 16.8 is a high-resolution photograph of the solar surface. The visible surface is highly mottled, or **granulated,** with regions of bright and dark gas known as *granules*. Each

bright granule measures about 1000 km across—comparable in size to a continent on Earth—and has a lifetime of between 5 and 10 minutes. Together, several million granules constitute the top layer of the convection zone, immediately below the photosphere.

Each granule forms the topmost part of a solar convection cell. Spectroscopic observation within and around the bright regions shows direct evidence for the upward motion of gas as it "boils" up from within—evidence that convection really does occur just below the photosphere. Spectral lines detected from the bright granules appear slightly bluer than normal, indicating Doppler-shifted matter approaching us at about 1 km/s. ∞ (Sec. 3.5) Spectroscopes focused on the darker portions of the granulated photosphere show the same spectral lines to be redshifted, indicating matter moving away from us.

The variations in brightness of the granules result strictly from differences in temperature. The upwelling gas is hotter and therefore emits more radiation than the cooler, downward-moving gas. The adjacent bright and dark gases appear to contrast considerably, but in reality their temperature difference is less than about 500 K. Careful measurements also reveal a much larger-scale flow on the solar surface. **Supergranulation** is a flow pattern quite similar to granulation, except that supergranulation cells measure some 30,000 km across. As with granulation, material upwells at the center of the cells, flows across the surface, then sinks down again at the edges. Scientists suspect that supergranules are the imprint on the photosphere of a deeper tier of large convective cells, like those depicted in Figure 16.7.

CONCEPT CHECK

✔ What are the two distinct ways in which energy moves outward from the solar core to the photosphere?

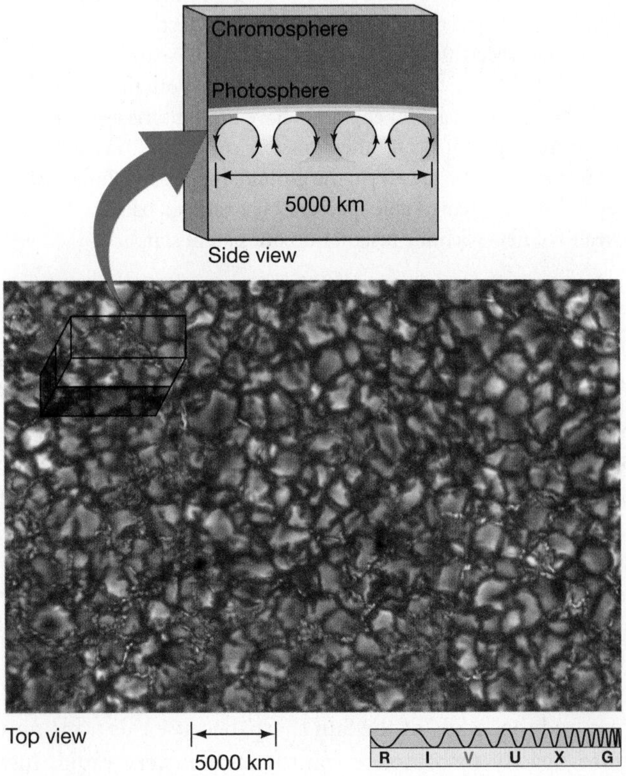

◀ **FIGURE 16.8 Solar Granulation** A photograph of the granulated solar photosphere, taken with the 1-m Swedish Solar Telescope looking directly down on the Sun's surface. Typical solar granules are comparable in size to Earth's continents. The bright portions of the image are regions where hot material is upwelling from below, as illustrated in Figure 16.7. The darker (redder) regions correspond to cooler gas that is sinking back down into the interior. The inset drawing shows a perpendicular cut through the solar surface. *(SST)*

16.3 The Sun's Atmosphere

Astronomers can glean an enormous amount of information about the Sun from an analysis of the absorption lines that arise in the photosphere and lower atmosphere. ∞ (Sec. 4.4) Figure 16.9 (see also Figure 4.4) is a detailed spectrum of the Sun spanning a range of wavelengths from 360 to 690 nm. Notice the intricate dark Fraunhofer absorption lines superposed on the background continuous spectrum.

Tens of thousands of spectral lines have been observed and cataloged in the solar spectrum. In all, some 67 elements have been identified in the Sun in various states of ionization and excitation. ∞ (Sec. 4.2) More elements probably exist there, but they are present in such small quantities that our instruments are simply not sensitive enough to detect them. Table 16.2 lists the 10 most common elements in the Sun. Notice that hydrogen is by far the most abundant element, followed by helium. This distribution is just what we saw on the jovian planets, and it is what we will find for the universe as a whole.

Solar Spectral Lines

As discussed in Chapter 4, spectral lines arise when electrons in atoms or ions make transitions between states of well-defined energies, emitting or absorbing photons of specific energies (i.e., wavelengths or colors) in the process. ∞ (Sec. 4.2) However, to explain the spectrum of the Sun (and,

TABLE 16.2 The Composition of the Sun		
Element	**Percentage of Total Number of Atoms**	**Percentage of Total Mass**
Hydrogen	91.2	71.0
Helium	8.7	27.1
Oxygen	0.078	0.97
Carbon	0.043	0.40
Nitrogen	0.0088	0.096
Silicon	0.0045	0.099
Magnesium	0.0038	0.076
Neon	0.0035	0.058
Iron	0.0030	0.14
Sulfur	0.0015	0.040

indeed, the spectra of all stars), we must slightly modify our earlier description of the formation of absorption lines. We explained these lines in terms of cool foreground gas intercepting light from a hot background source. In actuality, both the bright background and the dark absorption lines in Figure 16.9 form at roughly the same locations in the Sun—the solar photosphere and lower chromosphere. To understand how these lines are formed, consider again the solar energy emission process in a little more detail.

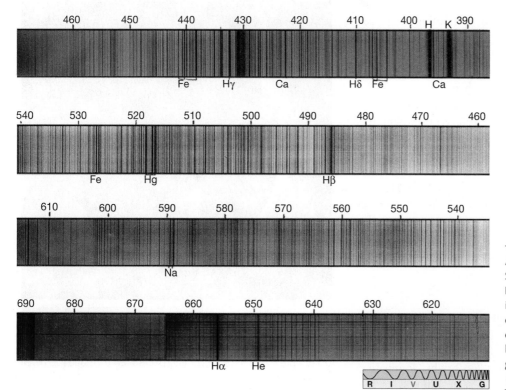

◄ FIGURE 16.9 **Solar Spectrum** A detailed visible spectrum of our Sun shows thousands of dark Fraunhofer (absorption) spectral lines indicating the presence of 67 different elements in various stages of excitation and ionization in the lower solar atmosphere. The numbers give wavelengths, in nanometers. (*Palomar Observatory/Caltech*)

Below the photosphere, the solar gas is sufficiently dense, and interactions among photons, electrons, and ions sufficiently common, that radiation cannot escape directly into space. In the solar atmosphere, however, the probability that a photon will escape without further interaction with matter depends on the photon's energy. Recall from Chapter 4 that an atom or ion can absorb a photon only if that photon's energy has just the right value to cause an electron to jump from one energy level to another. ∞ (Sec. 4.3) Hence, if the photon energy happens to correspond to some electronic transition in an atom or ion in the gas, then the photon may be absorbed again before it can travel very far—the more elements present of the type suitable for absorption, the lower the escape probability. Conversely, if the photon's energy does not coincide with any such transition, then the photon cannot interact further with the gas, and it leaves the Sun headed for interstellar spaces, or perhaps the detector of an astronomer on Earth.

Thus, when we look at the Sun, we are actually peering down into the solar atmosphere to a depth that depends on the wavelength of the light under study. Photons with wavelengths far from any absorption feature (i.e., having energies far from any atomic transition) are less likely to interact with matter as they travel through the solar gas and so come from deeper in the photosphere. However, photons with wavelengths near the centers of absorption lines are much more likely to be captured by an atom or ion and therefore escape mainly from higher (and cooler) levels. The lines are darker than their surroundings because the temperature where they form is lower than the 5800-K temperature at the base of the photosphere, where most of the continuous emission originates. (Recall that, by Stefan's law, the brightness of a radiating object depends on its temperature—the cooler the gas, the less energy it radiates.) ∞ (Sec. 3.4) Thus, the existence of Fraunhofer lines is direct evidence that the temperature in the Sun's atmosphere decreases with height above the photosphere.

Strictly speaking, spectral analysis allows us to draw conclusions only about the part of the Sun where the lines form—the photosphere and chromosphere. However, most astronomers think that, with the exception of the solar core (where nuclear reactions are steadily changing the composition—see Sec. 16.6), the data in Table 16.2 are representative of the entire Sun. That assumption is strongly supported by the excellent agreement between the standard solar model, which makes the same assumption, and helioseismological observations of the solar interior.

The Chromosphere

Above the photosphere lies the cooler chromosphere, the inner part of the solar atmosphere. This region emits very little light of its own and cannot be observed visually under normal conditions. The photosphere is just too bright, dominating the chromosphere's radiation. The relative dimness of the chromosphere results from its low density—large numbers of photons simply cannot be emitted by a tenuous gas containing very few atoms per unit volume. Still, although it is not normally seen, astronomers have long been aware of the chromosphere's existence. Figure 16.10 shows the Sun during an eclipse in which the photosphere—but not the chromosphere—is obscured by the Moon. The chromosphere's characteristic reddish hue is plainly visible. This coloration is due to the red Hα (hydrogen alpha) emission line of hydrogen, which dominates the chromospheric spectrum. ∞ *(More Precisely 4-1)*

The chromosphere is far from tranquil. Every few minutes, small solar storms erupt, expelling jets of hot matter known as *spicules* into the Sun's upper atmosphere (Figure 16.11). These long, thin spikes of matter leave the Sun's surface at typical speeds of about 100 km/s and reach several thousand kilometers above the photosphere. Spicules are not spread evenly across the solar surface. Instead, they cover only about 1 percent of the total area, tending to accumulate around the edges of supergranules. The Sun's magnetic field is also known to be somewhat stronger than average in those regions. Scientists speculate that the downward-moving material there tends to strengthen the solar magnetic field, and spicules are the result of magnetic disturbances in the Sun's churning outer layers.

The Transition Zone and the Corona

During the brief moments of an eclipse, if the Moon's angular size is large enough that both the photosphere and the chromosphere are blocked, the ghostly solar corona can be seen

R I V U X G

▲ FIGURE 16.10 **Solar Chromosphere** This photograph of a total solar eclipse shows the solar chromosphere a few thousand kilometers above the Sun's surface. *(G. Schneider)*

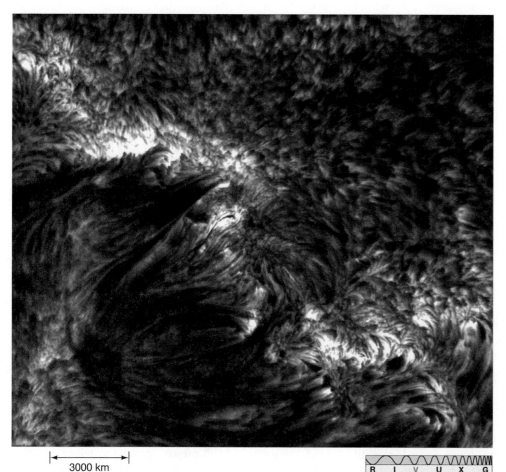

3000 km

◄ FIGURE 16.11 **Solar Spicules** Short-lived, narrow jets of gas that typically last mere minutes can be seen sprouting up from the solar chromosphere in this Hα image of the Sun. These so-called spicules are the thin, dark, spikelike regions. They appear dark against the face of the Sun because they are cooler than the underlying photosphere. *(SST)*

(Figure 16.12). With the photospheric light removed, the pattern of spectral lines changes dramatically. The intensities of the usual lines alter (suggesting changes in composition or temperature, or both), the spectrum shifts from absorption to emission, and an entirely new set of spectral lines suddenly appears. The shift from absorption to emission is entirely in accordance with Kirchhoff's laws, because we see the corona against the blackness of space, not against the bright continuous spectrum from the photosphere below. ∞ (Sec. 4.1)

These new coronal (and in some cases chromospheric) lines were first observed during eclipses in the 1920s. For years afterward, some researchers (for want of any better explanation) attributed them to a new nonterrestrial element, which they dubbed "coronium." We now recognize that these new spectral lines do not indicate any new kind of atom. Coronium does not exist. Rather, the new lines arise because atoms in the corona have lost several more electrons than atoms in the photosphere—that is, the coronal atoms are much more highly ionized. Therefore, their internal electronic structures, and hence their spectra, are quite different from the structure and spectra of atoms and ions in the photosphere. For example, astronomers have identified coronal lines corresponding to iron ions with as many as 13 of their normal 26 electrons missing. In the photosphere, most iron atoms have lost only 1 or 2 of their electrons.

▲ FIGURE 16.12 **Solar Corona** When both the photosphere and the chromosphere are obscured by the Moon during a solar eclipse, the faint corona becomes visible. This photograph clearly shows the emission of radiation from a relatively inactive solar corona. *(Bencho Angelov)*

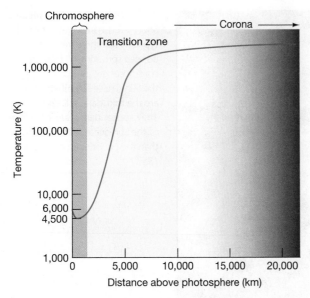

▲ FIGURE 16.13 Solar Atmospheric Temperature The change of gas temperature in the lower solar atmosphere is dramatic. The temperature, indicated by the blue line, reaches a minimum of 4500 K in the chromosphere and then rises sharply in the transition zone, finally leveling off at around 3 million K in the corona.

The cause of this extensive electron stripping is the high coronal *temperature*. The degree of ionization inferred from spectra observed during solar eclipses tells us that the temperature of the upper chromosphere exceeds that of the photosphere. Furthermore, the temperature of the solar corona, where even more ionization is seen, is higher still. Figure 16.13 shows how the temperature of the Sun's atmosphere varies with altitude. The temperature decreases to a minimum of about 4500 K some 500 km above the photosphere, after which it rises steadily. About 1500 km above the photosphere, in the transition zone, the temperature begins to rise rapidly, reaching more than 1 million K at an altitude of 10,000 km. Thereafter, in the corona, the temperature remains roughly constant at around 3 million K, although *SOHO* and other orbiting instruments have detected coronal "hot spots" having temperatures many times higher than this average value.

The cause of the rapid temperature rise is not fully understood. The temperature profile runs contrary to intuition: Moving away from a heat source, we would normally expect the heat to diminish, but this is not the case in the lower atmosphere of the Sun. The corona must have another energy source. Astronomers now think that magnetic disturbances in the solar photosphere are ultimately responsible for heating the corona (Section 16.5).

The Solar Wind

Electromagnetic radiation and fast-moving particles—mostly protons and electrons—escape from the Sun all the time. The radiation moves away from the photosphere at the speed of light, taking 8 minutes to reach Earth. The particles travel more slowly, although at the still considerable speed of about 500 km/s, reaching Earth in a few days. This constant stream of escaping solar particles is the solar wind.

The solar wind results from the high temperature of the corona. About 10 million km above the photosphere, the coronal gas is hot enough to escape the Sun's gravity, and it begins to flow outward into space. At the same time, the solar atmosphere is continuously replenished from below. If that were not the case, the corona would disappear in about a day. The Sun is, in effect, "evaporating"—constantly shedding mass through the solar wind. The wind is an extremely thin medium, however. Even though it carries away roughly 2 million tons of solar matter each second, less than 0.1 percent of the Sun's mass has been lost this way since the solar system formed 4.6 billion years ago.

CONCEPT CHECK

✔ Describe two ways in which the spectrum of the solar corona differs from that of the photosphere.

16.4 Solar Magnetism

The Sun has a powerful and complex magnetic field. Discovered in 1908 by American astronomer George Ellery Hale, the solar field still presents puzzles to scientists today. The structure of the Sun's magnetic field lines is crucial to understanding many aspects of the Sun's appearance and surface activity, yet the details of the field-line geometry, and even the mechanism responsible for generating and sustaining the entire solar field, remain subjects of intense research. Curiously, the keys to understanding many aspects of solar magnetism lie in a phenomenon first observed nearly three centuries before Hale's groundbreaking discovery.

Sunspots

Figure 16.14 is an optical photograph of the entire Sun, showing numerous dark blemishes on its surface. First studied in detail by Galileo around 1613, these "spots" provided one of the first clues that the Sun was not a perfect, unvarying creation, but rather a place of constant change. ∞ (Sec. 2.4) The dark areas are called **sunspots** and typically measure about 10,000 km across, approximately the size of Earth. As shown in the figure, they often occur in groups. At any given time, the Sun may have hundreds of sunspots, or it may have none at all.

Studies of sunspots show an **umbra,** or dark center, surrounded by a grayish **penumbra.** The close-up views in Figure 16.15 show each of these dark areas and the brighter undisturbed photosphere nearby. This gradation in darkness is really a gradual change in photospheric temperature—sunspots are simply *cooler* regions of the photospheric gas.

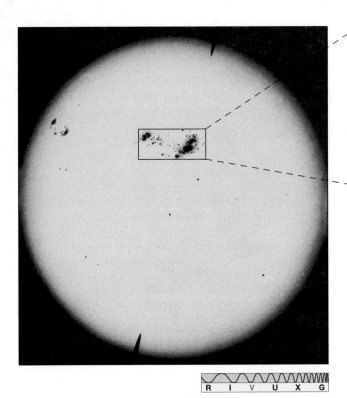

▲ **FIGURE 16.14 Sunspots** This photograph of the entire Sun, taken during a period of maximum solar activity, shows several groups of sunspots. The largest spots in the image are more than 20,000 km across, nearly twice the diameter of Earth. Typical sunspots are only about half that size. *(Palomar Observatory/Caltech)*

(a) |←——— 50,000 km ———→|

Penumbra

Umbra

(b) |←——— 10,000 km ———→|

R   I   V   U   X   G

▲ **FIGURE 16.15 Sunspots, Up Close** (a) An enlarged photograph of the largest pair of sunspots in Figure 16.14 shows how each spot consists of a cool, dark inner region called the umbra surrounded by a warmer, brighter region called the penumbra. The spots appear dark because they are slightly cooler than the surrounding photosphere. (b) A high-resolution image of a single typical sunspot—about the size of Earth—shows details of its structure as well as the surface granules surrounding it. *(Palomar Observatory/Caltech; SST/Royal Swedish Academy of Science)*

The temperature of the umbra is about 4500 K, compared with the penumbra's 5500 K. The spots, then, are certainly composed of hot gases. They seem dark only because they appear against an even brighter background (the 5800 K photosphere). If we could magically remove a sunspot from the Sun (or just block out the rest of the Sun's emission), the spot would glow brightly, just like any other hot object having a temperature of roughly 5000 K.

The Sun's Magnetic Field

What causes a sunspot? Why is it cooler than the surrounding photosphere? The answers to these questions are closely tied to the structure of the Sun's magnetic field. We saw in Chapter 4 that analysis of spectral lines can yield detailed information about the magnetic field at the location where the lines originate. ∞ (Sec. 4.5) Indeed, Hale's discovery of solar magnetism was made through observations of the Zeeman effect (broadening or splitting of spectral lines by a magnetic field) in Hα lines observed in sunspots. Most importantly, both the *strength* of the magnetic field and the *orientation* of a field line along the line of sight (toward or away from the observer) can be determined.

The magnetic field in a typical sunspot is about 1000 times greater than the field in neighboring, undisturbed photospheric regions (which is itself several times stronger than Earth's magnetic field). Furthermore, the field lines are not randomly oriented, but instead are directed roughly perpendicular to (out of or into) the Sun's surface. Scientists think that sunspots are cooler than their surroundings because these abnormally strong fields tend to block (or redirect) the convective flow of hot gas, which is normally toward the surface of the Sun.

The **polarity** of a sunspot simply indicates which way its magnetic field is directed relative to the solar surface. We conventionally label spots where field lines emerge from the interior as "S" and those where the lines dive below the photosphere as "N" (so field lines above the surface always run from S to N, as on Earth). Sunspots almost always come in pairs whose members lie at roughly the same latitude and have opposite magnetic polarities. Figure 16.16(a) illustrates

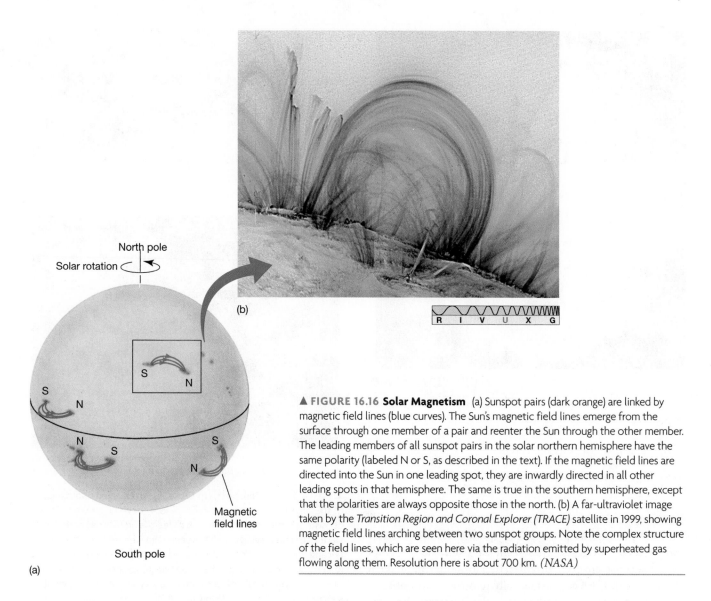

(b)

R   I   V   U   X   G

North pole

Solar rotation

South pole

Magnetic field lines

(a)

▲ FIGURE 16.16 **Solar Magnetism** (a) Sunspot pairs (dark orange) are linked by magnetic field lines (blue curves). The Sun's magnetic field lines emerge from the surface through one member of a pair and reenter the Sun through the other member. The leading members of all sunspot pairs in the solar northern hemisphere have the same polarity (labeled N or S, as described in the text). If the magnetic field lines are directed into the Sun in one leading spot, they are inwardly directed in all other leading spots in that hemisphere. The same is true in the southern hemisphere, except that the polarities are always opposite those in the north. (b) A far-ultraviolet image taken by the *Transition Region and Coronal Explorer (TRACE)* satellite in 1999, showing magnetic field lines arching between two sunspot groups. Note the complex structure of the field lines, which are seen here via the radiation emitted by superheated gas flowing along them. Resolution here is about 700 km. *(NASA)*

how magnetic field lines emerge from the solar interior through one member (S) of a sunspot pair, loop through the solar atmosphere, and then reenter the photosphere through the other member (N). As in Earth's magnetosphere, charged particles tend to follow the solar magnetic field lines. ∞ (Sec. 7.5) Figure 16.16(b) shows an actual image of solar magnetic loops, revealing high-temperature gas flowing along a complex network of magnetic field lines connecting two sunspot groups.

Despite the irregular appearance of the sunspots themselves, there is a great deal of order in the underlying solar field. *All* the sunspot pairs in the same solar hemisphere (north or south) at any instant have the *same* magnetic configuration. That is, if the leading spot (measured in the direction of the Sun's rotation) of one pair has N polarity, as shown in the figure, then all leading spots in that hemisphere have the same polarity. What's more, in the other hemisphere at the same time, all sunspot pairs have the

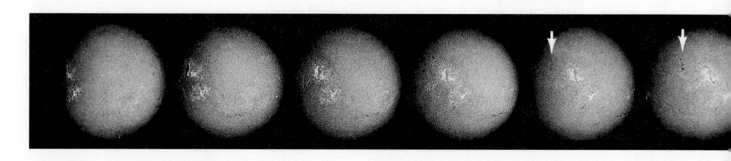

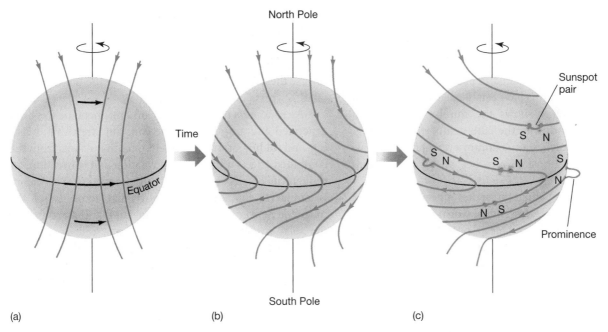

North Pole

Time

Sunspot pair

Equator

Prominence

South Pole

(a)   (b)   (c)

▲ **FIGURE 16.17 Solar Rotation** (a, b) The Sun's differential rotation wraps and distorts the solar magnetic field. (c) Occasionally, the field lines burst out of the surface and loop through the lower atmosphere, thereby creating a sunspot pair. The underlying pattern of the solar field lines explains the observed pattern of sunspot polarities. If the loop happens to occur on the edge of the Sun and is seen against the blackness of space, we see a phenomenon called a prominence, described in Section 16.5. (See Figure 16.21.)

opposite magnetic configuration (S polarity leading). To understand these regularities in sunspot polarities, we must look at the Sun's magnetic field in a little more detail.

The combination of differential rotation and convection radically affects the character of the Sun's magnetic field, which in turn plays a major role in determining the numbers and location of sunspots. As illustrated in Figure 16.17, the Sun's differential rotation distorts the solar magnetic field, "wrapping" it around the solar equator and eventually causing any originally north–south magnetic field to reorient itself in an east–west direction. At the same time, convection causes the magnetized gas to well up toward the surface, twisting and tangling the magnetic field pattern. In some places, the field

lines become kinked like a twisted garden hose, causing the field strength to increase. Occasionally, the field becomes so strong that it overwhelms the Sun's gravity, and a "tube" of field lines bursts out of the surface and loops through the lower atmosphere, forming a sunspot pair. The general east–west orientation of the underlying solar field accounts for the observed polarities of the resulting sunspot pairs in each hemisphere.

The Solar Cycle

Sunspots are not steady. Most change their size and shape, and all come and go. Figure 16.18 shows a time sequence in which a number of spots varied—sometimes growing,

▼ **FIGURE 16.18 Sunspot Rotation** The evolution of some sunspots and lower chromospheric activity over a period of 12 days. The sequence runs from left to right. An Hα filter was used to make these photographs, taken from the *Skylab* space station. An arrow follows one set of sunspots over the course of a week as they are carried around the Sun by its rotation. *(NASA)*

R I V U X G

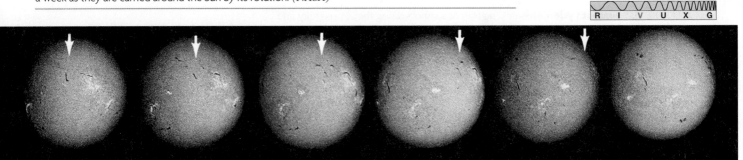

sometimes dissipating—over a period of several days. Individual spots may last anywhere from 1 to 100 days; a large group typically lasts 50 days. Not only do sunspots come and go with time, but their numbers and distribution across the face of the Sun also change fairly regularly. Centuries of observations have established a clear **sunspot cycle.** Figure 16.19(a) shows the number of sunspots observed each year during the 20th century. The average number of spots reaches a maximum every 11 or so years and then falls off almost to zero before the cycle begins afresh.

The latitudes at which sunspots appear vary as the sunspot cycle progresses. Individual sunspots do not move up or down in latitude, but new spots appear closer to the equator as older ones at higher latitudes fade away. Figure 16.19(b) is a plot of observed sunspot latitude as a function of time. At the start of each cycle, at *solar minimum,* only a few spots are seen, and these are generally confined to two narrow zones about 25° to 30° north and south of the solar equator. Approximately four years into the cycle, around *solar maximum,* the number of spots has increased markedly, and they are found within about 15° to 20° of the equator. Finally, by the end of the cycle, at solar minimum, the number has fallen again, and most sunspots lie within about 10° of the solar equator. The beginning of each new cycle appears to overlap the end of the last.

Complicating this picture further, the 11-year sunspot cycle is actually only half of a longer 22-year **solar cycle.**

During the first 11 years of the cycle, the leading spots of all the pairs in the northern hemisphere have the same polarity, while spots in the southern hemisphere have the opposite polarity (Figure 16.16). These polarities then reverse their signs for the next 11 years, so the full solar cycle takes 22 years.

Astronomers think that the Sun's magnetic field is both generated and amplified by the constant stretching, twisting, and folding of magnetic field lines that results from the combined effects of differential rotation and convection, although the details are still not well understood. The theory is similar to the "dynamo" theory that accounts for the magnetic fields of Earth and the jovian planets, except that the solar dynamo operates much faster and on a much larger scale. ∞ (Sec. 7.5) One prediction of this theory is that the Sun's magnetic field should rise to a maximum, then fall to zero, and reverse itself, more or less periodically, just as is observed. Solar surface activity, such as the sunspot cycle, simply follows the variations in the magnetic

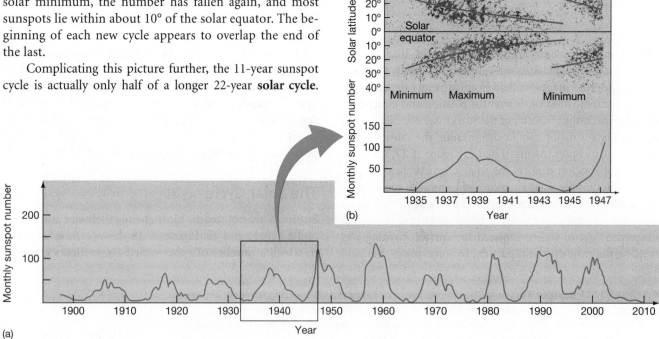

▲ **FIGURE 16.19 Sunspot Cycle** (a) Monthly number of sunspots during the 20th century, and beyond, showing yearly averages of the data to make long-term trends more evident. The (roughly) 11-year solar cycle is clearly visible. At the time of minimum solar activity, hardly any sunspots are seen. About 4 years later, at maximum solar activity, about 100 to 200 spots are observed per month. The most recent solar maximum occurred in 2001. (b) Sunspots cluster at high latitudes when solar activity is at a minimum. They appear at lower and lower latitudes as the number of sunspots peaks. They are again prominent near the Sun's equator as solar minimum is approached once more. The blue lines in the upper plot indicate how the "average" sunspot latitude varies over the course of the cycle.

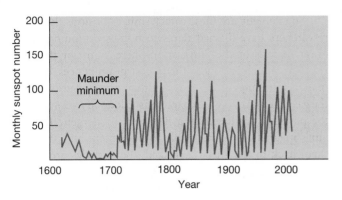

▲ FIGURE 16.20 **Maunder Minimum** Average number of sunspots occurring each month over the past four centuries. Note the absence of spots during the late 17th century.

field. The changing numbers of sunspots and their migration to lower latitudes are both consequences of the strengthening and eventual decay of the field lines as they become more and more tightly wrapped around the solar equator.

Figure 16.20 plots sunspot data extending back to the invention of the telescope. As can be seen, the 11-year "periodicity" of the solar sunspot cycle is far from regular. Not only does the period range from 7 to 15 years, but the sunspot cycle disappeared entirely over a number of years in the relatively recent past. The lengthy period of solar inactivity that extended from 1645 to 1715 is called the **Maunder minimum,** after the British astronomer who drew attention to these historical records. The corona was apparently also less prominent during total solar eclipses around that time, and Earth aurorae were sparse throughout the late 17th century. Lacking a complete understanding of the solar cycle, we cannot easily explain how it could shut down entirely. Most astronomers suspect changes in the Sun's convection zone or rotation pattern, but the specific causes of the Sun's century-long variations, as well as the details of the connection between solar activity and Earth's climate, remain a mystery (see *Discovery 16-2*).

CONCEPT CHECK

✔ What do observations of sunspot polarities tell us about the solar magnetic field?

16.5 The Active Sun

Most of the Sun's luminosity results from continuous emission from the photosphere. However, superimposed on this steady, predictable aspect of our star's energy output is a much more irregular component, characterized by explosive and unpredictable surface activity. Solar activity contributes little to the Sun's total luminosity and probably has no significant bearing on the evolution of the Sun, but it does affect us here on Earth. The size and duration of coronal holes are strongly influenced by the level of solar activity. Hence, so is the strength of the solar wind, and that in turn directly affects Earth's magnetosphere.

Active Regions

The photosphere surrounding a pair or group of sunspots can be a violent place, sometimes erupting explosively, spewing forth large quantities of energetic particles into the corona. The sites of these energetic events are known as **active regions.** Most groups of sunspots have active regions associated with them. Like all other aspects of solar activity, these phenomena tend to follow the solar cycle and are most frequent and violent around the time of solar maximum.

Figure 16.21 shows two large solar **prominences**— loops or sheets of glowing gas ejected from active regions on the solar surface, moving through the inner parts of the corona under the influence of the Sun's magnetic field. Magnetic instabilities in the strong fields found in and near sunspot groups may cause the prominences, although the details are not fully understood. The arching magnetic field lines in and around the active region are also easily seen (see also Figure 16.16b). The rapidly changing structure of the field lines and the fact that they can quickly transport mass and energy from one part of the solar surface to another, possibly tens of thousands of kilometers away, make the theoretical study of active regions an extraordinarily difficult task.

Quiescent prominences persist for days or even weeks, hovering high above the photosphere, suspended by the Sun's magnetic field. *Active prominences* come and go much more erratically, changing their appearance in a matter of hours or surging up from the solar photosphere and then immediately falling back on themselves. A typical solar prominence measures some 100,000 km in extent, nearly 10 times the diameter of planet Earth. Prominences as large as the one shown in Figure 16.21(a) (which traversed almost half a million kilometers of the solar surface) are less common and usually appear only at times of greatest solar activity. The largest prominences can release up to 10^{25} joules of energy, counting both particles and radiation—not much compared with the total solar luminosity of 4×10^{26} W, but still enormous by terrestrial standards. (All the power plants on Earth would take a billion years to produce that much energy.)

Flares are another type of solar activity observed low in the Sun's atmosphere near active regions. Also the result of magnetic instabilities, flares, like that shown in Figure 16.22, are even more violent (and even less well understood) than prominences. They often flash across a region of the Sun in minutes, releasing enormous amounts of energy as they go. Space-based observations indicate that X-ray and ultraviolet emissions are especially intense in the extremely compact hearts of flares, where temperatures can reach 100 million K.

So energetic are these cataclysmic explosions that some researchers have likened flares to bombs exploding in the lower regions of the Sun's atmosphere. A major flare can release as much energy as the largest prominences, but in a matter of minutes or hours rather than days or weeks. Unlike the gas that makes up the characteristic loop of a prominence, the particles produced by a flare are so energetic that the Sun's magnetic field is unable to hold them and shepherd them back to the surface. Instead, the particles are simply blasted into space by the violence of the explosion.

Figure 16.23 shows a **coronal mass ejection** from the Sun. Sometimes (but not always) associated with flares and prominences, these phenomena are giant magnetic "bubbles" of ionized gas that separate from the rest of the solar atmosphere and escape into interplanetary space. Such ejections occur about once per week at times of sunspot minimum, but up to two or three times per day at solar maximum. Carrying an enormous amount of energy, they can—if their fields are properly oriented—merge with Earth's magnetic field via a process known as *reconnection*, dumping some of their energy into the magnetosphere and potentially causing widespread communications and power disruptions on our planet (Figure 16.23b; see also *Discovery 16-2*).

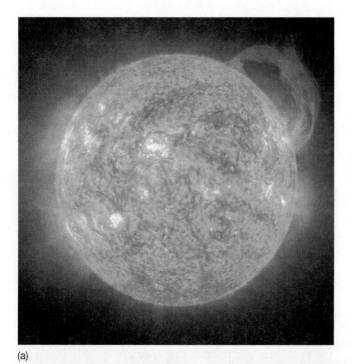

(a)

(b)

◀ FIGURE 16.21 **Solar Prominences** (a) This particularly large solar prominence was observed by ultraviolet detectors aboard the *SOHO* spacecraft in 2002. (b) Like a phoenix rising from the solar surface, this filament of hot gas measures more than 100,000 km in length. Earth could easily fit between its outstretched "arms." Dark regions in this *TRACE* image have temperatures less than 20,000 K; the brightest regions are about 1 million K. The ionized gas follows the solar magnetic field lines away from the Sun. Most of the gas will subsequently cool and fall back into the photosphere. (*NASA*)

R I V U X G

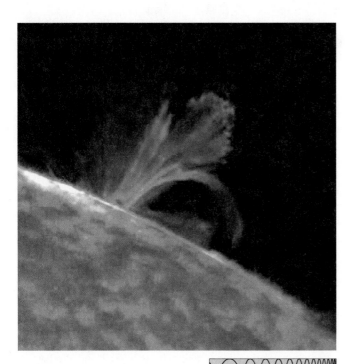

▲ **FIGURE 16.22 Solar Flare** Much more violent than a prominence, a solar flare is an explosion on the Sun's surface that sweeps across an active region in a matter of minutes, accelerating solar material to high speeds and blasting it into space. *(USAF)*

The Changing Solar Corona

Unlike the 5800 K photosphere, which emits most strongly in the visible part of the electromagnetic spectrum, the hot coronal gas radiates at much higher frequencies—primarily in the X-ray range. ∞ (Sec. 3.4) For this reason, X-ray telescopes have become important tools in the study of the solar corona. Figure 16.24(a) shows several X-ray images of the Sun. The full corona extends well beyond the regions shown, but the density of coronal particles emitting the radiation diminishes rapidly with distance from the Sun. The intensity of X-ray radiation farther out is too dim to be seen here.

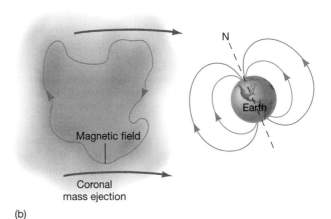

(b)

(c)

◀ **FIGURE 16.23 Coronal Mass Ejection** (a) A few times per week, on average, a giant magnetized "bubble" of solar material detaches itself from the Sun and rapidly escapes into space, as shown in this *SOHO* image taken in 2002. The circles are artifacts of an imaging system designed to block out the light from the Sun itself and exaggerate faint features at larger radii. (b) Should a coronal mass ejection encounter Earth with its magnetic field oriented opposite to our own, as illustrated, the field lines can join together as in part (c), allowing high-energy particles to enter and possibly severely disrupt our planet's magnetosphere, as well as human activities on and near the surface. By contrast, if the fields are oriented differently, the coronal mass ejection can slide by Earth with little effect. *(NASA/ESA)*

DISCOVERY 16-2

 Solar–Terrestrial Relations

Our Sun has often been worshipped as a god with power over human destinies. Obviously, the steady stream of solar energy arriving at our planet every day is essential to our lives, but over the past century there have also been repeated claims of a correlation between the Sun's activity and Earth's weather. Only recently, however, has the subject become scientifically respectable—that is, more natural than supernatural.

In fact, there do seem to be some correlations between the 22-year solar cycle (two sunspot cycles with oppositely directed magnetic fields) and periods of climatic dryness here on Earth. For example, near the start of the past eight cycles, there have been droughts in North America—at least within the middle and western plains from South Dakota to New Mexico. Another possible Sun–Earth connection is a link between solar activity and increased atmospheric circulation on our planet. As circulation increases, terrestrial storm systems deepen, extend over wider ranges of latitude, and carry more moisture. The relationship is complex and the subject controversial, because no one has yet shown any physical mechanism (other than the Sun's heat, which does not vary much during the solar cycle) that would allow solar activity to stir our terrestrial atmosphere. Without a better understanding of the physical mechanism involved, none of these effects can be incorporated into our weather-forecasting models.

Solar activity may also influence long-term climate on Earth. For example, the Maunder minimum (see Section 16.6) seems to correspond fairly well with the coldest years of the so-called Little Ice Age that chilled northern Europe and North America during the late 1600s. The accompanying "winter" scene actually captured one summer season in 17th-century Holland. How the active Sun and its abundance of sunspots may affect Earth's climate is a frontier problem in terrestrial climatology.

Measurements of the solar constant made over the past two decades indicate that the Sun's energy output varies with the solar cycle. Paradoxically, the Sun's luminosity is greatest when many dark sunspots cover its surface! Thus, the Maunder minimum does correspond to an extended period of lower-than-average solar emission. However, recent observed changes in the Sun's luminosity have been small—no more than 0.2 or 0.3 percent. It is not known by how much, if at all, the Sun's output declined during the Maunder minimum, nor how large a change would be needed to account for the alterations in climate that occurred.

One correlation that is definitely established, and also better understood, is that between solar activity and geomagnetic disturbances at Earth. The extra radiation and particles thrown off by flares or coronal mass ejections impinge on Earth's environment, overloading the Van Allen belts, thereby causing brilliant auroras in our atmosphere and degrading our communication networks. We are only beginning to understand how the radiation and particles emitted by solar phenomena also interfere with terrestrial radars, power networks, and other technological equipment. Some power outages on Earth are actually caused, not by increased customer demand or malfunctioning equipment, but by weather on the Sun!

We cannot yet predict just when and where solar flares or coronal mass ejections will occur. However, it would certainly be to our advantage to be able to do so, as that aspect of the active Sun affects our lives. This is a highly fertile area of astronomical research and one with clear terrestrial applications.

(Rijksmuseum, Amsterdam, Holland/The Bridgman Art Library)

In the mid-1970s, instruments aboard NASA's *Skylab* space station revealed that the solar wind escapes mostly through solar "windows" called **coronal holes.** The dark area moving from left to right in Figure 16.24(a), which shows more recent data from the Japanese *Yohkoh* X-ray solar observatory, represents a coronal hole. Not really holes, such structures are simply deficient in matter—vast regions of the Sun's atmosphere where the density is about

10 times lower than the already tenuous, normal corona. Note that the underlying solar photosphere looks black in these images because it is far too cool to emit X rays in any significant quantity.

Coronal holes are lacking in matter because the gas there is able to stream freely into space at high speeds, driven by disturbances the Sun's atmosphere and magnetic field. Figure 16.24(b) illustrates how, in coronal holes, the solar

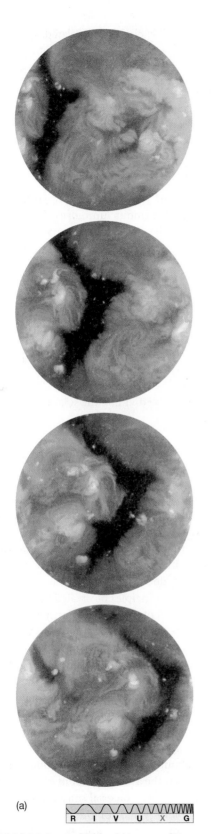

magnetic field lines extend from the surface far out into interplanetary space. Because charged particles tend to follow the field lines, they can escape, particularly from the Sun's polar regions, according to findings from *SOHO* and NASA's *Ulysses* spacecraft, which flew high above the ecliptic plane to explore the Sun's polar regions. In other regions of the corona, the solar magnetic field lines stay close to the Sun, keeping charged particles near the surface and inhibiting the outward flow of the solar wind (just as Earth's magnetic field tends to prevent the incoming solar wind from striking Earth), so the density remains relatively high. Because of the "open" field structure in coronal holes, flares and other magnetic activity (which, as we have seen, are related to magnetic loops near the solar photosphere) tend to be suppressed there.

The largest coronal holes, like that shown in Figure 16.24(a), can be hundreds of thousands of kilometers across and may survive for many months. Structures of this size are seen only a few times per decade. Smaller holes— perhaps only a few tens of thousand kilometers in size—are much more common, appearing every few hours.

Coronal holes appear to be an integral part of the process by which the Sun's large-scale field reverses and replenishes itself over the course of the solar cycle. Long-lived holes persist at the Sun's polar regions over much of the magnetic cycle, and the numbers and locations of other holes appear to change in step with solar activity. However, like many aspects of the solar magnetic field, the structure and evolution of coronal holes are not fully understood; they are currently the subject of intense research.

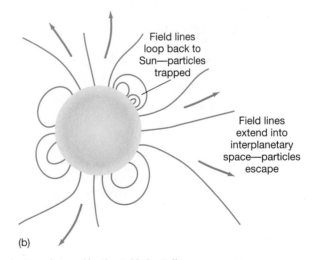

(a)

R I V U X G

(b)

▲ FIGURE 16.24 **Coronal Hole** (a) Images of X-ray emission from the Sun observed by the *Yohkoh* satellite. These frames were taken at roughly 2-day intervals, starting at the top. Note the dark, V-shaped coronal hole traveling from left to right, where the X-ray observations outline in dramatic detail the abnormally thin regions through which the high-speed solar wind streams forth. (b) Charged particles follow magnetic field lines (blue curves) that compete with gravity. When the field is trapped and loops back toward the photosphere, the particles are also trapped; otherwise, they can escape as part of the solar wind. (*ISAS/Lockheed Martin*)

Finally, the solar corona varies with the sunspot cycle. The photograph of the corona in Figure 16.12 shows the quiet Sun, at sunspot minimum. At such times, the corona is fairly regular in appearance and seems to surround the Sun more or less uniformly. Compare that image with Figure 16.25, which was taken in 1994 near a peak in the sunspot cycle. The active corona is much more irregular in appearance and extends farther from the solar surface. The "streamers" of coronal material pointing away from the Sun are characteristic of this phase.

Astronomers think that the corona is heated primarily by activity in the photosphere, which can inject large amounts of energy into the lower solar atmosphere. The myriad spicules and small-scale magnetic disturbances in the photosphere probably provide most of the energy needed to heat the corona. More extensive disturbances often move through the corona above an active site in the photosphere, distributing the energy throughout the coronal gas. Given this connection, it is hardly surprising that both the appearance of the corona and the strength of the solar wind are closely correlated with the solar cycle.

CONCEPT CHECK

✔ Why is solar activity important to life on Earth?

16.6 The Heart of the Sun

What powers the Sun? What forces are at work in the Sun's core to produce its enormous luminosity? By what process does the Sun shine, day after day, year after year, eon after eon? Answers to these questions are central to all astronomy. Without them, we can understand neither the physical existence of stars and galaxies in the universe nor the biological existence of life on Earth.

Solar Energy Production

In round numbers, the Sun's luminosity is 4×10^{26} W and its mass is 2×10^{30} kg. We can quantify how efficiently the Sun generates energy by dividing the solar luminosity by the solar mass:

$$\frac{\text{solar luminosity}}{\text{solar mass}} = 2 \times 10^{-4} \text{ W/kg}.$$

This simply means that, on average, every kilogram of solar material yields about 0.2 milliwatt of energy—0.0002 joule (J) of energy every second.

This is not much energy—a piece of burning wood generates about a million times more energy per unit mass per unit time than does our Sun, so the equivalent solar luminosity could (in principle) be created by a pile of burning logs comparable in mass to planet Earth. But there is one very important difference: The logs cannot continue to burn at this rate for billions of years.

To appreciate the magnitude of the energy generated by our Sun, we must consider not the ratio of the solar luminosity to the solar mass, but instead the total amount of energy generated by each gram of solar matter *over the entire lifetime of the Sun as a star.* This is easy to do. We simply multiply the rate at which the Sun generates energy by the age of the Sun, about 5 billion years. We obtain a value of 3×10^{13} J/kg. This is the average amount of energy radiated by every kilogram of solar material since the Sun formed. It represents a *minimum* value for the total energy radiated by the Sun, for more energy will be needed for every additional day the Sun shines. Should the Sun endure for another 5 billion years (as is predicted by theory), we would have to double this value.

Either way, this energy-to-mass ratio is very large. At least 60 trillion joules (on average) of energy must arise

▲ **FIGURE 16.25 Active Corona** Photograph of the solar corona during the July 1994 eclipse, near the peak of the sunspot cycle. At these times, the corona is much less regular and much more extended than at sunspot minimum (compare to Figure 16.12). Astronomers think that coronal heating is caused by surface activity on the Sun. The changing shape and size of the corona is the direct result of variations in prominence and flare activity over the course of the solar cycle. (*NCAR High Altitude Observatory*)

from every kilogram of solar matter to power the Sun throughout its lifetime. But the Sun's generation of energy is not explosive, releasing large amounts of energy in a short period. Instead, it is slow and steady, providing a uniform and long-lived rate of energy production. Only one known energy-generation mechanism can conceivably power the Sun in this way: **nuclear fusion**—the combining of light nuclei into heavier ones.

Nuclear Fusion

We can represent a typical fusion reaction symbolically as

nucleus 1 + nucleus 2 → nucleus 3 + energy.

For powering the Sun and other stars, the most important piece of this equation is the energy produced.

The essential point here is that, during a fusion reaction, the total mass *decreases*—the mass of nucleus 3 is less than the combined masses of nuclei 1 and 2. Where does this mass go? It is converted into energy in accordance with Einstein's famous equation of **mass-energy equivalence**

$$E = mc^2,$$

or

$$energy = mass \times (speed\ of\ light)^2.$$

This equation expresses the discovery, made by Albert Einstein at the beginning of the 20th century, that matter and energy are interchangeable—one can be converted into the other. To determine the amount of energy corresponding to a given mass, simply multiply the mass by the square of the speed of light (c in the equation). For example, the energy equivalent of 1 kg of matter is $1 \times (3 \times 10^8)^2$, or 9×10^{16} J. The speed of light is so large that even small amounts of mass translate into enormous amounts of energy.

The production of energy by a nuclear fusion reaction is an example of the **law of conservation of mass and energy,** which states that the *sum* of mass and energy (suitably converted into the same units, using Einstein's equation) must always remain *constant* during any physical process. *There are no known exceptions.* According to this law, an object can literally disappear, provided that some energy appears in its place. If magicians really made rabbits disappear, the result would be a flash of energy equaling the product of the rabbit's mass and the square of the speed of light—enough to destroy the magician, everyone in the audience, and probably all of the surrounding state as well! In the case of fusion reactions in the solar core, the energy is produced primarily in the form of electromagnetic radiation. The light we see coming from the Sun means that the Sun's mass must be slowly, but steadily, decreasing with time.

Charged Particle Interactions

All atomic nuclei are positively charged, so they repel one another. Furthermore, by the inverse-square law, the closer two nuclei come to one another, the greater is the repulsive force between them (Figure 16.26a). ∞ (Sec. 3.2) How, then, do nuclei—two protons, say—ever manage to fuse into anything heavier? The answer is that if they collide at high enough speeds, one proton can momentarily plow deep into the other, eventually coming within the exceedingly short range of the **strong nuclear force,** which binds

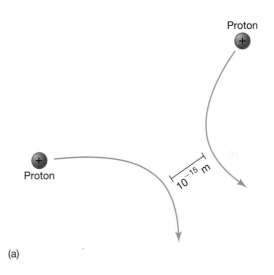

(a)

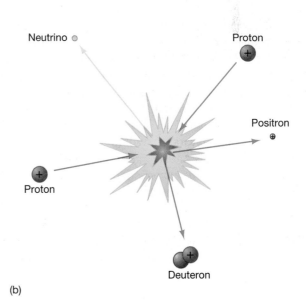

(b)

▲ FIGURE 16.26 **Proton Interactions** (a) Since like charges repel, two low-speed protons veer away from one another, never coming close enough for fusion to occur. (b) Sufficiently high-speed protons may succeed in overcoming their mutual repulsion, approaching close enough for the strong force to bind them together—in which case they collide violently, triggering nuclear fusion that ultimately powers the Sun.

nuclei together (see *More Precisely 16-1*). At distances less than about 10^{-15} m, the attraction of the nuclear force overwhelms the electromagnetic repulsion, and fusion occurs. Speeds in excess of a few hundred kilometers per second, corresponding to a gas temperature of 10^7 K or more, are needed to slam protons together fast enough to initiate fusion. Such conditions are found in the core of the Sun and at the centers of all stars.

The fusion of two protons is illustrated schematically in Figure 16.26(b). In effect, one of the protons turns into a neutron, creating new particles in the process, and combines with the other proton to form a **deuteron,** the nucleus of a special form of hydrogen called *deuterium.* Deuterium (also referred to as "heavy hydrogen") differs from ordinary hydrogen by virtue of an extra neutron in its nucleus. We can represent the reaction as follows:

proton + proton → deuteron + positron + neutrino.

The **positron** in this reaction is a positively charged electron. Its properties are identical to those of a normal, negatively charged electron, except for the positive charge. Scientists call the electron and the positron a "matter–antimatter pair"—the positron is said to be the *antiparticle* of the electron. The newly created positrons find themselves in the midst of a sea of electrons with which they interact immediately and violently. The particles and antiparticles annihilate (destroy) one another, producing pure energy in the form of gamma-ray photons.

The final product of the reaction is a particle known as a **neutrino,** a word derived from the Italian for "little neutral one." Neutrinos carry no electrical charge and are of very low mass—at most 1/100,000 the mass of an electron, which itself has only 1/2000 the mass of a proton. (The exact mass of the neutrino remains uncertain, although experimental evidence strongly suggests that it is not zero.) Neutrinos move at almost the speed of light and interact with hardly anything. They can penetrate, without stopping, several light-years of lead (a very dense material, widely used in terrestrial laboratories as an effective shield against radiation). Their interactions with matter are governed by the **weak nuclear force,** described in *More Precisely 16-1.* Despite their elusiveness, neutrinos can be detected with carefully constructed instruments. In the final section of this chapter we discuss some rudimentary neutrino "telescopes" and the important contribution they have made to solar astronomy.

Nuclei such as normal hydrogen and deuterium, containing the same number of protons, but different numbers of neutrons, represent different forms of the same element—they are known as **isotopes** of that element. Usually, there are about as many neutrons in a nucleus as protons, but the exact number of neutrons can vary, and most elements exist in a number of isotopic forms. To avoid confusion when talking about isotopes of the same element, nuclear physi-

cists attach a number to the symbol representing the element. This number indicates the total number of particles (protons plus neutrons) in the nucleus of an atom of the element. Thus, ordinary hydrogen is denoted by ^{1}H, deuterium by ^{2}H. Normal helium (two protons plus two neutrons) is ^{4}He (also referred to as helium-4), and so on. We will adopt this convention for the rest of the book.

The Proton–Proton Chain

The basic set of nuclear reactions powering the Sun (and the vast majority of all stars) is sketched in Figure 16.27. It is not a single reaction, but rather a sequence called the **proton–proton chain.** The following stages are shown in the figure.

I. First, two protons combine to form deuterium, as in Figure 16.26.

II. The resulting positrons annihilate with electrons, releasing energy in the form of gamma rays. The deuterons combine with protons to create an isotope of helium called helium-3 (containing only one neutron), releasing additional energy, again in the form of gamma-ray photons. Two of each of these sets of reactions are shown in Figure 16.27.

III. Finally, two helium-3 nuclei combine to produce helium-4, two protons, and still more gamma-ray energy.

Gargantuan quantities of protons are fused into helium by the proton–proton chain in the core of the Sun each and every second. The energy released ultimately becomes the sunlight that warms our planet.

Setting aside the temporary intermediate nuclei produced, we see that the *net* effect of the proton–proton chain is that four hydrogen nuclei (six protons consumed, two returned) combine to create one nucleus of the next lightest element, helium-4 (containing two protons and two neutrons, for a total mass of four), creating two neutrinos, and releasing energy in the form of gamma rays: ∞ (Sec. 4.2)

4 protons → helium-4 + 2 neutrinos + energy,

or

$$4\,(^1\text{H}) \rightarrow {}^4\text{He} + 2 \text{ neutrinos} + \text{energy}.$$

As discussed in more detail in *More Precisely 16-2,* to fuel the Sun's present energy output, hydrogen must be fused into helium in the core at a rate of 600 million tons per second—a lot of mass, but only a tiny fraction of the total amount available. The Sun will be able to sustain this rate of core burning for about another 5 billion years (see Chapter 20).

The Sun's nuclear energy is produced in the core in the form of gamma rays. However, as it passes through the cooler layers of the solar interior and photons are absorbed and reemitted, the radiation's blackbody spectrum steadily shifts toward lower and lower temperatures and, by Wien's law, the characteristic wavelength of the radiation increases.

ANIMATION/VIDEO Tritium-Helium Fusion

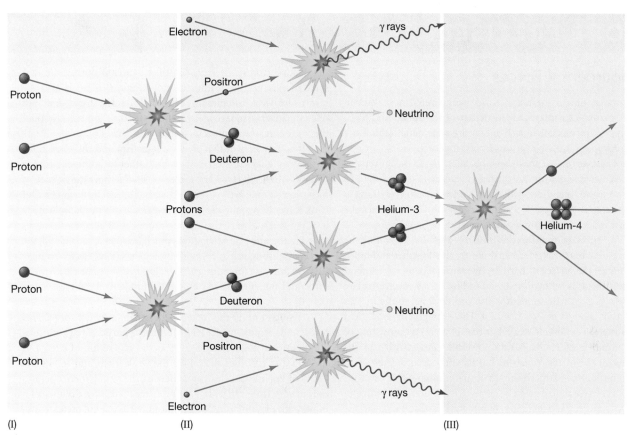

(I)   (II)   (III)

△ Interactive FIGURE 16.27 Solar Fusion In the proton–proton chain, a total of six protons (and two electrons) are converted into two protons, one helium-4 nucleus, and two neutrinos. The two leftover protons are available as fuel for new proton–proton reactions, so the net effect is that four protons are fused to form one helium-4 nucleus. Energy, in the form of gamma rays, is produced at each stage. (Most of the photons are omitted for clarity.) The three stages indicated correspond to reactions (I), (II), and (III) described in the text. This basic process powers the Sun, most stars, and ultimately life on Earth.

∞ (Sec. 2.4) The energy eventually leaves the photosphere mainly in the form of visible and infrared radiation. A comparable amount of energy is carried off by the neutrinos, which escape unhindered into space at almost the speed of light.

Other reaction sequences can produce the same end result as the proton–proton chain (see *More Precisely 20-1*). However, the sequence shown in Figure 16.27 is the simplest and produces almost 90 percent of the Sun's luminosity. Note how, at each stage of the chain, more massive, complex nuclei are created from simpler, lighter ones. We will see in Chapters 20 and 21 that not all stars are powered by hydrogen fusion. Nevertheless, *nuclear* fusion—the slow, but steady, transformation of light elements into heavier ones, creating energy in the process—is responsible for virtually *all* of the starlight we see.

CONCEPT CHECK

✔ Why does the fact that we see sunlight imply that the Sun's mass is slowly decreasing?

16.7 Observations of Solar Neutrinos

Theorists are quite sure that the proton-proton chain operates in the core of the Sun. However, because the gamma-ray energy created in the proton–proton chain is transformed into visible and infrared radiation by the time it emerges from the Sun, astronomers have no direct electromagnetic evidence of the core nuclear reactions. Instead, the *neutrinos* created in the proton–proton chain are our best bet for learning about conditions in the solar core. They travel cleanly out of the Sun, interacting with virtually nothing, and escape into space a few seconds after being created. Of course, the fact that they can pass through the entire Sun without interacting also makes neutrinos difficult to detect on Earth! Nevertheless, with knowledge of neutrino physics it is possible to construct neutrino detectors.

Over the past four decades, several experiments have been designed to detect solar neutrinos reaching Earth. Some detectors use large quantities of the elements chlorine

MORE PRECISELY 16-1

Fundamental Forces

Our studies of nuclear reactions have uncovered new ways in which matter can interact with matter at the subatomic level. Let's pause to consider in a slightly more systematic fashion the relationships among the various forces of nature.

As best we can tell, the behavior of all matter in the universe—from elementary particles to clusters of galaxies—is ruled by just four (or fewer) basic forces, which are *fundamental* to everything in the universe. In a sense, the search to understand the nature of the universe is the quest to understand the nature of these forces.

The **gravitational force** is probably the best known of the four. Gravity binds galaxies, stars, and planets together and holds humans on the surface of Earth. As we saw in Chapter 2, its magnitude decreases with distance according to an inverse-square law. ∞ (Sec. 2.7) Its strength is also proportional to the masses of each of the two objects involved. Thus, the gravitational field of an atom is extremely weak, but that of a galaxy, consisting of huge numbers of atoms, is very powerful. Gravity is by far the weakest of the forces of nature, but its effect accumulates as we move to larger and larger volumes of space, and nothing can cancel its attractive pull. As a result, gravity is the dominant force in the universe on all scales larger than that of Earth.

The **electromagnetic force** is another of nature's basic agents. Any particle having a net electric charge, such as an electron or a proton in an atom, exerts an electromagnetic force on any other charged particle. The everyday things we see around us are held together by this force. As with gravity, the strength of the electromagnetic force decreases with distance according to an inverse-square law. ∞ (Sec. 3.2) However, for subatomic particles, electromagnetism is much stronger than gravity. For example, the electromagnetic force between two protons exceeds their gravitational attraction by a factor of about 10^{36}. Unlike gravity, electromagnetic forces can repel (like charges) as well as attract (opposite charges). Positive and negative charges tend to neutralize each other, greatly diminishing their net electromagnetic influence. Above the microscopic level, most objects are in fact very close to being electrically neutral. Thus, except in unusual circumstances, the electromagnetic force is relatively unimportant on macroscopic scales.

A third fundamental force of nature is simply termed the **weak nuclear force.** This force is much weaker than electromagnetism, and its influence is somewhat more subtle. The weak nuclear force governs the emission of radiation from some radioactive atoms; the emission of a neutrino during the first stage of the proton–proton reaction (Figure 16.26) is also the result of a weak interaction. The weak nuclear force does not obey the inverse-square law. Its effective range is much less than the size of an atomic nucleus—about 10^{-18} m.

It is now known that electromagnetism and the weak force are not really separate forces at all, but rather two different aspects of a more basic **electroweak force.** At "low" temperatures, such as those found on Earth or even in stars, the electromagnetic and weak forces have quite distinct properties. However, as we will see in Chapter 27, at very high temperatures, such as those that prevailed in the universe when it was much less than a second old, the two are indistinguishable. Under those conditions, electromagnetism and the weak force are said to be "unified" into the electroweak force, and the universe has only three fundamental forces, rather than four.

Strongest of all the forces is the **strong nuclear force.** This force binds atomic nuclei and subnuclear particles (e.g., protons and neutrons) together and governs the generation of energy in the Sun and all other stars. Like the weak force, and unlike the forces of gravity and electromagnetism, the strong force operates only at very close range. It is unimportant outside a distance of a hundredth of a millionth of a millionth (10^{-15}) of a meter. However, within that range (e.g., in atomic nuclei), it binds particles with enormous strength. In fact, it is the range of the strong force that determines the typical sizes of atomic nuclei. Only when two protons are brought within about 10^{-15} m of one another can the attractive strong force overcome their electromagnetic repulsion. High-energy accelerator experiments suggest that, at very close quarters (less than 10^{-16} m), the strong force has a "hard" core where the attraction turns into repulsion. (This scale is too small to affect atomic nuclei, but it may be crucial in determining the physics of supernovae explosions—see Section 21.2.)

Very loosely speaking, we can say that the strong force is about 100 times stronger than electromagnetism, 1 million times stronger than the weak force, and 10^{38} times stronger than gravity. But not all particles are subject to all types of force. All particles interact through gravity because all have mass. However, only charged particles interact electromagnetically. Protons and neutrons are affected by the strong nuclear force, but electrons are not. Finally, under the right circumstances, the weak force can affect any type of subatomic particle, regardless of its charge.

or gallium, which happen to be slightly more likely than most to interact with neutrinos. The interactions turn chlorine nuclei into argon, or gallium into germanium. The new nuclei are radioactive, and the detection of the radiation from their decay signals the neutrino capture. Other detectors (two of which are shown in Figure 16.28) look for light produced when a high-energy neutrino occasionally collides with an electron in a water molecule, accelerating the elec-tron to almost the speed of light. As the high-speed electron moves through the water, it emits electromagnetic radiation, mainly in the ultraviolet part of the spectrum. At visible wavelengths, the water appears blue. Large photomultiplier tubes (light-amplification devices) detect the resultant faint glow that betrays the neutrino's passage. In all cases, the probability of a given neutrino interacting with matter in the detector is extraordinarily small—only 1 in 10^{15} of the

(a)

(b)

◀ **FIGURE 16.28 Neutrino Telescopes** (a) This swimming pool-sized "neutrino telescope" is buried beneath a mountain near Tokyo, Japan. Called Super Kamiokande, it is filled (in operation) with 50,000 tons of purified water, and contains 13,000 individual light detectors (some shown here being inspected by technicians) to sense the telltale signature—a brief burst of light—of a neutrino passing through the apparatus. (b) The Sudbury Neutrino Observatory (SNO), situated 2 km underground in Ontario, Canada. The SNO detector is similar in design to the Kamiokande device, but by using "heavy" water (with hydrogen replaced by deuterium) instead of ordinary water, and adding 2 tons of salt, it also becomes sensitive to other neutrino types. The device contains 10,000 light-sensitive detectors arranged on the inside of the large sphere shown here. *(ICRR, SNO)*

observed (and in fact have measured energies in the range predicted by the standard solar model), there is a real difference between the Sun's theoretical neutrino output and the number of neutrinos actually detected on Earth. The number of solar neutrinos reaching Earth is substantially less (by 50 to 70 percent) than the prediction of the standard solar model. This discrepancy is known as the **solar neutrino problem.**

How can we explain this clear disagreement between theory and observation? The broad agreement among several independent, well-designed, and thoroughly tested experiments implies to most scientists that experimental error is not the cause, and researchers are confident that the experimental results can be trusted. Indeed, the lead investigators of the two experiments shown in Figure 16.28 received the strongest possible scientific endorsement of their work in the form of the 2002 Nobel Prize. In that case, there are really only two possibilities: Either *solar neutrinos are not produced as frequently as we think,* or *not all of them make it to Earth.*

It is very unlikely that the resolution of the solar neutrino problem lies in the physics of the Sun's interior. For example, we might think of reducing the theoretical number of neutrinos by postulating a lower temperature in the solar core, but the nuclear reactions described in the previous section are just too well known, and the agreement between the standard solar model and helioseismological observations (Section 16.2) is far too close for conditions in the core to deviate much from the predictions of the model.

Instead, the answer involves the properties of the neutrinos themselves and has caused scientists to rethink some very fundamental concepts in particle physics. If neutrinos have even a minute amount of mass, theory indicates that it should be possible for them to change their properties—even to transform into other particles—during their 8-minute flight from the solar core to Earth through a process known as **neutrino oscillations.** In this picture, neutrinos are produced in the Sun at the rate required by the standard solar model, but some turn into something else—actually, other types of neutrinos—on their way to Earth and hence go undetected in

neutrinos passing through the apparatus is actually detected. Large amounts (tons) of target material and long-duration experiments (months or years) are needed to obtain accurate measurements.

The detectors' designs differ widely, they are sensitive to neutrinos of very different energies, and they disagree somewhat in the details of their results, but they all agree on one very important point: Although solar neutrinos are

MORE PRECISELY 16-2

Energy Generation in the Proton–Proton Chain

Let's look in a little more detail at the energy produced by fusion in the solar core and compare it with the energy needed to account for the Sun's luminosity. Using the notation presented in the text, the proton–proton chain may be compactly described by the following reactions:

proton fusion: $^1H + {}^1H \rightarrow {}^2H + \text{positron} + \text{neutrino}.$ (I)

deuterium fusion: $^2H + {}^1H \rightarrow {}^3He + \text{energy}.$ (II)

helium-3 fusion: $^3He + {}^3He \rightarrow {}^4He + {}^1H + {}^1H + \text{energy}.$ (III)

As discussed in the text and illustrated below, the net effect of the fusion process is that four protons combine to produce a nucleus of helium-4, in the process creating two neutrinos and two positrons (which are quickly converted into energy by annihilation with electrons).

We can calculate the total amount of energy released by accounting carefully for the total masses of the nuclei involved and applying Einstein's famous formula $E = mc^2$. This in turn allows us to relate the Sun's total luminosity to the consumption of hydrogen fuel in the core. Careful laboratory experiments have determined the masses of all the particles involved in the above reaction: The total mass of the protons is 6.6943×10^{-27} kg, the mass of the helium-4 nucleus is 6.6466×10^{-27} kg, and the neutrinos are virtually massless. We omit the positrons here—their masses will end up being counted as part of the total energy released. The difference between the total mass of the four protons and that of the final helium-4 nucleus, 0.0477×10^{-27} kg, is not great, but it is easily measurable.

Multiplying the vanished mass by the square of the speed of light yields 0.0477×10^{-27} kg $\times (3.00 \times 10^8 \text{ m/s})^2 = 4.28 \times 10^{-12}$ J. This is the energy produced in the form of radiation when 6.69×10^{-27} kg (the rounded-off mass of the four protons) of hydrogen fuses to helium. It follows that fusion of 1 kg of hydrogen generates $4.28 \times 10^{-12}/6.69 \times 10^{-27} = 6.40 \times 10^{14}$ J. Put another way, the process converts about 0.71 percent of the original mass into energy. A negligible fraction (actually, about 2 percent) of this energy is carried away by the neutrinos. The rest appears in the form of gamma rays and ultimately is radiated away from the solar photosphere—that is, it becomes the solar luminosity.

Thus we have established a direct connection between the Sun's energy output and the consumption of hydrogen in the core. The Sun's luminosity of 3.84×10^{26} W (see Table 16.1), or 3.84×10^{26} J/s (joules per second), implies a mass consumption rate of 3.84×10^{26} J/s$/6.40 \times 10^{14}$ J/kg $= 6.00 \times 10^{11}$ kg/s—600 million tons of hydrogen every second (1 ton = 1000 kg). A mass of 600 million tons sounds like a lot—the mass of a small mountain—but it represents only a few million million millionths of the total mass of the Sun. Put another way, of this 600 million tons, roughly 600 million tons/s $\times$ 0.0071 = 4.3 million tons per second of solar matter is converted into radiation—comparable to the mass carried away by the solar wind. Our parent star will be able to sustain this loss rate for a very long time.

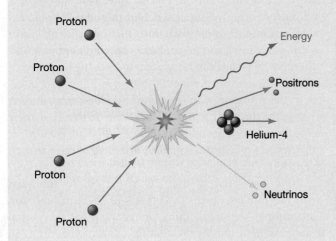

the experiments just described. (In the jargon of the field, the neutrinos are said to "oscillate" into other particles.)

In 1998 the Japanese group operating the Super Kamiokande detector shown in Figure 16.28(a) reported the first experimental evidence of neutrino oscillations (and hence of nonzero neutrino masses), although the observed oscillations did not involve neutrinos of the type produced in the Sun. In 2001, measurements made at the Sudbury Neutrino Observatory (SNO) in Ontario, Canada (Figure 16.28b), revealed strong evidence for the "other" neutrinos into which the Sun's neutrinos have been transformed. Further SNO observations with a modified detector, reported in 2002, confirmed the result. The total numbers of neutrinos observed are completely consistent with the standard solar model. The solar neutrino problem has been solved, the scientific method has proved itself again—and neutrino astronomy can claim its first major triumph!

PROCESS OF SCIENCE CHECK

✔ Using the solar neutrino problem as an example, discuss how scientific theory and observation evolve and adapt when they come into conflict.

CHAPTER REVIEW

SUMMARY

1 Our Sun is a **star (p. 386)**, a glowing ball of gas held together by its own gravity and powered by nuclear fusion at its center. The **photosphere (p. 386)** is the region at the Sun's surface from which virtually all the visible light is emitted. The main interior regions of the Sun are the **core (p. 386)**, where nuclear reactions generate energy; the **radiation zone (p. 386)**, where the energy travels outward in the form of electromagnetic radiation; and the **convection zone (p. 386)**, where the Sun's matter is in constant convective motion.

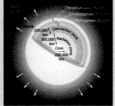

2 The amount of solar energy reaching a 1 m^2 at the top of Earth's atmosphere each second is a quantity known as the **solar constant (p. 386)**. The Sun's **luminosity (p. 387)** is the total amount of energy radiated from the solar surface per second. It is determined by multiplying the solar constant by the area of an imaginary sphere of radius 1 AU.

3 Much of our knowledge of the solar interior comes from mathematical models. The model that best fits the observed properties of the Sun is the **standard solar model (p. 388)**. **Helioseismology (p. 389)**—the study of vibrations of the solar surface caused by pressure waves in the interior—provides further insight into the Sun's structure. The effect of the solar convection zone can be seen on the surface in the form of **granulation (p. 392)** of the photosphere. Lower levels in the convection zone also leave their mark on the photosphere in the form of larger transient patterns called **supergranulation (p. 392)**.

4 Above the photosphere lies the **chromosphere (p. 386)**, the Sun's lower atmosphere. Most of the absorption lines seen in the solar spectrum are produced in the upper photosphere and the chromosphere. In the **transition zone (p. 386)** above the chromosphere, the temperature increases from a few thousand to around

a million kelvins. Above the transition zone is the Sun's thin, hot upper atmosphere, the solar **corona (p. 386)**. At a distance of about 15 solar radii, the gas in the corona is hot enough to escape the Sun's gravity, and the corona begins to flow outward as the **solar wind (p. 386)**.

5 **Sunspots (p. 396)** are Earth-sized regions on the solar surface that are a little cooler than the surrounding photosphere. They are regions of intense magnetism. Both the numbers and locations of sun spots vary in a roughly 11-year **sunspot cycle (p. 400)** as the Sun's magnetic field rises and falls. The overall direction of the field reverses from one sunspot cycle to the next. The 22-year cycle that results when the direction of the field is taken into account is called the **solar cycle (p. 400)**.

6 Solar activity tends to be concentrated in **active regions (p. 401)** associated with groups of sunspots. **Prominences (p. 401)** are looplike or sheetlike structures produced when hot gas ejected by activity on the solar surface interacts with the Sun's magnetic field. The more intense **flares (p. 401)** are violent surface explosions that blast particles and radiation into interplanetary space. **Coronal mass ejections (p. 402)** are huge blobs of magnetized gas escaping into interplanetary space. Most of the solar wind flows outward from low-density regions of the corona called **coronal holes (p. 404)**.

7 The Sun generates energy by converting hydrogen to helium in its core by the process of **nuclear fusion (p. 407)**. Nuclei are held together by the **strong nuclear force (p. 407)**. When four protons are converted to a helium nucleus in the **proton–proton chain (p. 408)**, some mass is lost. The **law of conservation of mass and energy (p. 407)** requires that this mass appear as energy, eventually resulting in the light we see. Very high temperatures are needed for fusion to occur.

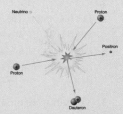

8 **Neutrinos (p. 408)** are nearly massless particles that are produced in the proton–proton chain and escape from the Sun. They interact via the **weak nuclear force (p. 408)**. Despite their elusiveness, it is possible to detect a small fraction of the neutrinos streaming from the Sun. Observations over several decades led to the **solar neutrino problem (p. 411)**—substantially fewer neutrinos are observed than are predicted by theory. The accepted explanation, supported by recent observational evidence, is that **neutrino oscillations (p. 411)** convert some neutrinos to other (undetected) particles en route from the Sun to Earth.

Mastering ASTRONOMY® *For instructor-assigned homework go to* **www.masteringastronomy.com**

Problems labeled **POS** explore the process of science | **VIS** problems focus on reading and interpreting visual information

REVIEW AND DISCUSSION

1. Name and briefly describe the main regions of the Sun.

2. How massive is the Sun, compared with Earth?

3. How hot is the solar surface? The solar core?

4. What is luminosity, and how is it measured in the case of the Sun?

5. What fuels the Sun's enormous energy output?

6. What are the ingredients and the end result of the proton–proton chain in the Sun?

7. Why is energy released in the proton–proton chain?

8. **POS**   How do scientists construct models of the Sun?

9. What is helioseismology, and what does it tell us about the Sun?

10. **POS**   How do observations of the Sun's surface tell us about conditions in the solar interior?

11. Describe how energy generated in the solar core eventually reaches Earth.

12. Why does the Sun appear to have a sharp edge?

13. **POS**   Give the history of "coronium," and tell how it increased our understanding of the Sun.

14. What is the solar wind?

15. Why do we say that the solar cycle is 22 years long?

16. What is the cause of sunspots, flares, and prominences?

17. Describe how coronal mass ejections may influence life on Earth.

18. Why are scientists so interested in solar neutrinos?

19. **POS**   What is the most likely solution to the solar neutrino problem?

20. What would we observe on Earth if the Sun's internal energy source suddenly shut off? How long do you think it might take—minutes, days, years, or millions of years—for the Sun's light to begin to fade? Repeat the question for solar neutrinos.

CONCEPTUAL SELF-TEST: MULTIPLE CHOICE

1. Compared with Earth's diameter, the Sun's diameter is about (**a**) the same; (**b**) ten times larger; (**c**) one hundred times larger; (**d**) one million times larger.

2. Overall, the Sun's average density is roughly the same as that of (**a**) rain clouds; (**b**) water; (**c**) silicate rocks; (**d**) iron–nickel meteorites.

3. The Sun spins on its axis roughly once each (**a**) hour; (**b**) day; (**c**) month; (**d**) year.

4. If astronomers lived on Venus instead of on Earth, the solar constant they measure would be (**a**) larger; (**b**) smaller; (**c**) the same.

5. The primary source of the Sun's energy is (**a**) fusion of light nuclei to make heavier ones; (**b**) fission of heavy nuclei into lighter ones; (**c**) the slow release of heat left over from the Sun's formation; (**d**) the solar magnetic field.

6. **VIS**   According to the standard model of the Sun (Figure 16.6), as the distance from the center increases, the density decreases

(**a**) at about the same rate as the temperature decreases; (**b**) faster than the temperature decreases; (**c**) more slowly than the temperature decreases; (**d**) but the temperature increases.

7. A typical solar granule is about the size of (**a**) a U.S. city; (**b**) a large U.S. state; (**c**) the Moon; (**d**) Earth.

8. As we move to greater and greater distances above the solar photosphere, the temperature in the Sun's atmosphere (**a**) steadily increases; (**b**) steadily decreases; (**c**) first decreases and then increases; (**d**) stays the same.

9. The time between successive sunspot maxima is about (**a**) a month; (**b**) a year; (**c**) a decade; (**d**) a century.

10. The solar neutrino problem is that (**a**) we detect more solar neutrinos than we expect; (**b**) we detect fewer solar neutrinos than we expect; (**c**) we detect the wrong type of neutrinos; (**d**) we can't detect solar neutrinos.

PROBLEMS

The number of dots preceding each Problem indicates its approximate level of difficulty.

1. • Use the reasoning presented in Section 16.1 to calculate the value of the "solar constant" (a) on Mercury at perihelion, (b) on Jupiter.

2. • Use Wien's law to determine the wavelength corresponding to the peak of the blackbody curve (a) in the core of the Sun, where the temperature is 10^7 K, (b) in the solar convection zone (10^5 K), and (c) just below the solar photo sphere (10^4 K). ∞ (Sec. 3.4). What form (visible, infrared, X ray, etc.) does the radiation take in each case?

3. •• The largest-amplitude solar pressure waves have periods of about 5 minutes and move at the speed of sound in the outer layers of the Sun, roughly 10 km/s. (a) How far does such a wave move during one wave period? (b) Approximately how many wavelengths are needed to completely encircle the Sun's equator? (c) Compare the wave period with the orbital period of an object moving just above the solar photosphere. ∞ *(More Precisely 2-2)*

4. • If convected solar material moves at 1 km/s, how long does it take to flow across the 1000-km expanse of a typical granule? Compare your answer with the roughly 10-minute lifetimes observed for most solar granules.

5. •• Use Stefan's law (flux $\propto T^4$, where T is the temperature in kelvins) to calculate how much less energy (as a fraction) is emitted per unit area of a 4500-K sunspot than from the surrounding 5800-K photosphere. ∞ (Sec. 3.4)

6. • The solar wind carries mass away from the Sun at a rate of about 2 million tons/s (1 ton = 1000 kg). At this rate, how long would it take for all of the Sun's mass to escape?

7. • Use the data presented in this chapter and in *More Precisely 8-1* to estimate the radius at which the speed of protons in the corona first exceeds the solar escape speed.

8. •• (a) Assuming constant luminosity, calculate how much equivalent mass (relative to the current mass of the Sun) the Sun has radiated into space in the 4.6 billion years since it formed. How much hydrogen has been consumed? (b) How long would it take the Sun to radiate its entire mass into space?

9. •• How long does it take for the Sun to convert one Earth mass of hydrogen into helium?

10. ••• The entire reaction sequence shown in Figure 16.27 generates 4.3×10^{-12} J of electromagnetic energy and releases two neutrinos. Assuming that neutrino oscillations transform half of these neutrinos into other particles by the time they travel 1 AU from the Sun, estimate the total number of solar neutrinos passing through Earth each second.

17

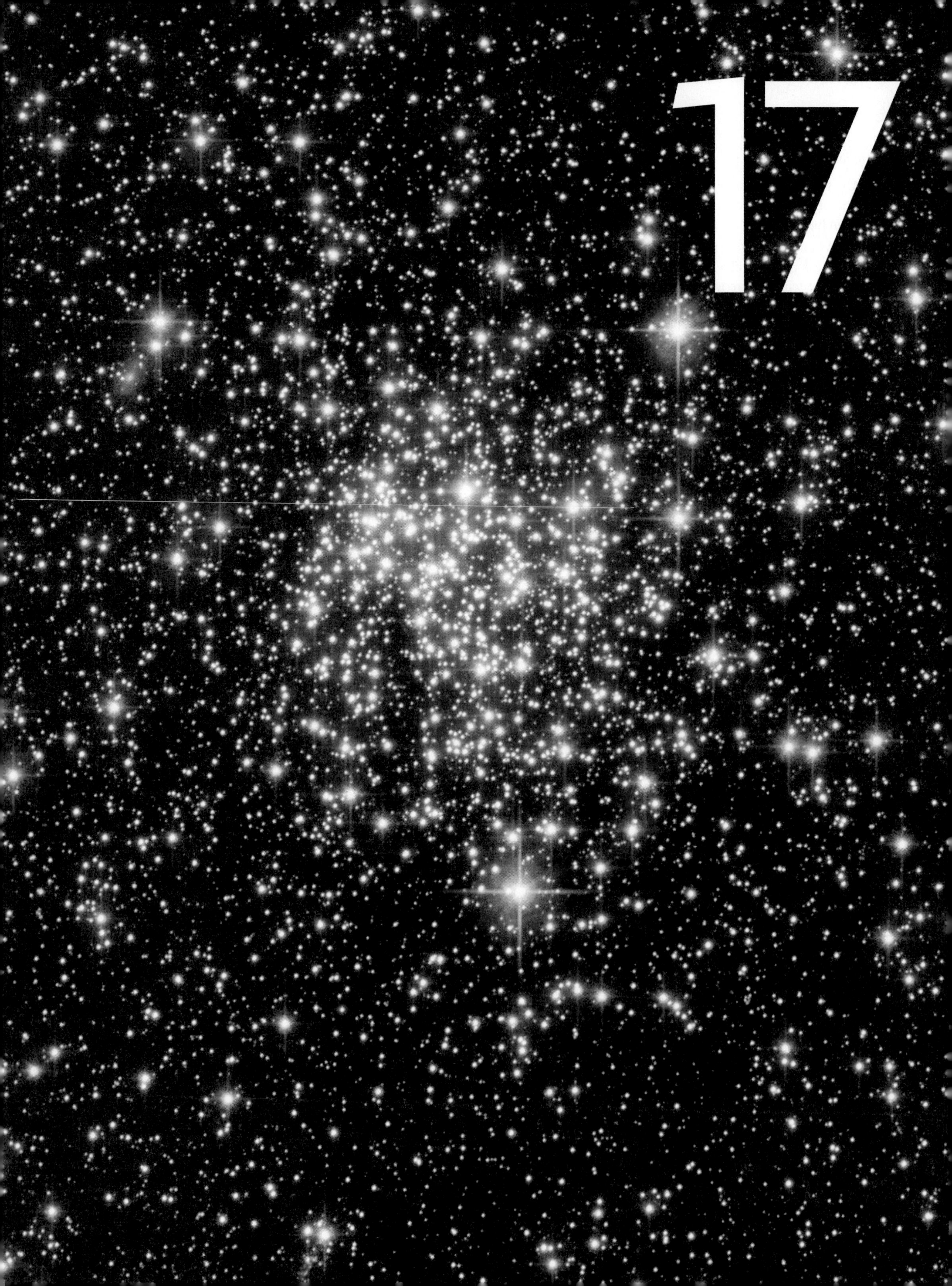

THE STARS

GIANTS, DWARFS, AND THE MAIN SEQUENCE

LEARNING GOALS

Studying this chapter will enable you to

1. Explain how stellar distances are determined.

2. Discuss the motions of the stars through space and how those motions are measured from Earth.

3. Distinguish between luminosity and apparent brightness, and explain how stellar luminosity is determined.

4. Explain the usefulness of classifying stars according to their colors, surface temperatures, and spectral characteristics.

5. Explain how physical laws are used to estimate stellar sizes.

6. Describe how a Hertzsprung–Russell diagram is used to identify stellar properties.

7. Explain how knowledge of a star's spectroscopic properties can lead to an estimate of its distance.

8. Explain how the masses of stars are measured and how mass is related to other stellar properties.

THE BIG PICTURE The total number of stars is virtually beyond our ability to count. Relatively few have been studied in detail. Yet by studying the light stars emit, astronomers have learned an enormous amount about their properties—their masses, luminosities, ages, and even their destinies.

Mastering ASTRONOMY®

Visit the Study Area in www.masteringastronomy.com for quizzes, animations, videos, interactive figures, and self-guided tutorials.

We have now studied Earth, the Moon, the solar system, and the Sun. To continue our inventory of the contents of the universe, we must move away from our local environment into the depths of space. In this chapter, we take a great leap in distance and consider stars in general. Our primary goal is to comprehend the nature of the stars that make up the constellations, as well as the myriad more distant stars we cannot perceive with our unaided eyes. Rather than studying their individual peculiarities, however, we will concentrate on determining the physical and chemical properties they share.

There is order in the legions of stars scattered across the sky. By cataloging and comparing their basic properties—luminosities, temperatures, composition, masses, and radii—astronomers gain new insights into how stars form and evolve. Like comparative planetology in the solar system, the study of the stars plays a vital role in furthering our understanding of the Galaxy and the universe we inhabit.

LEFT: The *Hubble Space Telescope recently imaged the magnificent star cluster NGC 265, a rich group of more than a thousand stars spanning some 65 light-years of space. Known as an open cluster, it is full of youthful stars shown here shining brightly (mostly) in blue light. This star field is part of the Small Magellanic Cloud, located nearly 200,000 light-years away in the southern sky. (ESA/NASA)*

17.1 The Solar Neighborhood

The Galaxy in which we live—the Milky Way—is an enormous collection of stars and interstellar matter held together by gravity. It contains more than 100 billion stars, spread throughout a volume of space nearly 100,000 light-years across, all orbiting the *galactic center* some 25,000 light-years from Earth (see Chapter 23). The Sun—and, in fact, every other star you see while gazing at the night sky—is part of this vast system. In this chapter we describe some of the distance-measurement techniques by which astronomers have extended their studies to larger and larger volumes of space, mapping out the distribution of stars on these vast scales.

As with the planets, knowing the distances to the stars is essential to determining many of their other properties. Looking deep into space within our Galaxy, astronomers can observe and study literally millions of individual stars. By observing other distant galaxies, we can statistically infer the properties of trillions more. All told, the observable universe probably contains several tens of sextillions (1 sextillion $= 10^{21}$) stars. Yet, remarkably, despite their incredible numbers, the essential properties of stars—their appearance in the sky, their births, lives, deaths, and even their interactions with their environment—can be understood in terms of just a few basic physical stellar quantities: luminosity (brightness), temperature (color), chemical composition, size, and mass. ∞ (Sec. 16.1)

The twin goals of measuring stellar distances and categorizing stellar properties have advanced hand in hand. As we will see, as more stellar distances become known, new insights into stellar properties are obtained, and these in turn present new techniques for distance measurement, applicable to even greater distances. In many ways, the story of how these techniques have evolved in tandem is the history of modern astronomy.

Stellar Parallax

Recall from Chapter 1 how surveyors and astronomers use *parallax* to measure distances to terrestrial and solar system objects. Parallax is the apparent shift of a foreground object relative to some distant background as the observer's point of view changes. ∞ (Sec. 1.6) To determine an object's parallax, we observe it from either end of some baseline and measure the angle through which the line of sight to the object shifts. In astronomical contexts, the angle is usually obtained by comparing photographs made from the two ends of the baseline.

As the distance to the object increases, the parallax becomes smaller and therefore harder to measure. Even the closest stars are so far away that no baseline on Earth is sufficient to allow an accurate determination of their distances. Their parallactic shifts, as seen from different points on Earth, are just too small. However, by observing a star *at different times of the year,* as shown in Figure 17.1 (see also Figure 1.30) and then comparing our observations, we can

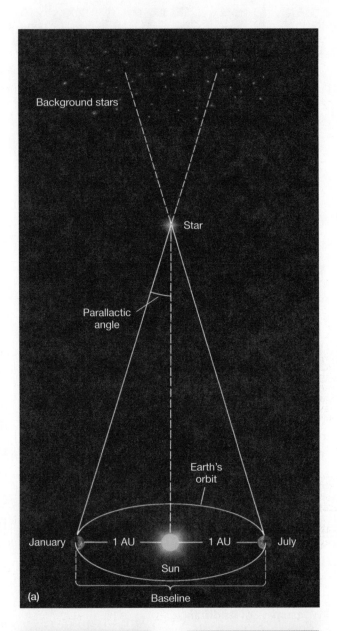

(b) As seen in January

As seen in July

Interactive FIGURE 17.1 Stellar Parallax (a) For observations of stars made 6 months apart, the baseline is twice the Earth–Sun distance, or 2 AU. Compare with Figure 1.30, which shows the same geometry, but on a much smaller scale. (b) The parallactic shift (exaggerated here, and indicated by the white arrow in this drawing) is usually measured photographically, as illustrated by the red star seen here. Images of the same region of the sky made at different times of the year are used to determine a star's apparent movement relative to background stars. This basic measurement is key to understanding the properties of stars throughout the universe.

extend our baseline to the diameter of Earth's orbit around the Sun, namely, 2 AU. Only with this enormously longer baseline do some stellar parallaxes become measurable. As indicated in the figure, a star's parallactic angle—or, more commonly, just its "parallax"—is conventionally defined to be *half* its apparent shift relative to the back ground as we move from one side of Earth's orbit to the other.

Because stellar parallaxes are so small, astronomers generally find it convenient to measure parallax in arc seconds rather than in degrees. If we ask at what distance a star must lie in order for its observed parallax to be exactly 1″, we get an answer of 206,265 AU, or 3.1×10^{16} m ∞ *(More Precisely 1-2)* Astronomers call this distance 1 **parsec** (1 pc), from "*par*allax in arc *sec*onds." Because parallax decreases as distance increases, we can relate a star's parallax to its distance by the following simple formula:

$$\text{distance (in parsecs)} = \frac{1}{\text{parallax (in arc seconds)}}.$$

Thus, a star with a measured parallax of 1″ lies at a distance of 1 pc from the Sun. The parsec is defined so as to make the conversion between distance and parallactic angle easy: An object with a parallax of 0.5″ lies at a distance of 1/0.5 = 2 pc, an object with a parallax of 0.1″ lies at 1/0.1 = 10 pc, and so on. One parsec is approximately equal to 3.3 light-years.

Our Nearest Neighbors

The closest star to Earth (besides the Sun) is called Proxima Centauri. This star is a member of a triple-star system (three separate stars orbiting one another, bound together by gravity) known as the Alpha Centauri complex. Proxima Centauri displays the largest known stellar parallax, 0.77″, which means that it is about 1/0.77 = 1.3 pc away—approximately 270,000 AU, or 4.3 light-years. That's the distance of the *nearest* star to Earth—almost 300,000 times the distance from Earth to the Sun! This is a fairly typical interstellar distance in the Milky Way Galaxy.

Vast distances can sometimes be grasped by means of analogies. Imagine Earth as a grain of sand orbiting a marble-sized Sun at a distance of 1 m. The nearest star, also a marble-sized object, is then more than 270 *kilometers* away. Except for the other planets in our solar system, themselves ranging in size from grains of

sand to millimeter-sized pellets and all lying within 50 m of the "Sun," nothing else of consequence exists in the 270 km separating the two stars. Such is the void of interstellar space.

The next nearest neighbor to the Sun beyond the Alpha Centauri system is called Barnard's star. Its parallax is 0.55″, so it lies at a distance of 1.8 pc, or 6.0 light-years—370 km in our model—from Earth. Figure 17.2 is a map of our nearest galactic neighbors—the 30 or so stars lying within 4 pc of Earth.

Ground-based images of stars are generally smeared out into a disk of radius 1″ or so by turbulence in Earth's atmosphere. ∞ (Sec. 5.4) However, astronomers have special equipment that can routinely measure stellar parallaxes of 0.03″ or less, corresponding to stars within about 30 pc (100 light-years) of Earth. Several thousand stars lie within this range, most of them of much lower luminosity than the Sun and invisible to the naked eye. High-resolution adaptive optics systems allow even more accurate measurements of stellar positions, extending the parallax range to over 100 pc in a few cases, although such measurements are not yet "routine." ∞ (Sec. 5.4)

Even greater precision can be achieved by placing instruments in space, above Earth's atmosphere. In the 1990s, data from the European *Hipparcos* satellite extended the

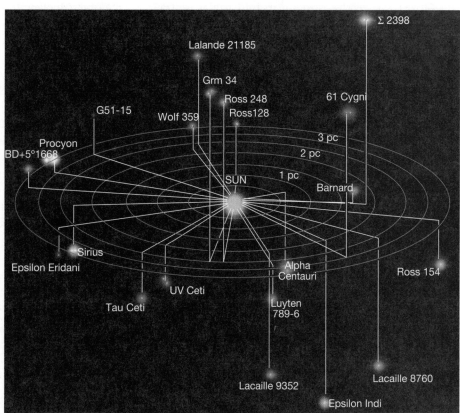

▲ **FIGURE 17.2 The Solar Neighborhood** A plot of the 30 closest stars to the Sun, projected so as to reveal their three-dimensional relationships. The circular gridlines represent distances from the Sun in the galactic plane; the vertical lines denote distances perpendicular to that plane. Notice that many stars are members of multiple-star systems. All lie within 4 pc (about 13 light-years) of Earth.

range of accurately measured parallaxes to well over 200 pc, encompassing nearly a million stars. Even so, the vast majority of stars in our Galaxy are far more distant. Following *Hipparcos*, the European Space Agency (ESA) has ambitious plans to greatly expand the scope of its stellar measurements. ESA's proposed *GAIA* project will have the astonishing range of 25,000 pc—spanning our entire Galaxy and encompassing roughly 1 billion stars! *GAIA* is currently scheduled for launch in 2012. In addition to mapping out the structure of the Milky Way Galaxy to unprecedented precision, this mission will allow astronomers to study in detail the properties of nearby stars of all masses and will also greatly expand our knowledge of extrasolar planetary systems. ∞ (Sec. 15.6) With these new data, in a time span of just three decades the fundamental stellar database upon which almost all of astronomy depends will have increased in size by a factor of a million. The results may be nothing short of revolutionary.

Stellar Motion

In addition to the apparent motion caused by parallax, stars have real spatial motion through the Galaxy. In Chapter 23, we will see how astronomers have measured the Sun's actual motion around the galactic center. However, *relative* to the Sun—that is, as seen by astronomers on Earth as we travel through space along with our parent star—stellar motion has two components. A star's *radial* velocity—along the line of sight—can be measured using the Doppler effect. ∞ (Sec. 3.5) For many nearby stars, their *transverse* velocity—perpendicular to our line of sight—can also be determined by careful monitoring of the star's position in the sky.

Figure 17.3 compares two photographs of the sky around Barnard's star. The photographs were made on the same day of the year, but 22 years apart. Note that the star, marked by the arrow, has moved during the 22-year interval shown: If the two images were superimposed, the images of the other stars in the field of view would coincide, but those of Barnard's star would not. Because Earth was at the same point in its orbit when these photographs were taken, the observed displacement is *not* the result of parallax caused by Earth's motion around the Sun. Instead, it indicates *real* spatial motion of Barnard's star relative to the Sun.

The annual movement of a star across the sky, as seen from Earth and corrected for parallax, is called **proper motion.** It describes the transverse component of a star's velocity relative to the Sun. (Both the star and the Sun are moving through space as they travel through the Galaxy; however, only their *relative* motion causes the star's position in the sky to change, as seen from Earth.) Like parallax, proper motion is measured in terms of angular displacement. Since the angles involved are typically very small, proper motion is usually expressed in arc seconds per year. Barnard's star moved 228″ in 22 years, so its proper motion is 228″/22 years, or 10.4″/yr.

A star's transverse velocity is easily calculated once its proper motion and its distance are known. At the distance of Barnard's star (1.8 pc), an angle of 10.4″ corresponds to a physical displacement of 0.000091 pc, or about 2.8 billion km. Barnard's star takes a year (3.2×10^7 s) to travel this distance, so its transverse velocity is 2.8 billion km/3.2×10^7 s, or 89 km/s. ∞ *(More Precisely 1-2)* Even though stars' transverse velocities are often quite large—tens or even hundreds of kilometers per second—their great distances from the Sun mean that their proper motion is small, and it usually takes many years for us to discern their movement across the sky. In all probability, every star in Figure 17.3 has some transverse motion relative to the Sun. However, only Barnard's star has proper motion large enough to be visible in these frames. In fact, Barnard's star has the largest known proper motion of any star. Only a few hundred stars have proper motions greater than 1″/yr.

Now consider the three-dimensional motion of our nearest neighbor, the Alpha Centauri system, sketched in Figure 17.4 in relation to our own solar system. Alpha Centauri's proper motion has been measured to be 3.7″/yr. At Alpha Centauri's distance of 1.35 pc, that measurement implies a transverse velocity of 24 km/s. We can determine the other component of motion—the radial velocity—by means of the Doppler effect. Spectral lines from Alpha Centauri are blueshifted by a tiny amount—about 0.0067 percent—allowing astronomers to measure the star system's radial velocity (relative to the Sun) as 300,000 km/s $\times$ 6.7 $\times$ 10^{-5} = 20 km/s toward us. ∞ (Sec. 3.5)

What is the true spatial motion of Alpha Centauri? Will this alien system collide with our own sometime in the future? The answer is no: Alpha Centauri's transverse velocity will steer it well clear of the Sun. We can combine the trans-

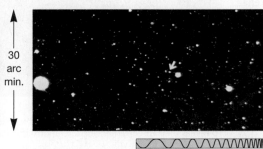

30 arc min.

R I V U X G

◀ **FIGURE 17.3 Proper Motion** A comparison of two photographic plates taken 22 years apart shows evidence of real spatial motion for Barnard's star (denoted by an arrow). *(Harvard College Observatory)*

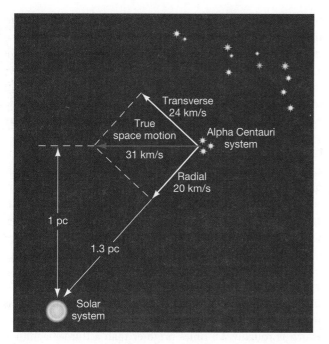

▲ **FIGURE 17.4 Real Spatial Motion** The motion of the Alpha Centauri star system drawn relative to our solar system. The transverse component of the velocity has been determined by observing the system's proper motion. The radial component is measured by using the Doppler shift of lines in Alpha Centauri's spectrum. The true spatial velocity, indicated by the red arrow, results from the combination of the two.

verse (24 km/s) and radial (20 km/s) velocities according to the Pythagorean theorem, as indicated in Figure 17.4. The total velocity is $\sqrt{24^2 + 20^2}$, or about 31 km/s, in the direction shown by the horizontal red arrow. As the figure indicates, Alpha Centauri will get no closer to us than about 1 pc, and that won't happen until 280 centuries from now.

CONCEPT CHECK

✔ Why can't astronomers use simultaneous observations from different parts of Earth's surface to determine stellar distances?

✔ Why are the spatial velocities of distant stars generally poorly known?

17.2 Luminosity and Apparent Brightness

Luminosity is an *intrinsic* property of a star—it does not depend in any way on the location or motion of the observer. Luminosity is sometimes referred to as the star's *absolute brightness*. However, when we look at a star, we see, not its luminosity, but rather its **apparent brightness**—the amount of energy striking a unit area of some light-sensitive surface or device (such as a charge-coupled device [CCD] chip or a human eye) per unit time. Apparent brightness is a measure, not of a star's luminosity, but of the *energy flux* (energy per unit area per unit time) produced by the star, as seen from Earth. A star's apparent brightness depends on our *distance* from the star. In this section, we discuss in more detail how these important quantities are related to one another.

Another Inverse-Square Law

Figure 17.5 shows light leaving a star and traveling through space. Moving outward, the radiation passes through imaginary spheres of increasing radius surrounding the source. The amount of radiation leaving the star per unit time—the star's luminosity—is constant, so the farther the light travels from the source, the less energy passes through each unit of area. Think of the energy as being spread out over an ever-larger area and therefore spread more thinly, or "diluted," as it expands into space.

Because the area of a sphere grows as the square of the radius, the energy per unit area—the star's apparent brightness, as seen by our eye or our telescope—is inversely proportional to the square of the distance from the star. Doubling the distance from a star makes it appear 2^2, or 4, times dimmer. Tripling the distance reduces the apparent brightness by a factor of 3^2, or 9, and so on.

Of course, the star's luminosity also affects its apparent brightness. Doubling the luminosity doubles the energy crossing any spherical shell surrounding the star and hence doubles the apparent brightness. We can therefore say that the apparent brightness of a star is directly proportional to the star's luminosity and inversely proportional to the square of its distance:

$$\text{apparent brightness (energy flux)} \propto \frac{\text{luminosity}}{\text{distance}^2}.$$

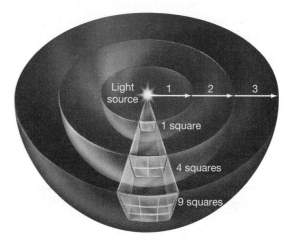

▲ **FIGURE 17.5 Inverse-Square Law** As light moves away from a source such as a star, it steadily dilutes while spreading over progressively larger surface areas (depicted here as sections of spherical shells). Thus, the amount of radiation received by a detector (the source's apparent brightness) varies inversely as the square of its distance from the source. Once the distance to a star is known, its luminosity—a fundamental stellar property—can then be determined.

ANIMATION/VIDEO The Inverse-Square Law

▲ **FIGURE 17.6 Luminosity** Two stars A and B of different luminosities can appear equally bright to an observer on Earth if the brighter star B is more distant than the fainter star A.

Thus, two identical stars can have the same apparent brightness if (and only if) they lie at the same distance from Earth.

However, as illustrated in Figure 17.6, two nonidentical stars can also have the same apparent brightness if the more luminous one lies farther away. A bright star (i.e., a star with large apparent brightness) is a powerful emitter of radiation (high luminosity), is near Earth, or both. Without additional information, we cannot distinguish between the effects of increasing luminosity and decreasing distance. Similarly, a faint star (a star with small apparent brightness) is a weak emitter (low luminosity), is far from Earth, or both.

Determining a star's luminosity is a twofold task. First, the astronomer must determine the star's apparent brightness by measuring the amount of energy detected through a telescope in a given amount of time. Second, the star's distance must be measured—parallax for nearby stars and by other means (to be discussed later) for more distant stars. The luminosity can then be found from the inverse-square law. This is basically the same reasoning we used earlier in Chapter 16, in our discussion of how astronomers measure solar luminosity. (In our new terminology, the solar constant is just the apparent brightness of the Sun.) ∞ (Sec. 16.1)

The Magnitude Scale

Instead of measuring apparent brightness in SI units (e.g., watts per square meter, the unit used for the solar constant in Chapter 16), astronomers often find it more convenient to work in terms of a construct called the **magnitude scale.** ∞ (Sec. 16.1) The scale dates from the second century B.C., when the Greek astronomer Hipparchus classified the naked-eye stars into six groups. The brightest stars were categorized as

first magnitude. The next brightest stars were labeled second magnitude, and so on, down to the faintest stars visible to the naked eye, which were classified as sixth magnitude. The range 1 (brightest) through 6 (faintest) spanned all the stars known to the ancients. Notice that magnitudes are really *rankings* in terms of apparent brightness (energy flux)—a *large* magnitude means a *faint* star. Just as "first rate" means "good" in everyday speech, "first magnitude" in astronomy means "bright."

When astronomers began using telescopes with sophisticated detectors to measure the light received from stars, they quickly discovered two important facts about the magnitude scale. First, the 1–6 magnitude range defined by Hipparchus spans about a factor of 100 in apparent brightness—a first-magnitude star is approximately 100 times brighter than a sixth-magnitude star. Second, the physiological characteristics of the human eye are such that each change in magnitude of 1 corresponds to a factor of about 2.5 in apparent brightness. In other words, to the human eye, a first-magnitude star is roughly 2.5 times brighter than a second-magnitude star, which is roughly 2.5 times brighter than a third-magnitude star, and so on. (By combining factors of 2.5, you can confirm that a first-magnitude star is indeed $(2.5)^5 \approx 100$ times brighter than a sixth-magnitude star.)

Modern astronomers have modified and extended the magnitude scale in a number of ways. First, we now *define* a change of 5 in the magnitude of an object (going from magnitude 1 to magnitude 6, say, or from magnitude 7 to magnitude 2) to correspond to *exactly* a factor of 100 in apparent brightness. Second, because we are really talking about apparent (rather than absolute) brightnesses, the numbers in Hipparchus's ranking system are called **apparent magnitudes.** Third, the scale is no longer limited to whole numbers: A star of apparent magnitude 4.5 is intermediate in apparent brightness between a star of apparent magnitude 4 and one of apparent magnitude 5. Finally, magnitudes outside the range from 1 to 6 are allowed: Very bright objects can have apparent magnitudes much less than 1, and very faint objects can have apparent magnitudes far greater than 6.

Figure 17.7 illustrates the apparent magnitudes of some astronomical objects, ranging from the Sun, at −26.7, to the faintest object detectable by the *Hubble* or *Keck* telescopes, an object having an apparent magnitude of 30—about as faint as a firefly seen from a distance equal to Earth's diameter. Note that this range in magnitudes corresponds to a very large factor (actually, of $10^{56.7/2.5} = 10^{22.7} \approx 5 \times 10^{22}$) in apparent brightness. Indeed, one of the main reasons that astronomers use this (admittedly rather intimidating) scale is that it allows them to compress a large spread in observed stellar properties into more "manageable" form.*

Apparent magnitude measures a star's apparent brightness when the star is seen at its actual distance from the Sun.

---

*Putting in the numbers, we can calculate that magnitude 1 corresponds to a flux of 1.1×10^{-8} W/m^2, magnitude 20 to 2.9×10^{-16} W/m^2, and so on, but astronomers find the "magnitude" versions more intuitive and much easier to remember.

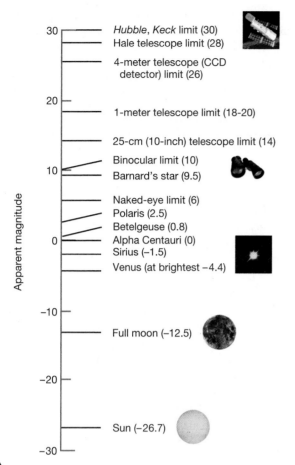

Interactive FIGURE 17.7 Apparent Magnitude This graph illustrates the apparent magnitudes of some astronomical objects and the limiting magnitudes (that is, the faintest magnitudes attainable) of some telescopes used to observe them. The original magnitude scale was defined so that the brightest stars in the night sky have magnitude 1, whereas the faintest stars visible to the naked eye have magnitude 6. The scale has since been extended to cover much brighter and much fainter objects. An increase of 1 in apparent magnitude corresponds to a decrease in apparent brightness by a factor of approximately 2.5.

To compare intrinsic, or absolute, properties of stars, however, astronomers imagine looking at all stars from a standard distance of 10 pc. There is no particular reason to use 10 pc—it is simply convenient. A star's **absolute magnitude** is its apparent magnitude when it is placed at a distance of 10 pc from the observer. Because the distance to the star is fixed in this definition, *absolute magnitude is a measure of a star's absolute brightness, or luminosity.*

We can use the earlier discussion of the inverse-square law to relate absolute and apparent magnitudes if the distance to the star is known. When a star farther than 10 pc away from us is moved to a point 10 pc away, its apparent brightness increases and hence its apparent magnitude decreases. Stars more than 10 pc from Earth therefore have apparent magnitudes that are greater than their absolute magnitudes. For example, if a star at a distance of 100 pc were moved to the standard 10-pc distance, its distance would decrease by a fac-

tor of 10, so (by the inverse-square law) its apparent brightness would *increase* by a factor of $10^2 = 100$. Its apparent magnitude (by definition) would therefore decrease by 5. In other words, at 100 pc distance, the star's apparent magnitude exceeds its absolute magnitude by 5.

For stars closer than 10 pc, the reverse is true. An extreme example is our Sun. Because of its proximity to Earth, it appears very bright and thus has a large negative apparent magnitude (Figure 17.7). However, the Sun's absolute magnitude is 4.83. If the Sun were moved to a distance of 10 pc from Earth, it would be only slightly brighter than the faintest stars visible to the naked eye in the night sky.

Knowledge of a star's apparent magnitude and distance allows us to compute its absolute magnitude (luminosity). Conversely, the numerical difference between a star's absolute and apparent magnitudes is a direct measure of the distance to the star. *More Precisely 17-1* presents more detail and some examples for the connection between absolute magnitude and luminosity and on the use of the magnitude scale in computing stellar luminosities and distances.

CONCEPT CHECK

✔ Two stars are observed to have the same apparent magnitude. On the basis of this information, what, if anything, can be said about their luminosities?

17.3 Stellar Temperatures

Looking at the night sky, you can tell at a glance which stars are hot and which are cool. In Figure 17.8, which shows the constellation Orion as it appears through a small telescope, the colors of the cool red star Betelgeuse (α) and the hot blue star Rigel (β) are clearly evident. Note that these colors are intrinsic properties of the stars and have *nothing* to do with Doppler redshifts or blueshifts. However, to obtain these stars' temperatures (3200 K for Betelgeuse and 11,000 K for Rigel), more precise observations are required. To make such measurements, astronomers turn to the radiation laws and the detailed properties of stellar spectra. ∞ (Secs. 3.4, 4.3)

Color and the Blackbody Curve

Astronomers can determine a star's surface temperature by measuring the star's apparent brightness (energy flux) at several frequencies and then matching the observations to the appropriate blackbody curve. ∞ (Sec. 3.4) In the case of the Sun, the theoretical curve that best fits the emission describes a 5800-K emitter. ∞ (Sec. 16.1) The same technique works for any other star, regardless of its distance from Earth.

Because the basic shape of the blackbody curve is so well understood, astronomers can estimate a star's temperature using as few as *two* measurements at selected wavelengths (which is fortunate, as detailed spectra of faint stars are often difficult and time consuming to obtain). This is

(a)

(b)

R I V U X G

◀ FIGURE 17.8 **Star Colors** (a) The different colors of the stars comprising the constellation Orion are easily distinguished in this photograph taken by a wide-field camera attached to a small telescope. The bright red star at the upper left is Betelgeuse (α); the bright blue-white star at the lower right is Rigel (β). (Compare with Figure 1.8.) The scale of the photograph is about 20° across. (b) An incredibly rich field of colorful stars, this time in the direction of the center of the Milky Way. Here, the field of view is just 2 arc minutes across—much smaller than in (a). The image is heavily populated with stars of many different temperatures. *(J Sanford/Astrostock-Sanford; NASA)*

covers the near ultraviolet, and infrared filters span longer wavelength parts of the spectrum.

Figure 17.9 shows how the B and V filters admit different amounts of light for objects radiating at different temperatures. In curve (a), corresponding to a very hot 30,000-K emitter, considerably more radiation (about 30 percent more) is received through the B filter than through the V filter, so this object looks brighter in B than in V. In curve (b), the temperature is 10,000 K, and the B and V fluxes are about the same. In the cool 3000-K curve (c), about five times more energy is received in the V range than in the B range, so the B image now is much fainter than the V image. In each case, it is possible to reconstruct the entire blackbody curve on the basis of only those two measurements, because no other blackbody curve can be drawn through both measured points. To the extent that a star's

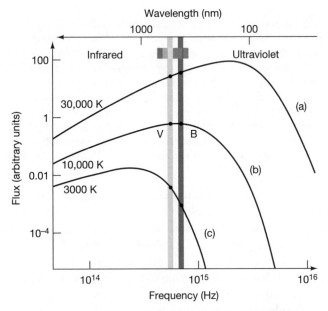

▲ FIGURE 17.9 **Blackbody Curves** The locations of the B (blue) and V (visual) filters, with blackbody curves for three different temperatures. Star (a) is very hot—30,000 K—so its B intensity is greater than its V intensity (as is actually the case for Rigel in Figure 17.8a). Star (b) has roughly equal B and V readings and so appears white. Its temperature is about 10,000 K. Star (c) is red; its V intensity greatly exceeds the B value, and its temperature is 3000 K (much as for Betelgeuse in Figure 17.8a). Thus, by using our knowledge of electromagnetic radiation, we can determine the temperatures of distant objects.

accomplished through the use of telescope filters that block out all radiation except that within specific wavelength ranges. For example, a B (blue) filter rejects all radiation except for a certain range from violet to blue light. Defined by international agreement to extend from 380 to 480 nm, this range corresponds to wavelengths to which photographic film happens to be most sensitive. Similarly, a V (visual) filter passes only radiation within the 490- to 590-nm range (green to yellow), corresponding to that part of the spectrum to which human eyes are particularly sensitive. Many other filters are also in routine use—a U (ultraviolet) filter

MORE PRECISELY 17-1

More on the Magnitude Scale

Let's recast our discussion of two important topics—stellar luminosity and the inverse-square law—in terms of magnitudes.

Absolute magnitude is equivalent to luminosity—an intrinsic property of a star. Given that the Sun's absolute magnitude is 4.83 (see Appendix 3, Table 6), we can construct a conversion chart (shown below) relating these two quantities. Since an increase in brightness by a factor of 100 corresponds to a decrease in magnitude by 5 units, it follows that a star with luminosity 100 times that of the Sun has absolute magnitude $4.83 - 5 = -0.17$, while a 0.01-solar luminosity star has absolute magnitude $4.83 + 5 = 9.83$. We can fill in the gaps by noting that 1 magnitude corresponds to a factor of $100^{1/5} \approx 2.512$; 2 magnitudes to $100^{2/5} \approx 6.310$, and so on. A factor of 10 in brightness corresponds to 2.5 magnitudes. You can use this chart to convert between solar luminosities and absolute magnitudes in many of the figures in this and later chapters.

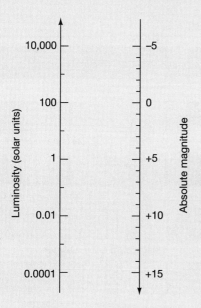

EXAMPLE 1 Let's calculate the luminosity (in solar units) of a star having absolute magnitude M (the conventional symbol for absolute magnitude, not to be confused with mass!). The star's absolute magnitude differs from that of the Sun by $(M - 4.38)$ magnitudes, so, in accordance with the reasoning just presented,

the luminosity L differs from the solar luminosity by a factor of $100^{-(M-4.83)/5}$, or $10^{-(M-4.83)/2.5}$. We can therefore write

$$L \text{ (solar units)} = 10^{-M-4.83)/2.5}.$$

Plugging in some numbers (taken from Appendix 3, Tables 5 and 6), we find that the Sun, with $M = 4.83$, of course has $L = 10^0 = 1$. Sirius A, with $M = 1.45$, has luminosity $10^{1.35} = 22$ solar units, Barnard's star, with $M = 13.24$, has luminosity $10^{-3.35} = 4.3 \times 10^{-4}$ solar units, Betelgeuse has $M = -5.14$ and luminosity 9700 Suns, and so on.

To cast the *inverse-square law* in these terms, recall that increasing the distance to a star by a factor of 10 decreases its apparent brightness by a factor of 100 (by the inverse-square law) and hence increases its apparent magnitude by 5 units. Increasing the distance by a factor of 100 increases the apparent magnitude by 10, and so on. Every increase in distance by a factor of 10 increases the apparent magnitude by 5. Since absolute magnitude is simply apparent magnitude at a distance of 10 pc, we can write

$$\text{apparent magnitude} - \text{absolute magnitude} = 5 \log_{10}\left(\frac{\text{distance}}{10 \text{ pc}}\right).$$

(The *logarithm* function—the LOG key on your calculator—is defined by the property that if $a = \log_{10}(b)$, then $b = 10^a$.) Even though it doesn't look much like it, this equation is precisely equivalent to the inverse-square law presented in the text! Note that for stars more than 10 pc from Earth, the apparent magnitude is greater than the absolute magnitude, while the reverse is true for stars closer than 10 pc.

EXAMPLE 2 The Sun, with an absolute magnitude of 4.83, seen from a distance of 100 pc, would have an apparent magnitude of $4.83 + 5 \log_{10}(100) = 14.83$, since $\log_{10}(100) = 2$. This is well below the threshold of visibility for binoculars or even a large amateur telescope (see Figure 17.7). We can also turn this around to illustrate how knowledge of a star's absolute and apparent magnitudes tells us its distance. The star Rigel Kentaurus (also known as Alpha Centauri) has absolute magnitude $+4.34$ and is observed to have apparent magnitude -0.01. The magnitude difference is -4.35, so its distance must therefore be $10 \text{ pc} \times 10^{-4.34/5} = 1.35$ pc, in agreement with the result (obtained by parallax) presented in the text.

spectrum is well approximated as a blackbody, measurements of the B and V fluxes are enough to specify the star's blackbody curve and thus yield its surface temperature.

Thus, astronomers can estimate a star's temperature simply by measuring and comparing the amount of light received through different colored filters. As discussed in Chapter 5, this type of non-spectral-line analysis using a standard set of filters is known as **photometry**. ∞ (Sec. 5.3) Table 17.1 lists, for several prominent stars, the surface

temperatures derived by photometric means, along with the color that would be perceived in the absence of filters.

Stellar Spectra

Color is a useful way to describe a star, but astronomers often use a more detailed scheme to classify stellar properties, incorporating additional knowledge of stellar physics obtained through spectroscopy. Figure 17.10 compares the

TABLE 17.1 Stellar Colors and Temperatures

Approximate Surface Temperature (K)	Color	Familiar Examples
30,000	Blue-violet	Mintaka (δ Orionis)
20,000	Blue	Rigel
10,000	White	Vega, Sirius
7000	Yellow-white	Canopus
6000	Yellow	Sun, Alpha Centauri
4000	Orange	Arcturus, Aldebaran
3000	Red	Betelgeuse, Barnard's star

spectra of several different stars, arranged in order of decreasing surface temperature (as determined from measurements of their colors). All the spectra extend from 400 to 650 nm, and each shows a series of dark absorption lines superimposed on a background of continuous color, like the spectrum of the Sun. ∞ (Sec. 16.3) However, the precise patterns of lines reveal many differences. Some stars display strong lines in the long-wavelength part of the spectrum (to the left in the figure). Other stars have their strongest lines at short wavelengths (to the right). Still others show strong absorption lines spread across the whole visible spectrum. What do these differences tell us?

Although spectral lines of many elements are present with widely varying strengths, the differences among the spectra in Figure 17.10 are not due to differences in composition. Detailed spectral analysis indicates that the seven stars shown have similar elemental abundances—all are more or less solar in makeup. ∞ (Sec. 16.3) Rather, as discussed in Chapter 4, the differences are due almost entirely to the stars' *temperatures*. ∞ (Sec. 4.5) The spectrum at the top of the figure is exactly what we would expect from a star with solar composition and a surface temperature of about 30,000 K, the second spectrum is what we would anticipate from a 20,000-K star, and so on, down to the 3000-K star at the bottom.

The main differences among the spectra in Figure 17.10 are as follows:

- Spectra of stars having surface temperatures exceeding 25,000 K usually show *strong* absorption lines of singly ionized helium (i.e., helium atoms that have lost one orbiting electron) and multiply ionized heavier elements, such as oxygen, nitrogen, and silicon (the latter lines are not shown in the figure). These strong lines are not seen in the spectra of cooler stars because only very hot stars can excite and ionize such tightly bound atoms.

- In contrast, the hydrogen absorption lines in the spectra of very hot stars are relatively *weak*. The reason is not a lack of hydrogen, which is by far the most abundant

element in all stars. At these high temperatures, however, much of the hydrogen is ionized, so there are few intact hydrogen atoms to produce strong spectral lines.

- Hydrogen lines are strongest in stars having intermediate surface temperatures of around 10,000 K. This temperature is just right for electrons to move frequently between hydrogen's second and higher orbitals, producing the characteristic visible hydrogen spectrum. ∞ (*More Precisely 4-1*) Lines of tightly bound atoms—for example, of helium and nitrogen—which need lots of energy for excitation, are rarely observed in the spectra of these stars, whereas lines from more loosely bound atoms—such as those of calcium and titanium—are relatively common.

- Hydrogen lines are again weak in stars with surface temperatures below about 4000 K, but now because the temperature is too low to boost many electrons out of the ground state. ∞ (Sec. 4.2) The most intense spectral lines in these stars are due to weakly excited heavy atoms;

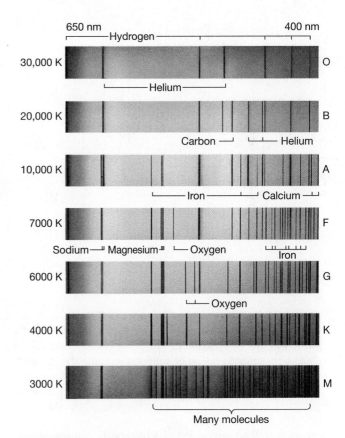

▲ **FIGURE 17.10 Stellar Spectra** Comparison of spectra observed for seven different stars having a range of surface temperatures. These are not actual spectra, which are messy and complex, but simplified artists' renderings illustrating a few spectral features. The spectra of the hottest stars, at the top, show lines of helium and multiply ionized heavy elements. In the coolest stars, at the bottom, helium lines are absent, but lines of neutral atoms and molecules are plentiful. At intermediate temperatures, hydrogen lines are strongest. All seven stars have about the same chemical composition.

no lines from ionized elements are seen. Temperatures in the coolest stars are low enough for molecules to survive, and many of the observed absorption lines are produced by molecules rather than by atoms. ∞ (Sec. 4.4)

Stellar spectra are the source of all the detailed information we have on stellar composition, and they do in fact reveal significant differences in composition among stars, particularly in the abundances of carbon, nitrogen, oxygen, and heavier elements. However, as we have just seen, these differences are *not* the primary reason for the different spectra that are observed. Instead, the main determinant of a star's spectral appearance is its temperature, and stellar spectroscopy is a powerful and precise tool for measuring this important stellar property.

Spectral Classification

Stellar spectra like those shown in Figure 17.10 were obtained for many stars well before the start of the 20th century as observatories around the world amassed spectra from stars in both hemispheres of the sky. Between 1880 and 1920, researchers correctly identified some of the observed spectral lines on the basis of comparisons between those lines and lines obtained in the laboratory. The researchers, though, had no firm understanding of how the lines were produced. Modern atomic theory had not yet been developed, so the correct interpretation of the line strengths, as just described, was impossible at the time.

Lacking a full understanding of how atoms produce spectra, early workers classified stars primarily according to their hydrogen-line intensities. They adopted an alphabetic A, B, C, D, E. . . scheme in which A stars, with the strongest hydrogen

lines, were thought to have more hydrogen than did B stars, and so on. The classification extended as far as the letter P.

In the 1920s, scientists began to understand the intricacies of atomic structure and the causes of spectral lines. Astronomers quickly realized that stars could be more meaningfully classified according to their surface temperature. Instead of adopting an entirely new scheme; however, they chose to shuffle the existing alphabetical categories—those based on the strengths of the hydrogen lines—into a new sequence based on temperature. In the modern scheme, the hottest stars are designated O, because they have very weak absorption lines of hydrogen and were classified toward the end of the original scheme. In order of decreasing temperature, the surviving letters now run O, B, A, F, G, K, M. (The other letter classes have been dropped.) These stellar designations are called **spectral classes** (or *spectral types*). Use the time-honored (and politically incorrect) mnemonic "**Oh, Be A F**ine **G**irl, **K**iss **Me**" to remember them in the correct order.*

Astronomers further subdivide each lettered spectral classification into 10 subdivisions, denoted by the numbers 0–9. By convention, the lower the number, the hotter is the star. For example, our Sun is classified as a G2 star (a little cooler than G1 and a little hotter than G3), Vega is a type A0, Barnard's star is M5, Betelgeuse is M2, and so on. Table 17.2

*Some astronomers have proposed the addition of two additional spectral classes—L and T—for low-mass, low-temperature stars whose odd spectra distinguish them from the M-class stars in the current scheme. For now, at least, the new classification has not been widely adopted. Astronomers are still uncertain whether these new objects are "true" stars, fusing hydrogen into helium in their cores, or whether they are "brown dwarfs" (see Chapter 20) that never achieved high enough central temperature for fusion to begin.

TABLE 17.2 Stellar Spectral Classes

Spectral Class	Approximate Surface Temperature (K)	Noteworthy Absorption Lines	Familiar Examples
O	30,000	Ionized helium strong; multiply ionized heavy elements; hydrogen faint	Mintaka (O9)
B	20,000	Neutral helium moderate; singly ionized heavy elements; hydrogen moderate	Rigel (B8)
A	10,000	Neutral helium very faint; singly ionized heavy elements; hydrogen strong	Vega (A0), Sirius (A1)
F	7000	Singly ionized heavy elements; neutral metals; hydrogen moderate	Canopus (F0)
G	6000	Singly ionized heavy elements; neutral metals; hydrogen relatively faint	Sun (G2), Alpha Centauri (G2)
K	4000	Singly ionized heavy elements; neutral metals strong; hydrogen faint	Arcturus (K2), Aldebaran (K5)
M	3000	Neutral atoms strong; molecules moderate; hydrogen very faint	Betelgeuse (M2), Barnard's star (M5)

lists the main properties of each stellar spectral class for the stars presented in Table 17.1.

We should not underestimate the importance of the early work in classifying stellar spectra. Even though the original classification was based on erroneous assumptions, the painstaking accumulation of large quantities of accurate data paved the way for rapid improvements in understanding once a theory came along that explained the observations.

CONCEPT CHECK

✔ Why does a star's spectral classification depend on its temperature?

17.4 Stellar Sizes

Most stars are unresolved points of light in the sky, even when viewed through the largest telescopes. Even so, astronomers can often make quite accurate determinations of their sizes.

Direct and Indirect Measurements

Some stars are big enough, bright enough, and close enough to allow us to measure their sizes *directly*. One well-known example is the bright red star Betelgeuse, a prominent member of the constellation Orion (Figure 17.8). In a technique known as *speckle interferometry,* many short-exposure images of a star, each too brief for Earth's turbulent atmosphere to smear it out into a disk, are combined to make a high-resolution map of the star's surface. ∞ (Sec. 5.4) In some cases, the results are detailed enough to allow a few surface features to be distinguished (Figure 17.11a). As shown in Figure 17.11(b), Betelgeuse is also (barely) large enough to be resolvable by the *Hubble Space Telescope* at short wavelengths. Steadily improving speckle and adaptive-optics techniques have allowed astronomers to construct very-high-resolution stellar images in a small number of cases. ∞ (Sec. 5.4)

Once a star's angular size has been measured, if its distance is also known, we can determine its radius by simple geometry. ∞ (Sec. 1.6) For example, with a distance of 130 pc and an angular diameter of up to 0.045″, Betelgeuse's maximum radius is 630 times that of the Sun. (We say "maximum radius" here because, as it happens, Betelgeuse is a *variable star*—its radius and luminosity vary somewhat irregularly, with a period of roughly 6 years.) All told, the sizes of perhaps a few dozen stars have been measured in this way.

Most stars are too distant or too small for such direct measurements to be made. Instead, their sizes must be inferred by indirect means, using the radiation laws. ∞ (Sec. 3.4) The radiation emitted by a star is governed by the Stefan–Boltzmann law, which states that the energy emitted *per unit area* per unit of time increases as the fourth power of the star's surface temperature. ∞ *(More Precisely 3-2)* To determine the star's luminosity, we must multiply by its surface area—large bodies radiate more energy than do small bodies at the same temperature. Because the surface area is proportional to the square of the radius, we have

$$\text{luminosity} \propto \text{radius}^2 \times \text{temperature}^4.$$

This **radius–luminosity–temperature relationship** is important because it demonstrates that knowledge of a star's luminosity and temperature can yield an estimate of the star's radius—an *indirect* determination of stellar size.

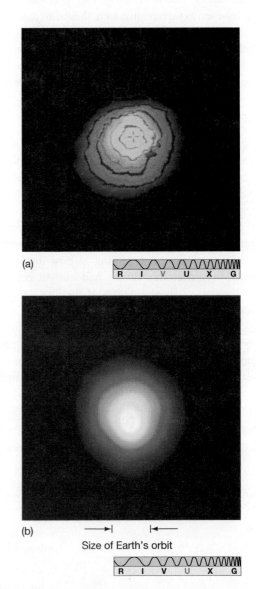

▲ FIGURE 17.11 Betelgeuse (a) The swollen star Betelgeuse (shown here in false color) is close enough for us to resolve its size directly, along with some surface features thought to be storms similar to those that occur on the Sun. Betelgeuse is such a huge star (about 600 times the size of the Sun) that its photosphere exceeds the size of Mars's orbit. Most of the surface features discernible here are larger than the entire Sun. (The dark lines are added contours of constant brightness, not part of the star itself.) (b) An ultraviolet view of Betelgeuse, in false color, as seen by a European camera onboard the *Hubble Space Telescope,* shows more clearly just how large this huge star is. *(NOAO; NASA/ESA)*

Giants and Dwarfs

Let's consider some examples to clarify these ideas. The star known as Aldebaran (the orange-red "eye of the bull" in the constellation Taurus) has a surface temperature of about 4000 K and a luminosity of 1.3×10^{29} W. Thus, its surface temperature is 0.7 times and its luminosity about 330 times, the corresponding quantities for our Sun. The radius–luminosity–temperature relationship (see *More Precisely 17-2*) then implies that the star's radius is almost 40 times the solar value. If our Sun were that large, its photosphere would extend halfway to the orbit of Mercury and, seen from Earth, would cover more than 20 degrees on the sky. A star as large as Aldebaran is known as a *giant*. More precisely,

MORE PRECISELY 17-2

Estimating Stellar Radii

We can combine the Stefan–Boltzmann law $F = \sigma T^4$ with the formula for the area of a sphere, $A = 4\pi R^2$, to obtain the relationship between a star's radius (R), luminosity (L), and temperature (T) described in the text:

$$L = 4\pi\sigma R^2 T^4,$$

or

$$\text{luminosity} \propto \text{radius}^2 \times \text{temperature}^4.$$

If we adopt convenient "solar" units, in which L is measured in solar luminosities (3.9×10^{26} W), R in solar radii (696,000 km), and T in units of the solar temperature (5800 K), we can eliminate the constant $4\pi\sigma$ and write this equation as

$$R^2 \text{ (in solar radii)} \times T^4 \text{ (in units of 5800 K)}.$$

As illustrated in the accompanying figure, both the radius and the temperature are important in determining the star's luminosity.

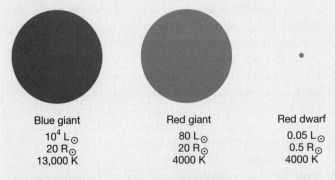

Blue giant	Red giant	Red dwarf
10^4 L$_\odot$	80 L$_\odot$	0.05 L$_\odot$
20 R$_\odot$	20 R$_\odot$	0.5 R$_\odot$
13,000 K	4000 K	4000 K

To compute the radius of a star from its luminosity and temperature, we rearrange terms so that the equation reads (in the same units)

$$R = \sqrt{L} / T^2.$$

This simple application of the radiation laws is the basis for almost every estimate of stellar size made in this text. Let's illustrate the process by computing the radii of two stars discussed in the text.

EXAMPLES In the solar units defined above, the star Aldebaran has luminosity $L = 1.3 \times 10^{29}$ W$/3.9 \times 10^{26}$ W $= 330$ units and temperature $T = 4000\ K/5800$ K $= 0.69$ unit. Thus, according to the equation, its radius is $R = \sqrt{330/0.69^2} = 18/0.48$ units $= 39$ solar radii—Aldebaran is a giant star. At the opposite extreme, Procyon B has $L = 2.3 \times 10^{23}$ W$/3.9 \times 10^{26}$ W $= 0.0006$ unit and $T = 8500\ K/5800 = 1.5$ units, so its radius is $R = \sqrt{0.0006/1.5^2} = 0.01$ times the radius of the Sun—Procyon B is a dwarf.

The following table lists luminosities, temperatures, and calculated radii (all in solar units) for some of the other stars mentioned in this chapter.

Star	Luminosity, L	Temperature, T	Radius, R
Sirius B	0.025	4.7	0.007
Barnard's star	0.0045	0.56	0.2
Sun	1	1	1
Sirius A	23	2.1	1.9
Vega	55	1.6	2.8
Arcturus	160	0.78	21
Rigel	63,000	1.9	70
Betelgeuse	36,000	0.55	630

Note that some of the luminosities shown in the table differ significantly from those in Appendix 6, Tables 5 and 6, and elsewhere in the text. This is because the values used elsewhere refer to visible light only, whereas the radiation laws (and the values in the table) refer to *total* luminosities. As we saw in Chapter 3, a star radiates its energy over a broad range of wavelengths that often extends well beyond the visible domain. ∞ (Sec. 3.4) A star like the Sun, whose emission happens to peak near the middle of the visible spectrum, emits most (roughly 80 percent) of its energy in the form of visible light. However, the cooler Aldebaran emits the bulk (about 75 percent) of its energy at infrared wavelengths, while the hot white dwarf Sirius B shines mainly in the ultraviolet—only about 10 percent of its energy is visible.

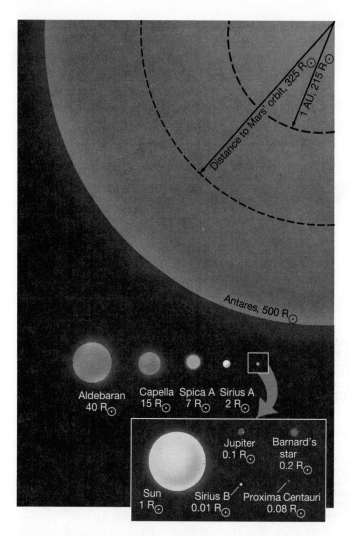

▲ **FIGURE 17.12 Stellar Sizes** Star sizes vary greatly. Shown here are the estimated sizes of several well-known stars, including a few of those discussed in this chapter. Only part of the red-giant star Antares can be shown on this scale, and the supergiant Betelgeuse would fill the entire page. (Here and in other figures, the symbol "⊙" stands for the Sun, so the symbol "R⊙" means "solar radius.")

giants are stars having radii between 10 and 100 times that of the Sun. Since any 4000-K object is reddish in color, Aldebaran is known as a **red giant.** Even larger stars, ranging up to 1000 solar radii in size, are known as **supergiants.** Betelgeuse is a prime example of a **red supergiant.**

Now consider Procyon B, a faint companion to Procyon A, one of the brightest stars in the night sky (see Appendix 3, Table 5). Procyon B's surface temperature is roughly 8500 K, about one and a half times that of the Sun. The star's total luminosity is 2.3×10^{25} W, about 0.0006 times the solar value. Again using the radius–luminosity–temperature relationship, we obtain a radius of 0.01 solar radii—slightly larger than that of Earth. Procyon B is hotter, but smaller and much less luminous, than our Sun. Such a star is known as a *dwarf.* In astronomy, the term **dwarf** refers to any star of radius comparable to or smaller than the Sun (including the

Sun itself). Because any 8500 K object glows white-hot, Procyon B is an example of a **white dwarf.**

The radii of the vast majority of stars (measured mostly with the radius–luminosity–temperature relationship) range from less than 0.01 to over 100 times the radius of the Sun. Figure 17.12 illustrates the estimated sizes of a few well-known stars.

CONCEPT CHECK

✔ Can we measure the radius of a star without knowing the star's distance from Earth?

17.5 The Hertzsprung–Russell Diagram

Astronomers use luminosity and surface temperature to classify stars in much the same way that height and weight serve to classify the bulk properties of human beings. We know that people's height and weight are well correlated: Tall people tend to weigh more than short ones. We might naturally wonder if the two basic stellar properties are also related in some way.

Figure 17.13 plots luminosity versus temperature for several well-known stars. Figures of this sort are called *Hertzsprung–Russell diagrams,* or **H–R diagrams,** after Danish astronomer Ejnar Hertzsprung and U.S. astronomer Henry Norris Russell, who independently pioneered the use of such plots in the second decade of the 20th century. The vertical scale, expressed in units of the solar luminosity (3.9×10^{26} W), extends over a large range, from 10^{-4} to 10^4; the Sun appears right in the middle of the luminosity range, at a luminosity of 1. Surface temperature is plotted on the horizontal axis, although in the unconventional sense of temperature increasing to the *left* (so that the spectral sequence O, B, A . . . reads from left to right). To change the horizontal scale so that temperature would increase conventionally to the right would play havoc with historical precedent.

As we have just seen, astronomers often use a star's *color* to measure its temperature. Indeed, the spectral classes plotted along the horizontal axis of Figure 17.13 are equivalent to the B/V color index. Also, because astronomers commonly express a star's luminosity as an absolute magnitude, stellar *magnitude* instead of stellar luminosity could be plotted on the vertical axis (see *More Precisely 17-1*). Many astronomers prefer to present their data in these more "observational" terms, and the diagrams corresponding to plots like Figure 17.13 are called **color-magnitude diagrams.** In this book, we will cast our discussion mainly in terms of the more "theoretical" quantities, temperature and luminosity, but realize that, for many purposes, color-magnitude and H–R diagrams represent pretty much the same thing.

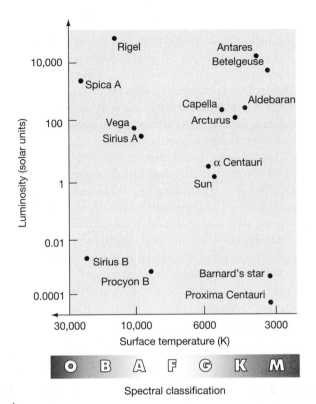

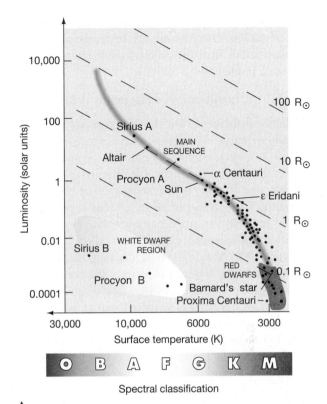

Interactive FIGURE 17.13 H–R Diagram of Well-Known Stars A plot of luminosity against surface temperature (or spectral class), known as an H–R diagram, is a useful way to compare stars. Plotted here are the data for some stars mentioned earlier in the text. The Sun, of course, has a luminosity of 1 solar unit. Its temperature, read off the bottom scale, is 5800 K—that of a G-type star. Similarly, the B-type star Rigel, at top left, has a temperature of about 11,000 K and a luminosity more than 10,000 times that of the Sun. The M-type star Proxima Centauri, at bottom right, has a temperature of about 3000 K and a luminosity less than $\frac{1}{10,000}$ that of the Sun. (See also Overlay 1 of the acetate insert.)

Interactive FIGURE 17.14 H–R Diagram of Nearby Stars Most stars have properties within the long, thin, shaded region of the H–R diagram known as the main sequence. The points plotted here are for stars lying within about 5 pc of the Sun. Each dashed diagonal line corresponds to a constant stellar radius, so that stellar size can be indicated on the same diagram as stellar luminosity and temperature. (Recall that the symbol "$R_\odot$" means "solar radius.") The main sequence is one of the most important correlations in all of astronomy; it allows us to expand our knowledge of stellar properties from the Sun's local neighborhood to the far reaches of our Galaxy.

The Main Sequence

The handful of stars plotted in Figure 17.13 gives little indication of any particular connection between stellar properties. However, as Hertzsprung and Russell plotted more and more stellar temperatures and luminosities, they found that a relationship does in fact exist: Stars are *not* uniformly scattered across the H–R diagram; instead, most are confined to a fairly well-defined band stretching diagonally from the top left (high temperature, high luminosity) to the bottom right (low temperature, low luminosity). In other words, cool stars tend to be faint (less luminous) and hot stars tend to be bright (more luminous). This band of stars spanning the H–R diagram is known as the **main sequence.**

Figure 17.14 shows a more systematic study of stellar properties, covering the 80 or so stars that lie within 5 pc of the Sun. As more points are included in the diagram, the main sequence "fills up," and the pattern becomes more

evident. The vast majority of stars in the immediate vicinity of the Sun lie on the main sequence.

The surface temperatures of main-sequence stars range from about 3000 K (spectral class M) to over 30,000 K (spectral class O). This relatively small temperature range—a difference of only a factor of 10—is determined mainly by the rates at which nuclear reactions occur in stellar cores. ∞ (Sec. 16.6) In contrast, the observed range in luminosities is very large, covering some eight orders of magnitude (i.e., a factor of 100 million), ranging from 10^{-4} to 10^4 times the luminosity of the Sun.

Using the radius–luminosity–temperature relationship (Section 17.4), astronomers find that stellar radii also vary along the main sequence. The faint, red M-type stars at the bottom right of the H–R diagram are only about one-tenth the size of the Sun, whereas the bright, blue O-type stars in the upper left are about 10 times larger than the Sun. The diagonal dashed lines in Figure 17.14 represent constant stellar radii, meaning that any star lying on a given line has the same radius, regardless of its luminosity or temperature. Along a constant-radius line, the radius–luminosity–temperature relationship implies that

$$\text{luminosity} \propto \text{temperature}^4.$$

By including such lines on our H–R diagrams, we can indicate stellar temperatures, luminosities, and radii on a single plot.

We see a very clear trend as we traverse the main sequence from top to bottom. At one end, the stars are large, hot, and bright. Because of their size and color, they are referred to as **blue giants.** The very largest are called **blue supergiants.** At the other end, stars are small, cool, and faint. They are known as **red dwarfs.** Our Sun lies right in the middle.

Figure 17.15 shows an H–R diagram for a different group of stars—the 100 stars of known distance having the greatest apparent brightness, as seen from Earth. Notice the much larger number of very luminous stars at the upper end of the main sequence than at the lower end. The reason for this excess of blue giants is simple: We can see very luminous stars a long way off. The stars shown in this figure are scattered through a much greater volume of space than those

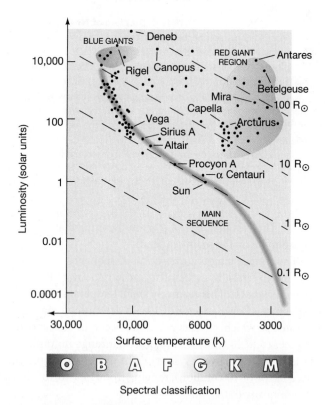

▲ FIGURE 17.15 **H–R Diagram of Brightest Stars** An H–R diagram for the 100 brightest stars in the sky is biased in favor of the most luminous stars—which appear toward the upper left—because we can see them more easily than we can the faintest stars. (Compare with Figure 17.14, which shows only the closest stars. Note that neither of these two figures is representative of a given region of space; rather, each is biased toward a different population of stars.)

depicted in Figure 17.14, but the sample is heavily biased toward the brightest objects. In fact, of the 20 brightest stars in the sky, only 6 lie within 10 pc of us; the rest are visible, despite their great distances, because of their high luminosities.

If very luminous blue giants are overrepresented in Figure 17.15, low-luminosity red dwarfs are surely underrepresented. In fact, no dwarfs appear on the diagram. This absence is not surprising because low-luminosity stars are difficult to observe from Earth. In the 1970s, astronomers began to realize that they had greatly underestimated the number of red dwarfs in our galaxy. As hinted at by the H–R diagram in Figure 17.14, which shows an unbiased sample of stars in the solar neighborhood, red dwarfs are actually the most common type of star in the sky. In fact, they probably account for upward of 80 percent of all stars in the universe. In contrast, O- and B-type supergiants are extremely rare, with only about 1 star in 10,000 falling into these categories.

White Dwarfs and Red Giants

Most stars lie on the main sequence. However, some of the points plotted in Figures 17.13 through 17.15 clearly do not. One such point in Figure 17.13 represents Procyon B, the white dwarf discussed earlier (Section 17.4), with surface temperature 8500 K and luminosity about 0.0006 times the solar value. A few more such faint, hot stars can be seen in Figure 17.14 in the bottom left-hand corner of the H–R diagram. This region, known as the **white-dwarf region,** is marked on Figure 17.14.

Also shown in Figure 17.13 is Aldebaran (discussed in Section 17.4), whose surface temperature is 4000 K and whose luminosity is some 300 times greater than the Sun's. Another point represents Betelgeuse (Alpha Orionis), the ninth-brightest star in the sky, a little cooler than Aldebaran, but more than 100 times brighter. The upper right-hand corner of the H–R diagram, where these stars lie (marked on Figure 17.15), is called the **red-giant region.** No red giants are found within 5 pc of the Sun (Figure 17.14), but many of the brightest stars seen in the sky are in fact red giants (Figure 17.15). Though relatively rare, red giants are so bright that they are visible to great distances. They form a third distinct class of stars on the H–R diagram, very different in their properties from both main-sequence stars and white dwarfs.

The *Hipparcos* mission (Section 17.1), in addition to determining hundreds of thousands of stellar parallaxes with unprecedented accuracy, also measured the colors and luminosities of more than 2 million stars. Figure 17.16 shows an H–R diagram based on a tiny portion of the enormous *Hipparcos* data set. The main-sequence and red-giant regions are clearly evident. Few white dwarfs appear, however, simply because the telescope was limited to observations of relatively bright objects—brighter than apparent magnitude 12. Almost no white dwarfs lie close enough to Earth that their magnitudes fall below this limit.

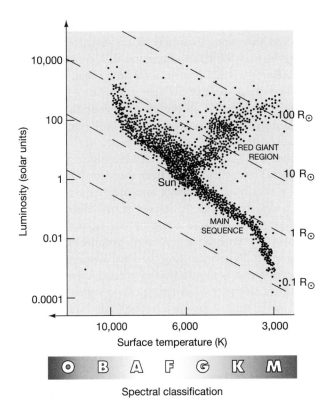

▲ FIGURE 17.16 *Hipparcos* **H–R Diagram** This simplified version of one of the most complete H–R diagrams ever compiled represents more than 20,000 data points, as measured by the European *Hipparcos* spacecraft for stars within a few hundred parsecs of the Sun.

About 90 percent of all stars in our solar neighborhood, and probably a similar percentage elsewhere in the universe, are main-sequence stars. About 9 percent of stars are white dwarfs, and 1 percent are red giants.

CONCEPT CHECK

✔ Only a tiny fraction of all stars are giants. Why, then, do giants account for so many of the brightest stars in the night sky?

17.6 Extending the Cosmic Distance Scale

We have already discussed the connections between luminosity, apparent brightness, and distance. Knowledge of a star's apparent brightness and its distance allows us to determine its luminosity using the inverse-square law. But we can also turn the problem around. If we somehow knew a star's luminosity and then measured its apparent brightness, the inverse-square law would tell us its distance from the Sun.

Spectroscopic Parallax

Most of us have a rough idea of the approximate intrinsic brightness (that is, the luminosity) of a typical traffic signal.

Suppose you are driving down an unfamiliar street and see a red traffic light in the distance. Your knowledge of the light's luminosity enables you immediately to make a mental estimate of its distance. A light that appears relatively dim (low apparent brightness) must be quite distant (assuming it's not just dirty). A bright one must be relatively close. Thus, *measurement of the apparent brightness of a light source, combined with some knowledge of its luminosity, can yield an estimate of its distance.*

For stars, the trick is to find an independent measure of the luminosity without knowing the distance. The H–R diagram can provide just that. For example, suppose we observe a star and determine its apparent magnitude to be 10. By itself, that doesn't tell us much—the star could equally well be faint and close, or bright and distant (Figure 17.6). But suppose we have some additional information: The star lies on the main sequence and has spectral type A0. Then we can read the star's luminosity off Figure 17.15. A main-sequence A0 star has a luminosity of approximately 100 solar units. According to *More Precisely 17-1,* this corresponds to an absolute magnitude of 0 and hence to a distance of 1000 pc.

This process of using stellar spectra to infer distances is called **spectroscopic parallax.*** The key steps are as follows:

1. We measure the star's apparent brightness and spectral type *without* knowing how far away it is.

2. Then we use the spectral type to estimate the star's luminosity.

3. Finally, we apply the inverse-square law to determine the distance to the star.

The existence of the main sequence allows us to make a connection between an easily measured quantity (spectral type) and the star's luminosity, which would otherwise be unknown. The term *spectroscopic parallax* refers to the specific process of using stellar spectra to infer luminosities and hence distances. However, as we will see in upcoming chapters, this essential logic (with a variety of different techniques replacing step 2) is used again and again as a means of distance measurement in astronomy. In practice, the "fuzziness" of the main sequence translates into a small (10–20 percent) uncertainty in the distance, but the basic idea remains valid.

In Chapter 2 we introduced the first "rung" on a ladder of distance-measurement techniques that will ultimately carry us to the edge of the observable universe. That rung is radar ranging on the inner planets. ∞ (Sec. 2.6) It establishes the scale of the solar system and defines the astronomical unit. In Section 17.1, we discussed a second rung in the cosmic distance ladder—stellar parallax—which is based on

---

*This unfortunate name is very misleading, as the method has mothing in common with stellar (geometric) parallax other than its use as a means of determining stellar distances.

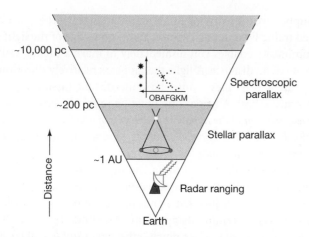

▲ **FIGURE 17.17 Stellar Distances** Knowledge of a star's luminosity and apparent brightness can yield an estimate of its distance. Astronomers use this third rung on the distance ladder, called spectroscopic parallax, to measure distances as far out as individual stars can be clearly discerned—several thousand parsecs.

the first, since Earth's orbit is the baseline. Now, having used the first two rungs to determine the distances and other physical properties of many nearby stars, we can employ that knowledge in turn to construct a third rung in the ladder—spectroscopic parallax. As illustrated schematically in Figure 17.17, this new rung expands our cosmic field of view still deeper into space.

Spectroscopic parallax can be used to determine stellar distances out to several thousand parsecs. Beyond that, spectra and colors of individual stars are difficult to obtain. The "standard" main sequence is obtained from H–R diagrams of stars whose distances can be measured by (geometric) parallax, so the method of spectroscopic parallax is calibrated by using nearby stars. Note that, in employing this method, we are assuming (without proof) that distant stars are basically similar to nearby stars and that *they fall on the same main sequence as nearby stars.* Only by making this assumption can we expand the boundaries of our distance-measurement techniques.

Of course, the main sequence is not really a line in the H–R diagram: It has some thickness. For example, the luminosities of main-sequence stars categorized as type A0 (Vega, for example) can actually range from about 30 to 100 times the luminosity of the Sun. The main reason for this range is the variation in stellar composition and age from place to place in our Galaxy. As a result, there is considerable uncertainty in the luminosity obtained by this method and hence a corresponding uncertainty in the distance of the star. Distances obtained by spectroscopic parallax are generally accurate to no better than about 25 percent.

Although this may not seem very accurate—a cross-country traveler in the United States would hardly be impressed to be told that the best estimate of the distance between Los Angeles and New York is somewhere between

3000 and 5000 km—it illustrates the point that, in astronomy, even something as simple as the distance to another star can be very difficult to measure. Still, an estimate with an uncertainty of ±25 percent is far better than no estimate at all. The deployment of the next generation of astrometry satellites (Section 17.1) promises to remedy this situation, combining radical improvements in both the range and accuracy of stellar parallax measurements.

Finally, realize that, because each rung in the distance ladder is calibrated using data from the lower rungs, changes made at *any* level will affect measurements made on *all* larger scales. As a result, the impact of new high-quality observations, such as those made by the *Hipparcos* mission (Section 17.1), extends far beyond the volume of space actually surveyed. By recalibrating the local foundations of the cosmic distance scale, *Hipparcos* caused astronomers to revise their estimates of distances on *all* scales—up to and including the scale of the universe itself. All distances quoted throughout this text reflect updated values based on *Hipparcos* data.

Luminosity Class

What if the star in question happens to be a red giant or a white dwarf and does not lie on the main sequence? Recall from Chapter 4 that detailed analysis of spectral *line widths* can provide information on the *density* of the gas where the line formed. ∞ (Sec. 4.5) The atmosphere of a red giant is much less dense than that of a main-sequence star, and this in turn is much less dense than the atmosphere of a white dwarf. Figure 17.18(b) and (c) illustrate the difference between the spectra of a main-sequence star and a red giant of the same spectral type.

Over the years, astronomers have developed a system for classifying stars according to the widths of their spectral lines. Because line width depends on density in the stellar photosphere, and because this density in turn is well correlated with luminosity, this stellar property is known as **luminosity class.** The standard luminosity classes are listed in Table 17.3 and shown on the H–R diagram in Figure 17.18(a). By determining a star's luminosity class,

TABLE 17.3 Stellar Luminosity Classes	
Class	**Description**
Ia	Bright supergiants
Ib	Supergiants
II	Bright giants
III	Giants
IV	Subgiants
V	Main-sequence stars and dwarfs

TABLE 17.4 Variation in Stellar Properties within a Spectral Class

Approximate Surface Temperature (K)	Luminosity (solar luminosities)	Radius (solar radii)	Object	Example
4900	0.3	0.8	K2V main-sequence star	ε Eridani
4500	110	21	K2III red giant	Arcturus
4300	4000	140	K2Ib red supergiant	ε Pegasi

astronomers can usually tell with a high degree of confidence what sort of object it is. Now we have a way of specifying a star's location in the diagram in terms of properties that are measurable by purely spectroscopic means: *Spectral type and luminosity class define a star on the H–R diagram just as surely as do temperature and luminosity.* The full specification of a star's spectral properties includes its luminosity class. For example, the Sun, a G2 main-sequence star, is of class G2V, the B8 blue supergiant Rigel is B8Ia, the red dwarf Barnard's star is M5V, the red supergiant Betelgeuse is M2Ia, and so on.

Consider, for example, a K2-type star (Table 17.4) with a surface temperature of approximately 4500 K. If the widths of the star's spectral lines tell us that it lies on the main sequence (i.e., it is a K2V star), then its luminosity is about 0.3 times the solar value. If the star's spectral lines are observed to be narrower than lines normally found in main-sequence stars, the star may be recognized as a K2III giant, with a luminosity 100 times that of the Sun (Figure 17.18a). If the lines are very narrow, the star might instead be classified as a K2Ib supergiant, brighter by a further factor of 40, at 4000 solar luminosities. In each case, knowledge of luminosity classes allows astronomers to identify the object and make an appropriate estimate of its luminosity and hence its distance.

CONCEPT CHECK

✔ Suppose astronomers discover that, due to a calibration error, *all* distances measured by geometric parallax are 10 percent larger than currently thought. What effect would this finding have on the "standard" main sequence used in spectroscopic parallax?

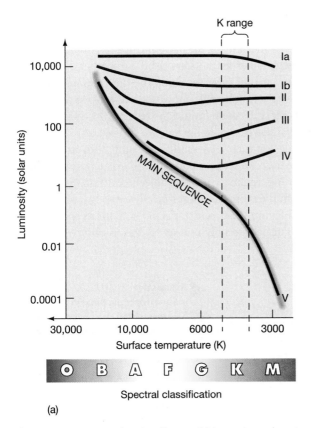

(a)

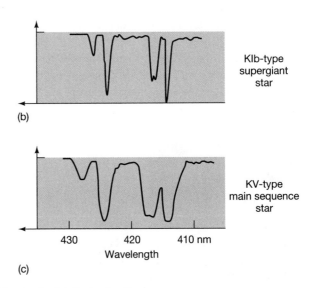

(b)

(c)

▲ **FIGURE 17.18 Luminosity Classes** (a) Approximate locations of the standard stellar luminosity classes in the H–R diagram. The widths of absorption lines also provide information on the density of a star's atmosphere. The denser atmosphere of a main-sequence K-type star has broader lines (c) than a giant star of the same spectral class (b).

17.7 Stellar Masses

What ultimately determines a star's position on the main sequence? The answer is its *mass* and its *composition.* Mass and composition are fundamental properties of any star. Together, they uniquely determine the star's internal structure, its external appearance, and even (as we will see in Chapter 20) its future evolution. The ability to measure these two key stellar properties is of the utmost importance if we are to understand how stars work. We have already seen how spectroscopy is used to determine composition. ∞ (Sec. 16.3) Now let's turn to the problem of finding a star's mass.

As with all other objects, we measure a star's mass by observing its gravitational influence on some nearby body—another star, perhaps, or a planet. If we know the distance between the two bodies, then we can use Newton's laws to calculate their masses. The extrasolar planetary systems that have recently been detected have not been studied well enough to provide independent stellar mass measurements, and we are a long way from placing our own spacecraft in orbit around other stars. ∞ (Sec. 15.5) Nevertheless, there are ways of determining stellar masses.

Binary Stars

Most stars are members of *multiple-star systems*—groups of two or more stars in orbit around one another. The majority of stars are found in **binary-star systems,** which consist of two stars in orbit about a common center of mass, held together by their mutual gravitational attraction. ∞ (Sec. 2.7) Other stars are members of triple, quadruple, or even more complex systems. The Sun is not part of a multiple-star system—if it has anything at all uncommon about it, it may be its lack of stellar companions.

Astronomers classify binary-star systems (or simply *binaries*) according to their appearance from Earth and the ease with which they can be observed. **Visual binaries** have widely separated members that are bright enough to be observed and monitored separately, as shown in Figure 17.19.

▲ FIGURE 17.19 **Visual Binary** The period and separation of a binary-star system can be observed directly if each star is clearly seen. At the left is an orbital diagram for the double star Kruger 60; at the right are actual photographs taken in some of the years indicated. *(Harvard College Observatory)*

The more common **spectroscopic binaries** are too distant to be resolved into separate stars, but they can be indirectly perceived by monitoring the back-and-forth Doppler shifts of their spectral lines as the stars orbit each other. Recall that motion toward an observer shifts the lines toward the blue end of the electromagnetic spectrum and motion away from the observer shifts them toward the red end. ∞ (Sec. 3.5) In a *double-line* spectroscopic binary, two distinct sets of spectral lines—one for each star—shift back and forth as the stars move. Because we see particular lines alternately approaching and receding, we know that the objects emitting the lines are in orbit. In the more common *single-line* systems, such as that shown in Figure 17.20, one star is too faint for its spectrum to be distinguished, so only one set of lines is observed to shift

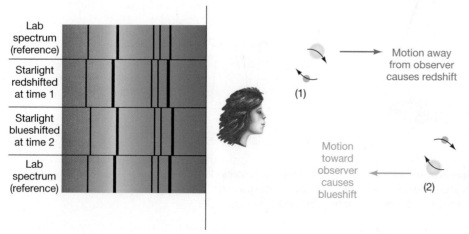

Interactive FIGURE 17.20 Spectroscopic Binary Properties of binary stars can be determined by measuring the periodic Doppler shift of one star relative to the other as they move in their orbits. Here we show a so-called single-line system, in which only one spectrum (from the brighter component) is visible. Reference spectra are shown at top and bottom at left. An observer would see the spectrum redshifted as the visible component moves away (top right) and blueshifted as it moves toward the observer (bottom right).

back and forth. The shifting means that the star that is observed must be in orbit around another star, even though the companion cannot be observed directly. (If this idea sounds familiar, it should—all of the extrasolar planetary systems discovered to date are extreme examples of single-line spectroscopic binaries.) ∞ (Sec. 15.5)

In the much rarer **eclipsing binaries,** the orbital plane of the pair of stars is almost edge-on to our line of sight. In this situation, depicted in Figure 17.21, we observe a periodic decrease in starlight as one star passes in front of (transits) the other. By studying the variation in the light from the binary system—the binary's **light curve**— astronomers can derive detailed information not only about the stars' orbits and masses, but also about their radii. Thus, eclipsing binaries provide an alternative means of measuring stellar radii that is *independent* of either the direct or the indirect methods described in Section 17.4.

For example, in the sequence shown in Figure 17.21, the maximum brightness (frames 1, 3, and 5) represents the combined brightnesses of the two stars, whereas the shallower minimum (frame 4) represents the brighter (larger) component only. These two pieces of information allow us to infer the individual brightnesses of the two stars. The deeper minima (frames 2 and 6) occur because the fainter red star partially blocks the light of the much brighter yellow star. The change in brightness tells us what fraction of the brighter star is obscured, and that in turn tells us the ratio of the *areas* of the two stars and hence (since area is proportional to radius squared) the ratio of their radii. If we also knew the components' orbital *speeds*—from Doppler measurements, say—then the widths of the minima and the time taken to go from minimum to maximum light would tell us the actual radii of the stars.

The preceding categories of binary-star systems are not mutually exclusive. For example, a single-line spectroscopic binary may also happen to be an eclipsing system. In that case, astronomers can use the eclipses to gain extra information about the fainter member of the pair. Occasionally, two unrelated stars just happen to lie close together in the sky,

even though they are actually widely separated. These *optical doubles* are just chance superpositions and carry no useful information about stellar properties.

Mass Determination

By observing the actual orbits of the stars, the back-and-forth motion of the spectral lines, or the dips in the light curve— whatever information is available—astronomers can measure the binary's orbital period. Observed periods span a broad range—from hours to centuries. How much additional information can be extracted depends on the type of binary involved.

If the distance to a *visual* binary is known, the semimajor axis of its orbit can be determined directly by simple geometry. Knowledge of the binary period and orbital semimajor axis is all we need to determine the combined mass of the component stars using the modified form of Kepler's third law. ∞ (Sec. 2.7) Since the orbits of both stars can be separately tracked, it is also possible to determine each of the individual stars' masses. Recall from Section 2.8 that, in any system of orbiting objects, each object orbits the common center of mass. Measuring the distance from each star to the center of mass of a visual binary yields the ratio of the stellar masses. Knowing both the sum of the masses and their ratio, we can then find the mass of each star.

For *spectroscopic* binaries, it is not possible to determine the semimajor axis directly. Doppler shift measurements give us information on the orbital velocities of the two stars, but only with regard to their radial components—that is, along the line of sight. As a result, we cannot determine the *inclination* of the orbit to our line of sight, and this imposes a limitation on how much information we can obtain—simply put, we cannot distinguish between a slow-moving binary seen edge-on and a fast-moving binary seen almost face-on (so that only a small component of the orbital motion is along the line of sight). We have already encountered this limitation in our study of extrasolar planets. ∞ (Sec. 15.6)

For a double-line spectroscopic system, individual radial velocities, and hence the ratio of the component masses, can be determined, but the uncertainty in the orbital inclination means that only lower limits on the individual masses can be obtained. For single-line systems, even less information is available, and only a fairly complicated relation between the component masses (known as the *mass function*) can be derived. However, if, as is often the case, the mass of the brighter star can be determined by other means (e.g., if the brighter star is recognized as a main-sequence star of a certain spectral class—see Figure 17.22), a lower limit can then be placed on the mass of the fainter, unseen star.

Finally, if a spectroscopic binary happens also to be an *eclipsing* system, then the uncertainty in the inclination is removed, as the binary is known to be edge-on or very nearly so. In that case, both masses can be determined for a

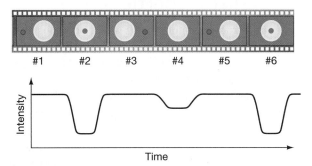

⚠ Interactive FIGURE 17.21 Eclipsing Binary If the two stars in a binary-star system happen to eclipse one another, additional **(MA)** information on their radii and masses can be obtained by observing the periodic decrease in starlight as one star passes in front of the other.

MORE PRECISELY 17-3

Measuring Stellar Masses in Binary Stars

As discussed in the text, most stars are members of binary systems—where two stars orbit one another, bound together by gravity. Here we describe—in an idealized case where the relevant orbital parameters are known—how we can use the observed orbital data, together with our knowledge of basic physics, to determine the masses of the component stars.

Consider the nearby visual binary system made up of the bright star Sirius A and its faint companion Sirius B, sketched in the accompanying figure. The binary's orbital period can be measured simply by watching the stars orbit one another, or alternatively by following the back-and-forth velocity wobbles of Sirius A due to its faint companion. It is almost exactly 50 years. The orbital semimajor axis can also be obtained by direct observation of the orbit, although in this case we must use some additional knowledge of Kepler's laws to correct for the binary's 46° inclination to the line of sight. ∞ (Sec. 2.5) It is 20 AU—a measured angular size of 7.5″ at a distance of 2.7 pc. ∞ *(More Precisely 1-2)* Once we know these two key orbital parameters, we can use the modified version of Kepler's third law to calculate the sum of the masses of the two stars. The result is $20^3/50^2 = 3.2$ times the mass of the Sun. ∞ (Sec. 2.8)

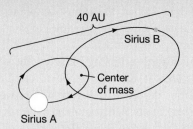

Further study of the orbit allows us to determine the individual stellar masses. Doppler observations show that Sirius A moves at approximately half the speed of its companion relative to their center of mass. ∞ (Secs. 2.8, 3.5) This implies that Sirius A must have twice the mass of Sirius B. It then follows that the masses of Sirius A and Sirius B are 2.1 and 1.1 solar masses, respectively.

Often the calculation of the masses of binary components is complicated by the fact that only partial information is available—we might only be able to see one star, or perhaps only spectroscopic velocity information is available (see Section 17.7). Nevertheless, this technique of combining elementary physical principles with detailed observations is how virtually every stellar mass quoted in this text has been determined.

---

double-line binary. For a single-line system, the mass function is simplified to the point where the mass of the unseen star is known if the mass of the brighter star can be found by other means (e.g., by recognizing it as a main-sequence star of known spectral type).

Despite all these qualifications and difficulties, the masses of individual component stars have been obtained for many nearby binary systems. *More Precisely 17-3* presents a simple example of how this is accomplished in practice.

17.8 Mass and Other Stellar Properties

We end our introduction to the stars with a brief look at how mass is correlated with the other stellar properties discussed in this chapter. Figure 17.22 is a schematic H–R diagram showing how stellar mass varies along the main sequence. There is a clear progression from low-mass red dwarfs to high-mass blue giants. With few exceptions, main-sequence stars range in mass from about 0.1 to 20 times the mass of the Sun. The hot O- and B-type stars are generally about 10 to 20 times more massive than our Sun. The coolest K- and M-type stars contain only a few tenths of a solar mass. The mass of a star at the time of its formation determines the star's location on the main sequence. Based on observations of stars within a few hundred light-years of the Sun, Figure 17.23 illustrates how the masses of main-sequence stars are distributed. Notice the huge fraction of low-mass stars, as

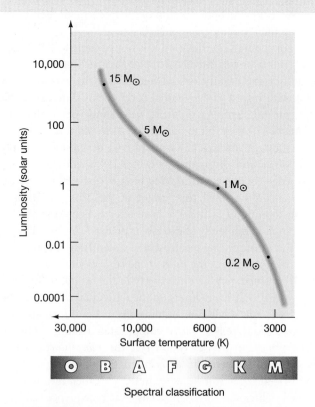

▲ FIGURE 17.22 **Stellar Masses** More than any other stellar property, mass determines a star's position on the main sequence. Low-mass stars are cool and faint; they lie at the bottom of the main sequence. Very massive stars are hot and bright; they lie at the top of the main sequence. (The symbol "M⊙" means "solar mass.") This connection between mass and luminosity is central to understanding how stars evolve in time.

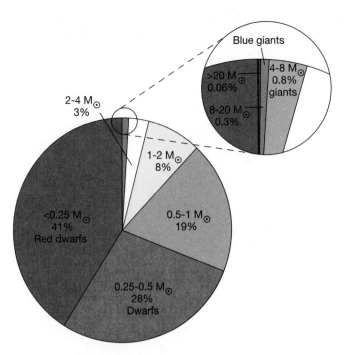

▲ **FIGURE 17.23 Stellar Mass Distribution** The range of masses of main-sequence stars, as determined from careful measurement of stars in the solar neighborhood.

well as the tiny fraction contributed by stars of more than a few solar masses.

Figure 17.24 illustrates how a main-sequence star's radius and luminosity depend on its mass. The two plots shown, of the *mass–radius* and *mass–luminosity* relations, are based on observations of binary-star systems. Along the main sequence,

both radius and luminosity increase with mass. As an approximate rule of thumb, we can say that radius increases proportionally to stellar mass, whereas luminosity increases much faster—almost as the *fourth power* of the mass (as indicated by the line in Figure 17.24b). Thus, a 2-solar-mass main-sequence star has a radius about twice that of the Sun and a luminosity of 16 (2^4) solar luminosities; a 0.2-solar-mass main-sequence star has a radius of roughly 0.2 solar radii and a luminosity of around 0.0016 (0.2^4) solar luminosity.

Table 17.5 compares some key properties of several well-known main-sequence stars, arranged in order of decreasing mass. Notice that the central temperature (obtained from mathematical models similar to those discussed in Chapter 16) differs relatively little from one star to another, compared with the large spread in stellar luminosities. ∞ (Sec. 16.2) The rapid rate of nuclear burning deep inside a star releases vast amounts of energy per unit time. How long can the fire continue to burn? We can estimate a main-sequence star's *lifetime* simply by dividing the amount of fuel available (the mass of the star) by the rate at which the fuel is being consumed (the star's luminosity):

$$\text{stellar lifetime} \propto \frac{\text{stellar mass}}{\text{stellar luminosity}}.$$

The mass–luminosity relation tells us that a star's luminosity is roughly proportional to the fourth power of its mass, so we can rewrite this expression to obtain, approximately,

$$\text{stellar lifetime} \propto \frac{1}{(\text{stellar mass})^3}.$$

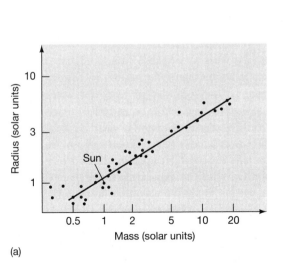

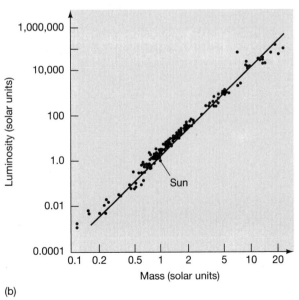

(a)   (b)

▲ **FIGURE 17.24 Stellar Radii and Luminosities** (a) Dependence of stellar radius on mass for main-sequence stars; actual measurements are plotted here. The radius increases nearly in proportion to the mass over much of the range (as indicated by the straight line drawn through the data points). (b) Dependence of main-sequence luminosity on mass. The luminosity increases roughly as the fourth power of the mass (indicated again by the straight line).

TABLE 17.5 Key Properties of Some Well-Known Main-Sequence Stars

Star	Spectral Type	Mass, M (Solar Masses)	Central Temperature (10^6 K)	Luminosity, L (Solar Luminosities)	Estimated Lifetime (M/L) (10^6 years)
Spica B*	B2V	6.8	25	800	90
Vega	A0V	2.6	21	50	500
Sirius A	A1V	2.1	20	22	1000
Alpha Centauri	G2V	1.1	17	1.6	7000
Sun	G2V	1.0	15	1.0	10,000
Proxima Centauri	M5V	0.1	0.6	0.00006	16,000,000

* The "star" Spica is, in fact, a binary system comprising a B1III giant primary (Spica A) and a B2V main-sequence secondary (Spica B).

The final column in Table 17.5 lists estimated lifetimes, based on the above proportionality and noting that the lifetime of the Sun (see Chapter 20) is about 10 billion years.

For example, the lifetime of a 10-solar-mass main-sequence O-type star is roughly $10/10^4 = 1/1000$ of the lifetime of the Sun, or about 10 million years. The nuclear reactions in such a massive star proceed so rapidly that its fuel is quickly depleted, despite its large mass. We can be sure that all the O- and B-type stars we now observe are quite young—less than a few tens of millions of years old. Massive stars older than that have already exhausted their fuel and no longer emit large amounts of energy. They have, in effect, died.

At the opposite end of the main sequence, the cooler K- and M-type stars have less mass than our Sun has. With their low core densities and temperatures, their proton–reactions churn away rather sluggishly, much more slowly than those in the Sun's core. The small energy release per unit time leads to low luminosities for these stars, so they have very long lifetimes. Many of the K- and M-type stars we now see in the night sky will shine on for at least another trillion years. The evolution of stars—large and small—is the subject of Chapters 20 and 21.

PROCESS OF SCIENCE CHECK

✔ How do we know the masses of stars that aren't components of binaries?

CHAPTER REVIEW

SUMMARY

1 The distances to the nearest stars can be measured by trigonometric parallax. A star with a parallax of 1 arc second (1″) is 1 **parsec (p. 419)**—about 3.3 light-years—away from Earth.

2 Stars have real motion through space as well as apparent motion as Earth orbits the Sun. A star's **proper motion (p. 420)**—its true motion across the sky—is a measure of the star's velocity perpendicular to our line of sight. The star's radial velocity—along the line of sight—is measured by the Doppler shift of the spectral lines emitted by the star.

3 The **apparent brightness (p. 421)** of a star is the rate at which energy from the star reaches a detector. Apparent brightness falls off as the inverse square of the distance. Optical astronomers use the **magnitude scale (p. 422)** to express and compare stellar brightnesses. The greater the magnitude, the fainter the star; a difference of five magnitudes corresponds to a factor of 100 in brightness. **Apparent magnitude (p. 422)** is a measure of apparent brightness. The **absolute magnitude (p. 423)** of a star is the apparent magnitude it would have if placed at a standard distance of 10 pc from the viewer. It is a measure of the star's luminosity.

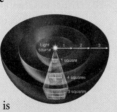

4 Astronomers often measure the temperatures of stars by measuring their brightnesses through two or more optical filters and then fitting a blackbody curve to the results. The measurement of the amount of starlight received through each member of a set of filters is called **photometry (p. 425)**. Spectroscopic observations of stars provide an accurate means of determining both stellar temperatures and stellar composition. Astronomers classify stars according to the absorption lines in their spectra. The standard stellar **spectral classes (p. 427)**, in order of decreasing temperature, are O, B, A, F, G, K, and M.

5 Only a few stars are large enough and close enough that their radii can be measured directly. The sizes of most stars are estimated indirectly through the **radius–luminosity–temperature re-**

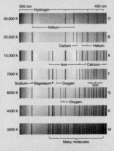

lationship (p. 428). Stars comparable in size to, or smaller than, the Sun are categorized as **dwarfs (p. 430)**, stars up to 100 times larger than the Sun are called **giants (p. 430)**, and stars more than 100 times larger than the Sun are known as **supergiants (p. 430)**. In addition to "normal" stars such as the Sun, two other important classes of star are **red giants (p. 430)** (and **red supergiants**) (p. 430), which are large, cool, and luminous, and **white dwarfs (p. 430)**, which are small, hot, and faint.

6 A plot of stellar luminosities versus stellar spectral classes (or temperatures) is called an **H–R diagram (p. 430)**, or a **color-magnitude diagram (p. 430)**. About 90 percent of all stars plotted on an H–R diagram lie on the **main sequence (p. 431)**, which stretches from hot, bright **blue supergiants (p. 432)** and **blue giants (p. 432)**, through intermediate stars such as the Sun, to cool, faint **red dwarfs (p. 432)**. Most main-sequence stars are red dwarfs; blue giants are quite rare. About 9 percent of stars are in the **white-dwarf region (p. 432)**, and the remaining 1 percent are in the **red-giant region (p. 432)**.

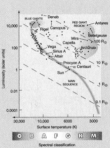

7 If a star is known to be on the main sequence, measurement of its spectral type allows its luminosity to be estimated and its distance to be measured. This method of determining distance, which is valid for stars up to several thousand parsecs from Earth, is called **spectroscopic parallax (p. 433)**. A star's **luminosity class (p. 434)** allows astronomers to distinguish main-sequence stars from giants and supergiants of the same spectral type.

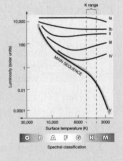

8 Most stars are not isolated in space, but instead orbit other stars in **binary-star systems (p. 436)**. In a **visual binary (p. 436)**, both stars can be seen and their orbit charted. In a **spectroscopic binary (p. 436)**, the stars cannot be resolved, but their orbital motion can be detected spec-

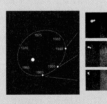

troscopically. In an **eclipsing binary (p. 437)**, the orbit is oriented in such a way that one star periodically passes in front of the other as seen from Earth and dims the light we receive. Studies of binary-star systems often allow stellar masses to be measured. The mass of a star determines the star's size, temperature, and brightness. Hot blue giants are much more massive than the Sun; cool red dwarfs are much less massive. High-mass stars burn their fuel rapidly and have much shorter lifetimes than the Sun. Low-mass stars consume their fuel slowly and may remain on the main sequence for trillions of years.

Mastering ASTRONOMY *For instructor-assigned homework go to* **www.masteringastronomy.com**

Problems labeled **POS** explore the process of science | **VIS** problems focus on reading and interpreting visual information

REVIEW AND DISCUSSION

1. How is parallax used to measure the distances to stars?

2. What is a parsec? Compare it with the astronomical unit.

3. Explain two ways in which a star's real motion through space translates into motion that is observable from Earth.

4. How do astronomers go about measuring stellar luminosities?

5. Describe how astronomers measure stellar radii.

6. Describe some characteristics of red-giant and white-dwarf stars.

7. What is the difference between absolute and apparent brightness?

8. How do astronomers measure stellar temperatures?

9. **POS** Briefly describe how stars are classified according to their spectral characteristics.

10. Why do some stars have very few hydrogen lines in their spectra?

11. What information is needed to plot a star on the H–R diagram?

12. What is the main sequence? What basic property of a star determines where it lies on the main sequence?

13. How are distances determined by spectroscopic parallax?

14. Why does the H–R diagram constructed from data on the brightest stars differ so much from the diagram constructed from data on the nearest stars?

15. Which stars are most common in our Galaxy? Why don't we see many of them in H–R diagrams?

16. Which stars are least common in our Galaxy?

17. **POS** How can stellar masses be determined by observing binary-star systems?

18. High-mass stars start off with much more fuel than low-mass stars. Why don't high-mass stars live longer?

19. **POS** In general, is it possible to determine the age of an individual star simply by noting its position on an H–R diagram? Explain.

20. Visual binaries and eclipsing binaries are relatively rare compared with spectroscopic binaries. Why is this?

CONCEPTUAL SELF-TEST: MULTIPLE CHOICE

1. **VIS** If Earth's orbit around the Sun were smaller, the parallactic angle to the star shown in Figure 17.1 ("Stellar Parallax") would be (**a**) smaller; (**b**) larger; (**c**) the same.

2. From a distance of 1 parsec, the angular size of Earth's orbit would be (**a**) 1 degree; (**b**) 2 degrees; (**c**) 1 arc minute; (**d**) 2 arc seconds.

3. According to the inverse-square law, if the distance to a light-bulb increases by a factor of 5, the bulb's apparent brightness (**a**) stays the same; (**b**) becomes 5 times less; (**c**) becomes 10 times less; (**d**) becomes 25 times less.

4. Compared with a star of absolute magnitude −2 at a distance of 100 pc, a star of absolute magnitude 5 at a distance of 10 pc will appear (**a**) brighter; (**b**) fainter; (**c**) to have the same brightness; (**d**) bluer.

5. **VIS** Pluto's apparent magnitude is approximately 14. According to Figure 17.7 ("Apparent Magnitude"), Pluto can be seen (**a**) with the naked eye on a dark night; (**b**) using binoculars; (**c**) using a 1-m telescope; (**d**) only with the *Hubble Space Telescope*.

6. Stars of spectral class M do not show strong lines of hydrogen in their spectra because (**a**) they contain very little hydrogen; (**b**) their surfaces are so cool that most hydrogen is in the ground state; (**c**) their surfaces are so hot that most hydrogen is ionized; (**d**) the hydrogen lines are swamped by even stronger lines of other elements.

7. Cool stars can be very luminous if they are very (**a**) small; (**b**) hot; (**c**) large; (**d**) close to our solar system.

8. **VIS** According to Figure 17.13 ("H–R Diagram of Prominent Stars"), Barnard's star must be (**a**) hotter; (**b**) larger; (**c**) closer to us; (**d**) bluer than Proxima Centauri.

9. **VIS** Figure 17.15 ("H–R Diagram of Bright Stars") shows Vega and Arcturus at approximately the same level on the vertical axis. This means that Arcturus must be (**a**) hotter than; (**b**) fainter than; (**c**) larger than; (**d**) of the same spectral class as Vega.

10. The mass of a star may be determined (**a**) by measuring its luminosity; (**b**) by determining its composition; (**c**) by measuring its Doppler shift; (**d**) by studying its orbit around a binary companion.

PROBLEMS

The number of dots preceding each Problem indicates its approximate level of difficulty.

1. • How far away is the star Spica, whose parallax is 0.012″? What would Spica's parallax be if it were measured from an observatory on Neptune's moon Triton as Neptune orbited the Sun?

2. •• A star lying 20 pc from the Sun has proper motion of 0.5″/yr. What is the star's transverse velocity? If the star's spectral lines are observed to be redshifted by 0.01 percent, calculate the magnitude of its three-dimensional velocity relative to the Sun.

3. • What is the luminosity of a star having three times the radius of the Sun and a surface temperature of 10,000 K?

4. • A certain star has a temperature twice that of the Sun and a luminosity 64 times greater than the solar value. What is the radius of the star, in solar units?

5. •• Two stars—A and B, with luminosities 0.5 and 4.5 times the luminosity of the Sun, respectively—are observed to have the same apparent brightness. Which star is more distant, and how much farther away is it than the other?

6. •• Calculate the solar energy flux (energy received per unit area per unit time), as seen from a distance of 10 pc from the Sun. Compare your answer with the solar constant at Earth.

7. •• Using the data shown in Figure 17.7, calculate the greatest distance at which a star like the Sun could be seen with (a) binoculars, (b) a typical 1-m telescope, (c) a 4-m telescope, and (d) the *Hubble Space Telescope*.

8. • A star has apparent magnitude 10.0 and absolute magnitude 2.5. How far away is it?

9. ••• Two stars in an eclipsing spectroscopic binary are observed to have an orbital period of 25 days. Further observations reveal that the orbit is circular, with a separation of 0.3 AU, and that one star is 1.5 times the mass of the other. What are the masses of the stars?

10. •• Given that the Sun's lifetime is about 10 billion years, estimate the life expectancy of (a) a 0.2-solar mass, 0.01-solar luminosity red dwarf, (b) a 3-solar mass, 30-solar luminosity star, (c) a 10-solar mass, 1000-solar luminosity blue giant.

18

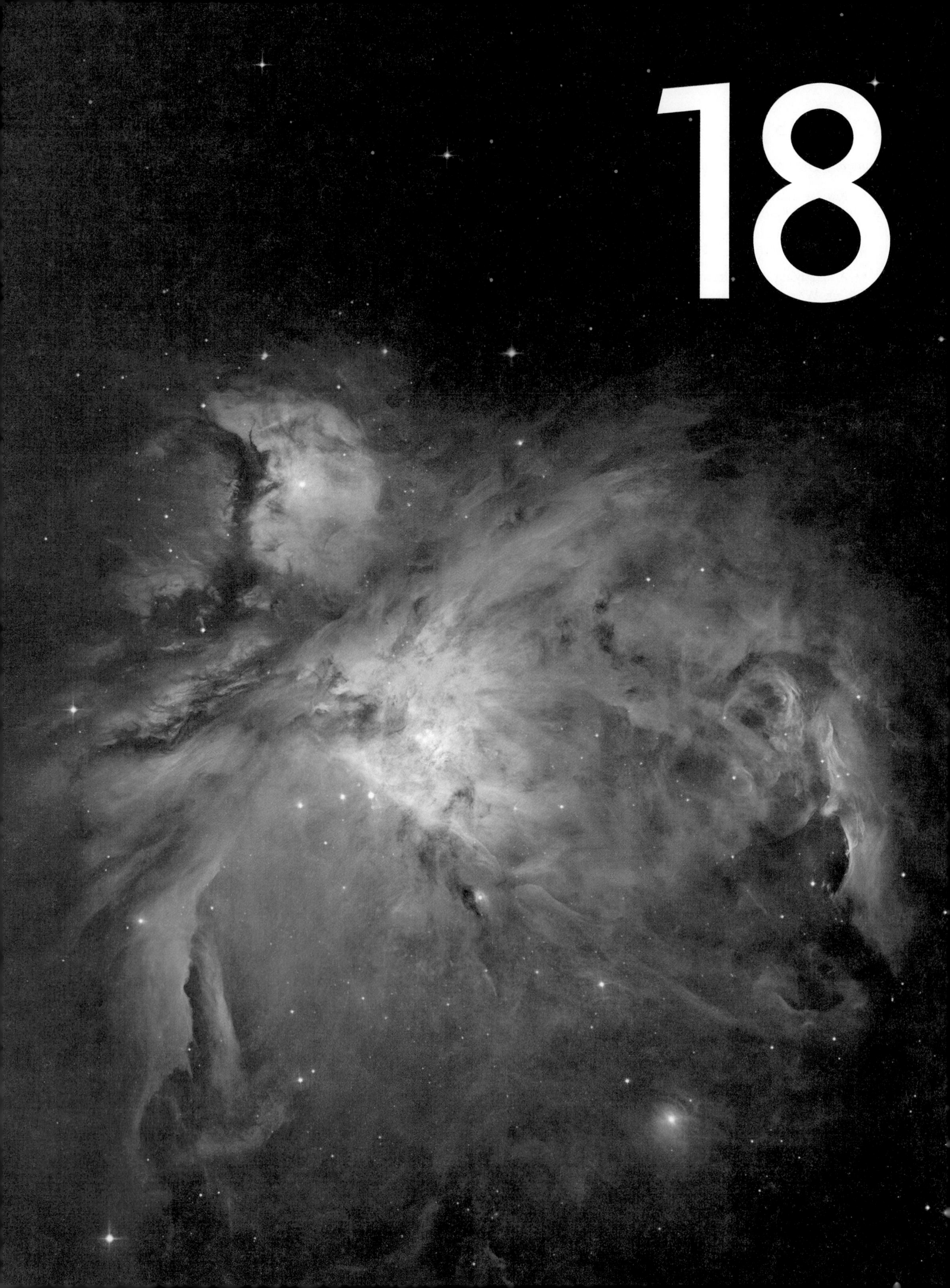

THE INTERSTELLAR MEDIUM

GAS AND DUST AMONG THE STARS

LEARNING GOALS

Studying this chapter will enable you to

1 Summarize the composition and physical properties of the interstellar medium.

2 Describe the characteristics of emission nebulae, and explain their significance in the life cycle of stars.

3 Discuss the properties of dark interstellar clouds.

4 Specify the radio techniques used to probe the nature of interstellar matter.

5 Discuss the nature and significance of interstellar molecules.

THE BIG PICTURE Interstellar space is the place both where stars are "born" and to which they return at "death." Rich in gas and dust, yet spread extraordinarily thinly throughout the vast, dark regions among the stars, normally dark interstellar matter glows brightly as gas clouds contract and form new stars.

Masting**ASTRONOMY**

Visit the Study Area in www.masteringastronomy.com for quizzes, animations, videos, interactive figures, and self-guided tutorials.

Stars and planets are not the only inhabitants of our Galaxy. The space around us harbors invisible matter throughout the dark voids between the stars. The density of this matter is extremely low—approximately a trillion trillion times less dense than matter in either stars or planets, far more tenuous than the best vacuum attainable on Earth. Only because the volume of interstellar space is so vast does its mass amount to anything at all.

So why bother to study this near-perfect vacuum? We do so for three important reasons. First, there is nearly as much mass in the "voids" among the stars as there is in the stars themselves. Second, interstellar space is the region out of which new stars are born. Third, interstellar space is also the region into which old stars expel their matter when they die. It is one of the most significant crossroads through which matter passes anywhere in our universe.

LEFT: *This remarkable image—actually a large mosaic of a billion bits of data stitched together from hundreds of smaller images—shows a classic star-forming region. The* Hubble Space Telescope *captured this optical view of the Orion Nebula, a stellar nursery lying roughly 1500 light-years from Earth, populated with thousands of young stars that have recently emerged from the loose matter comprising the surrounding nebulosity. Unprecedented detail has revealed much complexity in this nebula—but much insight as well. (STScI)*

18.1 Interstellar Matter

Figure 18.1 is a mosaic of photographs covering a much greater expanse of universal "real estate" than anything we have studied thus far. From our vantage point on Earth, the panoramic view shown here stretches all the way across the sky. On a clear night, it is visible to the naked eye as the Milky Way. In Chapter 23, we will come to recognize this band as the flattened disk, or *plane*, of our Galaxy.

The bright regions in this image are congregations of innumerable unresolved stars, merging together into a continuous blur at the resolution of the telescope. However, the dark areas are not simply "holes" in the stellar distribution. They are regions of space where *interstellar matter* obscures (blocks) the light from stars beyond, blocking from our view what would otherwise be a rather smooth distribution of bright starlight. Their very darkness means that they cannot easily be studied by the optical methods used to examine stellar matter. There is, quite simply, nothing to see!

Gas and Dust

From Figure 18.1 (see also Figure 18.4), it is evident that interstellar matter is distributed very unevenly throughout space. In some directions, the obscuring matter is largely absent, allowing astronomers to study objects literally billions of parsecs from the Sun. In other directions, there are small amounts of interstellar matter, so the obscuration is moderate, preventing us from seeing objects more than a few thousand parsecs away, but still allowing us to study nearby stars. Still other regions are so heavily obscured that starlight from even relatively nearby stars is completely absorbed before reaching Earth.

The matter among the stars is collectively termed the **interstellar medium.** It is made up of two components—*gas* and *dust*—intermixed throughout all space. The gas is made up mainly of individual atoms, of average size 10^{-10} m (0.1 nm) or so, and small molecules, no larger than about 10^{-9} m across. Interstellar dust is more complex, consisting of clumps of atoms and molecules—not unlike chalk dust or the microscopic particles that make up smoke, soot, or fog.

Apart from numerous narrow atomic and molecular absorption lines, the gas alone does not block radiation to any great extent. The obscuration that is evident in Figure 18.1 is caused by the dust. Light from distant stars cannot penetrate the densest accumulations of interstellar dust any more than a car's headlights can illuminate the road ahead in a thick fog.

Extinction and Reddening

We can use its effect on starlight to measure both the amount and the size of interstellar dust. As a rule of thumb, a beam of light can be absorbed or scattered only by particles having diameters comparable to or larger than the wavelength of the radiation involved. Thus, a range of dust particle sizes will tend to block shorter wavelengths most effectively. Furthermore, even for particles of a given size, the amount of obscuration (that is, absorption or scattering) produced by particles of a given size *increases* with *decreasing* wavelength. The size of a typical interstellar dust particle—or **dust grain**—is about 10^{-7} m (0.1 µm), comparable in size to the wavelength of visible light. Consequently, dusty regions of interstellar space are transparent to long-wavelength radio and infrared radiation, but opaque to shorter wavelength optical and ultraviolet radiation. The overall dimming of starlight by interstellar matter is called **extinction.**

Because the interstellar medium is more opaque to short-wavelength radiation than to radiation of longer wavelengths, light from distant stars is preferentially robbed of its higher

R I V U X G

▲ **FIGURE 18.1 Milky Way Mosaic** The Milky Way Galaxy photographed almost from horizon to horizon, thus spanning nearly 180°. This band contains high concentrations of stars, as well as interstellar gas and dust. The white box shows the field of view of Figure 18.4. *(Axel Mellinger)*

frequency ("blue") components. Hence, in addition to being generally diminished in brightness, stars also tend to appear redder than they really are. This effect, known as **reddening,** is conceptually similar to the process that produces spectacular red sunsets here on Earth. ∞ *(More Precisely 7-1)*

As illustrated in Figure 18.2(a), extinction and reddening change a star's apparent brightness and color. However, the patterns of absorption lines in the original stellar spectrum are still recognizable in the radiation reaching Earth, so the star's spectral class can be determined. Astronomers can use this fact to study the interstellar medium. From a main-sequence star's spectral and luminosity classes, astronomers learn the star's true luminosity and color. ∞ (Secs. 17.5, 17.6) They then measure the degree to which the starlight has been affected by extinction and reddening en route to Earth, and this, in turn, allows them to estimate both the numbers and the sizes of interstellar dust particles

along the line of sight to the star. By repeating these measurements for stars in many different directions and at many different distances from Earth, astronomers have built up a picture of the distribution and overall properties of the interstellar medium in the solar neighborhood.

Reddening can be seen very clearly in Figure 18.2(b), which shows a type of compact, dusty interstellar cloud called a *globule.* (We will discuss such clouds in more detail in Section 18.3.) The center of this cloud, called Barnard 68, is opaque to all optical wavelengths, so starlight cannot pass through it. However, near the edges, where there is less intervening cloud matter, some light does make it through. Notice how stars seen through the cloud are both dimmed and reddened relative to those seen directly. Figure 18.2(c) shows the same cloud in the infrared part of the spectrum. Much more of the radiation gets through, but even here reddening (or its infrared equivalent) can be seen.

ANIMATION/VIDEO Infrared View of Nebulae

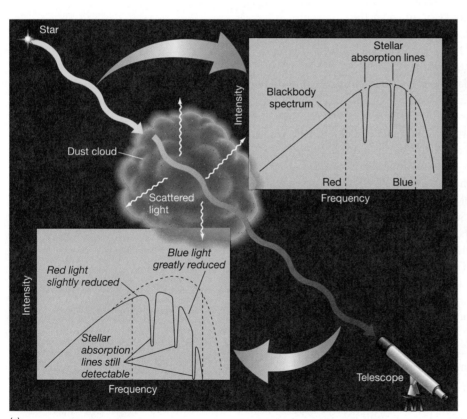

(b)

(c)

(a)

▲ **FIGURE 18.2 Reddening** (a) Starlight passing through a dusty region of space is both dimmed and reddened, but spectral lines are still recognizable in the light that reaches Earth. By recognizing stellar spectral features and inferring a star's intrinsic properties, astronomers can estimate the amount of obscuring dust along the line of sight. Note that this reddening has nothing to do with the Doppler effect—the frequencies of the lines are unchanged, although their intensities are reduced. (b) This dusty interstellar cloud, called Barnard 68, is opaque to visible light, except near its edges, where some light from background stars can be seen. Because blue light is more easily scattered or absorbed by dust than is red light, stars seen through the cloud appear red. The cloud spans about 0.2 pc and lies about 160 pc away. Frame (c) illustrates (in false color) how infrared radiation can penetrate Barnard 68, although it too is preferentially stripped of its shorter wavelengths. *(ESO)*

Overall Density

Gas and dust are found everywhere in interstellar space—no part of our Galaxy is truly devoid of matter. However, the density of the interstellar medium is extremely low. Overall, the gas averages roughly 10^6 atoms per cubic meter—just 1 atom per cubic centimeter—although there are large variations from place to place: Densities ranging from 10^4 to 10^9 atoms/m^3 have been found. Matter this diffuse is far less dense than the best vacuum—about 10^{10} molecules/m^3—ever attained in laboratories on Earth.

Interstellar dust is even rarer. On average, there is only one dust particle for every trillion or so atoms—just 10^{-6} dust particles per cubic meter, or 1000 per cubic *kilometer*. Some parts of interstellar space are so thinly populated that harvesting all the gas and dust in a region the size of Earth would yield barely enough matter to make a pair of dice.

How can such fantastically sparse matter diminish light radiation so effectively? The key is size—interstellar space is vast. The typical distance between stars (1 pc or so in the vicinity of the Sun) is much, much greater than the typical size of the stars themselves (around 10^{-7} pc). Stellar and planetary sizes pale in comparison to the vastness of interstellar space. Thus, matter can accumulate, regardless of how thinly it is spread. For example, an imaginary cylinder 1 m^2 in cross section and extending from Earth to Alpha Centauri would contain more than 10 billion billion dust particles. ∞ (Sec. 17.1) Over huge distances, dust particles accumulate slowly, but surely, to the point at which they can effectively block visible light and other short-wavelength radiation. Even though the density of matter is very low, interstellar space in the vicinity of the Sun contains about as much mass as exists in the form of stars.

Despite their rarity, dust particles make interstellar space a *relatively* dirty place. Earth's atmosphere, by comparison, is about a million times cleaner. Our air is tainted by only one dust particle for about every billion billion (10^{18}) atoms of atmospheric gas. If we could compress a typical parcel of interstellar space to equal the density of air on Earth, this parcel would contain enough dust to make a fog so thick that we would be unable to see our hand held at arm's length in front of us.

Composition

The composition of interstellar gas is reasonably well understood from spectroscopic studies of absorption lines formed when light from a distant star interacts with gas along the observer's line of sight (see Section 18.3). In most cases, the elemental abundances detected in interstellar gas mirror those found in other astronomical objects, such as the Sun, the stars, and the jovian planets. Most of the gas—about 90 percent by number—is atomic or molecular hydrogen; some 9 percent is helium, and the remaining 1 percent consists of heavier elements. The abundances of some heavy elements, such as carbon, oxygen, silicon, magnesium, and iron, are much lower in interstellar gas than in our solar system or in stars. The most likely explanation for this finding is that substantial quantities of these elements have been used to form the interstellar dust, taking them out of the gas and locking them up in a form that is much harder to observe.

In contrast to interstellar gas, the composition of interstellar dust is currently not very well known. We have some infrared evidence for silicates, graphite, and iron—the same elements that are underabundant in the gas—lending support to the theory that interstellar dust forms out of interstellar gas. The dust probably also contains some "dirty ice," a frozen mixture of ordinary water ice contaminated with trace amounts of ammonia, methane, and other chemical compounds. This composition is quite similar to that of cometary nuclei in our own solar system. ∞ (Sec. 14.2)

Dust Shape

Curiously, astronomers know the *shapes* of interstellar dust particles better than their composition. Although the minute atoms in the interstellar gas are basically spherical, the dust particles are not. Individual dust grains are apparently elongated or rodlike, as shown in Figure 18.3(a), although recent

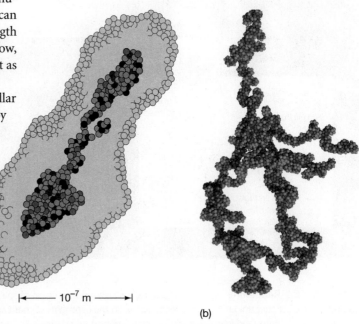

|←— 10^{-7} m —→|

(a)   (b)

▲ **FIGURE 18.3 Interstellar Dust** (a) A diagram of a typical interstellar dust particle, as inferred from polarization studies. The average size of such particles is only one ten-thousandth of a millimeter, yet space contains enough of them to obscure our view in certain directions. Part (b) shows the results of a computer simulation of how grains may grow as dust particles collide, stick, and fragment in interstellar space. The resulting grains are linear, or rodlike, on small scales, but may become tangled and twisted in complex ways on larger scales.

theoretical studies of how dust particles collide, stick, and break up suggest that their larger scale structure may be considerably more complex (Figure 18.3b).

Astronomers infer this elongated structure from the fact that the light emitted by stars is dimmed and partially **polarized,** or aligned, by the intervening dust. Recall from Chapter 3 that light consists of electromagnetic waves composed of vibrating electric and magnetic fields. ∞ (Sec. 3.2, Fig. 3.7) Normally, these waves are randomly oriented, and the radiation is said to be unpolarized. Stars emit unpolarized radiation from their photospheres. However, under the right conditions, the radiation can become polarized en route to Earth, with the electric fields all vibrating in roughly the same plane. One way in which this can happen is if the radiation interacts with an elongated dust grain, which tends to absorb electric waves vibrating parallel to its length.

Thus, if the light detected by our telescope is polarized, it is because some interstellar dust lies between the emitting object and Earth. Based on this reasoning, astronomers have determined not only that interstellar dust particles must be elongated in shape, but also that they tend to be *aligned* over large regions of space.

The alignment of the interstellar dust is the subject of ongoing research among astronomers. The current view, accepted by most, is that the dust particles are affected by a weak interstellar magnetic field, perhaps a million times weaker than Earth's field. Each dust particle responds to the field in much the same way that small iron filings are aligned by an ordinary bar magnet. Measurements of the blockage and polarization of starlight thus yield information about the size and shape of interstellar dust particles, as well as about magnetic fields in interstellar space.

CONCEPT CHECK

✔ If space is a near-perfect vacuum, how can there be enough dust in it to block starlight?

18.2 Emission Nebulae

Figure 18.4 shows a magnified view of the central part of Figure 18.1 (the region indicated by the rectangle in the earlier figure), in the general direction of the constellation Sagittarius. The field of view is mottled with stars and interstellar matter. The patchiness of the obscuration is evident. In addition, several large fuzzy patches of light are clearly visible. These fuzzy objects, labeled M8, M16, M17, and M20, correspond to the 8th, 16th, 17th, and 20th objects in a catalog compiled by Charles Messier, an 18th-century French astronomer.* Today they are known as **emission nebulae**—glowing clouds of hot interstellar matter. Figure 18.5 enlarges the left side of Figure 18.4, showing the nebulae more clearly.

---

*Messier was actually more concerned with making a list of celestial objects that might be confused with comets, his main astronomical interest. However, the catalog of 109 "Messier objects" is now regarded as a much more important contribution to astronomy than any comets Messier discovered.

◄ **FIGURE 18.4 Milky Way** A wide-angle photograph of a great swath of space in the direction of the center of our Galaxy, showing regions of brightness (vast fields of stars) as well as regions of darkness (where interstellar matter obscures the light from more distant stars). The field of view is roughly 30° across. The four nebulae discussed later in the chapter are labeled. (*Palomar/Caltech*)

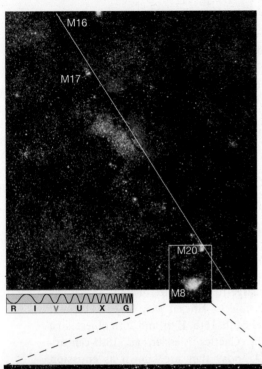

M16

M17

M20

M8

R I V U X G

◀ FIGURE 18.5 Galactic Plane A black-and-white photograph of a small portion (about 12° across) of the region of the sky shown in Figure 18.4, showing stars, gas, and dust, as well as several distinct fuzzy patches of light known as emission nebulae. The plane of the Milky Way is marked with a white diagonal line. (*Harvard College Observatory*)

R I V U X G

▲ FIGURE 18.6 M20–M8 Region A true-color enlargement of the bottom of Figure 18.5, showing M20 (top) and M8 (bottom) more clearly. The two nebulae are only a few degrees apart on the sky. (*P. Perkins*)

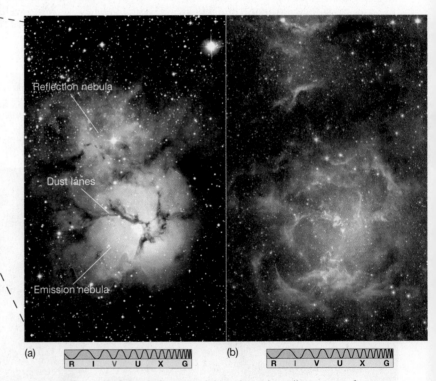

Reflection nebula

Dust lanes

Emission nebula

(a) R I V U X G   (b) R I V U X G

▲ FIGURE 18.7 Trifid Nebula Dotted throughout the Milky Way, star-forming regions stand out brightly against the surrounding darkness—if you look closely enough. (a) Further enlargement of the top of Figure 18.6, showing only M20 and its interstellar environment. The nebula itself (in red) is about 6 pc in diameter. It is often called the Trifid Nebula because of the dust lanes (in black) that trisect its midsection. The blue reflection nebula is unrelated to the red emission nebula; it is caused by starlight reflected from intervening dust particles. (b) A false-color infrared image taken by the *Spitzer Space Telescope* reveals bright regions of star-forming activity mostly in those lanes of dust. (*AURA; NASA*)

Observations of Emission Nebulae

Historically, astronomers have used the term **nebula** to refer to any "fuzzy" patch (bright or dark) on the sky—any region of space that was clearly distinguishable through a telescope, but not sharply defined, unlike a star or a planet. We now know that many (although not all) nebulae are clouds of interstellar dust and gas.

If a cloud happens to obscure stars lying behind it, we see it as a dark patch on a bright background, as in Figures 18.1, 18.2(b), and 18.4—a dark nebula. But if something within the cloud—a group of hot young stars, for example—causes it to glow, then we see a bright emission nebula instead. The method of spectroscopic parallax applied to stars that are visible within the emission nebulae shown in Figure 18.5 indicates that their distances from Earth range from 1200 pc (M8) to 1800 pc (M16). ∞ (Sec. 17.6) Thus, all four nebulae are near the limit of visibility for any object embedded in the dusty galactic plane. M16, at the top left, is approximately 1000 pc from M20, near the bottom.

We can gain a better appreciation of these nebulae by examining progressively smaller fields of view. Figure 18.6 is an enlargement of the region near the bottom of Figure 18.5, showing M20 at the top and M8 at the bottom, only a few degrees away. Figure 18.7 is an enlargement of the top of Figure 18.6, presenting a close-up of M20 and its immediate environment. The total area of the close-up view displayed measures some 10 pc across. Emission nebulae are among the most spectacular objects in the entire universe, yet they appear only as small, undistinguished patches of light when viewed in the larger context of the Milky Way, as in Figure 18.4. Perspective is crucial in astronomy.

The emission nebulae shown in Figures 18.5–18.7 are regions of glowing, ionized gas. At or near the center of each is at least one newly formed hot O- or B-type star producing huge amounts of ultraviolet light. As ultraviolet photons travel outward from the star, they ionize the surrounding gas. As electrons recombine with nuclei, they emit visible radiation, causing the gas to fluoresce, or glow. ∞ (Sec. 4.2) The reddish hue of these nebulae—and, in fact, of all emission nebulae—results when hydrogen atoms emit light in the red part of the visible spectrum. Specifically, it is caused by the emission of radiation at 656.3 nm—the Hα line discussed in Chapter 4. ∞ (More Precisely 4-1) Other elements in the nebula also emit radiation as their electrons recombine, but because hydrogen is so plentiful, its emission usually dominates.

Woven through the glowing nebular gas, and plainly visible in Figures 18.5–18.7, are lanes of dark, obscuring dust. These **dust lanes** are part of the nebulae and are not just unrelated dust clouds that happen to lie along our line of sight. The bluish region visible in Figures 18.6 and 18.7 immediately above M20 is another type of nebula unrelated to the red emission nebula itself. Called a **reflection nebula**, it is caused by starlight scattered from dust particles in interstellar clouds located just off the line of sight between Earth and the bright stars within M20. Reflection nebulae appear blue for much the same reason that Earth's daytime sky is blue: short-wavelength blue light is more easily scattered by interstellar matter back toward Earth and into our detectors. ∞ (More Precisely 7-1) Figure 18.8 sketches some of the key

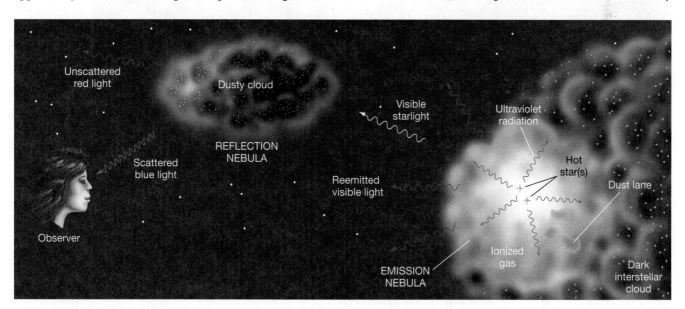

▲ **FIGURE 18.8 Nebular Structure** An emission nebula results when ultraviolet radiation from one or more hot stars ionizes part of an interstellar cloud. The nebula's reddish color is produced as electrons and protons recombine to form hydrogen atoms. Dust lanes may be seen if part of the parent cloud happens to obscure the emitting region. If some starlight happens to encounter another dusty cloud (or perhaps another part of the cloud harboring the emission nebula), some of the radiation, particularly at the shorter wavelength blue end of the spectrum, may be scattered back toward Earth, forming a reflection nebula.

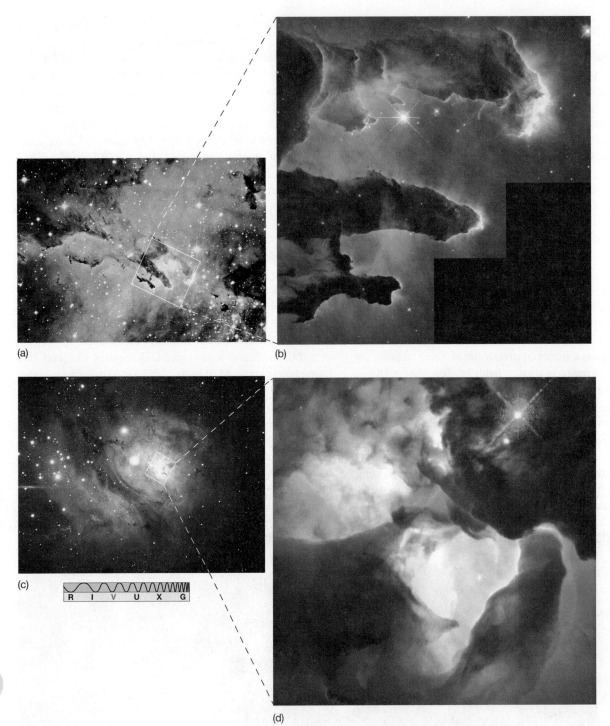

(a)

(b)

(c)

R I V U X G

(d)

▲ FIGURE 18.9 Emission Nebulae Enlargements of selected portions of Figure 18.5. (a) M16, the Eagle Nebula. (b) A *Hubble* image of huge pillars of cold gas and dust inside M16 shows delicate sculptures created by the action of stellar ultraviolet radiation on the original cloud. (c) M8, the Lagoon Nebula. (d) A high-resolution view of the core of M8, a region known as the Hourglass. Notice the irregular shape of the emitting regions, the characteristic red color of the light in the left frames, the bright stars within the gas, and the patches of obscuring dust. The insets at right are not shown in true color, rather the various colors accentuate observations at different wavelengths: Green represents emission from hydrogen atoms, red emission from singly ionized sulfur, and blue emission from doubly ionized oxygen. *(AURA; NASA)*

features of emission nebulae, illustrating the connection between the central stars, the nebula itself, and the surrounding interstellar medium.

Figure 18.9 shows enlargements of two of the nebulae visible in Figure 18.5. Notice again the hot, bright stars embedded within the glowing nebular gas and the predominant red coloration of the emitted radiation in parts (a) and (c). The relationship between the nebulae and their dust lanes is again evident in Figures 18.9(b) and (d), where regions of gas and dust are simultaneously silhouetted against background nebular emission and illuminated by foreground nebular stars.

The interaction between stars and gas is particularly striking in Figure 18.9(b). The three dark "pillars" visible in this spectacular *Hubble Space Telescope* image are part of the interstellar cloud from which the stars formed. The rest of the cloud in the vicinity of the new stars has already been heated and dispersed by their radiation in a process known as *photoevaporation*. The fuzz around the edges of the pillars, especially at the top right and center, is the result of this ongoing process. (See also an up-close view of another such pillar in M16 in the chapter-opening photo.) As photoevaporation continues, it eats away the less dense material first, leaving behind delicate sculptures composed of the denser parts of the original cloud, just as wind and water create spectacular structures in Earth's deserts and shores by eroding away the softest rock. The process is a dynamic one: The pillars will eventually be destroyed, but probably not for another hundred thousand or so years.

Spectroscopists often refer to the *ionization state* of an atom by attaching a roman numeral to the chemical symbol for the atom—I for the neutral (that is, not ionized) atom, II for a singly ionized atom (an atom missing one electron), III for a doubly ionized atom (one missing two electrons), and so on. Because emission nebulae are composed mainly of ionized hydrogen, they are often referred to as **HII regions.** Regions of space containing primarily neutral (atomic) hydrogen are known as **HI regions.**

Nebular Spectra

Most of the photons emitted by the recombination of electrons with atomic nuclei escape from the emission nebulae. Unlike the ultraviolet photons originally emitted by the embedded stars, these reemitted photons do not have enough energy to ionize the nebular gas, so they pass through the nebula relatively unhindered. Some eventually reach Earth. By studying these lower-energy photons, we can learn much about the detailed properties of emission nebulae.

Because at least one hot star resides near the center of every emission nebula, we might think that the combined spectrum of the star and the nebula would be hopelessly confused. In fact, they are not: We can easily distinguish nebular spectra from stellar spectra because the physical

conditions in stars and emission nebulae differ so greatly. In particular, emission nebulae are made of hot, thin gas that, as we saw in Chapter 4, yields detectable *emission* lines. ∞ (Sec. 4.1) When our spectroscope is trained on a star, we see a familiar stellar spectrum, consisting of a blackbody-like continuous spectrum and absorption lines, together with superimposed emission lines from the nebular gas. When no star appears in the field of view, only the emission lines are seen. Analyses of nebular spectra show compositions close to those derived from observations of the Sun and other stars and elsewhere in the interstellar medium: Hydrogen is about 90 percent abundant by number, followed by helium at about 9 percent; the heavier elements together make up the remaining 1 percent.

Unlike stars, nebulae are large enough for their actual sizes to be measurable by simple geometry. Coupling this information on size with estimates of the amount of matter along our line of sight (as revealed by the nebula's total emission of light), we can find the nebula's density. Generally, emission nebulae have only a few hundred particles, mostly protons and electrons, in each cubic centimeter—a density some 10^{22} times lower than that of a typical planet. Spectral-line widths imply that the gas atoms and ions have temperatures around 8000 K. ∞ (Sec. 4.5) Table 18.1 lists some vital statistics for each of the nebulae shown in Figure 18.5.

"Forbidden" Lines

When astronomers first studied the spectra of emission nebulae, they found many lines that did not correspond to anything observed in terrestrial laboratories. For example, in addition to the dominant red coloration just discussed, many nebulae emit light with a characteristic green color (see Figure 18.10). The greenish tint of portions of these nebulae puzzled astronomers in the early 20th century and defied explanation in terms of the properties of spectral lines known at the time, prompting speculation that the nebulae contained elements that were unknown on Earth. Some scientists even went so far as to invent the term "nebulium" for a supposed new element, much as the name helium came about when that element was first discovered in the Sun (recall also the fictitious element "coronium" from Chapter 16). ∞ (Sec. 16.3)

Later, with a fuller understanding of the workings of the atom, astronomers realized that these lines did in fact result from electron transitions within the atoms of familiar elements, but under unfamiliar conditions that were not reproducible in laboratories. Astronomers now understand that the greenish tint in Figure 18.10(b) and (c) is caused by a particular electron transition in doubly ionized oxygen. However, the structure of oxygen is such that an ion in the higher energy state for this transition tends to remain there for a very long time—many hours, in fact—before dropping

TABLE 18.1 Some Nebular Properties

Object	Approximate Distance (pc)	Average Diameter (pc)	Density (10^6 particles/m^3)	Mass (solar masses)	Temperature (K)
M8	1200	14	80	2600	7500
M16	1800	8	90	600	8000
M17	1500	7	120	500	8700
M20	1600	6	100	250	8200

back to the lower state and emitting a photon. Only if the ion is left undisturbed during this time, and not kicked into another energy state by a random interaction with another atom or molecule in the gas, will the transition actually occur and the photon be emitted.

In a terrestrial experiment, no atom or ion is left undisturbed for long. Even in a "low-density" laboratory gas, there are many trillions of particles per cubic meter, and each particle undergoes millions of collisions with other gas particles every second. The result is that an ion in the particular energy state that produces the peculiar green line in the nebular spectrum never has time to emit its photon in the lab—collisions kick it into some other state long before that occurs. For this reason, the line is usually called *forbidden,* even though it violates no law of physics; it simply occurs on Earth with such low probability that it is never seen.

In a typical emission nebula, the density is so low that collisions between particles are extremely rare. There is plenty of time for the excited ion to emit its photon, so the forbidden line is produced. Numerous forbidden lines are known in nebular spectra. These lines remind us once again that the environment in the interstellar medium is very different from conditions on Earth and warn us of the potential difficulties involved in extending our terrestrial

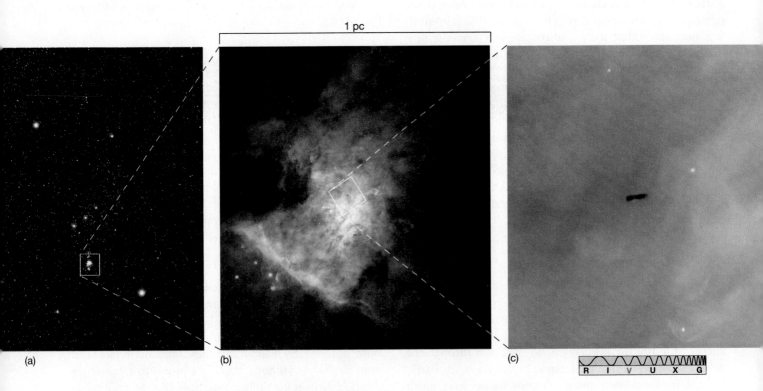

▲ **FIGURE 18.10 Orion Nebula** (a) Lying some 450 pc from Earth, the Orion Nebula (M42) is visible to the naked eye as the fuzzy middle "star" of Orion's sword. (b) Like all emission nebulae, the Orion Nebula consists of hot, glowing gas powered by a group of bright stars in the center. In addition to exhibiting red Hα emission, parts of the nebula show a slight greenish tint, caused by a so-called forbidden transition in ionized oxygen. (c) A high-resolution image shows rich detail in a region about 0.5 light-year across. Structural details are visible down to a level of 0.1", or 6 light-*hours*—a scale comparable to the dimensions of our solar system. *(NASA; ESO)*

experience from our laboratories to the study of interstellar space.

Some regions of interstellar space contain extremely dilute, even hotter gas than is found within emission nebulae. Ultraviolet observations by space-based instruments have found that these superheated interstellar "bubbles," making up the *intercloud medium,* may extend far into interstellar space beyond our local neighborhood and, conceivably, into the even vaster spaces among the galaxies. This high-temperature gas is probably the result of the violent expansion of debris from stars that exploded long ago. Somewhat like the Sun's faint corona, these regions are dark despite their high temperatures because the density of matter there is very low. ∞ (Sec. 16.3)

The Sun seems to reside in one such low-density region—a huge cavity called the "Local Bubble," sketched in Figure 18.11. The Local Bubble contains about 200,000 stars and extends for nearly 100 pc. It was probably carved out by multiple supernova explosions (see Chapters 20 and 21) that occurred several hundred thousand years ago in the Scorpius–Centaurus association, a rich cluster of bright young stars. Perhaps our hominid ancestors may have seen these ancient events—stellar catastrophes as bright as the full Moon—that now aid modern astronomers.

CONCEPT CHECK

✔ If emission nebulae are powered by ultraviolet radiation from very hot (blue-white) stars, why do they appear red?

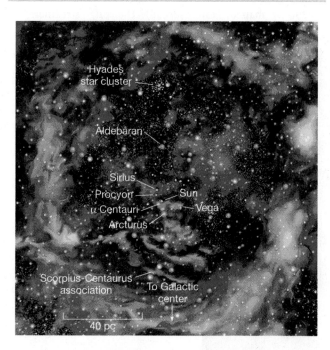

▲ FIGURE 18.11 **Local Bubble** The Sun resides in a vast low-density region of space that engulfs us nearly spherically. This cavity was likely caused by stellar explosions long ago, which then heated the nearby interstellar gas and expelled it well out of the solar neighborhood. Several prominent stars in our nighttime sky are plotted in this artist's conception, which depicts what the "bubble" might look like from afar.

18.3 Dark Dust Clouds

Emission nebulae and even the much larger interstellar bubbles are only small components of the interstellar medium. Most of space—in fact, more than 99 percent of it—is devoid of such regions and contains no stars. It is simply cold and dark. Look again at Figure 18.4, or just ponder the evening sky. The dark regions are by far the most representative of interstellar space. The average temperature of a typical dark region of interstellar matter is about 100 K. Compare this with 273 K, at which water freezes, and 0 K, at which atomic and molecular motions cease. ∞ *(More Precisely 3-1)*

Within these vast, dark interstellar voids lurks another type of astronomical object: the **dark dust cloud.** Dark dust clouds are even colder than their surroundings (with temperatures as low as a few tens of kelvins) and thousands or even millions of times denser. Along any given line of sight, their densities can range from 10^7 atoms/m^3 to more than 10^{12} atoms/m^3 (10^6 atoms/cm^3). Dark dust clouds are often called *dense* interstellar clouds by researchers, but we must recognize that even these densest interstellar regions are barely denser than the best vacuum achievable in terrestrial laboratories. Still, it is because their density is much larger than the average value of 10^6 atoms/m^3 in interstellar space that we can distinguish these clouds from the surrounding expanse of the interstellar medium.

Obscuration of Visible Light

Interstellar clouds bear little resemblance to terrestrial clouds. Most are much bigger than our solar system, and some are many parsecs across. (Yet even so, they make up no more than a few percent of the entire volume of interstellar space.) Despite their name, these clouds are made up primarily of gas, just like the rest of the interstellar medium. However, their absorption of starlight is due almost entirely to the dust they contain.

Figure 18.12(a) shows a region called L977, in the constellation Cygnus. L977 is a classic example of a dark dust cloud. The dense globule Barnard 68, shown in Figure 18.2(b), is another. Some early (18th-century) observers thought that these dark patches on the sky were simply empty regions of space that happened to contain no bright stars. However, by the late 19th century, astronomers had discounted this idea. They realized that seeing clear spaces among the stars would be like seeing clear tunnels between the trees in a huge forest, and it was extremely unlikely that as many tunnels would lead away from Earth as would be required to explain the observed dark regions.

Despite this realization, before the advent of radio astronomy astronomers had no direct means of studying clouds such as L977. Emitting no visible light, they are generally undetectable to the eye, except by the degree to which they dim starlight. However, as shown in Figure 18.12(b),

(a)

R I V U X G

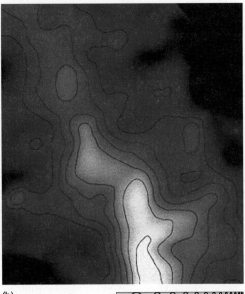

(b)

R I V U X G

◀ **FIGURE 18.12 Obscuration and Emission** (a) At optical wavelengths, this dark dust cloud (known as L977) can be seen only by its obscuration of background stars. (b) At radio wavelengths, it emits strongly in the CO molecular line, with the most intense radiation coming from the densest part of the cloud. *(C. and E. Lada)*

the cloud's radio emission—in this case from carbon monoxide (CO) molecules contained within its volume—outlines the cloud clearly at radio wavelengths, providing an indispensible tool for the study of such objects. We will return to the subject of molecular emission from interstellar clouds in Section 18.5.

Figure 18.13 is a spectacular wide-field image of another dark dust cloud. Taking its name from a neighboring star system, Rho Ophiuchus, this dust cloud resides relatively

nearby—about 170 pc from the Sun—making it one of the most intensely studied regions of star formation in the Milky Way. Pockets of heavy blackness mark regions where the dust and gas are especially concentrated and the light from the background stars is completely obscured. Measuring several parsecs across, the Ophiuchus cloud is only a tiny part of the grand mosaic shown in Figure 18.1. The cloud clearly is far from spherical. Indeed most interstellar clouds are very irregularly shaped. Note especially the long "streamers" of

Reflection nebula

Dust cloud

Antares

M4

R I V U X G

◀ **FIGURE 18.13 Dark Dust Cloud** The Ophiuchus dark dust cloud resides only 170 pc away, surrounded by colorful stars and nebulae that are actually small illuminated parts of a much bigger, and invisible, molecular cloud engulfing much of the 6-degree-wide region shown. Many stages of star formation can be seen in this spectacular four-image mosaic. The dark cloud itself is "visible" only because it blocks light coming from stars behind it. Notice the cloud's irregular shape, and especially its long "streamers" at upper left. The bright, giant star Antares, the (much more distant) star cluster M4, and a nearby blue reflection nebula are also noted. *(R. Gendler)*

(relatively) dense dust and gas at upper left. By contrast, the bright patches within the dark regions are foreground objects—emission nebulae and groups of bright stars. Some of them are part of the cloud itself, where newly formed stars near the edge of the cloud have created "hot spots" in the cold, dark gas. Others have no connection to the cloud and just happen to lie along the line of sight.

Dark and dusty interstellar clouds are sprinkled throughout our Galaxy. We can study them at optical wavelengths only if they happen to block the light emitted by more distant stars or nebulae. The dark outline of the L997 cloud in Figure 18.12(a) and the dust lanes visible in Figures 18.7 and 18.9 are good examples of this obscuration. Figure 18.14 shows another well-known, and particularly striking, example of such a cloud—the Horsehead Nebula in Orion. This curiously shaped finger of gas and dust projects out from the much larger dark cloud (called L1630) that fills the bottom half of the image and stands out clearly against the red glow of a background emission nebula. For reference, the stars and bright emission nebulae lie in front of the dark cloud; the red glow that silhouettes the Horsehead lies behind and above it.

Absorption Spectra

Astronomers first became aware of the true extent of dark interstellar clouds in the 1930s, as they studied the optical spectra of distant stars. The gas in such a cloud absorbs some of the stellar radiation in a manner that depends on the cloud's own temperature, density, and elemental abundance.

The absorption lines thus produced contain information about dark interstellar matter, just as stellar absorption lines reveal the properties of stars. ⌁ (Sec. 4.1)

Because the interstellar absorption lines are produced by cold, low-density gas, astronomers can easily distinguish them from the much broader absorption lines formed in stars' hot lower atmospheres. ⌁ (Sec. 4.5) Figure 18.15(a) illustrates how light from a star may pass through several interstellar clouds on its way to Earth. These clouds need not be close to the star, and, indeed, they usually are not. Each absorbs some of the stellar radiation in a manner that depends on its own temperature, density, velocity, and elemental abundance. Figure 18.15(b) depicts part of a typical spectrum produced in this way.

The narrow absorption lines contain information about dark interstellar clouds, just as stellar absorption lines reveal the properties of stars and nebular emission lines tell us about conditions in hot nebulae. By studying these lines, astronomers can probe the cold depths of interstellar space. In most cases, the elemental abundances detected in interstellar clouds mirror those found in other astronomical objects—perhaps not surprising, since (as we will see in Chapter 19) interstellar clouds are the regions that spawn emission nebulae and stars.

PROCESS OF SCIENCE CHECK

✔ How do astronomers use optical observations to probe the properties of dark dust clouds?

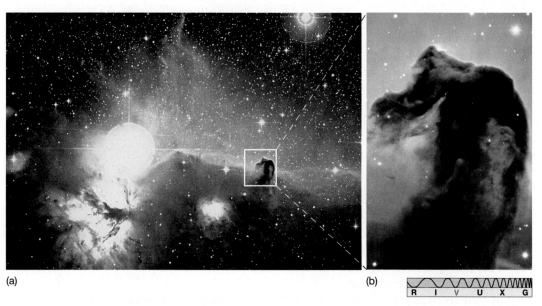

(a)   (b)

R I V U X G

▲ FIGURE 18.14 Horsehead Nebula (a) Located in the constellation Orion, not far from the Orion Nebula, this Horsehead Nebula is a striking example of a dark dust cloud, silhouetted against the bright background of an emission nebula. (b) A stunning image of the Horsehead, taken at highest resolution by the Very Large Telescope (VLT) in Chile. ⌁ (Sec. 5.2) The "neck" of the horse is about 0.25 pc across. This nebular region is roughly 1500 pc from Earth. (*Royal Observatory of Belgium; ESO*)

ANIMATION/VIDEO Horsehead Nebula

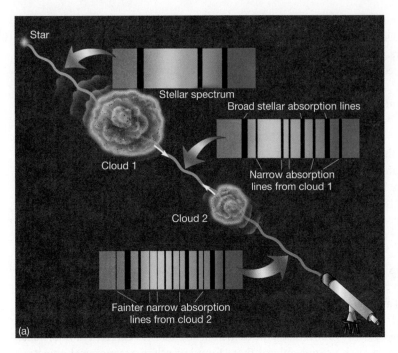

(a)

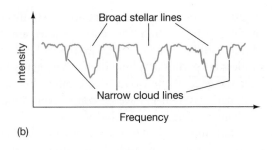

(b)

◀ FIGURE 18.15 **Absorption by Interstellar Clouds**
(a) Simplified diagram of some interstellar clouds between a hot star and Earth. Optical observations might show an absorption spectrum like that traced in (b). The wide, intense lines are formed in the star's hot atmosphere; narrower, weaker lines arise from the cold interstellar clouds. The smaller the cloud, the weaker are the lines. The redshifts or blueshifts of the narrow absorption lines provide information on cloud velocities. The widths of all the spectral lines depicted here are greatly exaggerated for the sake of clarity.

18.4 21-Centimeter Radiation

A basic difficulty with the optical technique just described is that we can examine interstellar clouds only along the line of sight to a distant star. To form an absorption line, a background source must provide radiation to absorb. The need to see stars through clouds also restricts this approach to relatively local regions, within a few thousand parsecs of Earth. Beyond that distance, stars are completely obscured, and optical observations are impossible. As we have seen, infrared observations provide a means of viewing the emission from some clouds, but they do not completely solve the problem because only the denser, dustier clouds emit enough infrared radiation for astronomers to study them in that part of the spectrum.

To probe interstellar space more thoroughly, we need a more general, more versatile observational method—one that does not rely on conveniently located stars and nebulae. In short, we need a way to detect cold, neutral interstellar matter anywhere in space through its *own* radiation. This may sound impossible, but such an observational technique does in fact exist. The method relies on low-energy *radio* emissions produced by the interstellar gas itself.

Electron Spin

Recall that a hydrogen atom has one electron orbiting a single-proton nucleus. Besides its orbital motion around the central proton, the electron also has some rotational motion—that is, *spin*—about its own axis. The proton also spins. This model is analogous to a planetary system in which, in addition to the orbital motion of a planet about a central star, both the planet (electron) and the star (proton) rotate about their own axes. But bear in mind the crucial difference between planetary and atomic systems: A planet orbiting the Sun is free to move in any orbit and spin at any rate, but within an atom, all physical quantities, such as energy, momentum, and angular momentum (spin), are *quantized*—they are permitted to take on only specific, distinct values. ∞ (Sec. 4.2)

The laws of physics dictate that there are exactly two possible spin configurations for a hydrogen atom in its ground state. The electron and proton can rotate in the same direction, with their spin axes parallel, or they can rotate with their axes antiparallel (i.e., parallel, but oppositely oriented). Figure 18.16 shows these two configurations. The antiparallel configuration has slightly less energy than the parallel state.

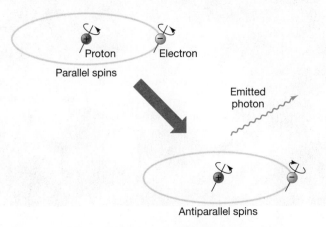

▲ FIGURE 18.16 **Hydrogen 21-cm Emission** A ground-state hydrogen atom changing from a higher-energy state (electron and proton spinning in the same direction) to a lower-energy state (spinning in opposite directions). The emitted photon carries away an energy equal to the energy difference between the two spin states.

Radio Emission

All matter in the universe tends to achieve its lowest possible energy state, and interstellar gas is no exception. A slightly excited hydrogen atom with the electron and proton spinning in the same direction eventually drops down to the less energetic, opposite-spin state as the electron suddenly and spontaneously reverses its spin. As with any other such change, the transition from a high-energy state to a low-energy state releases a photon with energy equal to the energy difference between the two levels.

Because that energy difference is very small, the energy of the emitted photon is very low. Consequently, the wavelength of the radiation is rather long—in fact, it is 21.1 cm, roughly the width of this book. That wavelength lies in the radio portion of the electromagnetic spectrum. Researchers refer to the spectral line that results from this hydrogen spin-flip process as **21-centimeter radiation.** This spectral line provides a vital probe into any region of the universe containing atomic hydrogen gas. Figure 18.17 shows typical

spectral profiles of 21-cm radio signals observed from several different regions of space. These tracings are the characteristic signatures of cold, atomic hydrogen in our Galaxy. Needing no visible starlight to help calibrate their signals, radio astronomers can observe *any* interstellar region that contains enough hydrogen gas to produce a detectable signal. Even the low-density regions between the dark clouds can be studied.

As can be seen in the figure, actual 21-cm lines are quite jagged and irregular, somewhat like nebular emission lines in appearance. The irregularities arise because there are usually numerous clumps of interstellar gas along any given line of sight, each with its own density, temperature, radial velocity, and internal motion. Thus, the intensity, width, and Doppler shift of the resultant 21-cm line vary from place to place. ∞ (Sec. 4.5) All these different lines are superimposed in the signal we eventually receive at Earth, and sophisticated computer analysis is generally required to disentangle them. The "average" figures quoted earlier for the temperatures (100 K) and densities (10^6 atoms/m^3) of the regions between the dark dust clouds are based on 21-cm measurements. Observations of the dark clouds themselves using 21-cm radiation yield densities and temperatures in good agreement with those obtained by optical spectroscopy.

All interstellar atomic hydrogen emits 21-cm radiation. But if all atoms eventually fall into their lowest-energy configuration, then why isn't all the hydrogen in the Galaxy in the lower energy state by now? Why do we see 21-cm radiation today? The answer is that the energy difference between the two states is comparable to the energy of a typical atom at a temperature of 100 K or so. As a result, atomic collisions in the interstellar medium are energetic enough to boost the electron into the higher energy configuration and so maintain comparable numbers of hydrogen atoms in either state. At any instant, any sample of interstellar hydrogen will contain many atoms in the upper level, so conditions will always be favorable for 21-cm radiation to be emitted.

Of great importance, the wavelength of this characteristic radiation is much larger than the typical size of interstellar dust particles. Accordingly, 21-cm radiation reaches Earth completely unscattered by interstellar debris. The opportunity to observe interstellar space well beyond a few thousand parsecs, and in directions lacking background stars, makes 21-cm observations among the most important and useful in all astronomy. We will see in Chapters 23 through 25 how such observations are indispensable in allowing astronomers to map out the large-scale structure of our own and other galaxies.

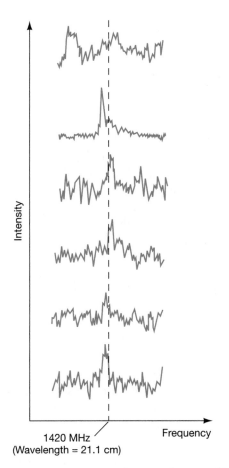

▲ **FIGURE 18.17** **21-cm Lines** Typical 21-cm radio spectral lines observed from several different regions of interstellar space. The peaks do not all occur at a wavelength of exactly 21.1 cm, corresponding to a frequency of 1420 MHz, because the gas in our Galaxy is moving with respect to Earth.

CONCEPT CHECK

✔ Why is 21-cm radiation so useful as a probe of galactic structure?

18.5 Interstellar Molecules

In some particularly cold (typically, 20-K) interstellar regions, densities can reach as high as 10^{12} particles/m^3. Until the late 1970s, astronomers regarded these regions simply as abnormally dense interstellar clouds, but it is now recognized that they belong to an entirely new class of interstellar matter. The gas particles in these regions are not in atomic form at all; they are molecules. Because of the predominance of molecules in these dense interstellar regions, they are known as **molecular clouds.** They literally dwarf even the largest emission nebulae, which were previously thought to be the most massive residents of interstellar space.

Molecular Spectral Lines

As noted in Chapter 4, much like atoms, molecules can become excited through collisions or by absorbing radiation. ∞ (Sec. 4.4) Furthermore, again like atoms, molecules eventually return to their ground states, emitting radiation in the process. The energy states of molecules are much more complex than those of atoms, however. Once more like atoms, molecules can undergo internal electron transitions, but unlike atoms, they can also rotate and vibrate. They do so in specific ways, obeying the laws of quantum physics. Figure 18.18 depicts a simple molecule rotating rapidly—that is, a molecule in an excited rotational state. After a length of time that depends on its internal makeup, the molecule relaxes back to a slower rotational rate (a state of lower energy). This change causes a photon to be emitted, carrying an energy equal to the energy difference between the two rotational states involved. The energy differences between these states are generally very small, so the emitted radiation is usually in the radio range.

We are fortunate that molecules emit radio radiation, because they are invariably found in the densest and dustiest parts of interstellar space. These are regions where the absorption of shorter wavelength radiation is enough to prohibit the use of ultraviolet, optical, and most infrared techniques that

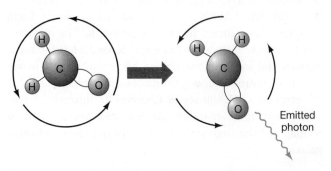

▲ **FIGURE 18.18 Molecular Emission** As a molecule changes from a rapid rotation (left) to a slower rotation (right), a photon is emitted that can be detected with a radio telescope. Depicted here is the formaldehyde molecule, H_2CO. The lengths of the curved arrows are proportional to the spin rate of the molecule.

might ordinarily detect changes in the energy states of the molecules. Only low-frequency radio radiation can escape.

Why are molecules found only in the densest and darkest of the interstellar clouds? One possible reason is that the dust serves to protect the fragile molecules from the normally harsh interstellar environment—the same absorption that prevents high-frequency radiation from getting out to our detectors also prevents it from getting in to destroy the molecules. Another possibility is that the dust acts as a catalyst that helps form the molecules. The grains provide both a place where atoms can stick together and react and a means of dissipating any heat associated with the reaction, which might otherwise destroy the newly formed molecules. Probably the dust plays both roles; the close association between dust grains and molecules in dense interstellar clouds argues strongly in favor of this view, although the details are still being debated.

Molecular Tracers

In mapping molecular clouds, radio astronomers are faced with a problem. Molecular hydrogen (H_2) is by far the most common constituent of these clouds, but unfortunately, despite its abundance, this molecule does not emit or absorb radio radiation. Rather, it emits only short-wavelength ultraviolet radiation, so it cannot easily be used as a probe of cloud structure. Nor are 21-cm observations helpful—they are sensitive only to *atomic* hydrogen, not to the *molecular* form of the gas. Theorists had expected H_2 to abound in these dense, cold pockets of interstellar space, but proof of its existence was hard to obtain. Only when spacecraft measured the ultraviolet spectra of a few stars located near the edges of some dense clouds was the presence of molecular hydrogen confirmed.

With hydrogen effectively ruled out as a probe of molecular clouds, astronomers must use observations of other molecules to study the dark interiors of these dusty regions. Molecules such as carbon monoxide (CO), hydrogen cyanide (HCN), ammonia (NH_3), water (H_2O), methyl alcohol (CH_3OH), formaldehyde (H_2CO), and about 150 others, some quite complex, are now known to exist in interstellar space.* These molecules are found only in very small quantities—they are generally 1 million to 1 billion times less abundant than H_2—but they are important as *tracers* of a cloud's structure and physical properties. They are produced by chemical reactions within molecular clouds. When we observe them, we know that the regions under study must also contain high densities of molecular hydrogen, dust, and other important constituents.

Some remarkably complex organic molecules, including formaldehyde (H_2CO), ethyl alcohol (CH_3CH_2OH), methylamine (CH_3NH_2), and formic acid (H_2CO_2), have been found in the densest of the dark interstellar clouds. Their presence has fueled speculation about the origins of life, both on Earth and in the interstellar medium—especially since the report (still unconfirmed) by radio astronomers in the mid-1990s of evidence that glycine (NH_2CH_2COOH), one of the key amino acids that form the large protein molecules in living cells, may also be present in interstellar space.

The rotational properties of different molecules often make them suitable as probes of regions with different physical properties. Formaldehyde may provide the most useful information on one region, carbon monoxide on another, and water on yet another, depending on the densities and temperatures of the regions involved. The data obtained thereby equip astronomers with a sophisticated spectroscopic "toolbox" for studying the interstellar medium.

For example, Figure 18.19 shows some of the sites where formaldehyde molecules have been detected near M20. At practically every dark area sampled between M16 and M8, the formaldehyde molecule is present in surprisingly large abundance (although it is still far less common than H_2). Analyses of spectral lines at many locations along the 12°-wide swath shown in Figure 18.5 indicate that the temperature and density are much the same in all the molecular clouds studied (50 K and 10^{11} molecules/m^3, on average). Figure 18.20 shows a contour map of the distribution of formaldehyde molecules in the immediate vicinity of the M20 nebula. After radio spectral lines of formaldehyde were observed at various locations, contours connecting regions of similar abundance were drawn. Notice that the amount of formaldehyde (and, we assume, the

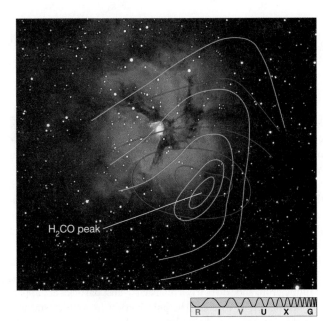

▲ **FIGURE 18.20 Molecules Near M20** Optical and infrared observations of star-forming regions produce more striking images, but radio studies yield vital information unobtainable at other wavelengths. This contour map of formaldehyde near the M20 nebula shows how that type of molecular gas is especially abundant in the darkest interstellar regions. Other kinds of molecules are similarly distributed. The contour values increase from the outside to the inside, so the maximum density of formaldehyde lies just to the bottom right of the visible nebula. The red and green contours outline the intensity of formaldehyde lines at different rotational frequencies. The nebula itself is about 6 pc across. (*Background image: AURA*)

amount of hydrogen) peaks in a dark region well away from the visible nebula.

Radio maps of interstellar gas and infrared maps of interstellar dust reveal that molecular clouds do not exist as distinct objects in space. Rather, they make up huge **molecular cloud complexes,** typically up to 50 pc across and containing enough gas to make a million stars like our Sun. About a thousand such giant complexes are currently known in our Galaxy.

The discovery of many interstellar molecules in the 1970s forced astronomers to rethink and reobserve interstellar space. In doing so, they realized that this active and vital domain is far from the void suspected by theorists not long before. As we will see in Chapter 19, regions of space once thought to contain nothing more than galactic "garbage"—the cool, tenuous darkness among the stars—now play a critical role in our understanding of stars and the interstellar medium from which they are born.

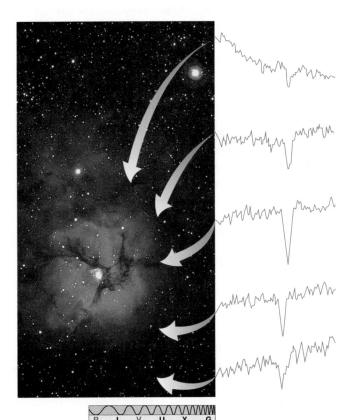

▲ **FIGURE 18.19 M20's Environment** Spectra indicate that formaldehyde molecules exist in the extended environment (arrows) around M20. The lines, which are formed by the absorption of background radiation, are most intense both in the dark dust lanes trisecting the nebula and in the dark regions beyond the nebula. (*Background image: AURA*)

PROCESS OF SCIENCE CHECK

✔ In mapping molecular clouds, why do astronomers use observations of "minority" molecules such as carbon monoxide and formaldehyde when these molecules constitute only a tiny fraction of the total number of molecules in interstellar space?

CHAPTER REVIEW

SUMMARY

1 The **interstellar medium (p. 446)** occupies the space among the stars. It is  made up of cold (less than 100 K) gas, mostly atomic or molecular hydrogen and helium, and **dust grains (p. 446)**. Interstellar dust is highly effective at blocking our view of distant stars, even though the density of the interstellar medium is very low. The spatial distribution of interstellar matter is patchy. The general diminution of starlight by dust is called **extinction (p. 446)**. In addition, the dust preferentially absorbs short-wavelength radiation, leading to a distinct **reddening (p. 447)** of light passing through interstellar clouds. Interstellar dust is thought to be composed of silicates, graphite, iron, and "dirty ice." Interstellar dust particles are apparently elongated or rodlike. The **polarization (p. 449)** of starlight provides a means of studying these particles.

2 A **nebula (p. 451)** is a general term for any fuzzy bright or dark patch on the sky. **Emission nebulae (p. 449)** are extended clouds of hot, glowing interstellar gas. Associated with star formation, they result when hot O- and B-type stars heat and ionize their surroundings. Studies of the emission lines produced by excited 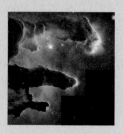 nebular atoms allow astronomers to measure the properties of nebulae. Nebulae are often crossed by dark **dust lanes (p. 451)**, part of the larger cloud from which they formed.

3 **Dark dust clouds (p. 455)** are cold, irregularly shaped regions in the interstellar medium whose constituent dust diminishes or completely obscures the light from background stars. The interstellar medium also contains many cold, dark **molecular clouds** **(p. 460)**, which are cool and dense enough that much of the gas exists in molecular form. Dust within these clouds probably both protects the molecules and acts as a catalyst to help them form. Often, several molecular clouds are found close to one another, forming 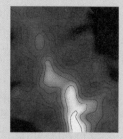 an enormous **molecular cloud complex (p. 461)** millions of times more massive than the Sun.

4 Cold, dark regions of interstellar space containing atomic hydrogen can be observed in the radio spectrum via the **21-centimeter radiation (p. 459)** produced when the electron in an atom of hydrogen reverses its spin, changing its energy slightly in the process. Molecular  clouds are observed mainly through the radio radiation emitted by the molecules they contain. Radio waves are not appreciably absorbed by the interstellar medium, so astronomers making observations at these wavelengths can often "see" to great distances.

5 Hydrogen is by far the most common constituent of molecular clouds, but molecular hydrogen happens to be very hard to observe. Astronomers usually study these clouds through observations of other "tracer" molecules that are less common, but much easier to detect. Many 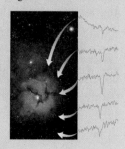 complex molecules have been identified in these regions.

Mastering ASTRONOMY *For instructor-assigned homework go to* **www.masteringastronomy.com**

Problems labeled **POS** explore the process of science | **VIS** problems focus on reading and interpreting visual information

REVIEW AND DISCUSSION

1. Give a brief description of the interstellar medium.

2. What is the composition of interstellar gas? What about interstellar dust?

3. Why is interstellar dust so much more effective than interstellar gas at absorbing starlight?

4. How dense, on average, is interstellar matter?

5. How is interstellar matter distributed throughout space?

6. What are some methods that astronomers use to study interstellar dust?

7. What is an emission nebula?

8. What is photoevaporation, and how does it change the structure and appearance of an emission nebula?

9. Why are some spectral lines that are observed in emission nebulae not normally seen in laboratories on Earth?

10. What is the Local Bubble? How did it form?

11. **POS** Describe some ways in which we can "see" a dark interstellar cloud.

12. Give a brief description of a dark dust cloud.

13. What is 21-cm radiation? With what element is it associated?

14. Why is 21-cm radiation useful to astronomers?

15. How does a molecular cloud differ from other interstellar matter?

16. **POS** Why can't astronomers use observations of hydrogen to explore the structure of molecular cloud complexes?

17. **POS** How do astronomers explore the structure of molecular cloud complexes?

18. If our Sun were surrounded by a cloud of gas, would this cloud be an emission nebula? Why or why not?

19. Compare the reddening of stars by interstellar dust with the reddening of the setting Sun.

20. What does the polarization of starlight tell us about the interstellar medium?

CONCEPTUAL SELF-TEST: MULTIPLE CHOICE

1. The chemical composition of the interstellar medium is basically similar to that of (**a**) the Sun; (**b**) Earth; (**c**) Venus; (**d**) Mars.

2. The density of atoms in the interstellar medium is most similar to (**a**) wildfire smoke; (**b**) dark rain clouds; (**c**) deep ocean water; (**d**) the interior of a TV tube.

3. Of the following objects, the one that shines most like an emission nebula shines is (**a**) a regular incandescent lightbulb with a filament; (**b**) a red hot ember from a campfire; (**c**) a glowing fluorescent light tube; (**d**) a star like the Sun.

4. Stars interact with emission nebulae by (**a**) exiting their atoms enough to emit light; (**b**) illuminating them like an advertising billboard; (**c**) causing them to contract; (**d**) heating them so they explode.

5. A dark interstellar globule is about the same size as (**a**) a cloud in Earth's atmosphere; (**b**) the entire planet Earth; (**c**) a star like the Sun; (**d**) the Oort cloud.

6. **VIS** The Ophiuchi cloud, shown in Figure 18.13 ("Dark Dust Cloud"), is dark because (**a**) there are no stars in this region; (**b**) the stars in this region are young and faint; (**c**) starlight from behind the cloud does not penetrate the cloud; (**d**) the region is too cold to sustain stellar fusion.

7. If a proton and an electron within a hydrogen atom initially have parallel spins, then change to have antiparallel spins, the atom must (**a**) absorb energy; (**b**) emit energy; (**c**) become hotter; (**d**) become larger.

8. Of the following telescopes, the one best suited to observing dark dust clouds is (**a**) an X-ray telescope; (**b**) a large visible-light telescope; (**c**) an orbiting ultraviolet telescope; (**d**) a radio telescope.

9. Of the following, the largest interstellar clouds are (**a**) molecular clouds; (**b**) dark dust clouds; (**c**) emission nebulae; (**d**) globules.

10. Molecular clouds are routinely studied using spectral lines from all but which of the following? (**a**) Molecular hydrogen; (**b**) Carbon monoxide; (**c**) Formaldehyde; (**d**) Water.

PROBLEMS

The number of dots preceding each Problem indicates its approximate level of difficulty.

1. • The average density of interstellar gas within the "Local Bubble" is much lower than the value mentioned in the text—in fact, it is roughly 10^3 hydrogen atoms/m^3. Given that the mass of a hydrogen atom is 1.7×1^{-27} kg, calculate the total mass of interstellar matter contained within a bubble volume equal in size to planet Earth.

2. • Assume the same average density as in the previous question, and calculate the total mass of interstellar hydrogen contained within a cylinder of cross-sectional area 1 m^2, extending from Earth to Alpha Centauri.

3. •• Given the average density of interstellar matter stated in Section 18.1, calculate how large a volume of space would have to be compressed to make a cubic meter of gas equal in density to air on Earth (1.2 kg/m^3).

4. • Interstellar extinction is sometimes measured in magnitudes per kiloparsec (1 kpc = 1000 pc). Light from a star 1500 pc away is observed to be diminished in intensity by a factor of 20 over and above the effect of the inverse-square law. What is the average interstellar extinction, in mag/kpc, along the line of sight?

5. •• A beam of light shining through a dense molecular cloud is diminished in intensity by a factor of 2 for every 5 pc it travels. By how many magnitudes is the light from a background star dimmed if the total thickness of the cloud is 60 pc?

6. •• A star of apparent magnitude 10 lies 500 pc from Earth. If interstellar absorption results in an average extinction of 2 mag/kpc, calculate the star's absolute magnitude and luminosity.

7. •• Spectroscopic observations of a certain star reveal it to be a B2II giant, with absolute magnitude −6. ∞ (Secs. 17.2, 17.6) The star's apparent magnitude is 14. Neglecting the effects of interstellar extinction, calculate the distance to the star. If the star's true distance is known (by other means) to be 5000 pc, calculate the average extinction, in mag/kpc, along the line of sight. ∞ (*More Precisely 17-1*)

8. • To carry enough energy to ionize a hydrogen atom, a photon must have a wavelength of less than 9.12×10^{-8} m (91.2 nm). Using Wien's law, calculate the temperature a star must have for the peak wavelength of its blackbody curve to equal this value. ∞ (Sec. 3.4)

9. •• Estimate the escape speeds near the edges of the four emission nebulae listed in Table 18.1, and compare them with the average speeds of hydrogen nuclei in those nebulae. ∞ (*More Precisely 8-1*) Do you think it is possible that the nebulae are held together by their own gravity?

10. •• If a group of interstellar clouds along the line of sight have radial velocities in the range 75 km/s (receding) to 50 km/s (approaching), calculate the range of frequencies and wavelengths over which the 21.1-cm (1420-MHz) line of hydrogen will be observed. ∞ (Sec. 3.5)

19

STAR FORMATION

A TRAUMATIC BIRTH

LEARNING GOALS

Studying this chapter will enable you to

1 Summarize the sequence of events leading to the formation of a star like our Sun.

2 Explain how the formation of a star depends on its mass.

3 Describe some of the observational evidence supporting the modern theory of star formation.

4 Explain the nature of interstellar shock waves, and discuss their possible role in the formation of stars.

5 Explain why stars form in clusters, and distinguish between open and globular star clusters.

THE BIG PICTURE Few issues in astronomy are more basic than knowing how stars form. Stars are the most numerous and obvious residents of the nighttime sky, and astronomers want to know in detail how they originate and evolve, and how they affect their surroundings.

W e now move from the interstellar medium—the gas and dust among the stars—back to the stars themselves. The next four chapters discuss the formation and evolution of stars. We have already seen that stars must evolve as they consume their fuel supply, and we have extensive observational evidence of stars at many different evolutionary stages. With the help of these observations, astronomers have developed an understanding of stellar evolution—the complex changes undergone by stars as they form, mature, grow old, and die.

We begin by studying how interstellar clouds of gas and dust are transformed into the myriad stars we see in the night sky. As we will see, the process is far from gentle—stellar nurseries are scenes of violent outbursts, interstellar shock waves, even actual collisions, as prestellar fragments grow in mass and compete for resources in a newborn cluster. The Sun and planet Earth are survivors of a similarly violent environment, some 4.5 billion years ago.

LEFT: *In this combined visible-infrared image captured by the new wide-field camera on the Hubble Space Telescope, we see a highly detailed view of the largest stellar nursery in our local galactic neighborhood. Called R136 and lying some 170,000 light-years away, this 100 light-year-wide region harbors a rich collection of thousands of young blue stars still embedded in the glowing reddish gas from which they formed a few million years ago. (STScI)*

MasteringASTRONOMY

Visit the Study Area in www.masteringastronomy.com for quizzes, animations, videos, interactive figures, and self-guided tutorials.

19.1 Star-Forming Regions

Our universe is constantly renewing itself. Literally billions of stars have been born, lived out their lives, and died since our Galaxy formed. We do not see this activity when we gaze at the nighttime sky, because the time scales on which stars play out this cosmic drama are enormously long by human standards. Even the shortest-lived O-type stars survive for millions of years. ∞ (Sec. 17.8) Nevertheless, we have plenty of evidence for ongoing stellar evolution throughout the cosmos.

Young Stars in the Universe

Our Sun, and probably most of the stars in our immediate cosmic neighborhood, formed billions of years ago. ∞ (Sec. 15.2) However, we know that many relatively nearby stars are much younger than this. The magnificent emission nebulae discussed in Chapter 18 and the ultraluminous, short-lived stars that power them are direct proof that star formation is a continuing process. ∞ (Sec. 18.2) The hottest stars in these regions must have formed less than a few million years ago—the blink of an eye, in cosmic terms—and there is no reason to suppose that galactic star formation has recently and abruptly ceased! Stars are forming all across the Milky Way, even as you read this.

In fact, star-forming regions are observed in many regions of the universe far beyond our own Galaxy. Figure 19.1 shows one of the largest such regions discovered to date. It lies in a galaxy quite similar to our own, about 1 million parsecs away. Almost 500 pc across, this vast stellar nursery dwarfs any emission nebula known in our Galaxy. Conceivably, the Milky Way may contain similarly large nebulae. If they exist, they are obscured by so much interstellar matter in our Galactic plane that we cannot see them. But whatever their size—large or small—emission nebulae are the birthplaces of all the stars in our night sky.

How and where do stars form? What factors determine the masses, luminosities, and spatial distribution of stars in our Galaxy and beyond? The association of bright emission nebulae with much larger dark dust clouds provides the key. ∞ (Sec. 18.3)

Simply put, star formation begins when part of the interstellar medium—one of those cold dark clouds—starts to collapse under its own weight. The cloud fragment heats up as it shrinks, and eventually its center becomes hot enough for nuclear fusion to begin. At that point, the contraction stops and a star is born.

R I V U X G

▲ FIGURE 19.1 **Extragalactic Star Formation** The giant star-forming region at the right, called NGC 604, is roughly 500 pc across. It is found in the nearby galaxy M33, displayed at the left on the much larger scale of 40,000 pc across. *(R. Gendler; NASA)*

Gravity and Heat

What determines which interstellar clouds collapse? For that matter, since all clouds exert a gravitational pull, why didn't they all collapse long ago? To answer these questions and understand the processes leading to the stars we see, we must explore in a little more detail the factors that compete with gravity in determining a cloud's fate. By far the most important of these is the random motion of atoms—or *heat*. *More Precisely 19-1* discusses some other factors that influence—and complicate—the star formation process.

MORE PRECISELY 19-1

Competition in Star Formation

In the text we describe star formation in terms of a competition between gravity, which tends to make interstellar clouds collapse, and heat, which tends to oppose it. In fact, the interstellar medium is a lot more complicated than our simple picture suggests, and heat is not the only factor that tends to oppose gravitational contraction. Two other important factors affecting star formation are *rotation* and *magnetism*.

Rotation—that is, spin—can compete with gravity's inward pull. As we saw in Chapter 6, a contracting cloud having even a small spin tends to develop a bulge around its midsection. ∞ (Sec. 6.7) As the cloud contracts, it must spin faster (to conserve its angular momentum), so the bulge grows and material on the edge tends to fly off into space. (Consider the analogy of mud flung from a rapidly rotating bicycle wheel.) Eventually, as in Figure 6.15, the cloud forms a flattened, rotating disk.

For material to remain part of the cloud and not be spun off into space, a force must be applied—in this case, the force of gravity. The more rapid the rotation, the greater is the tendency for the gas to escape, and the greater is the gravitational force needed to retain it. It is in this sense that we can regard rotation as opposing the inward pull of gravity. Should the rotation of a contracting gas cloud overpower gravity, the cloud would simply disperse. Thus, more mass is needed for a rapidly rotating interstellar cloud to contract to form a star than is needed for a cloud that does not rotate at all.

Magnetism can also hinder a cloud's contraction. Just as Earth, the outer planets, and the Sun all have some magnetism, magnetic fields permeate most interstellar clouds. As a cloud contracts, it heats up, and atomic encounters become violent enough to (partly) ionize the gas. As we noted in Chapter 7 in discussing Earth's Van Allen belts, and in Chapter 16 in discussing activity on the Sun, magnetic fields can exert electromagnetic control over charged particles. ∞ (Secs. 7.5, 16.5) In effect, the particles tend to become "tied" to the magnetic field—they are free to move *along* the field lines, but are inhibited from moving *perpendicular* to them.

As sketched in the accompanying figure, magnetism can hinder the contraction of an interstellar gas cloud, causing it to contract in a distorted way. Because the charged particles and the magnetic field are linked, the field itself follows the contraction of the cloud. The charged particles literally pull the magnetic field toward the cloud's center in the direction perpendicular to the field lines (shown in red). The three frames trace the evolution, top to bottom, of a slowly contracting interstellar cloud having some magnetism. The dashed lines within the cloud represent the regions where the field lines are distorted and compressed as the cloud shrinks. As the field lines are compressed, the magnetic field strength increases. In this way, the strength of the magnetism in a cloud can become much larger than that normally permeating general interstellar space. The primitive solar nebula may have contained a strong magnetic field created in just that manner.

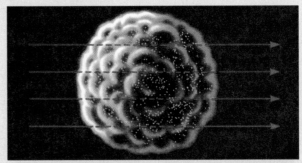

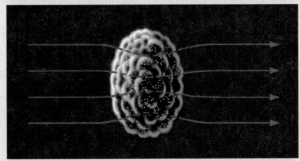

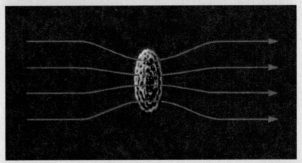

Theory suggests that even small quantities of rotation or magnetism can compete quite effectively with gravity and can greatly alter the evolution of a typical gas cloud. Unfortunately, the interplay of these factors is not well understood—both can lead to highly complex behavior as a cloud contracts, and the combination of the two is extremely difficult to study theoretically. In this chapter, we will try to gain an appreciation for the broad outlines of the star-formation process by neglecting these two complicating factors. Bear in mind, however, that both are probably important in determining the details.

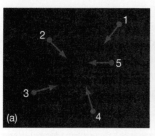
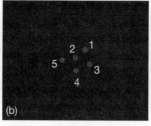
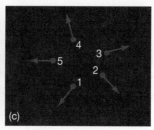

◀ **FIGURE 19.2 Atomic Motions**
The motions of a few atoms within an interstellar cloud are influenced by gravity so slightly that the atoms' paths are hardly changed (a) before, (b) during, and (c) after an accidental, random encounter.

We have already seen numerous instances of the competition between heat and gravity. ∞ *(More Precisely 8-1)* The temperature of a gas is simply a measure of the average speed of the atoms or molecules in it, so the higher the temperature, the greater the average speed of the molecules, and hence the higher the pressure of the gas. This is the main reason that the Sun and other stars don't collapse: The outward pressure of their heated gases exactly balances gravity's inward pull. ∞ (Fig. 16.2)

Consider a small portion of a large cloud of interstellar gas. Concentrate first on just a few atoms, as shown in Figure 19.2. Even though the cloud's temperature is very low, each atom still has some random motion because of the cloud's heat. ∞ *(More Precisely 3-1)* Each atom is also influenced by the gravitational attraction of all its neighbors. The gravitational force is not large, however, because the mass of each atom is so small. When a few atoms accidentally cluster for an instant, as shown in Figure 19.2(b), their combined gravity is insufficient to bind them into a lasting, distinct clump of matter. This accidental cluster will disperse as quickly as it forms. The effect of heat is much stronger than the effect of gravity.

Now consider a larger group of atoms. Imagine, for example, 50, 100, 1000—even a million—atoms, each gravitationally pulling on all the others. With increased mass, the force of gravity is now stronger than before. Will this many atoms exert a combined gravitational attraction strong enough to prevent the clump from dispersing again? The answer—at least under the conditions found in interstellar space—is still no. The gravitational attraction of even this mass of atoms is still far too weak to overcome the effect of heat.

How many atoms must be accumulated in order for their collective pull of gravity to prevent them from dispersing back into interstellar space? The answer, even for a typical cool (100 K) cloud, is a truly huge number. Nearly 10^{57} atoms are required—much more than the 10^{25} grains of sand on all the beaches of the world and even more than the 10^{51} elementary particles that constitute all the atomic nuclei in our entire planet. There is simply nothing on Earth comparable to a star.

Modeling Star Formation

The next two sections describe the currently accepted theoretical view of star formation, derived in large part from numerical experiments performed on high-speed computers. The results are mathematical predictions of a multifaceted problem incorporating gravity, heat, nuclear reaction rates, elemental abundances, and other physical conditions specifying the state of contracting interstellar clouds (see *More Precisely 19-1*).

Scientific theories always develop in response to experimental or observational data, and theories of star formation are no exception. ∞ (Sec. 1.2) The theory of star formation has evolved to explain innumerable observations of stars and star-forming regions. However, the phenomenology in this case is so complex and diverse that it is helpful to have a theoretical framework to "connect the dots" between phenomena that might otherwise appear unrelated. Accordingly, we present the theory first and then discuss how and where the observational data fit into and support the theoretical picture.

CONCEPT CHECK

✔ What basic competitive process controls star formation?

19.2 The Formation of Stars Like the Sun

Star formation begins when gravity begins to dominate over heat, causing a cloud to lose its equilibrium and start contracting. Only after the cloud has undergone radical changes in its internal structure is equilibrium finally restored.

In the process of becoming a main-sequence star like the Sun, an interstellar cloud goes through seven basic evolutionary stages, as listed in Table 19.1. The stages are characterized by varying central temperatures, surface temperatures, central densities, and radii of the prestellar object. They trace its progress from a cold, dark interstellar cloud to a hot, bright star. The numbers given in the table and in the following discussion are valid *only* for stars of approximately the same mass as that of the Sun. In the next section, we will relax this restriction and consider the formation of stars with masses different from that of the sun.

Note again the time scales involved in these stages—even the shortest spans a thousand human generations. Astronomers have not gained this insight by watching a single cloud or group of clouds evolve from start to finish. Rather, they combine theory and observation to refine a still evolving mathematical model of how stars form.

TABLE 19.1 Prestellar Evolution of a Solar-Type Star

Stage	Approximate Time to Next Stage (yr)	Central Temperature (K)	Surface Temperature (K)	Central Density (particles/m³)	Diameter* (km)	Object
1	2×10^6	10	10	10^9	10^{14}	Interstellar cloud
2	3×10^4	100	10	10^{12}	10^{12}	Cloud fragment Cloud fragment/protostar
3	10^5	10,000	100	10^{18}	10^{10}	
4	10^6	1,000,000	3000	10^{24}	10^8	Protostar
5	10^7	5,000,000	4000	10^{28}	10^7	Protostar
6	3×10^7	10,000,000	4500	10^{31}	2×10^6	Star
7	10^{10}	15,000,000	6000	10^{32}	1.5×10^6	Main-sequence star

* Round numbers; for comparison, recall that the diameter of the Sun is 1.4×10^6 km, whereas that of the solar system is roughly 1.5×10^{10} km.

Stage 1: An Interstellar Cloud

The first stage in the star-formation process is a dense interstellar cloud—the core of a dark dust cloud or perhaps a molecular cloud. These clouds are truly vast, sometimes spanning tens of parsecs (10^{14}–10^{15} km) across. Typical temperatures are about 10 K throughout, with a density of perhaps 10^9 particles/m³. Stage-1 clouds contain thousands of times the mass of the Sun, mainly in the form of cold atomic and molecular gas. (The dust in a stage-1 cloud both cools the cloud as it contracts and plays a crucial role in planet formation, but it constitutes a negligible fraction of the total mass of the cloud.) ∞ (Sec. 15.2)

Despite their low internal temperatures, most observed dark interstellar clouds seem to have enough internal pressure to support themselves against the force of gravity. ∞ *(More Precisely 8-1)* However, if such a cloud is to be the birthplace of stars, it must become unstable, start to collapse under its own gravity, and eventually break up into smaller pieces. Most astronomers think that the process of star formation is triggered when some external event, such as the shock of a nearby stellar explosion or the pressure wave produced when a nearby O- or B-type star forms and ionizes its surroundings, squeezes a cloud beyond the point where pressure can resist gravity's inward pull. ∞ (Secs. 16.2, 17.5) Or perhaps the cloud's supporting magnetic field leaks away as charged particles slowly drift across the confining field lines, leaving the gas unable to support its own weight *(More Precisely 19-1)*.

Whatever the cause, theory suggests that once the collapse begins, fragmentation into smaller and smaller clumps of matter naturally follows, as gravitational instabilities continue to operate in the gas. As illustrated in Figure 19.3, a typical cloud can break up into tens, hundreds, or even thousands, of fragments, each imitating the shrinking behavior of the parent cloud and contracting ever faster. The whole process, from a single stable cloud to many collapsing fragments, takes a few million years.

In this way, depending on the precise conditions under which fragmentation takes place, an interstellar cloud can produce either a few dozen stars, each much larger than our Sun, or a whole cluster of hundreds of stars, each comparable to or smaller than our Sun. There is little evidence of stars born in isolation, one star from one cloud. Most stars—perhaps even all stars—appear to originate as members of multiple systems or large groups of stars. The Sun, which is now found alone and isolated in space, probably escaped from the larger system in which it formed, perhaps after an encounter with another star or some much larger object (such as a molecular cloud).

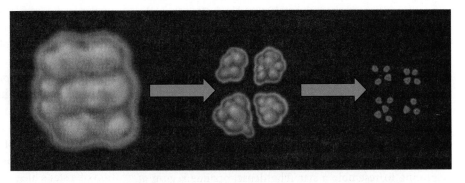

▲ **FIGURE 19.3 Cloud Fragmentation** As an interstellar cloud contracts, gravitational instabilities cause it to fragment into smaller pieces. The pieces themselves continue to fall inward and fragment, eventually forming many tens or hundreds of individual stars.

Stage 2: A Collapsing Cloud Fragment

The second stage in our evolutionary scenario represents the physical conditions in just one of the many fragments that develop in a typical interstellar cloud. A fragment destined to form a star like the Sun contains between one and two solar masses of material at this stage. Estimated to span a few hundredths of a parsec across, this fuzzy, gaseous blob is still about 100 times the size of our solar system. Its central density by this time is roughly 10^{12} particles/m^3.

Even though the fragment has shrunk substantially, its average temperature is not much different from that of the original cloud. The reason is that the gas constantly radiates large amounts of energy into space. The material of the fragment is so thin that photons produced within it easily escape without being reabsorbed by the cloud, so virtually all the energy released in the collapse is radiated away and does not cause any significant increase in temperature. Only at the center, where the radiation must traverse the greatest amount of material to escape, is there any appreciable temperature rise. The gas there may be as warm as 100 K by this stage. For the most part, however, the fragment stays cold as it shrinks.

The process of continued fragmentation is eventually stopped by the increasing density within the shrinking cloud. As stage-2 fragments continue to contract, they eventually become so dense that radiation cannot get out of the cloud easily. The trapped radiation then causes the temperature to rise, the pressure to increase, and the fragmentation to cease.

Stage 3: Fragmentation Ceases

By the start of stage 3, several tens of thousands of years after it first began contracting, a typical stage-2 fragment has shrunk to roughly the size of our solar system (still 10,000 times the size of our Sun). The density in the inner regions has just become high enough that the gas is opaque to the radiation it emits, so the core of the fragment begins to heat up considerably, as noted in Table 19.1. The central temperature has reached about 10,000 K—hotter than the hottest steel furnace on Earth. However, the temperature in the fragment's outer parts has not increased much. The gas there is still able to radiate its energy into space and so remains cool. The density increases much faster in the center of the fragment than near the edge, so the outside is both cooler and thinner than the interior. By this time, the central density is approximately 10^{18} particles/m^3 (still only 10^{-9} kg/m^3 or so).

For the first time, our contracting cloud fragment is beginning to resemble a star. The dense, opaque region at the center is called a **protostar**—an embryonic object at the dawn of star birth. The protostar's mass grows as more and more material rains down on it from the surrounding, still shrinking, fragment. However, the protostar's radius continues to decrease because pressure is still unable to overcome the relentless pull of gravity. After stage 3, we can distinguish a "surface" on the protostar—its *photosphere*. Inside the photosphere, the protostellar material is opaque to the radiation it emits.* From here on, the surface temperatures listed in Table 19.1 refer to the photosphere of the collapsing fragment and not to its low-density "periphery," where radiation can easily escape and the temperature remains low.

Stage 4: A Protostar

As the protostar evolves, it shrinks, its density grows, and its temperature rises, both in the core and at the photosphere. Some 100,000 years after the fragment began to form, it reaches stage 4, where its center seethes at about 1,000,000 K. Electrons and protons ripped from atoms whiz around at hundreds of kilometers per second, yet the temperature is still well short of the 10^7 K needed to ignite the proton–proton nuclear reactions that fuse hydrogen into helium. ∞ (Sec. 16.6) Still much larger than the Sun, our gassy heap is now about the size of Mercury's orbit. Heated by the material falling on it from above, it now has a surface temperature of a few thousand kelvins.

Knowing the protostar's radius and surface temperature, we can calculate its luminosity. Surprisingly, it turns out to be several thousand times the luminosity of the Sun. Even though the protostar has a surface temperature only about half that of the Sun, it is hundreds of times larger, making its total luminosity very large indeed—in fact, much greater than the luminosity of most main-sequence stars. Because nuclear reactions have not yet begun, the protostar's luminosity is due entirely to the release of gravitational energy as the protostar continues to shrink and material from the surrounding fragment continues to fall onto its surface.

By the time stage 4 is reached, our protostar's physical properties can be plotted on the Hertzsprung–Russell (H–R) diagram, as shown in Figure 19.4. Recall that an H–R diagram is a plot of two key stellar properties: surface temperature (increasing to the left) and luminosity (increasing upward). ∞ (Sec. 17.5) The luminosity scale in the figure is expressed in terms of the solar luminosity (4×10^{26} W). Our G2-type Sun is plotted at a temperature of 6000 K and a luminosity of 1 unit. As before, the dashed diagonal lines in the H–R diagram represent an object's radius, allowing us to follow the changes in the protostar's size as it evolves. At each phase of the star's evolution, its surface temperature and luminosity can be represented by a point on the diagram. The motion of that point as the star evolves is known as the star's **evolutionary track.** It is a graphical representation of a star's life.

*Note that this is the same definition of "surface" that we used for the Sun in Chapter 16. ∞ (Sec. 16.1)

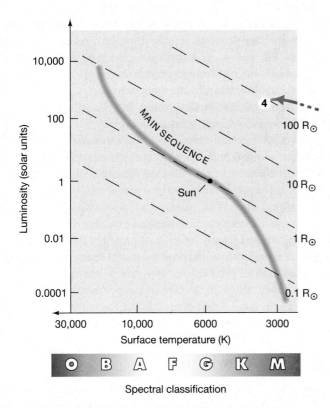

▲ FIGURE 19.4 Protostar on the H–R Diagram The red arrow indicates the approximate evolutionary track followed by an interstellar cloud fragment before reaching the end of the Kelvin–Helmholtz contraction phase as a stage-4 protostar. The boldface numbers on this and subsequent H–R plots refer to the prestellar evolutionary stages listed in Table 19.1 and described in the text. Recall from Chapter 17 that "$R_\odot$" denotes the radius of the Sun.

The red track in Figure 19.4 depicts the approximate path followed by our interstellar cloud fragment since it became a protostar at stage 3 (which itself lies off the right-hand edge of the figure). This early evolutionary track is known as the *Kelvin–Helmholtz contraction phase,* after the two European physicists (Lord Kelvin and Hermann von Helmholtz) who first studied the subject.

Figure 19.5 is an artist's sketch of an interstellar gas cloud proceeding along the evolutionary path outlined so far. As the stage-3 fragment contracts, it spins faster (to conserve angular momentum) and flattens into a rotating **protostellar disk** perhaps 100 AU in diameter, surrounding the central stage-4 protostar. ∞ *(More Precisely 6-1)* Recall that we first saw this process in Chapter 6, where we referred to the disk as the *solar nebula.* ∞ (Sec. 6.7) If the star is ultimately going to have a planetary system, by stage 4 that process is already well underway. ∞ (Sec. 15.2) However, regardless of whether planets actually form, astronomers think that protostellar disks are common—the vast majority of protostars (perhaps all) are accompanied by disks at this stage of their evolution.

Our protostar is still not in equilibrium. Even though its temperature is now so high that outward-directed pressure has become a powerful countervailing influence against gravity's continued inward pull, the balance is not yet perfect. The protostar's internal heat gradually diffuses out from the hot center to the cooler surface, where it is radiated away into space. As a result, the overall contraction slows, but it does not stop completely. From our perspective on Earth, this is quite fortunate: If the heated gas were somehow able to counteract gravity completely before the star reached the temperature and density needed to start nuclear burning in its core, the protostar would simply radiate away its heat and never become a true star. The night sky would be abundant in faint protostars, but completely lacking in the genuine article. Of course, there would be no Sun either, so it is unlikely that we, or any other intelligent life-form, would exist to appreciate these astronomical subtleties.

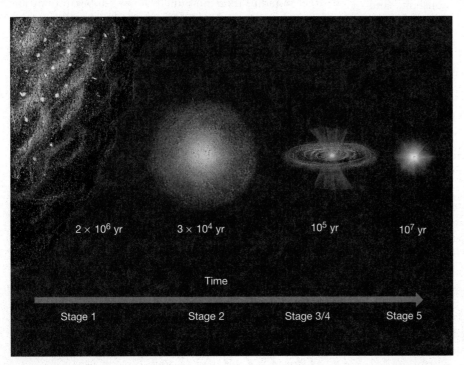

◄ FIGURE 19.5 Interstellar Cloud Evolution Artist's conception of the changes in an interstellar cloud during the early evolutionary stages outlined in Table 19.1. (Not drawn to scale.) Shown are a stage-1 interstellar cloud; a stage-2 fragment; a smaller, hotter stage-4 fragment with jets; and a stage-5 protostar. The duration of each stage, in years, is also indicated.

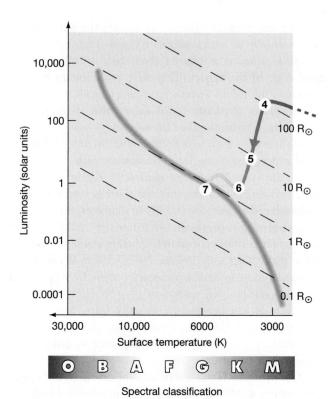

Interactive FIGURE 19.6 Newborn Star on the H–R Diagram The changes in a protostar's observed properties are shown by the path of decreasing luminosity, from stage 4 to stage 6, often called the Hayashi track. At stage 7, the newborn star has arrived on the main sequence.

After stage 4, the protostar on the H–R diagram moves down (toward lower luminosity) and slightly to the left (toward higher temperature), as shown in Figure 19.6. Its surface temperature remains almost constant, and it becomes less luminous as it shrinks. This portion of our protostar's evolutionary path, running from point 4 to point 6 in Figure 19.6, is often called the *Hayashi track*, after C. Hayashi, a 20th-century Japanese astrophysicist whose groundbreaking work in the 1960s on the evolution of pre-main-sequence stars still provides the theoretical basis for all studies of star formation.

Protostars on the Hayashi track often exhibit violent surface activity during this phase of their evolution, resulting in extremely strong protostellar winds, much denser than the solar wind that flows from our own Sun. As mentioned previously, this portion of the protostar's evolution is often called the **T Tauri** phase, after T Tauri, the first "star" (actually protostar) to be observed in that stage of prestellar development. ∞ (Sec. 15.3)

Stage 5: Protostellar Evolution

By stage 5 on the Hayashi track, the protostar approaches the main sequence. It has shrunk to about 10 times the size of the Sun, its surface temperature is about 4000 K, and its

luminosity has fallen to about 10 times the solar value. At this point, the central temperature has reached about 5,000,000 K. The gas is completely ionized by now, but the protons still do not have enough thermal energy to overcome their mutual electromagnetic repulsion and enter the realm of the nuclear binding force. ∞ (Sec. 16.6) The core is still too cool for nuclear fusion to begin.

Events proceed more slowly as the protostar approaches the main sequence. The initial contraction and fragmentation of the interstellar cloud occurred quite rapidly, but by stage 5, as the protostar nears the status of a full-fledged star, its evolution slows. The cause of this slowdown is heat: Even gravity must struggle to compress a hot object. The contraction is governed largely by the rate at which the protostar's internal energy can be radiated away into space. The greater this radiation of internal energy—that is, the more rapidly energy moves through the star to escape from its surface—the faster the contraction occurs. As the luminosity decreases, so, too, does the rate of contraction.

Stage 6: A Newborn Star

Some 10 million years after its first appearance, the protostar finally becomes a true star. By the bottom of the Hayashi track, at stage 6, when our roughly 1-solar-mass object has shrunk to a radius of about 1,000,000 km, the contraction has raised the central temperature to 10,000,000 K, enough to ignite nuclear burning. Protons begin fusing into helium nuclei in the core, and a star is born. As shown in Figure 19.6, the star's surface temperature at this point is about 4500 K, still a little cooler than the Sun. Even though the newly formed star is slightly larger in radius than our Sun, its lower temperature means that its luminosity is somewhat less than (actually, about two-thirds of) the solar value.

Stage 7: The Main Sequence at Last

Over the next 30 million years or so, the stage-6 star contracts a little more. In making this slight adjustment, the star's central density rises to about 10^{32} particles/m^3 (more conveniently expressed as 10^5 kg/m^3), the central temperature increases to 15,000,000 K, and the surface temperature reaches 6000 K. By stage 7, the star finally arrives at the main sequence, just about where our Sun now resides. Pressure and gravity are finally balanced, and the rate at which nuclear energy is generated in the core exactly matches the rate at which energy is radiated from the surface.

The evolutionary events just described occur over the course of some 40 to 50 million years. Although this is a long time by human standards, it is still less than 1 percent of the Sun's lifetime on the main sequence. Once an object begins fusing hydrogen in its core and establishes a "gravity-in, pressure-out" equilibrium, it is destined to burn steadily for a very long time. The star's location on the H–R diagram—that is, its surface temperature and luminosity—will remain virtually unchanged for the next 10 billion years.

CONCEPT CHECK

✔ What distinguishes a collapsing cloud from a protostar and a protostar from a star?

19.3 Stars of Other Masses

The numerical values and the evolutionary track just described are valid only for the case of a 1-solar-mass star. The temperatures, densities, and radii of prestellar objects of other masses exhibit similar trends, but the numbers and the tracks differ, in some cases considerably. Perhaps not surprisingly, the most massive fragments within interstellar clouds tend to produce the most massive protostars and, eventually, the most massive stars. Similarly, low-mass fragments give rise to low-mass stars. Whatever the mass, the end point of the prestellar evolutionary track is the main sequence.

The Zero-Age Main Sequence

Figure 19.7 compares the theoretical pre-main-sequence track taken by the Sun with the corresponding evolutionary tracks of a 0.3-solar-mass star and a 3-solar-mass star. All three tracks traverse the H–R diagram in the same general manner, but cloud fragments that eventually form stars more massive than the Sun approach the main sequence along a higher track on the diagram, whereas those destined to form less massive stars take a lower track. The *time* required for an interstellar cloud to become a main-sequence star also depends strongly on its mass. The most massive cloud fragments heat up to the required 10 million K and become O-type stars in a mere million years, roughly $\frac{1}{50}$ the time taken by the Sun. The opposite is the case for prestellar objects having masses less than that of our Sun. A typical M-type star, for example, requires nearly a billion years to form.

A star is considered to have reached the main sequence when hydrogen burning begins in its core and the star's properties settle down to stable values. The main-sequence line thus predicted by theory is called the **zero-age main sequence** (or ZAMS, for short). The fact that the theoretically derived zero-age main sequence agrees very well with the actual main sequences observed for stars in the vicinity of the Sun and in more distant star clusters (see Section 19.6) provides strong support for the modern theory of star formation and stellar structure. ∞ (Sec. 1.2)

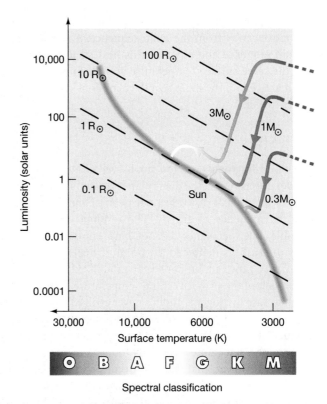

▲ FIGURE 19.7 **Prestellar Evolutionary Tracks** Some pre-main-sequence evolutionary paths for stars more massive and less massive than our Sun.

If all gas clouds contained precisely the same elements in exactly the same proportions, mass would be the sole determinant of a newborn star's location on the H–R diagram and the zero-age main sequence would be a well-defined line rather than a broad band. However, the composition of a star affects its internal structure (mainly by changing the opacity of its outer layers), and this in turn affects both the star's temperature and its luminosity on the main sequence. Stars with more heavy elements tend to be cooler and slightly less luminous than stars that have the same mass, but contain fewer heavy elements. As a result, differences in composition between stars "blur" the zero-age main sequence into the broad band we observe.

It is important to realize that *the main sequence is itself not an evolutionary track—stars do not evolve along it.* Rather, it is just a "way station" on the H–R diagram where stars stop and spend most of their lives—low-mass stars at the bottom, high-mass stars at the top. Once on the main sequence, a star stays in essentially the same location on the H–R diagram during its whole time as a stage-7 object. (In other words, a star that arrives on the main sequence as, say, a G-type star can never "work its way up" to become a B- or an O-type main-sequence blue supergiant or move down to become an M-type red dwarf.) As we will see in

Chapter 20, the next stage of stellar evolution occurs when a star moves away from the main sequence. A star leaving the main sequence and entering this next stage has pretty much the same surface temperature and luminosity it had when it arrived on the main sequence millions (or billions) of years earlier.

Failed Stars

Some cloud fragments are too small ever to become stars. Consider, for example, the giant planet Jupiter. It formed in the Sun's protostellar disk (the solar nebula) and contracted under the influence of gravity. The resultant heat is still detectable, but the planet did not have enough mass for gravity to crush its matter to the point of nuclear ignition. ∞ (Sec. 11.3) Instead, Jupiter became stabilized by heat and rotation before the planet's central temperature became hot enough to fuse hydrogen—Jupiter never evolved beyond the protostar stage. If it, or any of the other jovian planets, had continued to accumulate gas from the solar nebula, it might have become a star (almost certainly to the detriment of life on Earth). However, that did not occur—virtually all the matter present during the formative stages of our solar system is now gone, swept away by the solar wind during the Sun's T Tauri phase. ∞ (Sec. 15.2)

Low-mass gas fragments simply lack the mass needed to initiate nuclear burning. Rather than turning into stars, they continue to cool, eventually becoming compact, dark "clinkers"—cold fragments of unburned matter—orbiting a star or moving alone through interstellar space. On the basis of theoretical modeling, astronomers think that the minimum mass of gas needed to generate core temperatures high enough to begin nuclear fusion is about 0.08 solar mass (80 times the mass of Jupiter). Our practical definition of a star requires that it shine via the energy released by nuclear fusion reactions in its core. Thus this mass of 0.08 times the mass of the Sun is a lower limit on the masses of all stars in the universe.

Vast numbers of "substellar" objects may well be scattered throughout the universe—fragments frozen in time somewhere along the Kelvin–Helmholtz contraction phase. Small, faint, and cool (and growing ever colder), they are known collectively as **brown dwarfs.** For reasons discussed in more detail in *Discovery 19-1*, researchers generally reserve the term *brown dwarf* to mean a low-mass prestellar fragment of more than about 12 Jupiter masses (so Jupiter itself is not a brown dwarf, by this definition). Anything smaller is simply called a *planet.*

Observationally, these faint, low-mass objects are difficult to study, be they planets or brown dwarfs associated with stars or interstellar cloud fragments far from any star (see *Discovery 19-1*). Current observations suggest that up to 100 billion cold, dim substellar objects may lurk in the depths of interstellar space—a number comparable to the total number of "real" stars in our Galaxy.

19.4 Observations of Cloud Fragments and Protostars

How can we verify the theoretical picture just outlined? The age of our entire civilization is much shorter than the time needed for a single interstellar cloud to contract and form a star. We can never observe individual objects proceed through the full panorama of star birth. However, we can do the next best thing: We can observe many different objects—interstellar clouds, protostars, and young stars approaching the main sequence—as they appear today at different stages of their evolutionary paths.

The various evolutionary stages just described draw on evidence from different parts of the electromagnetic spectrum, and each observation is like part of a jigsaw puzzle. ∞ (Sec. 3.3) When properly oriented relative to all the others, the pieces can be used to build up a picture of the full life cycle of a star.

Evidence of Cloud Contraction

Prestellar objects at stages 1 and 2 are not yet hot enough to emit much infrared radiation, and certainly no optical radiation arises from their dark, cool interiors. The best way to study the early stages of cloud contraction and fragmentation is to observe the radio emission from interstellar molecules within those clouds. Consider again M20, the splendid emission nebula studied in Chapter 18 ∞ (Sec. 18.2) The brilliant region of glowing, ionized gas shown in Figure 18.7 is not our main interest here, however; instead, the youthful O- and B-type stars that energize the nebula alert us to the general environment in which stars are forming. Emission nebulae are indicators of star birth.

The region surrounding M20 contains galactic matter that seems to be contracting. The presence of (optically) invisible gas there was illustrated in Figure 18.20, which showed a contour map of the abundance of the formaldehyde (H_2CO) molecule. Formaldehyde and many other molecules are widespread in the vicinity of the nebula, especially throughout the dusty regions below and to the right of the emission nebula itself. Further analysis of the observations suggests that this region of greatest molecular abundance is also contracting and fragmenting and is well on its way toward forming a star—or, more likely, a star cluster.

The interstellar clouds in and around M20 thus provide tentative evidence of three distinct phases of star

DISCOVERY 19-1

Observations of Brown Dwarfs

Cruelly put, brown dwarfs are stellar failures—objects that formed through the contraction and fragmentation of an interstellar cloud, just as stars do, but fell short of the critical mass of about 0.08 solar mass (80 times the mass of Jupiter) needed to start hydrogen fusion in their cores. Interstellar space could contain huge numbers of these dim objects.

Although hundreds of brown dwarfs are now known, detecting them is no easy task, as they are small, cool, and hence very faint. ∞ (Sec. 3.4) We can detect stars by means of telescopes, and we can infer the presence of interstellar atoms and molecules by spectroscopic analysis, but astronomical objects of intermediate size outside our solar system remain hard to see. One place astronomers have looked is in binary-star systems, using many of the same techniques they employ in the search for extrasolar planets. ∞ (Sec. 15.5) The accompanying images show two binary-star systems containing brown dwarf candidates (marked by arrows). Note in each case how much fainter the brown dwarf is than its companion. Very high resolution is usually needed to separate the two.

The first (left) is an image of Gliese 623, which was originally identified as a binary system because of its variations in radial velocity. ∞ (Secs. 15.5, 17.7) From the binary's measured orbital separation and period, the mass of the faint companion appears to be approximately 0.1 solar mass—very close to the limit for a brown dwarf, although astronomers still aren't certain of its exact mass. ∞ (Sec. 2.8) The "rings" in the image are instrumental artifacts. The second image shows the binary-star system Gliese 229. These two objects are 7″ apart; the fainter "star" has a luminosity only a few millionths that of the Sun and an estimated mass about 50 times that of Jupiter. (The diagonal streak in the image is caused by an overexposure of the brighter star in the CCD chip used to record it.)

Actually, the dividing line between brown dwarfs and Jupiter-like planets is not completely clear-cut, especially given the varied properties of the many extrasolar planetary systems now known. ∞ (Sec. 15.6) Researchers distinguish between "stellar" objects (stars and brown dwarfs), which form within their own contracting cloud fragment as described in the text, and planets, which form in the nebular disk around a larger parent. For definiteness, many draw the dividing line at about 12 times the mass of Jupiter. Above that mass (but below 80 Jupiter masses), although core

temperatures never become high enough for hydrogen fusion to occur, a contracting fragment will experience a brief phase of *deuterium fusion,* as the core becomes hot enough for any deuterium nuclei that are present in the original cloud to combine. The phase ends once the deuterium is consumed, and the fragment's "nuclear" lifetime is over. Below 12 Jupiter masses, no nuclear fusion of any kind is expected. The art rendering in this box compares the sizes of some stars, brown dwarfs, and planets.

Infrared and spectroscopic studies offer other ways of searching for brown dwarfs, especially those that are not in binaries. Infrared observations are particularly effective because brown dwarfs emit most of their radiation in that part of the spectrum, whereas true stars tend to be brightest at the near-infrared and optical ranges. The final image below captures a star cluster just north of the Orion Nebula taken by the *Spitzer Space Telescope.* The bright objects are stars, but many of the faint specks are brown dwarf candidates. Researchers estimate that some 10–15 percent of the "stars" in Orion are actually brown dwarfs.

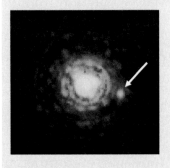

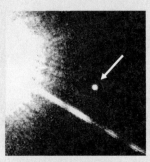

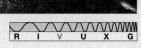

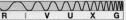

ANIMATION/VIDEO Binary Brown Dwarfs

(MA)

formation, as shown in Figure 19.8. The huge, dark molecular cloud surrounding the visible nebula is the stage-1 cloud. Both its density and its temperature are low—about 10^8 particles/m^3 and 20 K, respectively. Greater densities and temperatures typify smaller regions within this large cloud. The totally obscured regions labeled A and B, where the molecular emission of radio energy is strongest, are such denser, warmer fragments. Here, the total gas density is observed to be at least 10^9 particles/m^3, and the temperature is about 100 K. The Doppler shifts of the radio lines observed in the vicinity of region B imply that this portion of M20, labeled "contracting fragment" in the figure, is infalling. Recent infrared observations (Figure 19.8c) reveal the candidate protostars themselves, identified by the warmth of their growing embryos tucked inside. Less than a light-year across, the region has a total mass over a thousand times the mass of the Sun—considerably more than the mass of M20 itself. The region lies somewhere between stages 1 and 2 of Table 19.1.

The third star-formation phase shown in Figure 19.8 is M20 itself. The glowing region of ionized gas results directly from a massive O-type star that formed there within the past million years or so. Because the central star is already fully formed, this final phase corresponds to stage 6 or 7 of our evolutionary scenario.

Evidence of Cloud Fragments

Other parts of our Milky Way Galaxy provide sketchy evidence for prestellar objects in stages 3 through 5. The Orion complex, shown in Figure 19.9, is one such region. Lit from within by several O-type stars, the bright Orion Nebula is partly surrounded by a vast molecular cloud that extends even beyond the roughly 3×6-pc region bounded by the photograph in Figure 19.9(b).

The Orion molecular cloud harbors several smaller sites of intense radiation emitted by molecules deep within the core of the cloud fragment. Their extent, shown in

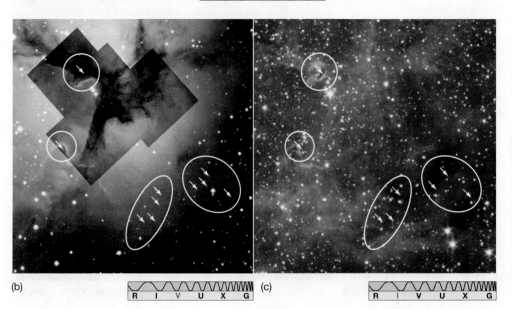

◀ FIGURE 19.8 **Star Formation Phases** (a) The M20 region shows observational evidence for three broad phases in the birth of a star. The parent cloud is stage 1 of Table 19.1. The region labeled "contracting fragment" likely lies between stages 1 and 2. Finally, the emission nebula (M20 itself) results from the formation of one or more massive stars (stages 6 and 7). (b) A close-up (including *Hubble* inlays) of the area near region B outlines (in drawn ovals) especially dense knots of dusty matter. (c) A *Spitzer Telescope* infrared image of the same scene reveals those cores thought to be stellar embryos (arrows). *(AURA; NASA)*

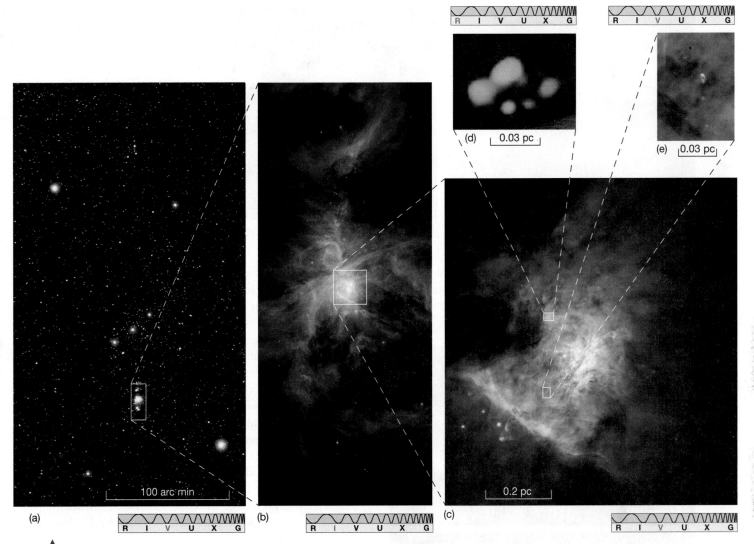

(d) | 0.03 pc |

(e) | 0.03 pc |

(a)

(b)

(c)

100 arc min

0.2 pc

Interactive FIGURE 19.9 Orion Nebula, Up Close The Orion Nebula has long been a treasure trove of information about star formation. (a) The constellation Orion, with the region around its famous emission nebula marked by a rectangle. The Orion Nebula is the middle "star" of Orion's sword (see Figure 1.8). (b) Enlargement of the framed region in part (a), suggesting how the nebula is partly surrounded by a vast molecular cloud. Various parts of this cloud are probably fragmenting and contracting, with even smaller sites forming protostars. The three frames at the right show some of the evidence for those protostars: (c) nearly real-color visible image of embedded nebular "knots" thought to harbor protostars, (d) false-color radio image of some intensely emitting molecular sites, and (e) high-resolution image of one of many young stars surrounded by disks of gas and dust where planets might ultimately form. (*Astrostock-Sanford; SST; CfA; NASA*)

Figures 19.9(d) and (e), measures about 10^{10} km, or $\frac{1}{1000}$ of a light-year, about the diameter of our solar system. Their density is about 10^{15} particles/m^3, much denser than the surrounding cloud. Although the temperature of these smaller regions cannot be estimated reliably, many researchers regard the regions as objects well on their way to stage 3. We cannot determine whether those regions will eventually form stars like the Sun, but it does seem certain that the intensely emitting objects in them are on the threshold of becoming protostars.

Evidence of Protostars

In the hunt for, and study of, objects at more advanced stages of star formation, radio techniques become less useful, because stages 4, 5, and 6 have increasingly higher temperatures. By Wien's law, their emission shifts toward shorter wavelengths, so these objects shine most strongly in the infrared. ∞ (Sec. 3.4) One particularly bright infrared emitter, known as the *Becklin–Neugebauer* object, was detected in the core of the Orion molecular cloud in

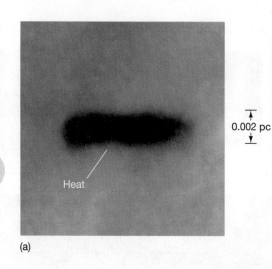

(a)

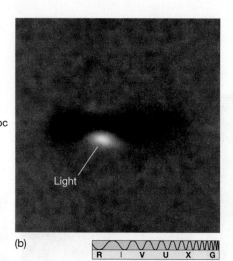

(b)

0.002 pc

Heat

Light

R   I   V   U   X   G

◄ FIGURE 19.10 **Protostars** (a) An edge-on infrared image of a planetary system-sized dusty disk in the Orion region, showing heat and light emerging from its center. On the basis of its temperature and luminosity, this unnamed source appears to be a low-mass protostar on the Hayashi track (around stage 5) in the H–R diagram. (b) An optical, face-on image of a slightly more advanced circumstellar disk surrounding an embedded protostar in Orion. *(NASA)*

the 1970s. Its luminosity is around a thousand times the luminosity of the Sun. Most astronomers agree that this warm, dense blob is a high-mass protostar, probably in or around stage 4.

Until the *Infrared Astronomy Satellite (IRAS)* was launched in the early 1980s, astronomers were aware of giant stars forming only in clouds far away. ∞ (Sec. 5.7) But *IRAS* showed that many such stars are forming much closer to home, and some of these protostars have masses comparable to that of our Sun. Figure 19.10 shows two examples of low-mass protostars, both spotted by *HST* in a rich star-forming region in Orion. Their infrared heat signatures are those expected of an object on the Hayashi track, at around stage 5.

The energy sources for some infrared objects seem to be luminous hot stars that are hidden from optical view by surrounding dark clouds. Apparently, these stars are already so hot that they emit large amounts of ultraviolet radiation, which is mostly absorbed by "cocoons" of dust surrounding them. The absorbed energy is then reemitted by the dust as

Disk

Jet

Protostar

Jet

(a)

(b)

Spherical wind

(c)

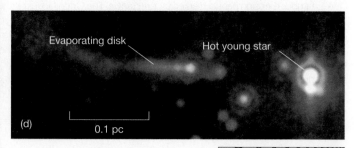

Evaporating disk

Hot young star

0.1 pc

(d)

R   I   V   U   X   G

▲ FIGURE 19.11 **Protostellar Wind** (a) The nebular disk around a protostar can be the site of intense heating and strong outflows, forming a bipolar jet perpendicular to the disk. (b) As the disk is blown away by the wind, the jets fan out, eventually (c) merging into a spherical wind. In contrast to this art, part (d) is an actual infrared image of a hot young star (at right) whose powerful winds are ripping away the disk (at left) surrounding a Sun-like star (at center). This system is located about 750 pc away in the star-forming cloud IC 1396. *(SST)*

infrared radiation. These bright infrared sources are known as *cocoon nebulae.* Two considerations support the idea that the hot stars heating the dust have only recently ignited: (1) The dust cocoons are predicted to disperse quite rapidly once their central stars form, and (2) they are invariably found in the dense cores of molecular clouds. The central stars probably lie near stage 6.

Protostellar Winds

Protostars often exhibit strong winds. Radio and infrared observations of hydrogen and carbon monoxide molecules in the Orion molecular cloud have revealed gas expanding outward at velocities approaching 100 km/s. High-resolution interferometric observations have disclosed expanding knots of water emission within the same star-forming region and have linked the strong winds to the protostars themselves. ∞ (Sec. 5.6) These winds may be related to the violent surface activity associated with many protostars.

As mentioned earlier, a young protostar may be embedded in an extensive protostellar disk of nebular material in which planets are forming. ∞ (Sec. 15.2) Strong heating within the turbulent disk and a powerful protostellar wind combine to produce a **bipolar flow,** expelling two "jets" of matter in the directions per pendicular to the disk, as illustrated by the art in Figure 19.11(a)–(c). As the protostellar wind gradually destroys the disk, blowing it away into space,

the outflow widens until, with the disk gone, the wind flows away from the star equally in all directions, as is approximately shown by actual infrared imagery in Figure 19.11(d). Figure 19.12 shows the emission from an especially clear bipolar flow, along with an artist's conception of the system producing it.

These outflows can be very energetic. Figure 19.13 shows a portion of the Orion molecular cloud, south of the Orion Nebula, where a newborn star is seen still surrounded by a bright nebula, its turbulent wind spreading out into the interstellar medium. Below the star (enlarged in the inset) are twin jets known as HH1 and HH2. ("HH" stands for *Herbig–Haro,* the investigators who first cataloged such objects.) Formed in another (unseen) protostellar disk—the protostar itself is still hidden within the dusty cloud fragment from which it formed—these jets have traveled outward for almost half a light-year before colliding with interstellar matter. More Herbig–Haro objects can be seen in the upper-right portion of the figure. One of them, the oddly shaped "waterfall," may be due to an earlier outflow from the same protostar responsible for the existence of HH1 and HH2.

PROCESS OF SCIENCE CHECK

✔ How can a "snapshot" of the universe today test our theories of the evolution of individual objects?

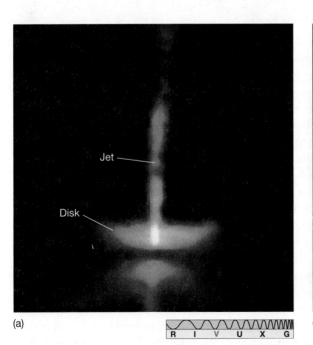

(a)

R I V U X G

(b)

▲ **FIGURE 19.12 Bipolar Jets** (a) This remarkable image shows two jets emanating from the young star system HH30, the result of infalling matter being expelled from an embryonic star near the center. The system is viewed roughly edge-on to the disk. (b) An artist's idealized conception of a young star system, showing two jets flowing perpendicular to the disk of gas and dust rotating around the star. *(NASA; D. Berry)*

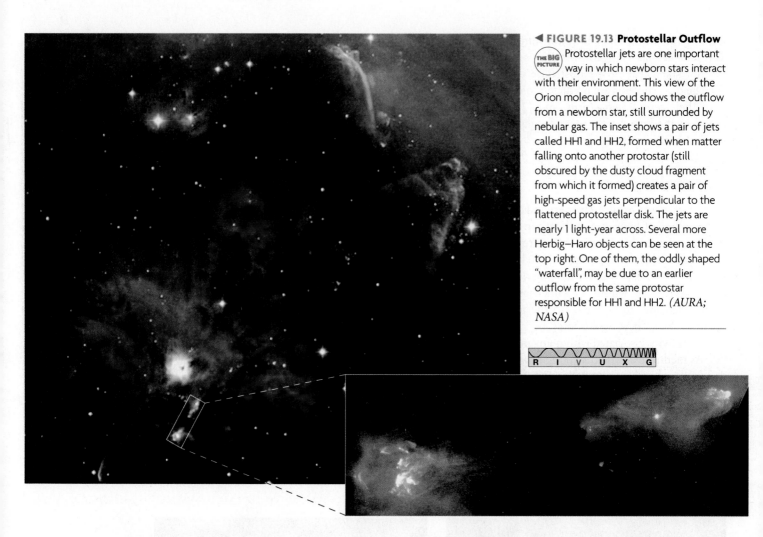

◄ FIGURE 19.13 **Protostellar Outflow** Protostellar jets are one important way in which newborn stars interact with their environment. This view of the Orion molecular cloud shows the outflow from a newborn star, still surrounded by nebular gas. The inset shows a pair of jets called HH1 and HH2, formed when matter falling onto another protostar (still obscured by the dusty cloud fragment from which it formed) creates a pair of high-speed gas jets perpendicular to the flattened protostellar disk. The jets are nearly 1 light-year across. Several more Herbig–Haro objects can be seen at the top right. One of them, the oddly shaped "waterfall", may be due to an earlier outflow from the same protostar responsible for HH1 and HH2. *(AURA; NASA)*

R I V U X G

19.5 Shock Waves and Star Formation

The subject of star formation is really much more complicated than the preceding discussion suggests. Interstellar space is populated with many kinds of clouds, fragments, protostars, stars, and nebulae, all interacting in a complex fashion and each type of object affecting the behavior of all the other types. For example, the presence of an emission nebula in or near a molecular cloud probably influences the evolution of the entire region. We can easily imagine expanding waves of matter driven outward by the high temperatures and pressures in the nebula. As the waves crash into the surrounding molecular cloud, interstellar gas tends to pile up and become compressed. Such a shell of gas, rushing rapidly through space, known as a **shock wave,** can push ordinarily thin matter into dense sheets, just as a plow pushes snow.

Many astronomers regard the passage of a shock wave through interstellar matter as the triggering mechanism needed to initiate star formation in a galaxy. Calculations show that when a shock wave encounters an interstellar cloud, it races around the thinner exterior of the cloud more rapidly than it can penetrate the cloud's thicker interior. Thus, shock waves do not blast a cloud from only one direction, but effectively squeeze it from many directions. Atomic bomb tests have experimentally demonstrated this squeezing: Shock waves created in the blast tend to surround buildings, causing them to be blown together (imploded) rather than apart (exploded). The "contracting fragment" in Figure 19.8 may well have been triggered by the shock wave from the M20 nebula. Note the correspondence between the shock-compressed region at the lower right and the high-density molecular gas revealed by radio studies (Figure 18.19). Once shock waves have begun compressing an interstellar cloud, natural gravitational instabilities take over, dividing the cloud into the fragments that eventually form stars.

Emission nebulae are by no means the only generators of interstellar shock waves. At least four other driving forces are available: the relatively gentle deaths of old stars in the form of planetary nebulae (to be discussed in Chapter 20); the much more violent ends of certain stars in supernova explosions (Chapter 21); the spiral-arm waves that plow through the Milky Way (Chapter 23); and interactions between galaxies (Chapter 24). Supernovae are by far the most energetic, and probably also the most efficient, means of piling up matter into dense clumps. However, they are relatively few and far between, so the other mechanisms may be more important overall in triggering star formation.

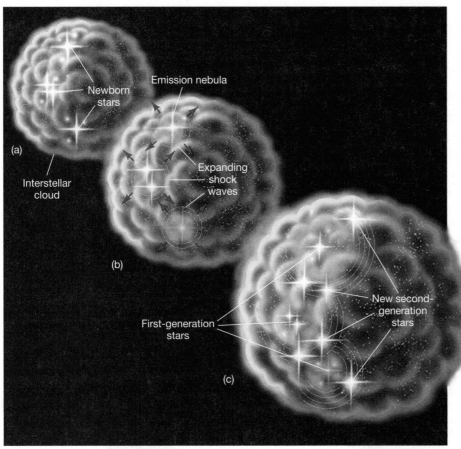

◄ FIGURE 19.14 **Generations of Star Formation** (a) Star birth and (b) shock waves lead to (c) more star births and more shock waves in a continuous cycle of star formation in many areas of our Galaxy. As in a chain reaction, old stars trigger the formation of new stars ever deeper into an interstellar cloud.

This picture of shock-induced star formation is complicated by the fact that O- and B-type stars form quickly, live briefly, and die explosively. These massive stars, themselves perhaps born of a passing shock wave, may in turn create new shock waves, either through the expanding nebular gas produced by their births or through their explosive deaths. The new shock waves can produce "second-generation" stars, which in turn will explode and give rise to still more shock waves, and so on. As depicted in Figure 19.14, star formation resembles a chain reaction. Other, lighter stars are also formed in the process, of course, but they are largely "along for the ride." It is the O- and B-type stars that drive the star-formation wave through the cloud.

Although the evidence is somewhat circumstantial, the presence of young (and thus fast-forming) O- and B-type stars in the vicinity of supernova remnants does suggest that the birth of stars is often initiated by the violent, explosive deaths of others.

Observational evidence lends some support to this chain-reaction picture. Groups of stars nearest molecular clouds do indeed appear to be the youngest, whereas those farther away seem to be older. Figure 19.15 shows an *HST* image of a star-forming region in the galaxy NGC 4214, which lies some 13 million light-years from Earth. A series of bright emission nebulae, powered by hot young stars, can be seen, suggesting that a wave of star formation recently swept across the region, triggering the sequence seen here.

◄ FIGURE 19.15 **A Wave of Star Formation?** A group of star-forming regions in the galaxy NGC 4214, possibly representing several generations in a chain of star formation. *(NASA)*

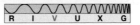

19.6 Star Clusters

The end result of the collapse of a cloud is a group of stars, all formed from the same parent cloud and lying in the same region of space. Such a collection of stars is called a **star cluster.** Figure 19.16 shows a spectacular view of a newborn star cluster and (part of) the interstellar cloud from which it came.

Because all the stars formed at the same time out of the same cloud of interstellar gas and under the same environmental conditions, clusters are near-ideal "laboratories" for stellar studies—not in the sense that astronomers can perform experiments on them, but because the properties of the stars are very tightly constrained. The only factor distinguishing one star from another in the same cluster is mass, so theoretical models of star formation and evolution can be compared with reality without the complications introduced by the broad spreads in age, chemical composition, and place of origin found when we consider all stars in our galactic neighborhood.

Clusters and Associations

Figure 19.17(a) shows a small star cluster called the Pleiades, or Seven Sisters, a well-known naked-eye object in the constellation Taurus, lying about 120 pc from Earth. This type of loose, irregular cluster, found mainly in the plane of the Milky Way (see Figure 18.4), is called an **open cluster.** Open clusters typically contain from a few hundred to a few tens of thousands of stars and are a few parsecs across.

Figure 19.17(b) shows the H–R diagram of stars in the Pleiades. The cluster contains stars in almost all parts of the main sequence—only the very brightest main-sequence stars are missing. (The brightest six or seven stars in the diagram have just left the main sequence, as will be discussed in Chapter 20.) Thus, even though we have no direct evidence of the cluster's birth, we can estimate its age as less than about 100 million years, the lifetime of a main-sequence B-type star. ∞ (Sec. 17.8) If all the stars in the cluster formed at the same time, then the red stars must be young, too. The wisps of leftover gas evident in the photograph are further evidence of the cluster's relative youth. In addition, the system is abundant in heavy elements that (as we will see) could have been created only within the cores of many generations of ancient stars long since perished.

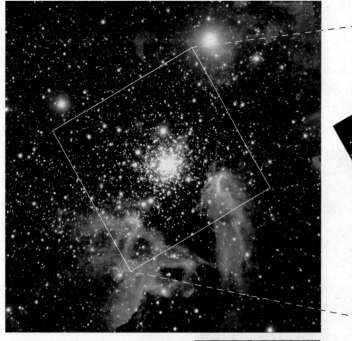

R I V U X G

▲ **FIGURE 19.16 Newborn Cluster** The star cluster NGC 3603 and a portion of the larger molecular cloud in which it formed. The cluster contains about 2000 bright stars and lies some 6000 pc from Earth. The field of view shown here spans about 20 light-years. The cluster is a few million years old, and the intense radiation from its most massive stars has cleared a cavity in the cloud several light-years across. The inset shows the central area more clearly, including the most massive star in the region (called Sher 25, above and to the left of the cluster), which is already near the end of its lifetime, having ejected part of its outer layers and formed a ring of gas. Many low-mass stars, less massive than the Sun, can also be seen. (*ESO; NASA*)

Less massive, but more extended, clusters are known as **associations.** These clusters typically contain no more than a few hundred bright stars, but may span many tens of parsecs. Associations tend to be rich in very young stars. Those containing many pre-main-sequence T Tauri stars are known as *T associations,* whereas those with prominent O- and B-type stars, such as the Trapezium in Orion (see Figure 19.20a below), are called *OB associations.* It is quite likely that the main difference between associations and open clusters is simply the efficiency (as measured by the fraction of gas that eventually ends up in stars) with which stars formed from the parent cloud.

Figure 19.18(a) shows a very different type of star cluster, called a **globular cluster.** All globular clusters are roughly

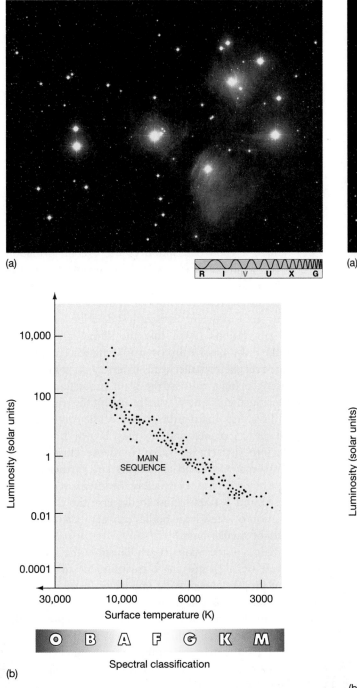

(a)

(b)

▲ FIGURE 19.17 **Open Cluster** (a) The Pleiades cluster (also known as the Seven Sisters or M45) lies about 120 pc from the Sun. The naked eye can see only six or seven of its brightest stars. (b) An H–R diagram for all the stars of this well-known open cluster. *(AURA)*

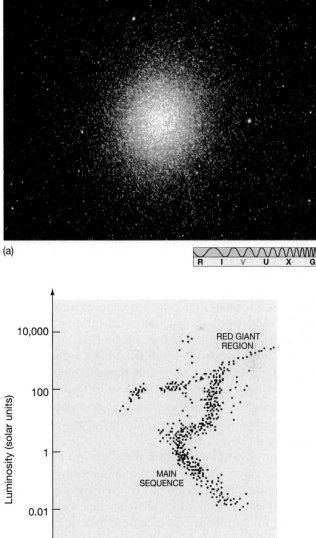

(a)

(b)

▲ FIGURE 19.18 **Globular Cluster** (a) The globular cluster Omega Centauri is approximately 5000 pc from Earth and spans some 40 pc in diameter. (b) A H–R diagram of some of its stars. *(J. Lodriguss)*

spherical (which accounts for their name), are generally found away from the Milky Way plane, and contain hundreds of thousands, and sometimes millions, of stars spread out over about 50 pc. Figure 19.18(b) is an H–R diagram of the cluster shown, which is called Omega Centauri. Notice the many differences between this H–R diagram and that of Figure 19.17(b)—globular clusters present a stellar environment very different from that of open clusters like the Pleiades. The distance to Omega Centauri cluster has been determined by a variation on the method of spectroscopic parallax, applied to the entire cluster rather than to individual stars. ∞ (Sec. 17.6) It lies about 5000 pc from Earth.

The most outstanding spectroscopic feature of globular clusters is their lack of upper-main-sequence stars. Astronomers in the 1920s and 1930s, working with instruments incapable of detecting stars fainter than about 1 solar luminosity at the distances of globular clusters, and having no theory of stellar evolution to guide them, were puzzled by the H–R diagrams they saw when they looked at the globular clusters. Indeed, a comparison of just the top halves of the diagrams (so that the lower main sequences cannot be seen) reveals few similarities between Figures 19.17(b) and 19.18(b).

Most globular clusters contain no main-sequence stars with masses greater than about 0.8 times the mass of the Sun. The more massive O- through F-type stars have long since exhausted their nuclear fuel and disappeared from the main sequence (in fact becoming the red giants and other luminous stars above the main sequence, as we will see in Chapter 20). ∞ (Sec. 17.8) From the theory of stellar evolution (Chapter 20), the A-type stars in Figure 19.18(b) are now known to be stars at much later stages in their evolution that just happen to be passing through the location of the upper main sequence. On the basis of these and other observations, astronomers estimate that most globular clusters are at least 10 billion years old—they contain the oldest known stars in our Galaxy.

Other observations confirm the great ages of globular clusters. For example, their spectra show few heavy elements, implying that these stars formed in the distant past, when heavy elements were much less abundant than they are today (Chapter 21). Astronomers speculate that the 150 or so globular clusters observed today are just the survivors of a much larger population of clusters that formed long ago.

Clusters and Nebulae

How many stars form in a cluster, and of what type are they? How much gas is left over? What does the collapsed cloud look like once star formation has run its course? At present, although the main stages in the formation of individual stars (stages 3–7) are becoming clearer, the answers to these more general questions (involving stages 1 and 2) are still sketchy. They await a more thorough understanding of the star-formation process.

In general, the more massive the collapsing region, the more stars are likely to form there. In addition, we know from H–R diagrams of observed stars that low-mass stars are much more common than high-mass ones. ∞ (Sec. 17.8) For every O- or B-type giant, hundreds or even thousands of G-, K-, and M-type dwarfs may form. The precise number of stars of any given mass or spectral type likely depends in a complex (and poorly understood) way upon conditions within the parent cloud. The same is true of the *efficiency* of star formation—the fraction of the total mass that actually finds its way into stars—which determines the amount of leftover material. However, if, as is usually the case, one or more O- or B-type stars form, their intense radiation and winds will cause the surrounding gas to disperse rapidly, leaving behind a young star cluster.

The Cluster Environment

In recent years, astronomers have come to realize that physical interactions—close encounters and even collisions—between protostars within a star cluster may be very important in determining the properties of the stars that eventually form. Supercomputer simulations of star-forming clouds suggest that, while the seven stages presented earlier (and listed in Table 19.1) remain a good description of the overall formation process, the sequence of events leading to a main-sequence star can be strongly influenced by events within the cluster itself. Figure 19.19 presents frames from one such simulation, illustrating some of the interactions just described.

The simulations reveal that the strong gravitational fields of the most massive protostars give them a competitive advantage over their smaller rivals in attracting gas from the surrounding nebula, causing the giant protostars to grow even faster. Such encounters usually disrupt the smaller protostellar disks, terminating the growth of the central protostars and ejecting planets and low-mass brown dwarfs from the disk into intracluster space. In dense clusters these interactions may even lead to mergers and further growth of massive objects. Thus, even before the intense radiation of newborn O and B stars begins to disperse the cluster gas, the formation of a few large bodies can significantly inhibit the growth of smaller ones.*

All these considerations clearly illustrate the important role played by a future star's environment in the star-formation process and provide important insight into why low-mass stars are so much more common than high-mass stars. The first few massive bodies to form tend to prevent the formation of additional high-mass stars by stealing their "raw material" and ultimately disrupting the environment in which other stars are growing. This tendency also helps explain the existence of brown dwarfs, by providing at least two natural

---

*Compare this picture of large objects dominating the accretion process at the expense of the smaller bodies around them with the standard view of planet formation presented in Chapter 15. ∞ (Sec. 15.1-3)

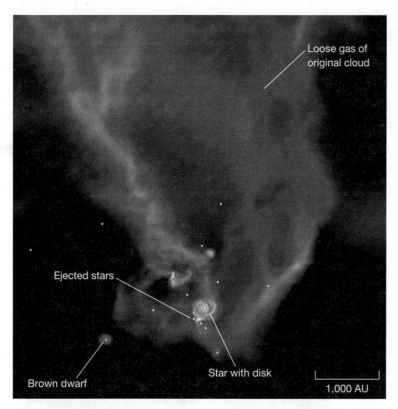

Loose gas of
original cloud

Ejected stars

Brown dwarf

Star with disk

1,000 AU

Interactive **FIGURE 19.19 Protostellar Collisions**
In the congested environment of a young cluster, star formation is a competitive and violent process. Large protostars may grow by "stealing" gas from smaller ones, and the extended disks surrounding most protostars can lead to collisions and even mergers. The cluster environment can play a crucial role in determining the types of stars that form. This frame from a supercomputer simulation shows a small star cluster emerging from an interstellar cloud that originally contained about 50 solar masses of material, distributed over a volume 1 light-year across. Remnants of the cloud are shown here in red. (*M. Bate, I. Bonnell, and V. Bromm*)

Nebula. The optical image in Figure 19.20(a) shows the Trapezium, the group of four bright stars responsible for ionizing the nebula; the infrared image in Figure 19.20(b) reveals an extensive cluster of stars within and behind the visible nebula. This remarkable infrared image shows many stages of star formation, most notably nearly 1000 new stars forming and giving rise to the speckled green fuzz created when jets of gas shoot out from those young stars and ram into the surrounding cloud.

ways (disk destruction and gas dispersion) in which star formation can stop before nuclear fusion begins in a growing stellar core. *Discovery 19-2* describes another system in which the gas-dispersal process may be almost complete.

Young star clusters are often shrouded in gas and dust, making them hard to see in visible light. However, infrared observations clearly demonstrate that stars really are found within star-forming regions. Figure 19.20 compares optical and infrared views of the central regions of the Orion

Cluster Lifetimes

Eventually, star clusters dissolve into individual stars. In some cases, the ejection of unused gas reduces a cluster's mass so much that it becomes gravitationally unbound and dissolves rapidly. In clusters that survive the early gas-loss phase, stellar encounters tend to eject the lightest stars from the cluster, just as the gravitational slingshot effect can propel spacecraft around the solar system. ∞ (*Discovery 6-1*) At the same time, the tidal gravitational field of the Milky Way Galaxy slowly strips outlying stars from the cluster. ∞ (Sec. 7.6) Occasional distant encounters with giant molecular clouds also tend to remove stars from a cluster. Even a near miss may disrupt the cluster entirely.

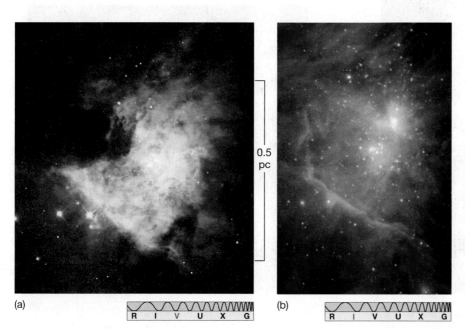

(a)

(b)

0.5 pc

R I V U X G

R I V U X G

Interactive **FIGURE 19.20 Star Formation in Orion** Some views of the central regions of the Orion Nebula. (a) A short-exposure visible-light image (observed with a filter that is transparent only to certain emission lines of oxygen) shows the nebula itself and four bright O-type stars known as the Trapezium, but few obvious other stars. (b) A *Spitzer Space Telescope* view of the same part of the nebula shows more clearly the irregular gas and dust amidst many young stars still embedded in that dust (see also Figure 5.28). (*Lick Observatory; NASA*)

DISCOVERY 19-2

Eta Carinae

At the heart of the Carina emission nebula (shown in the main figure at left) lies a remarkable object called Eta Carinae (object at right bottom). With an estimated mass of around 100 times the mass of the Sun and a luminosity of 5 million times the solar value, Eta Carinae is one of the most massive stars known. Formed probably only a few hundred thousand years ago, this star has had an explosive, though brief, life. In the mid-19th century, Eta Carinae produced an outburst that made it one of the brightest stars in the southern sky (even though it lies some 2200 pc away from Earth, a very long way compared to most of the bright stars visible in our night sky). During this "Great Eruption," which peaked in 1843, the star expelled more than 2 solar masses of material in less than a decade and released as much visible energy as a supernova explosion (see Chapter 21), yet it somehow survived the event.

The right images are close-ups of the most active part of the nebula: At top a *Chandra X-ray Telescope* image that gives a glimpse of the object's violence, and at bottom a *Hubble Telescope* image that was carefully processed to reveal fine detail, is the highest-resolution view of the explosion obtained to date. Dust lanes, tiny condensations in the outflowing material, and dark radial streaks of unknown origin all appear with exquisite clarity. The star itself is the white dot at the center of the image. The two ends of the "peanut" (at the top right and bottom left) are blobs of material ejected

in the 1843 outburst, now racing away from the star at hundreds of kilometers per second—perhaps enough to expel the surrounding nebular gas and convert the Carina Nebula into the Carina Cluster. Perpendicular to the line joining these two blobs is a thin disk of gas, also moving outward at high speed.

The details of the events leading to the Eta Carinae outburst are unclear. Quite possibly, such episodes of violent activity are the norm for supermassive stars. In 2005, astronomers discovered that Eta Carinae has a binary companion, an even hotter, but fainter star orbiting just 11 AU away—not much more than Eta Carinae's estimated radius of 5 AU. Many researchers suspect that the interaction between the two stars may have been responsible for the Great Eruption. However, although a few comparable outbursts have been observed in other galaxies, they are so rare that astronomers still do not know what constitutes "typical" behavior for such exotic objects.

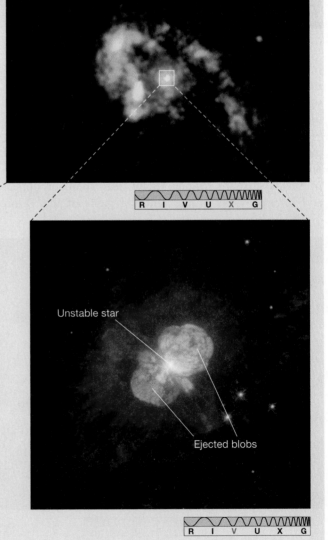

Unstable star

Ejected blobs

As a result of all these influences, most open clusters break up in a few hundred million years, although the actual lifetime depends on the cluster's mass. Loosely bound associations may survive for only a few tens of millions of years, whereas some very massive open clusters are known from their H–R diagrams to be almost 5 billion years old. In a sense, only when a star's parent cluster has completely dissolved is the star-formation process really complete. The road from a gas cloud to a single, isolated star like the Sun is long and tortuous indeed!

Take another look at the sky one clear, dark evening. Ponder all of the cosmic activity you have learned about as you peer upward at the stars. After studying this chapter, you may find that you have to modify your view of the night sky. Even the seemingly quiet nighttime darkness is dominated by continual change.

CONCEPT CHECK

✔ If stars in a cluster all form at the same time, how can some influence the formation of others?

CHAPTER REVIEW

SUMMARY

1 Stars form when an interstellar cloud collapses under its own gravity and breaks up into pieces comparable in mass to our Sun. A cold interstellar cloud may fragment into many smaller clumps of matter, from which stars eventually form. The evolution of the contracting cloud can be represented as an **evolutionary track (p. 470)** on the Hertzsprung–Russell diagram. As a collapsing prestellar fragment heats up and becomes denser, it eventually becomes a **protostar (p. 470)**—a warm, very luminous object that emits mainly infrared radiation. Eventually, a protostar's central temperature becomes high enough for hydrogen fusion to begin, and the protostar becomes a star.

2 For a star like the Sun, the whole formation process takes about 50 million years. More massive stars pass through similar formation stages, but much more rapidly. Stars less massive than the Sun take much longer to form. The **zero-age main sequence (p. 473)** is the region in the H–R diagram where stars lie when the formation process is over. Mass is the key property in determining a star's characteristics and life span. The most massive stars have the shortest formation times and main-sequence lifetimes. At the other extreme, some low-mass fragments never reach the point of nuclear ignition and become **brown dwarfs (p. 474)**.

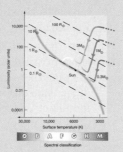

3 Many of the objects predicted by the theory of star formation have been observed in real astronomical objects. The dark interstellar regions near emission nebulae often provide evidence of cloud fragmentation and protostars. Radio telescopes are used in studying the early phases of cloud contraction and fragmentation; infrared observations allow us to see later stages of the process. Many

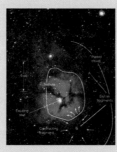

well-known emission nebulae, lit by several O- and B-type stars, are partially engulfed by molecular clouds, portions of which are fragmenting and contracting, with smaller sites forming protostars.

4 Protostars can produce powerful protostellar winds. These winds encounter less resistance in the directions perpendicular to a star's protostellar disk and often expel two jets of matter in the directions of the protostar's poles in a **bipolar flow (p. 479)**. The protostellar winds gradually destroy the disk, and eventually the wind flows away from the star equally in all directions. **Shock waves (p. 480)** are produced as young hot stars ionize the surrounding gas, forming emission nebulae. These shock waves can compress other interstellar clouds and trigger more star formation, possibly producing chain reactions of star formation in molecular cloud complexes.

5 A single collapsing and fragmenting cloud can give rise to hundreds or thousands of stars—a **star cluster (p. 482)**. The formation of the most massive stars may play an important role in suppressing the further formation of stars from lower-mass cluster members. **Open clusters (p. 482)**, with a few hundred to a few thousand stars, are found mostly in the plane of the Milky Way. They typically contain many bright blue stars, indicating that they formed relatively recently. **Globular clusters (p. 483)** are found mainly away from the Milky Way plane and may contain millions of stars. They include no main-sequence stars much

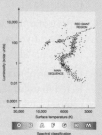

more massive than the Sun, indicating that they formed long ago. Infrared observations have revealed young star clusters or associations in several emission nebulae. Eventually, clusters break up into individual stars, although the entire process may take hundreds of millions or even billions of years.

Mastering**ASTRONOMY**® *For instructor-assigned homework go to **www.masteringastronomy.com***

Problems labeled **POS** explore the process of science | **VIS** problems focus on reading and interpreting visual information

REVIEW AND DISCUSSION

1. Briefly describe the basic chain of events leading to the formation of a star like the Sun.

2. What is the role of heat in the process of stellar birth?

3. What is the role of rotation in the process of stellar birth?

4. What is the role of magnetism in the process of stellar birth?

5. What is an evolutionary track?

6. Why do stars tend to form in groups?

7. Why does the evolution of a protostar slow down as the star approaches the main sequence?

8. In what ways do the formative stages of high-mass stars differ from those of stars like the Sun?

9. What are brown dwarfs?

10. What are T Tauri stars?

11. **POS** Stars live much longer than we do, so how do astronomers test the accuracy of theories of star formation?

12. At what evolutionary stages must astronomers use radio and infrared radiation to study prestellar objects? Why can't they use visible light?

13. Why has it been difficult until recently to demonstrate that stars and protostars actually exist within star-forming regions?

14. What is a shock wave? Of what significance are shock waves in star formation?

15. **POS** Explain the usefulness of the H–R diagram in studying the evolution of stars. Why can't evolutionary stages 1–3 be plotted on the diagram?

16. Compare the times necessary for the various stages in the formation of a star like the Sun. Why are some so short and others so long?

17. What do star clusters and associations have to do with star formation?

18. Compare and contrast the observed properties of open star clusters and globular star clusters.

19. **POS** How can we tell whether a star cluster is young or old?

20. In the formation of a star cluster with a wide range of stellar masses, is it possible for some stars to die out before others have finished forming? Do you think this will have any effect on the cluster's formation?

CONCEPTUAL SELF-TEST: MULTIPLE CHOICE

1. If a newly forming star has an excess of heat, then it will likely have (a) more gravity; (b) less gravity; (c) a slower contraction rate; (d) a rapid contraction rate.

2. The gravitational contraction of an interstellar cloud is primarily the result of its (a) mass; (b) composition; (c) diameter; (d) pressure.

3. The interstellar cloud from which our Sun formed was (a) slightly larger than the Sun; (b) comparable in size to Saturn's orbit; (c) comparable in mass to the solar system; (d) thousands of times more massive than the Sun.

4. A protostar that will eventually turn into a star like the Sun is significantly (a) smaller; (b) more luminous; (c) fainter; (d) less massive than the Sun.

5. Prestellar objects in which nuclear fusion never starts are referred to as (a) terrestrial planets; (b) brown dwarfs; (c) protostars; (d) globules.

6. The current theory of star formation is based upon (a) amassing evidence from many different regions of our Galaxy; (b) carefully studying the births of a few stars; (c) systematically measuring the masses and rotation rates of

interstellar clouds; (d) observations made primarily at short wavelengths.

7. **VIS** If the initial interstellar cloud in Figure 19.14 ("Generations of Star Formation") were much more massive, the result would be (a) the formation of more stars; (b) contraction of the cloud due to stronger gravitational attraction; (c) stars forming closer together; (d) stronger shock waves.

8. **VIS** One of the primary differences between the Pleiades cluster, shown in Figure 19.17(a), and Omega Centauri, shown in Figure 19.18(a), is that the Pleiades cluster is much (a) larger; (b) younger; (c) farther away; (d) denser.

9. **VIS** If the H–R diagram shown in Figure 19.18(b) ("Globular Cluster") were redrawn to illustrate a much younger cluster, the main-sequence turnoff would shift to (a) higher temperature; (b) higher pressure; (c) higher frequency; (d) a spectral classification of K or M.

10. A typical open cluster will dissolve in about the same amount of time as the time since (a) North America was first visited by Europeans; (b) dinosaurs walked on Earth; (c) Earth was formed; (d) the universe formed.

PROBLEMS

The number of dots preceding each Problem indicates its approximate level of difficulty.

1. •• In order for an interstellar gas cloud to contract, the average speed of its constituent particles must be less than half the cloud's escape speed. ∞ *(More Precisely 8-1)* Will a (spherical) molecular hydrogen cloud with a mass of 1000 solar masses, a radius of 10 pc, and a temperature of 10 K begin to collapse? Why or why not?

2. •• Under the same assumptions as in Problem 1, estimate the minimum mass needed to cause a 1000-K, 1-pc cloud to collapse.

3. •• Use the radius–luminosity–temperature relation to explain how a protostar's luminosity changes as it moves from stage 4 (temperature 3000 K, radius 2×10^8 km) to stage 6 (temperature 4500 K, radius 10^6 km). What is the change in absolute magnitude? ∞ (Sec. 17.2)

4. • A protostar on the Hayashi track evolves from a temperature of 3500 K and a luminosity 5000 times that of the Sun to a temperature of 5000 K and a luminosity of 3 solar units. What is the protostar's radius (a) at the start and (b) at the end of the evolution?

5. • What is the (approximate) absolute magnitude of a stage-5 protostar? (See Figure 19.6.)

6. •• Use the H–R diagrams in this chapter to estimate by what factor a 1000-solar-luminosity, 3000-K protostar is larger than a main-sequence star of the same luminosity.

7. • By how many magnitudes does a 3-solar-mass star decrease in brightness as it evolves from stage 4 to stage 6? (See Figure 19.7.)

8. •• As a simple model of the final stage of star formation, imagine that, between stages 6 and 7, a star's surface temperature increases with time at a constant rate, whereas the luminosity remains constant at the stage-7 level. The stage-7 radius is equal to the solar value. Using the temperatures given in Table 19.1, calculate the star's radius at a time exactly halfway between these two stages.

9. • What is the luminosity, in solar units, of a brown dwarf whose radius is 0.1 solar radius and whose surface temperature is 600 K (0.1 times that of the Sun)?

10. •• What is the maximum distance at which the brown dwarf in the previous problem could be observed by a telescope of limiting apparent magnitude (a) 18, (b) 30?

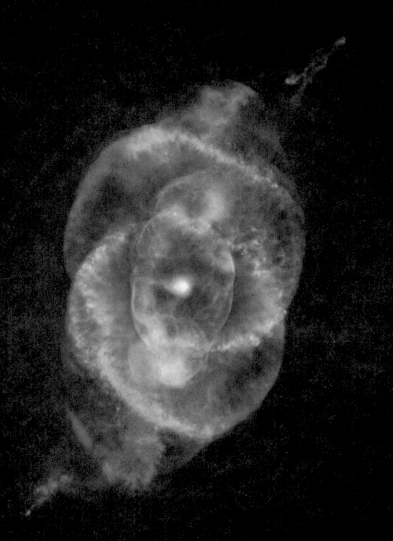

20

STELLAR EVOLUTION

THE LIFE AND DEATH OF A STAR

LEARNING GOALS

Studying this chapter will enable you to

1 Explain why stars evolve off the main sequence.

2 Outline the events that occur as a Sun-like star evolves from the main sequence to the giant branch.

3 Explain how the Sun will eventually come to fuse helium in its core, and describe what happens when that occurs.

4 Summarize the stages in the death of a typical low-mass star, and describe the resulting remnant.

5 Contrast the evolutionary histories of high-mass and low-mass stars.

6 Discuss the observations that help verify the theory of stellar evolution.

7 Explain how the evolution of stars in binary systems may differ from that of isolated stars.

THE BIG PICTURE No one has ever seen a single star move through all of its evolutionary stages. Like archaeologists who study bones and artifacts of different ages to learn more about the development of human culture, astronomers observe stars of many different ages, then assemble the data into a consistent model of how stars evolve over billions of years.

Mastering ASTRONOMY

Visit the Study Area in www.masteringastronomy.com for quizzes, animations, videos, interactive figures, and self-guided tutorials.

After reaching the main sequence, a newborn star changes little in outward appearance for more than 90 percent of its lifetime. However, at the end of that period, as the star begins to run out of fuel and die, its properties once again change greatly. Aging stars travel along evolutionary tracks that take them far from the main sequence as they end their lives. In this and the next two chapters, we will study the evolution of stars during and after their main-sequence burning stages.

We will find that the ultimate fate of a star depends primarily on its mass—although interactions with other stars can also play a decisive role—and that the final states of stars can be strange, indeed. By continually comparing theoretical calculations with detailed observations of stars and binaries of all types, astronomers have refined the theory of stellar evolution into a precise and powerful tool for understanding the universe.

LEFT: *This striking composite image combines visible light (colored red and purple) acquired by the* Hubble *telescope and X-ray radiation (blue) by the* Chandra *telescope. Known as NGC 6543, or informally as the Cat's Eye, this complex object is a planetary nebula—an old star (at center) shedding its outer layers over light-year dimensions as it ends its life.* (STScI/CXC)

20.1 Leaving the Main Sequence

Most stars spend most of their lives on the main sequence. A star like the Sun, for example, after spending a few tens of millions of years in its formative stages (1–6 in Chapter 19), resides on or near the main sequence (stage 7) for 10 billion years before turning into something else. ∞ (Sec. 19.2) That "something else" is the main topic of this chapter.

Observing Stellar Lifetimes

No one has ever witnessed (or ever will witness) the complete lifetime of any star, from birth to death. Stars take an enormously long time—millions, billions, and even trillions of years—to evolve. ∞ (Secs. 17.8, 19.2) Virtually all the low-mass stars that have ever formed still exist as stars. The coolest M-type stars—red dwarfs—consume their nuclear fuel so slowly that not one of them has yet left the main sequence. Some of them will shine steadily for a trillion years or more. By contrast, the most massive O- and B-type stars exhaust their fuel and leave the main sequence after only a few million years. Most of the high-mass stars that have ever existed perished long ago.

How then can we talk so confidently about what took place billions of years in the past and what will happen billions of years in the future? The answer is that, although we can never hope to follow the evolution of an individual star, we can observe billions of stars in the universe; that is enough to see examples of every possible stage of stellar development, thereby allowing us to test and refine our theoretical ideas. Just as anthropologists might piece together a picture of the human life cycle by studying snapshots of all the residents of a large city, so can astronomers construct a picture of stellar evolution by studying the myriad stars we see in the night sky.

Note that astronomers use the term "evolution" in this context to mean change *during the lifetime of an individual star.* Contrast this usage with the meaning of the term in biology, where it refers to changes in the characteristics of a *population* of plants or animals over many generations. In fact, as we will see in Chapter 21, populations of stars do evolve in the latter "biological" sense, as the overall composition of the interstellar medium (and hence of each new stellar generation) changes slowly over time due to nuclear fusion in stars. However, in astronomical parlance, "stellar evolution" always refers to changes that occur during a single stellar lifetime.

Structural Change

On the main sequence, a star slowly fuses hydrogen into helium in its core. This process of nuclear fusion is called **core hydrogen burning.** In Chapter 16, we saw how the proton–proton fusion chain powers the Sun. ∞ (Sec. 16.6) Here, by the way, is

another instance where astronomers use a fairly familiar term in a quite unfamiliar way: To astronomers "burning" always means nuclear fusion in a star's core and not the chemical reaction (such as the combustion of wood or gasoline in air) we would normally think of in everyday speech. Chemical burning does not directly affect atomic nuclei.

As discussed in Chapter 16 (see Figure 16.4), a main-sequence star is in a state of *hydrostatic equilibrium,* in which pressure's outward push exactly counteracts gravity's inward pull. ∞ (Sec. 16.2) This is a stable balance between gravity and pressure in which a small change in one always results in a small compensating change in the other. You should keep that figure in mind as you study the various stages of stellar evolution described next. Much of a star's complex behavior can be understood in these simple terms.

As the main-sequence star ages, its core temperature rises, and both its luminosity and radius increase. These changes are very slow, though—only a factor of three or four in luminosity over the Sun's entire 10-billion-year main-sequence lifetime, for example. As a result, the star's location on the H–R diagram remains almost unchanged during this phase (which is why we see the main sequence when we plot an H–R diagram for any reasonably large group of stars). ∞ (Sec. 17.5) Eventually, as the hydrogen in the core is consumed, the star's internal balance starts to shift, and both its internal structure and its outward appearance begin to change more rapidly: The star leaves the main sequence.

Once a star begins to move away from the main sequence, its days are numbered. The post-main-sequence stages of stellar evolution—the end of a star's life—depend critically on the star's mass. As a rule of thumb, we can say that low-mass stars die gently, whereas high-mass stars die catastrophically. The dividing line between these two very different outcomes lies around eight times the mass of the Sun, and in this chapter we will refer to stars of more than 8 solar masses as "high-mass" stars. Within both the "high-mass" and the "low-mass" (i.e., less than 8 solar masses) categories, there are substantial variations, some of which we will point out as we proceed.

Rather than dwelling on the many details, we will concentrate on a few representative evolutionary sequences. We begin by considering the evolution of a fairly low-mass star like the Sun. The stages described in the next few sections pertain to the Sun as it nears the end of its fusion cycle 5 billion years from now. The numbers continue the sequence begun in Chapter 19. In fact, most of the qualitative features of the discussion apply to any low-mass star, although the exact numbers vary considerably. Later, we will broaden our discussion to include all stars, large and small.

PROCESS OF SCIENCE CHECK

✔ How can astronomers "see" stars evolve in time?

20.2 Evolution of a Sun-like Star

The surface of a main-sequence star like the Sun occasionally erupts in flares and spots, but for the most part the star does not exhibit any sudden, large-scale changes in its properties. Its average surface temperature remains fairly constant, whereas its luminosity increases very slowly with time. The Sun has roughly the same surface temperature as it had when it formed nearly 5 billion years ago, even though it is some 30 percent brighter than it was at that time.

This state of affairs cannot continue indefinitely, however. Eventually, drastic changes occur in the star's interior structure. After approximately 10 billion years of steady core hydrogen burning, a Sun-like star begins to run out of fuel. The situation is a little like that of an automobile cruising effortlessly along a highway at a constant speed for many hours, only to have the engine suddenly cough and sputter as the gas gauge reaches empty. Unlike automobiles, though, stars are not easy to refuel.

Stage 8: The Subgiant Branch

As nuclear fusion proceeds, the composition of the star's interior changes as its hydrogen fuel is depleted. Figure 20.1 illustrates the increase in helium abundance and the corresponding decrease in hydrogen abundance that take place in the stellar core as the star ages. Three cases are shown: (a) the chemical composition of the original core, (b) the composition after 5 billion years, and (c) the composition after 10 billion years. Case (b) represents approximately the present state of our Sun.

The star's helium content increases fastest at the center, where temperatures are highest and the burning is fastest. The helium content also increases near the edge of the core, but more slowly because the burning rate is less rapid there. The inner, helium-rich region becomes larger and more deficient in hydrogen as the star continues to shine. Eventually, about 10 billion years after the star arrived on the main sequence (Figure 20.1c), hydrogen becomes depleted at the center, the nuclear fires there subside, and the location of principal burning moves to higher layers in the core. An inner core of nonburning pure helium starts to grow.

Without nuclear burning to maintain it, the outward-pushing gas pressure weakens in the helium inner core. However, the inward pull of gravity does not. Once the outward push against gravity is relaxed—even a little—structural changes in the star become inevitable. As the hydrogen is consumed, the inner core begins to contract. When all the hydrogen at the center is gone, the process accelerates.

If more heat could be generated, then the core might regain its equilibrium. For example, if helium in the core were to begin fusing into some heavier element, then energy would be created as a by-product of helium burning, and the necessary gas pressure would be reestablished. But the

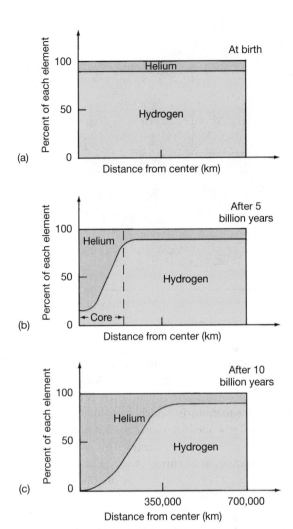

▲ FIGURE 20.1 **Solar Composition Change** Theoretical estimates of the changes in a Sun-like star's composition. Hydrogen (yellow) and helium (blue) abundances are shown (a) at birth, just as the star arrives on the zero-age main sequence; (b) after 5 billion years; and (c) after 10 billion years. At stage (b) only a few percent of the star's total mass has been converted from hydrogen into helium. This change speeds up as the nuclear burning rate increases with time.

helium at the center cannot burn—not yet, anyway. Despite its high temperature, the core is far too cold to fuse helium into anything heavier.

Recall from Chapter 16 that a minimum temperature of about 10^7 K is needed to fuse hydrogen into helium. Only above that temperature do colliding hydrogen nuclei (i.e., protons) have enough speed to overwhelm the repulsive electromagnetic force between them. ∞ (Sec. 16.6) Because helium nuclei, with two protons each, carry a greater positive charge, their electromagnetic repulsion is larger, and even higher temperatures are needed to cause them to fuse—at least 10^8 K. A core composed of helium at 10^7 K thus cannot generate energy through fusion.

The shrinkage of the helium core releases gravitational energy, driving up the central temperature and heating the

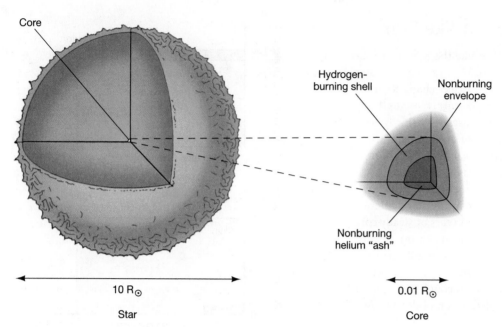

Core

Hydrogen-burning shell

Nonburning envelope

Nonburning helium "ash"

10 R$_\odot$

Star

0.01 R$_\odot$

Core

◄ **FIGURE 20.2 Hydrogen-Shell Burning** As a star's core converts more and more of its hydrogen into helium, the hydrogen in the shell surrounding the nonburning helium "ash" burns ever more violently. By the time shown here (a little after stage 8 in Table 20.1), the core has shrunk to a few tens of thousands of kilometers in diameter, whereas the star's photosphere is ten times the star's original size.

overlying burning layers. The higher temperatures—now well over 10^7 K (but still less than 10^8 K)—cause hydrogen nuclei to fuse even more rapidly than before. Figure 20.2 depicts this situation, in which hydrogen is burning at a furious rate in a shell surrounding the nonburning inner core of helium "ash" in the center. This phase is known as the **hydrogen-shell-burning** stage. The hydrogen shell generates energy faster than did the original main-sequence star's hydrogen-burning core, and the shell's energy production continues to increase as the helium core continues to shrink. Strange as it may seem, the star's response to the disappearance of the fire at its center is to get brighter!

Table 20.1 summarizes the key stages through which a solar-mass star evolves. The table is a continuation of

Table 19.1, except that the density units have been changed from particles per cubic meter to the more convenient kilograms per cubic meter and sizes are expressed as radii rather than diameters. The numbers in the "Stage" column refer to the evolutionary stages noted in the figures and discussed in the text.

After a lengthy stay on the main sequence, the star's temperature and luminosity are once again beginning to change, and we can trace these changes via the star's evolutionary track on the H–R diagram. ∞ (Sec. 19.2) Figure 20.3 shows the star's path away from the main sequence, labeled as stage 7. The star first evolves to the right on the diagram, its surface temperature dropping whereas its luminosity increases only slightly. By stage 8, the star's radius has increased to about

TABLE 20.1 Evolution of a Sun-like Star

Stage	Approximate Time to Next Stage (Yr)	Central Temperature (10^6 K)	Surface Temperature (K)	Central Density (kg/m^3)	Radius (km)	Radius (solar radii)	Object
7	10^{10}	15	6000	10^5	7×10^5	1	Main-sequence star
8	10^8	50	4000	10^7	2×10^6	3	Subgiant branch
9	10^5	100	4000	10^8	7×10^7	100	Helium flash
10	5×10^7	200	5000	10^7	7×10^6	10	Horizontal branch
11	10^4	250	4000	10^8	4×10^8	500	Asymptotic-giant branch
12	10^5	300	100,000	10^{10}	10^4	0.01	Carbon core
	—		3000	10^{-17}	7×10^8	1000	Planetary nebula*
13	—	100	50,000	10^{10}	10^4	0.01	White dwarf
14	—	Close to 0	Close to 0	10^{10}	10^4	0.01	Black dwarf

* Values refer to the envelope.

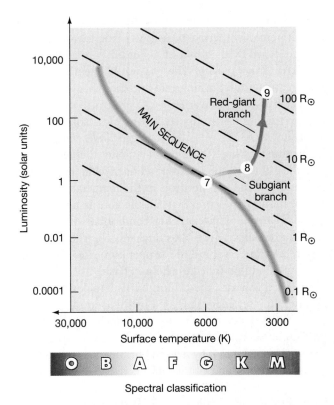

▲ **FIGURE 20.3 Red Giant on the H–R Diagram** As its helium core shrinks and its outer envelope expands, the star leaves the main sequence (stage 7). At stage 8, the star is well on its way to becoming a red giant. The star continues to brighten and grow as it ascends the red-giant branch to stage 9. As noted in Chapter 17, the dashed diagonal lines are lines of constant radius, allowing us to gauge the changes in the size of the star.

three times the radius of the Sun. The star at this stage is called a **subgiant.** Its roughly horizontal path from its main-sequence location (stage 7) to stage 8 on the figure is called the **subgiant branch.**

Stage 9: The Red-Giant Branch

Our aging star is now far from the main sequence and no longer in stable equilibrium. The helium core is unbalanced and shrinking. The rest of the core is also unbalanced, fusing hydrogen into helium at an ever-increasing rate. The gas pressure produced by this enhanced hydrogen burning causes the star's nonburning outer layers to increase in radius. Not even gravity can stop this inexorable change. While the core is shrinking and heating up, the overlying layers are expanding and cooling. The star is on its way to becoming a red giant. The transformation from normal main-sequence star to elderly red giant takes about 100 million years.

By stage 8, the star's surface temperature has fallen to the point at which much of the interior is opaque to the radiation from within. Beyond this point, convection carries the core's enormous energy output to the surface. One consequence of

that convection is that the star's surface temperature remains nearly constant between stages 8 and 9. The almost vertical path followed by the star between those stages is known as the **red-giant branch** of the H–R diagram. By stage 9, hydrogen shell burning in the still-shrinking core is proceeding so ferociously that the giant's luminosity is many hundreds of times the solar value. Its radius by this time is around 100 solar radii.

The red giant is huge—about the size of Mercury's orbit. In contrast, its helium core is surprisingly small—only about a thousandth the size of the entire star, making the core just a few times larger than Earth. The central density is enormous: Continued shrinkage of the red giant's core has compacted its helium gas to approximately 10^8 kg/m^3. Contrast this value with the 10^{-3} kg/m^3 in the giant's outermost layers, with the 5000 kg/m^3 average density of Earth, and with the 150,000 kg/m^3 in the present core of the Sun. About 25 percent of the mass of the entire star is packed into its planet-sized core.

A familiar example of a low-mass star in the red-giant phase is the KIII giant Arcturus (see Figure 17.15), one of the brightest stars in the sky. Its mass is about 1.5 times that of the Sun. Currently in the hydrogen-shell-burning stage and ascending the red-giant branch, Arcturus has a radius some 21 times that of the Sun and emits about 160 times more energy than the Sun, much of it in the infrared part of the spectrum.

Stage 10: Helium Fusion

Should the unbalanced state of a red-giant star continue, the core would eventually collapse, and the rest of the star would slowly drift into space. The forces and pressures at work inside a red giant would literally tear it apart. In fact, for stars less than about one-quarter the mass of the Sun, that is precisely what will eventually happen (in a few hundred billion years—see Section 20.3).

However, for a star like the Sun, this simultaneous shrinking and expanding does not continue indefinitely. A few hundred million years after a solar-mass star leaves the main sequence, something else happens: *Helium begins to burn in the core.* By the time the central density has risen to about 10^8 kg/m^3 (at stage 9), the temperature has reached the 10^8 K needed for helium to fuse into carbon, and the central fires reignite.

The reaction that transforms helium into carbon occurs in two steps. First, two helium nuclei come together to form a nucleus of beryllium-8 (^{8}Be), a highly unstable isotope that would normally break up into two helium nuclei in about 10^{-12} s. However, at the high densities found in the core of a red giant, it is possible that the beryllium-8 nucleus will encounter another helium nucleus before breakup occurs, fusing with the helium nucleus to form carbon-12 (^{12}C). This is the second step of the helium-burning reaction. In part, it is because of the electrostatic repulsion between beryllium-8 (containing four protons) and helium-4 (containing two) that the temperature must reach 10^8 K before that step can take place.

Symbolically, we can represent this next stage of stellar fusion as follows:

$$^4\text{He} + {}^4\text{He} \rightarrow {}^8\text{Be} + \text{energy},$$
$$^8\text{Be} + {}^4\text{He} \rightarrow {}^{12}\text{C} + \text{energy}.$$

Helium-4 nuclei are traditionally known as *alpha particles*. The term dates from the early days of nuclear physics, when the true nature of these particles, emitted by many radioactive materials, was unknown. Because three alpha particles are required to get from helium-4 to carbon-12, the foregoing reaction is usually called the **triple-alpha process.**

The Helium Flash

For stars comparable in mass to the Sun, a complication arises when helium fusion begins. At the high densities found in the core, the gas has entered a new state of matter whose properties are governed by the laws of quantum mechanics (the branch of physics describing the behavior of matter on subatomic scales), rather than by those of classical physics. ∞ (Sec. 4.2)

Up to now, we have been concerned primarily with the nuclei—protons, alpha particles, and so on—that make up virtually all the star's mass and that participate in the reactions that generate its energy. However, the star contains another important constituent: a vast sea of electrons stripped from their parent nuclei by the ferocious heat in the stellar interior. At this stage in our story, these electrons play an important role in determining the star's evolution.

Under the conditions found in the stage-9 red-giant core, a rule of quantum mechanics known as the *Pauli exclusion principle* (after Wolfgang Pauli, one of the founding fathers of quantum physics) prohibits the electrons in the core from being squeezed too close together. In effect, the exclusion principle tells us that we can think of the electrons as tiny rigid spheres that can be squeezed relatively easily up to the point of contact, but that become virtually incompressible thereafter. In the language of quantum mechanics, this condition is known as *electron degeneracy;* the pressure associated with the contact of the tiny electron spheres is called **electron degeneracy pressure.*** It has nothing to do with the *thermal pressure* (due to the star's heat) that we have been studying up to now. In fact, in our red-giant core, the pressure resisting the force of gravity is supplied almost entirely by degenerate electrons. Hardly any of the core's support results from "normal" thermal pressure, and this fact has dramatic consequences once the helium begins to burn.

Under normal ("nondegenerate") circumstances, the core could react to, and accommodate, the onset of helium burning, but in the core's degenerate state, the burning

becomes unstable, with literally explosive consequences. In a star supported by thermal pressure, the increase in temperature produced by the onset of helium fusion would lead to an increase in pressure. The gas would then expand and cool, reducing the burning rate and reestablishing equilibrium, just as discussed earlier.

In the electron-supported core of a solar-mass red giant, however, the pressure is largely *independent* of the temperature. When burning starts and the temperature increases, there is no corresponding rise in pressure, no expansion of the gas, no drop in the temperature, and no stabilization of the core. Instead, the core is unable to respond to the rapidly changing conditions within it. The pressure remains more or less unchanged as the nuclear reaction rates increase, and the temperature rises rapidly in a runaway condition called the **helium flash.**

For a few hours, the helium burns ferociously. Eventually, the flood of energy released by this period of runaway fusion heats the core to the point at which normal thermal pressure once again dominates. Finally able to react to the energy dumped into it by helium burning, the core expands, its density drops, and equilibrium is restored as the inward pull of gravity and the outward push of gas pressure come back into balance. The core, now stable, begins to fuse helium into carbon at temperatures well above 10^8 K.

The helium flash terminates the giant star's ascent of the red-giant branch of the H–R diagram. Yet, despite the violent ignition of helium in the core, the flash does *not* increase the star's luminosity. On the contrary, the energy released in the helium flash expands and cools the core and ultimately results in a *reduction* in the energy output. On the H–R diagram, the star jumps from stage 9 to stage 10, a stable state with steady helium burning in the core. As indicated in Figure 20.4, the surface temperature is now higher than it was on the red-giant branch, but the luminosity is considerably less than at the helium flash. This adjustment in the star's properties occurs quite quickly—in about 100,000 years.

At stage 10, our star is now stably burning helium in its core and fusing hydrogen in a shell surrounding it. The star resides in a well-defined region of the H–R diagram known as the **horizontal branch,** where core-helium-burning stars remain for a time before resuming their journey around the H–R diagram. The star's specific position within this region is determined mostly by its mass—not its original mass, but whatever mass remains after its ascent of the red-giant branch. The two masses differ because, during the red-giant stage, strong stellar winds eject large amounts of matter from a star's surface. As much as 20 to 30 percent of the original stellar mass may escape during that period. It so happens that more massive stars have lower surface temperatures at this stage, but all stars have roughly the same luminosity after the helium flash. As a result, stage-10 stars tend to lie along a horizontal line on the H–R diagram, with more massive stars to the right and less massive ones to the left.

The term refers to an idealized condition in which all particles (electrons in this case) are in their lowest possible energy states. As a result, the star cannot be compressed into a more compact configuration.

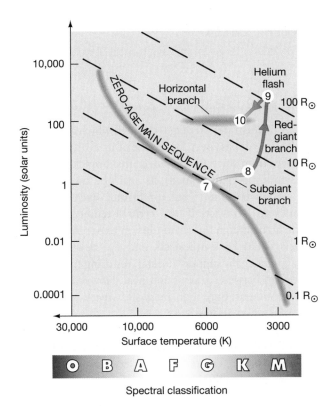

▲ FIGURE 20.4 Horizontal Branch A large increase in luminosity occurs as a star ascends the red-giant branch, ending in the helium flash. The star then settles down into another equilibrium state at stage 10, on the horizontal branch.

Stage 11: Back to the Giant Branch

The nuclear reactions in our star's helium core burn on, but not for long. Whatever helium exists in the core is rapidly consumed, and the dying star once again ascends the giant branch.

The triple-alpha helium-to-carbon fusion reaction—like the proton–proton and CNO-cycle hydrogen-to-helium reactions before it—proceeds at a rate that increases rapidly with temperature. At the extremely high temperatures found in the horizontal-branch core, the helium fuel doesn't last long—no more than a few tens of millions of years after the initial flash.

As helium fuses to carbon, a new carbon-rich inner core begins to form, and phenomena similar to those that took place during the earlier buildup of helium recur. Helium becomes depleted at the center of the star, and eventually fusion ceases there. The nonburning carbon core shrinks in size—even as its mass increases due to helium fusion—and heats up as gravity pulls it inward, causing the hydrogen- and helium-burning rates in the overlying lay of the core to increase. As illustrated in Figure 20.5, the star now contains a contracting carbon core surrounded by a helium-burning shell, which is in turn surrounded by a hydrogen-burning shell. The outer envelope of the star—the nonburning layers surrounding the core—expands, much as it did earlier during the first red-giant stage. By the time it reaches stage 11 in Figure 20.6, the star has become a swollen red giant for the second time.

To distinguish the second ascent of the giant branch from the first, the star's track during the second phase is often referred to as the **asymptotic-giant branch.*** The burning rates in the shells around the carbon core are much

*This rather intimidating term is borrowed from mathematics. An asymptote to a curve is a second curve that approaches ever closer to the first as the two are extended to infinity. Theoretically, if the star remained intact, the asymptotic-giant branch would approach the red-giant branch from the left as the luminosity increased and would effectively merge with the red-giant branch near the top of Figure 20.6. However, as we will see in Section 20.3, a Sun-like star will not live long enough for that to occur.

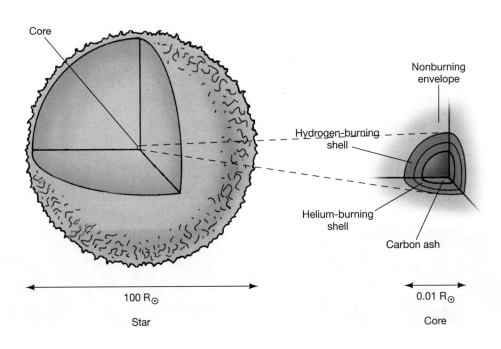

◀ FIGURE 20.5 Helium-Shell Burning Within a few million years after the onset of helium burning (stage 9), carbon ash accumulates in the star's inner core. Above this core, hydrogen and helium are still burning in concentric shells.

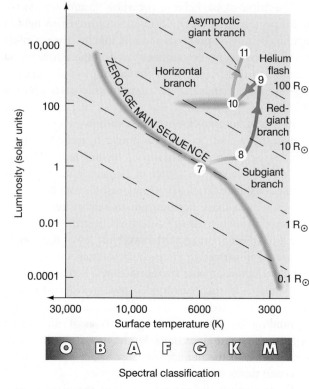

▲ **FIGURE 20.6 Reascending the Red-Giant Branch** A carbon-core star reenters the giant region of the H–R diagram—this time on a track called the *asymptotic-giant branch* (stage 11)—for the same reason it evolved there the first time around: Lack of nuclear fusion at the center causes the core to contract and the overlying layers to expand.

fiercer this time around, and the star's radius and luminosity increase to values even greater than those reached at the helium flash on the first ascent. The carbon core grows in mass as more and more carbon is produced in the helium-burning shell above it, but continues to shrink in radius, driving the hydrogen-burning and helium-burning shells to higher and higher temperatures and luminosities.

▼ **FIGURE 20.7 G-Type Star Evolution** Artist's conception of the relative sizes and changing colors of a normal G-type star (such as our Sun) during its formative stages, on the main sequence, and while passing through the red-giant and white-dwarf stages. At maximum swelling, the red giant is approximately 70 times the size of its main-sequence parent; the core of the giant is about $\frac{1}{15}$th the main-sequence size and would be barely discernible if this figure were drawn exactly to scale. The duration of time spent in the various stages—protostar, main-sequence star, red giant, and white dwarf—is roughly proportional to the lengths shown in this imaginary trek through space. The star's brief stay on the horizontal and asymptotic-giant branches are not shown.

20.3 The Death of a Low-Mass Star

Figure 20.7 illustrates the stages through which a G-type star like the Sun will pass over the course of its evolution. As our star moves from stage 10 (the horizontal branch) to stage 11 (the asymptotic-giant branch), its envelope swells, while its inner carbon core, too cool for further nuclear burning, continues to contract. If the central temperature could become high enough for carbon fusion to occur, still heavier products could be synthesized, and the newly generated energy might again support the star, restoring for a time the equilibrium between gravity and heat. However, as we will see in a moment, only high-mass stars reach temperatures high enough for this to occur.

For solar-mass stars, the central temperature never reaches the 600 million K needed for a new round of nuclear reactions to occur. The red giant is very close to the end of its nuclear-burning lifetime.

The Fires Go Out

Before the carbon core can attain the incredibly high temperatures needed for carbon ignition, its density reaches a point beyond which it cannot be compressed further. At about 10^{10} kg/m³, the electrons in the core once again become degenerate, the contraction of the core ceases, and the core's temperature stops rising. This stage (stage 12 in Table 20.1) represents the maximum compression that the star can achieve—there is simply not enough matter in the overlying layers to bear down any harder.

The core density at this stage is extraordinarily high. A single cubic centimeter of core matter would weigh 1000 kg on Earth—a ton of matter compressed into a volume about the size of a grape! Yet, despite the extreme compression of

Protostar   Main-sequence G-type star

Stage 4   Stage 7

the core, the central temperature is "only" about 300 million K. Some oxygen is formed via reactions between carbon and helium at the inner edge of the helium-burning shell—that is,

$$^{12}C + {}^{4}He \rightarrow {}^{16}O + energy.$$

However, collisions among nuclei are neither frequent nor violent enough to create any heavier elements. For all practical purposes, the central fires go out once carbon has formed.

Stage 12: A Planetary Nebula

Our aged stage-12 star is now in quite a predicament. Its inner carbon core no longer generates energy. The outer-core shells continue to burn hydrogen and helium, and as more and more of the inner core reaches its final, high-density state, the nuclear burning increases in intensity. Meanwhile, the envelope continues to expand and cool, reaching a maximum radius of about 300 times that of the Sun—big enough to engulf the planet Mars.

Around this time, the burning becomes quite unstable. The helium-burning shell is subject to a series of explosive *helium-shell flashes,* caused by the enormous pressure in the helium-burning shell and the extreme sensitivity of the triple-alpha burning rate to small changes in temperature. The flashes produce large fluctuations in the intensity of the radiation reaching the star's outermost layers, causing those layers to pulsate violently as the envelope repeatedly is heated, expands, cools, and then contracts (Figure 20.8). The amplitude of the pulsations grows as the temperature of the core continues to increase and the nuclear burning intensifies in the surrounding shells.

Compounding the star's problems is the increasing instability of its surface layers. Around the peak of each pulsation, the surface temperature drops below the point at which electrons can recombine with nuclei to form atoms. ∞ (Sec. 4.2) Each recombination produces additional photons, giving the gas a little extra outward "push" and causing some of it to escape. Thus, driven by increasingly intense radiation from within, and accelerated by instabilities in both the core and the outer layers, virtually all of the star's envelope is ejected into space in less than a few million years at a speed of a few tens of kilometers per second.

In time, a rather unusual-looking object results. The "star" now consists of two distinct parts, both of which constitute

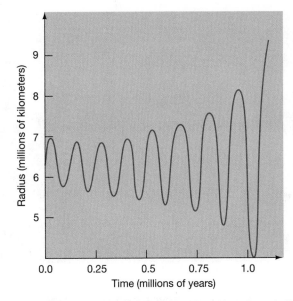

▲ **FIGURE 20.8 Red-Giant Instability** Buffeted by helium-shell flashes from within, and subject to the destabilizing influence of recombination, the outer layers of a red giant become unstable and enter into a series of growing pulsations. Eventually, the envelope is ejected and forms a planetary nebula.

stage 12 of Table 20.1. At the center is a small, well-defined core of mostly carbon ash. Hot, dense, and still very luminous, only the outermost layers of this core still fuse helium into carbon and oxygen. Well beyond the core lies an expanding cloud of dust and cool gas—the ejected envelope of the giant—spread over a volume roughly the size of our solar system.

As the core exhausts its last remaining fuel, it contracts and heats up, moving to the left in the H–R diagram. Eventually, it becomes so hot that its ultraviolet radiation ionizes the inner parts of the surrounding cloud, producing a spectacular display called a **planetary nebula**. Some well-known examples are shown in Figures 20.9 and 20.10. In all, more than 1500 planetary nebulae are known in our Galaxy. The word *planetary* here is misleading, for these objects have no association with planets. The name originated in the 18th century, when, viewed at poor resolution through small telescopes, these shells of glowing gas looked to some astronomers like the circular disks of planets in our solar system.

Note that the mechanism by which planetary nebulae shine is basically the same as that powering the emission nebulae we studied earlier: ionizing radiation from a hot star

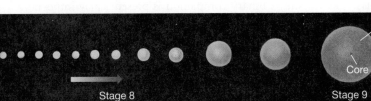
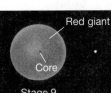

Red giant

White dwarf

Core

Stage 8

Stage 9

(a)

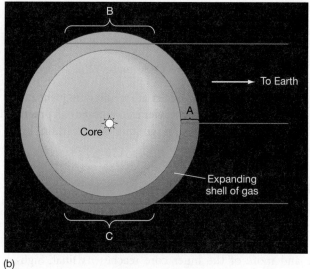

(b)

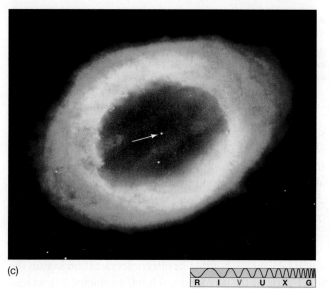

ANIMATION/VIDEO Helix Nebula Formation

(c)

R I V U X G

◀ **FIGURE 20.9 Ejected Envelope** A planetary nebula is an extended region of glowing gas surrounding an intensely hot central star (marked with an arrow here). The small, dense star is the core of a former red giant. The gas is what remains of the giant's envelope, now ejected into space. (a) Abell 39, some 2100 pc away, is a classic planetary nebula shedding a spherical shell of gas about 1.5 pc across. (b) The brightened appearance around the edge of Abell 39 is caused by the thinness of the shell of glowing gas around the central core. Very little gas exists along the line of sight between the observer and the central star (path A), so that part of the shell is invisible. Near the edge of the shell, however, more gas exists along the line of sight (paths B and C), so the observer sees a glowing ring. (c) The Ring Nebula, perhaps the most famous of all planetary nebulae at 1500 pc away and 0.5 pc across, is too small and dim to be seen with the naked eye. Astronomers once thought its appearance could be explained in much the same way as that of Abell 39. However, it now seems that the Ring really is ring shaped! Researchers are still unsure as to why a spherical star should eject a ring of material during its final days. *(AURA; NASA)*

embedded in a cool gas cloud. ∞ (Sec. 18.2) However, recognize that these two classes of object have very different origins and represent completely separate phases of stellar evolution. The emission nebulae discussed in Chapter 18 are signposts of recent stellar birth. Planetary nebulae, by contrast, indicate impending stellar death.

Astronomers once thought that the escaping giant envelope would be more or less spherical in shape, completely surrounding the core in three dimensions, just as it had while still part of the star. Figure 20.9(a) shows an example where this may well in fact be the case. The "ring" of this planetary nebula is in reality a three-dimensional shell of glowing gas—its halo-shaped appearance is only an illusion. As illustrated in Figure 20.9(b), the nebula looks brighter near the edges simply because there is more emitting gas along the line of sight there, creating the illusion of a bright ring.

However, such cases now seem to be in the minority. There is growing evidence that, for reasons not yet fully understood, the final stages of red-giant mass loss are often decidedly *non*spherical. For example, the famous Ring Nebula shown in Figure 20.9(c) may well actually *be* a ring, and not just our view of a glowing spherical shell, and many planetary nebulae are much more complex than that. As illustrated in Figure 20.10, planetary nebulae may exhibit jetlike and other irregular structures. Apparently, both the details of the gas-ejection process and the star's environment (such as whether a binary companion is present) play important roles in determining the nebula's shape and appearance.

The central star fades and eventually cools, and the expanding gas cloud becomes more and more diffuse, eventually dispersing into interstellar space. After just a few tens of thousands of years, the glowing planetary nebula disappears from view. As the cloud rejoins the interstellar medium, it

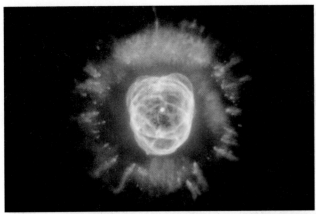

(a)

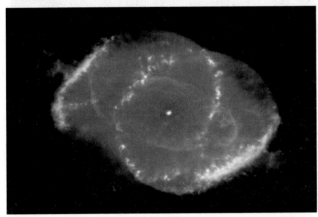

(b)

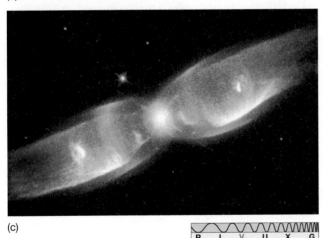

(c)

R I V U X G

envelope by convection during the star's final years, to enrich the interstellar medium when the giant envelope escapes. The evolution of low-mass stars is the source of virtually all the carbon-rich dust observed throughout the plane of our own and other galaxies. ∞ (Sec. 18.1)

Stage 13: A White Dwarf

The carbon core—the stellar remnant at the center of the planetary nebula—continues to evolve. Formerly concealed by the atmosphere of the red-giant star, the core becomes visible as the envelope recedes. Several tens of thousands of years are needed for the core to appear from behind the veil of expanding gas. The core is very small. By the time the envelope is ejected as a planetary nebula, the core has shrunk to about the size of Earth. (In some cases, it may be even smaller than our planet.) Its mass is about half the mass of the Sun. Shining only by stored heat, not by nuclear reactions, this small "star" has a white-hot surface when it first becomes visible, although it appears dim because of its small size. The core's temperature and size give rise to its new name: **white dwarf.** This is stage 13 of Table 20.1. The approximate path followed by the star on the H–R diagram as it evolves from stage-11 red giant to stage-13 white dwarf is shown in Figure 20.11.

Not all white-dwarf stars are found as the cores of planetary nebulae: Several hundred have been discovered "naked" in our Galaxy, their envelopes expelled to invisibility (or perhaps stripped away by a binary companion—to be discussed shortly) long ago. Figure 20.12 shows an example of a white dwarf, Sirius B, that happens to lie particularly close to Earth; it is the faint binary companion of the much brighter and better known Sirius A. ∞ *(More Precisely 17-2)* Some properties of Sirius B are listed in Table 20.2. With more than the mass of the Sun packed into a volume smaller than Earth, Sirius B has a density about a million times greater than anything familiar to us in the solar system. In fact, Sirius B has an unusually high mass for a white dwarf—it is thought to be the evolutionary product of a star roughly four times the mass of the Sun. *Discovery 20-1* discusses another possible peculiarity of Sirius B's evolution.

Hubble Space Telescope (HST) observations of nearby globular clusters have revealed the white-dwarf sequences

plays a vital role in the evolution of our Galaxy. During the final stages of the red giant's life, nuclear reactions between carbon and unburned helium in the core create oxygen and, in some cases, even heavier elements, such as neon and magnesium. Some of these reactions also release neutrons, which, carrying no electrical charge, have no electrostatic barrier to overcome and hence can interact with existing nuclei to form still heavier elements (see Chapter 21). All of these elements—helium, carbon, oxygen, and heavier ones—are "dredged up" from the depths of the core into the

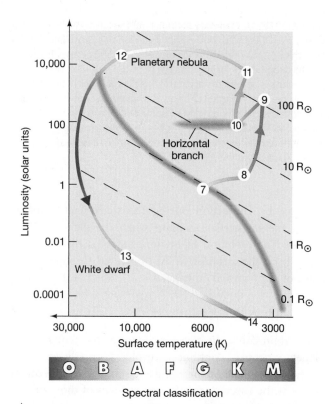

Interactive FIGURE 20.11 White Dwarf on the H–R Diagram
A star's passage from the horizontal branch (stage 10) to the white-dwarf stage (stage 13) by way of the asymptotic-giant branch creates an evolutionary path that cuts across the entire H–R diagram. This diagram illustrates the entire evolutionary lifetime of a typical low-mass star, and provides a basis for detailed comparison between theory and observations.

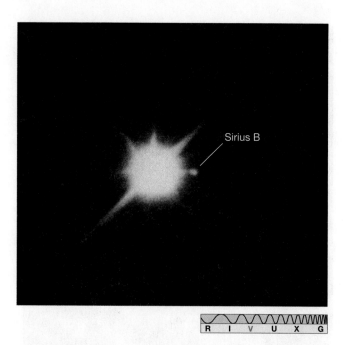

▲ **FIGURE 20.12 Sirius Binary System** Sirius B (the speck of light to the right of the much larger and brighter star Sirius A) is a white dwarf star, a companion to Sirius A. The "spikes" on the image of Sirius A are not real; they are caused by the support struts of the telescope. *(Palomar Observatory)*

long predicted by theory, but previously too faint to detect at such large distances. Figure 20.13(a) shows a ground-based view of the globular cluster M4, lying 1700 pc from Earth. Part (b) of the figure shows an *HST* closeup of a small portion of the cluster, revealing dozens of white dwarfs among the cluster's much brighter main sequence, red-giant, and horizontal-branch stars. When plotted on an H–R diagram (see Figure 20.14), the white dwarfs fall nicely along the path indicated in Figure 20.11.

Not all white dwarfs are composed of carbon and oxygen. As mentioned earlier, theory predicts that very low-mass stars (less than about one-quarter the mass of the Sun) will never reach the point of helium fusion. Instead, the core of such a star will become supported by electron degeneracy pressure before its central temperature reaches the 100 million K needed to start the triple-alpha process. The interiors of such stars are completely convective, ensuring that fresh hydrogen continually mixes from the envelope into the core. As a result, unlike the case of the Sun illustrated in Figure 20.2, a nonburning helium inner core never appears, and eventually *all* of the star's hydrogen is converted to helium, forming a *helium white dwarf*.

The time needed for this kind of transformation to occur is very long—hundreds of billions of years—so no helium white dwarfs have ever actually formed in this way. ∞ (Sec. 17.8) However, if a solar-mass star is a member of a binary system, it is possible for its envelope to be stripped away during the red-giant stage by the gravitational pull of its companion (see Section 20.6), exposing the helium core and terminating the star's evolution before helium fusion can begin. Several such low-mass helium white dwarfs have in fact been detected in binary systems.

Finally, in stars somewhat more massive than the Sun (close to the 8-solar-mass limit on "low-mass" stars at the time the carbon core forms), temperatures in the core may become high enough that an additional reaction,

$$^{16}O + {}^4He \rightarrow {}^{20}Ne + energy,$$

can occur, ultimately leading to the formation of a rare *neon–oxygen white dwarf*.

TABLE 20.2 Sirius B, a Nearby White Dwarf	
Mass	1.1 solar masses
Radius	0.0073 solar radius (5100 km)
Luminosity (total)	0.0025 solar luminosity (10^{24} W)
Surface temperature	27,000 K
Average density	$3.9 \times 10^9 \, kg/m^3$

DISCOVERY 20-1

Learning Astronomy from History

Sirius A, the brighter of the two objects shown in Figure 20.12, appears twice as luminous as any other visible star, excluding the Sun. Its absolute brightness is not very great, but because it is not very far from us (less than 3 pc away), its apparent brightness is very large. ∞ (Sec. 17.2) Sirius has been prominent in the nighttime sky since the beginning of recorded history. Cuneiform texts of the ancient Babylonians refer to the star as far back as 1000 B.C., and historians know that the star strongly influenced the agriculture and religion of the Egyptians of 3000 B.C.

Even though a star's evolution takes such a long time, we might have a chance to detect a slight change in Sirius because the recorded observations of this star go back several thousand years. The chances for success are improved in this case because Sirius A is so bright that even the naked-eye observations of the ancients should be reasonably accurate. Interestingly, recorded history does suggest that Sirius A has changed in appearance, but the observations are confusing. Every piece of information about Sirius recorded between the years 100 B.C. and A.D. 200 claims that this star was *red*. (No earlier records of its color are known.) In contrast, modern observations now show it to be white or bluish white—definitely *not* red.

If these reports are accurate, then Sirius has apparently changed from red to blue white in the intervening years. According to the theory of stellar evolution, however, no star should be able to change its color in this way in that short a time. Such a color change should take at least several tens of thousands of years and perhaps a lot longer. It should also leave some evidence of its occurrence.

Astronomers have offered several explanations for the rather sudden change in Sirius A, including the suggestions that (1) some ancient observers were wrong and other scribes copied their mistaken writings; (2) a galactic dust cloud passed between Sirius A and Earth some 2000 years ago, reddening the star much as Earth's dusty atmosphere often reddens our Sun at dusk; and (3) the companion to Sirius A, Sirius B, was a red giant and the dominant star of this double-star system 2000 years ago, but has since expelled its planetary nebular shell to reveal the white-dwarf star that we now observe.

Each of these explanations presents problems. How could the color of the sky's brightest star have been incorrectly recorded for hundreds of years? Where is the intervening galactic cloud now? Where is the shell of the former red giant? We are left with the uneasy feeling that the sky's brightest star doesn't fit particularly well into the currently accepted scenario of stellar evolution.

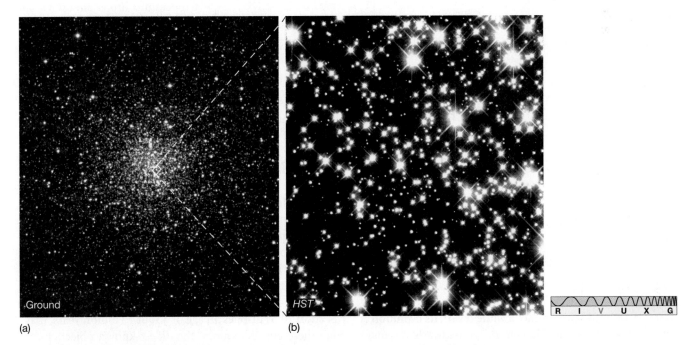

(a)   (b)

▲ **FIGURE 20.13 Distant White Dwarfs** (a) The globular cluster M4, as seen through a large ground-based telescope at Kitt Peak National Observatory in Arizona (see also Figure 18.13). At 1700 pc away, M4 is the closest globular cluster to us; it spans some 16 pc. (b) A peek at M4's "suburbs" by the *Hubble Space Telescope* shows nearly a hundred white dwarfs within a small 0.2 square-parsec region. *(AURA; NASA)*

(a)

◀ FIGURE 20.14 **Global Cluster H–R Diagram** (a) The globular cluster M80, some 8 kpc away. (b) Combined H–R diagram, based on ground- and space-based observations, for several globular clusters similar in overall composition to M80. The various evolutionary stages predicted by theory and depicted schematically in Figure 20.11 are clearly visible. Note also the blue stragglers—main-sequence stars that appear to have been "left behind" as other stars evolved into giants. They are probably the result of merging binary systems or actual collisions between stars of lower mass in this remarkably dense stellar system. (See also Figure 20.19.) The inset shows the H–R diagram of another globular cluster, NGC 2808, revealing that the main sequence is actually made up of three distinct sequences, increasing in helium content from right to left, and suggesting multiple generations of star formation shortly after the cluster formed. (*NASA; data courtesy W.E. Harris*)

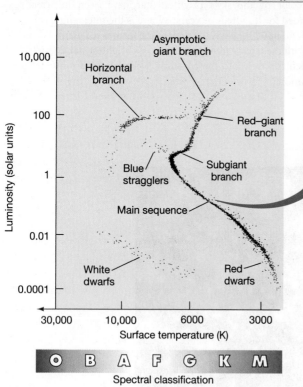

(b)

Stage 14: A Black Dwarf

Once an isolated star becomes a white dwarf, its evolution is over. (As we will see in Chapter 21, white dwarfs in binary systems may undergo further activity.) The isolated white dwarf continues to cool and dim with time, following the white–yellow–red track near the bottom of the H–R diagram of Figure 20.11 and eventually becoming a **black dwarf**—a

cold, dense, burned-out ember in space. This is stage 14 of Table 20.1, the graveyard of stars.

The cooling dwarf does not shrink much as it fades away. Even though its heat is leaking away into space, gravity does not compress it further. At the enormously high densities in the star (from the white-dwarf stage on), the resistance of electrons to being squeezed together—the same electron degeneracy that prevailed in the red-giant core around the time of the helium flash—supports the star, even as its temperature drops almost to absolute zero. As the dwarf cools, it remains about the size of Earth.

Comparing Theory with Reality

All the H–R diagrams and evolutionary tracks presented so far are theoretical constructs based largely on computer models of the interior workings of stars. Before continuing our study of stellar evolution, let's take a moment to compare our models with actual observations. Figure 20.14(a) shows the beautiful globular cluster M80, which lies about 8000 pc from Earth. Figure 20.14(b) shows a composite H–R diagram recently constructed by using the stars of a number of other globular clusters of roughly the same age and composition as M80. The diagram spans the entire range of stellar luminosities, from bright red giants to faint red and white dwarfs. Fitting theoretical models of the main-sequence, giant, and horizontal branches (see Section 20.5) implies an age of about 12 billion years, making these clusters among the oldest-known objects in the Milky Way Galaxy and, as such, key indicators of conditions in the early universe.

The great age of this cluster means that stars more massive than about 0.8 solar mass have already evolved beyond the red-giant stage, becoming mainly white dwarfs. The H–R diagram for this cluster can therefore be compared directly with Figure 20.11, as the red-giant, horizontal-branch, and asymptotic-giant-branch stars are all of roughly 1 solar mass. The similarity between theory and observation is striking: Stars in each of the evolutionary stages 7–13 can be seen in numbers consistent with the theoretical models. (See also the acetate inset in this chapter.) Astronomers place great confidence in the theory of stellar evolution precisely because its predictions are so often found to be in excellent agreement with plots of real stars.

Note that the points in Figure 20.14(b) are actually shifted a little to the left relative to Figure 20.11. This is because of differences in composition between stars such as the Sun and stars in globular clusters. For reasons to be discussed more fully in Chapter 21, the old globular cluster stars contain much lower concentrations of "heavy" elements (astronomical jargon for anything more massive than helium). One result of this relative paucity of these elements is that the interiors and atmospheres of those stars tend to be slightly more transparent to radiation from within, allowing the energy to escape more easily and making the stars slightly smaller and hotter than solar-type stars of the same mass.

The objects labeled as **blue stragglers** in Figure 20.14(b) appear at first sight to contradict the theory just described. They are observed in many star clusters, lying on the main sequence, but in locations suggesting that they should have evolved into white dwarfs long ago, given the parent cluster's age. They are main-sequence stars, but they did not form when the cluster did. Instead, they formed much more recently, through *mergers* of lower mass stars—so recently, in fact, that they have not yet had time to evolve into giants.

In some cases, these mergers are the result of stellar evolution in binary systems, as the component stars evolved, grew, and came into contact (see Section 20.6). In others, the mergers are thought to be the result of actual *collisions* between stars. The core of M80 contains a huge number of stars packed into a relatively small volume. For example, a sphere of radius 2 pc centered on the Sun contains exactly four stars, including the Sun itself. ∞ (Sec. 17.1) At the center of M80, the same 2-pc sphere would contain more than 10 *million* stars—our night sky would be ablaze with thousands of objects brighter than Venus! The dense central cores of globular clusters are among the few places in the entire universe where stellar collisions are likely to occur.

Recent high-precision measurements from *HST* have revealed another, and as yet incompletely resolved, mystery about globular clusters that may change significantly our view of how these systems formed. The inset to Figure 20.14 shows the H–R diagram of the cluster NGC 2808, revealing *three* distinct main sequences previously undetected in ground-based observations. The stars in the three sequences contain different amounts of helium, carbon, and nitrogen and are thought to be the result of distinct episodes of star formation occurring over the course of about 100 million years, the heavier elements produced by each stellar generation contributing to the composition of the next. Astronomers are uncertain just how this occurred, but it appears to have been quite common, as high-resolution studies of many globular clusters now reveal similar multiple populations.

CONCEPT CHECK

✔ Why does fusion cease in the core of a low-mass star?

20.4 Evolution of Stars More Massive than the Sun

High-mass stars evolve much faster than their low-mass counterparts. The more massive a star, the more ravenous is its fuel consumption and the shorter is its main-sequence lifetime. The Sun will spend a total of some 10 billion years on the main sequence, but a 5-solar-mass B-type star will remain there for only a hundred million years. A 10-solar-mass O-type star will depart in just 20 million years or so. This trend toward much faster evolution for more massive stars continues even after the star leaves the main sequence.

All evolutionary changes happen much more rapidly for high-mass stars because their larger mass and stronger gravity generate more heat, speeding up *all* phases of stellar evolution. In fact, helium fusion proceeds so quickly that the high-mass star has a very different evolutionary track. Its envelope swells and cools as the star becomes a *supergiant*.

Red Supergiants

Stars leave the main sequence for one basic reason: They run out of hydrogen in their cores. As a result, the early stages of stellar evolution beyond the main sequence are qualitatively the same in all cases: Main-sequence hydrogen burning in the core (stage 7) eventually gives way to the formation of a non-burning, collapsing helium core surrounded by a hydrogen-burning shell (stages 8 and 9). A high-mass star leaves the main sequence on its journey toward the red-giant region with an internal structure quite similar to that of its low-mass cousin. Thereafter, however, their evolutionary tracks diverge.

Figure 20.15 compares the post-main-sequence evolution of three stars, respectively, having masses 1, 4, and 10 times the mass of the Sun. Note that, whereas stars like the Sun ascend the red-giant branch almost vertically, stars of higher mass move nearly horizontally across the H–R diagram after leaving the upper main sequence. Their luminosities stay roughly constant as their radii increase and their surface temperatures drop.

In stars having more than about 2.5 times the mass of the Sun, helium burning begins smoothly and stably, *not*

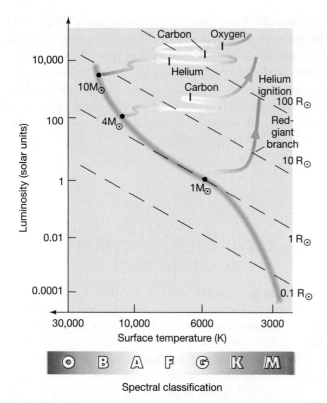

▲ **FIGURE 20.15 High-Mass Evolutionary Tracks** Evolutionary tracks for stars of 1, 4, and 10 solar masses (shown only up to helium ignition in the low-mass case). Stars with masses comparable to that of the Sun ascend the giant branch almost vertically, whereas higher-mass stars move roughly horizontally across the H–R diagram from the main sequence into the red-giant region. The most massive stars experience smooth transitions into each new burning stage. No helium flash occurs for stars more massive than about 2.5 solar masses. Some points are labeled with the element that has just started to fuse in the inner core.

explosively—there is no helium flash. Calculations indicate that the more massive a star, the lower is its core density when the temperature reaches the 10^8 K necessary for helium ignition, and the smaller is the contribution to the pressure from degenerate electrons. As a result, above 2.5 solar masses, the unstable core conditions described earlier do not occur. The 4-solar-mass red giant in Figure 20.15 remains a red giant as helium starts to fuse into carbon. There is no sudden jump to the horizontal branch and no subsequent reascent of the giant branch. Instead, the star loops smoothly back and forth near the top of the H–R diagram.

A much more important divergence occurs at approximately 8 solar masses—the dividing line between high and low mass mentioned in Section 20.1. A low-mass star never achieves the 600 million K needed to fuse carbon nuclei, so it ends its life as a carbon–oxygen (or possibly neon–oxygen) white dwarf. A high-mass star, however, can fuse not only hydrogen and helium, but also carbon, oxygen, and even heavier elements as its inner core continues to contract and its central temperature continues to rise. The rate of burning accelerates as the core evolves.

Evolution proceeds so rapidly in the 10-solar-mass star of Figure 20.15 that the star doesn't even reach the red-giant region before helium fusion begins. The star achieves a central temperature of 10^8 K while it is still quite close to the main sequence. As each element is burned to depletion at the center, the core contracts and heats up, and fusion starts again. A new inner core forms, contracts again, heats again, and so on. The star's evolutionary track continues smoothly across the supergiant region of the H–R diagram, seemingly unaffected by each new phase of burning. The star's radius increases as its surface temperature drops, so the star swells to become a red supergiant. ∞ (Sec. 17.4)

With heavier and heavier elements forming at an ever-increasing rate, the high-mass star shown in Figure 20.15 is very close to the end of its life. We will discuss the evolution and ultimate fate of such a star in more detail in the next chapter, but suffice it to say here that it is destined to die in a violent supernova—a catastrophic explosion releasing energy that will most likely literally blow the star to pieces—soon after carbon and oxygen begin to fuse in its core. High-mass stars evolve so rapidly that, for most practical observational purposes, they explode and die shortly after leaving the main sequence.

A good example of a post-main-sequence blue supergiant is the bright star Rigel in the constellation Orion. With a radius some 70 times that of the Sun and a total luminosity of more than 60,000 solar luminosities, Rigel is thought to have had an original mass about 17 times that of the Sun, although a strong stellar wind has probably carried away a significant fraction of its mass since it formed. Although still near the main sequence, Rigel is probably already fusing helium into carbon in its core.

Perhaps the best-known red supergiant is Betelgeuse (shown in Figures 17.8 and 17.11), also in Orion, and Rigel's rival for the title of brightest star in the constellation. Its luminosity is roughly 10^4 times that of the Sun in visible light and perhaps four times that in the infrared. Astronomers think that Betelgeuse is currently fusing helium into carbon and oxygen in its core, but its eventual fate is uncertain. As best we can tell, the star's mass at formation was between 12 and 17 times the mass of the Sun. However, like Rigel and many other supergiants, Betelgeuse has a strong stellar wind and is known to be surrounded by a huge shell of dust of its own making (see *Discovery 20-2*). It also pulsates, varying in radius by about 60 percent. The pulsations and strong wind may be related to the huge spots observed on the star's surface (Figure 17.11). Together, they suggest that Betelgeuse has lost a lot of mass since it formed, but just how much remains uncertain.

The End of the Road

Protostars and stars evolve because gravity always tends to cause a nonburning stellar core to contract and heat up. The contraction continues until it is halted either by electron

DISCOVERY 20-2

Mass Loss from Giant Stars

Astronomers now know that stars of all spectral types are active and have stellar winds. Consider the highly luminous, hot, blue O- and B-type stars, which have by far the strongest winds. Satellite and rocket observations have shown that their wind speeds may reach 3000 km/s. The result is a yearly mass loss sometimes exceeding 10^{-6} solar mass per year. Over the relatively short span of 1 million years, these stars blow a tenth of their total mass—more than an entire solar mass of material—into space. The powerful stellar winds, driven directly by the pressure of the intense ultraviolet radiation emitted by the stars themselves, hollow out vast cavities in the interstellar gas.

The black-and-white photograph here shows the supergiant star AG Carinae—50 times more massive than the Sun and a million times brighter—shedding its outer atmosphere. The star is shown puffing out vast clouds of gas and dust. (The star, at the center, is intentionally obscured to show the surrounding faint nebula more clearly; the bright vertical line is also an artifact—an effect of the optical system used to hide the star.)

The four-part accompanying *Hubble* image captures another stellar outburst in the second half of the year 2002, during which a star brightened more than a half-million times our Sun's luminosity. This star, with the tongue-twisting name V838 Monocerotis, is a highly variable (and poorly understood) red supergiant about 20,000 light-years distant. Actually, what we are seeing here is not matter being expelled outward as fast as the images imply; rather, a burst of light—often called a "light echo"—is illuminating shells of gas and dust now surrounding the star, but that had been shed long ago. For scale, the rightmost image is about 7 light-years across.

Observations made with radio, infrared, and optical telescopes have shown that luminous cool stars (e.g., K- and M-type red giants) also lose mass at rates comparable to those at which luminous hot stars lose mass. Red-giant wind velocities, however, are much lower, averaging merely 30 km/s. They carry roughly as much mass into space as do O-type stellar winds, because their densities are generally much greater. Also, because luminous red stars are inherently cool objects (with surface temperatures of only about 3000 K), they emit virtually no ultraviolet radiation, so the mechanism driving the winds must differ from that driving the winds of luminous hot stars. We can only surmise that gas turbulence, magnetic fields, or both in the atmospheres of the red giants are somehow responsible. The surface conditions in red giants are in some ways similar to those in T Tauri protostars, which are also known to exhibit strong winds. Possibly the same basic mechanism—violent surface activity—is responsible for both kinds of winds.

Unlike winds from hot stars, winds from these cool stars are rich in dust particles and molecules. Nearly all stars eventually evolve into red giants, so such winds are a major source of new gas and dust to interstellar space and also provide a vital link between the cycle of star formation and the evolution of the interstellar medium.

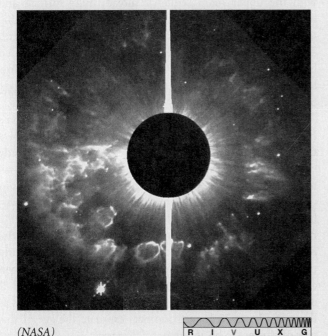

(NASA)

R I V U X G

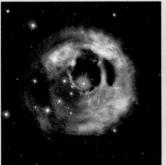

(NASA)

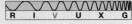

R I V U X G

ANIMATION/VIDEO Light Echo

MA

TABLE 20.3 End Points of Evolution for Stars of Different Masses

Initial Mass (Solar Masses)	Final State
less than 0.08	(hydrogen) brown dwarf
0.08–0.25	helium white dwarf
0.25–8	carbon–oxygen white dwarf
8–12 (approx.)*	neon–oxygen white dwarf
greater than 12*	supernova (Chapter 21)

* *Precise numbers depend on the (poorly known) amount of mass lost while the star is on, and after it leaves, the main sequence.*

degeneracy pressure or by the onset of a new round of nuclear fusion. In the latter case, a new nonburning core builds up, and the process repeats. The more massive the star, the more repetitions occur before the star finally dies. Table 20.3 lists some possible outcomes of stellar evolution for stars of different masses. For completeness, *brown dwarfs*—the end product of low-mass protostars unable even to fuse hydrogen in their cores—are included in the list ∞ (Sec. 19.3)

Note that our earlier dividing line of 8 solar masses between "low mass" and "high mass" really refers to the mass at the time the carbon core forms. Since very luminous stars often have strong stellar winds (*Discovery 20-2*), main-sequence stars as massive as 10 to 12 times the mass of the Sun may still manage to avoid going supernova. Unfortunately, we do not know exactly how much mass either Rigel or Betelgeuse has lost, so we cannot yet tell whether they are above or below the threshold for becoming a supernova. Either might explode or instead become a neon–oxygen white dwarf, but for now we can't say which. We may just have to wait and see—in a million years or so we will know for sure!

CONCEPT CHECK

✔ What is the essential evolutionary difference between high-mass and low-mass stars?

20.5 Observing Stellar Evolution in Star Clusters

Star clusters provide excellent test sites for the theory of stellar evolution. Every star in a given cluster formed at the same time, from the same interstellar cloud, and with virtually the same composition. Only the mass varies from one star to another, thus allowing us to check the accuracy of our theoretical models in a very straightforward way. Having studied the evolutionary tracks of individual stars in some detail, let's now consider how their collective appearance changes in time.

In Chapter 19, we saw how astronomers estimate the ages of star clusters by determining which of their stars have already left the main sequence. ∞ (Sec. 19.6) In fact, the main-sequence lifetimes that go into those age measurements represent only a tiny fraction of the data obtained from theoretical models of stellar evolution. Starting from the zero-age main sequence, astronomers can predict exactly how a newborn cluster should look at any subsequent time—which stars are on the main sequence, which are becoming giants, and which have already burned themselves out. Although we cannot see into the interiors of stars to test our models, we can compare stars' outward appearances with theoretical predictions. The agreement—in detail—between theory and observation is remarkably good.

The Evolving Cluster H–R Diagram

We begin our study shortly after the cluster's formation, with the upper main sequence already fully formed and burning steadily and with stars of lower mass just beginning to arrive on the main sequence, as shown in Figure 20.16(a). The appearance of the cluster at this early stage is dominated by its most massive stars: the bright blue supergiants. Now let's follow the cluster forward in time and see how it evolves by using an H–R diagram.

Figure 20.16(b) shows the appearance of our cluster's H–R diagram after 10 million years. The most massive O-type stars have left the main sequence. Most have already exploded and vanished, as just discussed, but one or two may still be visible as red supergiants. The remaining stars in the cluster are largely unchanged in appearance—their evolution is slow enough that little happens to them in such a relatively short period. The cluster's H–R diagram shows the main sequence slightly cut off, along with a rather poorly defined red-giant region. Figure 20.17 shows the twin open clusters h and chi Persei, along with their combined H–R diagram. Comparing Figure 20.17(b) with such diagrams as those in Figure 20.16, astronomers estimate the age of this pair of clusters to be about 10 million years.

After 100 million years (Figure 20.16c), stars brighter than type B5 or so (about 4–5 solar masses) have left the main sequence, and a few more red supergiants are visible. By this time, most of the cluster's low-mass stars have finally arrived on the main sequence, although the dimmest M-type stars may still be in their contraction phase. The appearance of the cluster is now dominated by bright B-type main-sequence stars and brighter red supergiants.

At any time during the cluster's evolution, the original main sequence is intact up to some well-defined stellar mass, corresponding to the stars that are just leaving the main sequence at that instant. We can imagine the main sequence being "peeled away" from the top down, with fainter and fainter stars turning off and heading for the giant branch as time goes on. Astronomers refer to the high-luminosity end of the observed main sequence as the **main-sequence turnoff.**

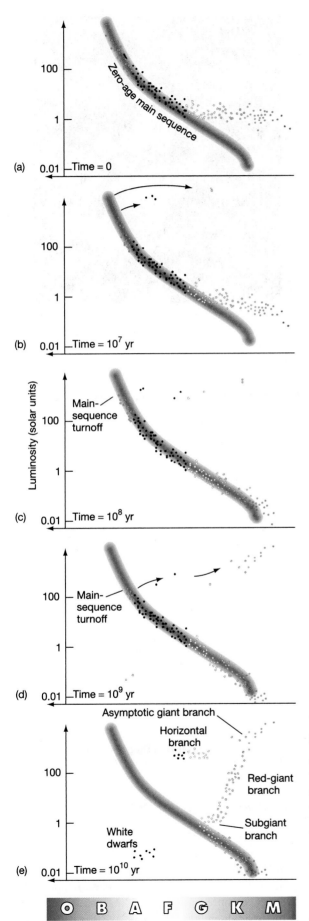

The mass of a star that is just evolving off the main sequence at any moment is known as the *turnoff mass*.

At 1 billion years, the main-sequence turnoff mass is around 2 solar masses, corresponding roughly to spectral type A2. The subgiant and giant branches associated with the evolution of low-mass stars are just becoming visible, as indicated in Figure 20.16(d). The formation of the lower main sequence is now complete. In addition, the first white dwarfs have just appeared, although they are often too faint to be observed at the distances of most clusters. Figure 20.18 shows the Hyades open cluster and its H–R diagram, which appears to lie between Figures 20.16(c) and 20.16(d), suggesting that the cluster's age is about 600 million years.

At 10 billion years, the turnoff point has reached solar-mass stars of spectral type G2. The subgiant and giant branches are now clearly discernible (see Figure 20.16e), and the horizontal and asymptotic-giant branches appear as distinct regions in the H–R diagram. Many white dwarfs are also present in the cluster. Although stars in all these evolutionary stages are also present in the 1-billion-year-old cluster shown in Figure 20.16(d), they are few in number then—typically only a few percent of the total number of stars in the cluster. Also, because they evolve so rapidly, these high-mass stars spend very little time in the various regions. Low-mass stars are much more numerous and evolve more slowly, so more of them spend more time in any given region of the H–R diagram, allowing their evolutionary tracks to be more easily discerned.

Figure 20.19 shows the globular cluster 47 Tucanae. By carefully adjusting their theoretical models until the cluster's main sequence, subgiant, red-giant, and horizontal branches are all well matched, astronomers have determined the age of 47 Tucanae to be between 10 and 12 billion years, a little older than our hypothetical cluster in Figure 20.16(e). In fact, globular-cluster ages determined in this way show a remarkably small spread: All the globular clusters in our Galaxy appear to have formed between about 10 and 12 billion years ago.

◀ **FIGURE 20.16 Cluster Evolution on the H–R Diagram** The properties of star clusters make them nearly ideal test-beds for the theory of stellar evolution. Here we show the changing H–R diagram of a hypothetical star cluster. (a) Initially, stars on the upper main sequence are already burning steadily while the lower main sequence is still forming. (b) At 10^7 years, O-type stars have already left the main sequence (as indicated by the arrows), and a few red giants are visible. (c) By 10^8 years, stars of spectral type B have evolved off the main sequence. More red giants are visible, and the lower main sequence is almost fully formed. (d) At 10^9 years, the main sequence is cut off at about spectral type A. The subgiant and red-giant branches are just becoming evident, and the formation of the lower main sequence is complete. A few white dwarfs may be present. (e) At 10^{10} years, only stars less massive than the Sun still remain on the main sequence. The cluster's subgiant, red-giant, horizontal, and asymptotic-giant branches are all discernible. Many white dwarfs have now formed.

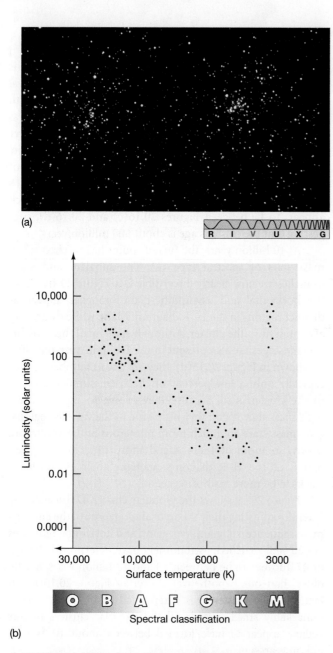

(b)

▲ FIGURE 20.17 **Newborn Cluster H–R Diagram** (a) The "double cluster" h and chi Persei, two open clusters that apparently formed at the same time, possibly even orbiting one another. (b) The H–R diagram of the pair indicates that the stars are very young—probably only about 10 million years old. Even so, the most massive stars have already left the main sequence. *(AURA)*

The Theory of Stellar Evolution

The modern theory of the lives and deaths of stars is one more excellent example of the scientific method in action. ∞ (Sec. 1.2) Faced with a huge volume of observational data, with little or no theory to organize or explain it, astronomers in the late 19th and early 20th centuries painstakingly classified and categorized the properties of the stars they observed. ∞ (Sec. 17.5) During the first half of the 20th century, as quantum mechanics began to yield detailed explanations of the behavior of

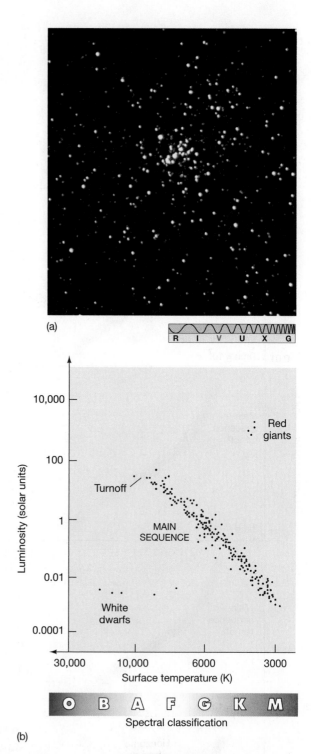

(b)

▲ FIGURE 20.18 **Young Cluster H–R Diagram** (a) The Hyades cluster, a relatively young group of stars visible to the naked eye, is found 46 pc away in the constellation Taurus. (b) The H–R diagram for this cluster is cut off at about spectral type A, implying an age of about 600 million years. A few massive stars have already become white dwarfs. *(AURA)*

light and matter on subatomic scales, theoretical explanations of many key stellar properties emerged. ∞ (Sec. 4.2) Since the 1950s, a truly comprehensive theory has emerged, tying together the basic disciplines of atomic and nuclear physics,

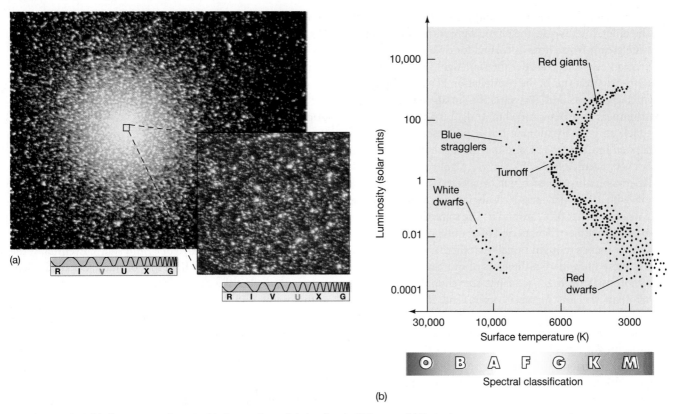

▲ FIGURE 20.19 **Old Cluster H–R Diagram**   (a) The southern globular cluster 47 Tucanae. (b) Fitting its main-sequence turnoff and its giant and horizontal branches to theoretical models gives 47 Tucanae an age of between 12 and 14 billion years, making it one of the oldest-known objects in the Milky Way Galaxy. The inset is a high-resolution ultraviolet image of 47 Tucanae's core region, taken with the *Hubble Space Telescope* and showing many blue stragglers—massive stars lying on the main sequence above the turnoff point, resulting perhaps from the merging of binary-star systems. (See also Figure 20.14.) The points representing white dwarfs, some red dwarfs, and blue stragglers have been added to the original data set, based on *Hubble* observations of this and other clusters. The white-dwarf data are for the cluster M4 (Figure 20.13). Data on the faintest main-sequence stars shown were obtained from ground-based observations. The thickness of the lower main sequence is due almost entirely to observational limitations, which make it difficult to determine accurately the apparent brightnesses and colors of low-luminosity stars. *(ESO; NASA)*

electromagnetism, thermodynamics, and gravitation into a coherent whole. Theory and observation have proceeded hand in hand, each refining and validating the details of the other as astronomers continue to hone their understanding.

Stellar evolution is one of the great success stories of astrophysics. Like all good scientific theories, it makes definite testable predictions about the universe while remaining flexible enough to incorporate new discoveries as they occur. At the start of the 20th century, some scientists despaired of ever knowing even the compositions of the stars, let alone why they shine and how they change. Today, the theory of stellar evolution is a cornerstone of modern astronomy. Its predictions extend our understanding of the cosmos literally to the limits of the observable universe.

PROCESS OF SCIENCE CHECK

✔ Why are observations of star clusters so important to the theory of stellar evolution?

20.6 Stellar Evolution in Binary Systems

We have noted that most stars in our Galaxy are not isolated objects, but are actually members of binary-star systems. However, our discussion of stellar evolution has so far focused exclusively on isolated stars. This narrow focus prompts us to ask how membership in a binary-star system changes the evolutionary tracks we have just described. Indeed, because nuclear burning occurs deep in a star's core, does the presence of a stellar companion have any significant effect at all? Perhaps not surprisingly, the answer depends on the distance between the two stars in question.

For a binary system whose component stars are very widely separated—that is, the distance between the stars is greater than perhaps a thousand stellar radii—the two stars evolve more or less independently of one another, each following the track appropriate to an isolated star of its particular mass. However, if the two stars are closer, then the gravitational pull of one may strongly influence the envelope of the other. In that case, the physical properties of both may deviate greatly from those calculated for isolated single stars.

As an example, consider the star Algol (Beta Persei, the second-brightest star in the constellation Perseus). By studying its spectrum and the variation in its light intensity, astronomers have determined that Algol is actually a binary (in fact, an eclipsing double-lined spectroscopic binary, as described in Chapter 17), and they have measured its properties very accurately. ∞ (Sec. 17.7) Algol consists of a 3.7-solar-mass main-sequence star of spectral type B8 (a blue giant) with a 0.8-solar-mass red-subgiant companion moving in a nearly circular orbit around it. The stars are 4 million km apart and have an orbital period of about 3 days.

A moment's thought reveals that there is something odd about these findings. On the basis of our earlier discussion, the more massive main-sequence star should have evolved *faster* than the less massive component. If the two stars formed at the same time (as is assumed to be the case), there should be no way that the 0.8-solar-mass star could be approaching the giant stage first. Either our theory of stellar evolution is seriously in error, or something has modified the evolution of the Algol system. Fortunately for theorists, the latter is the case.

As sketched in Figure 20.20, each star in a binary system is surrounded by its own teardrop-shaped "zone of influence," inside of which its gravitational pull dominates the effects of both the other star and the overall rotation of the binary. Any matter within that region "belongs" to the star and cannot easily flow onto the other component or out of the system. Outside the two regions, it is possible for gas to flow toward either star relatively easily. The two teardrop-shaped regions are called **Roche lobes,** after Edouard Roche, the French mathematician who first studied the binary-system problem in the 19th century and whose work we have already encountered in the context of planetary rings. ∞ (Sec. 12.4) The Roche lobes of the two stars meet at a point on the line joining them—the inner Lagrangian point (L₁), which we saw in Chapter 14 in discussing asteroid motions in the solar system. ∞ (Sec. 14.1) This Lagrangian point is a place where the gravitational pulls of the two stars exactly balance the rotation of the binary system. The greater the mass of one component, the larger is its Roche lobe and the farther from its center (and the closer to the other star) is the Lagrangian point.

Astronomers think that Algol started off as a *detached* binary, with both com-

ponents lying well within their respective Roche lobes. For reference, let us label the component that is now the 0.8-solar-mass subgiant as star 1 and the 3.7-solar-mass main-sequence star as star 2. Initially, star 1 was the more massive of the two, having perhaps three times the mass of the Sun. It thus turned off the main sequence first. Star 2 was originally a less massive star, perhaps comparable in mass to the Sun. As star 1 ascended the giant branch, it overflowed its Roche lobe and gas began to flow onto star 2. This transfer of matter had the effect of reducing the mass of star 1 and increasing that of star 2, which in turn caused the Roche lobe of star 1 to shrink as its gravity decreased. As a result, the rate at which star 1 overflowed its Roche lobe increased, and a period of unstable *rapid mass transfer* ensued, transporting most of star 1's envelope onto star 2. Eventually, the mass of star 1 became less than that of star 2. Detailed calculations show that the rate of mass transfer dropped sharply at that point, and the stars entered the relatively stable state we see today. These changes in Algol's components are illustrated in Figure 20.21.

Being part of a binary system has radically altered the evolution of both stars in the Algol system. The original high-mass star 1 is now a low-mass red subgiant, whereas the roughly solar mass star 2 is now a massive blue-giant main-sequence star. The removal of mass from the envelope of star 1 may prevent it from ever reaching the helium flash. Instead, its naked core may eventually be left behind as a *helium white dwarf.* In a few tens of millions of years, star 2 will itself begin to ascend the giant branch and fill its own Roche lobe. If star 1 is still a subgiant or a giant at that time, a contact binary system will result. If, instead, star 1 has by

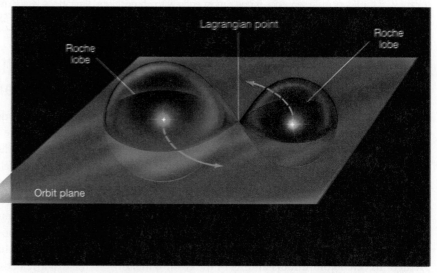

▲ FIGURE 20.20 **Stellar Roche Lobes** Each star in a binary system can be pictured as being surrounded by a "zone of influence," or Roche lobe, inside of which matter may be thought of as being "part" of that star. The two teardrop-shaped Roche lobes meet at the Lagrangian point between the two stars. Outside the Roche lobes, matter may flow onto either star with relative ease.

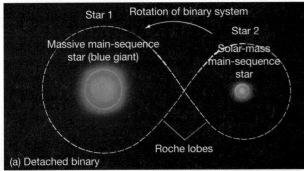

Star 1 Rotation of binary system

Massive main-sequence
star (blue giant)

Star 2

Solar-mass
main-sequence
star

Roche lobes

(a) Detached binary

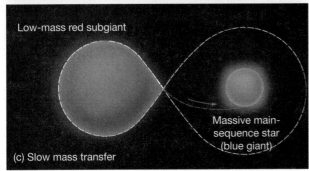

Red giant

Roche lobe

Intermediate-mass
main-sequence star

(b) Rapid mass transfer

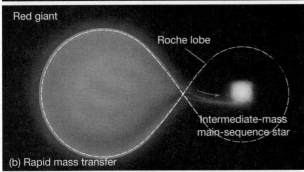

Low-mass red subgiant

Massive main-
sequence star
(blue giant)

(c) Slow mass transfer

◀ **FIGURE 20.21 Algol Evolution** The evolution of the binary star Algol. (a) Initially, Algol was probably a detached binary made up of two main-sequence stars: a relatively massive blue giant and a less massive companion similar to the Sun. (b) As the more massive component (star 1) left the main sequence, it expanded to fill, and eventually overflow, its Roche lobe, transferring large amounts of matter onto its smaller companion (star 2). (c) Today, star 2 is the more massive of the two, but it is on the main sequence. Star 1 is still in the subgiant phase and fills its Roche lobe, causing a steady stream of matter to pour onto its companion.

then become a white dwarf, a new mass-transfer period—with matter streaming from star 2 back onto star 1—will begin. In that case (as we will see in Chapter 21), Algol may have a very active and violent future in store for it.

Just as molecules exhibit few of the physical or chemical properties of their constituent atoms, binaries can display types of behavior that are quite different from the behavior of either of their component stars. The Algol system is a fairly simple example of binary evolution, yet it gives us an idea of the sorts of complications that can arise when two stars evolve interdependently. We will return to the subject in the next two chapters, when we continue our discussion of stellar evolution and the strange states of matter that may ensue.

CONCEPT CHECK

✔ Why is it important to understand the evolution of binary stars?

CHAPTER REVIEW

SUMMARY

1 Stars spend most of their lives on the main sequence, in the **core-hydrogen-burning** (p. 492) phase of stellar evolution, stably fusing hydrogen into helium at their centers. Stars leave the main sequence when the hydrogen in their cores is exhausted. The Sun is about halfway through its main-sequence lifetime and will reach this stage about 5 billion years from now. Low-mass stars evolve much more slowly than the Sun, and high-mass stars evolve much faster.

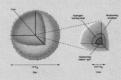

2 When the central nuclear fires in the interior of a solar-mass star cease, the helium in the star's core is still too cool to fuse into anything heavier. With no internal energy source, the helium core is unable to support itself against its own gravity and begins to

shrink. At this stage, the star is in the **hydrogen-shell-burning** (p. 494) phase, in which the nonburning helium at the center is surrounded by a layer of burning hydrogen. The energy released by the contracting helium core heats the hydrogen-burning shell, greatly increasing the nuclear reaction rates there. As a result, the star becomes much brighter, while the envelope expands and cools. A low-mass star like the Sun moves off the main sequence on the H–R diagram first along the **subgiant branch** (p. 495) and then almost vertically up the **red-giant branch** (p. 495).

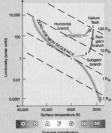

3 Eventually, the contracting core of a Sun-like star reaches the point at which helium begins to fuse into carbon, but conditions at the onset of helium burning are such that the electrons in the core have become degenerate—they can be thought of as tiny, hard spheres that, once brought into contact, present stiff resistance to being compressed any further. This **electron degeneracy pressure (p. 496)** makes the core unable to "react" to the new energy source, and helium burning begins violently in a **helium flash (p. 496).** The flash expands the core and reduces the star's luminosity, sending the star onto the **horizontal branch (p. 496)** of the H–R diagram. The star now has a core of burning helium surrounded by a shell of burning hydrogen.

4 As helium burns in the core, it forms an inner core of nonburning carbon. The carbon core shrinks and heats the overlying burning layers, andthe star once again becomes a red giant, even more luminous than before. It reenters the red-giant region of the H–R diagram along the **asymptotic-giant branch (p. 497).**

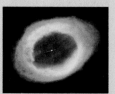

The core of a low-mass star never becomes hot enough to fuse carbon. Such a star continues to ascend the asymptotic-giant branch until its envelope is ejected into space as a **planetary nebula (p. 499).** At that point, the core becomes visible as a hot, faint, and extremely dense white dwarf, whereas the planetary nebula diffuses into space, carrying helium and some carbon into the interstellar medium. The white dwarf cools and fades, eventually becoming a cold **black dwarf (p. 504).**

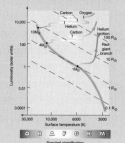

5 High-mass stars evolve more rapidly than low-mass stars because larger mass results in higher central temperature. High-mass stars never initiate a helium flash, and they attain central temperatures high enough to fuse carbon. These stars become red supergiants, forming heavier and heavier elements in their cores at an increasingly rapid pace, and eventually die explosively.

6 The theory of stellar evolution can be tested by observing star clusters, all of whose stars formed at the same time. As time goes by, the most massive stars leave the main sequence first, then the intermediate-mass stars, and so on. At any instant, no stars with masses above the cluster's **main-sequence turnoff (p. 508)** mass remain on the main sequence. Stars below this mass have not yet evolved into giants and so still lie on the main sequence. By comparing a particular cluster's main-sequence turnoff mass with theoretical predictions, astronomers can measure the age of the cluster.

7 Stars in binary systems can evolve quite differently from isolated stars because of interactions with their companions. Each star is surrounded by a teardrop-shaped **Roche lobe (p. 512),** which defines the region of space within which matter "belongs" to the star. As a binary star evolves into the giant phase, it may overflow its Roche lobe, and gas flows from the giant onto its companion. Stellar evolution in binaries can produce states that are not achievable in single stars. In a sufficiently wide binary, both stars evolve as though they were isolated.

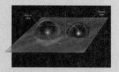

Mastering**ASTRONOMY**   *For instructor-assigned homework go to **www.masteringastronomy.com***

Problems labeled **POS** explore the process of science | **VIS** problems focus on reading and interpreting visual information

REVIEW AND DISCUSSION

1. Why don't stars live forever? Which types of stars live the longest?

2. **POS**   How do astronomers test the theory of stellar evolution?

3. How long can a star like the Sun keep burning hydrogen in its core?

4. Why is the depletion of hydrogen in the core of a star such an important event?

5. What makes an ordinary star become a red giant?

6. Roughly how big (in AU) will the Sun become when it enters the red-giant phase?

7. How long does it take for a star like the Sun to evolve from the main sequence to the top of the red-giant branch?

8. Do all stars eventually fuse helium in their cores?

9. What is a helium flash?

10. Describe an important way in which winds from red-giant stars are linked to the interstellar medium.

11. How do the late evolutionary stages of high-mass stars differ from those of low-mass stars?

12. What is the internal structure of a star on the asymptotic-giant branch?

13. What is a planetary nebula? Why do many planetary nebulae appear as rings?

14. What are white dwarfs? What is their ultimate fate?

15. Can you think of a way in which a helium white dwarf might exist today?

16. Why are white dwarfs hard to observe?

17. Do many black dwarfs exist in our Galaxy?

18. **POS**   How can astronomers measure the age of a star cluster?

19. What are the Roche lobes of a binary system?

20. Why is it odd that the binary system Algol consists of a low-mass red giant orbiting a high-mass main-sequence star? How did Algol come to be in this configuration?

CONCEPTUAL SELF-TEST: MULTIPLE CHOICE

1. A star will evolve "off the main sequence" when it uses up (a) all of its hydrogen; (b) half of its hydrogen; (c) most of the hydrogen in the core; (d) all of its gas.

2. On the main sequence, massive stars (a) conserve their hydrogen fuel by burning helium; (b) burn their hydrogen fuel more rapidly than the Sun; (c) burn their fuel more slowly than the Sun; (d) evolve into stars like the Sun.

3. Compared to other stars on the H–R diagram, red-giant stars are so named because they are (a) cooler; (b) fainter; (c) denser; (d) younger.

4. When the Sun is on the red-giant branch, it will be found at the (a) upper left; (b) upper right; (c) lower right; (d) lower left of the H–R diagram.

5. After the core of a Sun-like star starts to fuse helium on the horizontal branch, the core becomes (a) hotter; (b) cooler; (c) larger; (d) dimmer with time.

6. **VIS** If the evolutionary track in Overlay 3, showing a Sun-like star, were instead illustrating a significantly more massive star, its starting point (stage 7) would be (a) up and to the right; (b) down and to the left; (c) up and to the left; (d) down and to the right.

7. A white dwarf is supported by the pressure of tightly packed (a) electrons; (b) protons; (c) neutrons; (d) photons.

8. **VIS** When the Sun leaves the main sequence, in Figure 20.3, "Red Giant on the H–R Diagram," it will become (a) hotter; (b) brighter; (c) more massive; (d) younger.

9. A star like the Sun will end up as a (a) blue giant; (b) white dwarf; (c) binary star; (d) red dwarf.

10. Compared to the Sun, stars plotted near the bottom left of the H–R diagram are much (a) younger; (b) more massive; (c) brighter; (d) denser.

PROBLEMS

The number of dots preceding each Problem indicates its approximate level of difficulty.

1. • The Sun will leave the main sequence when roughly 10 percent of its hydrogen has been fused into helium. Using the data given in Section 16.5 and Table 16.2, calculate the total amount of mass destroyed (i.e., converted into energy) and the total energy released by the fusion of that amount of matter.

2. • Use the radius–luminosity–temperature relation to calculate the radius of a red supergiant with temperature 3000 K (half the solar value) and total luminosity 10,000 times that of the Sun. ∞ (Sec. 17.4) How many planets of our solar system would this star engulf?

3. • What would be the luminosity of the Sun if its surface temperature were 3000 K and its radius were a. 1 AU, b. 5 AU?

4. • Use the radius–luminosity–temperature relation to calculate the radius of a 12,000-K (twice the temperature of the Sun), 0.0004-solar-luminosity white dwarf.

5. •• The Sun will reside on the main sequence for 10^{10} years. If the luminosity of a main-sequence star is proportional to the fourth power of the star's mass, what is the mass of a star that is just now leaving the main sequence in a cluster that formed a. 400 million years ago, b. 2 billion years ago?

6. •• A main-sequence star at a distance of 20 pc is barely visible through a certain telescope. The star subsequently ascends the giant branch, during which time its temperature drops by a factor of three and its radius increases a hundredfold. What is the new maximum distance at which the star would still be visible in the same telescope?

7. • A Sun-like star goes through a rapid change in luminosity between stages 8 and 9, when its luminosity increases by about a factor of 100 in 10^5 years. On average, how rapidly does the star's absolute magnitude change, in magnitudes per year? Do you think this change would be noticeable in a distant star within a human lifetime?

8. • Calculate the average density of a red-giant core of 0.25 solar mass and radius 15,000 km. Compare your answer with the average density of the giant's envelope, if it has a 0.5 solar mass and its radius is 0.5 AU. Compare each of the two densities with the central density of the Sun. ∞ (Sec. 16.2)

9. • A 15-solar-mass blue supergiant with a surface temperature of 20,000 K becomes a red supergiant with the same total luminosity and a temperature of 4000 K. By what factor does its radius change?

10. •• The radius of Betelgeuse varies by about 60 percent within a period of 3 years. If the star's surface temperature remains constant, by how much does its absolute magnitude change during this time?

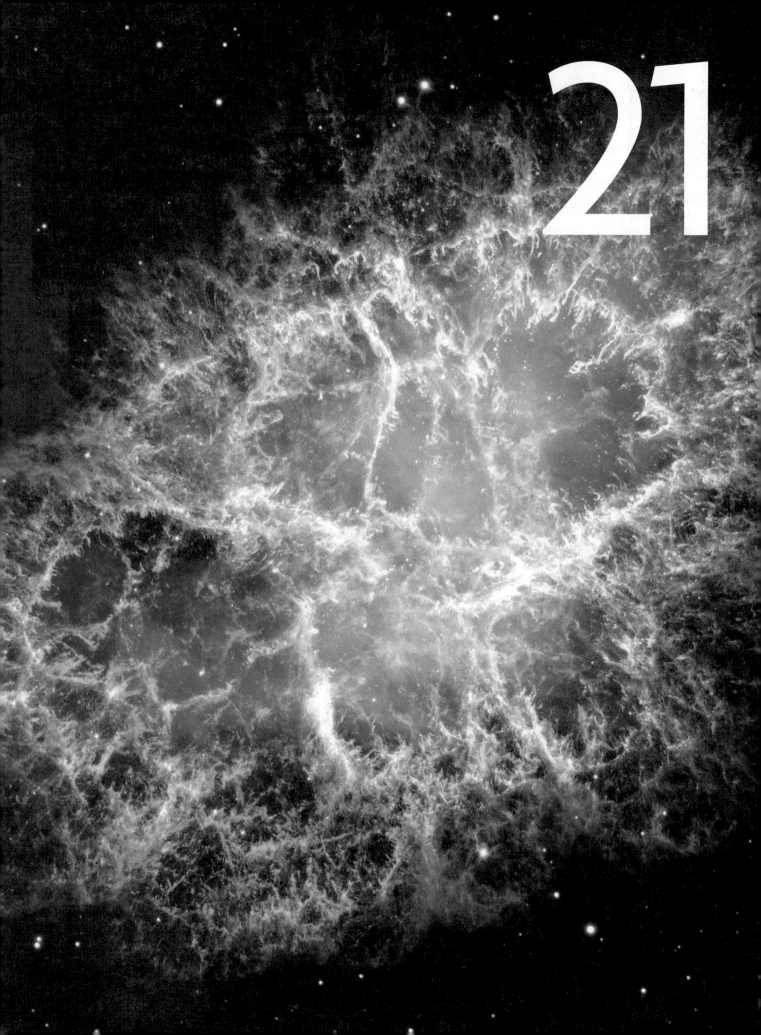

21

STELLAR EXPLOSIONS

NOVAE, SUPERNOVAE, AND THE FORMATION OF THE ELEMENTS

LEARNING GOALS

Studying this chapter will enable you to

1 Explain how white dwarfs in binary-star systems can become explosively active.

2 Summarize the sequence of events leading to the violent death of a massive star.

3 Describe the two types of supernovae, and explain how each is produced.

4 Describe the observational evidence for the occurrence of supernovae in our Galaxy.

5 Explain the origin of elements heavier than helium, and discuss the significance of these elements for the study of stellar evolution.

6 Outline how the universe continually recycles matter through stars and the interstellar medium.

What fate awaits a star when it runs out of fuel? For a low-mass star, the white-dwarf stage is not necessarily the end of the road: The potential exists for further violent activity if a binary companion can provide additional fuel. High-mass stars—whether they are or are not members of binaries—are also destined to die explosively, releasing vast amounts of energy, creating many heavy elements, and scattering the debris throughout interstellar space.

These cataclysmic explosions may trigger the formation of new stars, continuing the cycle of stellar birth and death. In this chapter, we will study in more detail both the processes responsible for the explosions and the mechanisms that create the elements from which we ourselves are made.

THE BIG PICTURE There is something philosophically intriguing about the idea that the deaths of some stars cause the birth of others, at the same time forming many of the elements that make up our own world. There is also something fantastic about the idea that we are made of stardust—yet it happens to be true.

LEFT: *All elements heavier than iron were created in supernovae—violent stellar explosions that mark the deaths of massive stars. Supernovae have been observed in many locations across the sky, often in galaxies far from our own. This billion-bit mosaic of several images from the* Hubble Space Telescope *shows a much closer example—the Crab Nebula, the debris field of a massive star that was actually seen in the sky about a thousand years ago as it blew itself to smithereens. (STScI)*

21.1 Life after Death for White Dwarfs

Although most stars shine steadily day after day and year after year, some change dramatically in brightness over very short periods of time. One type of star, called a **nova** (plural: *novae*), may increase enormously in brightness—by a factor of 10,000 or more—in a matter of days and then slowly return to its initial luminosity over a period of weeks or months. The word *nova* means "new" in Latin, and to early observers these stars did indeed seem new because they appeared suddenly in the night sky. Astronomers now recognize that a nova is not a new star at all. It is instead a white dwarf—a normally very faint star—undergoing an explosion on its surface that results in a rapid, temporary increase in the star's luminosity.

Figures 21.1(a) and (b) illustrate the brightening of a typical nova. Figure 21.1(c) shows a nova light curve, demonstrating how the luminosity rises dramatically in a matter of days and then fades slowly back to normal over the course of several months. On average, two or three novae are observed each year. Astronomers also know of many *recurrent novae*—stars that have been observed to "go nova" several times over the course of a few decades.

What could cause such an explosion on a faint, dead star? The energy involved is far too great to be explained by flares or other surface activity, and as we saw in the previous chapter, there is no nuclear activity in the dwarf's interior. ∞ (Sec. 20.3) To understand what happens, we must consider again the fate of a low-mass star after it enters the white-dwarf phase.

We noted in Chapter 20 that the white-dwarf stage represents the end point of a star's evolution. Subsequently, the star simply cools, eventually becoming a black dwarf—a burned-out ember in interstellar space. This scenario is quite correct for an *isolated* star, such as our Sun. However, should the star be part of a *binary* system, an important new possibility exists. If the distance between the two stars is small enough, then the dwarf's tidal gravitational field can pull matter—primarily hydrogen and helium—away from the surface of its main-sequence or giant companion. ∞ (Sec. 7.6) The system then becomes a mass-transferring

binary, similar to those discussed in Chapter 20. A stream of gas leaves the companion through the inner (L₁) Lagrangian point and flows onto the dwarf. ∞ (Sec. 20.6)

Because of the binary's rotation and the white dwarf's small size, material leaving the companion does not fall directly onto the dwarf, as indicated in Figure 20.21. Instead, such material "misses" the compact star, loops around behind it, and goes into orbit around it, forming a swirling, flattened disk of matter called an **accretion disk** (shown in Figure 21.2). Due to the effects of viscosity (i.e., friction) within the gas, the orbiting matter in the disk drifts gradually inward, its temperature increasing steadily as it spirals down onto the dwarf's surface. The inner part of the accretion

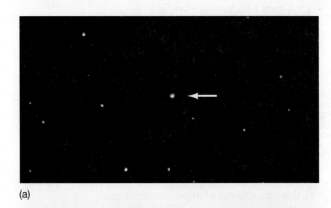

(a)

(b)

R I V U X G

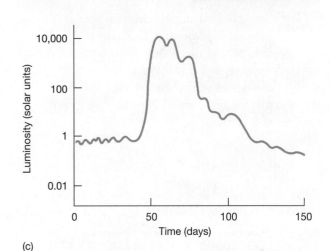

(c)

▶ **FIGURE 21.1 Nova** A nova is a star that suddenly increases enormously in brightness, then slowly fades back to its original luminosity. Novae are the result of explosions on the surfaces of faint white-dwarf stars caused by matter falling onto their surfaces from the atmosphere of larger binary companions. Shown is Nova Herculis 1934 in (a) March 1935 and (b) May 1935, after brightening by a factor of 60,000. (c) The light curve of a typical nova. The rapid rise and slow decline in the light received from the star, as well as the maximum brightness attained, are in good agreement with the explanation of the nova as a nuclear flash on a white-dwarf's surface. *(UC/Lick Observatory)*

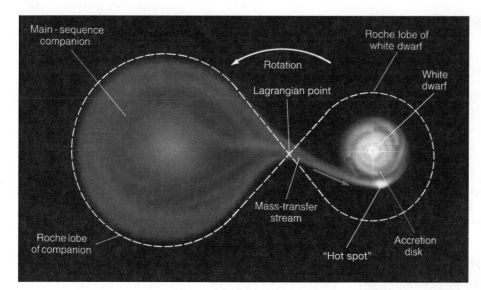

◄ FIGURE 21.2 **Close Binary System**
If a white dwarf in a semidetached binary system is close enough to its companion (in this case, a main-sequence star), its gravitational field can tear matter from the companion's surface. Compare Figure 20.21, but note that, unlike the scenario shown in that earlier figure, the matter does not fall directly onto the white dwarf's surface. Instead, it forms an accretion disk of gas spiraling down onto the dwarf.

disk becomes so hot that it radiates strongly in the visible, the ultraviolet, and even the X-ray portions of the electromagnetic spectrum. In many systems, the disk outshines the white dwarf itself and is the main source of the light emitted between nova outbursts. X rays from the hot disk are routinely observed in many galactic novae. The point at which the infalling stream of matter strikes the accretion disk often forms a turbulent "hot spot," causing detectable fluctuations in the light emitted by the binary system.

The "stolen" gas becomes hotter and denser as it builds up on the white dwarf's surface. Eventually, its temperature exceeds 10^7 K, and the hydrogen ignites, fusing into helium at a furious rate. (Figure 21.3a–d illustrates the sequence of events.) This surface-burning stage is as brief as it is violent: The star suddenly flares up in luminosity and then fades away as some of the fuel is exhausted and the remainder is blown off into space. If the event happens to be visible from Earth, we see a nova. Figure 21.4 shows two novae apparently caught in the act of expelling mass from their surfaces. A nova's decline in brightness results from the expansion and cooling of the white dwarf's surface layers as they are blown into space. Studies of the details of the brightness curve associated with a nova provide astronomers with a wealth of information about both the dwarf and its binary companion.

A nova represents one way in which a star in a binary system can extend its "active lifetime" well into the white-dwarf stage. Recurrent novae can, in principle, repeat their violent outbursts many dozens, if not hundreds, of times. But even more extreme possibilities exist at the end of stellar evolution. Vastly more energetic events may be in store, given the right circumstances.

CONCEPT CHECK

✔ Will the Sun ever become a nova?

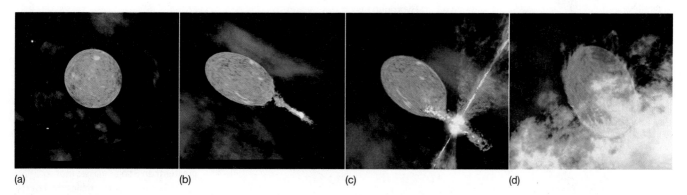

(a)   (b)   (c)   (d)

▲ FIGURE 21.3 **Nova Explosion** In this artist's conception, a white-dwarf star (actually faraway at upper left) orbits a cool red giant (a). As the dwarf swings around in an elliptical orbit, coming closer to the giant, material accretes from the giant to the dwarf and accumulates on the white dwarf's surface (b and c). The dwarf star then ignites in hydrogen fusion as a nova outburst (d). *(D. Berry)*

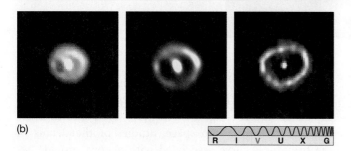

(a)

(b)

| R | I | V | U | X | G |

▲ **FIGURE 21.4 Nova Matter Ejection** (a) The ejection of material from a star's surface can clearly be seen in this image of Nova Persei, taken some 50 years after it suddenly brightened by a factor of 40,000 in 1901. This corresponds approximately to Figure 21.3(d). (b) Nova Cygni, imaged here with a European camera on the *Hubble Space Telescope,* erupted in 1992. At left, more than a year after the blast, a rapidly billowing bubble is seen; at right, 7 months after that, the shell continued to expand and distort. The images are fuzzy because the object is more than 10,000 light-years away. *(Palomar Observatory; ESA)*

21.2 The End of a High-Mass Star

A low-mass star—a star with a mass of less than about 8 solar masses—never becomes hot enough to burn carbon in its core. It ends its life as a carbon–oxygen (or possibly neon–oxygen) white dwarf. ∞ (Sec. 20.3) A high-mass star, however, can fuse not just hydrogen and helium, but also carbon, oxygen, and even heavier elements as its inner core continues to contract and its central temperature continues to rise. ∞ (Sec. 20.4) The burning rate accelerates as the core evolves. Can anything stop this runaway process? Is there a stable "white-dwarf-like" state at the end of the evolution of a high-mass star? What is the ultimate fate of such a star? To answer these questions, we must look more carefully at fusion in massive stars.

Fusion of Heavy Elements

Figure 21.5 is a cutaway diagram of the interior of a highly evolved star of large mass. Note the numerous layers in which various nuclei burn. As the temperature increases with depth, the ash of each burning stage becomes the fuel for the next stage. At the relatively cool periphery of the core, hydrogen fuses into helium. In the intermediate layers, shells of helium, carbon, and oxygen burn to form heavier nuclei. Deeper down reside neon, magnesium, silicon, and other heavy nuclei, all produced by nuclear fusion in the layers overlying the core. (Recall that, to astronomers, a "heavy" element is anything more massive than helium.) The core itself is composed of iron. We will study the key reactions in this burning chain in more detail later in the chapter.

As each element is burned to depletion at the center, the core contracts, heats up, and starts to fuse the ash of the previous burning stage. A new inner core forms, contracts again,

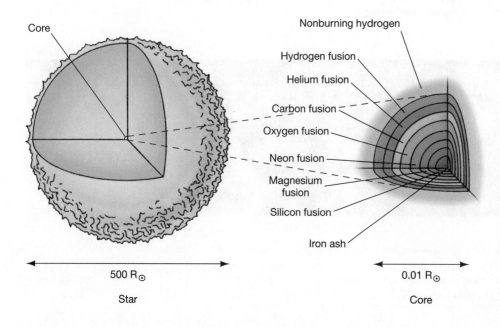

Core

Nonburning hydrogen
Hydrogen fusion
Helium fusion
Carbon fusion
Oxygen fusion
Neon fusion
Magnesium fusion
Silicon fusion
Iron ash

◀ **Interactive FIGURE 21.5 Heavy-Element Fusion**

(MA) Cutaway diagram of the interior of a highly evolved star of mass greater than 8 solar masses (not to scale). The interior resembles the layers of an onion, with shells of progressively heavier elements burning at smaller and smaller radii and at higher and higher temperatures.

500 R$_\odot$

Star

0.01 R$_\odot$

Core

heats again, and so on. Through each period of stability and instability, the star's central temperature increases, the nuclear reactions speed up, and the newly released energy supports the star for ever-shorter periods of time. For example, in round numbers, a star 20 times more massive than the Sun burns hydrogen for 10 million years, helium for 1 million years, carbon for a thousand years, oxygen for a year, and silicon for a week. Its iron core grows for less than a day.

Collapse of the Iron Core

Once the inner core begins to change into iron, our high-mass star is in trouble. As illustrated in Figure 21.6, iron is the *most stable* element there is. To understand the figure, imagine fusing four protons to form helium-4. According to the figure, the mass per particle of a helium-4 nucleus is less than the mass of a proton, so mass is lost and (in accordance with the law of conservation of mass and energy) energy is released. ∞ (Sec. 16.6) Similarly, combining three helium-4 nuclei to form carbon results in a net loss of mass, again releasing energy. In other words, the left side of the figure shows how light elements can fuse to release energy. The right side of the figure shows the opposite process, known as **fission.** Here, combining nuclei will increase the total mass per particle and hence absorb energy, so fusion can't occur.

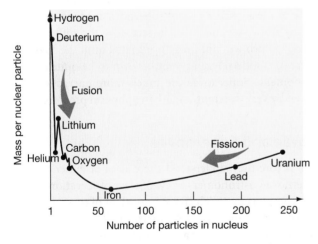

▲ **FIGURE 21.6 Nuclear Masses** This graph shows how the masses (per nuclear particle—proton or neutron) of most known nuclei vary with nuclear mass. It contains a lot of information about nuclear structure and stability. When light nuclei fuse (left side of the figure), the mass per particle decreases and energy is released. ∞ (Sec. 16.6) Similarly, when heavy nuclei split apart (right side), the total mass again decreases and energy is again released. The nucleus with the smallest mass per nuclear particle—the most stable element—is iron. It can be neither fused nor split to release energy. Nuclear burning in stars starts at hydrogen and moves through progressively heavier and heavier elements, all the way to iron. But fusion cannot form elements heavier than iron; those elements are created in the violent explosion that inevitably follows the appearance of iron in the star's core.

However, splitting a heavy nucleus (such as uranium, or plutonium, which lies just off the right edge of the figure) into lighter nuclei does release energy—this is how nuclear reactors and atomic bombs work.

Iron lies at the dividing line between these two types of behavior—at the lowest point of the curve in the figure. Iron nuclei are so compact that energy cannot be extracted either by combining them into heavier elements or by splitting them into lighter ones. In effect, iron plays the role of a fire extinguisher, damping the inferno in the stellar core. With the appearance of substantial quantities of iron, the central fires cease for the last time, and the star's internal support begins to dwindle. The star's foundation is destroyed, and its equilibrium is gone forever. Even though the temperature in the iron core has reached several billion kelvins by this stage, the enormous inward gravitational pull of matter ensures catastrophe in the very near future. Gravity overwhelms the pressure of the hot gas, and the star implodes, falling in on itself.

The core temperature rises to nearly 10 billion K. According to Wien's law, at that temperature individual photons have tremendously high energies—enough to split iron into lighter nuclei and then to break those lighter nuclei apart until only protons and neutrons remain. ∞ (Sec. 4.2) This process is known as *photo disintegration* of the heavy elements in the core. In less than a second, the collapsing core undoes all the effects of nuclear fusion that occurred during the previous 10 million years! But to split iron and lighter nuclei into smaller pieces requires a lot of energy (Figure 21.6, moving from iron to the left). After all, this splitting is just the opposite of the fusion reactions that generated the star's energy during earlier times. Photodisintegration *absorbs* some of the core's thermal energy—in other words, it cools the core and thus reduces the pressure there. As nuclei are destroyed, the core of the star becomes even less able to support itself against its own gravity. The collapse accelerates.

Now the core consists entirely of simple elementary particles—electrons, protons, neutrons, and photons—at enormously high densities, and it is still shrinking. As the density of the core continues to rise, the protons and electrons are crushed together, forming neutrons and neutrinos:

$$p + e \rightarrow n + \text{neutrino.}$$

This process is sometimes called the *neutronization* of the core. Recall from our discussion in Chapter 16 that the neutrino is an extremely elusive particle that hardly interacts at all with matter. ∞ (Sec. 16.6) Even though the central density by this time may have reached 10^{12} kg/m^3 or more, most of the neutrinos produced by neutronization pass through the core as if it weren't there. They escape into space, carrying away energy as they go, further reducing the core's pressure support.

◀ **FIGURE 21.7 Supernova 1987A**
A supernova called SN 1987A (arrow) was exploding near this nebula (called 30 Doradus) at the moment the photograph on the right was taken. The photograph on the left is the normal appearance of the star field. (See *Discovery 21-1*.) *(AURA)*

Supernova Explosion

The disappearance of the electrons and the escape of the neutrinos make matters even worse for the core's stability. There is now nothing to prevent it from collapsing all the way to the point at which the neutrons come into contact with one another, at the incredible density of about 10^{15} kg/m^3. At this point, the neutrons in the shrinking core offer rapidly increasing resistance to further compression, producing enormous pressures that finally slow the core's gravitational collapse. By the time the collapse is actually halted, however, the core has overshot its point of equilibrium and may reach densities as high as 10^{17} or 10^{18} kg/m^3 before turning around and beginning to reexpand. Like a fast-moving ball hitting a brick wall and bouncing back, the core becomes compressed, stops, and then rebounds—with a vengeance!

The events just described do not take long. Only about a second elapses from the start of the collapse to the "bounce" at nuclear densities. At that point, the core rebounds. An enormously energetic shock wave sweeps through the star at high speed, blasting all the overlying layers—including all the heavy elements just formed outside the iron inner core—into space. Although computer models are still somewhat inconclusive, and the details of how the shock reaches the surface and destroys the star remain uncertain, the end result is not: The star explodes, in one of the most energetic events known in the universe (see Figure 21.7). For a period of a few days, the exploding star may rival in brightness the entire galaxy in which it resides. This spectacular death rattle of a high-mass star is known as a **core-collapse supernova.**

CONCEPT CHECK

✔ Why does the iron core of a high-mass star collapse?

21.3 Supernovae

Let's compare a supernova with a nova. Like a nova, a **supernova** is a star that suddenly increases dramatically in brightness and then slowly dims again, eventually fading from view. In its unexploded state, a star that will become a supernova is known as the supernova's *progenitor*. In some cases, supernovae light curves can appear quite similar to those of novae, and a distant supernova can look a lot like a nearby nova—so much so, in fact, that the difference between the two was not fully appreciated until the 1920s. But novae and supernovae are now known to be quite different phenomena. Supernovae are much more energetic events, driven by very different underlying physical processes.*

Novae and Supernovae

Well before they understood the causes of either novae or supernovae, astronomers knew of clear observational differences between them. The most important of these differences is that a supernova is more than a million times brighter than a nova. A supernova produces a burst of light billions of times brighter than the Sun, reaching that level of brightness within just a few hours after the start of the outburst. The total amount of electromagnetic energy radiated by a supernova during the few months it takes to brighten and fade away is roughly 10^{43} J—nearly as much energy as the

---

*Note that, in discussing novae and supernovae, astronomers tend to blur the distinction between the observed event (the sudden appearance and brightening of an object in the sky), the process responsible for the event (a violent explosion in or on a star), and the object itself (the star itself is called a nova or a supernova, as the case may be). The two terms can have any of the three meanings, depending on the context.

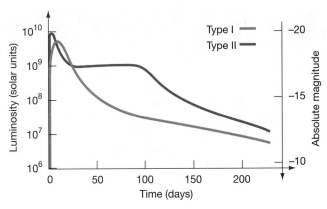

▲ **FIGURE 21.8 Supernova Light Curves** The light curves of typical Type I and Type II supernovae. In both cases, the maximum luminosity can sometimes reach that of a billion suns, but there are characteristic differences in the falloff of the luminosity after the initial peak. Type I light curves somewhat resemble those of novae (see Figure 21.1), but the total release of energy is much larger. Type II curves have a characteristic plateau during the declining phase.

Sun will radiate during its *entire* 10^{10}-year lifetime! (Enormous as this energy is, however, it pales in comparison with the energy emitted in the form of neutrinos, which may be 100 times greater.)

A second important difference is that the same star may become a nova many times, but a star can become a supernova only once. This fact was unexplained before astronomers knew the precise nature of novae and supernovae, but it is easily understood now that we understand how and why these explosions occur. The nova accretion–explosion cycle described earlier can take place over and over again, but a supernova destroys the star involved, with no possibility of a repeat performance.

In addition to the distinction between novae and supernovae, there are also important observational differences *among* supernovae. Some supernovae contain very little hydrogen, according to their spectra, whereas others contain a lot. Also, the light curves of the hydrogen-poor supernovae are qualitatively different from those of the hydrogen-rich ones. On the basis of these observations, astronomers divide supernovae into two classes, known simply as Type I and Type II. **Type I supernovae,** the hydrogen-poor kind, have a light curve somewhat similar in shape to that of typical novae; **Type II supernovae,** whose spectra show lots of hydrogen, usually have a characteristic "plateau" in the light curve a few months after the maximum (see Figure 21.8). Observed supernovae are divided roughly equally between these two categories.

Carbon-Detonation Supernovae

What is responsible for these differences among supernovae? Is there more than one way in which a supernova explosion can occur? The answer is yes. To understand the alternative supernova mechanism, we must return to the processes that cause novae and consider the long-term consequences of their accretion–explosion cycle.

Novae eject matter from a white dwarf's surface, but they do not necessarily expel or burn all the material that has accumulated since the last outburst. In other words, there is a tendency for the dwarf's mass to increase slowly with each new nova cycle. As its mass grows and the internal pressure required to support its weight rises, the white dwarf can enter into a new period of instability—with disastrous consequences.

Recall that a white dwarf is held up not by thermal pressure (heat), but by the degeneracy pressure of electrons that have been squeezed so close together that they have effectively come into contact with one another. ∞ (Sec. 20.3) However, there is a limit to the pressure that these electrons can exert. Consequently, there is a limit to the mass of a white dwarf, above which electrons cannot provide the pressure needed to support the star. Detailed calculations show that the maximum mass of a white dwarf is about 1.4 solar masses, a mass often called the *Chandrasekhar mass,* after the Indian-American astronomer Subramanyan Chandrasekhar, whose work in theoretical astrophysics earned him a Nobel Prize in physics in 1983.

If an accreting white dwarf exceeds the Chandrasekhar mass, the pressure of the degenerate electrons in its interior becomes unable to withstand the pull of gravity, and the star immediately starts to collapse. Its internal temperature rapidly rises to the point at which carbon can fuse into heavier elements. Carbon fusion begins everywhere throughout the white dwarf almost simultaneously, and the entire star explodes in another type of supernova—a so-called **carbon-detonation supernova**—comparable in violence to the "implosion" supernova associated with the death of a high-mass star, but born of a very different cause. In an alternative and (many astronomers think) possibly more common scenario, two white dwarfs in a binary system may collide and merge to form a massive, unstable star. The end result is the same: a carbon-detonation supernova.

We can now understand the differences between Type I and Type II supernovae. The explosion resulting from the detonation of a carbon white dwarf, the descendant of a low-mass star, is a supernova of Type I. Because this conflagration stems from a system containing virtually no hydrogen, we can readily see why the spectrum of a Type I supernova shows little evidence of that element. The appearance of the light curve (as we will soon see) results almost entirely from the radioactive decay of unstable heavy elements produced in the explosion itself.

The implosion–explosion of the core of a massive star, described earlier, produces a Type II supernova. Detailed computer models indicate that the characteristic shape of the Type II light curve is just what would be expected from the expansion and cooling of the star's outer envelope as it is

DISCOVERY 21-1

Supernova 1987A

In 1987, astronomers were treated to a spectacular supernova in the Large Magellanic Cloud (LMC), a small satellite galaxy orbiting our own (see Section 24.2). Observers in Chile first saw the explosion on February 24, and within a few hours, nearly all Southern Hemisphere telescopes and every available orbiting spacecraft were focused on the object. It was officially named SN 1987A. (The SN stands for "supernova," 1987 gives the year, and A identifies the supernova as the first seen that year.) This was one of the most dramatic changes observed in the universe in nearly 400 years. A 15-solar-mass B-type supergiant star with the catalog name SK-69°202 exploded and outshone all the other stars in the LMC combined for a few weeks, as shown in the "before" and "after" images of Figure 21.7.

Because the LMC is relatively close to Earth and because the explosion was detected so soon after it occurred, SN 1987A has provided astronomers with a wealth of detailed information on supernovae, allowing them to make key comparisons between theoretical models and observational reality. By and large, the theory of stellar evolution described in the text has held up very well. Still, SN 1987A did hold some surprises.

According to its hydrogen-rich spectrum, the supernova was of Type II—the core-collapse type—as expected for a high-mass parent star such as SK-69°202. But according to Figure 20.15 (which was computed for stars in our own Galaxy), the parent star should have been a red supergiant at the time of the explosion—not a blue supergiant, as was actually observed. This unexpected finding caused theorists to scramble in search of an explanation. It now seems that, relative to young stars in the Milky Way, the parent star's envelope was deficient in heavy elements. This deficiency had little effect on the evolution of the core and on the supernova explosion, but it did change the star's evolutionary track on the H–R diagram. Unlike a Milky Way star with the same mass, SK-69°202 shrank and looped back toward the main sequence once helium ignited in its core. Following the ignition of carbon, the star, with a surface temperature of around 20,000 K, had just begun to return to the right on the H–R diagram when the rapid chain of events leading to the supernova occurred.

The shape of the light curve of SN 1987A, shown in the first figure, also differed somewhat from the "standard" Type II shape (see Figure 21.8). The peak brightness was less than the expected value. For a few days after its initial detection, the supernova faded as it expanded and cooled rapidly. After about a week, the surface temperature had dropped to about 5000 K, at which point electrons and protons near the expanding surface recombined into atomic hydrogen, making the surface layers less opaque and

allowing more radiation from the interior to leak out. As a result, the supernova brightened rapidly as it grew. The temperature of the expanding layers reached a peak in late May, by which point the radius of the expanding photosphere was about 2×10^{10} km —a little larger than our solar system. Subsequently, the photosphere cooled as it expanded, and the luminosity dropped as the internal supply of heat from the explosion dissipated into space.

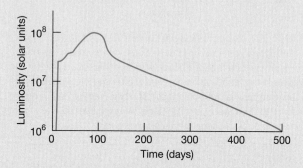

Much of the preceding description would apply equally well to a Type II supernova in our own Galaxy. The differences between the SN 1987A light curve shown here and the Type II light curve in Figure 21.8 are mainly the result of the (relatively) small size of SN 1987A's parent star. The peak luminosity of SN 1987A was less than that of a "normal" Type II supernova because SK-69°202 was small and quite tightly bound by gravity. A lot of the energy emitted in the form of visible radiation (and evident in Figure 21.8) was used up in expanding SN 1987A's stellar envelope, so far less was left over to be radiated into space. Thus, SN 1987A's luminosity during the first few months was lower than expected, and the early peak evident in the figure did not occur. The peak in the SN 1987A light curve at about 80 days actually corresponds to the plateau in the Type II light curve in Figure 21.8.

About 20 hours before the supernova was detected optically, a brief (13-second) burst of neutrinos was simultaneously recorded by underground detectors in Japan and the United States. ∞ (Sec. 16.7) As discussed in the text, the neutrinos are predicted to arise when electrons and protons in the star's collapsing core merge to form neutrons. The neutrinos preceded the light because they escaped during the collapse, whereas the first light of the explosion was emitted only after the supernova shock had plowed through the body of the star to the surface. In fact, theoretical models consistent with these observations suggest that vastly more energy was emitted in the form of neutrinos than in any other form. The supernova's neutrino luminosity was many tens of thousands of times greater than its optical energy output.

---

blown into space by the shock wave sweeping up from below. The expanding material consists mainly of unburned gas—hydrogen and helium—so it is not surprising that those elements are strongly represented in the supernova's observed spectrum. (See *Discovery 21-1* for an account of a

well-studied Type II supernova that confirmed many basic theoretical predictions, while also forcing astronomers to revise the details of their models.)

Figure 21.9 summarizes the processes responsible for the two different types of supernovae. We emphasize that,

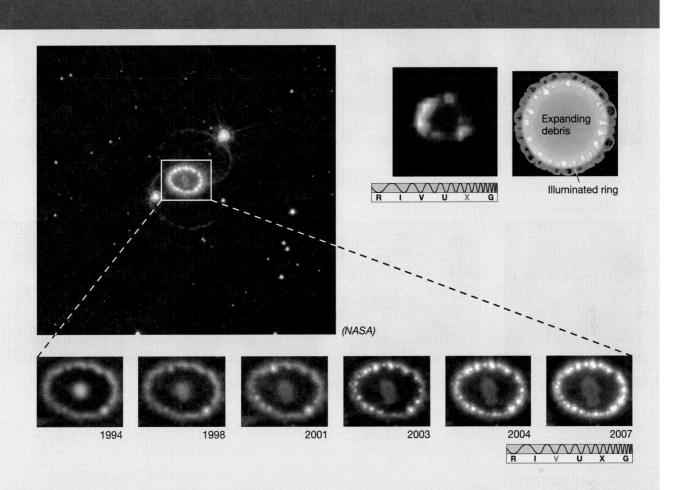

(NASA)

1994   1998   2001   2003   2004   2007

Expanding debris

Illuminated ring

R I V U X G

Despite some unresolved details in SN 1987A's behavior, the detection of the neutrino pulse is considered to be a brilliant confirmation of theory. This singular event—the detection of neutrinos—may well herald a new age of astronomy. For the first time, astronomers have received information from a specific body beyond the solar system by radiation outside the electromagnetic spectrum.

Theory predicts that the expanding remnant of SN 1987A is now on the verge of being resolvable by optical telescopes. The accompanying photographs show the barely resolved remnant (at center) surrounded by a much larger shell of glowing gas (in yellow). Scientists reason that the progenitor of the supernova expelled this shell during its red-giant phase, some 40,000 years before the explosion. The image we see results from the initial flash of ultraviolet light from the supernova hitting the ring and causing it to glow brightly. As the debris from the explosion itself strikes the ring, it has become a temporary, but intense, source of X rays. The

2000 *Chandra* X-ray image and diagram at top right show the fastest-moving ejecta impacting the irregular inner edge of the ring, forming the small (1000 AU in diameter) glowing regions on its left side. The six insets to the main image clearly show the ring "lighting up" as the shock wave from the explosion reaches it.

These images also show core debris (purple) moving outward toward the ring. The six insets show material cooling and becoming fainter as it expands at nearly 3000 km/s. The main image also revealed, to everyone's surprise, two additional faint rings that might be caused by radiation sweeping across an hourglass-shaped bubble of gas, itself perhaps the result of a nonspherical "bipolar" stellar wind from the progenitor star before the supernova occurred.

Buoyed by the success of stellar-evolution theory and armed with firm theoretical predictions of what should happen next, astronomers eagerly await future developments in the story of this remarkable object.

despite the similarity in the total amounts of energy involved, Type I and Type II supernovae are unrelated to one another. They occur in stars of very different types, under very different circumstances. All high-mass stars become Type II (core-collapse) supernovae, but only a tiny fraction of low-mass stars evolve into white dwarfs that ultimately explode as Type I (carbon-detonation) supernovae. However, there are far more low-mass stars than high-mass stars, so, by a remarkable coincidence, the two types of supernova occur at roughly the same rate.

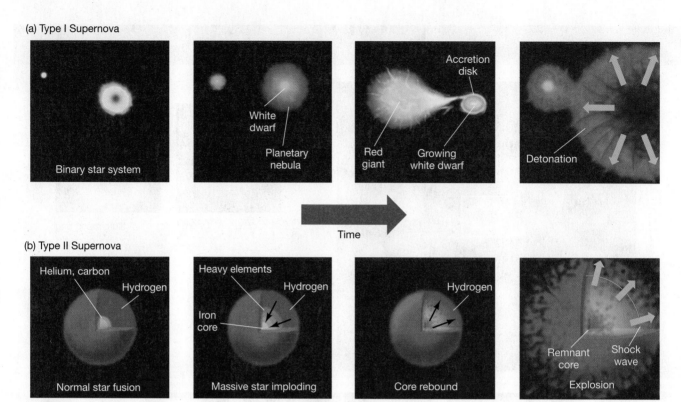

(a) Type I Supernova

Binary star system

White dwarf

Planetary nebula

Accretion disk

Red giant

Growing white dwarf

Detonation

Time

(b) Type II Supernova

Helium, carbon

Hydrogen

Normal star fusion

Heavy elements

Hydrogen

Iron core

Massive star imploding

Hydrogen

Core rebound

Remnant core

Shock wave

Explosion

▲ **FIGURE 21.9** **Two Types of Supernova** Type I and Type II supernovae have different causes. These sequences depict the evolutionary history of each type. (a) A Type I supernova usually results when a carbon-rich white dwarf pulls matter onto itself from a nearby red-giant or main-sequence companion. (b) A Type II supernova occurs when the core of a high-mass star collapses and then rebounds in a catastrophic explosion.

Supernova Remnants

We have plenty of evidence that supernovae have occurred in our Galaxy. Occasionally, the explosions themselves are visible from Earth. In many other cases, we can detect their glowing remains, or **supernova remnants.** One of the best-studied supernova remnants is known as the Crab Nebula, shown in Figure 21.10. The Crab has greatly dimmed now, but the original explosion in the year A.D. 1054 was so brilliant that manuscripts of ancient Chinese and Middle Eastern astronomers claim that its brightness greatly exceeded that of Venus and—according to some (possibly exaggerated) accounts—even rivaled that of the Moon. For nearly a month, this exploded star reportedly could be seen in broad daylight. Native Americans also left engravings of the event in the rocks of what is now the southwestern United States.

The Crab Nebula certainly has the appearance of exploded debris. Even today, the knots and filaments give a strong indication of past violence. In fact, astronomers have proof that this matter was ejected from some central explosion. Doppler-shifted spectral lines indicate that the nebula—the envelope of the high-mass star that exploded to create this Type II supernova—is expanding into space at several thousand kilometers per second. A vivid illustration of the phenomenon is provided by Figure 21.11, which was made by superimposing a positive image of the Crab Nebula taken in 1960 and a negative image

taken in 1974. If the gas were not in motion, the positive and negative images would overlap perfectly, but they do not. The gas moved outward in the intervening 14 years. Tracing the motion backward in time, astronomers have found that the explosion must have occurred about nine centuries ago, consistent with the Chinese observations.

The nighttime sky harbors many relics of stars that blew up long ago. Figure 21.12 is another example. It shows the Vela supernova remnant, whose expansion velocities imply that its central star exploded around 9000 B.C. The remnant lies only 500 pc away from Earth. Given its proximity, the Vela supernova may have been as bright as the Moon for several months. We can only speculate what impact such a bright supernova might have had on the myths, religions, and cultures of Stone Age humans when it first appeared in the sky.

Although hundreds of supernovae have been observed in other galaxies during the 20th century, no astronomer using modern equipment has ever observed a supernova in our own Galaxy. A viewable Milky Way star has not exploded since Galileo first turned his telescope to the heavens almost four centuries ago. The last supernovae observed in our Galaxy, by Tycho in 1572 and Kepler (and others) in 1604, caused a worldwide sensation in Renaissance times. The sudden appearance and subsequent fading of these very bright objects helped shatter the Aristotelian idea of an unchanging universe.

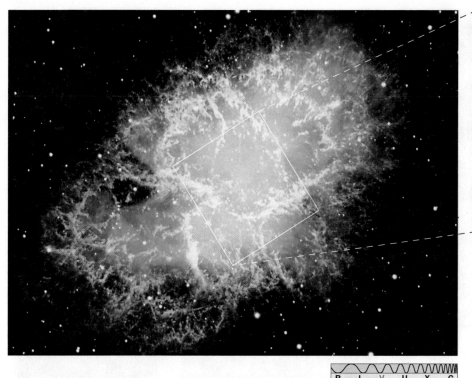
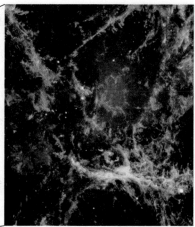

R I V U X G

Interactive FIGURE 21.10 Crab Supernova Remnant This remnant of an ancient Type II supernova is called the Crab Nebula (or M1 in the Messier catalog). It resides about 1800 pc from Earth and has an angular diameter about one-fifth that of the full Moon. Its debris is scattered over a region about 2 pc. Chinese astronomers observed this supernova explosion in A.D. 1054. The main image was taken with the Very Large Telescope in Chile, the inset by the *Hubble* telescope in orbit. Supernova remnants like the Crab offer astronomers vital information not only on the physics of supernovae, but also on how heavy elements formed in stars are dispersed back into interstellar space. *(ESA; NASA)*

Blast center

R I V U X G

◄ **FIGURE 21.11 The Crab in Motion** Positive and negative photographs of the Crab Nebula taken 14 years apart do not superimpose exactly, indicating that the gaseous filaments are still moving away from the site of the explosion. The positive image in glowing white was taken first, and then the black (negative) filaments were overlaid later—hence the black (but still glowing) outlying debris is farther from the center of the blast. The scale is roughly the same as in Figure 21.10. *(Harvard College Observatory)*

▲ **FIGURE 21.12 Vela Supernova Remnant** The glowing gases of the Vela supernova remnant are spread across 6° of the sky. The inset shows more clearly some of the details of the nebula's extensive filamentary structure. (The long diagonal streak was caused by the passage of an Earth-orbiting satellite while the photo was being exposed.) (*D. Malin/AAT*)

On the basis of stellar evolutionary theory, astronomers calculate that an observable supernova ought to occur in our Galaxy every 100 years or so. Even at a distance of several kiloparsecs, a supernova would (temporarily) outshine Venus, the brightest planet in our sky, so it seems unlikely that astronomers could have missed any since the last one nearly four centuries ago. Our part of the Milky Way seems long overdue for a supernova. However, a truly nearby supernova—within a few hundred parsecs, say—would be a very rare event, occurring only every 100,000 years or so. Humanity may be destined to see all supernovae from a distance.

PROCESS OF SCIENCE CHECK

✔ How did astronomers know, even before the mechanisms were understood, that there were at least two distinct physical processes at work in creating supernovae?

21.4 Formation of the Elements

Up to now, we have studied nuclear reactions mainly for their role in stellar energy generation. Now let's consider them again, but this time as the processes responsible for creating much of the world in which we live. The evolution of the elements,

combining nuclear physics with astronomy, is a complex subject and a very important problem in modern astronomy.

Types of Matter

We currently know of 115 different elements, ranging from the simplest—hydrogen, containing one proton—to the most complex, first reported in 2004 and known for now as ununpentium which has 115 protons and 184 neutrons in its nucleus (see Appendix 3, Table 2). In 1999, researchers claimed the discovery of elements 116 and 118, but the experimental findings have not been replicated, and these elements are not "officially" recognized.) All elements exist in different *isotopes*, each having the same number of protons, but a different number of neutrons. We often think of the most common or stable isotope as being the "normal" form of an element. Some elements, and many isotopes, are radioactively unstable, meaning that they eventually decay into other, more stable, nuclei.

The 81 stable elements found on Earth make up the overwhelming bulk of matter in the universe. In addition, 10 radioactive elements—including radon and uranium—also occur naturally on our planet. Even though the half-lives (the time required for half the nuclei to decay into something else) of these elements are very long (typically, millions or even billions of years), their slow, but steady, decay over the

4.5 billion years since the solar system formed means that they are scarce on Earth, in meteorites, and in lunar samples. ∞ (*More Precisely 7-2*, Sec. 14.4) They are not observed in stars—there is just too little of them to produce detectable spectral lines.

Besides these 10 naturally occurring radioactive elements, 19 more radioactive elements have been artificially produced under special conditions in nuclear laboratories on Earth. The debris collected after nuclear weapons tests also contains traces of some of these elements. Unlike the naturally occurring radioactive elements, the artificial ones decay into other elements quite quickly (in much less than a million years). Consequently, they, too, are extremely rare in nature. Two other elements round out our list: Promethium is a stable element that is found on our planet only as a by-product of nuclear laboratory experiments; technetium is an unstable element that is found in stars, but does not exist on Earth—any technetium that existed in our planet at its formation decayed long ago.

Abundance of Matter

How and where did all these elements form? Were they always present in the universe, or were they created after the universe formed? Since the 1950s, astronomers have come to realize that the hydrogen and most of the helium in the universe are *primordial*—that is, these elements date from the very earliest times (see Chapter 27). All other elements in our universe result from **stellar nucleosynthesis**—that is, they were formed by nuclear fusion in the hearts of stars.

To test this idea, we must consider not just the different kinds of elements and isotopes, but also their observed abundances, graphed in Figure 21.13. The curve shown is derived largely from spectroscopic studies of stars, including the Sun. The essence of the figure is summarized in Table 21.1, which combines all the known elements into eight groups based on the total numbers of nuclear particles (protons and neutrons)

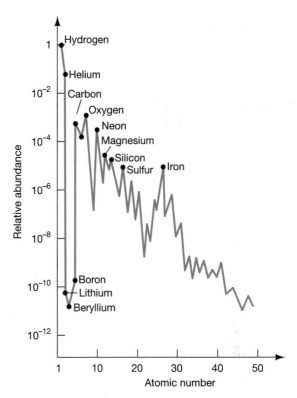

▲ **FIGURE 21.13 Elemental Abundance** A summary of the cosmic abundances of the elements and their isotopes, expressed relative to the abundance of hydrogen. The horizontal axis shows each of the listed elements' atomic number—the number of protons in the nucleus. Notice how many common terrestrial elements are found on "peaks" of the distribution, surrounded by elements that are tens or hundreds of times less abundant. Notice also the large peak around the element iron. The reasons for the peaks are discussed in the text.

that they contain. (All isotopes of all elements are included in both the table and the figure, although only a few elements are marked by dots and labeled in the figure.) Any theory proposed for the creation of the elements must reproduce these observed abundances. The most obvious feature is that heavy

TABLE 21.1 Cosmic Abundances of the Elements	
Elemental Group of Particles	**Percent Abundance by Number***
Hydrogen (1 nuclear particle)	90
Helium (4 nuclear particles)	9
Lithium group (7–11 nuclear particles)	0.000001
Carbon group (12–20 nuclear particles)	0.2
Silicon group (23–48 nuclear particles)	0.01
Iron group (50–62 nuclear particles)	0.01
Middle-weight group (63–100 nuclear particles)	0.00000001
Heaviest-weight group (over 100 nuclear particles)	0.000000001

The total does not equal 100 percent because of uncertainties in the abundance of helium. All isotopes of all elements are included.

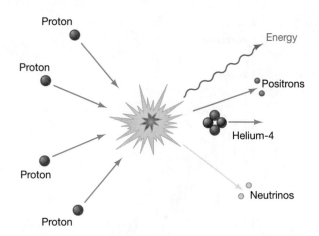

▲ **FIGURE 21.14 Proton Fusion** Diagram of the basic proton–proton hydrogen-burning reaction. Four protons combine to form a nucleus of helium-4, releasing energy in the process.

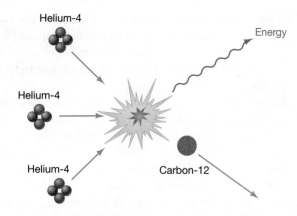

▲ **FIGURE 21.15 Helium Fusion** Diagram of the basic triple-alpha helium-burning reaction occurring in post-main-sequence stars. Three helium-4 nuclei combine to form carbon-12.

elements are generally much less abundant than lighter elements. However, the many peaks and troughs evident in the figure also represent important constraints.

Hydrogen and Helium Burning

Let's begin by reviewing the reactions leading to the production of heavy elements at various stages of stellar evolution. Look again at Figure 21.6 as we discuss the reactions involved. Stellar nucleosynthesis begins with the proton–proton chain studied in Chapter 16. ∞ (Sec. 16.6) Provided that the temperature is high enough—at least 10 million K—a series of nuclear reactions occurs, ultimately forming a nucleus of ordinary helium (^{4}He) from four protons (^{1}H):

$$4(^1\text{H}) \rightarrow {}^4\text{He} + 2 \text{ positrons} + 2 \text{ neutrinos} + \text{energy.}$$

Recall that the positrons immediately interact with nearby free electrons, producing high-energy gamma rays through matter–antimatter annihilation. The neutrinos rapidly escape, carrying energy with them, but playing no direct role in nucleosynthesis. The existence of these reactions has been directly confirmed in nuclear experiments conducted in laboratories around the world during recent decades. In massive stars, an alternate sequence of reactions called the CNO cycle, involving nuclei of carbon, nitrogen, and oxygen, may greatly accelerate the hydrogen-burning process, but the basic four-protons-to-one-helium-nucleus reaction, illustrated in Figure 21.14, is unchanged.

As helium builds up in the core of a star, the burning ceases, and the core contracts and heats up. When the temperature exceeds about 100 million K, helium nuclei can overcome their mutual electrical repulsion, leading to the *triple-alpha reaction,* which we discussed in Chapter 20 ∞ (Sec. 20.2):

$$3(^4\text{He}) \rightarrow {}^{12}\text{C} + \text{energy.}$$

The net result of this reaction is that three helium-4 nuclei are combined into one carbon-12 nucleus (Figure 21.15), releasing energy in the process.

Carbon Burning and Helium Capture

At higher and higher temperatures, heavier and heavier nuclei can gain enough energy to overcome the electrical repulsion between them. At about 600 million K (reached only in the cores of stars much more massive than the Sun), carbon nuclei can fuse to form magnesium, as depicted in Figure 21.16(a):

$$^{12}\text{C} + {}^{12}\text{C} \rightarrow {}^{24}\text{Mg} + \text{energy.}$$

However, because of the rapidly mounting nuclear charges—that is, the increasing number of protons in the

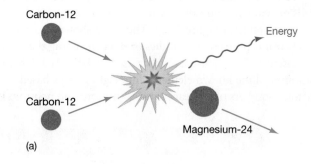

(a)

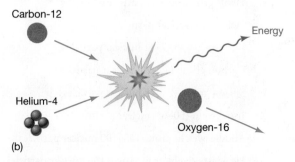

(b)

▲ **FIGURE 21.16 Carbon Fusion** Carbon can form heavier elements (a) by fusion with other carbon nuclei or, more commonly, (b) by fusion with a helium nucleus.

nuclei—fusion reactions between any nuclei larger than carbon require such high temperatures that they are actually quite uncommon in stars. The formation of most heavier elements occurs by way of an easier path. For example, the repulsive force between two carbon nuclei is three times greater than the force between a nucleus of carbon and one of helium. Thus, carbon–helium fusion occurs at a lower temperature than that at which carbon–carbon fusion occurs. As we saw in Section 20.3, at temperatures above 200 million K, a carbon-12 nucleus colliding with a helium-4 nucleus can produce oxygen-16:

$$^{12}\text{C} + {}^{4}\text{He} \rightarrow {}^{16}\text{O} + \text{energy}.$$

If any helium-4 is present, this reaction, shown in Figure 21.16(b), is much more likely to occur than the carbon–carbon reaction.

Similarly, the oxygen-16 thus produced may fuse with other oxygen-16 nuclei at a temperature of about 1 billion K to form sulfur-32:

$$^{16}\text{O} + {}^{16}\text{O} \rightarrow {}^{32}\text{S} + \text{energy}.$$

However, it is much more probable that an oxygen-16 nucleus will capture a helium-4 nucleus (if one is available) to form neon-20:

$$^{16}\text{O} + {}^{4}\text{He} \rightarrow {}^{20}\text{Ne} + \text{energy}.$$

The second reaction is more likely because it occurs at a lower temperature than that necessary for oxygen–oxygen fusion.

Thus, as the star evolves, heavier elements tend to form through **helium capture** rather than by fusion of like nuclei. As a result, elements with nuclear masses of 4 units (i.e., helium itself), 12 units (carbon), 16 units (oxygen), 20 units (neon), 24 units (magnesium), and 28 units (silicon) stand out as prominent peaks in Figure 21.13, our chart of cosmic abundances. Each element is built by combining the preceding element and a helium-4 nucleus as the star evolves.

Helium capture is by no means the only type of nuclear reaction occurring in evolved stars. As nuclei of many different kinds accumulate, a great variety of reactions become possible. In some, protons and neutrons are freed from their parent nuclei and are absorbed by others, forming new nuclei with masses intermediate between those formed by helium capture. Laboratory studies confirm that common nuclei, such as fluorine-19, sodium-23, phosphorus-31, and many others, are created in this way. However, their abundances are not as great as those produced directly by helium capture, simply because the helium-capture reactions are much more common in stars. For this reason, many of these elements (those with masses not divisible by four, the mass of a helium nucleus) are found in the troughs of Figure 21.13.

Iron Formation

Around the time silicon-28 appears in the core of a star, a competitive struggle begins between the continued capture of helium to produce even heavier nuclei and the tendency of more complex nuclei to break down into simpler ones. The cause of this breakdown is heat. By now, the star's core temperature has reached the unimaginably large value of 3 billion K, and the gamma rays associated with that temperature have enough energy to break a nucleus apart, as illustrated in Figure 21.17(a). This is the same process of photodisintegration that will ultimately accelerate the star's iron core in its final collapse toward a Type II supernova.

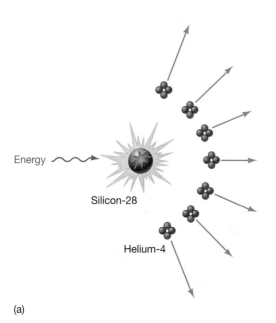

Energy

Silicon-28

Helium-4

(a)

◀ **FIGURE 21.17 Alpha Process** (a) At high temperatures, heavy nuclei (such as silicon, shown here) can be broken apart into helium nuclei by high-energy photons. (b) Other nuclei can capture the helium nuclei—or alpha particles—thus produced, forming heavier elements by the so-called alpha process. This process continues all the way to the formation of nickel-56 (in the iron group).

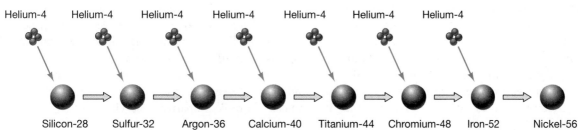

| Helium-4 | Helium-4 | Helium-4 | Helium-4 | Helium-4 | Helium-4 | Helium-4 |

(b)   Silicon-28   Sulfur-32   Argon-36   Calcium-40   Titanium-44   Chromium-48   Iron-52   Nickel-56

Under the intense heat, some silicon-28 nuclei break apart into seven helium-4 nuclei. Other nearby nuclei that have not yet photodisintegrated may capture some or all of these helium-4 nuclei, leading to the formation of still heavier elements (Figure 21.17b). The process of photodisintegration provides raw material that allows helium capture to proceed to greater masses. Photodisintegration continues, with some heavy nuclei being destroyed and others increasing in mass. In succession, the star forms sulfur-32, argon-36, calcium-40, titanium-44, chromium-48, iron-52, and nickel-56. The chain of reactions building from silicon-28 up to nickel-56 is

$$^{28}\text{Si} + 7(^{4}\text{He}) \rightarrow {}^{56}\text{Ni} + \text{energy}.$$

This two-step process—photodisintegration followed by the direct capture of some or all of the resulting helium-4 nuclei (or alpha particles)—is often called the *alpha process.*

Nickel-56 is unstable, decaying rapidly first into cobalt-56 and then into a stable iron-56 nucleus. Any unstable nucleus will continue to decay until stability is achieved, and iron-56 is the most stable of all nuclei (Figure 21.6). Thus, the alpha process leads inevitably to the buildup of iron in the stellar core.

Another way of describing Figure 21.6 is to say that iron's 26 protons and 30 neutrons are bound together more strongly than the particles in any other nucleus. Iron is said to have the greatest *nuclear binding energy* of any element—more energy per particle is required to break up (unbind) an iron-56 nucleus than the nucleus of any other element. This enhanced stability of iron explains why some of the heavier nuclei in the iron group are more abundant than many lighter nuclei (see Table 21.1 and Figure 21.13): Nuclei tend to "accumulate" near iron as stars evolve.

Making Elements Beyond Iron

If the alpha process stops at iron, how did heavier elements, such as copper, zinc, and gold, form? To form them, some nuclear process other than helium capture must have been involved. That other process is **neutron capture**: the formation of heavier nuclei by the absorption of neutrons.

Deep in the interiors of highly evolved stars, conditions are ripe for neutron capture to occur. Neutrons are produced as "by-products" of many nuclear reactions, so there are many of them present to interact with iron and other nuclei. Neutrons have no charge, so there is no repulsive barrier for them to overcome in combining with positively charged nuclei. As more and more neutrons join a nucleus, its mass continues to grow.

Adding neutrons to a nucleus—iron, for example—does not change the element. Rather, a more massive isotope of the same element is produced. Eventually, however, so many neutrons have been added to the nucleus that it becomes unstable and then decays radioactively to form a stable nucleus of some other element. The neutron-capture process then continues. For example, an iron-56 nucleus can capture a single neutron to form a relatively stable isotope, iron-57:

$$^{56}\text{Fe} + \text{n} \rightarrow {}^{57}\text{Fe}.$$

This reaction may be followed by another neutron capture:

$$^{57}\text{Fe} + \text{n} \rightarrow {}^{58}\text{Fe}.$$

Thus, another relatively stable isotope, iron-58, is produced, and this isotope can capture yet another neutron to produce an even heavier isotope of iron:

$$^{58}\text{Fe} + \text{n} \rightarrow {}^{59}\text{Fe}.$$

Iron-59 is known from laboratory experiments to be radioactively unstable. It decays in about a month into cobalt-59, which is stable. The neutron-capture process then resumes: Cobalt-59 captures a neutron to form the unstable cobalt-60, which in turn decays to nickel-60, and so on.

Each successive capture of a neutron by a nucleus typically takes about a year, so most unstable nuclei have plenty of time to decay before the next neutron comes along. Researchers usually refer to this "slow" neutron-capture mechanism as the *s-process.* It is the origin of the copper and silver in the coins in our pockets, the lead in our car batteries, and the gold (and the zirconium) in the rings on our fingers. As mentioned earlier, similar slow neutron-capture processes involving nuclei of lower mass are responsible for many of the elements intermediate between those formed by helium capture. These reactions are thought to be particularly important during the late (asymptotic-giant branch) stages of low-mass stars. ∞ (Sec. 20.3)

Making the Heaviest Elements

The s-process explains the synthesis of stable nuclei up to, and including, bismuth-209, the heaviest-known nonradioactive nucleus, but it cannot account for the heaviest nuclei, such as thorium-232, uranium-238, or plutonium-242. Any attempt to form elements heavier than bismuth-209 by slow neutron capture fails because the new nuclei decay back to bismuth as fast as they form. Accordingly, there must be yet another nuclear mechanism that produces the very heaviest nuclei. This process is called the *r-process* (where r stands for "rapid," in contrast to the "slow" s-process just described). The r-process operates very quickly, occurring (we think) literally during the supernova explosion that signals the death of a massive star.

During the first 15 minutes of the supernova blast, the number of free neutrons increases dramatically as heavy nuclei are broken apart by the violence of the explosion.

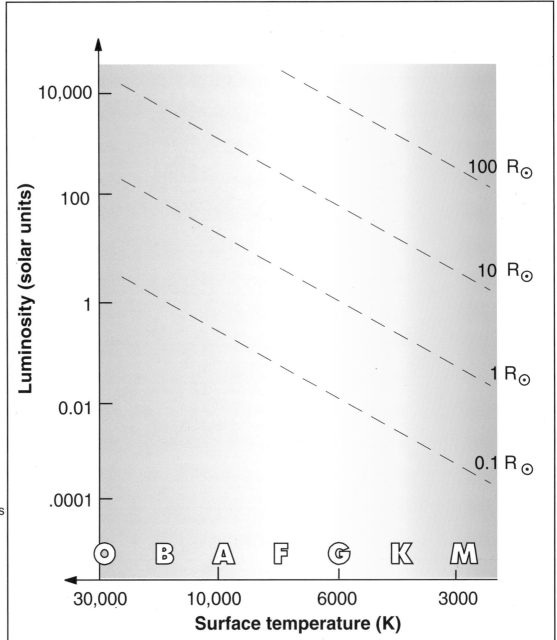

The H–R diagram plots stars by luminosity (vertical axis) and temperature, or spectral class (horizontal axis). The dashed diagonal lines are lines of constant radius.

Unlike the s-process, which stops when it runs out of stable nuclei, the neutron-capture rate during the supernova is so great that even unstable nuclei can capture many neutrons before they have time to decay. Jamming neutrons into light- and middleweight nuclei, the r-process is responsible for the creation of the heaviest-known elements. The heaviest of the heavy elements, then, are actually born *after* their parent stars have died. However, because the time available for synthesizing these heaviest nuclei is so brief, they never become very abundant. Elements heavier than iron (see Table 21.1) are a billion times less common than hydrogen and helium.

Observational Evidence for Stellar Nucleosynthesis

The modern picture of the formation of the elements involves many different types of nuclear reactions occurring at many different stages of stellar evolution, from main-sequence stars all the way to supernovae. Elements of the periodic table from hydrogen to iron are built first by fusion and then by alpha capture, with proton and neutron capture filling in the gaps. Elements beyond iron form by neutron capture and radioactive decay. Ultimately, these elements are ejected into interstellar space as the stars in which they form reach the ends of their lives.

Scientific theories must continually be tested and validated by experiment and observation, and the theory of stellar nucleosynthesis is no exception. ∞ (Sec. 1.2) Yet almost all of the nuclear processes just described take place deep in the hearts of stars, hidden from our view, and the stars responsible for the heavy elements we see today are all long gone. How, then, can we be sure that the sequences of events presented here actually occurred (and are still occurring today)? The answer is that the theory of stellar nucleosynthesis makes many detailed predictions about the numbers and types of elements formed in stars, affording astronomers ample opportunity to observe and test its consequences. We are reassured of the theory's basic soundness by three particularly convincing pieces of evidence.

First, the rates at which various nuclei are captured and the rates at which they decay are known from laboratory experiments. When these rates are incorporated into detailed computer models of the nuclear processes occurring in stars and supernovae, the resulting elemental abundances agree extremely well, point by point, with the observational data presented in Figure 21.13 and Table 21.1. The match is remarkably good for elements up through iron and is still fairly close for heavier nuclei. Although the reasoning is indirect, the agreement between theory and observation is so striking that most astronomers regard it as very strong evidence in support of the entire theory of stellar evolution and nucleosynthesis.

Second, the presence of one particular nucleus—technetium-99—provides direct evidence that heavy elements really do form in the cores of stars. Laboratory measurements show that the technetium nucleus has a radioactive half-life of about 200,000 years, a very short time, astronomically speaking. No one has ever found even traces of naturally occurring technetium on Earth, because it all decayed long ago. The observed presence of technetium in the spectra of many red-giant stars implies that it must have been synthesized in their cores through neutron capture—the only known way in which technetium can form—within the past few hundred thousand years and then transported by convection to the surface. Otherwise, we would not observe it. Many astronomers consider the spectroscopic evidence for technetium as proof that the s-process really does operate in evolved stars.

Third, the study of typical light curves from Type I supernovae indicates that radioactive nuclei form as a result of the explosion. Figure 21.18(a) (see also Figure 21.8) displays the dramatic rise in luminosity at the moment of explosion and the characteristic slower decrease in brightness. Depending on the initial mass of the exploded star, the luminosity takes from several months to many years to decrease to its original value, but the *shape* of the decay curve is nearly the same for all exploded stars. These curves have two distinct features: After the initial peak, the luminosity declines rapidly; then it decreases at a slower rate. This abrupt change in the rate of luminosity decay invariably occurs about 2 months after the explosion, regardless of the intensity of the outburst.

We can explain the two-stage decline of the luminosity curve in Figure 21.18(a) in terms of the radioactive decay of unstable nuclei, notably nickel-56 and its decay product cobalt-56, produced in abundance during the early moments of the supernova. From theoretical models of the explosion, we can calculate the amounts of these elements expected to form, and we know their half-lives from laboratory experiments. Because each radioactive decay produces a known amount of energy, we can then determine how the light emitted by these unstable elements should vary in time. The result is in very good agreement with the observed light curve in Figure 21.18(b)—the luminosity of a Type I supernova is entirely consistent with the decay of about 0.6 solar mass of nickel-56. More direct evidence for the presence of these unstable nuclei was first obtained in the 1970s, when a gamma-ray spectral feature of decaying cobalt-56 was identified in a supernova observed in a distant galaxy.

CONCEPT CHECK

✔ Why are the elements carbon, oxygen, neon, and magnesium, whose masses are multiples of four, as well as the element iron, so common on Earth?

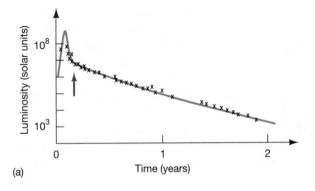

(a)

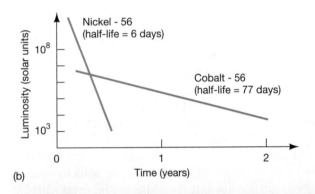

(b)

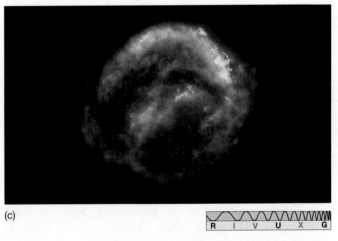

(c)

▲ FIGURE 21.18 **Supernova Energy Emission** (a) The light curve of a Type I supernova, showing not only the dramatic increase and slow decrease in luminosity, but also the characteristic change in the rate of decay about 2 months after the explosion (the time indicated by the arrow). This particular supernova occurred in the faraway galaxy IC 4182 in 1938. The crosses are the actual observations of the supernova's light. (b) Theoretical calculations of the light emitted by the radioactive decay of nickel-56 and cobalt-56 produce a light curve similar to those actually observed in real supernova explosions, lending strong support to the theory of stellar nucleosynthesis. (c) This is a composite image of the remains of the Kepler supernova—the last such object seen to explode in our Milky Way Galaxy, in 1604. The bubble-shaped debris field resides about 3000 pc away, is 4 pc across, and is expanding at 2000 km/s. X-ray data are blue and green; infrared data are red; optical data are yellow. (*NASA*)

21.5 The Cycle of Stellar Evolution

The theory of stellar nucleosynthesis can naturally account for the observed differences in the abundances of heavy elements between the old globular-cluster stars and stars now forming in our Galaxy. ⚭ (Sec. 20.5) Even though an evolved star continuously creates new heavy elements in its interior, changes in the star's composition are confined largely to the core, and the star's spectrum gives little indication of events within its core. Convection may carry some reaction products (such as the technetium observed in many red giants) from the core into the envelope, but the outer layers largely retain the star's original composition. Only at the end of the star's life are its newly created elements released and scattered into space.

Thus, the spectra of the *youngest* stars show the *most* heavy elements, because each new generation of stars increases the concentration of these elements in the interstellar clouds from which the next generation forms. Accordingly, the photosphere of a recently formed star contains a much greater abundance of heavy elements than that of a star that formed long ago. Knowledge of stellar evolution allows astronomers to estimate the ages of stars from purely spectroscopic studies, even when the stars are isolated and are not members of any cluster. ⚭ (Sec. 20.5) In the last three chapters, we have seen all the ingredients that make up the complete cycle of star formation and evolution in our Galaxy. Let's briefly summarize that process, which is illustrated in Figure 21.19:

1. *Stars form* when part of an interstellar cloud is compressed beyond the point at which it can support itself against its own gravity. The cloud collapses and fragments, forming a cluster of stars. The hottest stars heat and ionize the surrounding gas, sending shock waves through the surrounding cloud, modifying the formation of lower-mass stars, and possibly triggering new rounds of star formation. ⚭ (Sec. 19.6)

2. Within the cluster, *stars evolve*. The most massive stars evolve fastest, creating the heaviest elements in their cores and spewing them forth into the interstellar medium in supernovae. Lower-mass stars take longer to evolve, but they, too, can create heavy elements and contribute significantly to the "seeding" of interstellar space when they shed their envelopes as planetary nebulae. Roughly speaking, low-mass stars are responsible for most of the carbon, nitrogen, and oxygen that make life on Earth possible. High-mass stars produced the iron and silicon that make up Earth itself, as well as the heavier elements on which much of our technology is based.

3. The creation and *explosive dispersal* of newly formed elements are accompanied by further shock waves, whose

Interstellar medium

Supernova + heavy elements

Star formation

Stellar evolution

◀ **Interactive FIGURE 21.19 Stellar Recycling** The cycle of star formation and evolution continuously replenishes our Galaxy with new heavy elements and provides the driving force for the creation of new generations of stars. Clockwise from the top are an interstellar cloud (Barnard 68), a star-forming region (RCW 38), a massive star ejecting a "bubble" and about to explode (NGC 7635), and a supernova remnant and its heavy-element debris (N49). Earth and humanity are both direct consequences of this truly grand cosmic cycle. *(ESO; NASA)*

passage through the interstellar medium simultaneously enriches the medium and compresses it into further star formation. Each generation of stars increases the concentration of heavy elements in the interstellar clouds from which the next generation forms. As a result, recently formed stars contain a much greater abundance of heavy elements than do stars that formed long ago.

In this way, although some material is used up in each cycle—turned into energy or locked up in low-mass stars—the galaxy continuously recycles its matter. Each new round of formation creates stars with more heavy elements than

the preceding generation had. From the old globular clusters, which are observed to be deficient in heavy elements relative to the Sun, to the young open clusters, containing much larger amounts of these elements, we observe this enrichment process in action. Our Sun is the product of many such cycles. We ourselves are another. Without the elements synthesized in the hearts of stars, neither Earth nor the life it harbors would exist.

CONCEPT CHECK

✔ Why is stellar evolution important to life on Earth?

CHAPTER REVIEW

SUMMARY

1 A **nova (p. 518)** is a star that suddenly increases greatly in brightness, then slowly fades back to its normal appearance over a period of months. It is the result of a white dwarf in a binary system

drawing hydrogen-rich material from its companion. The gas spirals inward in an **accretion disk (p. 518)** and builds up on the white-dwarf's surface, eventually becoming hot and dense enough for the hydrogen to burn explosively, temporarily causing a large increase in the dwarf's luminosity.

2 Stars more massive than about 8 solar masses form heavier and heavier elements in their cores, at a more and more rapid pace. As they do so, their cores form a layered structure consisting 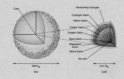 of burning shells of successively heavier elements. The process stops at iron, whose nuclei can neither be fused together nor split to produce energy. As a star's iron core grows in mass, it eventually becomes unable to support itself against gravity and begins to collapse. At the high temperatures produced during the collapse, iron nuclei are broken down into protons and neutrons. The protons combine with electrons to form more neutrons. Eventually, when the core becomes so dense that the neutrons are effectively brought into physical contact with one another, the collapse stops and the core rebounds, sending a violent shock wave out through the rest of the star. The star explodes in a **core-collapse supernova (p. 522)**.

3 Astronomers classify **supernovae (p. 522)** into two broad categories: Type I and Type II. These classes differ by their light curves and their composition. **Type I supernovae (p. 523)** are hydrogen poor and have a light curve similar in shape to that of a nova. **Type II supernovae (p. 523)** are hydrogen rich and have a characteristic plateau in the light curve a  few months after maximum. A Type II supernova is a core-collapse supernova. A Type I supernova occurs when a carbon–oxygen white dwarf in a binary system gains mass, collapses, and explodes as its carbon ignites. This type of supernova is called a **carbon-detonation supernova (p. 523)**.

4 Theory predicts that a supernova visible from Earth should occur within our Galaxy about once a century, although none has been observed in the last 400 years. We can see evidence of a past super nova in the form of a **supernova remnant (p. 526)**—a shell of exploded debris surrounding the site of the explosion and expanding into space at a speed of thousands of kilometers per second.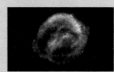

5 All elements heavier than helium are formed by **stellar nucleosynthesis (p. 529)**—the production of new elements by nuclear reactions in the cores of evolved stars. Elements heavier than car-  bon tend to form by **helium capture (p. 531)**, rather than by the fusion of more massive nuclei. At high enough core temperatures, photodisintegration breaks apart some heavy nuclei, providing helium-4 nuclei for the synthesis of even more massive elements, up to iron. Elements beyond iron form by **neutron capture (p. 532)** in the cores of evolved stars. During a supernova, rapid neutron capture occurs, producing the heaviest nuclei of all. Comparisons between theoretical predictions of element production and observations of element abundances in stars and supernovae provide strong support for the theory of stellar nucleosynthesis.

6 The processes of star formation, evolution, and explosion form a cycle that constantly enriches the interstellar medium with heavy elements and sows the seeds of new generations of stars. Without the elements produced in supernovae, life on Earth would be impossible.

Mastering**ASTRONOMY** *For instructor-assigned homework go to **www.masteringastronomy.com***

Problems labeled **POS** explore the process of science | **VIS** problems focus on reading and interpreting visual information

REVIEW AND DISCUSSION

1. Under what circumstances will a binary star produce a nova?

2. What is an accretion disk, and how does one form?

3. What is a light curve? How can it be used to identify a nova or a supernova?

4. Why does the core of a massive star collapse?

5. How do photodisintegration and neutronization contribute to the demise of a massive star?

6. What occurs in a massive star to cause it to explode?

7. What are the observational differences between Type I and Type II supernovae?

8. What is the Chandrasekhar mass, and what does it have to do with supernovae?

9. How do the mechanisms responsible for Type I and Type II supernovae explain their observed differences?

10. Roughly how often would we expect a supernova to occur in our own Galaxy? How often would we expect to *see* a galactic supernova?

11. **POS** What evidence is there that many supernovae have occurred in our Galaxy?

12. **POS** How can astronomers estimate the age of an isolated star?

13. What proof do astronomers have that heavy elements are formed in stars?

14. As a star evolves, why do heavier elements tend to form by helium capture rather than by fusion of like nuclei?

15. Why do the cores of massive stars evolve into iron and not heavier elements?

16. How are nuclei heavier than iron formed?

17. What is the r-process? When and where does it occur?

18. Why was supernova 1987A so important?

19. **POS** Why are neutrino detectors important to the study of supernovae?

20. Describe the role played by supernovae in "recycling" galactic matter.

CONCEPTUAL SELF-TEST: MULTIPLE CHOICE

1. A white dwarf can dramatically increase in brightness only if it (**a**) has another star nearby; (**b**) can avoid nuclear fusion in its core; (**c**) is spinning very rapidly; (**d**) is descended from a very massive star.

2. A nova differs from a supernova in that the nova (**a**) can occur only once; (**b**) is much more luminous; (**c**) involves only high-mass stars; (**d**) is much less luminous.

3. Which of the following stars will become hot enough to form elements heavier than oxygen? (**a**) A star that is half the mass of the Sun. (**b**) A star having the same mass as the Sun. (**c**) A star that is twice as massive as the Sun. (**d**) A star that is eight times more massive than the Sun.

4. A massive star becomes a supernova when it (**a**) collides with a stellar companion; (**b**) forms iron in its core; (**c**) suddenly increases in surface temperature; (**d**) suddenly increases in mass.

5. **VIS** Figure 21.8 ("Supernova Light Curves") indicates that a supernova whose luminosity declines steadily in time is most likely associated with a star that is (**a**) without a binary companion; (**b**) more than eight times the mass of the Sun; (**c**) on the main sequence; (**d**) comparable in mass to the Sun.

6. An observable supernova should occur in our Galaxy about once every (**a**) year; (**b**) decade; (**c**) century; (**d**) millennium.

7. Which one of the following *does not* provide evidence that supernovae have occurred in our Galaxy? (**a**) The rapid expansion and filamentary structure of the Crab Nebula. (**b**) Historical records from China and Europe. (**c**) The existence of binary stars in our Galaxy. (**d**) The existence of iron on Earth.

8. Nuclear fusion in the Sun will (**a**) never create elements heavier than helium; (**b**) create elements up to and including oxygen; (**c**) create all elements up to and including iron; (**d**) create some elements heavier than iron.

9. Most of the carbon in our bodies originated in (**a**) the core of the Sun; (**b**) the core of a red-giant star; (**c**) a supernova; (**d**) a nearby galaxy.

10. The silver atoms found in jewelry originated in (**a**) the core of the Sun; (**b**) the core of a red-giant star; (**c**) a supernova; (**d**) a nearby galaxy.

PROBLEMS

The number of dots preceding each Problem indicates its approximate level of difficulty.

1. ••• Estimate how close a 0.5-solar-mass white dwarf must come to the center of a 2-solar-mass subgiant with radius 10 times that of the Sun in order for the white dwarf's tidal field to strip matter from the companion's surface.

2. • Calculate the orbital speed of matter in an accretion disk just above the surface of a 0.6-solar-mass, 15,000-km-diameter white dwarf.

3. • A certain telescope can just detect the Sun at a distance of 10,000 pc. What is the apparent magnitude of the Sun at this distance? (For convenience, take the Sun's absolute magnitude to be 5.) What is the maximum distance at which the telescope can detect a nova having a peak luminosity of 10^5 solar luminosities?

4. • Repeat the previous calculation for a supernova having a peak luminosity 10^{10} times that of the Sun. What would be the apparent magnitude of the explosion if it occurred at a distance of 10,000 Mpc? Would it be detectable by any existing telescope?

5. •• At what distance would a supernova of absolute magnitude −20 look as bright as the Sun? As the Moon? Would you expect a supernova to occur that close to us?

6. • A (hypothetical) supernova at a distance of 150 pc has an absolute magnitude of −20. Compare its apparent magnitude with that of (a) the full Moon and (b) Venus at its brightest (see Figure 17.7). Would you expect a supernova to occur this close to us?

7. • A supernova's energy is often compared to the total energy output of the Sun over its lifetime. Using the Sun's current energy output, calculate its total energy output, assuming that the sun has a 10^{10} year main-sequence lifetime. How does this compare with the energy released by a supernova?

8. •• The *Hubble Space Telescope* is observing a distant Type I supernova with peak apparent magnitude 24. Using the light curve in Figure 21.8, estimate how long after the peak brightness the supernova will become too faint to be seen.

9. • The Crab Nebula is now about 1 pc in radius. If it was observed to explode in A.D. 1054, roughly how fast is it expanding? (Assume a constant expansion rate. Is that a reasonable assumption?)

10. •• Suppose that stars form in our Galaxy at an average rate of 10 per year. Suppose also that all stars greater than 8 solar masses explode as supernovae. Use Figure 17.23 to estimate the rate of Type II supernovae in our Galaxy.

22

NEUTRON STARS AND BLACK HOLES

STRANGE STATES OF MATTER

LEARNING GOALS

Studying this chapter will enable you to

1 Describe the properties of neutron stars, and explain how these strange objects are formed.

2 Explain the nature and origin of pulsars, and account for their characteristic radiation.

3 List and explain some of the observable properties of neutron-star binary systems.

4 Discuss the basic characteristics of gamma-ray bursts and some theoretical attempts to explain them.

5 Describe how black holes are formed, and discuss their effects on matter and radiation in their vicinity.

6 Describe Einstein's theories of relativity, and discuss how they relate to neutron stars and black holes.

7 Relate the phenomena that occur near black holes to the warping of space around them.

8 Discuss the difficulties in observing black holes, and explain some of the ways in which a black hole might be detected.

THE BIG PICTURE Neutron stars and black holes are among the most exotic objects in the universe. They are the end of the road for massive stars, and their bizarre properties boggle the imagination. Yet theory and observation seem to agree that, fantastic or not, they are real.

Mastering**ASTRONOMY**

Visit the Study Area in www.masteringastronomy.com for quizzes, animations, videos, interactive figures, and self-guided tutorials.

Our study of stellar evolution has led us to some very unusual and unexpected objects. Red giants, white dwarfs, and supernovae surely represent extreme states of matter completely unfamiliar to us here on Earth. Yet stellar evolution—and in particular, its end point, the death of a star—can have even more bizarre consequences. The strangest states of all result from the catastrophic implosion–explosion of stars much more massive than our Sun.

The almost unimaginable violence of a supernova may bring into being objects so extreme in their behavior that they require us to reconsider some of our most hallowed laws of physics. They open up a science fiction writer's dream of fantastic phenomena. They may even one day force scientists to construct a whole new theory of the universe.

LEFT: *This stunning image is actually a composite of three images taken by telescopes in orbit: optical light (in yellow) observed with* Hubble, *X-ray radiation (blue and green) with* Chandra, *and infrared radiation (red) with* Spitzer. *This object is known as Cassiopeia A, the remnant of a supernova whose radiation first reached Earth about 300 years ago. The small turquoise dot at the center may be a neutron star created in the blast, the sole survivor of the explosion. (NASA)*

22.1 Neutron Stars

In Chapter 21 we saw how some stars can explode violently as *supernovae*, scattering debris across large regions of interstellar space. What remains after a supernova? Is the entire progenitor (parent) star blown to bits and dispersed throughout interstellar space, or does some portion of it survive?

Stellar Remnants

For a Type I (carbon-detonation) supernova, most astronomers regard it as quite unlikely that any central remnant is left after the explosion. The entire star is shattered by the blast. However, for a Type II supernova, involving the implosion and subsequent rebound of a massive star's iron core, theoretical calculations indicate that part of the star may survive. ∞ (Sec. 21.2) The explosion destroys the parent star, but it may leave a tiny ultracompressed **remnant** at its center—all that remains of a star's inner core after stellar evolution has ceased. A white dwarf, the dense end-point of the evolution of a low-mass star, is another example of a stellar remnant.* ∞ (Sec. 20.3) Even by the high-density standards of a white dwarf, though, the matter within this severely compacted core is in a very strange state, unlike anything we are ever likely to find (or create) on Earth.

Recall from Chapter 21 that during the moment of implosion of a massive star—just prior to the supernova itself—the electrons in the core violently smash into the protons there, forming neutrons and neutrinos. ∞ (Sec. 21.2) The neutrinos leave the scene at (or nearly at) the speed of light, accelerating the collapse of the neutron core, which continues to contract until its particles come into contact. At that point, the central portion of the core rebounds, creating a powerful shock wave that races outward through the star, expelling matter violently into space.

The key point here is that the shock wave does not start at the very center of the collapsing core. The innermost part of the core—the region that "bounces"—remains intact as the shock wave it causes destroys the rest of the star. After the violence of the supernova has subsided, this ball of neutrons is all that is left. Researchers colloquially call this core remnant a **neutron star,** although it is not a star in any true sense of the word because all of its nuclear reactions have ceased forever.

Neutron-Star Properties

Neutron stars are extremely small and very massive. Composed purely of neutrons packed together in a tight ball about 20 km across, a typical neutron star is not much bigger than a small asteroid or a terrestrial city (see Figure 22.1), yet

---

These remnants are small and compact—no larger than Earth in the case of a white dwarf and far smaller still for a neutron star. They should not be confused with supernova remnants: glowing clouds of debris scattered across many parsecs of interstellar space. ∞ (Sec. 21.3)

▲ **FIGURE 22.1 Neutron Star** Neutron stars are not much larger than many of Earth's major cities. In this fanciful comparison, a typical neutron star sits alongside Manhattan Island. *(NASA)*

its mass is greater than that of the Sun. With so much mass squeezed into such a small volume, neutron stars are incredibly dense. Their average density can reach 10^{17} or even 10^{18} kg/m^3, nearly a billion times denser than a white dwarf. (For comparison, the density of a normal atomic nucleus is about 3×10^{17} kg/m^3.) A single thimbleful of neutron-star material would weigh 100 million tons—about as much as a good-sized terrestrial mountain. In a sense, we can think of a neutron star as a single enormous nucleus, with an atomic mass of around 10^{57}! At these densities, neutrons resist further packing in very much the same way as electrons do (at much lower densities) in a white dwarf—this **neutron degeneracy pressure** supports the neutron star.

Neutron stars are solid objects. Provided that a sufficiently cool one could be found, you might even imagine standing on it. However, doing so would not be easy, as a neutron star's gravity is extremely powerful. A 70-kg (150-pound) human would weigh the Earth equivalent of about 10 trillion kg (10 billion tons). The severe pull of a neutron star's gravity would flatten you much thinner than this piece of paper!

In addition to large mass and small size, newly formed neutron stars have two other very important properties. First, they *rotate* extremely rapidly, with periods measured in fractions of a second. This is a direct result of the law of conservation of angular momentum, which tells us that any rotating body must spin faster as it shrinks. ∞ *(More Precisely 6-1)* Even if the core of the progenitor star were initially rotating quite slowly (once every couple of weeks, say, as is observed in many upper main-sequence stars), it would be spinning a few times per second by the time it had reached a diameter of 20 km.

Second, newborn neutron stars have very strong *magnetic fields.* The original field of the progenitor star is amplified by the collapse of the core because the contracting material squeezes the magnetic field lines closer together, creating a magnetic field trillions of times stronger than Earth's.

In time, theory indicates, our neutron star will spin more and more slowly as it radiates its energy into space, and its magnetic field will diminish. However, for a few million years after its birth, these two properties combine to provide the primary means by which this strange object can be detected and studied.

CONCEPT CHECK
✔ Are all supernovae expected to lead to neutron stars?

22.2 Pulsars

Can we be sure that objects as strange as neutron stars really exist? The answer is a confident yes. The first observation of a neutron star occurred in 1967, when Jocelyn Bell, a graduate student at Cambridge University, made a surprising discovery. She observed an astronomical object emitting radio radiation in the form of rapid *pulses.* Each pulse consisted of an 0.01-second burst of radiation, after which there was nothing. Then, 1.34 s later, another pulse would arrive. The interval between the pulses was astonishingly uniform—so accurate, in fact, that the repeated emissions could be used as a precise clock. Figure 22.2 is a recording of part of the radio radiation from the pulsating object Bell discovered.

More than 1500 of these pulsating objects are now known in the Milky Way Galaxy. They are called **pulsars.** Each pulsar has its own characteristic pulse period and duration. In some cases, the pulse periods are so stable that they are by far the most accurate natural clocks known in the universe—more accurate even than the best atomic clocks on Earth. In some cases, the period is predicted to change by only a few seconds in a million years. The best current model describes a pulsar as a compact, spinning neutron star that periodically flashes radiation toward Earth.

The Lighthouse Model

When Bell made her discovery in 1967, she did not know what she was looking at. Indeed, no one at the time knew what a pulsar was. The explanation of pulsars as spinning neutron stars won Bell's thesis advisor, Antony Hewish, a share of the 1974 Nobel Prize in physics. Hewish reasoned that the only physical mechanism consistent with such precisely timed pulsations is a small rotating source of radiation. Only rotation can cause the high degree of regularity of the observed pulses, and only a small object can account for the sharpness of each pulse. Radiation emitted from different regions of an object larger than a few tens of kilometers across would arrive at Earth at slightly different times, blurring the pulse profile.

Figure 22.3 outlines the important features of this pulsar model. Two "hot spots" on the surface of a neutron star, or in the magnetosphere just above the surface, continuously emit radiation in a narrow "searchlight" pattern. These spots are most likely localized regions near the neutron-star's magnetic poles, where charged particles, accelerated to extremely high energies by the star's rotating magnetic field, emit radiation along the star's magnetic axis. The hot spots radiate more or less steadily, and the resulting beams sweep through space like a revolving lighthouse beacon, as the neutron star rotates. Indeed, this pulsar model is often known as the **lighthouse model.** If the neutron star happens to be oriented such that the beam sweeps across Earth, we see the star as a pulsar. The beams are observed as a series of rapid pulses— each time one of the beams flashes past Earth, a pulse is seen. The period of the pulses is the star's rotation period.

A few pulsars are definitely associated with supernova remnants, although not all such remnants have a detectable pulsar within them. Figure 22.4(c) shows a pair of optical photographs of the Crab pulsar, at the center of the Crab supernova remnant (Figures 22.4a and b). ∞ (Sec. 21.3) In the left frame, the pulsar is off; in the right frame, it is on. The rapid variation in the pulsar's light, with a pulse period of about 33 milliseconds, is shown in Figure 22.4(d). The Crab also pulses in the radio and X-ray parts of the spectrum. By observing the speed and direction of the Crab's ejected matter, astronomers have worked backward to pinpoint the location in space at

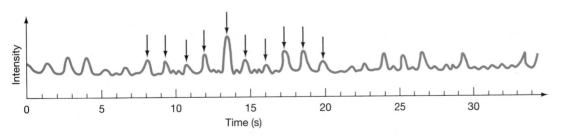

▲ **FIGURE 22.2 Pulsar Radiation** Pulsars emit periodic bursts of radiation. This recording shows the regular change in the intensity of the radio radiation emitted by the first such object discovered, known as CP 1919. Some of the object's pulses are marked by arrows.

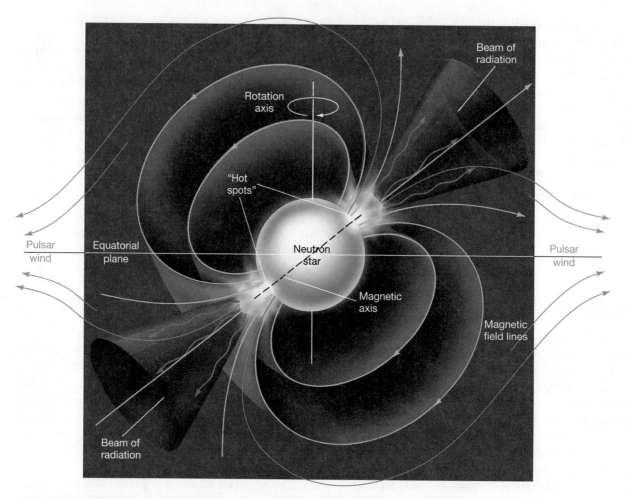

▲ **FIGURE 22.3 Pulsar Model** This diagram of the "lighthouse model" of neutron-star emission accounts for many of the observed properties of pulsars. Charged particles, accelerated by the magnetism of the neutron star, flow along the magnetic field lines, producing radiation that beams outward. At greater distances from the star, the field lines channel these particles into a high-speed outflow in the star's equatorial plane, forming a pulsar wind. The beam sweeps across the sky as the neutron star rotates. If it happens to intersect Earth, we see a pulsar—much like a lighthouse beacon.

which the explosion must have occurred and where the supernova core remnant should be located. ∞ (Sec. 21.3) It corresponds to the location of the pulsar. This is all that remains of the once massive star whose supernova was observed in 1054.

As indicated in Figure 22.3, the neutron star's strong magnetic field and rapid rotation channel high-energy particles from near the star's surface into the surrounding nebula (cf. the expanding envelope of the 1054 supernova—Figure 22.4a). The result is an energetic *pulsar wind* that flows outward at almost the speed of light, primarily in the star's equatorial plane. As it slams into the nebula, the wind heats the gas to very high temperatures. Figure 22.4(b) shows this process in action in the Crab—the combined *Hubble/Chandra* image reveals rings of hot X-ray-emitting gas moving rapidly away from the pulsar. Also visible in the image is a jet of hot gas (*not* the beam of radiation from the pulsar) escaping perpendicular to the equatorial plane. Eventually, the energy from the pulsar wind is deposited into

the Crab nebula, where it is radiated into space by the nebular gas, powering the spectacular display we see from Earth. ∞ (Fig. 21.10)

Most pulsars emit pulses in the form of radio radiation, but some (like the Crab) have been observed to pulse in the visible, X-ray, and gamma-ray parts of the spectrum as well. Figure 22.5 shows the Crab and the nearby Geminga pulsar in gamma rays. Geminga is unusual in that, although it pulsates strongly in gamma rays, it is barely detectable in visible light and not at all at radio wavelengths. Whatever types of radiation are produced, these electromagnetic flashes at different frequencies all occur at regular, repeated intervals, as we would expect, since they arise from the same object. However, pulses at different frequencies do not necessarily all occur at the same instant in the pulse cycle. The periods of most pulsars are quite short, ranging from about 0.03 s to 0.3 s (that is, flashing between 3 and 30 times per second). The human eye is insensitive to such rapid flashes, making it

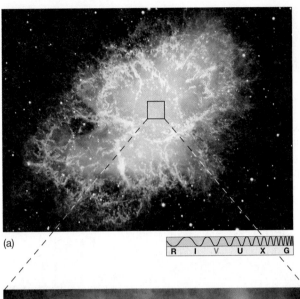

(a)

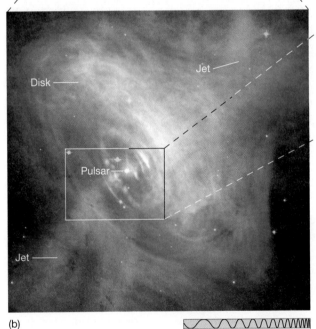

(b)

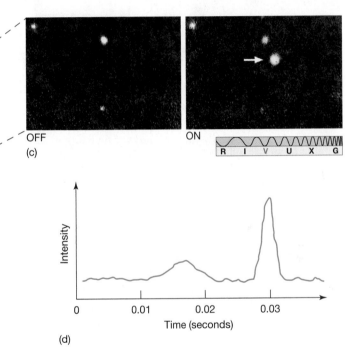

(c)   OFF   ON

(d)

◄ FIGURE 22.4 **Crab Pulsar** Despite their extreme properties, neutron stars offer the only working explanation of pulsars, and their association with supernova remnants ties them squarely to the theory of stellar evolution. Here, in the core of the Crab Nebula (a), the Crab pulsar (c) blinks on and off about 30 times each second. In this pair of optical images, the pulsing can be clearly seen. In the right frame, the pulsar (arrow) is on; in the left frame, it is off. (b) This more recent *Chandra* X-ray image of the Crab, superimposed on a *Hubble* optical image, shows the central pulsar, as well as rings of hot X-ray-emitting gas in the equatorial plane, driven by the pulsar wind and moving rapidly outward. Also visible in the image is a jet of hot gas (*not* the beam of radiation from the pulsar) escaping perpendicular to the equatorial plane. (d) The light curve shows the main pulse and its precursor. (The latter is probably related to the beam directed away from Earth.) *(ESO; NASA)*

impossible to observe the flickering of a pulsar by eye, even with a large telescope. Fortunately, instruments can record pulsations of light that the human eye cannot detect.

Most known pulsars are observed (usually by Doppler measurements) to have high speeds—much greater than the typical speeds of stars in our Galaxy. The most likely explanation for these anomalously high speeds is that neutron stars may receive substantial "kicks" due to asymmetries in the supernovae in which they formed. Such asymmetries, which are predicted by theory, are generally not very pronounced, but if the supernova's enormous energy is channeled even slightly in one direction, the newborn neutron star can recoil in the opposite direction with a speed of many tens or even hundreds of kilometers per second. Thus, observations of pulsar velocities give theorists additional insight into the detailed physics of supernovae.

Neutron Stars and Pulsars

All pulsars are neutron stars, but not all neutron stars are observed as pulsars, for two reasons. First, the two ingredients that make the neutron star pulse—rapid rotation and a strong magnetic field—both diminish with time, so the pulses gradually weaken and become less frequent. Theory indicates that, within a few tens of millions of years, the beam weakens and the pulses all but stop. Second, even a young, bright neutron star is not necessarily detectable as a pulsar from our vantage point on Earth. The pulsar beam depicted in Figure 22.3 is relatively narrow—perhaps as little as a few degrees across in some cases. Only if the neutron star happens to be oriented in just the right way do we actually see pulses. When we see those pulses from Earth, we call the body a pulsar. Note that we are using the term

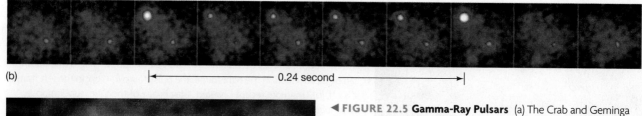

(b) |← — — — — — — 0.24 second — — — — — — →|

Geminga pulsar

Crab pulsar

(a)

R I V U X G

◄ **FIGURE 22.5 Gamma-Ray Pulsars** (a) The Crab and Geminga pulsars, which happen to lie fairly close (about 4.5 degrees) to one another in the sky. Unlike the Crab, Geminga is barely visible at optical wavelengths and undetectable in the radio region of the spectrum. (b) Sequence of *Compton Gamma-Ray Observatory* images showing Geminga's 0.24-s pulse period. (The Crab's 33-millesecond period is too rapid to be resolved by the detector.) (*NASA*)

"pulsar" here to mean the pulsing object we observe if the beam crosses Earth. However, many astronomers use the word more generically to mean *any* young neutron star producing beams of radiation as in Figure 22.3. Such an object will be a pulsar as seen from some directions—just not necessarily ours!

Given our current knowledge of star formation, stellar evolution, and neutron stars, our observations of pulsars are consistent with the ideas that (1) *every* high-mass star dies in a supernova explosion, (2) most supernovae leave a neutron star behind (a few result in black holes, as discussed in a moment), and (3) *all* young neutron stars emit beams of radiation, just like the pulsars we actually detect. A few pulsars are definitely associated with supernova remnants, clearly establishing those pulsars' explosive origin. On the basis of estimates of the rate at which massive stars have formed over the lifetime of the Milky Way, astronomers reason that, for every pulsar we know of, there must be several hundred thousand more neutron stars moving unseen somewhere in our Galaxy. Some formed relatively recently—less than a few million years ago—and simply happen not to be beaming their energy toward Earth. However, the vast majority are old, their youthful pulsar phase long past.

Neutron stars (and black holes too) were predicted by theory long before they were actually observed, although their extreme properties made many scientists doubt that they would ever be found in nature. The fact that we now have strong observational evidence, not just for their existence but also for the vitally important roles they play in many areas of high-energy astrophysics, is yet another testament to the fundamental soundness of the theory of stellar evolution.

CONCEPT CHECK

✔ Why don't we see pulsars at the centers of all supernova remnants?

22.3 Neutron-Star Binaries

We noted in Chapter 17 that most stars are not single, but instead are members of binary systems. ∞ (Sec. 17.7) Although many pulsars are known to be isolated (i.e., not part of any binary), at least some do have binary companions, and the same is true of neutron stars in general (even the ones not seen as pulsars). One important consequence of this pairing is that the masses of some neutron stars have been determined quite accurately. All the measured masses are fairly close to 1.4 times the mass of the Sun—the Chandrasekhar mass of the stellar core that collapsed to form the neutron-star remnant.

X-Ray Sources

The late 1970s saw several important discoveries about neutron stars in binary-star systems. Numerous X-ray sources were found near the central regions of our Galaxy and also near the centers of a few rich star clusters. Some of these sources, known as **X-ray bursters,** emit much of their energy in violent eruptions, each thousands of times more luminous than our Sun, but lasting only a few seconds. A typical burst is shown in Figure 22.6.

This X-ray emission arises on or near neutron stars that are members of binary systems. Matter torn from the surface of the (main-sequence or giant) companion by the neutron

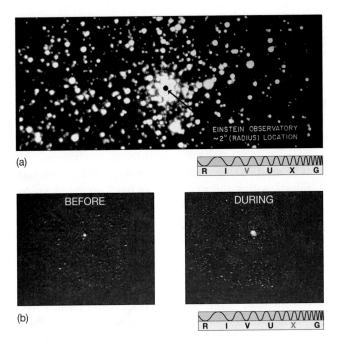

(a)

R I V U X G

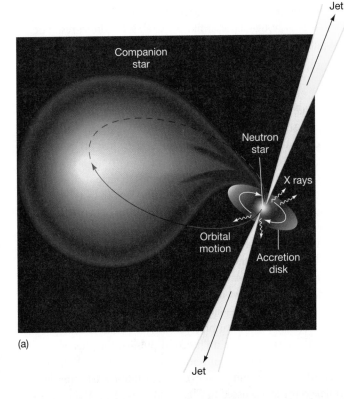

BEFORE

DURING

(b)

R I V U X G

Companion star

Jet

Neutron star

X rays

Orbital motion

Accretion disk

Jet

(a)

▶ **FIGURE 22.6 X-Ray Burster** An X-ray burster produces a sudden, intense flash of X rays, followed by a period of relative inactivity lasting as long as several hours. Then another burst occurs. The bursts are thought to be caused by explosive nuclear burning on the surface of an accreting neutron star, similar to the explosions on a white dwarf that give rise to novae. (a) An optical photograph of the globular star cluster Terzan 2, showing a 2" dot at the center where the X-ray bursts originate. (b) X-ray images taken before and during the outburst. The most intense X rays correspond to the position of the black dot shown in (a). *(SAO)*

star's strong gravitational pull accumulates on the neutron star's surface. As in the case of white-dwarf accretion (see Chapter 21), the material does not fall directly onto the surface. Instead, as illustrated in Figure 22.7(a), it forms an accretion disk (compare with Figure 21.2, which depicts the white-dwarf equivalent). ∞ (Sec. 21.1) The gas goes into a tight orbit around the neutron star and then spirals slowly inward. The inner portions of the accretion disk become extremely hot, releasing a steady stream of X rays.

As gas builds up on the neutron star's surface, its temperature rises due to the pressure of overlying material. Soon the temperature becomes hot enough to fuse hydrogen. The result is a sudden period of rapid nuclear burning that releases a huge amount of energy in a brief, but intense, flash of X rays—an *X-ray burst*. After several hours of renewed accumulation, a fresh layer of matter produces the next burst. Thus, an X-ray burst is much like a nova on a white dwarf, but occurring on a far more violent scale because of the neutron star's much stronger gravity. ∞ (Sec. 21.1)

Not all the infalling gas makes it onto the neutron star's surface, however; in at least one case—an object known as SS 433,* lying roughly 5000 pc from Earth—we have direct

The name simply identifies the object as the 433rd entry in a particular catalog of stars with strong optical emission lines.

▶ **FIGURE 22.7 X-Ray Emission** (a) Matter flows from a normal star toward a compact neutron-star companion and falls toward the surface in an accretion disk. As the gas spirals inward under the neutron star's intense gravity, it heats up, becoming so hot that it emits X rays. In at least one instance—the peculiar object SS 433—some material may be ejected in the form of two high-speed jets of gas. (b) False-color radiographs of SS 433, made at monthly intervals (left to right), show the jets moving outward and the central source rotating under the gravitational influence of the companion star. *(NRAO)*

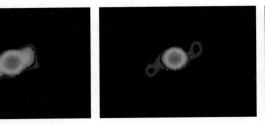
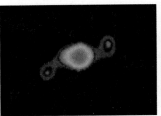
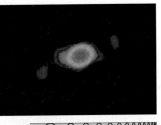

(b)

R I V U X G

observational evidence that some material is instead shot completely out of the system at enormously high speeds. SS 433 expels more than one Earth mass of material every year in the form of two oppositely directed narrow jets moving roughly perpendicular to the disk. Observations of the Doppler shifts of optical emission lines produced within the jets themselves imply speeds of almost 80,000 km/s—more than 25 percent of the speed of light! As the jets interact with the interstellar medium, they emit radio radiation, as shown in Figure 22.7(b).

Jets of this sort are apparently quite common in astronomical systems in which an accretion disk surrounds a compact object (such as a neutron star or a black hole). They are thought to be produced by the intense radiation and magnetic fields near the inner edge of the disk, although the details of their formation are still uncertain. Once again, note that these jets are *not* the "lighthouse" beams of radiation from the neutron star itself, shown in Figure 22.3, that can result in a pulsar, nor are they associated with a pulsar wind, as in Figure 22.4(b).

Since the discovery of SS 433, roughly a dozen stellar-mass objects with comparable properties have been discovered in our Galaxy, and we will see examples of similar phenomena on much larger scales in later chapters. Indeed, the current terminology for "stellar-scale" objects like SS 433—**microquasar**—derives from their much more energetic galactic counterparts (called *quasars*; see Sec. 24.4). SS 433 has been particularly important in the study of microquasars because we can actually observe both its disk and its jets, instead of simply having to assume their existence, as we do in more distant cosmic objects.

Millisecond Pulsars

In the mid-1980s an important new category of pulsars was found: a class of very rapidly rotating objects called **millisecond pulsars.** Some 250 are currently known in the Milky Way Galaxy. These objects spin hundreds of times per second (i.e., their pulse period is a few milliseconds). This speed is about as fast as a typical neutron star can spin without flying apart. In some cases, the star's equator is moving at more than 20 percent of the speed of light, a speed that suggests a phenomenon bordering on the incredible: a cosmic object of kilometer dimensions, more massive than our Sun, spinning almost at breakup speed and making nearly a thousand complete revolutions *every second!* Yet the observations and their interpretation leave little room for doubt.

The story of these remarkable objects is further complicated because many of them are found in globular clusters. This is odd, since globular clusters are known to be very old—10 billion years, at least. ∞ (Sec. 20.5) Yet, Type II supernovae (the kind that create neutron stars) are associated with massive stars that explode within a few tens of *millions* of years after their formation, and no stars have formed in any

globular cluster since the cluster itself came into being. Thus, no new neutron star has been produced in a globular cluster in a very long time. But the pulsar produced by a supernova is expected to slow down in only a few million years, and after 10 billion years its rotation should have all but ceased. Thus, the rapid rotation of the pulsars found in globular clusters cannot be a relic of their birth. Instead, these objects must have been "spun up"—that is, had their rotation rates increased—by some other, much more recent, mechanism.

The most likely explanation for the high rotation rate of pulsars is that the neutron star has been spun up by drawing in matter from a companion star. As matter spirals down onto the star's surface in an accretion disk, it provides a "push" that makes the neutron star spin faster (see Figure 22.8). Theoretical calculations indicate that this process can spin the star up to breakup speed in about a hundred million years. This general picture is supported by the finding that, of the 150 or so millisecond pulsars seen in globular clusters, roughly half are known to be members of binary systems. The remaining solo millisecond pulsars were probably formed when an encounter with another star ejected the pulsar from the binary or when the pulsar's own intense radiation destroyed its companion.

Thus, although a pulsar like the Crab is the direct result of a supernova, millisecond pulsars are the product of a two-stage process. First, the neutron star was formed in an ancient supernova, billions of years ago. Second, through a relatively recent interaction with a binary companion, the neutron star then achieved the rapid spin that we observe today. Once again, we see how members of a binary system can evolve in ways quite different from the manner in which single stars evolve. Notice that the scenario of accretion onto a neutron star from a binary companion is the same scenario that we just used to explain the existence of X-ray bursters. In fact, the two phenomena are closely linked. Many X-ray bursters may be on their way to becoming millisecond pulsars, and many millisecond pulsars are X-ray sources, powered by the trickle of material still falling onto them from their binary companions.

Figure 22.9 shows the globular cluster 47 Tucanae, together with a *Chandra* image of its core showing no fewer than 108 X-ray sources—about 10 times the number that had been known in the cluster prior to *Chandra*'s launch. Roughly half of these sources are millisecond pulsars; the cluster also contains two or three "conventional" neutron-star binaries. Most of the remaining sources are white-dwarf binaries, similar to those discussed in Chapter 21. ∞ (Sec. 21.1)

The way in which a neutron star can become a member of a binary system is the subject of active research, because the violence of a supernova explosion would be expected to blow the binary apart in many cases. Only if the supernova progenitor lost a lot of mass before the explosion would the binary system be likely to survive. Alternatively, by interacting with an existing binary and displacing one of its components, a neutron star may become part of a binary system *after* it is formed, as

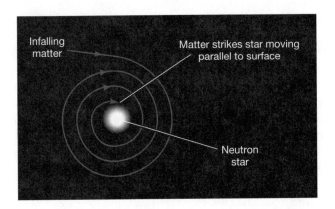

▲ **FIGURE 22.8 Millisecond Pulsar** Gas from a companion star spirals down onto the surface of a neutron star. As the infalling matter strikes the star, it moves almost parallel to the surface, so it tends to make the star spin faster. Eventually, this process can result in a millisecond pulsar—a neutron star spinning at the incredible rate of hundreds of revolutions per second.

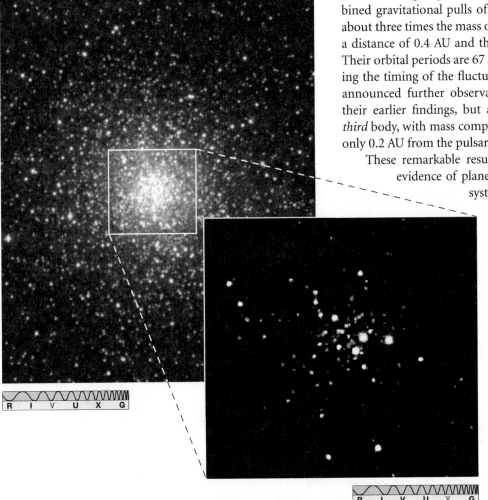

▲ **FIGURE 22.9 Cluster X-Ray Binaries** The dense core of the old globular cluster 47 Tucanae harbors more than 100 separate X-ray sources (shown in the *Chandra* image at the right). More than half of these are thought to be binary millisecond pulsars, still accreting small amounts of gas from their companions after an earlier period of mass transfer spun them up to millisecond speeds. *(ESO; NASA)*

depicted in Figure 22.10. Astronomers are eagerly searching the skies for more millisecond pulsars to test their ideas.

Pulsar Planets

Radio astronomers can capitalize on the precision with which pulsar signals repeat themselves to make extremely accurate measurements of pulsar motion. In January 1992, radio astronomers at the Arecibo Observatory found that the pulse period of a recently discovered millisecond pulsar lying some 500 pc from Earth varied in an unexpected, but quite regular, way. Careful analysis of the data has revealed that the period fluctuates on two distinct time scales—one of 67 days, the other of 98 days. The changes in the pulse period are small—less than one part in 10^7—but repeated observations have confirmed their reality.

The leading explanation for these fluctuations holds that they are caused by the Doppler effect as the pulsar wobbles back and forth in space. But what causes the wobble? The Arecibo group thinks that it is the result of the combined gravitational pulls of not one, but *two*, planets, each about three times the mass of Earth! One orbits the pulsar at a distance of 0.4 AU and the other at a distance of 0.5 AU. Their orbital periods are 67 and 98 days, respectively, matching the timing of the fluctuations. In April 1994, the group announced further observations that not only confirmed their earlier findings, but also revealed the presence of a *third* body, with mass comparable to Earth's Moon, orbiting only 0.2 AU from the pulsar.

These remarkable results constituted the first definite evidence of planet-sized bodies outside our solar system. A few other millisecond pulsars have since been found with similar behavior. However, it is unlikely that any of these planets formed in the same way as our own. Any planetary system orbiting the pulsar's progenitor star was almost certainly destroyed in the supernova explosion that created the pulsar. As a result, scientists are still unsure about how these planets came into being. One possibility involves the binary companion that provided the matter necessary to spin the pulsar up to millisecond speeds. Possibly, the pulsar's intense radiation and strong gravity destroyed the companion and then spread its matter out into a disk (a little like the solar nebula) in whose cool outer regions the planets might have condensed.

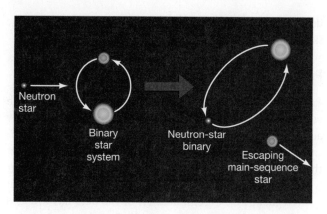

▲ **FIGURE 22.10 Binary Exchange** A neutron star can encounter a binary made up of two low-mass stars, ejecting one of them and taking its place. This mechanism provides a means of forming a binary system with a neutron-star component (which may later evolve into a millisecond pulsar) without having to explain how the binary survived the supernova explosion that formed the neutron star.

Astronomers have been searching for decades for planets orbiting main-sequence stars like our Sun, on the assumption that planets are a natural by-product of star formation. ∞ (Sec. 6.7) As we have seen, these searches have now identified many extrasolar planets, although the planetary systems discovered to date don't look much like our own—only a few planets even comparable in mass to Earth have so far been detected. ∞ (Sec. 15.6) It is ironic that the first Earth-sized planets to be found outside the solar system orbit a dead star and have little or nothing in common with our own world!

CONCEPT CHECK

✔ What is the connection between X-ray sources and millisecond pulsars?

22.4 Gamma-Ray Bursts

Discovered serendipitously in the late 1960s by military satellites looking for violators of the Nuclear Test Ban Treaty and first made public in the 1970s, **gamma-ray bursts** consist of bright, irregular flashes of gamma rays typically lasting only a few seconds (Figure 22.11a). Until the 1990s, it was thought that gamma-ray bursts were basically "scaled-up" versions of X-ray bursters in which matter accreted from the binary companion was subjected to even more violent nuclear burning, accompanied by the release of the more energetic gamma rays. However, this is not the case.

Distances and Luminosities

Figure 22.11(b) shows an all-sky plot of the positions of 2704 bursts detected by the *Compton Gamma-Ray Observatory (CGRO)* during its nine-year operational lifetime. ∞ (Sec. 5.7) On average, *CGRO* detected gamma-ray bursts at the rate of about one a day. Note that the bursts are distributed uniformly across the sky (their distribution is said to be *isotropic*), rather than being confined to the relatively narrow band of the Milky Way (cf. Figure 5.36). The bursts seemingly never repeat at the same location, show no obvious clustering, and appear unaligned with any known large-scale structure, near or far.

The isotropy of the *CGRO* data convinced most astronomers that the bursts do not originate within our own Galaxy, as had once been assumed, but instead are produced far beyond the Milky Way—at so-called *cosmological distances,* comparable to the scale of the universe itself. However, although *CGRO* detected thousands of gamma-ray bursts, it was unable to determine the distance to any of them. The reason for this brings us back to a fundamental issue in astronomy: the difficulties involved in measuring distances in the universe. ∞ (Sec. 17.6)

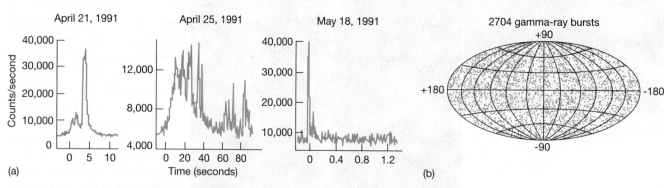

▲ **FIGURE 22.11 Gamma-Ray Bursts** (a) Plots of intensity versus time (in seconds) for three gamma-ray bursts. Note the substantial differences between them. Some bursts are irregular and spiky, whereas others are much more smoothly varying. Whether this wide variation in the burst appearance means that more than one physical process is at work is presently unknown. (b) Positions on the sky of all the gamma-ray bursts detected by *Compton Observatory* during its nearly 9-year operating lifetime. The bursts appear to be distributed isotropically (uniformly) across the entire sky. The plane of the Milky Way Galaxy runs horizontally across the center of the map, which is also the direction to the center of our Galaxy. *(NASA)*

Gamma-ray observations in and of themselves do not provide enough information to tell us how far away a burst is. Rather, in order to determine the distance, astronomers must somehow associate the burst with some other object in the sky whose distance can be measured by other means. Such objects are referred to as burst *counterparts,* and the techniques for studying them usually involve observations in the optical or X-ray parts of the electromagnetic spectrum. Unfortunately, as we saw in Chapter 5, gamma rays are far too penetrating to be focused by conventional optics. ∞ (Sec. 5.7) As a result, *CGRO*'s burst positions were uncertain by several degrees, so optical or X-ray telescopes had to scan very large regions of the sky as they looked for the counterpart. Compounding this problem, the "afterglows" of a burst at X-ray or optical wavelengths generally fade quite rapidly, severely limiting the time available to complete the search.

The first direct measurement of the distance to a gamma-ray burst was made on May 8, 1997, thanks to a combination of gamma-ray, X-ray, and optical observations of the gamma-ray burst GRB 970508. (The number simply indicates the date of the burst's detection.) The Italian–Dutch *BeppoSAX* satellite recorded the burst in both the gamma- and X-ray regions of the spectrum. The importance of the X-ray observations is that they allowed a much more accurate determination of the burst's location in the sky—in fact, to within a few arc minutes. That was enough for ground-based optical astronomers to look for and find GRB 970508's optical afterglow. For the first time, astronomers had observed an optical counterpart to a gamma-ray burst.

The counterpart's optical spectrum, obtained with the Keck telescope, revealed a very important piece of information. Several absorption lines of iron and magnesium were identified, but they were redshifted by almost a factor of two in wavelength. Such redshifts, as we will see in Chapter 24, are the result of the expansion of the universe, and they are clear proof—a "smoking gun," if you will—that at least this gamma-ray burst, and presumably all others, really did occur at cosmological distances. The events responsible for GRB 970508 occurred more than 2 *billion* parsecs from Earth.

Figure 22.12 shows optical images of another gamma-ray burst, GRB 971214, along with the "host" galaxy in which it resided. According to the redshift of lines observed in the galaxy's spectrum, this burst was almost

5 billion pc away. By the time the *Hubble* image (b) was taken (about 2 months after the Keck image and 4 months after the initial burst of gamma rays), the afterglow had faded, but a faint image of a host galaxy remained.

In all, more than 200 optical or X-ray afterglows of gamma-ray bursts have been detected, and several dozen distances are known. All are very large, implying that the bursts must be extremely energetic, since otherwise they wouldn't be detectable by our equipment. If we assume that the gamma rays are emitted equally in all directions (a big assumption!), then we can calculate the total energy emitted from the fraction we see, using the inverse-square law. ∞ (Sec. 17.2) We find that each burst apparently generates more energy—and in some cases hundreds of times more energy—than a typical supernova explosion, all in a matter of seconds! According to this calculation, GRB 971214 was the most violent explosion ever observed in the universe.

Finding the counterpart to a gamma-ray burst requires an accurate measurement of the burst's location (to limit the region of the sky that must be scanned) and fast communication (so that other telescopes can start to look before the afterglow fades). The most successful searches for burst counterparts have been carried out by satellites combining gamma-ray detectors with X-ray and/or optical telescopes. An example is NASA's *Swift* mission, in operation since 2005. *Swift* combines a wide-angle gamma-ray detector (to monitor as much of the sky as possible) with two telescopes: one X-ray and one

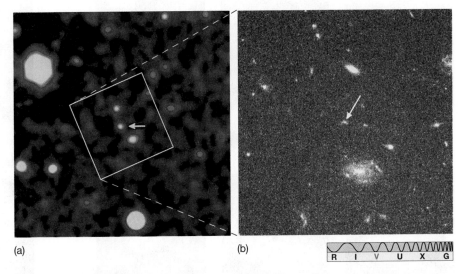

(a) (b)

R I V U X G

▲ **FIGURE 22.12 Gamma-Ray Burst Counterparts** Optical images of the gamma-ray burst GRB 971214 (the numbers simply mean December 14, 1997, the day the burst was recorded). Image (a), taken with the Keck telescope, shows the visible afterglow of the gamma-ray source (arrow) to be quite bright, comparable to two other prominent sources in the overlaid box. A spectrum of the afterglow showed it to be highly redshifted, placing it near the limits of the observable universe, almost 5 billion parsecs away. By the time the *Hubble* image (b) was taken (about 2 months after the Keck image and 4 months after the initial burst of gamma rays was detected by the Italian–Dutch *BeppoSax* satellite), the afterglow had faded, but a faint image of a host galaxy remains. The different colors result from the use of different filters to observe the light. *(Keck; NASA)*

optical/ultraviolet instrument. The gamma-ray detector spots the burst and determines its direction to an accuracy of about 4 arc minutes. Within seconds, the on-board computer automatically points the X-ray and optical telescopes toward the burst and relays its coordinates to other telescopes, in space and on the ground. *Swift* locates burst counterparts at the rate of about 1 per week, and has played a pivotal role in advancing our understanding of these violent phenomena.

What Causes the Bursts?

Not only are gamma-ray burst sources extremely energetic, they are also very *small*. The millisecond flickering in the bursts implies that whatever their origin, all of their energy must come from a volume no larger than a *few hundred kilometers* across. The reasoning is as follows: If the emitting region were, say, 300,000 km—1 light-second—across, even an instantaneous change in intensity at the source would be smeared out over a time interval of 1 s as seen from Earth, because light from the far side of the object would take 1 s longer to reach us than light from the near side. For the gamma-ray variation not to be blurred by the light-travel time, the source cannot be more than 1 light-millisecond, or just 300 km, in diameter.

Most theoretical attempts to explain gamma-ray bursts picture the burst as a *relativistic fireball*—an expanding region, probably a jet, of superhot gas radiating furiously in the gamma-ray part of the spectrum. (The term "relativistic" here means that particles are moving at nearly the speed of light and that Einstein's theory of relativity is needed to describe them—see Section 22.6.) The complex burst structure and afterglows are produced as the fireball expands, cools, and interacts with its surroundings.

If the gamma-ray burst is emitted as a jet, then its total energy may be reduced to more "manageable" levels, because the bright flash we see is representative of only a small fraction of the sky. This may make the apparently huge luminosity of the December 1997 burst somewhat easier to explain. As an analogy, consider a handheld laser pointer of the sort commonly used in talks and lectures. It radiates only a few milliwatts of power, much less than a household lightbulb, but it appears enormously bright if you happen to look directly into the beam. (*Don't* do this, by the way!) The laser beam is bright because all of its energy is concentrated in almost a single direction, instead of being radiated isotropically into space.

Two leading models for the energy source have emerged, as sketched in Figure 22.13. The first (Figure 22.13a) is the "true" end point of a binary-star system—the *merger* of the

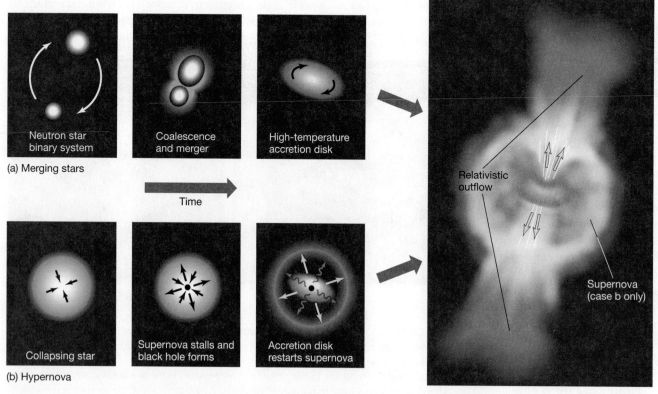

(a) Merging stars

Neutron star binary system · Coalescence and merger · High-temperature accretion disk

Time

(b) Hypernova

Collapsing star · Supernova stalls and black hole forms · Accretion disk restarts supernova

Relativistic outflow

Supernova (case b only)

▲ FIGURE 22.13 **Gamma-Ray Burst Models** Neutron stars and black holes are intimately connected with the most violent phenomena in the universe. Two models have been proposed to explain gamma-ray bursts. Part (a) depicts the merger of two neutron stars; part (b) shows the core collapse, implosion, and stalled supernova, or *hypernova,* of a single massive star. Both models predict a relativistic fireball, perhaps releasing energy in the form of jets, as shown.

component stars. Suppose that both members of the binary evolve to become neutron stars. As the system continues to evolve, gravitational radiation (see *Discovery 22-2*) is released, and the two ultradense stars spiral in toward each other. Once they are within a few kilometers of one another, coalescence is inevitable. Such a merger will likely produce a violent explosion comparable in energy to that generated by a supernova and perhaps energetic enough to explain the flashes of gamma rays we observe. The overall rotation of the binary system may channel the energy into a high-speed, high-temperature jet.

The second model (Figure 22.13b), sometimes called a *hypernova,* is a "failed" supernova—but what a failure! In this picture, a very massive star undergoes core collapse much as described earlier for a Type II supernova, but instead of forming a neutron star, the core collapses to a black hole (see Section 22.5). ∞ (Sec. 21.2) At the same time, the blast wave racing outward through the star stalls. Instead of being blown to pieces, the inner part of the star begins to implode, forming an accretion disk around the black hole and creating a relativistic jet. The jet punches its way out of the star, producing a gamma-ray burst as it slams into the surrounding shells of gas expelled from the star during the final stages of its nuclear-burning lifetime. ∞ (*Discovery 20-2*) At the same time, intense radiation from the accretion disk may restart the stalled supernova, blasting what remains of the star into space.

The idea of a relativistic fireball has become widely accepted among workers in this branch of astrophysics. Many researchers regard it as likely that the energy is released as a jet and that the most intense gamma-ray bursts occur when the jet is directed toward us ("looking into the laser"). Which of the two models just described is correct? Most experts in the field would now say that the answer is probably *both*. The hypernova model predicts bursts of relatively long duration and is the leading explanation of the "long" gamma-ray bursts lasting more than about 2 seconds. Detailed calculations indicate that the neutron star merger model can naturally account for the short bursts.

Thanks to the "rapid response network" provided by *Swift* and other instruments, astronomers now have detailed observations of many afterglows of both types of burst. Both the spectra and the light curves of the long-burst afterglows are precisely what astronomers would expect from a supernova produced by a very massive (25-solar mass) star. ∞ (Sec. 21.3) The rapidly fading X-ray afterglows from the short bursts are consistent with the predictions of the merger scenario. It seems that there are, in fact, two distinct classes of gamma-ray bursts, and that both of the models depicted in Figure 22.13 contribute to the total.

CONCEPT CHECK

✔ What are gamma-ray bursts, and why have they posed such a challenge to current theory?

22.5 Black Holes

Table 22.1 lists the properties of some of the dense stellar remnants we have encountered in this text. Brown, white, and black dwarfs are held up by electron degeneracy—the resistance of tightly packed electrons to further compression. ∞ (Secs. 20.2, 20.3) The much denser neutron stars, as we have just seen, are supported by a similar mechanism involving neutrons instead. Squeezed together, the neutrons in a neutron star form a hard ball of matter that not even gravity can compress further. Or can it? Is it possible that, given enough matter packed into a small enough volume, the collective pull of gravity can eventually crush any opposing pressure? Can gravity continue to compress a massive star into an object the size of a planet, a city, a pinhead—even smaller? The answer, apparently, is yes.

The Final Stage of Stellar Evolution

Although the precise figure is uncertain, mainly because the behavior of matter at very high densities is not well enough understood, most researchers concur that the mass of a neutron star cannot exceed about 3 solar masses. That is the neutron-star equivalent of the white dwarf's Chandrasekhar mass limit discussed in the previous chapter. ∞ (Sec. 21.3) Above this limit, not even tightly packed neutrons can withstand the star's gravitational pull. In fact, we know of *no* force that can counteract gravity once neutron degeneracy pressure is overwhelmed. If enough material is left behind after a supernova such that the central core exceeds the 3-solar-mass limit, gravity wins the battle with pressure once and for all, and the star's central core collapses forever. Stellar evolution theory indicates that this is the fate of any star whose main-sequence mass exceeds about 25 times the mass of the Sun.

The limit of 3 solar masses is uncertain, in part because it ignores the effects of magnetism and rotation, both of which are surely present in the cores of evolved stars. Because these effects can compete with gravity, they influence stellar evolution. ∞ (Sec. 19.1) In addition, we do not know precisely how the basic laws of physics might change in regions of very dense matter that is both rapidly spinning and strongly magnetized. Generally speaking, theorists expect that the neutron-star mass limit increases when magnetism and rotation are included because even larger amounts of mass will then be needed for gravity to compress stellar cores into neutron stars or black holes, but the amount of the increase is not currently known.

As the stellar core shrinks, the gravitational pull in its vicinity eventually becomes so great that even light itself is unable to escape. The resultant object therefore emits no light, no radiation, and no information whatsoever. Astronomers call this bizarre end point of stellar evolution, in which a massive core remnant collapses in on itself and vanishes forever, a **black hole.**

TABLE 22.1 Properties of Stellar Remnants

Remnant	Typical mass (solar masses)	Typical radius (km)	Typical density (kg/m³)	Support	Context (section)
brown dwarf	less than 0.08	70,000	10^5	electron degeneracy	H fusion never started (19.3)
white dwarf	less than 1.4	10,000	10^9	electron degeneracy	stellar core after fusion stops at C/O (20.3)
black dwarf	less than 1.4	10,000	10^9	electron degeneracy	"cold" white dwarf (20.3)
neutron star	1.4–3 (approx.)	10	10^{18}	neutron degeneracy	remnant of a core collapse supernova (22.1)
black hole	more than 3	10	infinite at the center	none	remnant of a core collapse supernova with massive progenitor (22.5)

Escape Speed

Newtonian mechanics—up to now our reliable and indispensable tool in understanding the universe—cannot adequately describe conditions in or near black holes. ∞ (Sec. 2.8) To comprehend these collapsed objects, we must turn instead to the modern theory of gravity: Einstein's *general theory of relativity,* discussed in Section 22.6. Still, we can usefully discuss some aspects of these strange bodies in more or less Newtonian terms. Let's consider again the familiar Newtonian concept of escape speed—the speed needed for one object to escape from the gravitational pull of another—supplemented by two key facts from relativity: (1) Nothing can travel faster than the speed of light, and (2) all things, *including light,* are attracted by gravity.

A body's escape speed is proportional to the square root of the body's mass divided by the square root of its radius. ∞ (Sec. 2.8) Earth's radius is 6400 km, and the escape speed from Earth's surface is just over 11 km/s. Now consider a hypothetical experiment in which Earth is squeezed on all sides by a gigantic vise. As our planet shrinks under the pressure, its mass remains the same, but its escape speed increases because the planet's radius is decreasing. For example, suppose Earth were compressed to one-fourth its present size. Then the proportionality mentioned in the first sentence of this paragraph predicts that our planet's escape speed would double. Consequently, any object escaping from this hypothetically compressed Earth would need a speed of at least 22 km/s to do so.

Imagine compressing Earth some more. Squeeze it by an additional factor of, say, a thousand, making its radius hardly more than a kilometer. Now a speed of about 630 km/s would be needed to escape from the planet's gravitational pull. Compress Earth still further, and the escape speed continues to rise. If our hypothetical vise were to squeeze Earth hard enough to crush its radius to about a centimeter, then the speed needed to escape the planet's

surface would reach 300,000 km/s. But this is no ordinary speed—it is the speed of light, the fastest speed allowed by the laws of physics as we currently know them.

Thus, if, by some fantastic means, the entire planet Earth could be compressed to less than the size of a grape, the escape speed would exceed the speed of light. However, because nothing can in fact exceed that speed, the compelling conclusion is that *nothing—absolutely nothing—could escape from the surface of such a compressed body.*

Black Hole Properties

The origin of the term *black hole* now becomes clear: No form of radiation—radio waves, visible light, X rays, indeed, photons of any wavelength—would be able to escape the intense gravity of our grape-sized Earth. With no photons leaving, our planet would be invisible and uncommunicative. No signal of any sort could be sent to the universe beyond. For all practical purposes, such a supercompact Earth could be said to have disappeared from the universe! Only its gravitational field would remain behind, betraying the presence of its mass, now shrunk to a point.

The "one-way" nature of the flow of energy and matter into a black hole means that almost all information about the matter falling into it—gas, stars, spaceships, or people— is lost. Only a few scraps survive. In fact, we now know that, regardless of the composition, structure, or history of the objects that formed the hole, only three physical properties can be measured from the outside: the hole's *mass, charge,* and *angular momentum.* All other information is lost once matter enters the hole. Thus, *just three numbers* are required to completely describe a black hole's outward appearance and interaction with the rest of the universe.

In this chapter, we will concentrate on black holes that formed from nonrotating, electrically neutral matter. Such objects are *completely specified* once their masses are known.

The Event Horizon

Astronomers have a special name for the critical radius at which the escape speed from an object would equal the speed of light and within which the object could no longer be seen. It is the **Schwarzschild radius,** after Karl Schwarzschild, the German scientist who first studied its properties. The Schwarzschild radius of any object is simply proportional to the object's mass. For Earth, the Schwarzschild radius is 1 cm; for Jupiter, which is about 300 Earth masses, it is approximately 3 m; for the Sun, with a mass of 300,000 Earth masses, it is 3 km. For a 3-solar-mass stellar core remnant, the Schwarzschild radius is about 9 km. As a convenient rule of thumb, the Schwarzschild radius of an object is simply 3 km, multiplied by the object's mass, measured in solar masses. Every object has a Schwarzschild radius; it is the radius to which the object would have to be compressed for it to become a black hole. Put another way, a black hole is an object that happens to lie within its own Schwarzschild radius.

The surface of an imaginary sphere with radius equal to the Schwarzschild radius and centered on a collapsing star is called the **event horizon.** It defines the region within which no event can ever be seen, heard, or known by anyone outside. Even though there is no matter of any sort associated with it, we can think of the event horizon as the "surface" of a black hole.

A 1.4-solar-mass neutron star has a radius of about 10 km and a Schwarzschild radius of 4.2 km. If we were to keep increasing the star's mass, the star's Schwarzschild radius would grow, although its actual physical radius would not. In fact, the radius of a neutron star *decreases* slightly with increasing mass. By the time the neutron star's mass exceeded about 3 solar masses, it would lie just within its own event horizon, and it would collapse of its own accord. It would not stop shrinking at the Schwarzschild radius: The event horizon is not a physical boundary of any kind—just a communications barrier. The remnant would shrink right past the Schwarzschild radius to ever-diminishing size on its way to being crushed to a point.

Thus, provided that at least 3 solar masses of material remain behind after a supernova explosion, the remnant core will collapse catastrophically, diving below the event horizon in less than a second. The core simply "winks out," disappearing and becoming a small dark region from which nothing can escape—a literal black hole in space. Theory indicates that this is the likely fate of stars having more than about 20 to 25 times the mass of the Sun.

22.6 Einstein's Theories of Relativity

The objects we have been studying in this and the last few chapters have taken us far beyond the scope of Newtonian mechanics and gravitation discussed in Chapter 2. ∞ (Sec. 2.8) Now, in the face of extreme states of matter, speeds comparable to that of light, and gravitational fields so intense that not even light can escape, these "workhorse" theories must give way to more refined tools. These tools are the theories of *special* and *general relativity.*

Special Relativity

By the latter part of the 19th century, physicists were well aware of the special status of the speed of light, *c*. It was, they knew, the speed at which all electromagnetic waves traveled, and, as best they could tell, it represented an upper limit on the speeds of *all* known particles. Scientists struggled without success to construct a theory of mechanics and radiation in which *c* was a natural speed limit.

In 1887, a fundamental experiment carried out by the American physicists A. A. Michelson and E. W. Morley compounded theorists' problems further by demonstrating another important and unique aspect of light: The measured speed of a beam of light is *independent* of the motion of either the observer or the source (see *Discovery 22-1*). No matter what our motion may be relative to the source of the radiation, we always measure precisely the same value for *c*: 299,792.458 km/s.

A moment's thought tells us that this is a decidedly nonintuitive statement. For example, if we were traveling in a car moving at 100 km/h and we fired a bullet forward with a speed of 1000 km/h relative to the car, an observer standing at the side of the road would see the bullet pass by at $100 + 1000 = 1100$ km/h, as illustrated in Figure 22.14(a). However, the Michelson–Morley experiment tells us that if we were traveling in a rocket ship at one-tenth the speed of light, 0.1*c*, and we shone a searchlight beam ahead of us (Figure 22.14b), an outside observer would measure the speed of the beam not as 1.1*c*, as the example of the bullet would suggest, but as *c*. The rules that apply to particles moving at or near the speed of light are different from those we are used to in everyday life.

The **special theory of relativity** (or just *special relativity*) was proposed by Einstein in 1905 to deal with the preferred status of the speed of light. The theory is the mathematical framework that allows us to extend the familiar laws of physics from low speeds (i.e., speeds much less than *c*, which are often referred to as *nonrelativistic*) to very high (or *relativistic*) speeds, comparable to *c*. The essential features of the theory as follows:

1. The speed of light, *c*, is the maximum possible speed in the universe, and all observers measure the same value for *c*, regardless of their motion. Einstein broadened this statement into the *principle of relativity*: The basic laws of physics are *the same* to all unaccelerated observers.

2. There is no absolute frame of reference in the universe; that is, there is no "preferred" observer relative to whom all other velocities can be measured. Put another way, there is no way to tell who is moving and who is not. Instead, only relative velocities between observers matter (hence the term "relativity").

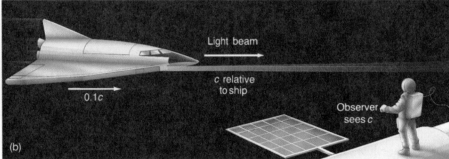

▲ **FIGURE 22.14** **Speed of Light** (a) A bullet fired from a speeding car would be measured by an outside observer to have a speed equal to the sum of the speeds of the car and of the bullet. (b) A beam of light shining forward from a high-speed spacecraft would still be observed to have speed *c*, regardless of the speed of the spacecraft. The speed of light is thus independent of the speed of the source or of the observer.

3.   Neither space nor time can be considered independently of one another. Rather, they are each components of a single entity: **spacetime.** There is no absolute, universal time—observers' clocks tick at different rates, depending on the observers' motions relative to one another.

Special relativity is equivalent to Newtonian mechanics in describing objects that move much more slowly than the speed of light, but it differs greatly in its predictions at relativistic velocities. (See *Discovery 22-1*.) Yet, despite their often nonintuitive nature, all of the theory's predictions have been repeatedly verified to a high degree of accuracy. Today, special relativity lies at the heart of modern science. No scientist seriously doubts its validity.

General Relativity

Einstein's special theory of relativity is cast in terms of frames of reference ("observers") moving at *constant* speeds with respect to one another. In constructing his theory, Einstein rewrote the laws of motion expounded by Newton more than two centuries previously. ∞ (Sec. 2.7) But Newton's other great legacy—the theory of gravitation—does not deal with observers moving at constant relative velocities. Rather, gravity causes observers to *accelerate* relative to one another, making for a much more complex mathematical problem. Fitting gravity into special relativity took Einstein another decade. The result once again overturned scientists' conception of the universe.

In 1915, Einstein illustrated the connection between special relativity and gravity with the following famous "thought experiment." Imagine that you are enclosed in an elevator with no windows, so that you cannot directly observe the outside world, and the elevator is floating in space. You are weightless. Now suppose that you begin to feel the floor press up against your feet. Weight has apparently returned. There are two possible explanations for this, as shown in Figure 22.15. A large mass could have come nearby, and you are feeling its downward gravitational attraction (Figure 22.15a), *or* the elevator has begun to accelerate upward, and the force you feel is that exerted by the elevator as it accelerates

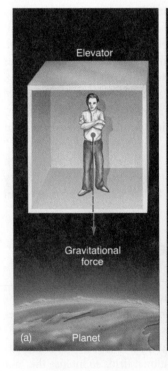

▶ **FIGURE 22.15** **Einstein's Elevator** A passenger in a closed elevator floating in space feels weight return—he feels a force exerted on his feet by the elevator floor. Einstein reasoned that no experiment conducted entirely within the elevator can tell the passenger whether the force is (a) due to the gravity of a nearby massive object or (b) caused by the acceleration of the elevator itself.

DISCOVERY 22-1

Special Relativity

In 1887 the Michelson–Morley experiment attempted to determine Earth's motion relative to the "absolute" space through which light supposedly moved. As illustrated in the first figure, Michelson and Morley expected the measured speed of a beam of light to change as their equipment moved due to Earth's rotation and orbit around the Sun—faster when the beam was moving opposite to Earth's motion (to the left in the figure) and slower when Earth was "catching up" on it (to the right). In fact, they measured precisely the *same* speed of light for any orientation of their apparatus. This meant either that Earth was not moving through space—which conflicts with the fact that we see stellar parallax—or that Newtonian thinking and human intuition somehow go awry when light is involved. Far from measuring the properties of absolute space, the Michelson–Morley experiment ultimately demolished the entire concept—and with it, the 19th-century view of the universe.

With the special theory of relativity, Einstein explained the Michelson–Morley experiment and elevated the speed of light to the status of a constant of nature. He rewrote the laws of mechanics to reflect that new fact and thereby opened the door to a flood of new physics and a much deeper understanding of the universe. But many commonsense ideas had to be abandoned in the process and replaced with some decidedly less intuitive concepts.

Imagine that you are an observer watching a rocket ship fly past at relative velocity v and that the craft is close enough for you to make detailed observations inside its cabin. If v is much less than the speed of light, c, you would see nothing out of the ordinary—special relativity is consistent with familiar Newtonian mechanics at low velocities. As the ship's velocity increases, however, you begin to notice that it appears to *contract* in the direction in which it is moving. A meterstick on board, identical at launch to the one in your laboratory, is now shorter than its twin. This is called **Lorentz contraction** (or Lorentz–Fitzgerald contraction). The graph shows the stick's measured length aboard the moving ship: At low speeds (bottom) the meterstick measures 1 meter, but at high speeds (top) the stick is shortened considerably. A meterstick moving at 90 percent of the speed of light would shrink to a little less than half a meter.

At the same time, the ship's clock, synchronized prior to launch with your own, now ticks *more slowly*. This phenomenon, known as *time dilation*, has been observed many times in laboratory experiments in which fast-moving radioactive particles are observed to decay more slowly than if they were at rest in the lab. Their internal clocks—their half-lives—are slowed by their rapid motion. ∞ *(More Precisely 7-2)* Although no material particle can actually reach the speed of light, Einstein's theory implies that, as approaches c, the measured length of the meterstick will shrink to nearly zero and the clock will slow to a virtual stop.

Of course, from the point of view of an astronaut on board the spaceship, *you* are the one moving rapidly. Seen from the ship, *you* appear to be compressed in the direction of motion, and *your* clock runs slowly! How can this be? The answer is that, in relativity, the familiar concept of *simultaneity*—the idea that two events happen "at the same time"—is no longer well defined, but depends on the observer.

When measuring the length of the moving meterstick, you note the positions of the two ends *at the same time, according to your clock*. But those two events—the two measurements—do *not* occur at the same time as seen by the astronaut on the spaceship. From her viewpoint, your measurement of the leading end of the meterstick occurs *before* your measurement of the trailing end, resulting in the Lorentz contraction you observe. A similar argument applies to measurements of time, such as the period between two clock ticks. Time dilation occurs because the measurements occur at the same location and different times in one frame, but at *different places and times* in the other.

Further experimentation would show that the *mass* of the rocket ship also rises as the ship accelerates, becoming nearly infinite as the ship's speed approaches that of light. Finally, perhaps the best-known prediction of special relativity is that the rocket ship's *energy* and mass are proportional to one another, connected by the famous equation $E = mc^2$.

Einstein's revolutionary ideas required physicists to abandon some long-held, cherished, and "obvious" facts about the universe. Perhaps not surprisingly, they encountered initial opposition from many of Einstein's colleagues, but the gain in scientific understanding soon overcame the price in unfamiliarity. Within just a few years, special relativity had become almost universally accepted, and Einstein was on his way to becoming the best-known scientist on the planet.

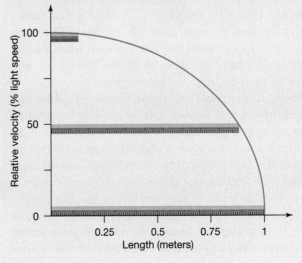

you at the same rate (Figure 22.15b). The crux of Einstein's argument is this: There is *no* experiment that you can perform within the elevator (without looking outside) that will let you distinguish between these two possibilities.

Thus, Einstein reasoned, there is no way to tell the difference between a gravitational field and an accelerated frame of reference (such as the rising elevator in the thought experiment). This statement is known more formally as the **equivalence principle.** Using it, Einstein set about incorporating gravity into special relativity as a general acceleration of all particles. However, he found that another major modification to the theory of special relativity had to be made. As we have just seen, a central concept in relativity is the notion that space and time are not separate quantities, but instead must be treated as a single entity, spacetime. To incorporate the effects of gravity, the mathematics forced Einstein to the unavoidable conclusion that spacetime had to be *curved.* The resulting theory, the result of including gravity within the framework of special relativity, is called **general relativity.**

The central concept of general relativity is this: Matter—all matter—tends to "warp" or curve space in its vicinity. Objects such as planets and stars react to this warping by changing their paths. In the Newtonian view of gravity, particles move on curved trajectories because they are acted upon by a gravitational force. ∞ (Sec. 2.7) In Einsteinian relativity, those same particles move on curved trajectories because they are falling freely through space, following the curvature of spacetime produced by some nearby massive object. The more the mass, the greater is the warping. Thus, in general relativity, there is no such thing as a "gravitational force" in the Newtonian sense. Objects move as they do because they follow the curvature of spacetime, which is determined by the amount of matter present. Stated more loosely, as summed up by the reknowned physicist John Archibald Wheeler, "Spacetime tells matter how to move, and matter tells spacetime how to curve."

Some props may help you visualize these ideas. Bear in mind, however, that these props are not real, but only tools to help you grasp some exceedingly strange concepts. Imagine a pool table with the tabletop made of a thin rubber sheet rather than the usual hard felt. As Figure 22.16 suggests, such a rubber sheet becomes distorted when a heavy weight (e.g., a rock) is placed on it. The heavier the rock (Figure 22.16a), the larger is the distortion.

Trying to play pool on this table, you would quickly find that balls passing near the rock were deflected by the curvature of the tabletop (Figure 22.16b). The pool balls are not attracted to the rock in any way; rather, they respond to the curvature of the sheet produced by the rock's presence. In much the same way, anything that moves through space—matter *or radiation*—is deflected by the curvature of spacetime near a star. For example, Earth's orbital path is the trajectory that results as our planet falls freely in the relatively gentle curvature of space created by our Sun. When the curvature is small

(i.e., gravity is weak), both Einstein and Newton predict the same orbit—the one we observe. However, as the gravitating mass increases, the two theories begin to diverge.

Curved Space and Black Holes

Modern notions about black holes rest squarely on the general theory of relativity. Although white dwarfs and (to a lesser extent) neutron stars can be adequately described by the classical Newtonian theory of gravity, only the modern Einsteinian theory of relativity can properly account for the bizarre physical properties of black holes.

In Figure 22.16, we saw how the distortion of space (the rubber sheet in our analogy) increases as the mass of the object causing the distortion increases. In these terms, a *black hole* is a region of space where the gravitational field becomes overwhelming and the curvature of space extreme. At the event horizon itself, the curvature is so great that

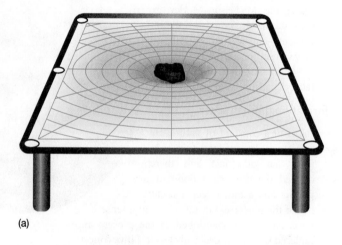

(a)

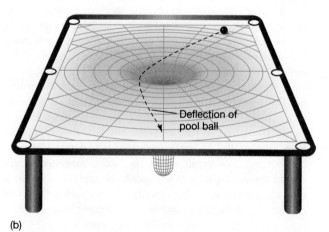

(b)

▲ **Interactive FIGURE 22.16 Curved Space** (a) A pool table made of a thin rubber sheet sags when a weight is placed on it.
(MA) Likewise, space is bent, or warped, in the vicinity of any massive object. (b) As the weight increases, so does the warping of space. A ball shown rolling across the table is deflected by the curvature of the surface, in much the same way as a planet's curved orbit is determined by the curvature of spacetime produced by the Sun.

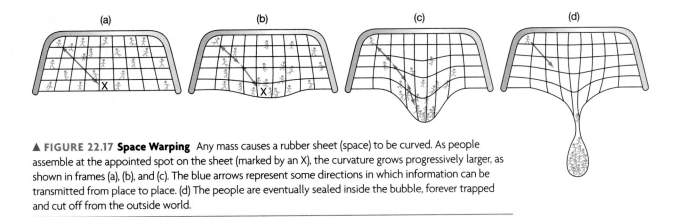

▲ FIGURE 22.17 **Space Warping** Any mass causes a rubber sheet (space) to be curved. As people assemble at the appointed spot on the sheet (marked by an X), the curvature grows progressively larger, as shown in frames (a), (b), and (c). The blue arrows represent some directions in which information can be transmitted from place to place. (d) The people are eventually sealed inside the bubble, forever trapped and cut off from the outside world.

space "folds over" on itself, causing objects within to become trapped and disappear.

Let's consider another analogy. Imagine a large extended family of people living on a huge rubber sheet—a sort of gigantic trampoline. Deciding to hold a reunion, they converge on a given place at a given time. As shown in Figure 22.17, one person remains behind, not wishing to attend. She keeps in touch with her relatives by means of "message balls" rolled out to her (and back from her) along the surface of the sheet. These message balls are the analog of radiation carrying information through space.

As the people converge, the rubber sheet sags more and more. Their accumulating mass creates an increasing amount of space curvature. The message balls can still reach the lone person far away in nearly flat space, but they arrive less frequently as the sheet becomes more and more warped and stretched—as shown in Figures 22.17(b) and (c)—and the balls have to climb out of a deeper and deeper well. Finally, when enough people have arrived at the appointed spot, the mass becomes too great for the rubber to support them. As illustrated in Figure 22.17(d), the sheet pinches off into a "bubble," compressing the people into oblivion and severing their communications with the lone survivor outside. This final stage represents the formation of an event horizon around the reunion party.

Right up to the end—the pinching off of the bubble— two-way communication is possible. Message balls can reach the outside from within (but at a slower and slower rate as the rubber stretches), and messages from outside can get in without difficulty. However, once the event horizon (the bubble) forms, balls from the outside can still fall in, but they can no longer be sent back out to the person left behind, no matter how fast they are rolled. They cannot make it past the "lip" of the bubble in Figure 22.17(d). This analogy (very) roughly depicts how a black hole warps space completely around on itself, isolating its interior from the rest of the universe. The essential ideas—the slowing down and eventual cessation of outward-going signals and the one-way nature of the event horizon once it forms—all have parallels in the case of stellar black holes.

PROCESS OF SCIENCE CHECK

✔ How do Newton's and Einstein's theories differ in their descriptions of gravity?

22.7 Space Travel Near Black Holes

Black holes are *not* cosmic vacuum cleaners. They don't cruise around interstellar space, sucking up everything in sight. The orbit of an object near a black hole is basically the same as its orbit near a star of the same mass. Only if the object happens to pass within a few Schwarzschild radii (perhaps 50 or 100 km for a typical 5- to 10-solar-mass black hole formed in a supernova) of the event horizon is there any significant difference between its actual orbit and the one predicted by Newtonian gravity and described by Kepler's laws. Of course, if some matter does happen to fall into a black hole—if the object's orbit happens to take it too close to the event horizon—it will be unable to get out. Black holes are like turnstiles, permitting matter to flow in only one direction: inward.

Because a black hole will accrete at least a little material from its surroundings, its mass, and hence also the radius of its event horizon, tends to increase slowly over time.

Tidal Forces

Matter flowing into a black hole is subject to great tidal stress. An unfortunate person falling feet first into a solar-mass black hole would find himself stretched enormously in height and squeezed unmercifully laterally. He would be torn apart even before he reached the event horizon, for the pull of gravity would be much stronger at his feet (which are closer to the hole) than at his head. The tidal forces at work in and near a black hole are the same basic phenomenon that is responsible for ocean tides on Earth and the spectacular volcanoes on Io. The only difference is that the tidal forces near a black hole are far stronger than any other force we know in the solar system.

As illustrated (with some artistic license) in Figure 22.18, a similar fate awaits any kind of matter falling into a black hole. Whatever falls in—gas, people, space probes—is vertically stretched and horizontally squeezed and accelerated to high speeds in the process. The net result of all this stretching and squeezing is numerous, and violent collisions among the torn-up debris cause a great deal of frictional heating of the infalling matter. Material is simultaneously torn apart and heated to high temperatures as it plunges into the hole.

So efficient is the heating that, before reaching the hole's event horizon, matter falling into the hole emits radiation of its own accord. For a black hole of mass comparable to the Sun, the energy is expected to be emitted in the form of X rays. In effect, the gravitational energy of matter outside the black hole is converted into heat as that matter falls toward the hole. Thus, contrary to what we might expect from an object whose defining property is that nothing can escape from it, the region surrounding a black hole is expected to be a *source* of energy. Of course, once the hot matter falls below the event horizon, its radiation is no longer detectable—it never leaves the hole.

Approaching the Event Horizon

One safe way to study a black hole would be to go into orbit around it well beyond the disruptive influence of the hole's strong tidal forces. After all, Earth and the other planets of our solar system orbit the Sun without falling into it and

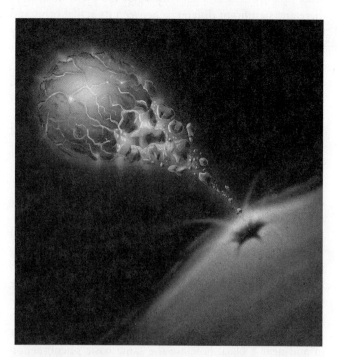

▲ FIGURE 22.18 **Black-Hole Heating** Any matter falling into the clutches of a black hole will become severely distorted and heated. This sketch shows an imaginary planet being pulled apart by a black hole's gravitational tides.

without being torn apart. The gravity field around a black hole is basically no different; however, even from a stable circular orbit, a close investigation of the hole would be unsafe for humans. Endurance tests conducted on astronauts of the United States and the former Soviet Union indicate that the human body cannot withstand stress greater than about 10–20 times the pull of gravity on Earth's surface. This breaking point would occur about 3000 km from a 10-solar-mass black hole (which, recall, would have a 30-km event horizon). Closer than that, the tidal effect of the hole would tear a human body apart.

Let's instead send an imaginary indestructible astronaut—a mechanical robot, say—in a probe toward the center of the hole. Watching from a safe distance in our orbiting spacecraft, we can then examine the nature of space and time near the hole. Our robot will be a useful explorer of theoretical ideas, at least down to the event horizon. After that boundary is crossed, there is no way for the robot to return any information about its findings.

Suppose, for example, our robot has an accurate clock and a light source of known frequency mounted on it. From our safe vantage point far outside the event horizon, we could use telescopes to read the clock and measure the frequency of the light we receive. What might we discover? We would find that the light from the robot would become more and more redshifted as the robot neared the event horizon. Even if the robot used rocket engines to remain motionless, a redshift would still be detected. The redshift is not caused by motion of the light source, nor is it the result of the Doppler effect arising as the robot falls into the hole. Rather, it is a redshift induced by the black hole's gravitational field, predicted by Einstein's general theory of relativity and known as **gravitational redshift.**

We can explain gravitational redshift as follows: According to general relativity, photons are attracted by gravity. As a result, in order to escape from a source of gravity, photons must expend some energy. They have to do work to get out of the gravitational field. They don't slow down at all—photons always move at the speed of light—they just lose energy. Because a photon's energy is proportional to the frequency of its radiation, light that loses energy must have its frequency reduced (or, equivalently, its wavelength lengthened). In other words, as illustrated in Figure 22.19, radiation coming from the vicinity of a massive object will be redshifted to a degree depending on the strength of the object's gravitational field.

As photons traveled from the robot's light source to the orbiting spacecraft, they would become gravitationally redshifted. From our standpoint on the orbiting spacecraft, a green light, say, would become yellow and then red as the robot astronaut neared the black hole. From the robot's perspective, the light would remain green. As the robot got closer to the event horizon, the radiation from its light source would become undetectable with optical telescopes. The radiation reaching us in the orbiting spacecraft would by then be lengthened so much that infrared and then radio telescopes would be needed to detect it. When the robot probe got closer still to the event horizon, the radiation it emitted as visible light would be shifted to wavelengths even longer than conventional radio waves by the time it reached us.

Light emitted *from the event horizon itself* would be gravitationally redshifted to infinitely long wavelengths. In other words, each photon would use all its energy trying to escape from the edge of the hole. What was once light (on the robot) would have no energy left upon its arrival at the safely orbiting spacecraft. Theoretically, this radiation would reach us—still moving at the speed of light—but with zero energy. Thus, the light radiation originally emitted would be redshifted beyond our perception.

Now, what about the robot's clock? Assuming that we could read it, what time would it tell? Would there be any observable change in the rate at which the clock ticked as it moved deeper into the hole's gravitational field? From the safely orbiting spacecraft, we would find that any clock close to the hole would appear to tick more *slowly* than an equivalent

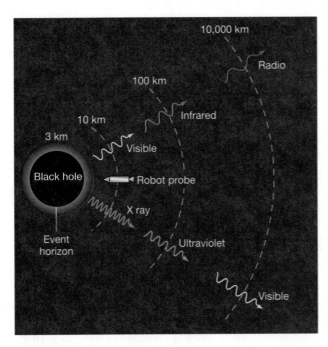

▲ **FIGURE 22.19 Gravitational Redshift** As a photon escapes from the strong gravitational field close to a black hole, it must expend energy to overcome the hole's gravity. This energy does not come from a change in the speed at which the photon travels. (That speed is always 300,000 km/s, even under these extreme conditions.) Rather, the photon "gives up" energy by increasing its wavelength. Thus, the photon's frequency changes, and the photon is (red)shifted into a less energetic region of the spectrum. This figure shows the effect on two beams of radiation, one of visible light and one of X rays, emitted from a space probe as it nears the event horizon of a 1-solar-mass black hole. (Note that the light is *not* coming from within the black hole itself.) The beams are shifted to longer and longer wavelengths as they move farther from the event horizon.

clock on board the spacecraft. The closer the clock came to the hole, the slower it would appear to run. On reaching the event horizon, the clock would seem to stop altogether. It would be as if the robot astronaut had found immortality! All action would become virtually frozen in time. Consequently, an external observer would never actually witness an infalling astronaut sink below the event horizon. Such a process would appear to take forever.

This apparent slowing down of the robot's clock is known as **time dilation.** It is another clear prediction of general relativity and in fact is closely related to the gravitational redshift. To see this connection, imagine that we use our light source as a clock, with the passage of (say) a wave crest constituting a "tick." The clock thus ticks at the frequency of the radiation. As the wave is redshifted, the frequency drops, and fewer wave crests pass the distant observer each second—the clock appears to slow down. This thought experiment demonstrates that the redshift of the radiation and the slowing of the clock are one and the same.

From the point of view of the indestructible robot, however, relativity theory predicts no strange effects at all. To the infalling robot, the light source hasn't reddened, and the clock keeps perfect time. In the robot's frame of reference, everything is normal. Nothing prohibits it from coming within the Schwarzschild radius of the hole. No law of physics constrains an object from passing through an event horizon. There is no barrier at the event horizon and no sudden lurch as it is crossed; it is only an imaginary boundary in space. Travelers passing through the event horizon of a sufficiently massive hole (such as might lurk in the heart of our own Galaxy, as we will see) might not even know it—at least until they tried to get out!

The gravitational fields of most astronomical objects are far too weak to produce any significant gravitational redshift, although in many cases the effect can still be measured. Delicate laboratory experiments on Earth and on satellites in near-Earth orbits have succeeded in detecting the tiny gravitational redshift produced by even our own planet's weak gravity. Sunlight is redshifted by only about a thousandth of a nanometer. A few white-dwarf stars do show some significant gravitational reddening of their emitted light, however. Their radii are much smaller than that of our Sun, so their surface gravity is very much stronger than the Sun's. Neutron stars should show a substantial shift in their radiation, but it is difficult to disentangle the effects of gravity, magnetism, and environment on the signals we observe.

Deep Down Inside

No doubt you are wondering what lies within the event horizon of a black hole. The answer is simple: No one really knows. However, the question is of great interest to theorists, as it raises some fundamental issues that lie at the forefront of modern physics.

Can an entire star simply shrink to a point and vanish? General relativity predicts that, without some agent to compete with gravity, the core remnant of a high-mass star will collapse all the way to a point at which both its density and its gravitational field become infinite. Such a point is called a **singularity.** We should not take this prediction of infinite density too literally, however. Singularities are not physical—rather, they always signal the breakdown of the theory producing them. In other words, the present laws of physics are simply inadequate to describe the final moments of a star's collapse.

As it stands today, the theory of gravity is incomplete because it does not incorporate a proper (i.e., a quantum mechanical) description of matter on very small scales. As our collapsing stellar core shrinks to smaller and smaller radii, we eventually lose our ability even to describe, let alone predict, its behavior. Perhaps matter trapped in a black hole never actually reaches a singularity. Perhaps it just approaches this bizarre state in a manner that we will someday understand as the subject of *quantum gravity*—the merger of general relativity with quantum mechanics—develops.

Having said that, we can at least estimate how small the core can get before current theory fails. It turns out that by the time that stage is reached, the core is already much smaller than any elementary particle. Thus, although a complete description of the end point of stellar collapse may well require a major overhaul of the laws of physics, for all practical purposes the prediction of collapse to a point is valid. Even if a new theory somehow succeeds in doing away with the central singularity, it is unlikely that the external appearance of the hole or the existence of its event horizon will change. Any modifications to general relativity are expected to occur only on submicroscopic scales, not on the macroscopic (kilometer-sized) scale of the Schwarzschild radius.

Singularities are places where the rules break down, and some very strange things may occur near them. Many possibilities have been envisaged—gateways into other universes, time travel, the creation of new states of matter— but none of them has been proved, and certainly none of them has ever been observed. Because these regions are places where science fails, their presence causes serious problems for many of our cherished laws of physics, from causality (the idea that cause should precede effect, which runs into immediate and severe problems if time travel is possible) to energy conservation (which is violated if material can hop from one universe to another through a black hole). It is currently unclear whether the removal of the central singularity by some future all-encompassing theory would necessarily also eliminate all of these problematic side effects.

Disturbed by the possibility of such chaos in science, some researchers have even proposed a "principle of cosmic censorship": Nature always hides *any* singularity, such as

that found at the center of a black hole, inside an event horizon. In that case, even though physics fails, its breakdown cannot affect us outside, so we are safely insulated from any effects the singularity may have. What would happen if we one day found a so-called *naked singularity* somewhere—a singularity uncloaked by an event horizon? Would relativity theory still hold there? For now, we just don't know.

What sense are we to make of black holes? Do black holes and all the strange phenomena that occur in and around them really exist? The basis for understanding these weird objects is the relativistic concept that mass warps spacetime—which has already been found to be a good representation of reality, at least for the weak gravitational fields produced by stars and planets (see *More Precisely 22-1* and *Discovery 22-2*). The larger the concentration of mass, the greater is the spacetime warping and, apparently, the stranger are the observational consequences. These consequences are part and parcel of general relativity, and black holes are one of its most striking predictions. As long as general relativity stands as the correct theory of gravity in the universe, black holes are real.

CONCEPT CHECK

✔ Why would you never actually witness an infalling object crossing the event horizon of a black hole?

22.8 Observational Evidence for Black Holes

Theoretical ideas aside, is there any observational evidence for black holes? Can we prove that these strange invisible objects really do exist?

Stellar Transits?

One way in which we might think we would detect a black hole is if we observed it transiting (passing in front of) a star. Unfortunately, such an event would be extremely hard to see. The approximately 12,000-km-diameter planet Venus is barely noticeable when it transits the Sun, so a 10-km-wide object moving across the image of a faraway star would be completely invisible with either current equipment or any equipment available in the foreseeable future.

Actually, we are even worse off than the previous paragraph suggests. Suppose we were close enough to the star to resolve the disk of the transiting black hole. Then the observable effect would not be a black dot superimposed on a bright background; instead, the background starlight would be deflected as it passed the black hole on its way to Earth, as indicated in Figure 22.20. The effect is the same as the bending of distant starlight around the edge of the Sun, a phenomenon that has been repeatedly measured during solar eclipses

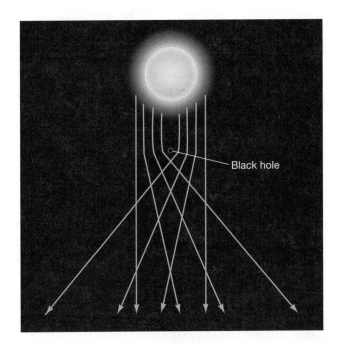

▲ **FIGURE 22.20 Gravitational Light Deflection** The gravitational bending of light around the edges of a small, massive black hole makes it impossible to observe the hole as a black dot superimposed against the bright background of its stellar companion.

Black hole

throughout the last several decades (see *More Precisely 22-1*). With a black hole, much larger deflections would occur. As a result, the image of a black hole in front of a bright companion star would show not a neat, well-defined black dot, but rather a fuzzy image virtually impossible to resolve, even from nearby.

Black Holes in Binary Systems

A much better way to find black holes is to look for their effects on other objects. Our Galaxy harbors many binary-star systems in which only one object can be seen. Recall from our study of binary-star systems in Section 17.7 that we need to observe the motion of only one star to infer the existence of an unseen companion and measure some of its properties. In the majority of cases, the invisible companion is simply small and dim, nothing more than an M-type star hidden in the glare of an O- or B-type partner or perhaps

shrouded by dust or other debris, making it invisible to even the best available equipment. In either case, the invisible object is *not* a black hole.

A few close binary systems, however, have peculiarities suggesting that one of their members may be a black hole. Some of the most interesting observations, made during the 1970s and 1980s by Earth-orbiting satellites, revealed binary systems in which the invisible member emits large amounts of X rays. The mass of the emitting object is measured as several solar masses, so we know that it is not simply a small, dim star. Nor is it likely that visible radiation from the X-ray source is obscured by dusty circumstellar debris—in the cases of interest, intense radiation from the binary components would have dispersed the debris into interstellar space long ago.

One particular binary system drawing much attention lies in the constellation Cygnus. Figure 22.21(a) shows the area of the sky in Cygnus where astronomers have reasonably good evidence for a black hole. The rectangle outlines the celestial system of interest, some 2000 pc from Earth. The black-hole candidate is an X-ray source called Cygnus X-1, studied in detail by the *Uhuru* satellite in the early

▶ **FIGURE 22.21 Cygnus X-1** (a) The brightest star in this photograph is a member of a binary system whose unseen companion, called Cygnus X-1, is a leading black hole candidate. (b) An X-ray image of field of view outlined by the rectangle in part (a). Of course, X rays cannot be seen directly; X rays emitted by the Cygnus X-1 source were captured in space by an X-ray detector aboard a satellite, changed into radio signals for transmission to the ground, and changed again into electronic signals that were then viewed on a video screen, from which this picture was taken. (*Harvard-Smithsonian Center for Astrophysics*)

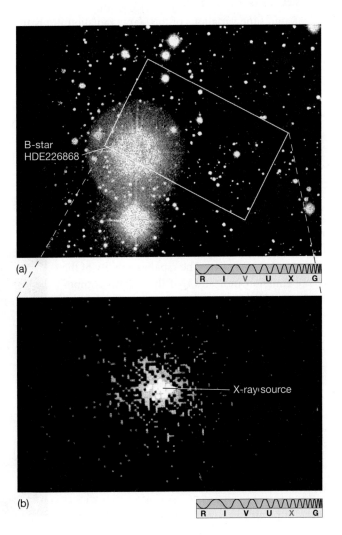

B-star
HDE226868

(a)

R I V U X G

X-ray source

(b)

R I V U X G

1970s. The main observational features of this binary system are as follows:

1. The visible companion of the X-ray source—a blue B-type supergiant with the catalog name HDE 226868—was identified a few years after Cygnus X-1 was discovered. Assuming that the companion lies on the main sequence, we know that its mass must be around 25 times the mass of the Sun.

2. Spectroscopic observations indicate that the binary system has an orbital period of 5.6 days. Combining this information with further spectroscopic measurements of the visible component's orbital speed, astronomers estimate the total mass of the system to be around 35 solar masses, implying that Cygnus X-1 has a mass about 10 times the mass of the Sun. ∞ (Sec. 17.7)

3. Other detailed studies of Doppler-shifted spectral lines suggest that hot gas is flowing from the bright star toward an unseen companion. ∞ (Sec. 4.5)

4. X-ray radiation emitted from the immediate neighborhood of Cygnus X-1 implies the presence of very high temperature gas, perhaps as hot as several million kelvins (see Figure 22.21b).

5. Rapid time variations of this X-ray radiation imply that the size of the X-ray-emitting region of Cygnus X-1 must be extremely small—in fact, less than a few hundred kilometers across. The reasoning is basically the same as in the discussion of gamma-ray bursts in Section 22.4: X rays from Cygnus X-1 have been observed to vary in intensity on time scales as short as a millisecond. For this variation not to be blurred by the travel time of light across the source, Cygnus X-1 cannot be more than 1 light-millisecond, or 300 km, in diameter.

These properties suggest that the invisible X-ray-emitting companion could be a black hole. The X-ray-emitting region is likely an accretion disk formed as matter drawn from the visible star spirals down onto the unseen component. The rapid variability of the X-ray emission indicates that the unseen component must be compact—a neutron star or a black hole. The mass limit of the dark component argues for the latter, for a neutron star's mass cannot exceed about 3 solar masses. Figure 22.22 is

an artist's conception of this intriguing object. Note that most of the gas drawn from the visible star ends up in a doughnut-shaped accretion disk of matter. As the gas flows toward the black hole, it becomes superheated and emits the X rays we observe just before they are trapped forever below the event horizon.

A few other black hole candidates are known. For example, the third X-ray source discovered in the Large Magellanic Cloud—LMC X-3—is an invisible object that, like Cygnus X-1, orbits a bright companion star. Reasoning similar to that applied to Cygnus X-1 suggests that LMC X-3 has 10 times more mass than the Sun—too massive to be anything but a black hole. Similarly, the X-ray binary system A0620–00 contains an invisible compact object of mass 3.8 times that of the Sun. In total, perhaps two dozen known objects in or near our Galaxy may be black holes. Cygnus X-1, LMC X-3, and A0620–00 have the strongest claims.

Black Holes in Galaxies

Perhaps the strongest evidence for black holes comes not from binary systems in our own Galaxy, but from observations of the *centers* of many galaxies, including our own. Using high-resolution observations at wavelengths ranging from radio to ultraviolet, astronomers have found that stars and gas near the centers of many galaxies are moving extremely rapidly, orbiting some very massive, unseen object. Masses inferred from Newton's laws range from millions to billions of times the mass of the Sun. ∞ (*More Precisely 2-2*)

The intense energy emission from the centers of these galaxies and the short-timescale fluctuations in that emission suggest the presence of massive, compact objects. In addition, as in the *radio galaxy* shown in Figure 22.23, these objects are also observed to have extended *jets* reminiscent of—but vastly larger than—those associated with neutron stars and black holes. The leading (and at present, the only)

▶ **Interactive FIGURE 22.22 Black-Hole Binary**
(MA) Observations of X-ray binary systems provide strong evidence for stellar-mass black holes. This is an artist's conception of a binary system (THE BIG PICTURE) containing a large, bright, visible star and an invisible, X-ray-emitting black hole. This painting is based on data obtained from detailed observations of Cygnus X-1. (*L. Chaisson*)

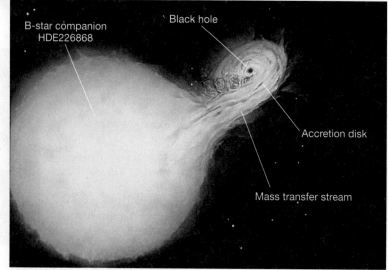

B-star companion HDE226868

Black hole

Accretion disk

Mass transfer stream

ANIMATION/VIDEO Black Hole Geometry

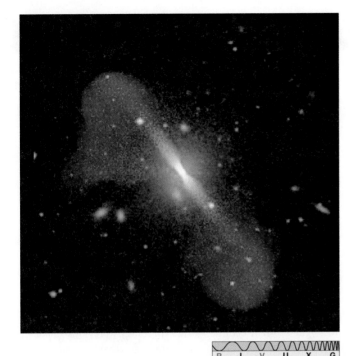

▲ FIGURE 22.23 **Active Galaxy** Many galaxies are thought to harbor massive black holes at their centers. Indeed, the best evidence for black holes anywhere in the universe is found in galactic nuclei. Shown here in false color is the galaxy 3C296. Blue color shows the distribution of stars in the central elliptical galaxy; red shows huge jets of radio emission extending 500,000 light-years across. (See also the opening image of Chapter 24.) *(NRAO)*

explanation is that these energetic objects are powered by huge central **supermassive black holes** accreting stars and gas from their surroundings. The radio-emitting jets are rooted in the parsec-sized accretion disks surrounding the black holes themselves. We will return to the observations, and the question of how these black holes might have formed, in Chapters 23–25.

Thus, astronomers know of "stellar-mass" black holes, comparable in mass to the Sun, and supermassive black holes of millions or billions of solar masses. The former are the result of stellar evolution, as discussed in the last few chapters; the latter have grown in the centers of galaxies, as we will see in Part 4 of the text. Is there anything in between? In 2000, X-ray astronomers reported the first evidence for the long-sought, but elusive, missing link between these two classes of black holes. Figure 22.24 shows an unusual-looking galaxy called M82, currently the site of an intense and widespread burst of star formation (see Chapter 24); the red plumes are winds of hot gas escaping from numerous star-forming sites in the otherwise quiescent part of the galaxy (shown in blue). The inset shows a *Chandra* image of the innermost few thousand parsecs of M82, revealing a number of bright X-ray sources close to—but *not* at—the center of the galaxy. Their spectra and X-ray luminosities suggest that some may be accreting compact objects with masses ranging from 100 to almost 1000 times the mass of the Sun. If confirmed, they will be the first *intermediate-mass black holes* ever observed.

Too large to be remnants of normal stars and too small to warrant the "supermassive" label, these objects present a puzzle to astronomers. Where did they come from? One possible origin is suggested by follow-up infrared observations from the Subaru and Keck telescopes on Mauna Kea, indicating that some of the X-ray sources are apparently associated with dense, young star clusters. ∞ (Secs. 5.2, 19.6) Theorists speculate that collisions between high-mass main-sequence stars in the congested cores of such clusters could lead to the runaway growth of extremely massive and highly unstable stars, which could then collapse to form intermediate-mass

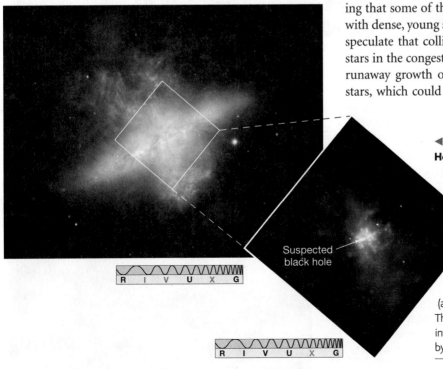

Suspected black hole

◀ FIGURE 22.24 **Intermediate-Mass Black Holes?** X-ray observations (inset below) of the interior of the starburst galaxy M82 (at top, about 100,000 light-years across and 12 million light-years away) reveal a collection of bright sources thought to be the result of matter accreting onto intermediate-mass black holes. The black holes are probably young, have masses between 100 and 1000 times the mass of the Sun, and lie relatively far (about 600 light-years) from the center of M82. The brightest (and possibly most massive) intermediate-mass black hole candidate is marked by an arrow. *(Subaru; NASA)*

MORE PRECISELY 22-1

Tests of General Relativity

Special relativity is the most thoroughly tested and most accurately verified theory in the history of science. General relativity, however, is on somewhat less-firm experimental ground.

The problem with verifying general relativity is that its effects on Earth and in the solar system—the places where we can most easily perform tests—are very small. Just as special relativity produces major departures from Newtonian mechanics only when velocities approach the speed of light, general relativity predicts large departures from Newtonian gravity only when extremely *strong* gravitational fields are involved—effect, when orbital speeds and escape velocities become relativistic.

We will encounter other experimental and observational tests of general relativity elsewhere in this chapter. (See *Discovery 22-2.*) Here, we consider just two "classical" tests of the theory—the deflection of light by the Sun and the effect of relativity on the orbit of Mercury. These tests are solar system observations that helped ensure the acceptance of Einstein's theory. Later, more accurate measurements confirmed and strengthened the test results. Bear in mind, however, that there are currently no tests of general relativity in the "strong-field" regime—that part of the theory that predicts black holes, for example—so the full theory has never been experimentally tested. Scientists hope that the experiments described in *Discovery 22-2* will be able to test that part of the theory.

At the heart of general relativity is the premise that everything, including light, is affected by gravity because of the curvature of spacetime. Shortly after he published his theory in 1915, Einstein noted that light from a star should be deflected by a measurable amount as it passes the Sun. According to the theory, the closer to the Sun the light comes, the more it is deflected. Thus, the maximum deflection should occur for a ray that just grazes the solar surface. Einstein calculated that the deflection angle should be 1.75"—a small, but detectable, amount. Of course, it is normally impossible to see stars close to the Sun. However, during a solar eclipse, when the Moon blocks the Sun's light, the observation becomes possible, as illustrated in the first (highly exaggerated) figure.

In 1919, a team of observers led by the British astronomer Sir Arthur Eddington succeeded in measuring the deflection of starlight during an eclipse. The results were in excellent agreement with the prediction of general relativity. Virtually overnight Einstein became world famous. His previous major

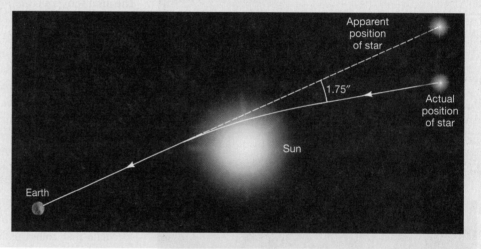

black holes. Figure 22.25 shows the current best candidate for a "nearby" star cluster harboring an intermediate-mass black hole—the globular cluster G1 in the Andromeda Galaxy. ∞ (Sec. 3.1) The peculiar orbits of stars near the center of this cluster suggest a black hole of mass 20,000 times that of the Sun, and observations of the cluster in both radio and X rays are consistent with theoretical expectations of the emission from such a massive object in the cluster's core. However, both theory and observations remain controversial, and further systematic studies of many other galaxies will be needed to resolve the issue.

▶ FIGURE 22.25 **Black-Hole Host?** Astronomers have found that stars near the center of the massive globular cluster G1 do not move as expected if the cluster's mass is distributed like its light. The observations suggest that an intermediate-mass black hole may reside at the cluster's center. *(NASA)*

accomplishments notwithstanding, this single prediction assured him a permanent position as the best-known scientist on Earth! The high-precision *Hipparcos* satellite observed shifts in the apparent positions of many stars, even those whose line of sight is far from the Sun. ∞ (Sec. 17.1) The shifts are exactly as predicted by Einstein's theory.

EXAMPLE According to general relativity, a beam of light passing an object of mass M at distance R is deflected through an angle (in radians) of $4GM/Rc^2$, where $G = 6.67 \times 10^{-11}$ N m²/kg² is the gravitational constant and $c = 3.00 \times 10^8$ m/s is the speed of light. Putting in the numbers for the Sun, we obtain $M = 1.99 \times 10^{30}$ kg and $R = 696,000$ km, and remembering that 1 radian = 57.3°; we can calculate the deflection to be $(4 \times 6.67 \times 10^{-11} \times 1.99 \times 10^{30})/(6.96 \times 10^8 \times [3.00 \times 10^8]^2) \times 57.3$ (degrees per radian) $\times 3600$ (arc seconds per degree) = 1.75", as previously stated. ∞ *(More Precisely 1-2)* In more convenient units, we can write

$$\text{deflection (are seconds)} = 1.75 \frac{M \,(\text{solar masses})}{R \,(\text{solar radii})}$$

Note that the deflection is proportional to the mass M and inversely proportional to the distance R. Thus, Earth, with mass $M = 3.0 \times 10^{-6}$ and radius $R = 9.2 \times 10^{-3}$ solar units, would produce a deflection of just 0.57 milli-arc second (thousandths of an arc second), whereas a white dwarf such as Sirius B, with $M = 1.1$ and $R = 0.0073$ in the same units would deflect the beam by 4.4 arc minutes. ∞ (Sec. 20.3) (Neutron stars and black holes produce even greater effects, but the preceding simple formula is valid only when the deflection is small—less than a few degrees.)

The second prediction of general relativity testable within the solar system is that planetary orbits deviate slightly from the perfect ellipses of Kepler's laws. Again, the effect is greatest where gravity is strongest—that is, closest to the Sun. Thus, the largest relativistic effects are found in the orbit of Mercury. Relativity predicts that Mercury's orbit is not a closed ellipse. Instead, its orbit should rotate slowly, as shown in the second (again exaggerated) diagram. The amount of rotation is very small—only 43" per century—but Mercury's orbit is so well charted that even this tiny effect is measurable.

In fact, the observed rotation rate is 540" per century, much greater than that predicted by relativity. However, when other (nonrelativistic) gravitational influences—primarily the perturbations due to the other planets—are taken into account, the rotation is in complete agreement with the foregoing prediction.

Do Black Holes Exist?

You may have noticed that the identification of an object as a black hole really proceeds by elimination. Loosely stated, the argument goes as follows: "Object X is compact and very massive. We don't know of anything else that can be that small and that massive. Therefore, object X is a black hole." For the very massive compact objects observed (or inferred to be) in the centers of galaxies, the absence of viable alternatives means that the black-hole hypothesis has become widely accepted among astronomers. However, Cygnus X-1 and the other suspected stellar-mass black holes in binary systems all have masses relatively close to the dividing line separating neutron stars from black holes. Given the present uncertainties in both observation and theory, might they conceivably be merely dim, dense neutron stars and not black holes at all?

Most astronomers do not think this likely but it highlights a problem. It is difficult to unambiguously distinguish a 10-solar-mass black hole from, say, a 10-solar-mass neutron star (if one could somehow exist). Both objects would affect a companion star's orbit in the same way; both would tear mass from its surface, and both would form an accretion disk around themselves that would emit intense X rays (although many researchers think that the accretion disks may differ sufficiently in detail that the nature of the central object might be identifiable from observations).

We have stressed throughout this text that scientific theories unsupported by observational or experimental evidence are destined not to survive. ∞ (Sec. 1.2) Black holes are a clear prediction of Einstein's general theory of relativity, which is widely regarded as the correct description of gravity in the presence of strong fields and orbital speeds comparable to the speed of light. But we have also seen that general relativity

DISCOVERY 22-2

Gravity Waves: A New Window on the Universe

Electromagnetic waves are common everyday phenomena. They involve periodic changes in the strengths of electric and magnetic fields. ∞ (Sec. 3.2) Electromagnetic waves move through space and transport energy. Any accelerating charged particle, such as an electron in a broadcasting antenna or on the surface of a star, generates electromagnetic waves.

The modern theory of gravity—Einstein's theory of relativity—also predicts waves that move through space. A *gravity wave* is the gravitational counterpart of an electromagnetic wave. *Gravitational radiation* results from changes in the strength of a gravitational field. Any time an object of any mass accelerates, a gravity wave should be emitted at the speed of light. The wave should produce small distortions in the space through which it passes. Gravity is an exceedingly weak force compared with electromagnetism, so these distortions are expected to be very small—in fact, much smaller than the diameter of an atomic nucleus. ∞ (More Precisely 16-1) Yet many researchers think that these tiny distortions are measurable.

The objects most likely to produce gravity waves detectable on Earth are close binary systems containing black holes, neutron stars, or white dwarfs. As these massive components orbit one another, their acceleration results in rapidly changing gravitational fields and the emission of gravitational radiation. As energy escapes in the form of gravity waves, the two objects spiral toward one another, orbiting more rapidly and emitting even more gravitational radiation. As we saw in Section 22.4, neutron-star mergers may well also be the origin of some gamma-ray bursts, so gravitational radiation might provide an alternative means of studying these violent and mysterious phenomena.

Such a slow but steady decay in the orbit of a binary system has in fact been detected. In 1974, radio astronomer Joseph Taylor and his student Russell Hulse at the University of Massachusetts discovered an unusual binary system. Both components are neutron stars, and one is observable from Earth as a pulsar. This system has become known as the *binary pulsar*. Measurements of the periodic Doppler shift of the pulsar's radiation prove that its orbit is shrinking at exactly the rate predicted by relativity theory if the energy were being carried off by gravity waves. The two neutron stars should merge in an energetic burst of gravitational radiation and gamma rays in less than 300 million years (although most of the radiation will be emitted during the last few seconds). Even though the waves themselves have not been detected, the binary pulsar is regarded by most astronomers as very strong evidence in favor of general relativity. Taylor and Hulse received the 1993 Nobel Prize in physics for their discovery. The first accompanying figure illustrates the scale of, and predicted orbital changes in, the binary pulsar's orbit.

In 2004, radio astronomers announced the discovery of a *double-pulsar* binary system with an even shorter period than the binary pulsar, implying stronger relativistic effects and a shorter merger time—about 85 million years. Because both components

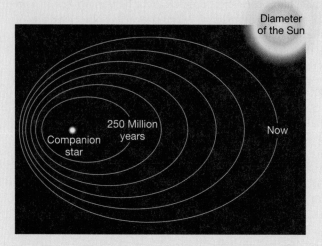

are pulsars, and since the system, by pure luck, also happens to be seen almost exactly edge on by observers on Earth, leading to eclipses, this system has provided a wealth of detailed information on both neutron stars and gravitational physics.

The second figure shows part of an ambitious gravity-wave observatory called *LIGO*—short for *Laser Interferometric Gravity-Wave Observatory*—which became operational in 2003. Twin detectors, one (shown here) in Hanford, Washington, the other in Livingston, Louisiana, use interference between two laser beams to measure the tiny distortions of space produced by gravitational radiation in the lengths of the 4-km-long arms should a gravity wave pass by. ∞ (Discovery 3-1) The instrument is in theory capable of detecting gravity waves from many galactic and extragalactic sources, although, so far, no gravity waves have been detected. NASA has ambitious plans to launch a space-based interferometer millions of kilometers across, sensitive to the long-wavelength gravitational waves produced by supermassive black holes in the hearts of galaxies across the universe. However, its status is currently uncertain, and no launch date has been set.

NSF

has been tested most thoroughly in situations where gravity is weak and velocities are relatively low, and not at all under the extreme conditions expected near a black hole. So we can legitimately ask, "Do we have any unambiguous evidence that the massive, compact objects just described really are black holes?"

The short answer—at least, if measurements of mass and size alone are insufficient to convince you of a black hole's reality—is no. Detailed measurements of black-hole properties are hard to make and even harder to interpret. Black holes tend to live in very messy astrophysical environments. Astronomers have uncovered tantalizing hints of black-hole event horizons (rather than hard neutron star surfaces) in several systems, but none have as yet proved conclusive. However, as technology continues to improve, we can expect many more such observations, with increasing precision, as astronomers test a key prediction of Einstein's theory.

So have black holes really been discovered? Despite the uncertainties, the answer is probably yes. Skepticism is healthy in science, but only the most stubborn astronomers (and some do exist!) would take serious issue with the many lines of theoretical reasoning that support the case for black holes. The crucial role played by black holes in the theories of stellar evolution, gamma-ray bursts, and (as we will see in Chapters 24 and 25) the structure and evolution of galaxies is a clear indication of their widespread acceptance in astronomy.

Can we guarantee that future modifications to the theory of compact objects will not invalidate some or all of our arguments? No, but similar statements could be made in many areas of astronomy—indeed, about any theory in any area of science. We conclude that, strange as they are, black holes have been detected, both in our Galaxy and beyond. Perhaps someday future generations of space travelers will visit Cygnus X-1 or the center of our Galaxy and (carefully!) test these conclusions firsthand. Until then, we will have to continue to rely on improving theoretical models and observational techniques to guide our discussions of the mysterious objects known as black holes.

PROCESS OF SCIENCE CHECK

✔ How do astronomers "see" black holes?

CHAPTER REVIEW

SUMMARY

1 A core-collapse supernova may leave behind a **remnant (p. 540)**—an ultra-compressed ball of material called a **neutron star (p. 540).** Neutron stars are extremely dense and, at formation, are predicted to be very hot, strongly magnetized, and rapidly rotating. They cool down, lose much of their magnetism, and slow down as they age.

2 According to the **lighthouse model (p. 541)**, neutron stars, because they are magnetized and rotating, send regular bursts of electromagnetic energy into space. The beams are produced by charged particles confined by the strong magnetic fields. When we can see the beams from Earth, we call the source a **pulsar (p. 541)**. The pulse period is the rotation period of the neutron star. Because the pulse energy is beamed into space and because neutron stars slow down as they radiate energy into space, not all neutron stars are seen as pulsars.

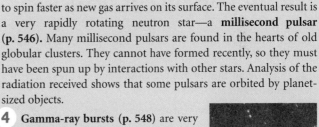

3 A neutron star in a close binary system can draw matter from its companion, forming an accretion disk. The material in the disk heats up before it reaches the neutron star, making the disk a strong source of X rays. As gas builds up on the star's surface, the star eventually becomes hot enough to fuse hydrogen. When hydrogen burning starts, it does so explosively, and an **X-ray burster (p. 544)** results. The rapid rota-

tion of the inner part of the accretion disk causes the neutron star to spin faster as new gas arrives on its surface. The eventual result is a very rapidly rotating neutron star—a **millisecond pulsar (p. 546)**. Many millisecond pulsars are found in the hearts of old globular clusters. They cannot have formed recently, so they must have been spun up by interactions with other stars. Analysis of the radiation received shows that some pulsars are orbited by planet-sized objects.

4 **Gamma-ray bursts (p. 548)** are very energetic flashes of gamma rays observed about once per day, distributed uniformly over the entire sky. In some cases, their distances have been measured, placing them far away from us and implying that they are extremely luminous. The leading theoretical models for these explosions postulate the violent merger of neutron stars in a distant binary system or the recollapse and subsequent violent explosion following a "failed" supernova in a very massive star.

5 Einstein's special theory of relativity deals with the behavior of particles moving at speeds comparable to the speed of light. It agrees with Newton's theory at low velocities, but makes many very different predictions for high-speed motion. All of its predictions have been repeatedly

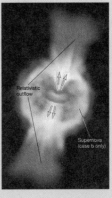

verified by experiment. The modern replacement for Newtonian gravity is Einstein's **general theory of relativity (p. 556)**, which describes gravity in terms of the warping, or bending, of **spacetime (p. 554)** by the presence of mass. The more mass, the greater the warping. All particles—including photons—respond to that warping by moving along curved paths.

6 The upper limit on the mass of a neutron star is about 3 solar masses. Beyond that mass, the star can no longer support itself against its own gravity, and it collapses to form a **black hole (p. 551)**, a region of space from which nothing can escape. Very massive stars, after exploding as supernovae, form black holes rather than neutron stars. Conditions in and near black holes can only be described by general relativity. The radius at which the escape speed from a collapsing star equals the speed of light is called the **Schwarzschild radius (p. 553)**. The surface of an imaginary sphere, of radius equal to the Schwarzschild radius, surrounding a black hole is called the **event horizon (p. 553)**.

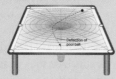

7 To a distant observer, light leaving a spaceship that is falling into a black hole would be subject to **gravitational redshift** (p. 558) as the light climbed out of the hole's intense gravitational field. At the same time, a clock on the spaceship would show **time dilation (p. 559)**—the clock would appear to slow down as the ship approached the event horizon. The observer would never see the ship reach the surface of the hole. Once within the event horizon, no known force can prevent a collapsing star from contracting all the way to a point-like **singularity (p. 560)**, at which point both the density and the gravitational field of the star become infinite. This prediction of relativity theory has yet to be proved. Singularities are places where the known laws of physics break down.

8 Once matter falls into a black hole, it can no longer communicate with the outside. However, on its way in, it can form an accretion disk and emit X rays, just as in the case of a neutron star. The best candidates for black holes are binary systems in which one component is a compact X-ray source. Cygnus X-1, a well-studied X-ray source in the constellation Cygnus, is a long-standing black-hole candidate. Studies of orbital motions imply that some binaries contain compact objects too massive to be neutron stars, leaving black holes as the only alternative. There is also strong evidence for **supermassive black holes (p. 563)** in the centers of many galaxies, including our own.

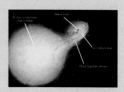

Mastering**ASTRONOMY** *For instructor-assigned homework go to www.masteringastronomy.com*

Problems labeled **POS** explore the process of science | **VIS** problems focus on reading and interpreting visual information

REVIEW AND DISCUSSION

1. How does the way in which a neutron star forms determine some of its most basic properties?

2. What would happen to a person standing on the surface of a neutron star?

3. Why aren't all neutron stars seen as pulsars?

4. Why do many neutron stars move at high speeds relative to their neighbors?

5. What are X-ray bursters?

6. What is the favored explanation for the rapid spin rates of millisecond pulsars?

7. Why do you think astronomers were surprised to find a pulsar with a planetary system?

8. Why do astronomers think that gamma-ray bursts are very distant and very energetic?

9. Describe two leading models for gamma-ray bursts.

10. What does it mean to say that the measured speed of a light beam is independent of the motion of the observer?

11. Use your knowledge of escape speed to explain why black holes are said to be "black."

12. According to special relativity, what is special about the speed of light?

13. **POS** Why is it so difficult to test the predictions of general relativity? Describe two tests of the theory.

14. What would happen to someone falling into a black hole?

15. What is an event horizon?

16. **POS** What is the principle of cosmic censorship? Do you think it is a sound scientific principle?

17. What makes Cygnus X-1 a good black-hole candidate?

18. **POS** What evidence is there for black holes much more massive than the Sun?

19. Imagine that you had the ability to travel at will through the Galaxy. Explain why you would discover many more neutron stars than those known to observers on Earth. Where would you be most likely to find these objects?

20. **POS** Do you think that planet-size objects discovered orbiting a pulsar should be called planets? Why or why not?

CONCEPTUAL SELF-TEST: MULTIPLE CHOICE

1. A neutron star is about the same size as (a) a school bus; (b) a U.S. city; (c) the Moon; (d) Earth.

2. A neutron star's immense gravitational attraction is due primarily to its small radius and (a) rapid rotation rate; (b) strong magnetic field; (c) large mass; (d) high temperature.

3. The most rapidly "blinking" pulsars are those that (a) spin fastest; (b) are oldest; (c) are most massive; (d) are hottest.

4. The X-ray emission from a neutron star in a binary system comes mainly from (a) the hot surface of the neutron star itself; (b) heated material in an accretion disk around the neutron star; (c) the neutron star's magnetic field; (d) the surface of the companion star.

5. **VIS**   According to Figure 22.11, gamma-ray bursts are observed to occur (a) mainly near the Sun; (b) throughout the Milky Way Galaxy; (c) approximately uniformly over the entire sky; (d) near pulsars.

6. Black holes result from stars having initial masses (a) less than the mass of the Sun; (b) between 1 and 2 times the mass

of the Sun; (c) up to 8 times the mass of the Sun; (d) more than 25 times the mass of the Sun.

7. If the Sun were magically to turn into a black hole of the same mass, (a) Earth would start to spiral inward; (b) Earth's orbit would remain unchanged; (c) Earth would fly off into space; (d) Earth would be torn apart.

8. **VIS**   According to the second figure in *Discovery 22-1*, a meterstick in a spaceship traveling at half the speed of light would appear to have a length of (a) 1 meter; (b) 0.87 meter; (c) 0.50 meter; (d) 0.15 meter.

9. The best place to search for black holes is in a region of space that (a) is dark and empty; (b) has recently lost some stars; (c) has strong X-ray emission; (d) is cooler than its surroundings.

10. The best evidence for supermassive black holes in the centers of galaxies is (a) the absence of stars there; (b) rapid gas motion and intense energy emission; (c) gravitational redshift of radiation emitted from near the center; (d) unknown visible and X-ray spectral lines.

PROBLEMS

The number of dots preceding each Problem indicates its approximate level of difficulty.

1. • The angular momentum of a solid body is proportional to the angular velocity of the body times the square of its radius. ∞ *(More Precisely 6-1)* Using the law of conservation of angular momentum, estimate how fast a collapsed stellar core would spin if its initial spin rate was 1 revolution per day and its radius decreased from 10,000 km to 10 km.

2. • What would your mass be if you were composed entirely of neutron-star material of density 3×10^{17} kg/m³? (Assume that your average density is 1000 kg/m³.) Compare your answer with the mass of a typical 10-km-diameter rocky asteroid.

3. • Calculate the surface gravitational acceleration and escape speed of a 1.4-solar-mass neutron star with a radius of 10 km. What would be the escape speed from a neutron star of the same mass and radius 4 km?

4. •• Use the radius–luminosity–temperature relation to calculate the luminosity of a 10-km-radius neutron star for temperatures of 10^5 K, 10^7 K, and 10^9 K. At what wavelengths does the star radiate most strongly in each case? What do you conclude about the detectability of neutron stars? Could the brightest of them be plotted on an H–R diagram?

5. •• A gamma-ray detector of area 0.5 m² in the midst of a gamma-ray burst records photons having total energy 10^{-8} joule. If the burst occurred 1000 Mpc away, calculate the total amount of energy it released (assuming that the energy was emitted isotropically), and compare your answer with the Sun's total energy output on the main sequence. How would this figure change if the burst occurred 10,000 pc away instead, in the halo of our Galaxy? What if it occurred within the Oort cloud of our own solar system, at a distance of 50,000 AU?

6. •• A gamma-ray burst 5000 Mpc away releases 10^{45} joules of energy isotropically in the form of gamma rays, each of energy

250 keV. ∞ *(More Precisely 4-1)* Some of the rays are detected by an instrument in Earth orbit with an effective collecting area of 0.75 m². How many gamma-ray photons strike the detector?

7. • A 10-km-radius neutron star is spinning 1000 times per second. Calculate the speed of a point on its equator, and compare your answer with the speed of light. (The equator is the circumference of a circle, and recall that the circumference is equal to $2\pi r$, where r is the diameter.) Also, calculate the orbital speed of a particle in a circular orbit just above the surface if the neutron star's mass is 1.4 times the mass of the Sun.

8. • Supermassive black holes are thought to exist in the centers of many galaxies. What would be the Schwarzschild radii of black holes of 1 million and 1 billion solar masses, respectively? How does the 1-million-solar-mass black hole compare in size with the Sun? How does the 1-billion-solar-mass black hole compare in size with the solar system?

9. •• Use the information presented in *More Precisely 22-1* to estimate the deflection of a beam of light that just grazes the surface of (a) the Moon, (b) Jupiter, and (c) Sirius B. (d) A future generation of space astrometry missions may be able to measure angles as small as 10^{-6} arcsec accurately. At what distance from the Sun would this deflection occur?

10. •• Calculate the tidal acceleration on a 2-m-tall human falling feet-first into a 1-solar-mass black hole; that is, compute the difference in the accelerations (forces per unit mass) on their head and their feet just as the feet cross the event horizon. ∞ *(Sec. 7.6)* Repeat the calculation for a 1-million-solar-mass black hole and for a 1-billion-solar-mass black hole (see question 8). Compare these accelerations with the acceleration due to gravity on Earth ($g = 9.8$ m/s²).

PART 4

GALAXIES AND COSMOLOGY

Andromeda Galaxy
as observed in 1890
(*J. Roberts*)

It is hard to imagine, but less than 100 years ago the Sun was considered the center of the universe. Earlier studies by Copernicus, Kepler, Galileo, and others had dethroned Earth from a central position, but the Sun itself, at the center of our solar system, was still assumed also to be the center of the Milky Way—which a century ago was identical to the universe. Our true place in the cosmos, and even the existence of countless other galaxies comparable to our own, were completely unknown.

Enter the American astronomer Harlow Shapley (1885–1972), who, by studying variable stars in globular clusters, was able to deduce the size and scale of the Milky Way Galaxy, as well as our position in it. His results, announced in 1918, showed not only that our extended home in space was immensely larger than had previously been realized—about 100,000 light-years across—but also that Earth resided in what he called the "galactic suburbs," now known to be about 25,000 light-years from the center of the Galaxy. Shapley demonstrated that our Sun is not central, unique, or special in any way. His work was a milestone in our understanding of our place in the universe, certainly one of the most important astronomical discoveries of the 20th century.

Ironically, Shapley's dramatic discovery of the increased size and scale of the Milky Way led him astray regarding another, even more profound, advance in our knowledge at that time: the realization that our Galaxy is only one of many galaxies in the universe. The sheer size of the Milky Way implied by his observations caused him to oppose the idea of a vastly larger universe—he found it hard to believe that there could be other, distant galaxies as huge as our own. Even among eminent scientists, personal biases can sometimes cloud scientific judgment.

The stage was set for a "Great Debate" that occurred at the National Academy of Sciences in Washington in 1920. At issue were the fuzzy "spiral nebulae" (which today we call galaxies): Were they close enough to be part of our own Milky Way, or were they sufficiently distant to be whole galaxies unto themselves? Shapley held that, given that his research had revised upward the size of the Milky Way, the spiral nebulae must be part of our own Galaxy. His opponent, Heber Curtis of California's Lick Observatory, while incorrectly rejecting the great size of our Galaxy, correctly argued that the spirals were remote aggregates of stars similar to the Milky Way. Both men presented other scientific arguments supporting their views (see Section 23.2), but both also let personal feelings affect their comprehension of our home Galaxy. With no objective measurements of the true distances to the nebulae, the debate ended in a draw.

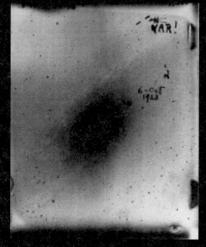

Edwin Hubble's discovery of variable stars in Andromeda (*Carnegie Institute*)

Harlow Shapley working at his rotating octagonal desk (*Harvard*)

Andromeda Galaxy as observed today *(R. Gendler)*

Shapley's rival, the Caltech astronomer Edwin Hubble (1884–1953), broke the stalemate just a few years later by using the premier optical telescope of the day, the 2.5-m (100-inch) reflector atop Mt. Wilson. He first resolved the Andromeda Nebula into individual stars and then carefully measured its variable stars, thereby proving that Andromeda was a genuine galaxy millions of light-years distant, well outside our Milky Way. Ironically, Hubble used the same basic technique that Shapley and his Harvard colleagues had pioneered. It was yet another milestone along the road extending the Copernican principle: Neither Earth nor the Sun is special in any way, and even the Galaxy in which we live is just one of myriad galaxies in a much, much larger cosmos.

Today, astronomers have extensively mapped the distri-

bution of variable stars in Andromeda. Curiously, we are still struggling to determine that galaxy's distance to better than 10 percent accuracy. Even in the decades-long lifetime of this textbook, Andromeda's quoted distance has fluctuated from about 2.2 to as much as 2.9 million light-years; in this edition, we have averaged the most recent measurements to arrive at a value of 2.5 million light-years. The correct value is important, for upon it rests a key rung in the so-called distance ladder. This cosmic yardstick is used to measure the ranges to billions of other, more distant galaxies and hence to gauge the vastly larger realm of the universe itself.

The Shapley-Curtis debate of yesteryear, together with our current struggles to pin down accurate distances to the truly faraway galaxies, constitute good case studies of how the scientific method actually works. Science is practiced by human beings, and scientists are no different from others who have strong emotions and personal values. Yet, over the course of time, and through much criticism and debate, scientific issues eventually gain a measure of objectivity. By demanding tests and proven facts, the scientific community gradually damps the subjectivity of individuals and arrives at a more objective view among a community of critical thinkers. Reasoned skepticism and repeated testing are hallmarks of the modern scientific method.

Hubble Ultra Deep Field *(STScI)*

571

23

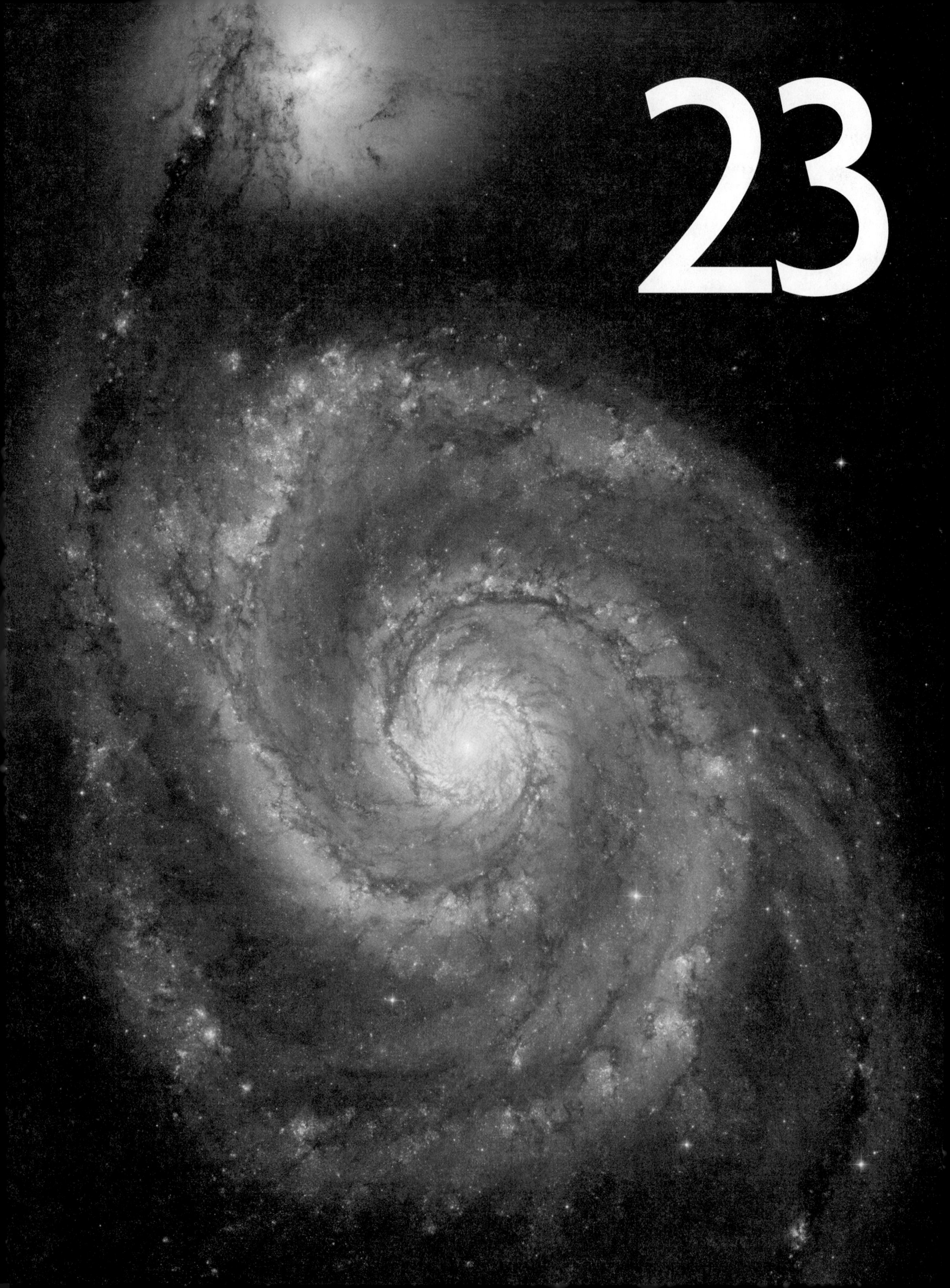

THE MILKY WAY GALAXY

A SPIRAL IN SPACE

LEARNING GOALS

Studying this chapter will enable you to

1. Describe the overall structure of the Milky Way Galaxy, and specify how the various regions differ from one another.

2. Explain the importance of variable stars in determining the size and shape of our Galaxy.

3. Describe the orbital paths of stars in different regions of the Galaxy, and explain how these motions are accounted for by our understanding of how the Galaxy formed.

4. Discuss some possible explanations for the existence of the spiral arms observed in our own and many other galaxies.

5. Explain what studies of Galactic rotation reveal about the size and mass of our Galaxy, and discuss the possible nature of dark matter.

6. Describe some of the phenomena observed at the center of our Galaxy.

THE BIG PICTURE The Milky Way is just one galaxy out of nearly a hundred billion in the observable universe, but for astronomers it plays much the same role among galaxies as the Sun does among stars. Our understanding of galaxies throughout the cosmos rests squarely on our knowledge of the size, structure, and dynamics of our own.

Mastering ASTRONOMY

Visit the Study Area in www.masteringastronomy.com for quizzes, animations, videos, interactive figures, and self-guided tutorials.

Looking up on a dark, clear night, we are struck by two aspects of the night sky. The first is that the individual stars we see are roughly uniformly distributed in all directions. They all lie relatively close to us, mapping out the local Galactic neighborhood within a few hundred parsecs of the Sun. But this is only a local impression. Ours is a rather provincial view. Beyond those nearby stars, the second thing we notice is a fuzzy band of light—the Milky Way—stretching across the heavens.

From the Northern Hemisphere, this band is most easily visible in the summertime, arcing high above the horizon. Its full extent forms a great circle that encompasses the entire celestial sphere. This is the insider's view of the galaxy in which we live—the blended light of countless distant stars. As we consider much larger volumes of space, on scales far, far greater than the distances between neighboring stars, a new level of organization becomes apparent as the large-scale structure of the Milky Way Galaxy is revealed.

LEFT: *Galaxies like this one, known as M51 or the Whirlpool, contain roughly a hundred billion stars, bound together by gravity. The graceful winding arms of this majestic spiral galaxy sweep across some 100,000 light-years of space; the whole object is 33 million light-years distant. Its size, shape, and mass approximate those of our own Galaxy, which has never been photographed in its entirety because we live inside it. If this were our Galaxy, the Sun would reside in one of its spiral arms, two-thirds of the way out from the center. (STScI)*

23.1 Our Parent Galaxy

A **galaxy** is a gargantuan collection of stellar and interstellar matter—stars, gas, dust, neutron stars, black holes—isolated in space and held together by its own gravity. Astronomers are aware of literally billions of galaxies beyond our own. The particular galaxy we happen to inhabit is known as the **Milky Way Galaxy,** or just *the Galaxy*, with a capital *G*.

Our Sun lies in a part of the Galaxy known as the **Galactic disk**—an immense, circular, flattened region containing most of our Galaxy's luminous stars and interstellar matter (and virtually everything we have studied so far in this book). Figure 23.1 illustrates how, viewed from within, the Galactic disk appears as a band of light stretching across our night sky, a band known as the *Milky Way*. As indicated in the figure, if we look in a direction away from the Galactic disk (red arrows), we see relatively few stars in our field of view. However, if our line of sight happens to lie within the disk (white and blue arrows), we see so many stars that their light merges into a continuous blur.

Paradoxically, although we can study individual stars and interstellar clouds that lie near the Sun in great detail, our location within the Galactic disk makes deciphering our Galaxy's large-scale structure from Earth a very difficult task—a little like trying to unravel the layout of paths, bushes, and trees in a city park without being able to leave one particular park bench. In some directions, the interpretation of what we see is ambiguous and inconclusive. In others, foreground objects completely obscure our view of what lies beyond, but we cannot move around them to get a better look. As a result, astronomers who study the Milky Way Galaxy are often guided in their efforts by comparisons with more distant, but much more easily observable, systems.

Figures 23.2 and 23.3 show three galaxies thought to resemble our own in overall structure. Figure 23.2 is the Andromeda Galaxy, the nearest major galaxy to the Milky Way Galaxy, lying nearly 800 kpc (roughly 2.5 million light-years) away. Andromeda's apparent elongated shape is a consequence of the angle at which we happen to view it. In fact, our Galaxy, like this one, consists of a circular galactic disk of matter that fattens to a **Galactic bulge** at the center. The disk and bulge are embedded in a roughly spherical ball of faint old stars known as the **Galactic halo.** These three basic galactic regions are indicated on the figure. (The halo stars are so faint that they cannot be discerned here.) Figures 23.3(a) and (b) show views of two other galaxies—one seen face-on, the other edge-on—that illustrate these points more clearly.

CONCEPT CHECK

✔ Why do we see the Milky Way as a band of light across the night sky?

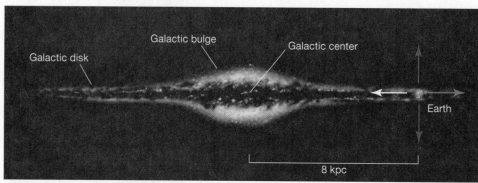

Galactic disk · Galactic bulge · Galactic center · Earth · 8 kpc

(a) Artist's view of Milky Way from afar

(b) Real image of Milky Way from inside

R I V U X G

◄ **FIGURE 23.1 Galactic Plane** (THE BIG PICTURE) The overall shape of our Galaxy can be inferred from some very basic observations. (a) Gazing from Earth toward the Galactic center (white arrow) in this artist's conception, we see myriad stars stacked up within the thin band of light known as the Milky Way. Looking in the opposite direction (blue arrow), we still see the Milky Way band, but now it is fainter because our Sun lies far from the Galactic center, so we see more stars when looking toward the center than when looking in the opposite direction. Looking perpendicular to the disk (red arrows), we see far fewer stars. (b) This is a real optical view of the sky around us. The white band is the disk of our Milky Way Galaxy—dimly seen with the naked eye from very dark locations on Earth. This is what one sees looking toward the Galactic center—in the direction of the white arrow in part (a). (*Lund Observatory*)

(a)

Galactic bulge

Galactic disk

Galactic halo

R I V U X G

(b)

(c)

▲ FIGURE 23.2 **Andromeda Structure** (a) The Andromeda Galaxy probably resembles fairly closely the overall layout of our own Milky Way Galaxy. The disk and bulge are clearly visible in this image, which is about 30,000 pc across. The faint stars of the halo, completely surrounding the disk and bulge, cannot be seen here. The white stars sprinkled all across this image are not part of Andromeda's halo; they are foreground stars in our own Galaxy, lying in the same region of the sky as Andromeda, but about a thousand times closer. (b) More detail within the inner parts of the galaxy. (c) The galaxy's peculiar—and still unexplained—double core; this inset covers a region only 15 pc across. *(R. Gendler; Palomar/Caltech; NASA)*

(a)

(b)

R I V U X G

▲ FIGURE 23.3 **Spiral Galaxies** (a) This galaxy, catalogued as M101 and seen nearly face-on, is somewhat similar in its overall structure to our own Milky Way Galaxy and Andromeda. (b) The galaxy NGC 4565 happens to be oriented in such a way that we see it edge-on, allowing us to make out its disk and central bulge. *(R. Gendler)*

23.2 Measuring the Milky Way

Before the 20th century, astronomers' conception of the cosmos differed markedly from the modern view. The fact that we live in just one of many enormous "islands" of matter separated by even larger tracts of apparently empty space was completely unknown, and the clear distinction between "our Galaxy" and "the universe" did not exist. The twin ideas that (1) *the Sun is not at the center of the Galaxy* and (2) *the Galaxy is not at the center of the universe* required both time and hard observational evidence before they gained widespread acceptance. The growth in our knowledge of our Galaxy, as well as the realization that there are many other distant galaxies similar to our own, has gone hand in hand with the development of the cosmic distance scale.

Star Counts

In the late 18th century, long before the distances to any stars were known, the English astronomer William Herschel tried to estimate the size and shape of our Galaxy simply by counting how many stars he could see in different directions in the sky. Assuming that all stars were of about equal brightness, he concluded that the Galaxy was a somewhat flattened, roughly disk-shaped collection of stars lying in the plane of the Milky Way, with the Sun at or near its *center* (Figure 23.4). Subsequent refinements to this approach led to much the same picture. Early in the 20th century, some astronomers went so far as to estimate the dimensions of this "Galaxy" as about 10 kpc in diameter by 2 kpc thick.

Today the Milky Way Galaxy is known to be several tens of kiloparsecs across, and the Sun is known to lie far from the center. How could the older picture have been so flawed? The answer is that the earlier observations were made in the visible part of the electromagnetic spectrum, and astronomers failed to take into account the (then-unknown) absorption of visible light by interstellar gas and dust. ∞ (Sec. 18.1) Only in the 1930s did astronomers begin to realize the true extent and importance of the interstellar medium.

Any objects in the Galactic disk more than a few kiloparsecs away from us are hidden from our view (in visible light) by the effects of interstellar dust. The apparent falloff in the density of stars with distance in the plane of the Milky Way is thus not a real thinning of their numbers in space, but simply a consequence of the murky environment in the Galactic disk. The long "fingers" in Herschel's map are directions in which the obscuration happens to be a little less severe than in others. However, because some obscuration occurs in all directions in the disk, the falloff is roughly similar no matter which way we look, so the Sun appears to be more or less at the center. The horizontal extent of Figure 23.4 corresponds approximately to the span of the blue and white arrows in Figure 23.1.

Radiation coming to us from above or below the plane of the Galaxy, where there is less gas and dust along the line of sight, arrives on Earth relatively unscathed. There is still some patchy obscuration, but the Sun happens to be located where the view out of the disk is largely unimpeded by nearby interstellar clouds.

Spiral Nebulae and Globular Clusters

We have just seen how astronomers' attempts to probe the Galactic disk by optical means are frustrated by the effects of the interstellar medium. However, looking in other directions, out of the Milky Way plane, we can see to much greater distances. During the first quarter of the 20th century, studies of the large-scale structure of our Galaxy focused on two particularly important classes of objects, both found mainly away from the Milky Way. The first is *globular clusters,* those tightly bound swarms of old, reddish stars we met in Chapter 19. ∞ (Sec. 19.6) About 150 are now known in our own Galaxy. The second class consisted of objects that were known at the time as *spiral nebulae.* Examples are shown in Figures 23.2(a) and 23.3(a). We know them today as **spiral galaxies,** comparable in size to our own.

Early 20th-century astronomers had no means of determining the distances to any of these objects. They are too far away to have any observable parallax, and with the technology of the day, main-sequence stars (after the discovery of the main sequence in 1911) could not be clearly identified and measured. For these reasons, neither of the techniques discussed in Chapter 17—trigonometric and spectroscopic parallax—was applicable. ∞ (Secs. 17.1, 17.6) As a result, even the most basic properties—size, mass, and stellar and interstellar content—of globular clusters and spiral nebulae were unknown. It was assumed that the globular clusters lay within our own Galaxy, which was thought at the time to be

▲ **FIGURE 23.4 Herschel's Galaxy Model** Eighteenth-century astronomer William Herschel constructed this "map" of the Galaxy by counting the numbers of stars he saw in different directions of the sky. Our Sun (marked by the large yellow dot) appears to lie near the center of the distribution. The long axis of the diagram lies in the plane of the Galactic disk.

relatively small (using the size estimates just mentioned). The locations of the spiral nebulae were much less clear.

Knowing the distance to an object is vitally important to understanding its true nature. As an example, consider again the Andromeda "nebula" (Figure 23.2). In the late 19th century, when improved telescopes and photographic techniques allowed astronomers to obtain images showing detail comparable to that in Figure 23.2(a), the newly released photographs caused great excitement among astronomers, who thought they were seeing the formation of a star from a swirling gaseous disk! Comparing Figure 23.2(a) with the figures in Chapter 15 (especially Figure 15.2b), we can perhaps understand how such a mistake could be made—*if* we thought that we were looking at a relatively close, star-sized object. Far from demonstrating that Andromeda was distant and large, the new observations seemed to confirm that it was just a small part of our own Galaxy.

Further observations soon made it clear that Andromeda was not a star-forming region. Andromeda's parallax is too small to measure, indicating that it must be at least several hundred parsecs from Earth, and, even at 100 pc—which we now know is vastly less than Andromeda's true distance—an object the size of the solar nebula would be impossible to resolve and simply would not look like Figure 23.2(a). (See Section 22.4 for another, more recent example of how distance measurements directly affect our theoretical understanding of observational data.)

During the first quarter of the 20th century, both the size of our Galaxy and the distances to the spiral nebulae were hotly debated in astronomical circles (see below, and also the discussion in the Part 4 Opener on p. 570). One school of thought maintained that the spiral nebulae were relatively small systems contained within our Galaxy. Other astronomers held that the spirals were much larger objects, lying far outside the Milky Way Galaxy and comparable to it in size. However, with no firm distance information, both arguments were inconclusive. Only with the discovery of a new distance-measurement technique—which we discuss next—was the issue finally settled in favor of the latter view. However, in the process, astronomers' conception of our own Galaxy changed radically and forever.

A New Yardstick

An important by-product of the laborious effort to catalog stars around the turn of the 20th century was the systematic study of **variable stars**—stars whose luminosity changes with time, some quite erratically, others more regularly. Only a small fraction of stars fall into this category, but those that do are of great astronomical significance.

We encountered several examples of variable stars in earlier chapters. Often, the variability is the result of membership in a binary system. In an eclipsing binary, for example, the total brightness varies because one star periodically blocks the light of the other. ∞ (Sec. 17.7) In novae and supernovae, binary membership has more violent consequences. ∞ (Sec. 21.3) These latter objects are called *cataclysmic variables,* because of their sudden, large changes in brightness.

In other instances, however, the variability is a basic trait of a star and is not dependent on its being a part of a binary system. We call such a star an *intrinsic variable.* A particularly important class of intrinsic variables is the **pulsating variable stars,** which vary cyclically in luminosity in very characteristic ways. Two types of pulsating **variable** stars that have played central roles in revealing both the true extent of our Galaxy and the distances to our galactic neighbors are the **RR Lyrae** and **Cepheid variables.** Following long-standing astronomical practice, the names come from the first star of each class to be discovered—in this case, the variable star labeled RR in the constellation Lyra and the variable star Delta Cephei, the fourth brightest star in the constellation Cepheus. Note, by the way, that these stars have *nothing* whatsoever to do with the pulsars discussed in the previous chapter! Pulsars are rapidly rotating neutron stars beaming energy into space as they spin; as we will see in a moment, pulsating variable stars are "normal" stars undergoing a temporary period of instability as they evolve. ∞ (Sec. 22.2)

RR Lyrae and Cepheid variable stars are recognizable by the characteristic shapes of their light curves. RR Lyrae stars all pulsate similarly (Figure 23.5a), with only small differences in period between them. Observed periods range from about 0.5 to 1 day. Cepheid variables also pulsate in distinctive ways (the regular "sawtooth" pattern in Figure 23.5b), but different Cepheids can have quite different pulsation periods, ranging from about 1 to 100 days. The period of any given RR Lyrae or Cepheid variable is, to a high degree of accuracy, the same from one cycle to the next. The key point is that pulsating variable stars can be recognized and identified *just by observing the variations in the light they emit.*

Why do Cepheids and RR Lyrae variables pulsate? The basic mechanism was first suggested by the British astrophysicist Sir Arthur Eddington in 1941. The structure of any star is determined in large part by how easily radiation can travel from the core to the photosphere—that is, by the *opacity* of the interior, the degree to which the gas hinders the passage of light through it. If the opacity rises, the radiation becomes trapped, the internal pressure increases, and the star "puffs up." If the opacity falls, radiation can escape more easily, and the star shrinks. According to theory, under certain circumstances a star can become unbalanced and enter a state in which the flow of radiation causes the opacity first to rise—making the star expand, cool, and diminish in luminosity—and then to fall, leading to the pulsations we observe.

The conditions necessary to cause pulsations are not found in main-sequence stars. Rather, they occur in evolved post-main-sequence stars as they pass through a region of the

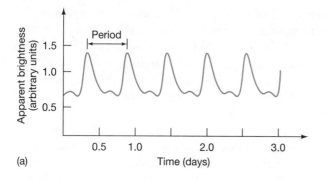

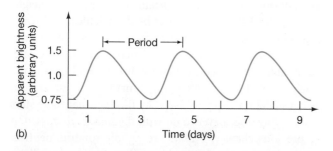

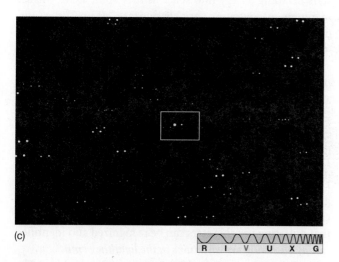

▲ FIGURE 23.5 **Variable Stars** (a) Light curve of the pulsating variable star RR Lyrae. All RR Lyrae-type variables have essentially similar light curves, with periods of less than a day. (b) The light curve of a Cepheid variable star called WW Cygni, having a period of several days. (c) This Cepheid is shown here (boxed) on successive nights, near its maximum and minimum brightness; two photos, one from each night, were superimposed and then slightly displaced. *(Harvard College Observatory)*

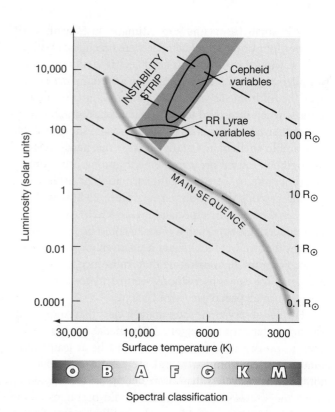

▲ FIGURE 23.6 **Variable Stars on the H–R Diagram** Pulsating variable stars are found in the instability strip of the H–R diagram. As a high-mass star evolves through the strip, it becomes a Cepheid variable. Low-mass horizontal-branch stars in the instability strip are RR Lyrae variables.

H–R diagram known as the *instability strip* (Figure 23.6). When a star's temperature and luminosity place it in this strip, the star becomes internally unstable. Both its temperature and its radius vary in a regular way, causing the pulsations we observe: For the reasons just described, as the star brightens, its radius shrinks and its surface becomes hotter; as its luminosity decreases, the star expands and cools. As we learned in Chapter 20, high-mass stars evolve across the upper part of the H–R diagram. When their evolutionary tracks take them into the in-

stability strip, they become Cepheid variables. ∞ (Sec. 20.4) RR Lyrae variables are low-mass horizontal-branch stars that lie within the lower portion of the instability strip. ∞ (Sec. 20.2) Thus, pulsating variables are normal stars passing through a brief—roughly million-year—phase of instability as a natural part of stellar evolution.

The Cosmic Distance Scale

The importance of these stars to Galactic astronomy lies in the fact that once we recognize a star as being of the RR Lyrae or Cepheid type, we can infer its luminosity, and that in turn allows us to measure its distance. The distance calculation is precisely the same as that presented in Chapter 17 during our discussion of spectroscopic parallax. ∞ (Sec. 17.6) Comparing the star's (known) luminosity with its (observed) apparent brightness yields an estimate of its distance by the inverse-square law: ∞ (Sec. 17.2)

$$\text{apparent brightness} \propto \frac{\text{luminosity}}{\text{distance}^2}.$$

In this way, astronomers can use pulsating variables as a means of determining distances, both within our own Galaxy and far beyond.

How do we infer a variable star's luminosity? For RR Lyrae stars, doing so is simple. As we saw in Chapter 20, all horizontal branch stars have basically the same luminosity (averaged over a complete pulsation cycle)—about 100 times that of the Sun. ∞ (Sec. 20.2) Thus, once a variable star is recognized as being of the RR Lyrae type, its luminosity is immediately known. For Cepheids, we make use of a close correlation between average luminosity and pulsation period, discovered in 1908 by Henrietta Leavitt of Harvard University (see *Discovery 23-1*) and known simply as the **period–luminosity relationship.** Cepheids that vary slowly—that is, that have long periods—have large luminosities; conversely, short-period Cepheids have low luminosities.

Figure 23.7 illustrates the period–luminosity relationship for Cepheids found within a thousand parsecs or so of Earth. Astronomers can plot such a diagram for relatively nearby stars because they can measure their distances by using stellar or spectroscopic parallax. Once the distances are known, the luminosities of those stars can be calculated. We know of no exceptions to the period–luminosity relationship, and it is consistent with theoretical calculations of pulsations in evolved stars. Consequently, we assume that it holds for all Cepheids, near and far. Thus, a simple measurement of a Cepheid variable's pulsation period immediately tells us its luminosity—we just read it off the plot in Figure 23.7. (The roughly constant luminosities of the RR Lyrae variables are also indicated in the figure.)

This distance-measurement technique works well, provided that the variable star can be clearly identified and its pulsation period measured. With Cepheids, the method allows astronomers to estimate distances out to about 25 million parsecs, enough to take us all the way to the nearest galaxies. The less-luminous RR Lyrae stars are not so easily seen as Cepheids, so their useful range is not as great. However, they are much more common, so, within their limited range, they are actually more useful than Cepheids.

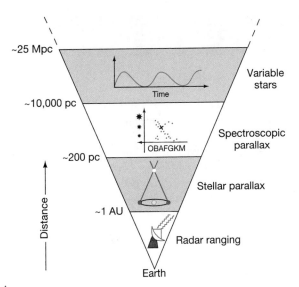

Interactive FIGURE 23.8 Variable Stars on Distance Ladder Application of the period–luminosity relationship of Cepheid variable stars allows us to determine distances out to about 25 Mpc with reasonable accuracy.

Figure 23.8 extends our cosmic distance ladder, begun in Chapter 2 with radar ranging in the solar system and expanded in Chapter 17 to include stellar and spectroscopic parallax, by adding variable stars as a fourth method of determining distance. Note that because the period–luminosity relationship is calibrated by using nearby stars, this latest rung inherits any and all uncertainties and errors present in the lower levels. Uncertainties also arise from the "scatter" of data points shown in Figure 23.7. Although the overall connection between period and luminosity is unmistakable, the individual data points do not quite lie on a straight line; instead, a range of possible luminosities corresponds to any measured period.

The Size and Shape of Our Galaxy

Many RR Lyrae variables are found in globular clusters. Early in the 20th century, the American astronomer Harlow Shapley used observations of variable stars to make two very important discoveries about the Galactic globular cluster system. First, he showed that most globular clusters reside at great distances—many thousands of parsecs—from the Sun. Second, by measuring the direction and distance of each cluster, he was able to determine the three-dimensional distribution of the clusters in space (Figure 23.9). In this way, Shapley demonstrated that the globular clusters map out a truly gigantic, and roughly *spherical*, volume of space, about 30 kpc across.* However, the center of the distribution lies nowhere near our Sun; rather, it is located nearly 8 kpc away from us, in the direction of the constellation Sagittarius.

The Galactic globular cluster system and the Galactic halo, of which it is a part, are somewhat flattened in the direction perpendicular to the disk, but the degree of flattening is quite uncertain. The halo is certainly much less flattened than the disk, however.

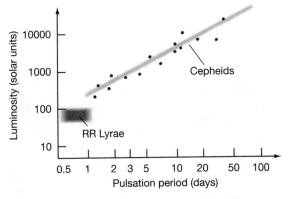

▲ **FIGURE 23.7 Period–Luminosity Plot** A plot of pulsation period versus average absolute brightness (that is, luminosity) for a group of Cepheid variable stars. The two properties are quite tightly correlated. The pulsation periods of some RR Lyrae variables are also shown.

DISCOVERY 23-1

Early "Computers"

A large portion of the early research in observational astronomy focused on monitoring stellar luminosities and analyzing stellar spectra. Much of this pioneering work was done using photographic methods. What is not so well known is that most of the labor was accomplished by women. Around the turn of the twentieth century, a few dozen dedicated women—assistants at the Harvard College Observatory—created an enormous database by observing, sorting, measuring, and cataloging photographic information that helped form the foundation of modern astronomy. Some of them went far beyond their duties in the lab to make several of the basic astronomical discoveries often taken for granted today.

The first photograph below taken in 1910, shows several of those women carefully examining star images and measuring variations in luminosity or wavelengths of spectral lines. In the cramped quarters of the Harvard Observatory, they inspected image after image to collect a vast body of data on millions of measurements of hundreds of thousands of stars. Note the plot of stellar luminosity changes pasted on the wall at the left. The pattern is so regular that it likely belongs to a Cepheid variable. Known as "computers" (there were no electronic devices then), these women were paid 25 cents an hour.

The second photograph, taken in 1913, shows a more formal portrait of another group of staff members, along with their director, E.C. Pickering. Though looking rather stern here, Pickering was often described as a true Victorian gentleman who championed a policy, unique at the time, of admitting women to the staff. Also prominent here (and symmetrically positioned to Pickering's left) is Annie Cannon, perhaps the most accomplished of the early group of women who, beginning in 1880, undertook a survey of the skies that lasted for more than half a century—work that netted Cannon the first Oxford honorary degree awarded to a woman.

The first major result of this work was a record of the brightnesses and spectra of tens of thousands of stars, published in 1890 under the direction of Williamina Fleming (seen standing in the photograph below). On the basis of this compilation,

several of these women made fundamental contributions to astronomy. In 1897 Antonia Maury (who is also pictured in the first photo at left rear) undertook the most detailed study of stellar spectra to that time, enabling Hertzsprung and Russell independently to develop what is now called the H–R diagram. In 1898 Annie Cannon proposed the spectral classification system (described in Chapter 17) that is now the international standard for categorizing stars. ∞ (Sec. 17.5) And in 1908 Henrietta Leavitt discovered the period–luminosity relationship for Cepheid variable stars, which later allowed Pickering's successor as director, Harlow Shapley (see the introductory essay for Part 4), to recognize our Sun's true position in the universe.

All was not work, however, and socializing was common among this generation of astronomers. The third photograph (below) shows a 1920s scene from a humorous play portraying life at the Observatory, starring (at center) the then youngest of the "lady computers," Cecilia Payne, who would go on to become one of the foremost astronomers of the 20th century (see the introductory essay for Part 3).

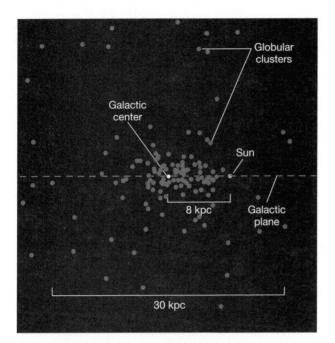

▲ **FIGURE 23.9 Globular Cluster Distribution** Our Sun does not coincide with the center of the very large collection of globular clusters (indicated by the pink dots). Instead, more globular clusters are found in one direction than in any other. The Sun resides closer to the edge of the collection, which measures roughly 30 kpc across. The globular clusters outline the true distribution of stars in the Galactic halo.

In a brilliant intellectual leap, Shapley realized that the distribution of globular clusters maps out the true extent of stars in the Milky Way Galaxy—the region that we now call the Galactic halo. The hub of this vast collection of matter, 8 kpc from the Sun, is the **Galactic center.** Figure 23.9 shows

the distribution, based on modern data, of the 138 globular clusters lying within 20 kpc of the center. As illustrated in Figure 23.10, we live in the "suburbs" of this huge ensemble—in the Galactic disk, the thin sheet of young stars, gas, and dust that cuts through the center of the halo. Since Shapley's time, astronomers have identified many individual stars—that is, stars not belonging to any globular cluster—within the Galactic halo.

Shapley's bold interpretation of the globular clusters as defining the overall distribution of stars in our Galaxy was an enormous step forward in human understanding of our place in the universe. Five hundred years ago, Earth was considered the center of all things. Copernicus argued otherwise, demoting our planet to an undistinguished location removed from the center of the solar system. In Shapley's time, as we have just seen, the prevailing view was that our Sun was the center not only of the Galaxy, but also of the universe. Shapley showed otherwise. With his observations of globular clusters, he simultaneously increased the size of our Galaxy by almost a factor of 10 over earlier estimates and banished our parent Sun to its periphery, virtually overnight!

The Shapley–Curtis Debate

Curiously, Shapley's dramatic revision of the size of the Milky Way Galaxy and our place in it only strengthened his erroneous opinion that the spiral nebulae were part of our Galaxy and that our Galaxy was essentially the entire universe. He regarded as beyond belief the idea that there could be other structures as large as our Galaxy. The scientific issues involved in understanding the nature of the spiral nebulae were clearly drawn in a famous 1920 debate between

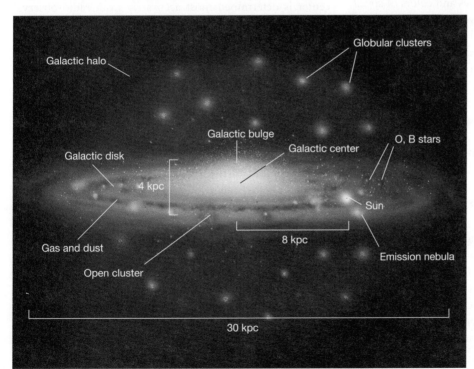

◀ **Interactive FIGURE 23.10 Stellar Populations in Our Galaxy** Based on observations of young stars and gas in the disk, and the old stars and globular clusters in the halo, astronomers have constructed a comprehensive picture of the structure of our Galaxy. This artist's conception of a (nearly) edge-on view of the Milky Way Galaxy shows schematically the distributions of young blue stars, open clusters, old red stars, and globular clusters. (The brightness and size of our Sun are greatly exaggerated for clarity.)

Shapley and Lick Observatory astronomer Heber Curtis. (See also p. 570.) We list here some key elements of the debate:

1. *Size of the Milky Way.* Shapley correctly asserted that the diameter of the Milky Way Galaxy was much larger than the "conventional" scale based on star counts, but then incorrectly concluded that similar-sized galaxies beyond our own could not exist. Curtis incorrectly accepted the smaller size for our Galaxy, but correctly argued that similar galaxies might exist beyond our own.

2. *Distribution of the nebulae on the sky.* Curtis noted that the observed spiral nebulae were generally found away from the plane of the Galaxy, and he suggested that our Galaxy had a "ring" of occulting material in its plane, like those observed in many edge-on spirals, preventing us from seeing the nebulae in the plane. Shapley simply had to accept the notion that spiral nebulae were, for some unknown reason, not found in the Galactic plane. Curtis was almost correct in this point. Note, however, that the effects of absorption by interstellar dust were completely unknown at the time. ∞ (Sec. 18.1)

3. *Observations of novae.* Shapley argued (correctly) that the observed apparent brightnesses of some "novae" seen in spiral nebulae implied enormous luminosities if the nebulae lay at large distances. ∞ (Secs. 17.2, 21.1) Curtis suggested (also correctly) that these anomalous events might be members of a second, much brighter, class of nova—today we call them *supernovae*. ∞ (Sec. 21.3)

4. *Brightness and spectra of the nebulae.* Shapley pointed out that the measured brightnesses and colors of spiral nebulae differed from what he would have expected to see if our Galaxy were observed from afar, suggesting that the nebulae were somehow fundamentally different from the Milky Way. Curtis had no answer. We know today that these differences exist because of interstellar absorption and reddening, which prevent astronomers from getting a comparable view of our own Galaxy, but all this was unknown at the time. ∞ (Sec. 18.1) Curtis did correctly note that spectral lines seen in spiral nebulae were generally consistent with the nebulae's being assemblages of large numbers of stars, supporting his argument that they were stellar systems comparable to our own Galaxy. ∞ (Sec. 4.2)

5. *Rotation of the nebulae.* Shapley cited published measurements of the angular rotation speeds of some spiral nebulae, which implied that the nebulae would have to be spinning faster than the speed of light if they were very distant and hence very large. ∞ (*More Precisely 1-2*) Curtis simply responded that the observations were in error. Curtis was right, but he couldn't prove it at the time.

We see that both men made some correct and some incorrect statements (or conclusions) about the problem. However, with the observations of the day, their disagreements could not be resolved, and the debate was inconclusive. But technology marches on, and just a few years later, in 1925, American astronomer Edwin Hubble reported that he had observed Cepheids in the Andromeda Galaxy and finally succeeded in measuring its distance. His work firmly established Andromeda as a separate galaxy lying far beyond our own, finally extending the Copernican principle to the Galaxy itself.

CONCEPT CHECK

✔ Can variable stars be used to map out the structure of the Galactic disk?

23.3 Galactic Structure

Based on optical, infrared, and radio observations of stars, gas, and dust, Figure 23.10 illustrates the different spatial distributions of the disk, bulge, and halo of the Milky Way Galaxy. The extent of the halo is based largely on optical observations of globular clusters and other halo stars. However, as we have seen, optical techniques can cover only a small portion of the dusty Galactic disk. Much of our knowledge of the structure of the disk on larger scales is based on radio observations, particularly of the 21-cm radio emission line produced by atomic hydrogen. ∞ (Sec. 18.4)

The center of the gas distribution coincides roughly with the center of the globular cluster system, lying about 8 kpc from the Sun. In fact, the location of the Galactic center is determined most accurately from radio observations of Galactic gas. The densities of both stars and gas in the disk decline quite rapidly beyond about 15 kpc from the Galactic center (although some radio-emitting gas has been observed out to at least 50 kpc).

The Spatial Distribution of Stars

Perpendicular to the Galactic plane, the disk in the vicinity of the Sun is relatively thin—"only" 300 pc thick, or about one hundredth of the 30-kpc Galactic diameter. Don't be fooled, though: Even if you could travel at the speed of light, it would take you a thousand years to traverse the thickness of the Galactic disk. The disk may be thin compared with the Galactic diameter, but it is huge by human standards.

Actually, the thickness of the Galactic disk depends on the kinds of objects measured. Young stars and interstellar gas are more tightly confined to the plane than are stars like the Sun, and solar-type stars in turn are more tightly confined than are older K- and M-type dwarfs. The reason for these differences is that stars form in interstellar clouds close to the plane of the disk, but then tend to drift out of the disk over time, mainly due to their interactions with other stars and molecular clouds. Thus, *as stars age, their abundance above and below the plane of the disk slowly increases.* Note that these considerations do not apply to the Galactic halo, whose ancient stars and globular clusters extend far above and below the Galactic plane. As we will see in a moment, the halo is a remnant of an early stage of our Galaxy's evolution and predates the formation of the disk.

Recently, improved observational techniques have revealed an intermediate category of Galactic stars, midway between the old halo and the younger disk, both in age and in spatial distribution. Consisting of stars with estimated ages in the range of 7–10 billion years, this *thick-disk* component of the Milky Way Galaxy measures some 2–3 kpc from top to bottom. Its thickness is too great to be explained by the slow drift just described. Like the halo, it appears to be a vestige of our Galaxy's distant past.

Also shown in Figure 23.10 is our Galaxy's central bulge, measuring roughly 6 kpc across in the plane of the Galactic disk by 4 kpc perpendicular to the plane. Obscuration by interstellar dust makes it difficult to study the detailed structure of the Galactic bulge in optical images of the Milky Way. (See, for example, Figure 18.4, which would clearly show a large portion of the bulge were it not for interstellar absorption.) However, at longer wavelengths, which are less affected by interstellar matter, a much clearer picture emerges (Figure 23.11; see also Figure 23.3b). Detailed measurements of the motion of gas and stars in and near the bulge imply that it is actually football shaped, about half as wide as it is long, with the long axis of the "football" lying in the Galactic plane. On the basis of these observations, astronomers speculate that the central part of our Galaxy may have a distinctly elongated, or barlike, appearance and that we may live in a galaxy of the "barred-spiral" type, as discussed further in Chapter 24.

Stellar Populations

Aside from their distributions in space, the three components of the Galaxy—disk, bulge, and halo—have several other properties that distinguish them from one another. First, the halo contains almost *no* gas or dust—just the opposite of the disk and bulge, in which interstellar matter is common. Second, there are clear differences in both *appearance* and *composition* among disk, bulge, and halo stars: Stars in the Galactic bulge and halo are found to be distinctly *redder* than stars found in the disk. Observations of other spiral galaxies also show this trend—the blue-white tint of the disk and the yellowish coloration of the bulge are evident in Figures 23.2(a) and 23.3(a).

All the bright, blue stars visible in our sky are part of the Galactic disk, as are the young, open star clusters and star-forming regions. In contrast, the cooler, redder stars—including those found in the old globular clusters—are more uniformly distributed throughout the disk, bulge, and halo. Galactic disks appear bluish because main-sequence O- and B-type blue supergiants are very much brighter than G-, K-, and M-type dwarfs, even though the dwarfs are present in far greater numbers.

The explanation for the marked difference in stellar content between disk and halo is this: Whereas the gas-rich Galactic disk is the site of ongoing star formation and so contains stars of all ages, all the stars in the Galactic halo are *old.* The absence of dust and gas in the halo means that no new stars are forming there, and star formation apparently ceased long ago—at least 10 billion years in the past, judging from the types of halo stars we now observe. (Recall from Chapter 20 that most globular clusters are thought to be between 10 and 12 billion years old.) ∞ (Sec. 20.5) The gas density is very high in the inner part of the Galactic bulge, making that region the site of vigorous ongoing star formation, and both very old and very young stars mingle there. The bulge's gas-poor outer regions have properties more similar to those of the halo.

Support for this picture comes from studies of the spectra of halo stars, which indicate that these stars are far less abundant in heavy elements (i.e., elements heavier than helium) than are nearby stars in the disk. In Chapter 21, we saw how each successive cycle of star formation and evolution enriches the interstellar medium with the products of stellar nucleosynthesis, leading to a steady increase in heavy elements with time. ∞ (Sec. 21.5) Thus, the scarcity of these elements in halo stars is consistent with the view that the halo formed long ago.

Astronomers often refer to young disk stars as *Population I* stars and to old halo stars as *Population II* stars. The idea of

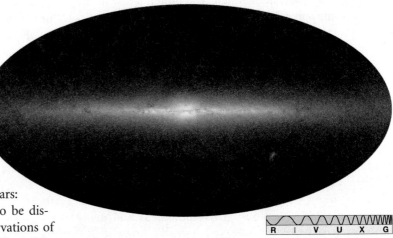

A Interactive **FIGURE 23.11 Infrared View of the Milky Way**
A wide-angle infrared image of the disk and bulge of the Milky Way Galaxy, as observed by the Two Micron All Sky Survey. Compare Figure 23.3(b). (*UMass/Caltech*)

two stellar "populations" dates from the 1940s, when the differences between disk and halo stars first became clear. The names are something of an oversimplification, as there is actually a continuous variation in stellar ages throughout the Milky Way Galaxy, not a simple division of stars into two distinct "young" and "old" categories. Nevertheless, the terminology is widely used.

Orbital Motion

Now let's turn our attention to the *dynamics* of the Milky Way Galaxy—that is, to the motion of the stars, dust, and gas it contains. Are the internal motions of our Galaxy's members disordered and random, or are they part of some gigantic "traffic pattern"? The answer depends on our perspective. The motion of stars and clouds we see on small scales (within a few tens of parsecs from the Sun) seems random, but on larger scales (hundreds or thousands of parsecs) the motion is much more orderly.

As we look around the Galactic disk in different directions, a clear pattern of motion emerges (Figure 23.12). Radiation received from stars and interstellar gas clouds in the upper-right and lower-left quadrants of the figure is generally *blueshifted*. At the same time, radiation from stars and gas sampled in the upper-left and lower-right quadrants tends to be *redshifted*. In other words, some regions of the Galaxy (those in the blueshifted directions) are approaching the Sun, whereas others (the redshifted ones) are receding from us. ∞ (Sec. 3.5)

Careful study of the positions and velocities of stars and gas clouds near the Sun leads us to two important conclusions about the motion of the Galactic disk. First, the entire disk is *rotating*—stars, gas, and dust all move in roughly circular paths around the Galactic center, their orbits governed by the Galaxy's gravitational pull. The orbital speed in the vicinity of the Sun is about 220 km/s. Thus, at the Sun's distance of 8 kpc from the Galactic center, material takes about 225 million years (an interval of time sometimes called 1 *Galactic year*) to complete one circuit.

Second, the Galactic rotation period depends on distance from the Galactic center, being shorter closer to the center and longer at greater distances. In other words, the Galactic disk rotates not as a solid object, but *differentially*. ∞ (Secs. 11.1, 16.1) Accurate measurements, made by the *Hipparcos* satellite, of stars within a few hundred parsecs of the Sun have proved particularly valuable in measuring these important Galactic properties. ∞ (Sec. 17.1) Similar *differential rotation* is observed in Andromeda and many other spiral galaxies.

This picture of orderly circular motion about the Galactic center applies only to the disk: Stars in the Galactic halo and bulge are not so well behaved. The old globular clusters in the halo and the faint, reddish individual stars in both the halo and the bulge do *not* share the disk's

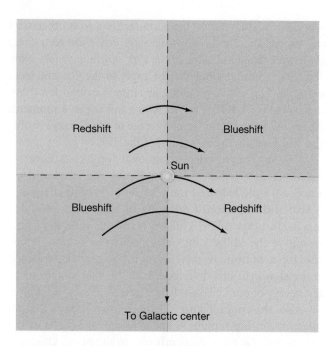

▲ **FIGURE 23.12 Orbital Motion in the Galactic Disk** Stars and interstellar clouds in the neighborhood of the Sun show systematic Doppler motions, implying that the disk of the Galaxy is spinning in a well-ordered way. These four Galactic quadrants are drawn to intersect not at the Galactic center, but at the Sun, the location from which our observations are made. Because the Sun orbits faster than stars and gas at larger radii, we are moving away from material at top left and gaining on that at top right, resulting in the Doppler shifts indicated. Similarly, stars and gas in the bottom left quadrant are gaining on us, while material at bottom right is pulling away. The curved arrows represent the angular speed of the disk material.

well-defined rotation. Instead, their orbital orientations are largely random.* Although these objects do orbit the Galactic center, they move in all directions, their paths filling an entire three-dimensional volume rather than a nearly two-dimensional disk.

Figure 23.13 contrasts the motion of bulge and halo stars with the much more regular orbits of stars in the Galactic disk. At any given distance from the Galactic center, bulge, or halo, stars move at speeds comparable to the disk's rotation speed at that radius, but in *all* directions, not just one. Their orbits carry these stars repeatedly through the plane of the disk and out the other side. (They don't collide with stars in the disk because interstellar distances are huge compared with the diameters of individual stars—a star or even an entire star cluster passes through the disk almost as though it weren't there—see Section 25.2.) Some well-known stars in the vicinity of the Sun—the bright giant Arcturus, for example—are actually halo

---

*Halo stars do, in fact, have some net rotation about the Galactic center, but the rotational component of their motion is overwhelmed by the larger random component. The motion of bulge stars also has a rotational component, larger than that of the halo, but still smaller than the random component of stellar motion in the bulge.

TABLE 23.1 Overall Properties of the Galactic Disk, Halo, and Bulge

Galactic Disk	Galactic Halo	Galactic Bulge
highly flattened	roughly spherical—mildly flattened	somewhat flattened and elongated in the plane of the disk ("football shaped")
contains both young and old stars	contains old stars only	contains young and old stars; more old stars at greater distances from the center
contains gas and dust	contains no gas and dust	contains gas and dust, especially in the inner regions
site of ongoing star formation	no star formation during the last 10 billion years	ongoing star formation in the inner regions
gas and stars move in circular orbits in the Galactic plane	stars have random orbits in three dimensions	stars have random orbits but some net rotation about the Galactic center
spiral arms	no obvious substructure	central regions probably elongated into a bar; ring of gas and dust near center
overall white coloration, with blue spiral arms	reddish in color	yellow-white

stars that are "just passing through" the disk on orbits that take them far above and below the Galactic plane.

Recently, astronomers have detected numerous *tidal streams* in the Galactic halo—groups of stars thought to be the remnants of globular clusters and even small satellite

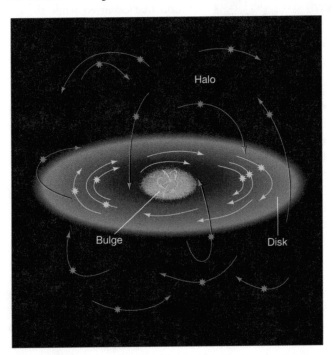

▲ **FIGURE 23.13 Stellar Orbits in Our Galaxy** Stars in the Galactic disk (blue curves) move in orderly, circular orbits about the Galactic center. In contrast, halo stars (orange curves) have orbits with largely random orientations and eccentricities. The orbit of a typical halo star takes it high above the Galactic disk, then down through the disk plane, and then out the other side and far below the disk. The orbital properties of bulge stars are intermediate between those of disk stars and those of halo stars.

galaxies (see Sec. 24.1) torn apart by our Galaxy's tidal field. Just as micrometeoroid swarms in our solar system follow the orbit of their disrupted parent comet long after the comet itself is gone, stars in a tidal stream are now spread out around the entire original orbit of their parent cluster or galaxy. ∞ (Sec. 14.4) We will discuss the processes responsible for them further in Chapter 25.

Table 23.1 compares some key properties of the three basic components of the Galaxy.

CONCEPT CHECK

✔ Why do astronomers regard the disk and the halo as different components of our Galaxy?

23.4 The Formation of the Milky Way

Is there some evolutionary scenario that can naturally account for the Galactic structure we see today? The answer is that there is, and it takes us all the way back to the birth of our Galaxy, more than 10 billion years ago. ∞ (Sec. 20.5) Not all the details are agreed upon by all astronomers, but the overall picture is now fairly widely accepted. For simplicity, we confine our discussion here to the Galactic disk and halo; in many ways, the bulge is intermediate in its properties between these two extremes. Figure 23.14 illustrates the current view of our Galaxy's evolution, starting (not unlike the star-formation scenario outlined in Chapter 19) from a contracting cloud of pregalactic gas. ∞ (Sec. 19.1)

When the first Galactic stars and globular clusters formed, the gas in our Galaxy had not yet accumulated into

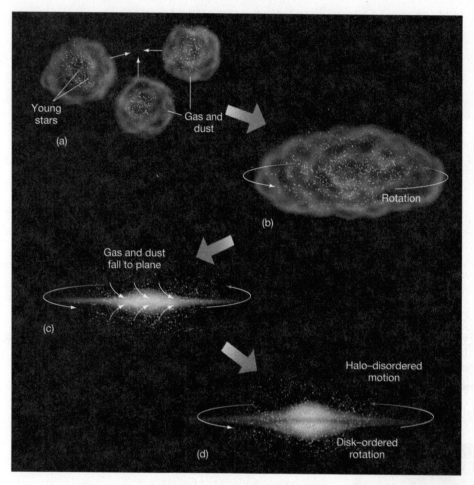

◀ **FIGURE 23.14 Milky Way Formation**

(THE BIG PICTURE) Theoretical scenarios for the formation of our Galaxy and others account for the disk and halo. (a) The Milky Way Galaxy possibly formed through the merger of several smaller systems. (b) Astronomers reason that early on, our Galaxy was irregularly shaped, with gas distributed throughout its volume. When stars formed during this stage, there was no preferred direction in which they moved and no preferred location in which they were found. Their orbits carried them throughout an extended three-dimensional volume surrounding the newborn Galaxy. (c) In time, the gas and dust fell to the Galactic plane and formed a spinning disk. The stars that had already formed were left behind in the halo. (d) New stars forming in the disk inherit its overall rotation and so orbit the Galactic center on ordered, circular orbits.

ful activity. The thick disk, with its intermediate-age stars, may represent an intermediate stage of star formation that occurred while the gas was still flattening into the plane.

Recent studies of the composition of stars in the Galactic disk suggest that the infall of halo gas is still going on today. The best available models of star formation and stellar nucleosynthesis predict that the fraction of heavy elements in disk stars should be significantly *greater* than is actually observed, unless the gas in the disk is steadily being "diluted" by relatively unevolved gas arriving from the halo at a rate of perhaps 5–10 solar masses per year. ∞ (Sec. 21.5) This may not sound like much mass, but accumulated over billions of years it actually amounts to a significant fraction of the total mass of the disk (see Section 23.6).

This theory also explains the randomly oriented orbits of the halo stars and the more ordered motion of the disk (Figure 23.14d). When the halo developed, the irregularly shaped Galaxy was rotating only very slowly, so there was no strongly preferred direction in which matter tended to move. As a result, halo stars were free to travel along nearly any path once they formed (or when their parent systems merged), leading to the random halo orbits we observe today. After the Galactic disk formed, however, stars that formed from its gas and dust inherited its rotational motion and so move on well-defined, circular orbits. Again, the thick disk's orbital properties are consistent with the idea that it formed while gas was still sinking to the Galaxy's midplane.

In principle, the structure of our Galaxy bears witness to the conditions that created it. In practice, however, the

a disk. Instead, it was spread out over an irregular and quite extended region of space, spanning many tens of kiloparsecs in all directions (Figure 23.14b). When the first stars formed, they were distributed throughout that volume. Their distribution today (the Galactic halo) reflects that fact—it is an imprint of their birth. Many astronomers think that the very first stars formed even earlier, in smaller systems that later merged to create our Galaxy (Figure 23.14a). Probably, many more stars were born during the mergers themselves, as interstellar gas clouds collided and began to collapse. ∞ (Sec. 19.5) Whatever the details, the present-day halo would look much the same in either case.

Since those early times, rotation has flattened the gas in our Galaxy into a relatively thin disk (Figure 23.14c). Physically, the process is similar to the flattening of the solar nebula during the formation of the solar system, as described in Chapters 6 and 15, except on a vastly larger scale. ∞ (Secs. 6.7, 15.2) Star formation in the halo ceased billions of years ago when the raw materials—the gas and dust—cooled and fell toward the Galactic plane. Ongoing star formation in the disk gives it its bluish tint, but the halo's short-lived bright blue stars have long since burned out, leaving only the long-lived red stars that give the halo its characteristic pinkish glow. The Galactic halo is ancient, whereas the disk is full of youth-

interpretation of the observations is made difficult by the sheer complexity of the system we inhabit and by the many competing physical processes that have modified its appearance since it formed. As a result, the early stages of the Milky Way are still quite poorly understood. We will return to the subject of galaxy formation in Chapters 24 and 25.

CONCEPT CHECK

✔ Why are there no young halo stars?

23.5 Galactic Spiral Arms

If we want to look beyond our immediate neighborhood and study the full extent of the Galactic disk, we cannot rely on optical observations, as interstellar absorption severely limits our vision. In the 1950s, astronomers developed a very important tool for exploring the distribution of gas in our Galaxy: spectroscopic radio astronomy.

Radio Maps of the Milky Way

The keys to observing Galactic interstellar gas are the 21-cm radio emission line produced by atomic hydrogen and the many radio molecular lines formed in molecular cloud complexes. ∞ (Sec. 18.4) Long-wavelength radio waves are largely unaffected by interstellar dust, so they travel more or less unimpeded through the Galactic disk, allowing us to "see" to great distances. Because hydrogen is by far the most abundant element in interstellar space, the 21-cm signals are strong enough that a large portion of the disk can be observed in this way. As noted in Chapter 18, observations of spectral lines from "tracer" molecules, such as carbon monoxide, allow us

to study the distribution of the densest interstellar clouds. ∞ (Sec. 18.5)

Earlier, we noted that observations of stars within several hundred parsecs of the Sun have allowed astronomers to measure the rotation rate of the Galaxy in the solar neighborhood. As indicated in Figure 23.15, in order to probe to greater distances, astronomers often turn to radio observations (illustrated here with 21-cm radiation), because long-wavelength radio waves are largely unaffected by interstellar dust, allowing astronomers to study virtually the entire Galactic disk. ∞ (Sec. 18.4)

However, the distances to the clouds emitting the radio radiation are often poorly known. How, then, can we determine just where in the disk a cloud lies? Astronomers accomplish this by using all available data, coupled with our knowledge of Newtonian mechanics, to construct a *mathematical model* of the rotation of stars and gas throughout the Galactic disk. ∞ (Sec. 2.8) Assuming circular orbits, the model allows us to turn a measured radial velocity into a distance along the line of sight. As in so many areas of astronomy, theory and observations complement one another: The data refine the theoretical model, whereas the model in turn provides the framework needed to interpret further observations. ∞ (Sec. 1.2)

Radio astronomers couple their observations with this Galactic model to turn their measurements into detailed information about the distribution of gas along the line of sight. Because of the differential rotation described in Section. 23.3, the measured velocity of a cloud depends on its *distance* from the Sun (Figure 23.15), and the Galactic model provides the connection between the two. Furthermore, the strength of the signal is a measure of the *density* of gas in the cloud—denser clouds contain more gas and emit more radiation. Thus, knowing direction, distance, and density,

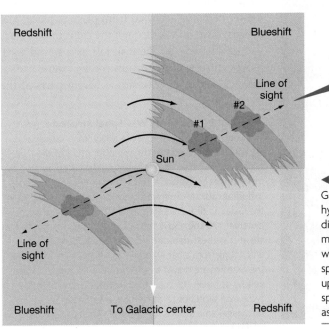

◄ **FIGURE 23.15 Gas in the Galactic Disk** Because the disk of our Galaxy is rotating differentially, 21-cm radio signals from different clumps of hydrogen matter along any given line of sight are Doppler shifted by different amounts, allowing both their densities and distances to be measured. As in Figure 23.12, the black arrows represent the angular speed at which gas orbits the Galactic center. The graph at right is a typical radio spectrum, showing how cloud #2, which is part of an outer spiral arm in the upper right quadrant, is blueshifted more than cloud #1, which is in an inner spiral arm. By repeating these observations in many different directions, astronomers map out the distribution of gas in our Galaxy.

astronomers can use observations along different lines of sight to map out the radio-emitting gas in our Galaxy.

Spiral Structure

Interstellar gas in the Galactic disk exhibits an organized pattern on a grand scale. Near the center, the gas in the disk fattens markedly in the Galactic bulge. Radio-emitting gas has been observed out to at least 50 kpc from the Galactic center. Over much of the inner 20 kpc or so of the disk, the gas is confined within about 100 pc of the Galactic plane. Beyond that distance, the gas distribution spreads out somewhat, to a thickness of several kiloparsecs, and shows definite signs of being "warped," possibly because of the gravitational influence of a pair of nearby galaxies (to be discussed in Chapter 24; see also Figure 23.16).

Radio studies provide perhaps the best direct evidence that we live in a spiral galaxy. Figure 23.16 is an artist's conception (based on observational data) of the appearance of our Galaxy as seen from far above the disk. The figure clearly shows our Galaxy's **spiral arms,** pinwheel-like structures originating close to the Galactic bulge and extending outward throughout much of the Galactic disk. Our Sun lies near the edge of one of these arms, which wraps around a large part of the disk. Notice, incidentally, the scale markers on Figures 23.9, 23.10, and 23.16: The Galactic globular-cluster distribution (Figure 23.9), the luminous stellar component of the disk (Figure 23.10), and the known spiral structure (Figure 23.16) all have roughly the *same* diameter—about 30 kpc. This scale is fairly typical of spiral galaxies observed elsewhere in the universe.

Survival of the Spiral Arms

The spiral arms in our Galaxy are made up of much more than just interstellar gas and dust. Studies of the Galactic disk within a kiloparsec or so of the Sun indicate that young stellar and prestellar objects—emission nebulae, O- and B-type stars, and recently formed open clusters—are also distributed in a spiral pattern that closely follows the distribution of interstellar clouds. The obvious conclusion is that the spiral arms are the part of the Galactic disk where star formation takes place. The brightness of the young stellar objects just listed is the main reason that the spiral arms of other galaxies are easily seen from afar (e.g., Figure 23.3a).

A central problem facing astronomers trying to understand spiral structure is how that structure persists over long periods. The basic issue is simple: Differential rotation makes it impossible for any large-scale structure "tied" to the disk material to survive. Figure 23.17 shows how a spiral pattern consisting always of the same group of stars and gas clouds would necessarily disappear within a few hundred million years. How, then, do the Galaxy's spiral arms retain their structure over long periods in spite of differential rotation?

A leading explanation for the existence of spiral arms holds that they are **spiral density waves**—coiled waves of gas compression that move through the Galactic disk, squeezing clouds of interstellar gas and triggering the process of star formation as they go. ∞ (Sec. 19.5) The spiral arms we observe are defined by the denser-than-normal clouds of gas the density waves create and by the new stars formed as a result of the spiral waves' passage.

This explanation of spiral structure avoids the problem of differential rotation, because the wave pattern is not tied to any particular piece of the Galactic disk. The spirals we see are merely patterns moving through the disk, not great masses of matter being transported from place to place. The density wave moves through the collection of stars and gas making up the disk just as a sound wave moves through air or an ocean wave passes through water, compressing different parts of the

30 kpc

◀ **FIGURE 23.16 Milky Way Spiral Structure** An

(THE BIG PICTURE) artist's conception of our Milky Way Galaxy seen face-on illustrates the spiral structure of the Galactic disk. This image is based on data accumulated by many teams of astronomers during the past few decades, including radio and infrared maps of stars, gas, and dust in the Galactic disk. Painted from the perspective of an observer 100 kpc above the Galactic plane, the spiral arms are at their best-determined positions, apparently emanating from a bar whose length is twice its width. All the features are drawn to scale (except for the oversized yellow dot near the top, which represents our Sun). The two small blotches to the left are dwarf galaxies, called the Magellanic Clouds. We study them in Chapter 24. *(Adapted from JPL)*

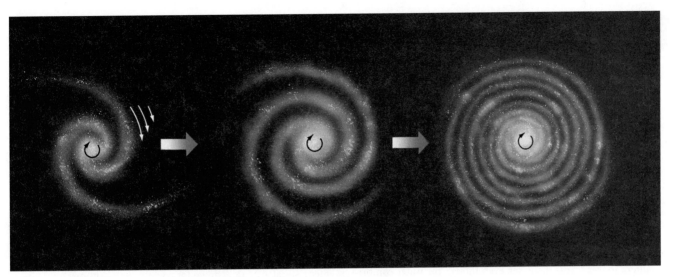

▲ Interactive **FIGURE 23.17 Differential Galactic Rotation** The disk of our Galaxy rotates differentially—stars close to the Galactic center take less time to orbit the center than those farther out. (The white arrows represent the angular speed of the disk; the big red arrows denote time evolution from left to right.) If spiral arms were somehow tied to the material of the Galactic disk, this differential rotation would cause the spiral pattern to wind up and disappear in a few hundred million years. Spiral arms would be too short-lived to be consistent with the numbers of spiral galaxies we observe today.

disk at different times. Even though the rotation rate of the disk material varies with distance from the Galactic center, the wave itself remains intact, defining the Galaxy's spiral arms.

In fact, over much of the visible portion of the Galactic disk (within about 15 kpc of the center), the spiral wave pattern is predicted to rotate *more slowly* than the stars and gas. Thus, as shown in Figure 23.18, Galactic material catches up with the wave, is temporarily slowed down and compressed as it passes through, and then continues on its way. (For a more down-to-earth example of an analogous process, see *Discovery 23-2*.)

As material enters the density wave from behind, the gas is compressed and forms stars. Dust lanes mark the regions of highest-density gas. The most prominent stars—the bright O- and B-type blue giants—live for only a short time, so young stellar associations, emission nebulae, and open clusters with long main sequences are found only within the arms, near their birth sites, just ahead of the dust lanes. The brightness of these young systems emphasizes the spiral structure. Further downstream, ahead of the spiral arms, we see mostly older stars and star clusters. These objects have had enough time since their formation to outdistance the wave and pull away from it. Over millions of years, their random individual motions, superimposed on their overall rotation around the Galactic center, distort and

▲ **FIGURE 23.18 Spiral Density Waves** Density-wave theory holds that the spiral arms seen in our own and many other galaxies are waves of gas compression and star formation moving through the material of the galactic disk. In the painting at right, gas motion is indicated by red arrows and arm motion by white arrows. Gas enters an arm from behind, is compressed, and forms stars. The spiral pattern is delineated by dust lanes, regions of high gas density, and newly formed O- and B-type stars. The inset at left shows the spiral galaxy NGC 1566, which displays many of the features just described. *(AURA)*

DISCOVERY 23-2

Density Waves

In the late 1960s, American astrophysicists C.C. Lin and Frank Shu proposed a way in which spiral arms in the Galaxy could persist for many Galactic rotations. They argued that the arms themselves contain no "permanent" matter. They should thus not be viewed as assemblages of stars, gas, and dust moving intact through the disk—those would quickly be destroyed by differential rotation. Instead, a spiral arm should be envisaged as a density wave—a wave of compression and expansion sweeping through the Galaxy.

A wave in water builds up material temporarily in some places (crests) and lets it down in others (troughs). The wave *pattern* moves across the water, even though the water comprising the peaks and troughs does not. ∞ (Sec. 3.1) Similarly, as the spiral density wave encounters galactic matter, the gas is compressed to form a region of slightly higher than normal density. Galactic material enters the wave, is temporarily slowed down and compressed as it passes through, and then continues on its way. The compression triggers the formation of new stars and nebulae. In this way, the spiral arms are formed and re-formed repeatedly, without disappearing completely. Lin and Shu showed that the process can in fact maintain a spiral pattern for very long periods.

The accompanying figure illustrates the formation of a density wave in a much more familiar context: a traffic jam on a highway, triggered by the presence of a repair crew moving slowly down the road. As cars approach the crew, they slow down temporarily. Then they speed up again as they pass the work site and continue on their way. The result, as might be reported by a high-flying traffic helicopter, is a region of high traffic density, concentrated around the location of the work crew and moving with it. An observer on the side of the road, however, sees that the jam never contains the same cars for very long. Cars constantly catch up to the bottleneck, move slowly through it, and then speed up again, only to be replaced by more cars arriving from behind.

The traffic jam is analogous to the region of high stellar density in a Galactic spiral arm. Just as the traffic density wave is not tied to any particular group of cars, the spiral arms are not attached to any particular piece of disk material. Stars and gas enter a spiral arm, slow down for a while, then continue on their orbits around the Galactic center. The result is a moving region of high stellar and gas density, involving different parts of the disk at different times. Notice also that, just as in our Galaxy, the wave moves more slowly than, and independently of, the overall traffic flow.

We can extend our traffic analogy a little further. Most drivers are well aware that the effects of such a tie-up can persist long after the road crew responsible for it has stopped work and gone home for the night. Similarly, spiral density waves can continue to move through the disk even after the disturbance that originally produced them has long since subsided. According to spiral density wave theory, that is precisely what has happened in the Milky Way. Some disturbance in the past produced the wave, which has been moving through the Galactic disk ever since.

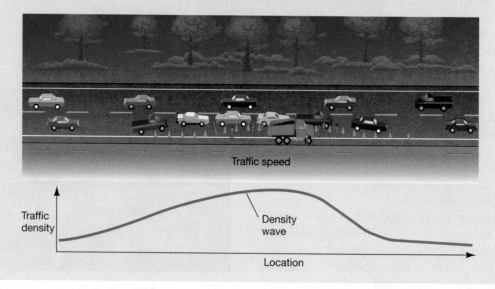

Traffic speed

Traffic density

Density wave

Location

eventually destroy their original spiral configuration, and they become part of the general disk population.

Note, incidentally, that although the spirals shown in Figure 23.18 have two arms each, astronomers are not certain how many arms make up the spiral structure in our own Galaxy (see Figure 23.16). The theory makes no strong predictions on this point.

An alternative possibility is that the formation of stars drives the waves, instead of the other way around. Imagine a row of newly formed massive stars somewhere in the disk. The emission nebula created when these stars form, and the supernovae when they die, send shock waves through the surrounding gas, triggering new star formation. ∞ (Secs. 19.5, 21.5) Thus, as illustrated in Figure 23.19(a), the formation

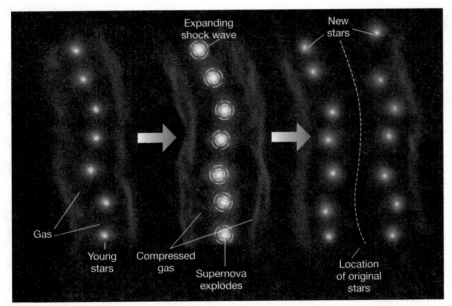

of one group of stars provides the mechanism for the creation of others. Computer simulations suggest that it is possible for the "wave" of star formation created in this manner to take on the form of a partial spiral and for the pattern to persist for some time. However, the process, sometimes known as **self-propagating star formation,** can produce only pieces of spirals, as are seen in some galaxies (Figure 23.19b). It apparently cannot produce the galaxy-wide spiral arms seen in other galaxies and present in our own. It may well be that there is more than one process at work in the spectacular spirals we see.

Origin of Spiral Structure

An important question (but one that unfortunately is not answered by either of the two theories just described) is Where do these spirals come from? What was responsible for generating the density wave in the first place or for creating the line of newborn stars whose evolution drives the advancing spiral arm? Scientists speculate that (1) the gravitational effects of our satellite galaxies (the Magellanic Clouds, to be discussed in Chapter 24), (2) instabilities in the gas near the Galactic bulge, or (3) the possible barlike asymmetry within the bulge itself may have had a big enough influence on the disk to get the process going.

The first possibility is supported by growing evidence that many other spiral galaxies seem to have been affected by gravitational interactions with neighboring systems in the relatively recent past (see Chapter 24). However, many astronomers still regard the other two possibilities as equally likely. For example, they point to *isolated* spirals, whose structure clearly cannot be the result of an external interaction. The fact is that we still don't know for sure how galaxies—including our own—acquire such beautiful spiral arms.

CONCEPT CHECK
✔ Why can't spiral arms simply be clouds of gas and young stars orbiting the Galactic center?

23.6 The Mass of the Milky Way Galaxy

We can measure our Galaxy's mass by studying the motions of gas clouds and stars in the Galactic disk. Recall from Chapter 2 that Newton's Law of Gravity (in the form of the modified version of Kepler's third law) connects the period, orbital size, and masses of any two objects in orbit around each other: ∞ (Sec. 2.8)

$$\text{total mass (solar masses)} = \frac{\text{orbital size (AU)}^3}{\text{orbital period (years)}^2}.$$

As we saw earlier, the distance from the Sun to the Galactic center is about 8 kpc, and the Sun's orbital period is 225 million years. Substituting these numbers into the preceding equation, we obtain a mass of $(8000 \times 206,000)^3/(225,000,000)^2$, or almost 9×10^{10} solar masses—90 *billion* times the mass of our Sun! The Milky Way Galaxy is truly enormous, in mass as well as in size.

But what mass have we just measured? When we performed the analogous calculation in the case of a planet orbiting the Sun, there was no ambiguity: The result of our calculation was the mass of the Sun. ∞ *(More Precisely 2-2)* However, the Galaxy's matter is not concentrated at the Galactic center (as the Sun's mass is concentrated at the center of the solar system); instead, Galactic matter is distributed over a large volume of space. Some of it lies inside the Sun's orbit (i.e., within 8 kpc of the Galactic center), and some lies outside, at large distances from both the Sun and the center of the

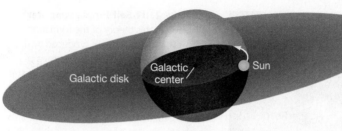

▲ FIGURE 23.20 **Weighing the Galaxy** The orbital speed of a star or gas cloud moving around the Galactic center is determined only by the mass of the Galaxy lying inside the orbit (within the gray-shaded sphere). Thus, to measure the Galaxy's total mass, we must look for objects orbiting at large distances from the center.

Galaxy. What portion of the Galaxy's mass controls the Sun's orbit? Isaac Newton answered this question three centuries ago: The Sun's orbital period is determined by the portion of the Galaxy that lies *within the orbit of the Sun* (Figure 23.20). This is the mass computed in the foregoing equation.

Galactic Rotation

The Sun's motion around the Galactic center tells us that the total Galactic mass within the Sun's orbit is about 90 billion solar masses, but it says nothing about the mass lying outside that orbit—that is, more than 8 kpc from the center. To determine the mass of the Galaxy on larger scales, we must measure the orbital motion of stars and gas at greater distances from the Galactic center. Astronomers have found that the most effective way to do this is to make radio observations of gas in the Galactic disk, because radio waves are relatively unaffected by interstellar absorption and allow us to probe to great distances, far beyond the Sun's orbit. On the basis of these studies, radio astronomers have determined our Galaxy's rotation rate at various distances from the Galactic center. The resultant plot of rotation speed versus distance from the center (Figure 23.21) is called the Galactic **rotation curve**.

Knowing the Galactic rotation curve, we can now repeat our earlier calculation to compute the total mass that lies within any given distance from the Galactic center. We find, for example, that the mass within about 15 kpc from the center—the volume defined by the globular clusters and the known spiral structure—is roughly 2×10^{11} solar masses, about twice the mass contained within the Sun's orbit. Does the distribution of matter in the Galaxy "cut off" beyond 15 kpc, where the luminosity drops off sharply? Surprisingly, it does not.

Newton's laws of motion predict that if all of the mass of the Galaxy were contained within the edge of the visible structure, then the orbital speed of stars and gas beyond 15 kpc would decrease with increasing distance from the Galactic center, just as the orbital speeds of the planets diminish as we move outward from the Sun. The dashed line in Figure 23.21 indicates what the rotation curve would look like in that case. However, the true rotation curve is quite different: Far from falling off at larger distances, it *rises* slightly, out to the limits of our measurement capabilities. This slight rise implies that the amount of mass contained within successively larger radii continues to grow beyond the orbit of the Sun, apparently out to a distance of at least 40 or 50 kpc.

According to the equation presented at the beginning of this section, the amount of mass within 40 kpc is approximately 6×10^{11} solar masses. Since 2×10^{11} solar masses lie within 15 kpc of the Galactic center, we have to conclude that at least twice as much mass lies *outside* the luminous part of our Galaxy—the part made up of stars, star clusters, and spiral arms—as lies inside!

Dark Matter

On the basis of these observations of the Galactic rotation curve, astronomers now regard the luminous portion of the Milky Way Galaxy—the region outlined by the globular clusters and by the spiral arms—as merely the "tip of the Galactic iceberg." Our Galaxy is in reality very much larger. The luminous region is surrounded by an extensive, invisible **dark halo,** which dwarfs the inner halo of stars and globular clusters and extends well beyond the 15-kpc radius once thought to represent the limit of our Galaxy. But what is the composition of this dark halo? We do not detect enough stars or interstellar matter to account for the mass that our computations tell us must be there. We are inescapably drawn to the conclusion

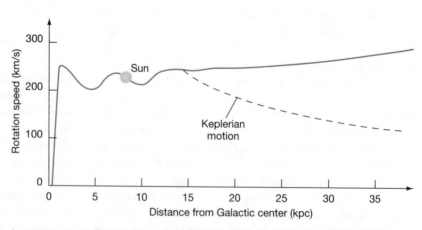

▲ FIGURE 23.21 **Galaxy Rotation Curve** The rotation curve for the Milky Way Galaxy plots rotation speed against distance from the Galactic center. We can use this curve to compute the mass of the Galaxy that lies within any given radius. The dashed curve is the rotation curve expected if the Galaxy "ended" abruptly at a radius of 15 kpc, the limit of most of the known spiral structure and the globular cluster distribution. The fact that the red curve does not follow this dashed line, but instead stays well above it, indicates that additional unseen matter must be beyond that radius.

that most of the mass in our Galaxy exists in the form of invisible **dark matter,** which we presently do not understand.

Note, incidentally, that even the "visible" portion of our Galaxy contains substantial amounts of dark matter. The total mass of stars and gas within 15 kpc of the center is estimated, from direct measurements of the luminosities of stars and radio emission from the interstellar medium, to be around 6×10^{10} solar masses. Most of this mass lies in the disk. Comparing this mass with that inferred from the Galactic rotation curve, we see that, even inside this luminous region, dark matter accounts for roughly two-thirds of the Galaxy's total mass.

The term *dark* here does not refer just to matter that is undetectable in visible light: The material has (so far) escaped detection at *all* wavelengths, from radio to gamma rays. Only by its gravitational pull do we know of its existence. Dark matter is not hydrogen gas (atomic or molecular), nor is it made up of ordinary stars. Given the amount of matter that must be accounted for, we would have been able to detect it with present-day instruments if it were in either of those forms. Its nature and its consequences for the evolution of galaxies and the universe are among the most important questions in astronomy today.

Many candidates have been suggested for this dark matter, although none is proven. Stellar-mass black holes may supply some of the unseen mass, but given that they are the evolutionary products of (relatively rare) massive stars, it is unlikely that there could be enough of them to hide large amounts of Galactic matter. ∞ (Sec. 22.8) Currently among the strongest "stellar" contenders are brown dwarfs—low-mass prestellar objects that never reached the point of core nuclear burning—white dwarfs, and faint, low-mass red dwarfs. ∞ (Secs. 19.3, 20.3) In the jargon of the field, these objects are collectively known as *MAssive Compact Halo Objects,* or MACHOs for short. In principle, they could exist in great numbers throughout the Galaxy, yet would be exceedingly hard to see because they are so faint.

Hubble Space Telescope observations of globular clusters seem to argue against at least the last of the three possibilities listed for MACHOs. Figure 23.22 shows a *Hubble* image of a relatively nearby globular cluster— one close enough that very faint red dwarfs could have been detected if any existed. The *Hubble* data suggest that there is a cutoff at about 0.2 solar mass, below which stars form much less frequently than had previously been supposed. As a result, stars with

very low mass may be unexpectedly rare, at least in the Galactic halo.

A radically different alternative is that the dark matter is made up of exotic *subatomic particles* that pervade the entire universe. In order to account for the properties of dark matter, these particles must have mass (to produce the observed gravitational effects), but also must interact hardly at all with "normal" matter (because otherwise we would be able to see them). One class of candidate particles satisfying these requirements has been dubbed *Weakly Interacting Massive Particles,* or WIMPs. Many theoretical astrophysicists think that such "dark-matter particles" could have been produced in abundance during the very earliest moments of our universe. If they survived to the present day, there might be enough of them to account for all the dark matter apparently out there. We will discuss this possibility and its far-reaching implications in more detail in Chapter 27. These ideas are hard to test, however, because these particles would necessarily be very difficult to detect. Several detection experiments on Earth have been attempted, so far without success.

A few astronomers have proposed a very different explanation for the "dark matter problem," suggesting that its resolution may lie not in the nature of dark matter, but rather in a modification to Newton's law of gravity that increases the gravitational force on very large (Galactic) scales, doing away with the need for dark matter in the first place. We must emphasize that the vast majority of scientists do *not* accept this view in any way. However, the very fact that it

R I V U X G

▲ **FIGURE 23.22 Missing Red Dwarfs** Sensitive visible-light observations with the *Hubble Space Telescope* have apparently ruled out faint red-dwarf stars as candidates for dark matter. The object shown here, the globular cluster 47 Tucanae, is one of many regions searched in the Milky Way. The inset, 0.4 pc on a side, is a high-resolution *Hubble* image of part of the cluster. The red dwarfs that would be expected if they existed in sufficient numbers to account for the dark matter in the Galaxy are not found. (The red stars that are seen are giants.) (*AAT; NASA*)

ANIMATION/VIDEO Rotating Globular Cluster

MA

has been proposed—and is being seriously discussed in scientific circles—is a testament to our current level of uncertainty. Dark matter is one of the great unsolved mysteries in astronomy today.

The Search for Stellar Dark Matter

Recently, researchers have obtained insight into the distribution of stellar dark matter by using a key element of Albert Einstein's theory of general relativity: the prediction that a beam of light can be deflected by a gravitational field, which has already been verified in the case of starlight that passes close to the Sun. ∞ (Sec. 22.6, *More Precisely 22-1*) The effect is small in the case of light grazing the Sun, but it has the potential for making distant and otherwise invisible stellar objects observable from Earth. Here's how.

Imagine looking at a distant star as a faint foreground object (a MACHO, such as a brown or white dwarf) happens to

cross your line of sight. As illustrated in Figure 23.23, the intervening object deflects a little more starlight than usual toward you, resulting in a temporary, but quite substantial, *brightening* of the distant star. In some ways, the effect is like the focusing of light by a lens, so the process is known as **gravitational lensing.** The foreground object is referred to as a *gravitational lens*. The amount of brightening and the duration of the effect depend on the mass, distance, and speed of the lensing object. Typically, the apparent brightness of the background star increases by a factor of two to five for a period of several weeks. Thus, even though the foreground object cannot be seen directly, its effect on the light of the background star makes it detectable. (In Chapter 25, we will encounter other instances of gravitational lensing in the universe, but on very much larger scales.)

Of course, stars are very small compared with the distance scale of the Galaxy, and the probability that one star will pass almost directly in front of another, as seen from Earth, is extremely low. But by observing millions of stars every few days

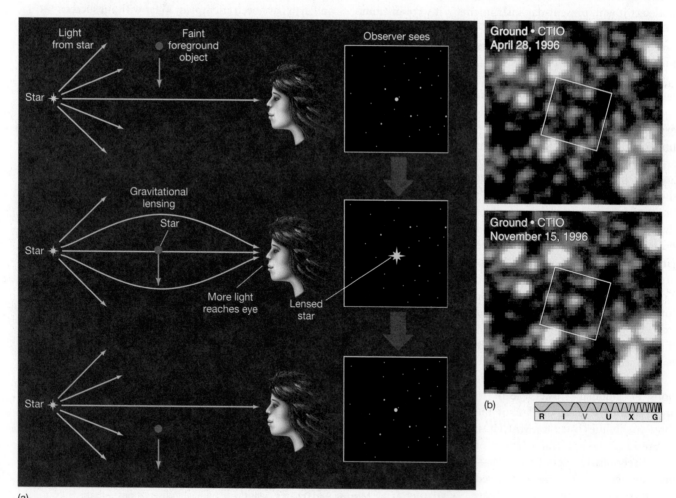

▲ **Interactive FIGURE 23.23 Gravitational Lensing** (a) Gravitational lensing by a faint foreground object (such as a brown dwarf) can temporarily cause a background star to brighten significantly, providing a means of detecting otherwise invisible stellar dark matter. (b) The brightening of a star during a lensing event, this one implying that a massive, but unseen, object passed in front of the unnamed star at the center of the two boxes imaged 6 months apart. *(AURA)*

over a period of years (using automated telescopes and high-speed computers to reduce the burden of coping with so much data), astronomers have been able to see enough of these events to let them estimate the amount of stellar dark matter in the Galactic halo. The technique represents an exciting new means of probing the structure of our Galaxy. The first lensing events were reported in late 1993. Subsequent observations are consistent with lensing by low-mass white dwarfs and suggest that such stars could account for a significant fraction—perhaps as much as 20 percent—but apparently *not* all, of the dark matter inferred from dynamical studies.

Bear in mind, though, that the identity of the dark matter is not necessarily an all-or-nothing proposition. It is perfectly conceivable—and, in fact, most astronomers think it likely—that more than one type of dark matter exists. For example, it is quite possible that most of the dark matter in the inner (visible) parts of galaxies is in the form of brown dwarfs and very low mass stars, whereas the dark matter farther out may be primarily in the form of exotic particles. We will return to this perplexing problem in later chapters, when we discuss some theories of how galaxies form and evolve, and how matter in the universe may have come into being.

PROCESS OF SCIENCE CHECK

✔ The nature of dark matter particles is unknown, yet most scientists regard these particles as the best solution to the dark matter problem. How do you think these statements square with the experimental scientific method presented in Section 1.2?

23.7 The Galactic Center

Theory predicts that the Galactic bulge should be densely populated with billions of stars, with the highest densities found closest to the Galactic center. However, we are unable to see this central region of our Galaxy—the interstellar medium in the Galactic disk shrouds what otherwise would be a stunning view. Figure 23.24 shows the optical view we do have of the part of the Milky Way toward the Galactic center, in the general direction of the constellation Sagittarius. Here, the Galactic plane is nearly vertical.

Observations at other wavelengths allow us to peer more deeply into the congested central regions of our Galaxy. The inset to Figure 23.24 is an adaptive-optics infrared image of the innermost parsec. ∞ (Sec. 5.4) It shows a dense central cluster containing roughly 1 million stars. That's a stellar density some 10 *million* times greater than in our solar neighborhood, high enough that stars must experience frequent close encounters and even collisions with one another.

Over the past two decades, combined radio, infrared, and X-ray observations have allowed astronomers to paint a detailed—and intriguing—picture of the Galactic center. They reveal complex structure on many scales, and violent activity in our Galaxy's core.

Galactic Activity

Figure 23.25(a) is an infrared view of a somewhat smaller part of Figure 23.24, and here the Galactic plane is horizontal. On

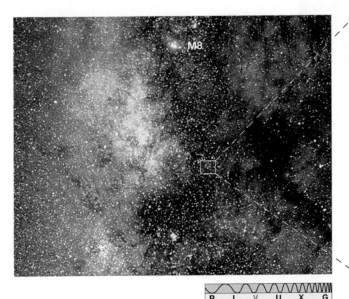

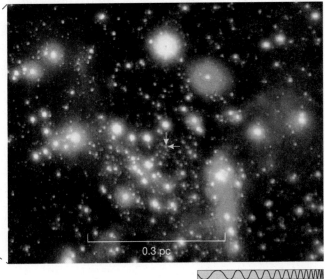

0.3 pc

R I V U X G

A Interactive FIGURE 23.24 Galactic Center Photograph of stellar and interstellar matter in the direction of the Galactic center. Because of heavy obscuration, even the largest optical telescopes can see no farther than one-tenth the distance to the center. To connect with previous figures, the M8 nebula can be seen at extreme top center. ∞ (Sec. 18.2) The field is roughly 10° vertically across and is a continuation of the bottom part of Figure 18.5. The overlaid box outlines the location of the center of our Galaxy. The inset at right shows an adaptive-optics infrared view of the dense stellar cluster surrounding the Galactic center, whose very core is indicated by the twin arrows. *(AURA; ESO)*

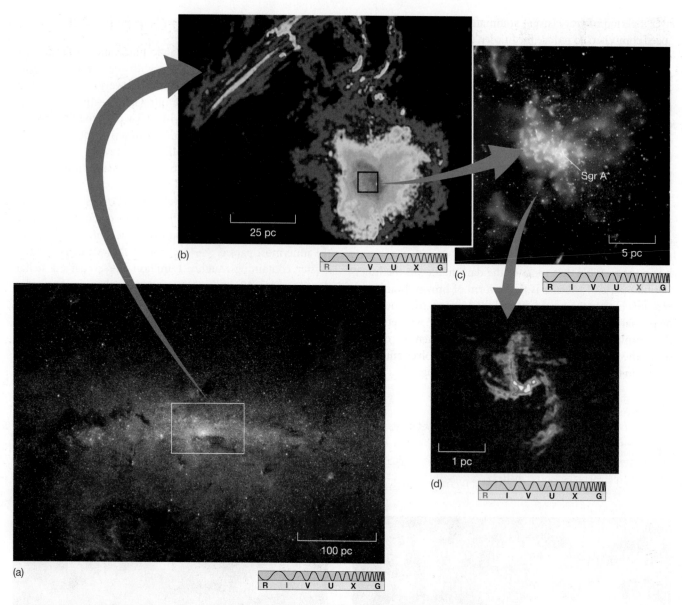

(b)

25 pc

R   I   V   U   X   G

(c)

Sgr A*

5 pc

R   I   V   U   X   G

(d)

1 pc

R   I   V   U   X   G

(a)

100 pc

R   I   V   U   X   G

▲ **FIGURE 23.25 Galactic Center Close-Up** Detailed observations of the center of our Galaxy reveal another important component of galactic structure. (a) An infrared image of part of the Galactic plane shows many bright stars packed into a relatively small volume surrounding the Galactic center (white box). The average density of matter in this boxed region is estimated to be about a million times that in the solar neighborhood. (b) The central portion of our Galaxy, as observed in the radio part of the spectrum. This image shows a region about 100 pc across surrounding the Galactic center (which lies within the bright blob at the bottom right). The long-wavelength radio emission cuts through the Galaxy's dust, providing a view of matter in the immediate vicinity of the Galaxy's center. (c) A recent *Chandra* image showing the relation of a hot supernova remnant (red) and Sgr A*, the suspected black hole at the very center of our Galaxy. (d) The spiral pattern of radio emission arising from Sagittarius A itself suggests a rotating ring of matter only a few parsecs across. All images are false-color, since they lie outside the visible spectrum. (*SST; NRAO; NASA*)

this scale, infrared radiation has been detected from what appear to be huge clouds rich in dust. In addition, radio observations indicate a ring of molecular gas nearly 400 pc across, containing hundreds of thousands of solar masses of material and rotating around the Galactic center at about 100 km/s. The origin of this ring is unclear, although researchers suspect that the gravitational influence of our Galaxy's elongated, ro-

tating bulge may well be involved, deflecting gas from farther out into the dense central regions.

Higher-resolution radio observations reveal further structure on smaller scales. Figure 23.25(b) shows a region called Sagittarius A. (The name simply means that it is the brightest radio source in the constellation Sagittarius.) It lies at the center of the boxed region in Figure 23.24 and

Figure 23.25(a)—and, we think, at the center of our Galaxy. On a scale of about 25 pc, extended filaments can be seen. Their presence suggests to many astronomers that strong *magnetic fields* operate in the vicinity of the center, creating structures similar in appearance to (but much larger than) those observed on the active Sun. ∞ (Sec. 16.5)

On even smaller scales (Figure 23.25c), *Chandra* observations indicate an extended region of hot X-ray-emitting gas, apparently associated with a supernova remnant, in addition to many other individual bright X-ray sources. And within that lies a rotating ring or disk of molecular gas only a few parsecs across, with streams of matter spiraling inward toward the center (shown, again in the radio, in Figure 23.25d). Note that the scale of this disk is comparable to that of the dense central cluster shown in the inset to Figure 23.24.

What could cause all this activity? An important clue comes from the Doppler broadening of infrared spectral lines emitted from the central swirling whirlpool of gas. ∞ (Sec. 4.5) The extent of the broadening indicates that the gas is moving very rapidly. In order to keep this gas in orbit, whatever is at the center must be extremely massive—more than a million solar masses. Given the twin requirements of large mass and small size, a leading contender is a *supermassive black hole*. ∞ (Sec. 22.8)

The hole itself is not the source of the energy, of course. Instead, the vast accretion disk of matter drawn toward the hole by the enormous gravity emits the energy as it falls in, just as we saw (on a much smaller scale) in Chapter 22 when we discussed X-ray emission from neutron stars and stellar-mass black holes. ∞ (Secs. 22.3, 22.8) The strong magnetic fields, thought to be generated within the accretion disk as matter spirals inward, may act as "particle accelerators," creating extremely high-energy particles detected on Earth as *cosmic rays*.

In the late 1990s, the *Compton Gamma Ray Observatory* found indirect evidence for a fountain of high-energy particles, possibly produced by violent processes close to the event horizon, gushing outward from the hole into the halo more than a thousand parsecs beyond the Galactic center. ∞ (Sec. 5.7) Astronomers have reason to suspect that similar events are occurring at the centers of many other galaxies.

The Central Black Hole

Astronomers have identified a candidate for the supermassive black hole at the Galactic center. At the very heart of Sagittarius A, is a remarkable object with the odd-sounding name **Sgr A*** (pronounced "saj ay star"). By the standards of the active galaxies to be studied in Chapter 24, this compact **Galactic nucleus** is not particularly energetic. Still, radio observations made during the past two decades, along with more recent X- and gamma-ray observations, suggest that it is nevertheless a pretty violent place. Its total energy output (at all wavelengths) is estimated to be 10^{33} W, more than a million times that of the Sun.

VLBI observations using radio telescopes arrayed from Hawaii to Massachusetts imply that Sgr A* cannot be much larger than 10 AU, and it is probably a good deal smaller than that. ∞ (Sec. 5.6) This size is consistent with the view that the energy source is a massive black hole. Figure 23.26 is perhaps the strongest evidence to date supporting the black-hole picture. It shows a high-resolution infrared image of the innermost 0.04 pc (or 8000 AU across) near the Galactic center, centered on Sgr A*. Using advanced adaptive-optics techniques on the Keck telescopes and the VLT, U.S. and European researchers have created the first-ever diffraction-limited (0.05" resolution) images of the region. ∞ (Sec. 5.4)

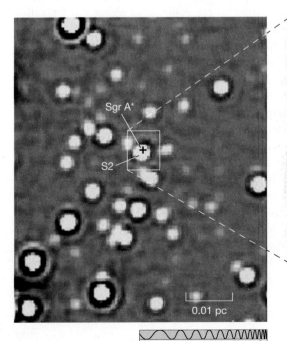

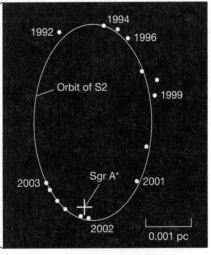

◀ **FIGURE 23.26 Orbits Near the Galactic Center** This extremely close-up map of the Galactic center (left) was obtained by infrared adaptive optics, resulting in an ultra-high-resolution image of the innermost 0.1 pc of the Milky Way. The resolution is high enough that the orbits of individual stars can be tracked with confidence around a still-unseen Sgr A* (marked with a cross). The inset shows the orbit of the innermost star in the frame, labeled S2, between 1992 and 2003. The solid line shows the best-fitting orbit for S2 around a black hole of 4 million solar masses, located at Sgr A*. *(ESO)*

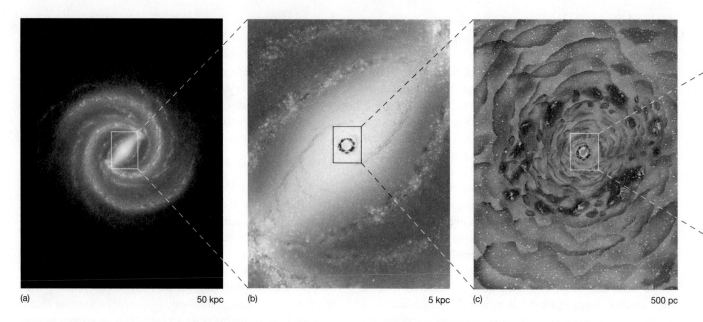

(a)   50 kpc   (b)   5 kpc   (c)   500 pc

▲ **FIGURE 23.27 Galactic Center Zoom** Six artist's conceptions, centered on the Galactic center and each increasing in resolution by a factor of 10. Frame (a) shows the same scene as Figure 23.16. Frame (f) is a rendition of a vast whirlpool within the innermost 0.5 parsec of our Galaxy. The data imaged in Figure 23.25 do not closely match these artistic renderings because the Figure 23.25 view is parallel to the Galactic disk—along the line of sight from the Sun to the Galactic center—whereas these six paintings portray a simplified view perpendicular to the disk, while progressively zooming down onto that disk. *(L. Chaisson)*

Remarkably, the image quality is good enough that the *proper motions* of several of the stars—their orbits around the Galactic center—can clearly be seen. The inset shows a series of observations of one of the brightest stars—called S2—over a 10-year period. The motion is consistent with an orbit around a massive object at the location of Sgr A*, in accordance with Newton's laws of motion. ∞ (Sec. 2.7). The solid curve on the figure shows the elliptical orbit that best fits the observations: a 15-year orbit with a semimajor axis of 950 AU, corresponding (from Kepler's third law, as modified by Newton) to a central mass of approximately 4 million solar masses. The small size of the central object is very clearly demonstrated by the motion of another star in the group (S16), whose extremely eccentric orbit brings it within just 45 AU of the center.

Other observations, using adaptive-optics infrared-imaging techniques, have revealed a bright source very close to Sgr A* that seems to vary with a 10-minute period. ∞ (Sec. 5.4) The source could be a hot spot on the accretion disk that circles the purported hole. Note that, even with the large mass just mentioned, if Sgr A* is a genuine black hole, the size of its event horizon is still only 0.02 AU. ∞ (Sec. 22.5) Such a small region, 8 kpc away, is currently unresolvable with any telescope now in existence.

Figure 23.27 places these findings into a simplified perspective. Each frame is centered on the Galaxy's core, and each increases in resolution by a factor of 10. Frame (a) renders the Galaxy's overall shape, as painted in Figure 23.16. This frame measures about 50 kpc across. Frame (b) spans a distance of 5 kpc from side to side and is nearly filled by the Galactic bar and the great sweep of the innermost spiral arm. Moving in to a 500-pc span, frame (c) depicts part of the 400-pc ring of matter mentioned earlier and some young dense star clusters, evidence of recent star formation near the Galactic center. The dark blobs represent giant molecular clouds, the pink patches emission nebulae associated with star formation within those clouds. In parts (b) and (c), the artist has peeled away the bright bulge, enabling us to "see" better into the central regions.

In Figure 23.27(d), at 50 pc, a pinkish (thin, warm) region of ionized gas surrounds the reddish (thicker, warmer) heart of the Galaxy, corresponding approximately to the images shown in Figure 23.25(b) and(c). The sources of the energy responsible for this vast ionized cloud are supernovae and activity in the Galactic center. The central cluster and the surrounding star-forming ring can also be seen. Figure 23.27(e), spanning 5 pc, depicts the central cluster and ring in more detail, along with the tilted, spinning whirlpool of hot (10^4 K) gas surrounding the center of our Galaxy. The innermost part of this gigantic whirlpool is shown in Figure 27.27(f), in which a swiftly spinning, white-hot disk of gas with temperatures in the millions of kelvins nearly engulfs

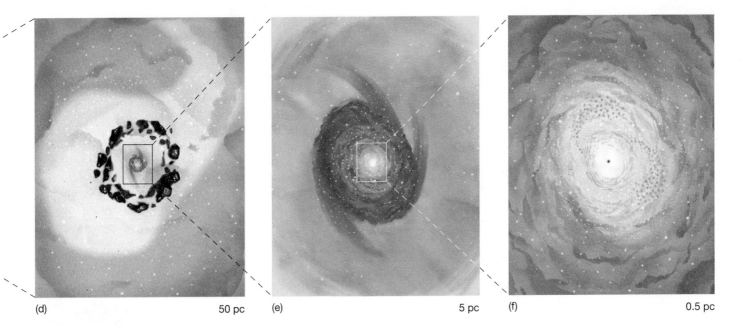

(d) 50 pc (e) 5 pc (f) 0.5 pc

the central black hole. Small star clusters within the central cluster are torn apart by the black hole's gravitational pull. The black hole and even the stellar orbits shown in Figure 23.26 are still too small to be pictured (even as minute dots) on this scale.

The last decade has seen an explosion in our knowledge of the innermost few parsecs of our Galaxy, and astronomers are working hard to decipher the clues hidden within its invisible radiation. Still, we are only now beginning to appreciate the full magnitude of this strange new realm deep in the heart of the Milky Way.

CONCEPT CHECK

✔ What is the most likely explanation for the energetic events observed at the Galactic center?

CHAPTER REVIEW

SUMMARY

1 A **galaxy (p. 574)** is a huge collection of stellar and interstellar matter isolated in space and bound together by its own gravity. Because we live within it, the **Galactic disk (p. 574)** of our own Milky Way Galaxy appears as a broad band of light across the sky, a band called the Milky Way. Near the center, the disk thickens into the **Galactic bulge (p. 574)**. The disk is surrounded by a roughly spherical **Galactic halo (p. 574)** of old stars and star clusters. Like many others visible in the sky, our Galaxy is a **spiral galaxy (p. 576)**.

2 The halo can be studied using **variable stars (p. 577)**, whose luminosity changes with time. **Pulsating variable stars (p. 577)** vary in brightness in a repetitive and predictable way. Of particular importance to astronomers are **RR Lyrae variables (p. 577)** and **Cepheid**

variables (p. 577). All RR Lyrae stars have roughly the same luminosity. For Cepheids, the luminosity can be determined using the **period–luminosity relationship (p. 579)**. Knowing the luminosity, astronomers can apply the inverse-square law to determine the distance. The brightest Cepheids can be seen at distances of millions of parsecs, extending the cosmic distance ladder well beyond our own Galaxy. In the early 20th century, Harlow Shapley used RR Lyrae stars to determine the distances to many of the Galaxy's globular clusters and found that they have a roughly spherical distribution in space, but the center of the sphere lies far from the Sun. The center of their distribution is close to the **Galactic center (p. 581)**, about 8 kpc away.

3 The Galactic halo lacks gas and dust, so no new stars are forming there. All halo stars are old. The gas-rich disk is the site of current star formation and contains many young stars. Stars in the halo and bulge move on largely random three-dimensional orbits that pass repeatedly through the plane of the disk, but have no preferred orientation. Stars and gas in the disk move on roughly circular orbits around the Galactic center. Halo stars appeared early on, before the Galactic disk took shape, when there was no preferred orientation for their orbits. After the disk formed, stars born there inherited its overall spin and so move on circular orbits in the Galactic plane.

4 Radio observations clearly reveal the extent of our Galaxy's **spiral arms (p. 588)**, regions of the densest interstellar gas where star formation is taking place. The spirals cannot be "tied" to the disk material, as the disk's differential rotation would have wound them up long ago. Instead, they may be **spiral density waves (p. 588)** that move through the disk, triggering star formation as they pass by. Alternatively, the spirals may arise from **self-propagating star formation (p. 591)**, when shock waves produced by the formation and evolution of one generation of stars trigger the formation of the next.

5 The Galactic **rotation curve (p. 592)** plots the orbital speed of matter in the disk versus distance from the Galactic

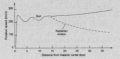

center. By applying Newton's laws of motion, astronomers can determine the mass of the Galaxy. They find that the Galactic mass continues to increase beyond the radius defined by the globular clusters and the spiral structure we observe. Our Galaxy, like many others, has an invisible **dark halo (p. 592)** containing far more mass than can be accounted for in the form of luminous matter. The **dark matter (p. 593)** making up these dark halos is of unknown composition. Leading candidates include low-mass stars and exotic subatomic particles. Recent attempts to detect stellar dark matter have used the fact that a faint foreground object can occasionally pass in front of a more distant star, deflecting the star's light and causing its apparent brightness to increase temporarily. This deflection is called **gravitational lensing (p. 594)**.

6 Astronomers working at infrared and radio wavelengths have uncovered evidence for energetic activity within a few parsecs of the Galactic center. The leading explanation is that a black hole roughly 4 million times more massive than the Sun resides there. The hole lies at the center of a dense star cluster containing millions of stars, which is in turn surrounded by a star-forming disk of molecular gas. The observed activity is thought to be powered by accretion onto the black hole, as well as by supernova explosions in the cluster surrounding it.

Mastering ASTRONOMY *For instructor-assigned homework go to **www.masteringastronomy.com***

Problems labeled **POS** explore the process of science | **VIS** problems focus on reading and interpreting visual information

REVIEW AND DISCUSSION

1. **POS** What evidence do we have that we live in a disk galaxy?

2. Why is it difficult to map out our Galaxy from our vantage point on Earth?

3. What are spiral nebulae? How did they get that name?

4. In what region of the Galaxy are globular clusters found?

5. How are Cepheid variables used in determining distances?

6. Roughly how far out into space (looking out of the Galactic disk) can we use Cepheids to measure distance?

7. What important discoveries were made early in the twentieth century by using RR Lyrae variables?

8. Of what use is radio astronomy in the study of Galactic structure?

9. Contrast the motions of disk and halo stars.

10. How do we know that the Milky Way Galaxy has spiral arms?

11. **POS** Explain why galactic spiral arms are thought to be regions of recent and ongoing star formation.

12. Describe what happens to interstellar gas as it passes through a spiral density wave.

13. What is self-propagating star formation?

14. What do the red stars in the Galactic halo tell us about the history of the Milky Way?

15. What does the rotation curve of our Galaxy tell us about the Galaxy's total mass?

16. **POS** What evidence is there for dark matter in the Galaxy?

17. Describe some candidates for Galactic dark matter.

18. What is gravitational lensing, and can astronomers use it to search for dark matter?

19. Why can't optical astronomers easily study the center of our Galaxy?

20. **POS** Describe some ways in which astronomers can observe the Galactic center.

CONCEPTUAL SELF-TEST: MULTIPLE CHOICE

1. Most of the bright stars in our Galaxy are located in the Galactic (**a**) center; (**b**) bulge; (**c**) halo; (**d**) disk.

2. **VIS** According to Figure 23.7 ("Period–Luminosity Plot"), a Cepheid variable star with luminosity 1000 times that of the Sun has a pulsation period of roughly (**a**) 1 day; (**b**) 3 days; (**c**) 10 days; (**d**) 50 days.

3. Globular clusters are found mainly (**a**) in the Galactic center; (**b**) in the Galactic disk; (**c**) in spiral arms; (**d**) in the Galactic halo.

4. Shapley measured the distances to globular clusters by using (**a**) trigonometric parallax; (**b**) a comparison of the absolute and apparent magnitudes of variable stars; (**c**) spectroscopic parallax; (**d**) radar ranging.

5. In the Milky Way Galaxy, our Sun is located (**a**) near the Galactic center; (**b**) about halfway out from the center; (**c**) at the outer edge; (**d**) in the halo.

6. A telescope searching for newly formed stars would make the most discoveries if it were pointed (**a**) directly away from the Galactic center; (**b**) perpendicular to the Galactic disk; (**c**) within a spiral arm; (**d**) between spiral arms.

7. The first stars that formed in the Milky Way now (**a**) have random orbits in the halo; (**b**) orbit in the Galactic plane; (**c**) orbit closest to the Galactic center; (**d**) orbit in the same direction as the Milky Way spins.

8. **VIS** Figure 23.21 ("Galaxy Rotation Curve") tells us that (**a**) the Galaxy rotates like a solid body; (**b**) far from the center, the Galaxy rotates more slowly than we would expect based on the light we see; (**c**) far from the center, the Galaxy rotates more rapidly than we would expect based on the light we see; (**d**) there is no matter beyond about 15 kpc from the Galactic center.

9. Most of the mass of the Milky Way exists in the form of (**a**) stars; (**b**) gas; (**c**) dust; (**d**) dark matter.

10. A black hole probably exists at the Galactic center because (**a**) stars near the center of the Milky Way are disappearing; (**b**) no stars can be seen in the vicinity of the Galactic center; (**c**) stars near the center of the Milky Way are orbiting some unseen object; (**d**) the Galaxy rotates faster than astronomers would expect.

PROBLEMS

The number of dots preceding each Problem indicates its approximate level of difficulty.

1. • Calculate the angular diameter of a prestellar nebula of radius 100 AU lying 100 pc from Earth. Compare this with the roughly 6° diameter of the Andromeda Galaxy (Figure 23.2a).

2. •• How close would the nebula in the previous question have to be in order to have the same angular diameter as Andromeda? Calculate the apparent magnitude of the central star if it had a luminosity 10 times that of the Sun.

3. • What is the greatest distance at which an RR Lyrae star of absolute magnitude 0 could be seen by a telescope capable of detecting objects as faint as 20th magnitude?

4. • A typical Cepheid variable is 100 times brighter than a typical RR Lyrae star. How much farther away than RR Lyrae stars can Cepheids be used as distance-measuring tools?

5. •• An astronomer looking through the *Hubble Space Telescope* can see a star with solar luminosity at a distance of 100,000 pc. The brightest Cepheids have luminosities 30,000 times greater than that of the Sun. Taking the Sun's absolute magnitude to be 5, calculate the absolute magnitudes of these bright Cepheids. Neglecting interstellar absorption, how far away can *HST* see them?

6. ••• What is the maximum distance at which *Hubble* could see the Cepheid in the previous question if it lay in the Galactic disk, with an average interstellar extinction of 2.5 magnitudes per kiloparsec?

7. •• Calculate the proper motion (in arc seconds per year) of a globular cluster with a transverse velocity (relative to the Sun) of 200 km/s and a distance of 3 kpc. Do you think that this motion is measurable?

8. •• Calculate the total mass of the Galaxy lying within 20 kpc of the Galactic center if the rotation speed at that radius is 240 km/s.

9. •• Using the data presented in Figure 23.21, estimate the distance from the Galactic center at which matter takes (a) 100 million years and (b) 500 million years to complete one orbit.

10. •• Material at an angular distance of 0.2" from the Galactic center is observed to have an orbital speed of 1200 km/s. If the Sun's distance to the Galactic center is 8 kpc, and the material's orbit is circular and is seen edge-on, calculate the radius of the orbit and the mass of the object around which the material is orbiting.

24

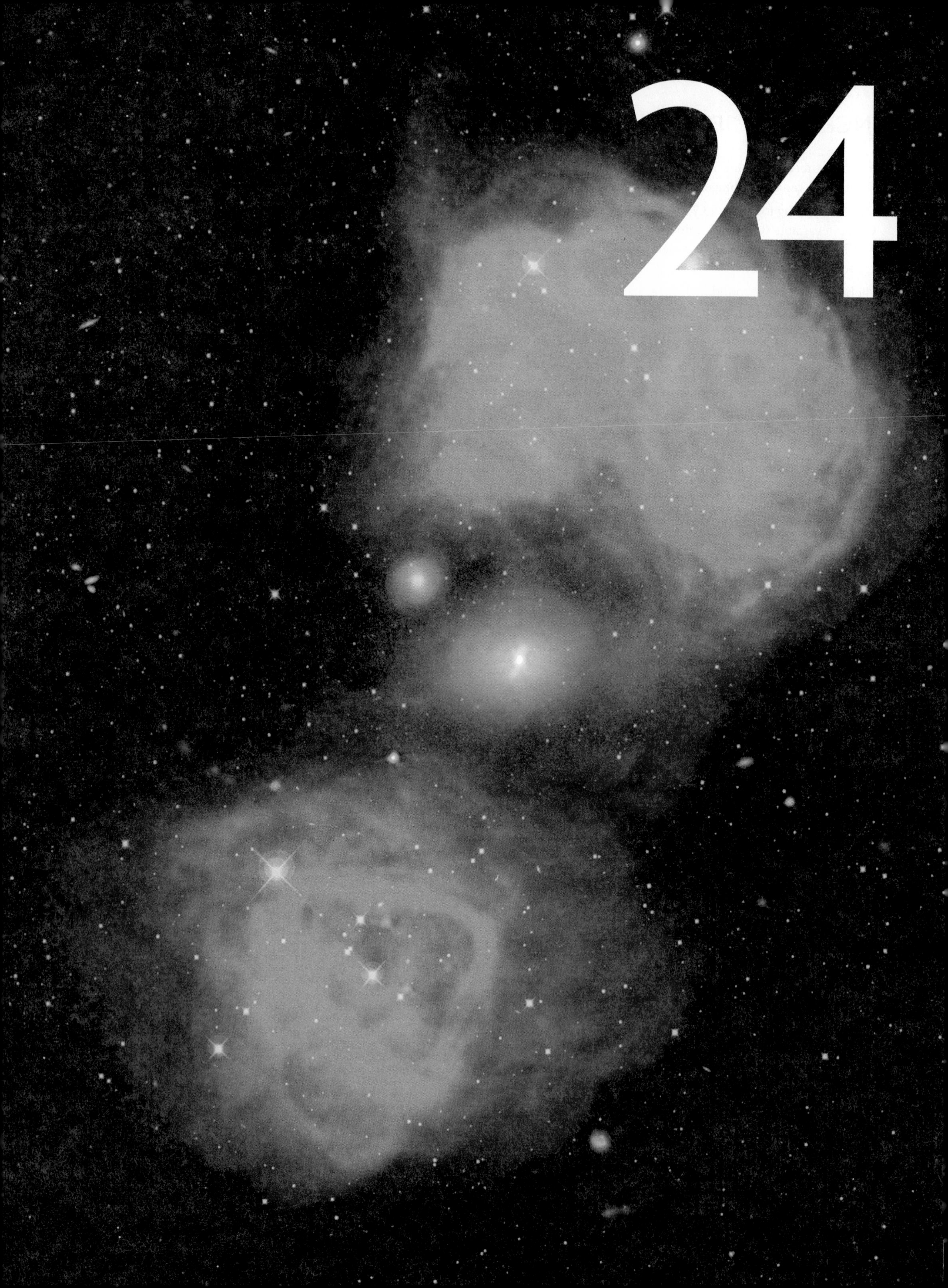

GALAXIES

BUILDING BLOCKS OF THE UNIVERSE

LEARNING GOALS

Studying this chapter will enable you to

1 Describe the basic properties of normal galaxies.

2 Discuss the distance-measurement techniques that enable astronomers to map the universe beyond the Milky Way.

3 Describe how galaxies are observed to clump into clusters.

4 State Hubble's law and explain how it is used to derive distances to the most remote objects in the observable universe.

5 Specify the basic differences between active and normal galaxies.

6 Describe some important features of active galaxies.

7 Explain what drives the central engine thought to power all active galaxies.

THE BIG PICTURE The light we see from most galaxies is fairly well explained as the cumulative output of billions of stars. But some "active" galaxies are different. They are much more luminous, and most of their energy comes from intensely bright central nuclei harboring supermassive black holes.

Visit the Study Area in www.masteringastronomy.com for quizzes, animations, videos, interactive figures, and self-guided tutorials.

As our field of view expands to truly cosmic scales, the focus of our studies shifts dramatically. Planets become inconsequential, stars themselves mere points of hydrogen consumption. Now entire galaxies become the "atoms" from which the universe is built—distant realms completely unknown to scientists just a century ago.

We know of literally millions of galaxies beyond our own. Most are smaller than the Milky Way, some comparable in size, a few much larger. Many are sites of explosive events far more energetic than anything ever witnessed in our own Galaxy. All are vast, gravitationally bound assemblages of stars, gas, dust, dark matter, and radiation separated from us by almost incomprehensibly large distances. The light we receive tonight from the most distant galaxies was emitted long before Earth existed. By studying the properties of galaxies, we gain insight into the history of our Galaxy and the universe in which we live.

LEFT: *This is a double image, superposing optical light (acquired by the Hubble Space Telescope in Earth orbit) on radio emission (captured by the Very Large Array in New Mexico). At center (in white) is a giant, visible elliptical galaxy cataloged as NGC 1316, which is probably devouring its small northern neighbor. The result is the complex radio emission (in orange), called Fornax A, spanning more than a million light-years of space. (NRAO/STScI)*

24.1 Hubble's Galaxy Classification

Figure 24.1 shows a vast expanse of space lying about 100 million pc from Earth. Almost every patch or point of light in this figure is a separate galaxy—hundreds can be seen in just this one photograph. Over the years, astronomers have accumulated similar images of many millions of galaxies. We begin our study of these enormous accumulations of matter simply by considering their appearance on the sky.

Seen through even a small telescope, images of galaxies look distinctly nonstellar. They have fuzzy edges, and many are quite elongated—not at all like the sharp, pointlike images normally associated with stars. Although it is difficult to tell from the photograph, some of the blobs of light in Figure 24.1 are spiral galaxies like the Milky Way Galaxy and Andromeda. Others, however, are definitely not spirals—no disks or spiral arms can be seen. Even when we take into account their different orientations in space, galaxies do *not* all look the same.

The American astronomer Edwin Hubble was the first to categorize galaxies in a comprehensive way. Working with the then recently completed 2.5-m optical telescope on Mount Wilson in California in 1924, he classified the galaxies he saw into four basic types—*spirals, barred spirals, ellipticals,* and *irregulars*—solely on the basis of their visual appearance. Many modifications and refinements have been incorporated over the years, but the basic **Hubble classification scheme** is still widely used today.

Spirals

We saw several examples of **spiral galaxies** in Chapter 23—for example, our own Milky Way Galaxy and our neighbor Andromeda. ∞ (Sec. 23.1) All galaxies of this type contain a flattened galactic disk in which spiral arms are found, a central galactic bulge with a dense nucleus, and an extended halo of faint, old stars. ∞ (Sec. 23.3) The stellar density (i.e., the number of stars per unit volume) is greatest in the *galactic nucleus,* at the center of the bulge. However, within this general description, spiral galaxies exhibit a wide variety of shapes, as illustrated in Figure 24.2.

In Hubble's scheme, a spiral galaxy is denoted by the letter S and classified as type a, b, or c according to the size of its central bulge. Type Sa galaxies have the largest bulges, Type Sc the smallest. The tightness of the spiral pattern is quite well correlated with the size of the bulge (although the correspondence is not perfect). Type Sa spiral galaxies tend to have tightly wrapped, almost circular, spiral arms, Type Sa galaxies typically have more open spiral arms, and Type Sc spirals often have a loose, poorly defined spiral structure. The arms also tend to become more "knotty," or clumped, in appearance as the spiral pattern becomes more open.

(a)

R I V U X G

▲ **FIGURE 24.1 Coma Cluster** (a) A collection of many galaxies, each consisting of hundreds of billions of stars. Called the Coma Cluster, this group of galaxies lies more than 100 million pc from Earth. (The blue spiked object at top right is a nearby star; virtually every other object in this image is a galaxy.) (b) A recent *Hubble Space Telescope* image of part of the cluster. *(AURA; NASA)*

(b)

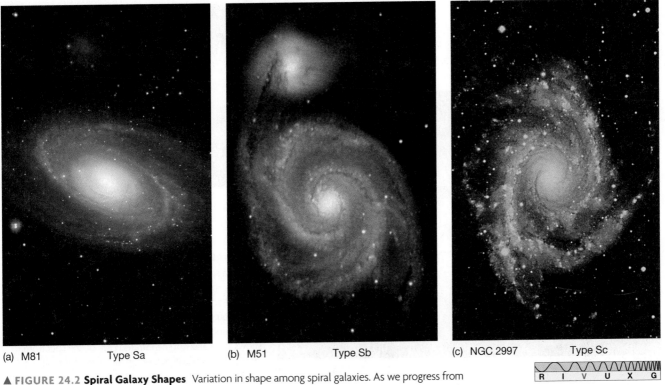

(a) M81   Type Sa   (b) M51   Type Sb   (c) NGC 2997   Type Sc

R	I	V	U	X	G

▲ **FIGURE 24.2 Spiral Galaxy Shapes** Variation in shape among spiral galaxies. As we progress from type Sa to Sb to Sc, the bulges become smaller and the spiral arms tend to become less tightly wound. *(R. Gendler; NOAO; D. Malin/AAT)*

The bulges and halos of spiral galaxies contain large numbers of reddish old stars and globular clusters, similar to those observed in our own Galaxy and in Andromeda. Most of the light from spirals, however, comes from A- through G-type stars in the galactic disk, giving these galaxies an overall whitish glow. We assume that thick disks exist, too, but their faintness makes this assumption hard to confirm— the thick disk in the Milky Way contributes only a percent or so of our Galaxy's total light. ∞ (Sec. 23.3)

Like the disk of the Milky Way, the flat disks of typical spiral galaxies are rich in gas and dust. Type Sc galaxies contain the most interstellar matter, Sa galaxies the least. The 21-cm radio radiation emitted by spirals betrays the presence of the gas, and obscuring dust lanes are clearly visible in many systems (see Figures 24.2b and c). ∞ (Sec. 18.4) Stars are forming within the spiral arms, which contain numerous emission nebulae and newly formed O- and B-type stars. ∞ (Secs. 18.2, 23.5) The arms appear bluish because of the presence of bright blue O- and B-type stars there. The photo of the Sc galaxy NGC 2997 shown in Figure 24.2(c) reveals the preponderance of interstellar gas, dust, and young blue stars tracing the spiral pattern particularly clearly. Spirals are not necessarily young galaxies, however: Like our own Galaxy, they are simply rich enough in interstellar gas to provide for continued stellar birth.

Most spirals are not seen face-on, as they are shown in Figure 24.2. Many are tilted with respect to our line of sight,

making their spiral structure hard to discern. However, we do not need to see spiral arms to classify a galaxy as a spiral. The presence of the disk, with its gas, dust, and newborn stars, is sufficient. For example, the galaxy shown in Figure 24.3 is classified as a spiral because of the clear line of obscuring dust seen along its midplane. (Incidentally, this relatively nearby galaxy was another of the "nebulae" figuring in the Shapley–Curtis debate discussed in Chapter 23. ∞ (Sec. 23.2) The visible dust lane was interpreted by Curtis as an obscuring "ring" of material, leading him to suggest that our Galactic plane might contain a similar feature.)

Barred Spirals

A variation of the spiral category in Hubble's classification scheme is the **barred-spiral galaxy**. Barred spirals differ from ordinary spirals mainly by the presence of an elongated "bar" of stellar and interstellar matter passing through the center and extending beyond the bulge, into the disk. The spiral arms project from near the ends of the bar rather than from the bulge (their origin in normal spirals). Barred spirals are designated by the letters SB and are subdivided, like the ordinary spirals, into categories SBa, SBb, and SBc, depending on the size of the bulge. Again, like ordinary spirals, the tightness of the spiral pattern is correlated with the size of the bulge. Figure 24.4 shows the variation among barred-spiral galaxies. In the case of the SBc category, it is

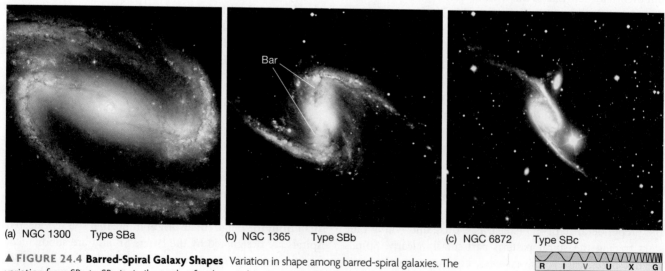

◄ **FIGURE 24.3 Sombrero Galaxy** The Sombrero Galaxy, a spiral system seen edge-on. Officially cataloged as M104, this galaxy has a dark band composed of interstellar gas and dust. The large size of this galaxy's central bulge marks it as type Sa, even though its spiral arms cannot be seen from our perspective. The inset shows this galaxy in the infrared part of the spectrum, highlighting its dust content in false-colored pink. *(NASA)*

often hard to tell where the bar ends and the spiral arms begin.

Frequently, astronomers cannot distinguish between spirals and barred spirals, especially when a galaxy happens to be oriented with its galactic plane nearly edge-on toward Earth, as in Figure 24.3. Because of the physical and chemical similarities of spiral and barred-spiral galaxies, some researchers do not even bother to distinguish between them. Others, however, regard the differences in their structures as very important, arguing that these differences suggest basic dissimilarities in the ways the two types of galaxies formed and evolved.

The discovery that the bulge of our own Galaxy is elongated suggests that the Milky Way may be a barred

(a) NGC 1300   Type SBa   (b) NGC 1365   Type SBb   (c) NGC 6872   Type SBc

▲ **FIGURE 24.4 Barred-Spiral Galaxy Shapes** Variation in shape among barred-spiral galaxies. The variation from SBa to SBc is similar to that for the spirals in Figure 24.2, except that now the spiral arms begin at either end of a bar through the galactic center. In frame (c), the bright star is a foreground object in our own Galaxy; the object at top center is another galaxy that is probably interacting with NGC 6872. *(NASA; AAT; ESO)*

spiral, of type SBb or SBc. However, the full extent of the bar remains uncertain. Some astronomers place our Galaxy in an intermediate category lying between Hubble's spiral and barred-spiral types. ∞ (Sec. 23.3)

Ellipticals

Unlike the spirals, **elliptical galaxies** have no spiral arms and, in most cases, no obvious galactic disk—in fact, other than possessing a dense central nucleus, they often exhibit little internal structure of any kind. As with spirals, the stellar density increases sharply in the central nucleus. Denoted by the letter E, these systems are subdivided according to how elliptical they appear on the sky. The most circular are designated E0, slightly flattened systems are labeled E1, and so on, all the way to the most elongated ellipticals, of type E7 (Figure 24.5).

Notice, by the way, that the Hubble type of an elliptical galaxy depends both on the galaxy's intrinsic three-dimensional shape *and* on its orientation relative to the line of sight. Consider, for example, a spherical galaxy, a cigar-shaped galaxy seen end on, and an M&M-shaped galaxy seen face on; all would appear circular in the sky. As a result, it is often difficult to decipher a galaxy's true shape from its visual appearance.

There is a large range in both the size and the number of stars contained in elliptical galaxies. The largest elliptical galaxies are much larger than our own Milky Way Galaxy. These *giant ellipticals* can range up to hundreds of kiloparsecs across and contain trillions of stars. At the other extreme, *dwarf ellipticals* may be as small as 1 kpc in diameter and contain fewer than a million stars. The significant observational differences between giant and dwarf ellipticals have led many astronomers to conclude that these galaxies are members of

separate classes, with quite different histories of formation and stellar content. The dwarfs are by far the most common type of ellipticals, outnumbering their brighter counterparts by about 10 to 1. However, most of the *mass* that exists in the form of elliptical galaxies is contained in the larger systems.

The absence of spiral arms is not the only difference between spirals and ellipticals: Most ellipticals also contain little or no cool gas and dust. The 21-cm radio emission from neutral hydrogen gas is, with few exceptions, completely absent, and no obscuring dust lanes are seen. In most cases, there is no evidence of young stars or ongoing star formation. Like the halo of our own Galaxy, ellipticals are made up mostly of old, reddish, low-mass stars. Also, as in the halo of our Galaxy, the orbits of stars in ellipticals are disordered, exhibiting little or no overall rotation; objects move in all directions, not in regular, circular paths as in our Galaxy's disk. Ellipticals differ from our Galaxy's halo in at least one important respect, however: X-ray observations reveal large amounts of very *hot* (several million kelvins) interstellar gas distributed throughout their interiors, often extending well beyond the visible portions of the galaxies (Figure 24.5a,b).

Some giant ellipticals are exceptions to many of the foregoing general statements about elliptical galaxies, as they have been found to contain disks of gas and dust in which stars are forming. Astronomers think that these systems may be the results of collisions among gas-rich galaxies (see Section 25.2). Indeed, galactic collisions may have played an important role in determining the appearance of many of the systems we observe today.

Intermediate between the E7 ellipticals and the Sa spirals in the Hubble classification is a class of galaxies that show evidence of a thin disk and a flattened bulge, but that contain no gas and no spiral arms. Two such objects are shown in

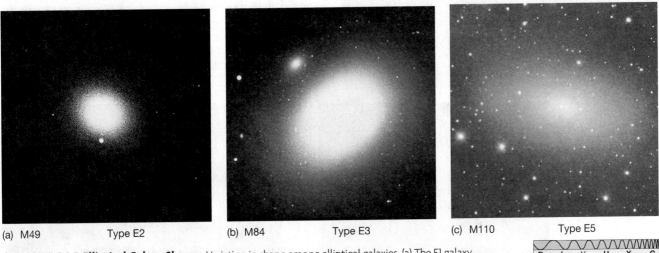

| (a) M49 | Type E2 | (b) M84 | Type E3 | (c) M110 | Type E5 |

▲ **FIGURE 24.5 Elliptical Galaxy Shapes**   Variation in shape among elliptical galaxies. (a) The E1 galaxy M49 is nearly circular in appearance. (b) M84 is a slightly more elongated elliptical galaxy, classified as E3. Both galaxies lack spiral structure, and neither shows evidence of cool interstellar dust or gas, although each has an extensive X-ray halo of hot gas that extends far beyond the visible portion of the galaxy. (c) M110 is a dwarf elliptical companion to the much larger Andromeda Galaxy. (*AURA; SAO; R. Gendler*)

R   I   V   U   X   G

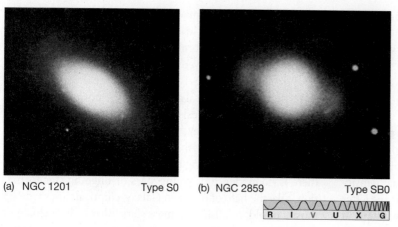

(a) NGC 1201   Type S0   (b) NGC 2859   Type SB0

R I V U X G

◄ **FIGURE 24.6 S0 Galaxies** (a) S0 (or lenticular) galaxies contain a disk and a bulge, but no interstellar gas and no spiral arms. They are in many respects intermediate between E7 ellipticals and Sa spirals in their properties. (b) SB0 galaxies are similar to S0 galaxies, except for a bar of stellar material extending beyond the central bulge. *(Palomar/Caltech)*

Figure 24.6. These galaxies are known as **S0 galaxies** if no bar is evident and **SB0 galaxies** if a bar is present. They are also known as *lenticular* galaxies, because of their lens-shaped appearance. They look a little like spirals whose dust and gas have been stripped away, leaving behind just a stellar disk. Observations in recent years have shown that many normal elliptical galaxies have faint disks within them, like the S0 galaxies. As with the S0s, the origin of these disks is uncertain, but some researchers suspect that S0s and ellipticals may be closely related.

Irregulars

The final class of galaxies identified by Hubble is a catch-all category—**irregular galaxies**—so named because their visual appearance does not allow us to place them into any of the other categories just discussed. Irregulars tend to be rich in interstellar matter and young, blue stars, but they lack any regular structure, such as well-defined spiral arms or central bulges. They are divided into two subclasses: Irr I galaxies and Irr II galaxies. The Irr I galaxies often look like misshapen spirals.

Irregular galaxies tend to be smaller than spirals, but somewhat larger than dwarf ellipticals. They typically contain between 10^8 and 10^{10} stars. The smallest such galaxies are called *dwarf irregulars*. As with elliptical galaxies, the dwarf type is the most common irregular. Dwarf ellipticals and dwarf irregulars occur in approximately equal numbers and together make up the vast majority of galaxies in the universe. They are often found close to a larger "parent" galaxy.

Figure 24.7 shows the **Magellanic Clouds,** a famous pair of Irr I galaxies that orbit the Milky Way Galaxy. They are shown to proper scale in Figure 23.16. Studies of Cepheid

◄ **FIGURE 24.7 Magellanic Clouds** The Magellanic Clouds are prominent features of the night sky in the Southern Hemisphere. Named for the 16th-century Portuguese explorer Ferdinand Magellan, whose around-the-world expedition first brought word of these fuzzy patches of light to Europe, they are dwarf irregular (Irr I) galaxies, gravitationally bound to our own Milky Way Galaxy. They orbit our Galaxy and accompany it on its trek through the cosmos. (a) The Clouds' relationship to one another in the southern sky. Both the Small (b) and the Large (c) Magellanic Cloud have distorted, irregular shapes, although some observers claim they can discern a single spiral arm in the Large Cloud. *(F. Espenak; Harvard Observatory)*

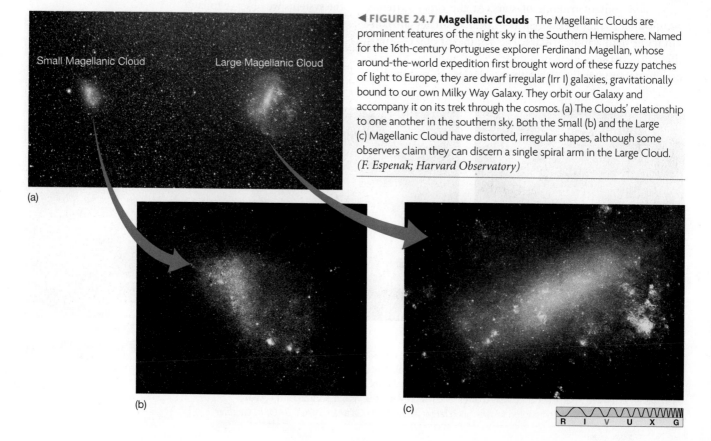

Small Magellanic Cloud   Large Magellanic Cloud

(a)

(b)

(c)

R I V U X G

variables within the Clouds show them to be approximately 50 kpc from the center of our Galaxy. ∞ (Sec. 23.2) The Large Cloud contains about 6 billion solar masses of material and is a few kiloparsecs across. Both Clouds contain lots of gas, dust, and blue stars (and the recent, well-documented supernova discussed in *Discovery 21-1*), indicating ongoing star formation. Both also contain many old stars and several old globular clusters, so we know that star formation has been going on in them for a very long time.

Radio studies hint at a possible bridge of hydrogen gas connecting the Milky Way to the Magellanic Clouds, although more observational data are still needed to establish this link beyond doubt. It is possible that the tidal force of the Milky Way tore a stream of gas from the Clouds the last time their orbits brought them close to our Galaxy. Of course, gravity works both ways, and many researchers reason that the forces exerted by the Clouds may in turn be responsible for distorting our Galaxy, warping and thickening the outer parts of the Galactic disk. ∞ (Sec. 23.5)

The much rarer Irr II galaxies (Figure 24.8), in addition to their irregular shape, have other peculiarities, often exhibiting a distinctly explosive or filamentary appearance. Their appearance once led astronomers to suspect that violent events had occurred within them. However, it now seems more likely that, in some (but probably not all) cases, we are seeing the result of a close encounter or collision between two previously "normal" systems.

The Hubble Sequence

Table 24.1 summarizes the basic characteristics of the various types of galaxies. When he first developed his classification scheme, Hubble arranged the galaxies into the "tuning fork" diagram shown in Figure 24.9. The variation in types across the diagram, from ellipticals to spirals to irregulars, is often referred to as the *Hubble sequence*.

Hubble's primary aim in creating this diagram was to indicate similarities in appearance among galaxies. However, he also regarded the tuning fork as an evolutionary sequence from left to right, with E0 ellipticals evolving into flatter ellipticals and S0 systems and ulti-mately forming disks and spiral arms. Indeed, Hubble's terminology referring to ellipticals as "early-type" and spirals as "late-type" galaxies is still widely used today. However, as far as modern astronomers can tell, there is no direct evolutionary connection of this sort along the Hubble sequence. Isolated normal galaxies do *not* evolve from one type to another. Spirals are not ellipticals that have grown arms, nor are ellipticals spirals that have somehow expelled their star-forming disks. Some astronomers do suspect that bars may be transient features and that barred-spiral galaxies may therefore evolve into ordinary spirals, but, in general, astronomers know of no simple parent–child relationship among Hubble types.

However, the key word in the previous paragraph is *isolated*. As described in Section 25.2, there is now strong

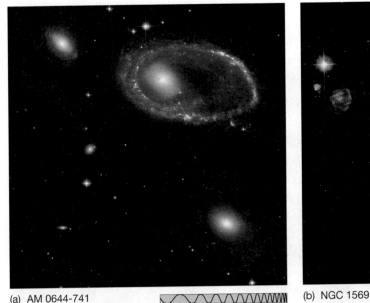

(a) AM 0644-741

(b) NGC 1569

▲ **FIGURE 24.8 Irregular Galaxy Shapes** Some irregular (Irr II) galaxies. (a) The strangely shaped galaxy AM 0644-741 is probably plunging headlong into a group of several other galaxies (two of them shown here), causing huge rearrangements of its stars, gas, and dust. (b) The galaxy NGC 1569 seems to show an explosive appearance, probably the result of a recent galaxywide burst of star formation. (*NASA*)

TABLE 24.1 Galaxy Properties by Type

	Spiral/Barred Spiral (S/SB)	Elliptical* (E)	Irregular (Irr)
Shape and structural properties	Highly flattened disk of stars and gas, containing spiral arms and thickening central bulge. Sa and SBa galaxies have the largest bulges, the least obvious spiral structure, and roughly spherical stellar halos. SB galaxies have an elongated central "bar" of stars and gas.	No disk. Stars smoothly distributed through an ellipsoidal volume ranging from nearly spherical (E0) to very flattened (E7) in shape. No obvious substructure other than a dense central nucleus.	No obvious structure. Irr II galaxies often have "explosive" appearances.
Stellar content	Disks contain both young and old stars; halos consist of old stars only.	Contain old stars only.	Contain both young and old stars.
Gas and dust	Disks contain substantial amounts of gas and dust; halos contain little of either.	Contain hot X-ray emitting gas, little or no cool gas and dust.	Very abundant in gas and dust.
Star formation	Ongoing star formation in spiral arms.	No significant star formation during the last 10 billion years.	Vigorous ongoing star formation.
Stellar motion	Gas and stars in disk move in circular orbits around the galactic center; halo stars have random orbits in three dimensions.	Stars have random orbits in three dimensions	Stars and gas have highly irregular orbits.

* As noted in the text, some giant ellipticals appear to be the result of collisions between gas-rich galaxies and are exceptions to many of the statements listed here.

observational evidence that collisions and tidal interactions *between* galaxies are commonplace and that these encounters are the main physical processes driving the evolution of galaxies. We will return to this important subject in Chapter 25.

CONCEPT CHECK

✔ In what ways are large spirals like the Milky Way and Andromeda not representative of galaxies as a whole?

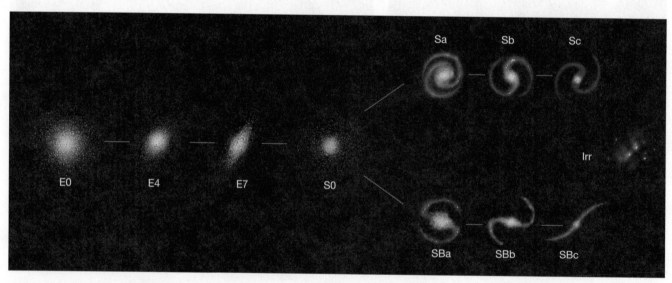

▲ **FIGURE 24.9 Galactic "Tuning Fork"** Hubble's tuning fork diagram, showing his basic galaxy classification scheme. The terminology is widely used today, and still defines our discussion of "normal" galaxies in the universe. Hubble's placement of the four basic types of galaxies—ellipticals, spirals, barred spirals, and irregulars—in the diagram is suggestive of evolution, but in fact the tuning fork has no physical meaning. As we will see in Chapter 25, galaxies do evolve, but not (in either direction) along the "Hubble sequence" defined by this figure.

24.2 The Distribution of Galaxies in Space

Now that we have seen some of their basic properties, let us ask how galaxies are spread through the expanse of the universe beyond the Milky Way. Galaxies are not distributed uniformly in space. Rather, they tend to clump into still larger agglomerations of matter. As we will see, this uneven distribution is crucial in determining both their appearance and their evolution. As always in astronomy, our understanding hinges on our ability to tell how far away an object lies. We therefore begin by looking more closely at the means used by astronomers to measure distances to galaxies.

Extending the Distance Scale

Astronomers estimate that some 40 billion galaxies as bright as (or brighter than) our own exist in the observable universe. Some reside close enough for the Cepheid variable technique to work—astronomers have detected and measured the periods of Cepheids in galaxies as far away as 25 Mpc (see Figure 24.10). ∞ (Sec. 23.2) However, some galaxies contain no Cepheid stars (can you think of some reasons that this might be?), and, in any case, most known galaxies lie much farther away than 25 Mpc. Cepheid variables in very distant galaxies simply cannot be observed well enough, even through the world's most sensitive telescopes,

to allow us to measure their apparent brightnesses and periods. To extend our distance-measurement ladder, therefore, we must find some new class of object to study. What individual objects are bright enough for us to observe at great distances?

One way in which researchers have tackled this problem is through observations of **standard candles**—easily recognizable astronomical objects whose luminosities are confidently known. The basic idea is very simple. Once an object is identified as a standard candle—by its appearance or by the shape of its light curve, say—its luminosity can be estimated. Comparison of the luminosity with the apparent brightness then gives the object's distance and, hence, the distance to the galaxy in which it resides. ∞ (Sec. 17.2) Note that, apart from the way in which the luminosity is determined, the Cepheid variable technique relies on identical reasoning. However, the term *standard candle* tends to be applied only to very bright objects.

To be most useful, a standard candle must (1) have a *well-defined luminosity,* so that the uncertainty in estimating its brightness is small, and (2) be *bright enough* to be seen at large distances. Over the years, astronomers have explored the use of many types of objects as standard candles—novae, emission nebulae, planetary nebulae, globular clusters, Type I (carbon-detonation) supernovae, and even entire galaxies have been employed. Not all have been equally useful, however: Some have larger intrinsic spreads in their luminosities than others, making them less reliable for measuring distances.

In recent years, planetary nebulae and Type I supernovae have proved particularly reliable as standard candles. ∞ (Secs. 20.3, 21.3) The latter have remarkably consistent peak luminosities and are very bright, allowing them to be identified and measured out to distances of many hundreds of megaparsecs. The small luminosity spread of Type I supernovae is a direct consequence of the circumstances in which these violent events occur. As discussed in Chapter 21, an accreting white dwarf explodes when it reaches the well-defined critical mass at which carbon fusion begins. ∞ (Sec. 21.3) The magnitude of the explosion is relatively insensitive to the details of how the white dwarf formed or how it subsequently reached critical mass, with the result that all such supernovae have

R I V U X G

▲ **FIGURE 24.10 Cepheid in Virgo** This sequence of six snapshots chronicles the periodic changes in a Cepheid variable star in the spiral galaxy M100, a member of the Virgo Cluster of galaxies. The Cepheid appears at the center of each inset, taken at the different times indicated during 1994. The star looks like a square because of the high magnification of the digital CCD camera—we are seeing individual pixels of the image. The 24th-magnitude star varies by about a factor of two in brightness every 7 weeks. *(NASA)*

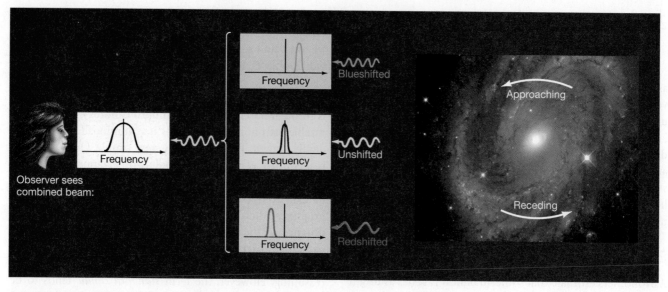

▲ FIGURE 24.11 **Galaxy Rotation** A galaxy's rotation causes some of the radiation it emits to be blueshifted and some to be redshifted (relative to what the emission would be from an unmoving source). From a distance, when the radiation from the galaxy is combined into a single beam and analyzed spectroscopically, the redshifted and blueshifted components combine to produce a broadening of the galaxy's spectral lines. The amount of broadening is a direct measure of the rotation speed of the galaxy, such as the one at the right, NGC 4603, about 100 million light-years away. *(NASA)*

quite similar properties.* Thus, when a Type I supernova is observed in a distant galaxy (we assume that it occurs *in the galaxy*, not in the foreground), astronomers can quickly obtain an accurate estimate of the galaxy's distance.

An important alternative to standard candles was discovered in the 1970s, when astronomers found a close correlation between the rotational speeds and the luminosities of spiral galaxies within a few tens of megaparsecs of the Milky Way Galaxy. Rotation speed is a measure of a spiral galaxy's total mass, so it is perhaps not surprising that this property should be related to luminosity. ∞ (Sec. 23.5) What *is* surprising, though, is how tight the correlation is. The **Tully-Fisher relation**, as it is now known (after its discoverers), allows us to obtain a remarkably accurate estimate of a spiral galaxy's luminosity simply by observing how fast the galaxy rotates. As usual, comparing the galaxy's (true) luminosity with its (observed) apparent brightness yields its distance.

To see how the method is used, imagine that we are looking edge-on at a distant spiral galaxy and observing one particular emission line, as illustrated in Figure 24.11. Radiation from the side of the galaxy where matter is generally approaching us is blueshifted by the Doppler effect. Radiation from the other side, which is receding from us, is redshifted by a similar amount. The overall effect is that line radiation from the galaxy is "smeared out," or broadened, by

Recall from Chapter 21 that a Type II supernova also occurs when a growing stellar core—this time at the center of a massive star—reaches a critical mass. ∞ (Sec. 21.2) However, the outward appearance of the explosion can be significantly modified by the amount of stellar material through which the blast wave must travel before it reaches the star's surface, resulting in a greater spread in observed luminosities. ∞ (Discovery 21-1)

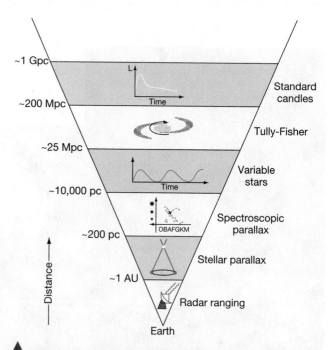

⚠ **Interactive** FIGURE 24.12 **Extragalactic Distance Ladder** An inverted pyramid summarizes the distance techniques used to study different realms of the universe. The techniques shown in the bottom four layers—radar ranging, stellar parallax, spectroscopic parallax, and variable stars—take us as far as the nearest galaxies. To go farther, we must use new techniques—the Tully-Fisher relation and the use of standard candles—based on distances determined by the four lowest techniques.

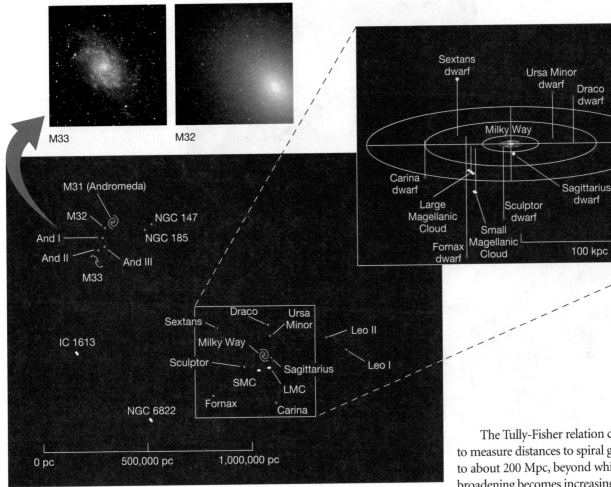

▲ **FIGURE 24.13 Local Group** The Local Group is made up of nearly 50 galaxies within approximately 1 Mpc of our Milky Way Galaxy. Only a few are spirals; most of the rest are dwarf-elliptical or irregular galaxies, only some of which are shown here. Spirals are colored blue, ellipticals pink, and irregulars white, all of them depicted approximately to scale. The inset map (top right) shows the Milky Way in relation to some of its satellite galaxies. The photographic insets (top left) show two well-known neighbors of the Andromeda Galaxy (M31): the spiral galaxy M33 and the dwarf-elliptical galaxy M32 (also visible in Figure 23.2a, a larger-scale view of the Andromeda system). *(M. BenDaniel; NASA)*

the galaxy's rotation. The faster the rotation, the greater the amount of broadening (see Figure 4.18 for the stellar equivalent). By measuring the amount of broadening, we can therefore determine the galaxy's rotation speed. Once we know that, the Tully-Fisher relation tells us the galaxy's luminosity.

The particular line normally used in these studies actually lies in the radio part of the spectrum. It is the 21-cm line of cold, neutral hydrogen in the galactic disk. ∞ (Sec. 18.4) This line is used in preference to optical lines because (1) optical radiation is strongly absorbed by dust in the disk under study and (2) the 21-cm line is normally very narrow, making the broadening easier to observe. In addition, astronomers often use *infrared,* rather than optical, luminosities, to avoid absorption problems caused by dust, both in our own Galaxy and in others.

The Tully-Fisher relation can be used to measure distances to spiral galaxies out to about 200 Mpc, beyond which the line broadening becomes increasingly difficult to measure accurately. A somewhat similar connection, relating line broadening to a galaxy's *diameter,* exists for elliptical galaxies. Once the galaxy's diameter and angular size are known, its distance can be computed from elementary geometry. ∞ *(More Precisely 1-2)* These methods bypass many of the standard candles often used by astronomers and so provide independent means of determining distances to faraway objects.

As indicated in Figure 24.12, standard candles and the Tully-Fisher relation form the fifth and sixth rungs of our cosmic distance ladder, introduced in Chapter 1 and expanded in Chapters 17 and 23. ∞ (Secs. 1.6, 17.1, 17.6, 23.2) In fact, they stand for perhaps a dozen or so related, but separate, techniques that astronomers have employed in their quest to map out the universe on large scales. Just as with the lower rungs, we calibrate the properties of these new techniques by using distances measured by more local means. In this way, the distance-measurement process "bootstraps" itself to greater and greater distances. However, at the same time, the errors and uncertainties in each step accumulate, so the distances to the farthest objects are the least well known.

ANIMATION/VIDEO The Evolution of Galaxies

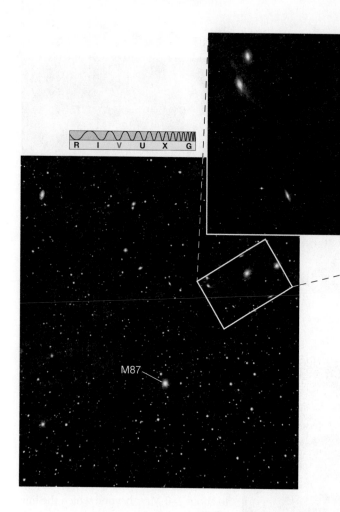

R I V U X G

M86

M87

◀ FIGURE 24.14 **Virgo Cluster**
The central region of the Virgo
Cluster of galaxies, about 17 Mpc
from Earth. Many large spiral and
elliptical galaxies can be seen. The
inset shows several galaxies
surrounding the giant elliptical
known as M86. An even bigger
elliptical galaxy, M87, noted at the
bottom, will be discussed later in
the chapter. (*M. BenDaniel;
AURA*)

Clusters of Galaxies

Figure 24.13 sketches the locations of all the known major astronomical objects within about 1 Mpc of the Milky Way. Our Galaxy appears with its dozen or so satellite galaxies—including the two Magellanic Clouds discussed earlier and a recently discovered companion (labeled "Sagittarius dwarf" in the figure) lying almost within our own Galactic plane. The Andromeda Galaxy, lying 800 kpc from us, is also shown, surrounded by satellites of its own. Two of Andromeda's galactic neighbors are shown in insets. M33 is a spiral, and M32 is a dwarf elliptical, easily seen in Figure 23.2(a) below and to the right of Andromeda's central bulge.

All told, nearly 50 galaxies are known to populate our Galaxy's neighborhood. Three of them (the Milky Way, Andromeda, and M33) are spirals; the remainder are dwarf irregulars and dwarf ellipticals. Together, these galaxies form the **Local Group**—a new level of structure in the universe above the scale of our Galaxy. As indicated in Figure 24.13, the Local Group's diameter is a little over 1 Mpc. The Milky Way Galaxy and Andromeda are by far its largest members, and most of the smaller galaxies are gravitationally bound to one or the other of them. The combined gravity of the galaxies in the Local Group binds them together, like stars in a star cluster, but on a millionfold larger scale. More generally, a collection of galaxies held together by their mutual gravitational attraction is called a **galaxy cluster.**

Moving beyond the Local Group, the next large concentration of galaxies we come to is the *Virgo Cluster* (Figure 24.14), named after the constellation in which it is found. Lying some 17 Mpc from the Milky Way, the Virgo Cluster does not contain a mere 50 galaxies, however. Rather, it houses more than 2500 galaxies, bound by gravity into a tightly knit group about 3 Mpc across.

Wherever we look in the universe, we find galaxies, and the majority of galaxies are members of groups or clusters of galaxies. In practice, the distinction between a "group" and "cluster" is mainly a matter of convention. Groups generally contain only a few bright galaxies (such as the Milky Way and Andromeda) and are quite irregular in shape, whereas large, "rich" clusters like Virgo may contain thousands of individual galaxies distributed fairly smoothly in space. The Coma cluster, shown in Figure 24.1 and lying approximately 100 Mpc away, is another example of a rich cluster. Figure 24.15 is a long-exposure photograph of a much more distant rich cluster, lying about 700 Mpc from Earth. A sizeable minority of galaxies (perhaps 20 to 30 percent) are not members of any group or cluster, but are apparently isolated systems, moving alone through intercluster space. (For simplicity, we will use the term "cluster" below to refer to any gravitationally bound collection of galaxies, large or small.)

We will return to the large-scale distribution of matter in the universe in Chapters 25 and 26.

PROCESS OF SCIENCE CHECK

✔ What are some of the problems astronomers encounter in measuring the distances to faraway galaxies?

R I V U X G

▲ **FIGURE 24.15 Distant Galaxy Cluster** The galaxy cluster Abell 1689 contains huge numbers of galaxies and resides roughly 2 billion light-years from Earth. Virtually every patch of light in this photograph is a separate galaxy. Thanks to the high resolution of the optics on board the *Hubble Space Telescope*, we can now discern, even at this great distance, spiral structure in some of the galaxies. We also see many galaxies colliding—some tearing matter from one another, others merging into single systems. *(NASA)*

24.3 Hubble's Law

Now that we have seen some basic properties of galaxies throughout the universe, let's turn our attention to the large-scale *motions* of galaxies and galaxy clusters. Within a galaxy cluster, individual galaxies move more or less randomly. You might expect that, on even larger scales, the clusters themselves would also have random, disordered motion—some clusters moving this way, some that. In fact, that is not the case: On the largest scales, galaxies and galaxy clusters alike move in a very *ordered* way.

Universal Recession

In 1917, the American astronomer Vesto M. Slipher, working under the direction of Percival Lowell, reported that virtually every spiral galaxy he observed had a redshifted spectrum—it was *receding* from our Galaxy. ⊙ (Sec. 3.5) It is now known that, except for a few nearby systems, *every* galaxy takes part in a general motion away from us in all directions. Individual galaxies that are not part of galaxy clusters are steadily receding. Galaxy clusters, too, have an overall recessional motion, although their individual member galaxies move randomly

with respect to one another. (Consider a jar full of fireflies that has been thrown into the air. The fireflies within the jar, like the galaxies within the cluster, have random motions due to their individual whims, but the jar as a whole, like the galaxy cluster, has some directed motion as well.)

Figure 24.16 shows the optical spectra of several galaxies, arranged in order of increasing distance from the Milky Way Galaxy. The spectra are redshifted, indicating that the associated galaxies are receding. Furthermore, the extent of the redshift increases progressively from top to bottom in the figure. There is a connection between Doppler shift and distance: The greater the distance, the greater the redshift. This trend holds for nearly all galaxies in the universe. (Two galaxies within our Local Group, including Andromeda, and a few galaxies in the Virgo Cluster display blueshifts and so are moving toward us, but this results from their local motions within their parent clusters—recall the fireflies in the jar.)

Figure 24.17(a) shows recessional velocity plotted against distance for the galaxies of Figure 24.16. Figure 24.17(b) is a similar plot for some more galaxies within about 1 billion parsecs of Earth. Plots like these were first made by Edwin Hubble in the 1920s and now bear his name: *Hubble diagrams.* The data points generally fall close to a straight line, indicating that the rate at which a galaxy recedes is *directly proportional* to its distance from us. This rule is called **Hubble's law.** We can construct such a diagram for any collection of galaxies, provided that we can determine their distances and velocities. The universal recession described by the Hubble diagram is sometimes called the *Hubble flow.*

The recessional motions of the galaxies prove that the cosmos is neither steady nor unchanging on the largest scales. The universe (actually, *space itself*—see Section 26.2) is expanding! However, let's be clear on just *what* is expanding and what is not. Hubble's law does not mean that humans, Earth, the solar system, or even individual galaxies and galaxy clusters are physically increasing in size. These groups of atoms, rocks, planets, stars, and galaxies are held together by their own internal forces and are not themselves getting bigger. Only the largest framework of the universe—the vast distances separating the galaxy clusters—is expanding.

To distinguish recessional redshift from redshifts caused by motion *within* an object—for example, galactic orbits within a cluster or explosive events in a galactic nucleus—the redshift resulting from the Hubble flow is called the **cosmological redshift.** Objects that lie so far away that they exhibit a large cosmological redshift are said to be at *cosmological distances*—distances comparable to the scale of the universe itself.

Hubble's law has some fairly dramatic implications. If nearly all galaxies show recessional velocity according to Hubble's law, then doesn't that mean that they all started their journey from a single point? If we could run time backward, wouldn't all the galaxies fly back to this one point, perhaps the site of some violent event in the remote past? The answer is

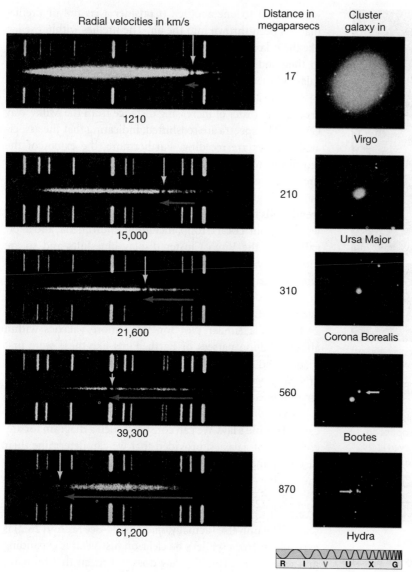

Radial velocities in km/s

Distance in megaparsecs

Cluster galaxy in

1210

17

Virgo

15,000

210

Ursa Major

21,600

310

Corona Borealis

39,300

560

Bootes

61,200

870

Hydra

R I V U X G

◀ **FIGURE 24.16 Galaxy Spectra** Optical spectra, shown at left, of several galaxies named on the right. Both the extent of the redshift (denoted by the horizontal red arrows) and the distance from the Milky Way Galaxy to each galaxy (numbers in center column) increase from top to bottom. The vertical yellow arrow in each spectrum highlights a particular spectral feature (a pair of dark absorption lines). The horizontal red arrows indicate how this feature shifts to longer wavelengths in spectra of more distant galaxies. The white lines at the top and bottom of each spectrum are laboratory references. *(Palomar/Caltech)*

per second per megaparsec, the most commonly used unit for H_0). Astronomers continually strive to refine the accuracy of the Hubble diagram and the resulting estimate of H_0, because Hubble's constant is one of the most fundamental quantities of nature; it specifies the rate of expansion of the entire cosmos.

Hubble's original value for H_0 was about 500 km/s/Mpc, far higher than the currently accepted value. This overestimate was due almost entirely to errors in the cosmic distance scale at the time, particularly the calibrations of Cepheid variables and standard candles. The measured value dropped rapidly as various observational errors were recognized and resolved and distance measurements became more reliable. Published estimates of H_0 entered the "modern" range (within, say, 20 percent of the current value) in roughly the mid-1960s.

During the last two decades of the 20th century, the various distance-measurement techniques then in use appeared to disagree fundamentally in their values for H_0. For example, infrared Tully-Fisher observations and *HST* studies of Cepheid variables as far away as the Virgo cluster seemed to produce systematically high results (70–80 km/s/Mpc). ∞ (Sec. 23.2) On the other hand, visible-light Tully-Fisher studies and techniques using standard candles, including Type I supernovae, tended to return lower values (50–65 km/s/Mpc).

Was this discrepancy the result of unknown errors in some (or all) of the approaches used? Apparently not. Astronomers worked hard to refine their methods still further, and now, early in the 21st century, the leading measurements of H_0 are all remarkably consistent with one another. We will adopt a rounded-off value of $H_0 = 70$ km/s/Mpc (a choice roughly in the middle of all recent results, and also consistent with some precise cosmological measurements to be discussed in Chapter 27) as the best current estimate of Hubble's constant for the remainder of the text.

yes—but not in the way you might expect! In Chapters 26 and 27, we will explore the ramifications of the Hubble flow for the past and future evolution of our universe. For now, however, we set aside its cosmic implications and use Hubble's law simply as a convenient distance-measuring tool.

Hubble's Constant

The constant of proportionality between recessional velocity and distance in Hubble's law is known as **Hubble's constant**, denoted by the symbol H_0. The data shown in Figure 24.17 then obey the equation

$$\text{recessional velocity} = H_0 \times \text{distance}.$$

The value of Hubble's constant is the slope of the straight line—recessional velocity divided by distance—in Figure 24.17(b). Reading the numbers off the graph, we get roughly 70,000 km/s divided by 1000 Mpc, or 70 km/s/Mpc (kilometers

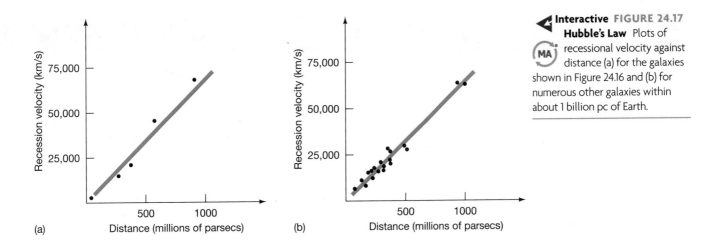

Interactive FIGURE 24.17
Hubble's Law Plots of recessional velocity against distance (a) for the galaxies shown in Figure 24.16 and (b) for numerous other galaxies within about 1 billion pc of Earth.

The Top of the Distance Ladder

Using Hubble's law, we can derive the distance to a remote object simply by measuring the object's recessional velocity and dividing by Hubble's constant. Hubble's law thus tops our inverted pyramid of distance-measurement techniques (Figure 24.18). This seventh method simply assumes that Hubble's law holds. If that assumption is correct, Hubble's law enables us to measure great distances in the universe—so long as we can obtain an object's spectrum, we can determine how far away it is. Notice, however, that the uncertainty in Hubble's constant translates directly into a similar uncertainty in all distances determined from Hubble's law.

Many redshifted objects have recessional motions that are a substantial fraction of the speed of light. The most distant objects thus far observed in the universe—some young galaxies and quasars (Section 24.4)—have redshifts (fractional increases in wavelength) of more than 7, meaning that their radiation has been stretched in wavelength not by just a few percent, as with most of the objects we have discussed, but *eightfold*. Their ultraviolet spectral lines are shifted all the way into the infrared part of the spectrum! *More Precisely 24-1* discusses in more detail the meaning and interpretation of such large redshifts, apparently implying recessional velocities comparable to the speed of light. According to Hubble's law, the objects that exhibit these redshifts lie almost 9000 Mpc away from us, as close to the limits of the observable universe as astronomers have yet been able to probe.

The speed of light is finite. It takes time for light—or, for that matter, any kind of radiation—to travel from one point in space to another. The radiation that we now see from these most distant objects originated long ago. Incredibly, that radiation was emitted almost 13 billion years ago (see Table 24.2), well before our planet, our Sun, and perhaps even our Galaxy came into being!

CONCEPT CHECK

✔ How does the use of Hubble's law differ from the other extragalactic distance-measurement techniques we have seen in this text?

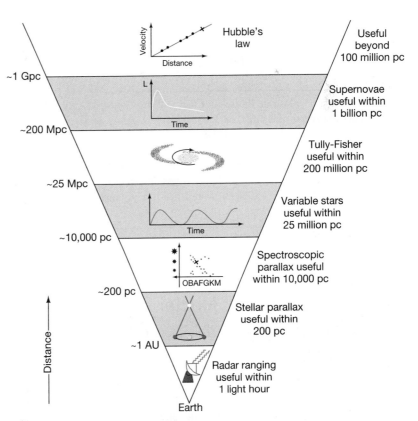

Interactive FIGURE 24.18 Cosmic Distance Ladder Hubble's law tops the hierarchy of distance-measurement techniques. It is used to find the distances of astronomical objects all the way out to the limits of the observable universe.

MORE PRECISELY 24-1

Relativistic Redshifts and Look-Back Time

In discussing very distant objects, astronomers usually talk about their *redshifts* rather than their distances. Indeed, it is common for researchers to speak of an event occurring "at" a certain redshift—meaning that the light received today from that event is redshifted by the specified amount. Of course, because of Hubble's law, redshift and distance are equivalent to one another. However, redshift is the preferred quantity because it is a directly observable property of an object, whereas distance is derived from redshift with the use of Hubble's constant, whose value is not accurately known. (In Chapter 26 we will see another, much more fundamental, reason why astronomers favor the use of redshift in studies of the cosmos.)

The redshift of a beam of light is, by definition, the *fractional* increase in the wavelength of the light resulting from the recessional motion of the source. ∞ (Sec. 3.5) Thus, a redshift of 1 corresponds to a *doubling* of the wavelength. From the formula for the Doppler shift given previously, the redshift of radiation received from a source moving away from us with speed v is given by

$$\text{redshift} = \frac{\text{observed wavelength} - \text{true wavelength}}{\text{true wavelength}}$$

$$= \frac{\text{recessional velocity, } v}{\text{speed of light, } c}.$$

EXAMPLE   Let's illustrate this relationship with two examples, rounding the speed of light, c, to 300,000 km/s. A galaxy at a distance of 100 Mpc has a recessional speed (by Hubble's law) of 70 km/s/Mpc × 100 Mpc = 7,000 km/s. Its redshift is therefore 7,000 km/s ÷ 300,000 km/s = 0.023. Conversely, an object that has a redshift of 0.05 has a recessional velocity of 0.05 × 300,000 km/s = 15,000 km/s and hence a distance of 15,000 km/s ÷ 70 km/s/Mpc = 210 Mpc.

Unfortunately, although the foregoing equation is correct for low speeds, it does not take into account the effects of relativity. As we saw in Chapter 22, the rules of everyday physics have to be modified when speeds begin to approach the speed of light. ∞ *(Discovery 22-1)* The formula for the Doppler shift is no exception. In particular, although the formula is valid for speeds much less than the speed of light, when $v = c$ the redshift is not unity, as the equation suggests, but is in fact *infinite*. That is, radiation received from an object moving away from us at nearly the speed of light is redshifted to almost infinite wavelength.

Thus, do not be alarmed to find that many galaxies and quasars have redshifts greater than unity. This does not mean that they are receding faster than light! It simply means that the preceding simple formula is not applicable. In fact, the real connection between redshift and distance is quite complex, requiring us to make key assumptions about the past history of the universe (see Chapter 26). In place of a formula, we can use Table 24.2, which presents a conversion chart relating redshift and distance. All of the values shown are based on reasonable assumptions and are usable even for large redshifts. We take Hubble's constant to be 70 km/s/Mpc and assume a flat universe in which matter (mostly dark) contributes just over one-quarter of the total density (see Section 26.6). The conversions in the table are used consistently throughout this text. The column headed "v/c" gives equivalent recessional velocities based on the Doppler effect, taking relativity properly into account. Even though this is *not* the correct interpretation of the redshift (see Section 26.2), we include it here for comparison, simply because it is so often quoted in the popular media.

Because the universe is expanding, the "distance" to a galaxy is not very well defined. Do we mean the distance to the galaxy when it emitted the light we see today, the present distance to the galaxy (as presented in the table, even though we do not see the galaxy as it is today), or some other, more appropriate measure? Largely because of this ambiguity, astronomers prefer to work in terms of a quantity known as the *look-back time* (shown in the last column of Table 24.2), which is simply how long ago an object emitted the radiation we see today. Astronomers talk frequently about redshifts and sometimes about look-back times, but they

24.4 Active Galactic Nuclei

The galaxies described in Section 24.1—those falling into the various Hubble classes—are generally referred to as **normal galaxies.** The majority of galaxies fall into this broad category. As we have seen, their luminosities range from a million or so times that of the Sun for dwarf ellipticals and irregulars to more than a trillion solar luminosities for the largest giant ellipticals known. For comparison, in round numbers, the luminosity of the Milky Way Galaxy is 2×10^{10} solar luminosities, or roughly 10^{37} W.

In these last two sections we focus our attention on "bright" galaxies, conventionally taken to mean galaxies with luminosities more than about 10^{10} times the solar value. In these terms, our Galaxy is bright, but not abnormally so.

Galactic Radiation

A substantial fraction of bright galaxies—perhaps as many as 40 percent—don't fit well into the "normal" category. Their *spectra* differ significantly from those of their normal cousins, and their *luminosities* can be extremely large. Known

hardly ever talk of distances to high-redshift objects (and *never* about recession velocities, despite what you hear on the news!). Bear in mind, however, that redshift is the only unambiguously measured quantity in this discussion. Statements about "derived" quantities, such as distances and look-back times, all require that we make specific assumptions about how the universe has evolved with time.

For nearby sources, the look-back time is numerically equal to the distance in light-years: The light we receive tonight from a galaxy at a distance of 100 million light-years was emitted 100 million years ago. However, for more distant objects, the look-back time and the present distance in light-years differ because of the expansion of the universe, and the divergence increases dramatically with increasing redshift.

As a simple analogy, imagine an ant crawling across the surface of an expanding balloon at a constant speed of 1 cm/s relative to the balloon's surface. After 10 seconds, the ant may think it has traveled a distance of 10 cm, but an outside observer with a tape measure will find that it is actually more than 10 cm from its starting point (measured along the surface of the balloon) because of the balloon's expansion. In exactly the same way, the present distance to a galaxy with a given redshift depends on how the universe expanded in the past. For example, a galaxy now located 15 billion light-years from Earth was much closer to us when it emitted the light we now see. Consequently, its light has taken considerably less than 15 billion years—in fact, about 10 billion years—to reach us.

TABLE 24.2 Redshift, Distance, and Look-Back Time				
Redshift	V/C	Present Distance		Look-Back Time
		(Mpc)	(10^6 light-years)	(millions of years)
0.000	0.000	0	0	0
0.010	0.010	42	137	137
0.025	0.025	105	343	338
0.050	0.049	209	682	665
0.100	0.095	413	1350	1290
0.200	0.180	809	2640	2410
0.250	0.220	999	3260	2920
0.500	0.385	1880	6140	5020
0.750	0.508	2650	8640	6570
1.000	0.600	3320	10,800	7730
1.500	0.724	4400	14,400	9320
2.000	0.800	5250	17,100	10,300
3.000	0.882	6460	21,100	11,500
4.000	0.923	7310	23,800	12,100
5.000	0.946	7940	25,900	12,500
6.000	0.960	8420	27,500	12,700
7.000	0.969	8820	28,800	12,900
8.000	0.976	9150	29,800	13,000
9.000	0.980	9430	30,700	13,100
10.000	0.984	9660	31,500	13,200
50.000	0.999	12,300	40,100	13,600
100.000	1.000	12,900	42,200	13,700
∞	1.000	14,600	47,500	13,700

collectively as **active galaxies,** they are of great interest to astronomers. The brightest among them are the most energetic objects known in the universe, and all may represent an important, if intermittent, phase of galactic evolution (see Section 25.4). At optical wavelengths, active galaxies often *look* like normal galaxies—familiar components such as disks, bulges, stars, and dark dust lanes can be identified. At other wavelengths, however, their unusual properties are much more apparent.

Most of a normal galaxy's energy is emitted in or near the visible portion of the electromagnetic spectrum, much like the radiation from stars. Indeed, to a large extent, the light we see from a normal galaxy *is* just the accumulated light of its many component stars (once the effects of interstellar dust are taken into account), each described approximately by a blackbody curve. ∞ (Sec. 3.4) By contrast, as illustrated schematically in Figure 24.19, the radiation from active galaxies does *not* peak in the visible. Most active galaxies do emit substantial amounts of visible radiation, but far more of their energy is emitted at invisible wavelengths, both longer and shorter than those in the visible range. Put another way, the radiation from active galaxies is *inconsistent* with what we

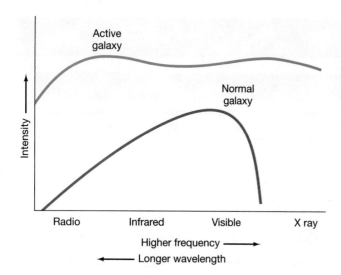

▲ **FIGURE 24.19 Galaxy Energy Spectra** The energy emitted by a normal galaxy differs significantly from that emitted by an active galaxy. This plot illustrates the general run of intensity for all galaxies of a particular type and does not represent any one individual galaxy.

▲ **FIGURE 24.20 Active Galaxy** This image of the galaxy NGC 7742 resembles a fried egg, with a ring of blue star-forming regions surrounding a very bright yellow core that spans about 1 kpc. An active galaxy, NGC 7742 combines star formation with intense emission from its central nucleus and lies roughly 24 Mpc away. (*NASA*)

would expect if it were the combined radiation of myriad stars. Their radiation is said to be *nonstellar*.

Many luminous galaxies with nonstellar emission are known to be *starburst galaxies*—previously normal systems currently characterized by widespread episodes of star formation, most likely as a result of interactions with a neighbor. The irregular galaxy M82 shown in Figure 24.8 is a prime example. We will study these important systems and their role in galaxy evolution in Chapter 25. For purposes of this text, however, we will use the term "active galaxy" to mean a system whose abnormal activity is related to violent events occurring in or near the *galactic nucleus*. Such systems are also known as **active galactic nuclei.**

Even with this restriction, there is still considerable variation in the properties of galaxies, and astronomers have identified and cataloged a bewildering array of systems falling into the "active" category. For example, Figure 24.20 shows an active galaxy exhibiting both nuclear activity and widespread star formation, with a blue-tinted ring of newborn stars surrounding an extended 1-kpc-wide core of intense emission. Rather than attempting to describe the entire "zoo" of active galaxies, we will instead discuss three basic species: the energetic *Seyfert galaxies* and *radio galaxies* and the even more luminous *quasars*. Although these objects all lie toward the "high-luminosity" end of the active range and represent perhaps only a few percent of the total number of active galaxies, their properties will allow us to identify and discuss features common to active galaxies in general.

The association of galactic activity with the central nucleus is reminiscent of the discussion in Chapter 23 of the center of the Milky Way. ∞ (Sec. 23.7) In our own Galaxy,

it seems clear that the activity in the nucleus is associated with the central *supermassive black hole,* whose presence is inferred from observations of stellar orbits in the innermost fraction of a parsec. As we will see, most astronomers think that basically the same thing is going on in the nuclei of active galaxies and that "normal" and "active" galaxies may differ principally in the degree to which the nonstellar nuclear component of the radiation outshines the light from the rest of the galaxy. This is a powerful unifying theme for understanding the evolution of galaxies, and we will return to it in Chapter 25. For the remainder of this chapter, we concentrate on describing the properties of active galaxies and the black holes that power them.

Seyfert Galaxies

In 1943, Carl Seyfert, an American optical astronomer studying spiral galaxies from Mount Wilson Observatory, discovered the type of active galaxy that now bears his name. **Seyfert galaxies** are a class of astronomical objects whose properties lie between those of normal galaxies and those of the most energetic active galaxies known.

Superficially, Seyferts resemble normal spiral galaxies (Figure 24.21a). Indeed, the stars in a Seyfert's galactic disk and spiral arms produce about the same amount of visible radiation as do the stars in a normal spiral galaxy. However, most of a Seyfert's energy is emitted from the galactic nucleus—the center of the overexposed white patch in the figure. The nucleus of a Seyfert galaxy is some 10,000 times brighter than the center of our own Galaxy. In fact, the

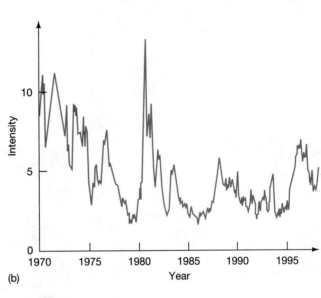

▲ FIGURE 24.21 **Seyfert Galaxy** (a) The Circinus galaxy, a Seyfert with a bright compact core, lies some 4 Mpc away. It is one of the closest active galaxies. (b) This graph illustrates the irregular variations in luminosity of the Seyfert galaxy 3C 84 over two decades. Because this Seyfert emits strongly in the radio part of the electromagnetic spectrum, these observations were made with large radio telescopes. The optical and X-ray luminosities vary as well. *(NASA; NRAO)*

brightest Seyfert nuclei are 10 times more energetic than the *entire* Milky Way.

Some Seyferts produce radiation spanning a broad range in wavelengths, from the infrared all the way through ultraviolet and even X rays. However, the majority (about 75 percent) emit most of their energy in the infrared. Scientists think that much of the high-energy radiation in these Seyferts is absorbed by dust in or near the nucleus and then reemitted as infrared radiation.

Seyfert spectral lines have many similarities to those observed toward the center of our own Galaxy. ∞ (Sec. 23.7) Some of the lines are very broad, most likely indicating rapid (5000 km/s or more) internal motion within the nuclei. ∞ (Sec. 4.5) However, not all of the lines are broad, and some Seyferts show no broad lines at all. In addition, their energy emission often varies in time (Figure 24.21b). A Seyfert's luminosity can double or halve within a fraction of a year. These rapid fluctuations in luminosity lead us to conclude that the source of energy emissions in Seyfert galaxies must be quite compact—simply put, as we saw in Chapter 22, an object cannot "flicker" in less time than radiation takes to cross it. ∞ (Sec. 22.4) The emitting region must therefore be less than 1 light-year across—an extraordinarily small region, considering the amount of energy emanating from it.

Together, the rapid time variability and large radio and infrared luminosities observed in Seyferts imply violent nonstellar activity in their nuclei. As mentioned earlier, this activity is likely similar in *nature* to processes occurring at the center of our own Galaxy, but its *magnitude* is thousands of times greater than the comparatively mild events within our own Galaxy's heart. ∞ (Sec. 23.7)

Radio Galaxies

As the name suggests, **radio galaxies** are active galaxies that emit large amounts of energy in the radio portion of the electromagnetic spectrum. They differ from Seyferts not only in the wavelengths at which they radiate, but also in both the appearance and the extent of their emitting regions.

Figure 24.22(a) shows the radio galaxy Centaurus A, which lies about 4 Mpc from Earth. Almost none of this galaxy's radio emission comes from a compact nucleus. Instead, the energy is released from two huge extended regions called **radio lobes**—roundish clouds of gas spanning about half a megaparsec and lying well beyond the visible galaxy.* Undetectable in visible light, the radio lobes of radio galaxies

*The term "visible galaxy" is commonly used to refer to those components of an active galaxy that emit visible "stellar" radiation, as opposed to the nonstellar and invisible "active" component of the galaxy's emission.

(a)

◄ FIGURE 24.22 **Centaurus A Radio Lobes** Radio galaxies are the largest and arguably the most impressive active galaxies. Centaurus A, shown here optically in (a), has giant radio-emitting lobes (b) extending a million parsecs or more beyond the central galaxy and emitting large amounts of invisible energy. This entire object is thought to be the result of a collision between two galaxies that took place about 500 million years ago. The lobes cannot be imaged in visible light and are mainly observable with radio telescopes; they are shown here in pastel false colors, with decreasing intensity from red to yellow to green to blue. The inset (at right) shows a *Chandra* image of one of the lobes up close, proving that at least the jets in the inner parts of the lobes do emit X rays. *(ESO; NRAO; SAO)*

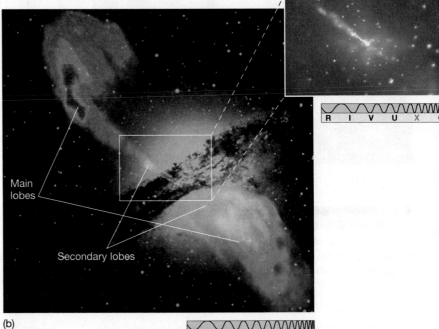

Main lobes

Secondary lobes

(b)

are truly enormous. From end to end, they typically span more than 10 times the size of the Milky Way Galaxy, comparable in scale to the entire Local Group.

Figure 24.22(b) shows the relationship between the galaxy's visible, radio, and X-ray emissions. In visible light, Centaurus A is apparently a large E2 galaxy some 500 kpc in diameter, bisected by an irregular band of dust. Centaurus A is a member of a small cluster of galaxies, and numerical simulations suggest that this peculiar galaxy is probably the result of a collision between an elliptical galaxy and a smaller spiral galaxy about 500 million years ago. In the crowded confines of a cluster, such collisions may be commonplace (see Section 25.2). The radio lobes are roughly symmetrically placed, jutting out from the center of the visible galaxy and roughly perpendicular to the dust lane, suggesting that they consist of material ejected in opposite directions from the galactic nucleus. This conclusion is strengthened by the presence of a pair of smaller secondary lobes closer to the visible galaxy and by the presence of a roughly 1-kpc-long jet of matter in the galactic center, all aligned with the main lobes (and marked in the figure).

If the material was ejected from the nucleus at close to the speed of light and has subsequently slowed, then Centaurus A's outer lobes were created a few hundred million years ago, quite possibly around the time of the collision thought to be responsible for the galaxy's odd optical appearance. The secondary lobes were expelled more recently. Apparently, some violent process at the center of Centaurus

A—most probably triggered by the collision—started up around that time and has been intermittently firing jets of matter out into intergalactic space ever since.

Centaurus A is a relatively low-luminosity source that happens to lie very close to us, astronomically speaking, making it particularly easy to study. Figure 24.23 shows a much more powerful emitter, called Cygnus A, lying roughly 250 Mpc from Earth. The high-resolution radio map in Figure 24.23(b) clearly shows two narrow, high-speed jets joining the radio lobes to the center of the visible galaxy (the dot at the center of the radio image). Notice that, as with Centaurus A, Cygnus A is a member of a small group of galaxies, and the optical image (Figure 24.23a) appears to show two galaxies colliding.

The radio lobes of the brightest radio galaxies (such as Cygnus A) emit roughly 10 times more energy than the Milky Way Galaxy does at all wavelengths, coincidentally about the same amount of energy emitted by the most luminous Seyfert nuclei. However, despite their names, radio galaxies actually radiate far more energy at shorter wavelengths. Their total energy output can be a hundred (or more) times greater than their radio emission. Most of this energy comes from the nucleus of the visible galaxy.

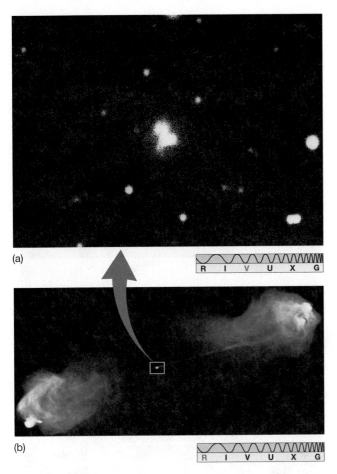

(a)

R I V U X G

(b)

R I V U X G

▲ **FIGURE 24.23 Cygnus A** (a) Cygnus A also appears to be two galaxies in collision. (b) On a much larger scale, it displays radio-emitting lobes on either side of the optical image. The optical galaxy in (a) is about the size of the small dot at the center of (b). Note the thin line of radio-emitting material joining the right lobe to the central galaxy. The distance from one lobe to the other is approximately a million light-years. *(NOAO; NRAO)*

With total luminosities up to a thousand times that of the Milky Way, bright radio galaxies are among the most energetic objects known in the universe. Their radio emission lets us study in detail the connection between the small-scale nucleus and the large-scale radio lobes.

Not all radio galaxies have obvious radio lobes. Figure 24.24 shows a *core-dominated* radio galaxy, most of whose energy is emitted from a small central nucleus (which radio astronomers refer to as the *core*) less than 1 pc across. Weaker radio emission comes from an extended region surrounding the nucleus. It is likely that all radio galaxies

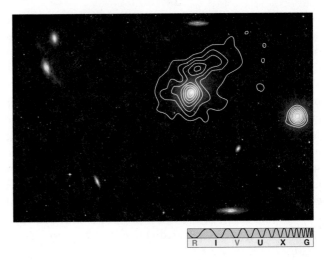

R I V U X G

▲ **FIGURE 24.24 Core-Dominated Radio Galaxy** As shown by this radio contour map of the radio galaxy M86, the radio emission comes from a bright central nucleus, which is surrounded by an extended, less-intense radio halo. The radio map is superimposed on an optical image of the galaxy and some of its neighbors, a wider-field version of which was shown previously in Figure 24.14. *(CfA)*

have jets and lobes, but what we observe depends on our perspective. As illustrated in Figure 24.25, when a radio galaxy is viewed from the side, we see the jets and lobes. However, if we view the jet almost head-on—in other words, looking *through* the lobe—we see a core-dominated system.

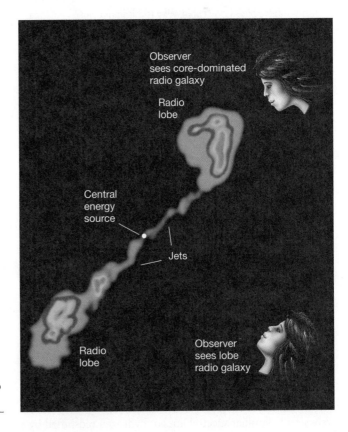

Observer sees core-dominated radio galaxy

Radio lobe

Central energy source

Jets

Radio lobe

Observer sees lobe radio galaxy

▶ **FIGURE 24.25 Radio Galaxy** A central energy source produces high-speed jets of matter that interact with intergalactic gas to form radio lobes. The system may appear to us as either radio lobes or a core-dominated radio galaxy, depending on our location with respect to the jets and lobes.

In fact, jets are a fairly common feature of active galaxies of all types. Figure 24.26 presents several images of the giant elliptical galaxy M87, a prominent member of the Virgo Cluster (Figure 24.14). A long-time exposure (Figure 24.26a) shows a large, fuzzy ball of light—a fairly normal-looking E1 galaxy about 100 kpc across. A shorter exposure of M87 (Figure 24.26b), capturing only the galaxy's bright inner regions, reveals a long (2 kpc) thin jet of matter ejected from the galactic center at nearly the speed of light. Computer enhancement shows that the jet is made up of a series of distinct "blobs" more or less evenly spaced along its length, suggesting that the material was ejected during bursts of activity. The jet has also been imaged in the radio, infrared (Figure 24.26c), and X-ray regions of the spectrum.

Our location with respect to a jet also affects the *type* of radiation we see. The theory of relativity tells us that radiation emitted by particles moving close to the speed of light is strongly concentrated, or beamed, in the direction of motion. ∞ *(Discovery 22-1)* As a result, if we happen to be directly in line with the beam, the radiation we receive is both very intense and Doppler shifted toward short wavelengths. ∞ (Sec. 3.5) The resulting object is called a **blazar** (Figure 5.35b). Much of the luminosity of the hundred or so known blazars is received in the form of X- or gamma rays.

CONCEPT CHECK

✔ The energy emission from an active galactic nucleus does not resemble a blackbody curve. Why is this important?

Quasars

In the early days of radio astronomy, many radio sources were detected for which no corresponding visible object was known. By 1960, several hundred such sources were listed in the *Third Cambridge Catalog,* and astronomers were scanning the skies in search of visible counterparts to these radio sources. Their job was made difficult both by the low resolution of the radio observations (which meant that the observers did not know exactly where to look) and by the faintness of the objects at visible wavelengths.

In 1960, astronomers detected what appeared to be a faint blue star at the location of the radio source 3C 48 (the 48th object on the third *Cambridge* list) and obtained its spectrum. Containing many unknown and unusually broad emission lines, the object's peculiar spectrum defied interpretation. 3C 48 remained a unique curiosity until 1962, when another similar-looking—and similarly mysterious— faint blue object with "odd" spectral lines was discovered and identified with the radio source 3C 273 (Figure 24.27).

The following year saw a breakthrough when astronomers realized that the strongest unknown lines in 3C 273's spectrum were simply familiar spectral lines of hydrogen redshifted by a

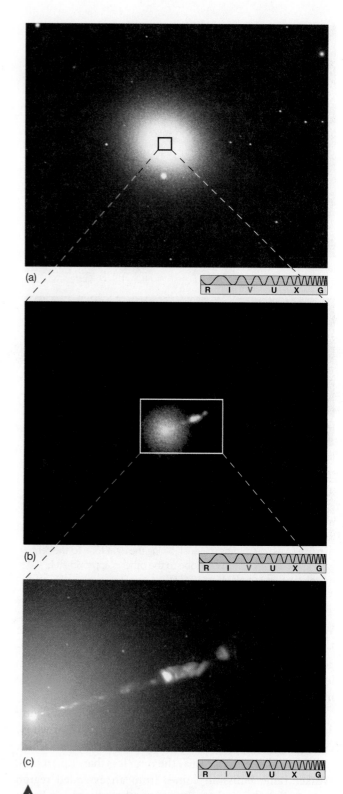

Interactive FIGURE 24.26 M87 Jet The giant elliptical galaxy M87 (also called Virgo A) is displayed here in a series of zooms. (a) A long optical exposure of the galaxy's halo and embedded central region. (b) A short optical exposure of its core and an intriguing jet of matter, on a somewhat smaller scale. (c) An infrared image of M87's jet, examined more closely compared with (b). The bright point at left in (c) marks the bright nucleus of the galaxy; the bright blob near the center of the image corresponds to the bright "knot" visible in the jet in (b). *(NOAO; NASA)*

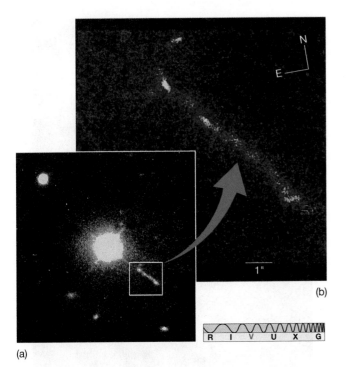

▲ **FIGURE 24.27 Quasar 3C 273** (a) The bright quasar 3C 273 displays a luminous jet of matter, but the main body of the quasar is starlike in appearance. (b) The jet extends for about 30 kpc and can be seen better in this high-resolution image. *(AURA)*

10^{40} W, comparable to 20 trillion Suns or a thousand Milky Way Galaxies. More generally, quasars range in luminosity from around 10^{38} W—about the same as the brightest Seyferts—up to nearly 10^{42} W. A value of 10^{40} W (comparable to the luminosity of a bright radio galaxy) is fairly typical.

Clearly not stars (because of their enormous luminosities), these objects became known as *quasi-stellar radio sources* ("quasi-stellar" means "starlike"), or **quasars.** (The name persists even though we now know that not all such highly redshifted, starlike objects are strong radio sources.) More than 100,000 quasars are now known, and the numbers are increasing rapidly as large-scale surveys probe deeper and deeper into space (see *Discovery 25-1*). The distance to the *closest* quasar is 240 Mpc, and the farthest lies more than 9000 Mpc away. Most quasars lie well over 1000 Mpc from Earth. Since light travels at a finite speed, these faraway objects represent the universe as it was in the distant past. The implication is that most quasars date back to much earlier periods of galaxy formation and evolution, rather than more recent times. The prevalence of these energetic objects at great distances tells us that the universe was once a much more violent place than it is today.

Quasars share many properties with Seyferts and radio galaxies. Their radiation is nonstellar and may vary irregularly in brightness over periods of months, weeks, days, or (in some cases) even hours, and some quasars show evidence of jets and extended emission features. Note the jet of luminous matter in 3C 273 (Figure 24.27), reminiscent of the jet in M87 and extending nearly 30 kpc from the center of the quasar. Figure 24.30 shows a quasar with radio lobes similar to those seen in Cygnus A (Figure 24.23b).

very unfamiliar amount—about 16 percent, corresponding to a recession velocity of 48,000 km/s! Figure 24.28 shows the spectrum of 3C 273. Some prominent emission lines and the extent of their redshift are marked on the diagram. Once the nature of the strange spectral lines was known, astronomers quickly found a similar explanation for the spectrum of 3C 48, whose 37 percent redshift implied that it was receding from Earth at the astonishing rate of almost one-third the speed of light!

Their huge speeds mean that neither of these two objects can be members of our Galaxy. In fact, their large redshifts indicate that they are very far away indeed. Applying Hubble's law (with our adopted value of the Hubble constant, $H_0 = 70$ km/s/Mpc), we obtain distances of 650 Mpc for 3C 273 and 1400 Mpc for 3C 48. (See again *More Precisely 24-1* for more information of how these distances are determined and what the large redshifts really mean.)

However, this explanation of the unusual spectra created an even deeper mystery. A simple calculation using the inverse-square law reveals that, despite their unimpressive optical appearance (see Figure 24.29), these faint "stars" are in fact among the brightest-known objects in the universe! Object 3C 273, for example, has a luminosity of about

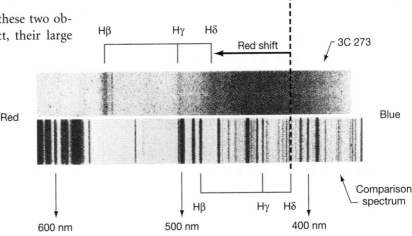

▲ **FIGURE 24.28 Quasar Spectrum** Optical spectrum of the distant quasar 3C 273. Notice both the redshift and the widths of the three hydrogen spectral lines marked as Hβ, Hγ, and Hδ. ∞ (Sec. 14.3) The redshift indicates the quasar's enormous distance. The width of the lines implies rapid internal motion within the quasar. *(Adapted from Palomar/Caltech)*

R I V U X G

▲ **FIGURE 24.29 Typical Quasar** Although quasars are the most luminous objects in the universe, they are often unimpressive in appearance. In this optical image, a distant quasar (marked by an arrow) is seen close (in the sky) to nearby normal stars. The quasar's much greater distance makes it appear fainter than the stars, but intrinsically it is much, much brighter. Often starlike in appearance, quasars are generally identified via their unusual nonstellar colors or spectra. *(SDSS)*

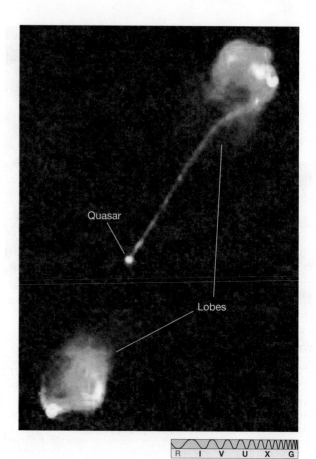

R I V U X G

▲ **FIGURE 24.30 Quasar Jets** Radio image of the quasar 3C 175, showing radio jets feeding faint radio lobes. The bright (white) central object is the quasar, some 3000 Mpc away. The lobes themselves span approximately a million light-years—comparable in size to the radio galaxies discussed earlier. *(NRAO)*

Quasars have been observed in all parts of the electromagnetic spectrum, although many emit most of their energy in the optical and infrared. About 10–15 percent of quasars (called "radio-loud" quasars) also emit substantial amounts of energy at radio wavelengths, presumably as a result of unresolved jets.

Astronomers once distinguished between active galaxies and quasars on the basis of their appearance, spectra, and distance from us, but today most astronomers think that quasars *are* in fact just the intensely bright nuclei of distant active galaxies lying too far away for the galaxies themselves to be seen. (Figure 25.19 presents *Hubble Space Telescope* observations of several relatively nearby quasars in which the surrounding galaxies are clearly visible.)

PROCESS OF SCIENCE CHECK

✔ How did the determination of quasar distances change astronomers' understanding of these objects?

24.5 The Central Engine of an Active Galaxy

The present consensus among astronomers is that, despite their differences in appearance and luminosity, Seyferts, radio galaxies, quasars—as well as "normal" galactic nuclei—share a common energy-generation mechanism.

As a class, active galactic nuclei have some or all of the following properties:

1. They have *high luminosities,* generally greater than the 10^{37} W characteristic of a bright normal galaxy.

2. Their energy emission is mostly *nonstellar*—it cannot be explained as the combined radiation of even trillions of stars.

3. Their energy output can be highly *variable,* implying that their energy is emitted from a small central nucleus much less than a parsec across.

4. They may exhibit *jets* and other signs of explosive activity.

5. Their optical spectra may show broad emission lines, indicating *rapid internal motion* within the energy-producing region.

6. Often the activity appears to be associated with *interactions* between galaxies.

The principal questions, then, are How can such vast quantities of energy arise from these relatively small regions of space? Why is the radiation nonstellar? and What is the origin of the jets and extended radio-emitting lobes? We first consider how the energy is *produced* and then turn to the question of how it is actually *emitted* into intergalactic space.

Energy Production

As illustrated in Figure 24.31, the leading model for the central engine of active galaxies is a scaled-up version of the process powering X-ray binaries in our own Galaxy and the activity in our Galactic center—accretion of gas onto a supermassive black hole, releasing huge amounts of energy as the matter sinks onto the central object. ∞ (Secs. 22.3, 22.8, 23.7) In order to power the brightest active galaxies, theory suggests that the black holes involved must be *billions* of times more massive than the Sun.

As with this model's smaller scale counterparts, the infalling gas forms an *accretion disk* and spirals down toward the black hole, heating up to high temperatures by friction within the disk and emitting large amounts of radiation as a result. In the case of an active galaxy, however, the origin of the accreted gas is not a binary companion, as it is in stellar X-ray sources, but entire stars and clouds of interstellar gas—most likely diverted into the galactic center by an encounter with another galaxy—that come too close to the hole and are torn apart by its strong gravity.

Accretion is extremely efficient at converting infalling mass (in the form of gas) into energy (in the form of electromagnetic radiation). Detailed calculations indicate that as much as 10 or 20 percent of the total mass–energy of the infalling matter can be radiated away before it crosses the hole's event horizon and is lost forever. ∞ (Sec. 22.5) Since the total mass–energy of a star like the Sun—the mass times the speed of light squared—is about 2×10^{47} J, it follows that the 10^{38}-W luminosity of a bright active galaxy can be accounted for by the consumption of "only" 1 solar mass of gas per decade by a billion-solar-mass black hole. More or less luminous active galaxies would require correspondingly more or less fuel. Thus, to power a 10^{40}-W quasar, which is 100 times brighter, the black hole simply consumes 100 times more fuel, or 10 stars per year. The central black hole of a 10^{36}-W Seyfert galaxy would devour only one Sun's worth of material every thousand years.

The small size of the emitting region is a direct consequence of the compact central black hole. Even a billion-solar-mass black hole has a radius of only 3×10^9 km, or 10^{-4} pc—about 20 AU—and theory suggests that the part of the accretion disk responsible for most of the emission would be much less than 1 pc across. ∞ (Sec. 22.5) Instabilities in the accretion disk can cause fluctuations in the energy released, leading to the variability observed in many objects. The broadening of the spectral lines seen in the nuclei of many active galaxies may result from the rapid orbital motion of the gas in the black hole's intense gravity.

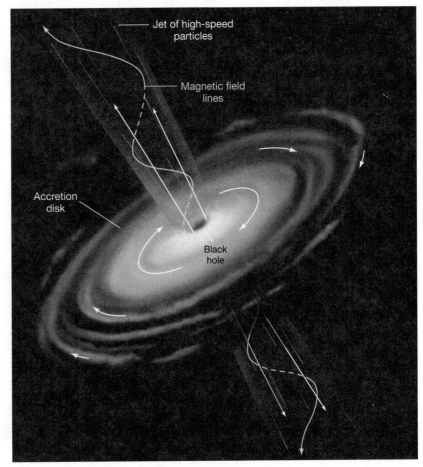

▲ FIGURE 24.31 Active Galactic Nucleus The leading theory for the energy source in active galactic nuclei holds that these objects are powered by material accreting onto a supermassive black hole. As matter spirals toward the hole, it heats up, producing large amounts of energy. At the same time, high-speed jets of gas can be ejected perpendicular to the accretion disk, forming the jets and lobes observed in many active objects. The jets transport magnetic fields generated in the disk out to the radio lobes, where they play a crucial role in producing the observed radiation.

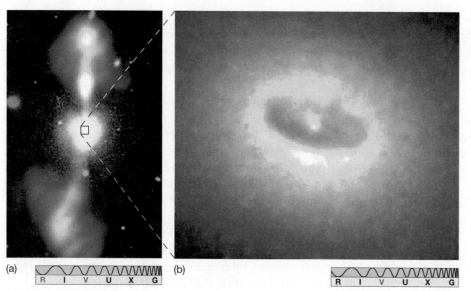

◀ FIGURE 24.32 **Giant Elliptical Galaxy** (a) A combined optical/radio image of the giant elliptical galaxy NGC 4261, in the Virgo Cluster, shows a white visible galaxy at center, from which blue-orange (false-color) radio lobes extend for about 60 kpc. (b) A close-up photograph of the galaxy's nucleus reveals a 100-pc-diameter disk surrounding a bright hub thought to harbor a black hole. *(NRAO; NASA)*

The jets shown in Figure 24.31 consist of material (mainly electrons and protons) blasted out into space—and completely out of the visible portion of the galaxy—from the inner regions of the disk. The details of how jets form remain uncertain, and many active galactic nuclei do not have them, but there is a growing consensus among theorists that jets are a fairly common feature of accretion flows, large and small. They are most likely formed by strong magnetic fields produced within the accretion disk itself.

These fields accelerate charged particles to nearly the speed of light and eject them parallel to the disk's rotation axis. Figure 24.32 shows a *Hubble Space Telescope* image of a disk of gas and dust at the core of the radio galaxy NGC 4261 in the Virgo Cluster. Consistent with the model just described, the disk is perpendicular to the huge jets emanating from the galaxy's center.

Figure 24.33 shows further evidence in favor of this model, in the form of imaging and spectroscopic data from

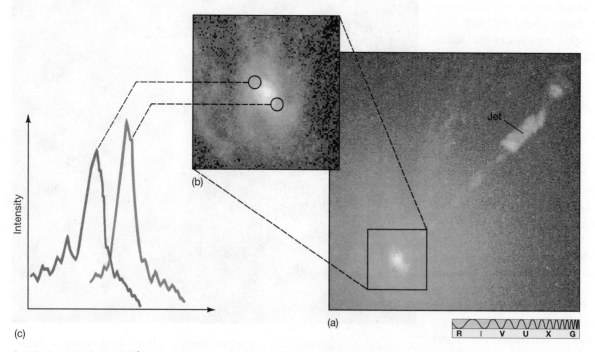

▲ **FIGURE 24.33** **M87 Disk** Recent images and spectra of M87 support the idea of a rapidly whirling accretion disk at the galaxy's heart. (a) An image of the central region of M87, similar to that shown in Figure 24.26(c), shows the galaxy's bright nucleus and jet. (b) A magnified view of the nucleus suggests a spiral swarm of stars, gas, and dust. (c) Spectral-line features observed on opposite sides of the nucleus show opposite Doppler shifts, implying that material on one side of the nucleus is coming toward us and material on the other side is moving away from us. The implication is that an accretion disk spins perpendicular to the jet and that at its center is a black hole having some 3 billion times the mass of the Sun. *(NASA)*

the center of M87, suggesting a rapidly rotating disk of matter orbiting the galaxy's center, again perpendicular to the jet. Measurements of the gas velocity on opposite sides of the disk indicate that the mass within a few parsecs of the center is approximately 3×10^9 solar masses; we assume that this is the mass of the central black hole. At M87's distance, *HST*'s resolution of 0.05 arc second cor responds to a scale of about 5 pc, so we are still far from seeing the (solar system–sized) central black hole itself, but the improved "circumstantial" evidence has convinced many astronomers of the basic correctness of the theory.

Energy Emission

Theory suggests that the radiation emitted by the hot accretion disk surrounding a supermassive black hole should span a broad range of wavelengths, from infrared through X rays, corresponding to the broad range of temperatures in the disk as the gas heats up. This accounts for the observed spectra of at least some active galactic nuclei. However, it also appears that in many cases the high-energy radiation emitted from the accretion disk is "reprocessed"—that is, absorbed and reemitted at longer wavelengths—by material beyond the nucleus before eventually reaching our detectors.

Researchers think that the most likely site of this reprocessing is a rather fat, donut-shaped ring of gas and dust surrounding the inner accretion disk where the energy is actually produced. As illustrated in Figure 24.34, if our line of sight to the black hole does not intersect this dusty donut, then we see the "bare" energy source emitting large amounts of high-energy radiation (with broad emission lines, since we can see the rapidly moving gas near the black hole). ∞ (Sec. 4.5) If the donut intervenes, we see instead large amounts of infrared radiation reradiated from the dust (and only narrow emission lines, from gas farther from the center). The structure of the donut itself is quite uncertain and may in reality bear little resemblance to the rather regular-looking ring shown in the figure. Many astronomers suspect that the absorbing region may actually be a dense *outflow* of gas driven from the accretion disk's outer edge by the intense radiation within.

A different reprocessing mechanism operates in many jets and radio lobes. This mechanism involves the *magnetic fields* possibly produced within the accretion

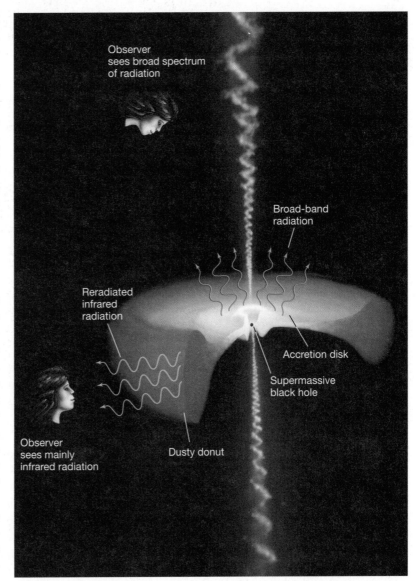

▲ **FIGURE 24.34 Dusty Donut** Remarkably, given the diversity of active galaxies, most researchers think that a single "generic" model can account for their observed properties. The accretion disk surrounding a massive black hole, drawn here with some artistic license, consists of hot gas at many different temperatures (hottest nearest the center). When viewed from above or below, the disk radiates a broad spectrum of electromagnetic energy extending into the X-ray band. However, the dusty infalling gas that ultimately powers the system is thought to form a rather fat, donut-shaped region outside the accretion disk (shown here in dull red), which effectively absorbs much of the high-energy radiation reaching it, re-emitting it mainly in the form of cooler, infrared radiation. Thus, when the accretion disk is viewed from the side, strong infrared emission is observed. For systems with jets (as shown here), the appearance of the jet, radiating mostly radio waves and X rays, also depends on the viewing angle. (See also Figure 24.25.) *(Adapted from D. Berry)*

disk and transported by the jets into intergalactic space (Figure 24.31). As sketched in Figure 24.35(a), whenever a charged particle (here an electron) encounters a magnetic field, the particle tends to spiral around the magnetic field lines. We have already encountered this idea in the

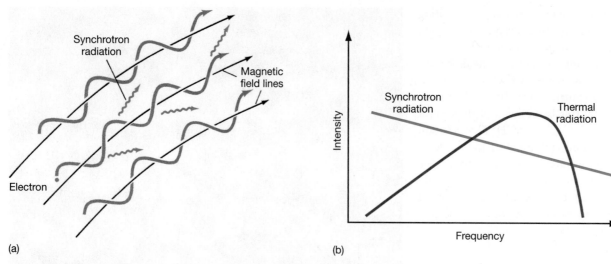

▲ FIGURE 24.35 **Synchrotron Radiation** (a) Charged particles, especially fast-moving electrons (red), emit synchrotron radiation (blue) while spiraling in a magnetic field (black). This process is not confined to active galaxies. It occurs on smaller scales as well, when charged particles interact with magnetism in Earth's Van Allen belts ∞ (Sec. 7.5), when charged matter arches above sunspots ∞ (Sec. 16.5), in the vicinity of neutron stars ∞ (Sec. 22.2), and at the center of our own Galaxy ∞ (Sec. 23.7). (b) Variation of the intensity of thermal and synchrotron (nonthermal) radiation with frequency. Thermal radiation, described by a blackbody curve, peaks at some frequency that depends on the temperature of the source. Nonthermal synchrotron radiation, by contrast, is more intense at low frequencies and is independent of the temperature of the emitting object.

discussions of Earth's magnetosphere and solar activity. ∞ (Secs. 7.5, 16.5)

As the particles whirl around, they emit electromagnetic radiation. ∞ (Sec. 3.2) The radiation produced in this way—called **synchrotron radiation,** after the type of particle accelerator in which it was first observed—is *nonthermal* in nature, meaning that there is no link between the emission and the temperature of the radiating object. Hence, the radiation is not described by a blackbody curve. Instead, its intensity decreases with increasing frequency, as shown in Figure 24.35(b). This is just what is needed to explain the overall spectrum of radiation from radio galaxies and radio-loud quasars. Observations of the radiation received from the jets and radio lobes of active galaxies are completely consistent with synchrotron radiation.

Eventually, the jet is slowed and stopped by the intergalactic medium, the flow becomes turbulent, and the magnetic field grows tangled. The result is a gigantic radio lobe emitting virtually all of its energy in the form of synchrotron radiation. Thus, even though the radio *emission* comes from an enormously extended volume of space that dwarfs the visible galaxy, the *source* of the energy is still the accretion disk— a billion billion times smaller in volume than the radio lobe— lying at the galactic center. The jets serve merely as a conduit that transports energy from the nucleus, where it is generated, into the lobes, where it is finally radiated into space.

The existence of the inner lobes of Centaurus A and the blobs in M87's jet imply that the formation of a jet may be an intermittent process (or, as in the case of the Seyferts discussed earlier, may not occur at all), and, as we have seen, there is also evidence to indicate that much, if not all, of the activity observed in nearby active galaxies has been sparked by recent interaction with a neighbor. Many nearby active galaxies (e.g., Centaurus A) appear to have been "caught in the act" of interacting with another galaxy, suggesting that the fuel supply can be "turned on" by a companion. The tidal forces involved divert gas and stars into the galactic nucleus, triggering an outburst that may last for many millions of years.

What do active galaxies look like between active outbursts? What other connections exist between them and the normal galaxies we see? To answer these important questions, we must delve more deeply into the subject of *galaxy evolution*, to which we turn in Chapter 25.

CONCEPT CHECK

✔ How does accretion onto a supermassive black hole power the energy emission from the extended radio lobes of a radio galaxy?

CHAPTER REVIEW

SUMMARY

1 The **Hubble classification scheme (p. 604)** divides galaxies into several classes, depending on their appearance. **Spiral galaxies (p. 604)** have flattened disks, central bulges, and spiral arms. Their halos consist of old stars, whereas the gas-rich disks are the sites of ongoing star formation. **Barred-spiral galaxies (p. 605)** contain an extended "bar" of material projecting beyond the central bulge. **Elliptical galaxies (p. 607)** have no disk and contain little or no cool gas or dust, although very hot interstellar gas is observed. In most cases, they consist entirely of old stars. They range in size from dwarf ellipticals, which are much less massive than the Milky Way Galaxy, to giant ellipticals, which may contain trillions of stars. **S0 and SB0 galaxies (p. 608)** are intermediate in their properties between ellipticals and spirals. **Irregular galaxies (p. 608)** are galaxies that do not fit into any of the other categories. Many are rich in gas and dust and are the sites of vigorous star formation.

2 Astronomers often use **standard candles (p. 611)** as distance-measuring tools. These are objects that are easily identifiable and whose luminosities lie within some reasonably well-defined range. Comparing luminosity and apparent brightness, astronomers determine distance with the use of the inverse-square law. An alternative approach is the **Tully-Fisher relation (p. 612)**, an empirical correlation between rotational velocity and luminosity in spiral galaxies.

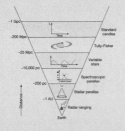

3 The Milky Way, Andromeda, and several other smaller galaxies form a small gravitationally bound collection of galaxies called the **Local Group (p. 614)**. **Galaxy clusters (p. 614)** consist of many galaxies orbiting one another, bound together by their own gravity. The nearest large galaxy cluster to the Local Group is the Virgo Cluster.

4 Distant galaxies are observed to be receding from the Milky Way at speeds proportional to their distances from us. This relationship is called **Hubble's law (p. 615)**. The constant of proportionality in the law is **Hubble's constant (p. 616)**. Its value is thought to be around 70 km/s/Mpc. Astronomers use Hubble's law to determine distances to the most remote objects in the universe. The redshift associated with the Hubble expansion is called the **cosmological redshift (p. 615)**.

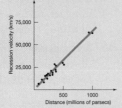

5 **Active galaxies (p. 619)** can be much more luminous than normal galaxies and have nonstellar spectra, emitting most of their energy outside the visible part of the electromagnetic spectrum. Often the nonstellar activity suggests rapid internal motion and is associated with a bright **active galactic nucleus (p. 620)**. Many active galaxies have high-speed, narrow jets of matter shooting out from their central nuclei. The jets transport energy from the nucleus, where it is generated, to enormous **radio lobes (p. 621)** lying far beyond the visible portion of the galaxy, where it is radiated into space. The jets often appear to be made up of distinct "blobs" of gas, suggesting that the process that generates the energy is intermittent.

6 A **Seyfert galaxy (p. 620)** looks like a normal spiral, but has an extremely bright central galactic nucleus. Spectral lines from Seyfert nuclei are very broad, indicating rapid internal motion, and the rapid variability in the luminosity of Seyferts implies that the source of the radiation is much less than 1 light-year across. **Radio galaxies (p. 621)** emit large amounts of energy in the radio part of the spectrum. The corresponding visible galaxy is usually elliptical. **Quasars (p. 625)**, or quasi-stellar objects, are the most luminous objects known. In visible light they appear starlike, and their spectra are usually substantially redshifted. All quasars are very distant, indicating that we see them as they were in the remote past.

7 The generally accepted explanation for the observed properties of all active galaxies is that their energy is generated by the accretion of galactic gas onto a supermassive (million- to billion-solar-mass) black hole lying in the galactic center. The small size of the accretion disk explains the compact extent of the emitting region, and the high-speed orbit of gas in the black hole's intense gravity accounts for the rapid motion that is observed. Typical luminosities of active galaxies require the consumption of about 1 solar mass of material every few years. Some of the infalling matter may be blasted out into space, producing magnetized jets that create and feed the galaxy's radio lobes. The accretion disk emits at a broad range of temperatures, producing a nonstellar spectrum. In addition, much of the radiation may be reprocessed into the infrared by a ring of dust surrounding the disk. On larger scales, charged particles spiraling around magnetic field lines produce **synchrotron radiation (p. 630)**, whose spectrum is consistent with the radio emission from radio galaxies and jets.

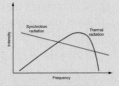

*For instructor-assigned homework go to **www.masteringastronomy.com***

Problems labeled **POS** explore the process of science　|　**VIS** problems focus on reading and interpreting visual information

REVIEW AND DISCUSSION

1. What distinguishes one type of spiral galaxy from another?

2. Describe some similarities and differences between elliptical galaxies and the halo of our own Galaxy.

3. Describe the four rungs in the distance-measurement ladder used to determine the distance to a galaxy lying 5 Mpc away.

4. Describe the contents of the Local Group. How much space does it occupy compared with the volume of the Milky Way?

5. What are standard candles, and why are they important to astronomy?

6. How is the Tully-Fisher relation used to measure distances to galaxies?

7. What is the Virgo Cluster?

8. What is Hubble's law?

9. **POS** How is Hubble's law used by astronomers to measure distances to galaxies?

10. What is the most likely range of values for Hubble's constant? What are the uncertainties in its value?

11. Name two basic differences between normal galaxies and active galaxies.

12. Are there any "nearby" active galaxies—within 50 Mpc of Earth, say?

13. Describe some of the basic properties of Seyfert galaxies.

14. **POS** What is the evidence that the radio lobes of some active galaxies consist of material ejected from the galaxy's center?

15. **POS** How do we know that the energy-emitting regions of many active galaxies must be very small?

16. What was it about the spectra of quasars that was so unexpected and surprising?

17. Why do astronomers prefer to speak in terms of redshifts rather than distances to faraway objects?

18. How do we know that quasars are extremely luminous?

19. Briefly describe the leading model for the central engine of an active galaxy.

20. **POS** How is the process of synchrotron emission related to observations of active galaxies?

CONCEPTUAL SELF-TEST: MULTIPLE CHOICE

1. Young stars in a galactic disk are (**a**) evenly distributed within and between spiral arms; (**b**) mostly found in the space between spiral arms; (**c**) mostly found in the spiral arms; (**d**) older than stars in the halo.

2. Astronomers classify elliptical galaxies by (**a**) the number of stars they contain; (**b**) their colors; (**c**) how flattened they appear; (**d**) their diameters.

3. Using the method of standard candles, we can, in principle, find the distance to a campfire if we know (**a**) the number of logs used; (**b**) the fire's temperature; (**c**) the length of time the fire has been burning; (**d**) the type of wood used in the fire.

4. **VIS** If the galaxy in Figure 24.11 ("Galaxy Rotation") were smaller and spinning more slowly, then, in order to represent it correctly, the figure should be redrawn to show (**a**) a greater blueshift; (**b**) a greater redshift; (**c**) a narrower combined line; (**d**) a larger combined amplitude.

5. Within 30 Mpc of the Sun, there are about (**a**) 3 galaxies; (**b**) 30 galaxies; (**c**) a few thousand galaxies; (**d**) a few million galaxies.

6. **VIS** According to Figure 24.17 ("Hubble's Law"), a galaxy 500 million parsecs away has a velocity of roughly (**a**) 25,000 km/s away from us; (**b**) 35,000 km/s toward us; (**c**) 35,000 km/s away from us; (**d**) 75,000 km/s toward us.

7. **VIS** According to Figure 24.19 ("Galaxy Energy Spectra"), active galaxies (**a**) emit most of their energy at long wavelengths; (**b**) emit very little energy at high frequencies; (**c**) emit large amounts of energy at all wavelengths; (**d**) emit most of their energy in the visible part of the spectrum.

8. If the light from a galaxy fluctuates in brightness very rapidly, the region producing the radiation must be (**a**) very large; (**b**) very small; (**c**) very hot; (**d**) rotating very rapidly.

9. Quasar spectra (**a**) are strongly redshifted; (**b**) show no spectral lines; (**c**) look like the spectra of stars; (**d**) contain emission lines from unknown elements.

10. Active galaxies are very luminous because they (**a**) are hot; (**b**) contain black holes in their cores; (**c**) are surrounded by hot gas; (**d**) emit jets.

PROBLEMS

The number of dots preceding each Problem indicates its approximate level of difficulty.

1. • A supernova of luminosity 1 billion times the luminosity of the Sun is used as a standard candle to measure the distance to a faraway galaxy. From Earth, the supernova appears as bright as the Sun would appear from a distance of 10 kpc. What is the distance to the galaxy?

2. •• A Cepheid variable star in the Virgo cluster has an absolute magnitude of −5 and is observed to have an apparent magnitude of 26.3. Use these figures to calculate the distance to the Virgo cluster.

3. • According to Hubble's law, with H_0 = 70 km/s/Mpc, what is the recessional velocity of a galaxy at a distance of 200 Mpc? How far away is a galaxy whose recessional velocity is 4000 km/s? How do these answers change if H_0 = 60 km/s/Mpc? If H_0 = 80 km/s/Mpc?

4. •• According to Hubble's law, with H_0 = 70 km/s/Mpc, how long will it take for the distance from the Milky Way Galaxy to the Virgo Cluster to double?

5. ••• Assuming Hubble's law with H_0 = 70 km/s/Mpc, what would be the angular diameter of a giant elliptical galaxy of actual diameter 80 kpc if its 656.3-nm Hα line is observed at 700 nm?

6. •• A certain quasar has a redshift of 0.25 and an apparent magnitude of 13. Using the data from Table 24.2, calculate the quasar's absolute magnitude and hence its luminosity. Compare the apparent brightness of the quasar, viewed from a distance of 10 pc, with that of the Sun as seen from Earth.

7. • What are the absolute magnitude and luminosity of a quasar with a redshift of 5 and an apparent magnitude of 22?

8. • On the basis of the data presented in the text, calculate the orbital speed of material orbiting at a distance of 0.5 pc from the center of M87.

9. •• Spectral lines from a Seyfert galaxy are observed to be redshifted by 0.5 percent and to have broadened emission lines indicating an orbital speed of 250 km/s at an angular distance of 0.10 from its center. Assuming circular orbits, use Kepler's laws to estimate the mass within this 0.10 radius. ∞ (Sec. 23.6)

10. • A quasar consumes 1 solar mass of material per year, converting 15 percent of it directly into energy. What is the quasar's luminosity, in solar units?

25

GALAXIES AND DARK MATTER

THE LARGE-SCALE STRUCTURE OF THE COSMOS

LEARNING GOALS

Studying this chapter will enable you to

1 Describe some of the methods used to determine the masses of distant galaxies.

2 Explain why astronomers think that most of the matter in the universe is invisible.

3 Discuss some theories of how galaxies form and evolve.

4 Explain the role of black holes and active galaxies in current theories of galactic evolution.

5 Summarize what is known about the large-scale distribution of galaxies in the universe.

6 Describe some techniques used by astronomers to probe the universe on very large scales.

THE BIG PICTURE Galaxies are among the grandest, most beautiful objects in the sky, and they dominate our view of deep space, but they represent just a tiny fraction of all matter in the cosmos. Vast quantities of unseen cosmic material—dark matter—actually account for most of the mass in the universe.

Visit the Study Area in www.masteringastronomy.com for quizzes, animations, videos, interactive figures, and self-guided tutorials.

On scales much larger than even the largest galaxy clusters, the dynamics of the universe itself becomes apparent, new levels of structure are revealed, and a humbling new reality emerges. We may be star stuff, the product of countless cycles of stellar evolution, but we are not the stuff of the cosmos. The universe in the large is composed of matter fundamentally different from the familiar atoms and molecules that make up our bodies, our planet, our star and Galaxy, and all the luminous matter we observe in the heavens. Only its gravity announces the presence of this strange kind of matter, providing the backdrop against which galaxies form and evolve.

By comparing and classifying the properties of galaxies near and far, astronomers have begun to understand their formation, dynamics, and evolution. By mapping out the distribution of those galaxies in space, astronomers trace out the immense realms of the universe. Points of light in the uncharted darkness, they remind us that our position in the universe is no more special than that of a boat adrift at sea.

LEFT: *Some galaxies are bright and splendid, like the five big ones in this image, known as Stephan's Quintet. Others are dim and distant, like several that appear smaller in the background. One member of this group, NGC 7320 (upper left), is actually seven times closer to Earth than the other four, which are in the process of colliding. Mergers and acquisitions are common among galaxies, but astronomers still don't fully understand how galaxies formed long ago. (STScI)*

25.1 Dark Matter in the Universe

In Chapter 23, we saw how measurements of the orbital velocities of stars and gas in our own Galaxy reveal the presence of an extensive *dark-matter halo* surrounding the galaxy we see. ∞ (Sec. 23.6) Do other galaxies have similar dark halos? And what evidence do we have for dark matter on larger scales? To answer these questions, we need a way to calculate the masses of galaxies and galaxy clusters, then compare those masses with the luminous matter we actually observe.

How can we measure the masses of such large systems? Surely, we can neither count all their stars nor estimate their interstellar content very well: Galaxies are just too complex for us to take a direct inventory of their material makeup. Instead, we must rely on indirect techniques. Despite their enormous sizes, galaxies and galaxy clusters obey the same physical laws as do the planets in our own solar system. To calculate galaxy masses, we turn as usual to Newton's law of gravity.

Masses of Galaxies and Galaxy Clusters

Astronomers can calculate the masses of some spiral galaxies by determining their *rotation curves,* which plot rotation speed versus distance from the galactic center, as illustrated in Figure 25.1. Rotation curves for a few nearby spirals are shown in Figure 25.1(b). The mass within any given radius then follows directly from Newton's laws. ∞ (Sec. 2.7) The rotation curves shown imply masses ranging from about 10^{11} to 5×10^{11} solar masses within about 25 kpc of the center—comparable to the results obtained for our own Galaxy using the same technique. ∞ (Sec. 23.6)

Distant galaxies are generally too far away for such detailed curves to be drawn. Nevertheless, by observing the broadening of their spectral lines—as discussed in Chapter 24 in the context of the Tully-Fisher relation—we can still measure the overall rotation speed of these galaxies. ∞ (Sec. 24.2) Estimating a galaxy's size then leads to an estimate of its mass. Similar techniques have been applied to ellipticals and irregulars. In general, the approach is useful for measuring the mass lying within about 50 kpc of a galaxy's center—the extent of the electromagnetic emission from stellar and interstellar material.

To probe farther from the centers of galaxies, astronomers turn to *binary* galaxy systems (Figure 25.2a), whose components may lie hundreds of kiloparsecs apart. The orbital period of such a system is typically billions of years, far too long for the orbit to be accurately measured. However, by estimating the period and semimajor axis from the available information—the line-of-sight velocities and the angular separation of the components—an approximate total mass can be derived. ∞ (*More Precisely 2-2*)

Galaxy masses obtained in this way are fairly uncertain. However, by combining many such measurements, astronomers can obtain quite reliable *statistical* information about galaxy masses. Most normal spirals (the Milky Way

(a)

R I V U X G

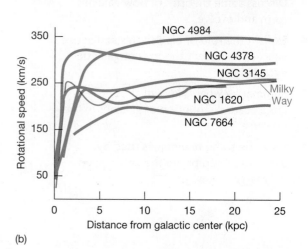

(b)

Interactive FIGURE 25.1 Galaxy Rotation Curves (a) By observing orbital velocities at different distances from the center of a disk galaxy, astronomers can plot a rotation curve for the galaxy. This is M64, the "Evil Eye" galaxy, some 5 Mpc distant. (b) Rotation curves for some nearby spiral galaxies, with the corresponding curve for our own Galaxy (from Figure 23.21) marked in red for comparison. These curves indicate masses of a few hundred billion times the mass of the Sun, much greater than can be accounted for by the observed stars and gas, providing strong evidence for dark matter in the universe. *(NASA)*

Galaxy included) and large ellipticals contain between 10^{11} and 10^{12} solar masses of material. Irregular galaxies often contain less mass, about 10^8 to 10^{10} times that of the Sun. Dwarf ellipticals and dwarf irregulars can contain as little as 10^6 or 10^7 solar masses of material.

We can use another statistical technique to derive the combined mass of all the galaxies within a galaxy cluster. As

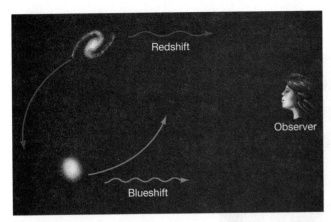

(a)

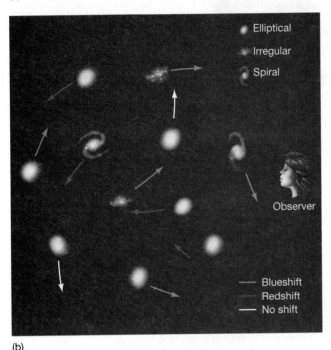

(b)

▲ **FIGURE 25.2 Galaxy Masses** (a) In a binary galaxy system, galaxy masses may be estimated by observing the orbit of one galaxy about the other. (b) The mass of a galaxy cluster may be estimated by observing the motion of many galaxies in the cluster and estimating how much mass is needed to prevent the cluster from flying apart.

Visible Matter and Dark Halos

The rotation curves of the spiral galaxies shown in Figure 25.1 remain flat (that is, do not decline, and even rise slightly) far beyond the galaxies' visible images, implying that those galaxies—and perhaps all spiral galaxies—contain large amounts of *dark matter,* in the form of invisible dark halos similar to that surrounding the Milky Way. ∞ (Sec. 23.6) Overall, spiral galaxies seem to contain from 3 to 10 times more mass than can be accounted for in the form of luminous matter. Studies of elliptical galaxies suggest similarly large dark halos surrounding these galaxies, too.

Astronomers find even greater discrepancies between visible light and total mass when they study galaxy clusters. Calculated cluster masses range from 10 to nearly 100 times the mass suggested by the light emitted by individual cluster galaxies. Put another way, a lot more mass is needed to bind galaxy clusters than we can see. Thus, the problem of dark matter exists not just in our own Galaxy, but also in other galaxies and, to an even greater degree, in galaxy clusters as well. In that case, we are compelled to accept the fact that *upwards of 90 percent of the matter in the universe is dark*—and not just in the visible portion of the spectrum. The mass goes undetected at *any* electromagnetic wavelength.

As discussed in Chapter 23, many possible explanations for the dark matter have been suggested, ranging from stellar remnants of various sorts to exotic subatomic particles. ∞ (Sec. 23.6) Whatever its nature, the dark matter in clusters apparently cannot simply be the accumulation of dark matter within individual galaxies. Even including the galaxies' dark halos, we still cannot account for all the dark matter in galaxy clusters. As we look on larger and larger scales, we find that a larger and larger fraction of the matter in the universe is dark.

Intracluster Gas

In addition to the luminous matter observed within the cluster galaxies themselves, astronomers also have evidence for large amounts of *intracluster gas*—superhot (more than 10 million K), diffuse intergalactic matter filling the space among the galaxies. Satellites orbiting above Earth's atmosphere have detected substantial amounts of X-ray radiation from many clusters. Figure 25.3 shows false-color X-ray images of one such system. The X-ray-emitting region is centered on, and comparable in size to, the visible cluster image.

Further evidence for intracluster gas can be found in the appearance of the radio lobes of some active galaxies. ∞ (Sec. 24.4) In some systems, known as *head–tail* radio galaxies, the lobes seem to form a "tail" behind the main part of the galaxy. For example, the lobes of radio galaxy NGC 1265, shown in Figure 25.4, appear to be "swept back" by some onrushing wind, and, indeed, this is the most likely explanation for the galaxy's appearance. If NGC 1265 were

depicted in Figure 25.2(b), each galaxy within a cluster moves relative to all other members of the cluster, and we can estimate the cluster's mass simply by asking how massive it must be in order to bind its galaxies gravitationally. For example, if we find that galaxies in a cluster are moving with an average speed of 1000 km/s and the cluster radius is 3 Mpc (both typical values), it follows from Newton's laws—*assuming* that the cluster is gravitationally bound—that the mass of the cluster must be around (3 Mpc) × (1000 km/s)^2/G ≈ 7 × 10^{14} solar masses. ∞ *(More Precisely 2-2)* Cluster masses obtained in this way generally lie in the range of 10^{14}–10^{15} solar masses. Notice that this calculation gives us no information whatsoever about the masses of individual galaxies. It tells us only about the *total* mass of the cluster.

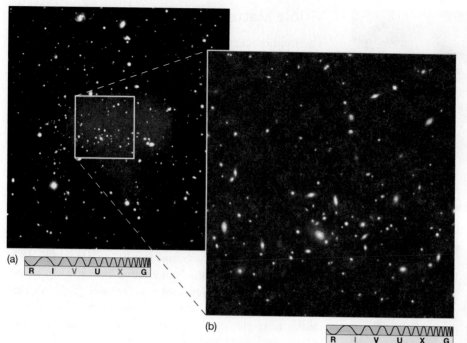

R I V U X G

(a)

(b)

R I V U X G

◀ **FIGURE 25.3 Galaxy Cluster X-Ray Emission** (a) Superposition of infrared and X-ray radiation from the distant galaxy cluster Abell 1835. The X rays are shown as a fuzzy, bluish cloud of hot gas filling the intracluster spaces among the galaxies. (b) A longer infrared exposure of the central region, showing the richness of this cluster, which spans about a million parsecs. Images like these demonstrate that the space between the galaxies within galaxy clusters is filled with superheated gas. *(NASA; AURA; ESA)*

from 10 to 100 times more mass in gas than exists in stars.

Why is the intracluster gas so hot? Simply because its particles are bound by gravity and hence are moving at speeds comparable to those of the galaxies in the cluster— 1000 km/s or so. Since temperature is just a measure of the speed at which the gas particles move, this speed translates (for protons) to a temperature of 40 million K. ∞ *(More Precisely 8-1)*

Where did the gas come from? There is so much of it that it could not have been expelled from the galaxies themselves. Instead, astronomers think that it is mainly *primordial*—gas that has been around since the universe formed and that never became part of a galaxy. However, the intracluster gas does contain some heavy elements—carbon, nitrogen, and so on—implying that at least some of it is material ejected from galaxies after enrichment by stellar evolution. ∞ *(Sec. 21.5)* Just how this occurred remains a mystery.

at rest, it would be just another double-lobe source, perhaps quite similar to Centaurus A (Figure 24.22). However, the galaxy is traveling through the intergalactic medium of its parent galaxy cluster (known as the Perseus Cluster), and the outflowing matter forming the lobes tends to be left behind as NGC 1265 moves.

How much gas do these observations reveal? At least as much matter—and, in most cases, significantly *more*—exists within clusters in the form of hot gas as is visible in the form of stars. This is a lot of material, but it still doesn't solve the dark-matter problem. To account for the total masses of galaxy clusters implied by dynamical studies, we would have to find

PROCESS OF SCIENCE CHECK

✔ What assumptions are we making when we infer the mass of a galaxy cluster from observations of the spectra of its constituent galaxies?

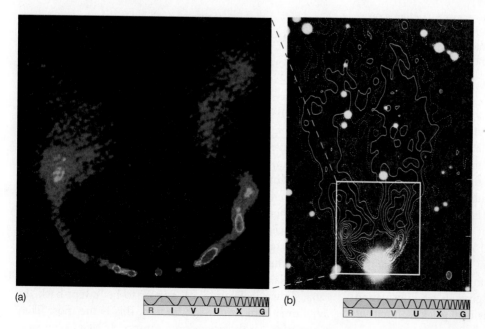

(a)

R I V U X G

(b)

R I V U X G

◀ **FIGURE 25.4 Head–Tail Radio Galaxy** (a) Radiograph, in false color, of the head–tail radio galaxy NGC 1265. (b) The same radio data, in contour form, superposed on the optical image of the galaxy. Astronomers reason that this object is moving rapidly through space, trailing a "tail" behind as it goes. *(NRAO; Palomar/Caltech)*

25.2 Galaxy Collisions

Contemplating the congested confines of a rich galaxy cluster (such as Virgo or Coma, with thousands of member galaxies orbiting within a few megaparsecs), we might expect that collisions among galaxies would be common. ∞ (Sec. 24.2) Gas particles collide in our atmosphere, and hockey players collide in the rink. So, do galaxies in clusters collide, too? The answer is yes, and this simple fact plays a pivotal role in our understanding of how galaxies evolve.

Figure 25.5 apparently shows the aftermath of a bull's-eye collision between a small galaxy (perhaps one of the two at the right, although that is by no means certain) and the larger galaxy at the left. The result is the "Cartwheel" galaxy, about 150 Mpc from Earth, its halo of young stars resembling a vast ripple in a pond. The ripple is most likely a density wave created by the passage of the smaller galaxy through the disk of the larger one. ∞ (Sec. 23.5) The disturbance is now spreading outward from the region of impact, creating new stars as it goes.

Figure 25.6 shows an example of a close encounter that hasn't (yet) led to an actual collision. Two spiral galaxies are apparently passing each other like majestic ships in the night. The larger and more massive galaxy on the left is called NGC 2207; the smaller one on the right is IC 2163. Analysis of this image suggests that IC 2163 is now swinging past NGC 2207 in a counterclockwise direction, having made a close approach some 40 million years ago. The two galaxies seem destined to undergo further close encounters, as IC 2163 apparently does not have enough energy to escape the gravitational pull of NGC 2207. Each time the two galaxies experience a close encounter, bursts of star formation erupt in both as their interstellar clouds of gas and dust are pushed, shoved, and shocked. In roughly a billion years, these two galaxies will probably merge into a single, massive galaxy.

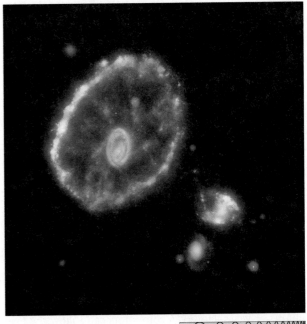

R I V U X G

▲ FIGURE 25.5 **Cosmic Cartwheel** The "Cartwheel" galaxy (upper left) appears to be the result of a collision that has led to an expanding ring of star formation moving outward through the galactic disk. The offending galaxy might be one of the smaller objects at right, or as implied by recent radio observations of loose, linking gas, it might have been another galaxy now farther away and beyond the field of view. This is a false-color composite image combining four spectral bands: infrared in red (from *Spitzer*), optical in green (from *Hubble*), ultraviolet in blue (from *Galex*), and X ray in purple (from *Chandra*). *(NASA)*

These examples illustrate how an interaction with another galaxy—a close encounter or an actual collision—can have dramatic consequences for a galaxy, especially its interstellar gas. The rapidly varying gravitational forces during the interaction compress the gas, often resulting in a galaxy-wide episode of star formation. The result is a **starburst galaxy,** a spectacular example of which is shown in Figure 25.7.

No human will ever witness an entire galaxy collision, for it lasts many

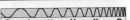

R I V U X G

◀ FIGURE 25.6 **Galaxy Encounter** This encounter between two spirals, NGC 2207 (left) and IC 2163, has already led to bursts of star formation in each. Eventually the two will merge, but not for a billion years or so. *(NASA)*

ANIMATION/VIDEO Galaxy Collision I

◀ **FIGURE 25.7 Starburst Galaxy** This interacting galaxy pair (IC 694, at the left, and NGC 3690) shows starbursts now under way in both galaxies—hence the bluish tint. Such intense, short-lived bursts probably last for no more than a few tens of millions of years—a small fraction of a typical galaxy's lifetime. *(W. Keel)*

shown in Figure 25.6, but the details of the original structure have been largely obliterated by the collision. Notice the similarity to the real image of NGC 4038/4039 (Figure 25.8a), the so-called Antennae galaxies, which show extended tails, as well as double galactic centers only a few hundred parsecs across. Star formation induced by the collision is clearly traced by the blue light from thousands of young, hot stars. The simulations indicate that, as with the galaxies in Figure 25.6, ultimately the two galaxies will merge into one.

Galaxies in clusters apparently collide quite often. Many collisions and near misses similar to those shown in the previous figures have been observed (see also Section 24.4), and a straightforward calculation reveals that, given the crowded conditions in even a modest cluster, close encounters are the norm rather than the exception. The reason is simple: The distance between adjacent galaxies in a cluster averages a few hundred thousand parsecs, which is not much greater (certainly less than five times more) than the size of a typical galaxy, including its extended dark halo. Galaxies simply do not have that much room to roam around without bumping into one another. Many researchers think that most galaxies

millions of years. However, computers can follow the event in a matter of hours. Simulations modeling in detail the gravitational interactions among stars and gas, and incorporating the best available models of gas dynamics, allow astronomers to better understand the effects of a collision on the galaxies involved and even estimate the eventual outcome of the interaction.

The particular calculation shown in Figure 25.8(b) began with two colliding spiral galaxies, not so different from those

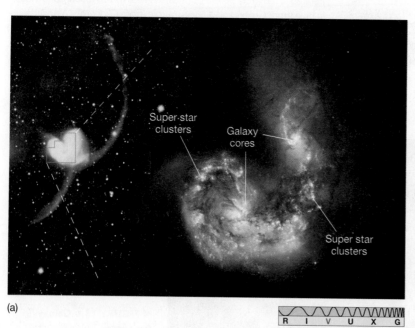

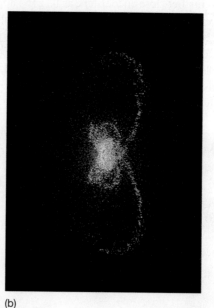

(a)

(b)

▲ **FIGURE 25.8 Galaxy Collision** (a) The "Antennae" galaxies collided a few tens of millions of years ago. The long tidal "tails" (black-and-white image at the left) mark their final plunge. Strings of young, bright "super star clusters" (color image at center) are the result of violent shock waves produced in the gas disks of the two colliding galaxies. (b) A computer simulation of the encounter shows many of the same features as the real thing, strengthening the case that we really are seeing a collision in progress. Such simulations also demonstrate the crucial role played by dark-matter halos in mediating galaxy interactions. *(AURA; NASA; J. Barnes)*

in most clusters have been strongly influenced by collisions, in some cases in the relatively recent past.

In the smaller groups, the galaxies' speeds are low enough that interacting galaxies tend to "stick together," and mergers, as shown in the computer simulation, are the most common outcome. In larger groups, galaxies move faster and tend to pass through one another without sticking. Either way, the encounters have substantial effects on the galaxies involved (see Section 25.3). If we wait long enough, we will have an opportunity to see for ourselves what a galaxy collision is really like: Our nearest large neighbor, the Andromeda Galaxy (Fig. 23.2), is currently approaching the Milky Way at a velocity of 120 km/s. In a few billion years it will collide with our Galaxy, and we will then be able to test astronomers' theories firsthand!

Curiously, although a collision may wreak havoc on the large-scale structure of the galaxies involved, it has essentially no effect on the individual stars they contain. The stars within each galaxy just glide past one another. In contrast to galaxies in the cluster, the stars in a galaxy are so small compared with the distances between them that when two galaxies collide, the star population merely doubles for a time, and the stars continue to have so much space that they do not run into each other. Collisions can rearrange the stellar and interstellar contents of each galaxy, often producing a spectacular burst of star formation that may be visible to enormous distances, but from the point of view of the stars, it's clear sailing.

CONCEPT CHECK

✔ What role do collisions play in the evolution of galaxies?

25.3 Galaxy Formation and Evolution

With Hubble's law as our guide to distances in the universe, and armed now with knowledge of the distribution of dark matter on galactic and larger scales, let's turn to the question of how galaxies came to be the way they are. Can we explain the different types of galaxy we see? Astronomers know of no simple evolutionary connections among the various categories in the Hubble classification scheme. ∞ (Sec. 24.1) To answer the question, we must understand how galaxies formed.

Unfortunately, compared with the theories of star formation and stellar evolution, the theory of galaxy formation and evolution is still very much in its infancy. Galaxies are far more complex than stars, they are harder to observe, and the observations are harder to interpret. In addition, we have only partial information on conditions in the universe during the formation process, quite unlike the corresponding situation for stars. ∞ (Sec. 18.3) Finally, and most important, stars almost never *collide* with one another, with the result that most single stars and binaries evolve in isolation. Galaxies, however, may suffer numerous collisions during their lives,

making it much harder to decipher their pasts. Indeed, collisions like those described in the previous section blur the distinction between formation and evolution to the point where it can be hard to separate one from the other.

Nevertheless, some general ideas have begun to gain widespread acceptance, and we can offer some insights into the processes responsible for the galaxies we see. We first describe a general scenario for how large galaxies form from smaller ones, then discuss how galaxies change in time due to both internal stellar evolution and external influences. Finally, we consider how the galaxy types in Hubble's classification might fit in to this broad picture.

Mergers and Acquisitions

The seeds of galaxy formation were sown in the very early universe, when small density fluctuations in the primordial matter began to grow (see Section 27.5). Our discussion here begins with these "pregalactic" blobs of gas already formed. The masses of the various fragments were quite small—only a few million solar masses, comparable to the masses of the smallest present-day dwarf galaxies, which may in fact be remnants of that early time. Most astronomers think that galaxies grew by repeated *merging* of smaller objects, as illustrated in Figure 25.9(a). Contrast this with the process of star formation, in which a large cloud fragments into smaller pieces that eventually become stars. ∞ (Sec. 19.2)

Theoretical evidence for this picture of **hierarchical merging** is provided by computer simulations of the early universe, which clearly show merging taking place. Further strong support comes from observations that galaxies at large redshifts (meaning that they are very distant and the light we see was emitted long ago) are distinctly smaller and more irregular than those found nearby. Figure 25.9(b) (see also Figure 25.10) shows some of these images, which include objects up to 5 *billion* parsecs away. The vague bluish patches are separate small galaxies, each containing only a few percent of the mass of the Milky Way Galaxy. Their irregular shape is thought to be the result of galaxy mergers; the bluish coloration comes from young stars that formed during the merging process.

Figure 25.9(c) shows more detailed views of some of the objects in Figure 25.9(b), all lying in the same region of space, about 1 Mpc across and almost 5000 Mpc from Earth. Each blob seems to contain several billion stars spread throughout a distorted spheroid about a kiloparsec across. Their decidedly bluish tint suggests that active star formation is already underway. We see them as they were nearly 10 billion years ago, a group of young galaxies possibly poised to merge into one or more larger objects.

Hierarchical merging provides the conceptual framework for all modern studies of galaxy evolution. It describes a process that began billions of years ago and continues (albeit at a greatly reduced rate) to the present day, as galaxies continue to collide and merge. By studying how galaxy

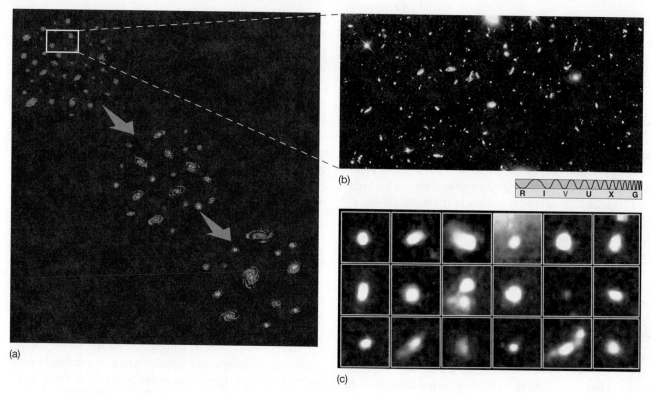

▲ **FIGURE 25.9 Galaxy Formation** (a) The present view of galaxy formation holds that large systems were built up from smaller ones through collisions and mergers, as shown schematically in this drawing. (b) This photograph, one of the deepest ever taken of the universe (i.e., looking at the faintest, most distant objects), provides "fossil evidence" for hundreds of galaxy shards and fragments, up to 5000 Mpc distant. (c) Enlargements of selected portions of (b) reveal rich (billion-star) "star clusters," all lying within a relatively small volume of space (about 1 Mpc across). Their proximity to one another suggests that we may be seeing a group of pregalactic fragments about to merge to form a galaxy. The events pictured took place about 10 billion years ago. *(NASA)*

properties vary with distance, and hence look-back time, astronomers try to piece together the merger history of the universe. ∞ *(More Precisely 24-1)*

Figure 25.10 is a remarkable image from the *Hubble Space Telescope* showing billions of years of galaxy evolution in a single tiny patch of the sky. The large, bright galaxies with easily discernible Hubble types are mostly (according to their redshifts) relatively nearby objects. They are seen here against a backdrop of small, faint, irregular galaxies lying much farther away. The size and appearance of these distant galaxies compared with those in the foreground strongly support the basic idea that galaxies grew by mergers and were smaller and less regular in the past.

Evolution and Interaction

Left alone, a galaxy will evolve slowly and fairly steadily as interstellar clouds of gas and dust are turned into new generations of stars and main-sequence stars evolve into giants and, ultimately, into compact remnants—white dwarfs, neutron stars, and black holes. The galaxy's overall color, composition, and appearance change in a more or less predictable way as the cycle of stellar evolution recycles and enriches the galaxy's interstellar matter. ∞ (Sec. 21.5) If the galaxy is an elliptical, lacking interstellar gas, it will tend to become fainter and redder in time as the more massive stars burn out and are not replaced. ∞ (Sec. 20.5) For a gas-rich galaxy, such as a spiral or an irregular, hot, bright stars will lend a bluish coloration to the overall light for as long as gas remains available to form them. As in our own Galaxy, the star-forming lifetime of a spiral disk may be prolonged by infall of fresh gas from the galaxy's surroundings. ∞ (Sec. 23.4)

But many—perhaps most—galaxies are *not* alone. They reside in small groups and clusters, and, as we have just seen, may *interact with other galaxies* repeatedly over extended periods of time. As described in the previous section, these interactions can rearrange a galaxy's internal structure, compressing interstellar gas, and triggering a sudden, intense burst of star formation. Encounters may also divert fuel to a central black hole, powering violent activity in some galactic nuclei. ∞ (Sec. 24.4) Thus, starbursts and nuclear activity are key indicators of interactions and mergers between galaxies.

Careful studies of starburst galaxies and active galactic nuclei indicate that most galactic encounters probably took place long ago—at redshifts greater than about 1, when the clusters were more compact and galaxy collisions were correspondingly more frequent. ∞ *(More Precisely 24-1)* We see the majority of these violent events as they unfolded roughly 10 billion years ago. Nevertheless, the galaxy interactions observed locally are extensions of this same basic process into the present day. We have ample evidence that galaxies evolved, and are still evolving, in response to external factors long after the first pregalactic fragments formed and merged.

Computer simulations have shown that the extensive dark-matter halos surrounding most, if not all, galaxies are crucial to galaxy interactions. The dark halos make the galaxies much larger than their optical appearance would suggest, making interactions and mergers all the more likely. Consider two galaxies approaching one another. As they orbit, the galaxies interact with each other's dark halos, slowing the galaxies' motion and stripping halo material by tidal forces. The halo matter is redistributed between the galaxies or is entirely lost from the system. In either case, the interaction changes the orbits of the galaxies, causing them to spiral toward one another and eventually to merge.

Types of Merger

The different types and masses of galaxies can lead to an almost bewildering variety of possible interactions. Here we consider just a few of the many possibilities.

If one galaxy of an interacting pair happens to have a much lower mass than the other, its interaction with the other's halo causes it to spiral inward, ultimately to be disrupted near the center of the larger system. This process is colloquially termed *galactic cannibalism* and may explain why supermassive galaxies are often found at the cores of rich galaxy clusters. Having "dined" on their companions, they now lie at the center of the cluster, waiting for more "food" to arrive. Figure 25.11 is a remarkable combination of images that has apparently captured this process at work in a distant cluster.

We also have examples of galactic cannibalism closer to home. The small Sagittarius dwarf galaxy (Figure 24.13) is already well on its way to suffering a similar fate at the center of the Milky Way, and theory indicates that the Magellanic Clouds (Figure 24.7) will eventually meet the same end. As illustrated in Figure 25.12, the "streams" of halo stars, all with similar orbits and composition, in the halo of the Milky Way Galaxy may well be the stellar remnants of such disruption in the past. ∞ (Sec. 23.3)

Now consider two interacting disk galaxies, one a little smaller than the other, but each having a mass comparable to the Milky Way Galaxy. As shown in the computer-generated frames of Figure 25.13, the smaller galaxy can distort the larger one substantially, causing spiral arms to appear where none existed before, triggering an extended episode of star formation. The entire event requires several hundred million years—a span of evolution that a supercomputer can model in

▲ FIGURE 25.10 **Hubble Deep Field** Numerous small, irregularly shaped young galaxies can be seen in this very deep optical image. Known as the Hubble Deep Field–North, this image, made with an exposure of approximately 100 hours, captured objects as faint as 30th magnitude. ∞ (Sec. 17.2) (As in Figure 25.9, "deep" in this context implies "faint," meaning that we are looking at objects far away and as they were long ago.) Redshift measurements (as denoted by the superposed values observed at the Keck Observatory in Hawaii) indicate that some of these galaxies lie well over 1000 Mpc from Earth. ∞ *(More Precisely 24-1)* The field of view is about 2 arc minutes across, or less than 1 percent of the area subtended by the full Moon. *(NASA; Keck)*

R I V U X G

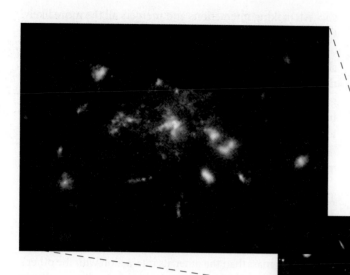

◄ **FIGURE 25.11 Galactic Cannibalism** Here we have a dramatic glimpse of a large and massive galaxy under assembly by the merging of smaller, lighter galaxies. This is the way most galaxies probably developed in the earlier universe—by means of a "bottom up" scenario that hierarchically built really big objects by merging star-rich building blocks. This image captures a formative process that occurred about 10 billion years ago, only a few billion years after the Big Bang. The bigger image highlights a region at upper left (in the white box) catalogued MRC 1138-262 and nicknamed the "Spiderweb" Galaxy. The inset shows more clearly dozens of small galaxies about to merge into a single huge object, in fact altogether one of the most massive galaxies known. (*NASA/ESA*)

R I V U X G

minutes. The final frame of the figure looks remarkably similar to the double galaxy shown in the opening photo for Chapter 23, and in fact, the simulation was constructed to mimic the sizes, shapes, and velocities in that binary galaxy system. The magnificent spiral galaxy is M51, popularly known as the Whirlpool Galaxy, about 10 Mpc from Earth. Its smaller companion is an irregular galaxy that may have drifted past M51 millions of years ago.

What if the colliding galaxies are comparable in size and mass? Computer simulations reveal that such a merger can destroy a spiral galaxy's disk, creating a galaxy-wide starburst episode. The violence of the merger and the effects of subsequent supernovae eject much of the remaining gas into intergalactic space, creating the hot intracluster gas noted in Section 24.1. Once the burst of star formation has subsided, the resulting object looks very much like an elliptical galaxy. The elliptical's hot X-ray halo is the last vestige of the original spiral's disk. The merging galaxies in Figures 25.7 and 25.8 may be examples of this phenomenon in progress.

Making the Hubble Sequence

If galaxies form and evolve by repeated mergers, can we account for the Hubble sequence and, specifically, differences between spirals and ellipticals? ∞ (Sec. 24.1) The details are still far from certain, but, remarkably, the answer now seems to be a qualified yes. Collisions and close encounters are random events and do *not* represent a "genuine" evolutionary connection between galaxies. Nevertheless, observations and computer simulations do suggest some plausible ways in which the observed Hubble types might have arisen, starting from a universe populated only by irregular, gas-rich galaxy fragments.

As we have just seen, the simulations reveal that "major" mergers—collisions between large galaxies of comparable

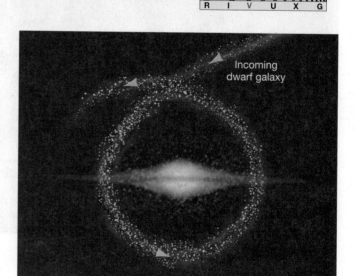

Incoming dwarf galaxy

▲ **FIGURE 25.12 Tidal Streams in the Milky Way** This illustration depicts the breakup and dispersion of the contents of an incoming star-rich galaxy companion captured by our Milky Way. Such streams of stars, all in similar orbits and having much the same composition, are found in the halo of our Galaxy today. Eventually, the smaller galaxy dissolves within the larger one, a case of the littler one being "digested," much as other dwarf companion galaxies were probably "consumed" by our Galaxy long ago.

Time ▬▬▬▶

▲ FIGURE 25.13 **Galaxy Interaction** Galaxies can change their shapes long after their formation. In this computer-generated sequence, two galaxies closely interact over several hundred million years. The smaller galaxy, in red, has gravitationally disrupted the larger galaxy, in blue, changing it into a spiral galaxy. Compare the result of this supercomputer simulation with the opening image of Chapter 23 (p. 572), which shows the spiral galaxy M51 and its small companion. (*J. Barnes & L. Hernquist*)

size—tend to destroy galactic disks, effectively turning spirals into ellipticals (Figure 25.14a). On the other hand, "minor" mergers, in which a small galaxy interacts with, and ultimately is absorbed by, a larger one, generally leave the larger galaxy intact, with more or less the same Hubble type as it had before the merger (Figure 25.14b). This is the most likely way for large spirals to grow—in particular, our own Galaxy probably formed in such a manner.

Supporting evidence for this general picture comes from observations that spiral galaxies are relatively rare in regions of high galaxy density, such as the central regions of rich galaxy clusters. These observations are consistent with the view that the fragile disks of spiral galaxies are easily destroyed by collisions, which are more common in dense galactic environments. Spirals also seem to be more common at larger redshifts (that is, in the past), implying that their numbers are decreasing with time, presumably also as the result of collisions. However, nothing in this area of astronomy is clear cut, and astronomers know of numerous isolated elliptical galaxies in low-density regions of the universe that are hard to explain as the result of mergers. In addition, the competition between infall, which acts to sustain galactic disks, and collisions, which tend to destroy them, remains poorly understood, as is the effect of activity in galactic nuclei, to be discussed in Section 25.4.

In principle, the starbursts associated with galaxy mergers leave their imprint on the star-formation history of the universe in a way that can be correlated with the properties of galaxies. As a result, studies of star formation in distant galaxies have become a very important way of testing and quantifying the details of the entire hierarchical merger scenario.

ANIMATION/VIDEO Galaxy Merger

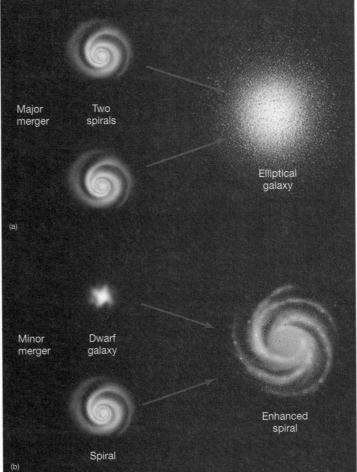

▲ FIGURE 25.14 **Galaxy Mergers** (a) When comparably sized galaxies come together, the result is probably an elliptically shaped galaxy, as their original arms and disks do not likely survive the encounter. (b) By contrast, if a large spiral absorbs a smaller companion, the probable result is merely a larger spiral, with much of its original geometry unchanged.

CONCEPT CHECK

✔ Other than scale, in what important ways does galaxy evolution differ from that of stars?

25.4 Black Holes in Galaxies

Now let's ask how quasars and active galaxies fit into the framework of galaxy evolution just described. The fact that quasars are more common at great distances from us demonstrates that they were much more prevalent in the past than they are today. ∞ (Sec. 24.4) Quasars have been observed with redshifts of up to 6.4 (the current record, as of early 2010), so the process must have started at least 13 billion years ago (see Table 24.2). However, most quasars have redshifts between 2 and 3, corresponding to an epoch some 2 billion years later. Most astronomers agree that quasars represent an early stage of galaxy evolution—an "adolescent" phase of development, prone to frequent flare-ups and "rebellion" before settling into more steady "adulthood." This view is reinforced by the fact that the same black hole energy-generation mechanism can account for the luminosities of quasars, active galaxies, and the central regions of normal galaxies like our own.

Black Hole Masses

In Chapter 24 we saw the standard model of active galactic nuclei accepted by most astronomers—accretion of gas onto a supermassive black hole. ∞ (Sec. 24.5) We also saw that a large fraction of all "bright" galaxies exhibit activity of some sort, even though in many cases it represents only a small fraction of the galaxy's total energy output. This suggests that these galaxies may also harbor central black holes, with the potential of far greater activity under the right circumstances. Our own Galaxy is a case in point. ∞ (Sec. 23.7) The 4-million-solar-mass black hole at the center of the Milky Way is not currently active, but if fresh fuel were supplied (say, by a star or molecular cloud coming too close to the hole's intense gravitational field), it might well become a (relatively weak) active galactic nucleus.

In fact, in recent years, astronomers have amassed evidence that many bright normal galaxies contain supermassive black holes at their centers. Figure 25.15 presents perhaps the most compelling evidence for such a black hole in a normal galaxy. It comes from a radio study of NGC 4258, a spiral galaxy about 6 Mpc away. Using the Very Long Baseline Array, a continent-wide interferometer comprising 10 radio telescopes, a U.S.–Japanese team has achieved an angular resolution hundreds of times better than that attainable with *HST*. ∞ (Sec. 5.6) The observations reveal a group of molecular clouds swirling in an organized fashion about the galaxy's center. Doppler measurements indicate a slightly warped, spinning disk centered precisely on the galaxy's heart. The rotation speeds imply the presence of more than 40 million solar masses packed into a region less than 0.2 pc across.

Similar evidence exists for supermassive black holes in the nuclei of several dozen bright galaxies—some normal, some active—within a few tens of megaparsecs of the Milky Way. Some observers would go so far as to say that in every case where a galaxy has been surveyed and a black hole *could*

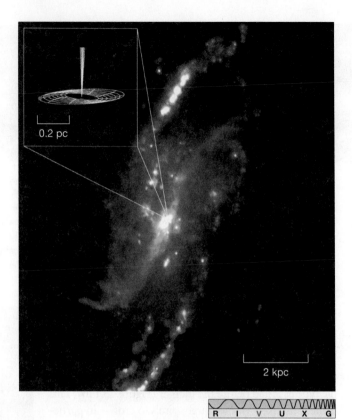

0.2 pc

2 kpc

R I V U X G

▲ **FIGURE 25.15 Galactic Black Hole** A network of radio telescopes has probed the core of the spiral galaxy NGC 4258, shown here in the light of mostly hydrogen emission. Within the innermost 0.2 pc (inset), observations of Doppler-shifted molecular clouds (designated by red, green, and blue dots) show that they obey Kepler's third law perfectly and have revealed a slightly warped disk of rotating gas (shown in the inset as an artist's conception). At the center of the disk presumably lurks a huge black hole. (*J. Moran*)

have been detected, given the resolution and the sensitivity of the observations, a black hole *has* in fact been found. It is a small step to the remarkable conclusion that *every bright galaxy—active or not—contains a central supermassive black hole.* This unifying principle connects our theories of normal and active galaxies in a very fundamental way.

Astronomers have also found that there is a correlation between the masses of the central black holes and the properties of the galaxy in which they reside. As illustrated in Figure 25.16, the largest black holes tend to be found in the most massive galaxies (as measured by the mass of the bulge). The reason for this correlation is not fully understood, but most astronomers take it to mean that, at the very least, the evolution of normal and active galaxies must be very closely connected, as we now discuss.

The Quasar Epoch

Where did the supermassive black holes in galaxies come from? To be honest, the answer is not yet known. The processes whereby the first billion-solar-mass black holes formed early in

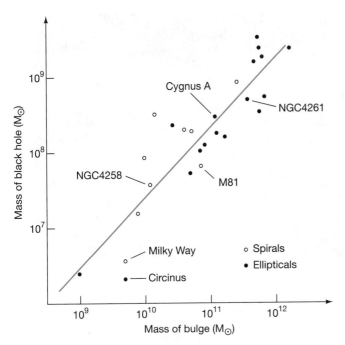

▲ FIGURE 25.16 Black-Hole Masses Careful observations of nearby normal and active galaxies reveal that the mass of the central black hole is well correlated with the mass of the galactic bulge. In this diagram, each point represents a different galaxy. The straight line is the best fit to the data, implying a black-hole mass of 1/200 the mass of the bulge. *(Data courtesy L. Ferrarese)*

the history of the universe are not fully understood. However, the accretion responsible for quasar energy emission also naturally accounts for the mass of the black hole—only a few percent of the infalling mass is converted to energy; the rest is trapped forever in the black hole once it crosses the event horizon. ∞ (Sec. 22.5) Simple estimates suggest that the accretion rates needed to power the quasars are generally consistent with black hole masses inferred by other means.

Since the brightest known quasars devour about a thousand solar masses of material every year, it is unlikely that they could maintain their luminosity for very long—even a million years would require a billion solar masses, enough to account for the most massive black holes known. ∞ (Sec. 24.4) This suggests that a typical quasar spends only a fairly short amount of time in its highly luminous phase—perhaps only a few million years in some cases—before running out of fuel. Thus, most quasars were relatively brief events that occurred long ago.

To make a quasar, we need a black hole and enough fuel to power it. Although fuel was abundant at early times in the universe's history in the form of gas and newly formed stars, black holes were not. They had yet to form, most probably by the same basic stellar evolutionary processes we saw in Chapter 21, although the details are again not well known. ∞ (Sec. 21.2) The building blocks of the supermassive black holes that would ultimately power the quasars may well have been relatively small black holes having masses

tens or perhaps a few hundreds of times the mass of the Sun, formed by the first generations of stars. These small black holes sank to the center of their still-forming parent galaxy and merged to form a single, more massive black hole.

As galaxies merged, so, too, did their central black holes, and eventually supermassive (1-million- to 1-billion-solar-mass) black holes existed in the centers of many young galaxies. Some supermassive black holes may have formed directly by the gravitational collapse of the dense central regions of a protogalactic fragment or perhaps by accretion or a rapid series of mergers in a particularly dense region of the universe. These events resulted in the earliest (redshift-6) quasars known, shining brightly 13 billion years ago. However, in most cases, all those mergers took time—roughly another 2 billion years. By then (at redshifts between 2 and 3, roughly 11 billion years ago), many supermassive black holes had formed, and there was still plenty of merger-driven fuel available to power them. This was the height of the "quasar epoch" in the universe.

Until recently, astronomers were confident that black holes would merge when their parent galaxies collided, but they had no direct evidence of the process—no image of two black holes "caught in the act." In 2002, the *Chandra* X-ray observatory discovered a binary black hole—two supermassive objects, each having a mass a few tens of millions of times that of the Sun—in the center of the ultraluminous starburst galaxy NGC 6240, itself the product of a galaxy merger some 30 million years ago. Figure 25.17 shows optical and X-ray views of the system. The black holes are the two blue-white objects near the center of the (false-color) X-ray image. Orbiting just 1000 pc apart, they are losing energy through interactions with stars and gas and are predicted to merge in about 400 million years. NGC 6240 lies just 120 Mpc from Earth, so we are far from seeing a quasar merger in the early universe. Nevertheless, astronomers think that events similar to this must have occurred countless times billions of years ago, as galaxies collided and quasars blazed.

Distant galaxies are generally much fainter than their bright quasar cores. As a result, until quite recently, astronomers were hard pressed to discern any galactic structure in quasar images. Since the mid-1990s, several groups of astronomers have used the *Hubble Space Telescope* to search for the "host" galaxies of moderately distant quasars. After removing the bright quasar core from the *HST* images and carefully analyzing the remnant light, the researchers have reported that, in every case studied—several dozen quasars so far—a host galaxy can be seen enveloping the quasar. Figure 25.18 shows some of the longest quasar exposures ever taken. Even without sophisticated computer processing, the hosts are clearly visible.

As we saw in Chapter 24, the connection between active galaxies and galaxy clusters is well established, and many relatively nearby quasars are also known to be members of clusters. ∞ (Sec. 24.4) The link is less clear-cut for the most distant quasars, however, simply because they are so far away

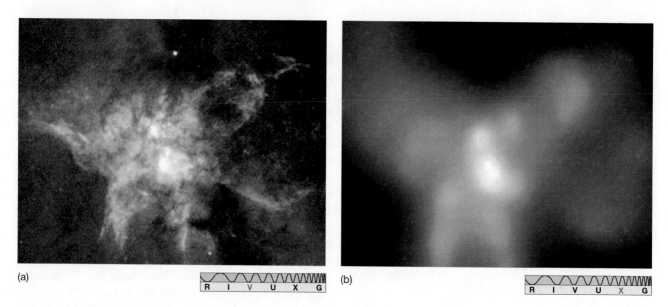

(a)

R I V U X G

(b)

R I V U X G

▲ **FIGURE 25.17 Binary Black Hole** These (a) optical *(Hubble)* and (b) X-ray *(Chandra)* images of the starburst galaxy NGC 6240 show two supermassive black holes (the blue-white objects near the center of the X-ray image) orbiting about 1 kpc apart. According to theoretical estimates, they will merge in about 400 million years, releasing an intense burst of gravitational radiation in the process. ∞ *(Discovery 22-2)* The colors of the optical image are real; the false colors in the X-ray image indicate a range of energies in the X-ray band. *(NASA)*

that other cluster members are very faint and extremely hard to see. However, as the number of known quasars continues to increase, evidence for quasar clustering (and presumably, therefore, for quasar membership in young galaxy clusters) mounts. Thus, as best we can tell, quasar activity is intimately related to interactions and collisions in young galaxy clusters.

This connection also suggests a possible way in which the growth of black holes might be tied to the growth of their parent galaxies. Many astronomers think that a process called **quasar feedback,** in which some fraction of the quasar's enormous energy output is absorbed by the surrounding galactic gas, might explain the correlation of black hole and bulge masses shown in Figure 25.16. In this picture, which is appealing but by no means certain, the absorbed energy expels the gas from the galaxy, simultaneously shutting down both galactic star formation and the quasar's own fuel supply.

Active and Normal Galaxies

Early on, frequent mergers may have replenished the quasar's fuel supply, extending its luminous lifetime. However, as the

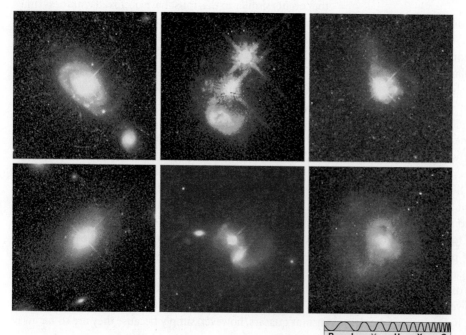

R I V U X G

◀ **FIGURE 25.18 Quasar Host Galaxies** These long-exposure *Hubble Space Telescope* images of distant quasars clearly show the young host galaxies in which the quasars reside, lending strong support to the view that quasars represent an early, highly luminous phase of galactic evolution. The quasar at the top left is the best example, having the catalog name PG0052 + 251 and residing roughly 700 Mpc from Earth. Note that several of the quasars appear to be associated with interacting galaxies, consistent with current theories of galaxy evolution. *(NASA)*

merger rate declined, these systems spent less and less of their time in the "quasar" phase. The rapid decline in the number of bright quasars roughly 10 billion years ago marks the end of the quasar epoch. Today, the number of quasars has dropped virtually to zero (recall that the nearest lies hundreds of megaparsecs away). ∞ (Sec. 24.4)

Large black holes do not simply vanish. If a galaxy contained a bright quasar 10 billion years ago, the black hole responsible for all that youthful activity must still be present in the center of the galaxy today. We see some of these black holes as active galaxies. The remainder reside dormant in normal galaxies all around us. In this view, *the difference between an active galaxy and a normal one is mainly a matter of fuel supply.* When the fuel runs out and a quasar shuts down, its central black hole remains behind, its energy output reduced to a relative trickle. The black holes at the hearts of normal galaxies are simply quiescent, awaiting another interaction to trigger a new active outburst. Occasionally, two nearby galaxies may interact with each other, causing a flood of new fuel to be directed toward the central black hole of one or both. The engine starts up for a while, giving rise to the nearby active galaxies—radio galaxies, Seyferts, and others—we observe.

Should this general picture be correct, it follows that many relatively nearby galaxies (but probably *not* our own Milky Way, whose central black hole is even now only a paltry 3–4 million solar masses) must once have been brilliant quasars. ∞ (Sec. 23.7) Perhaps some alien astronomer, thousands of megaparsecs away, is at this very moment observing the progenitor of M87 in the Virgo cluster—seeing it as it was billions of years ago—and is commenting on its enormous luminosity, nonstellar spectrum, and possibly its high-speed jets, and wondering what exotic physical process can account for its violent activity! ∞ (Sec. 24.4)

Finally, Figure 25.19 suggests some possible (but unproven!) evolutionary connections among quasars, active galaxies, and normal galaxies. If the largest black holes reside in the most massive galaxies, and also tend to power the

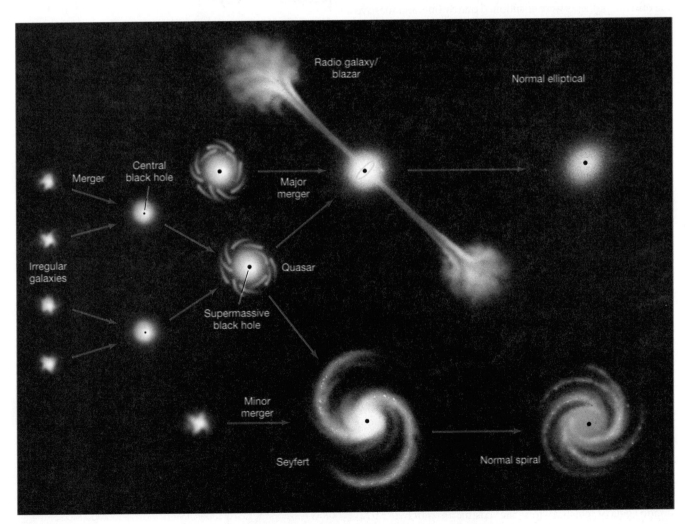

▲ **FIGURE 25.19 Galaxy Evolution** Some possible evolutionary sequences for galaxies, beginning with galaxy mergers leading to the highly luminous quasars, decreasing in violence through the radio and Seyfert galaxies, and ending with normal ellipticals and spirals. The central black holes that powered the early activity are still there at later times; they simply run out of fuel as time goes on.

brightest active galactic nuclei, then we would expect that the most luminous nuclei should reside in the largest galaxies, which probably came into being via "major" mergers of other large galaxies. Since the products of such mergers are elliptical galaxies, we have a plausible explanation of why the brightest active galaxies—the radio galaxies—should be associated with large ellipticals. ∞ (Sec. 24.4) Furthermore, the path to spiral galaxies would necessarily have entailed a series of mergers involving smaller galaxies, resulting in the less-violent Seyferts along the way.

Active Galaxies and the Scientific Method

When active galactic nuclei—especially quasars—were first discovered, their extreme properties defied conventional explanation. Initially, the idea of supermassive (million- to billion-solar-mass) black holes in galaxies was just one of several competing, and very different, hypotheses advanced to account for the enormous luminosities and small sizes of those baffling objects. However, as observational evidence mounted, the other hypotheses were abandoned one by one, and massive black holes in galactic nuclei became first the leading, and eventually the standard, theory of active galaxies.

As often happens in science, a theory once itself considered extreme is now the accepted explanation for these phenomena. Far from threatening the laws of physics, as some astronomers once feared, active galaxies are now an integral part of our understanding of how galaxies form and evolve. The synthesis of studies of normal and active galaxies, galaxy formation, and large-scale structure is one of the great triumphs of extragalactic astronomy.

CONCEPT CHECK

✔ Does *every* galaxy have the potential for activity?

25.5 The Universe on Large Scales

Many galaxies, including our own, are members of galaxy clusters—megaparsec-sized structures held together by their own gravity. ∞ (Sec. 24.2) Our own small cluster is called the Local Group. Figure 25.20 shows the locations of the Virgo cluster, the closest "large" large cluster, and several other well-defined clusters in our cosmic neighborhood. The region displayed is about 70 Mpc across. Each point in the figure represents an entire galaxy whose distance has been determined by one of the methods described in Chapter 24.

Clusters of Clusters

Does the universe have even greater groupings of matter, or do galaxy clusters top the cosmic hierarchy? Most astronomers have concluded that the galaxy clusters are themselves clustered, forming titanic agglomerations of matter known as **superclusters.**

Together, the galaxies and clusters in the vicinity of the Milky Way form the *Local Supercluster,* also known as the Virgo Supercluster. Aside from the Virgo Cluster itself, it contains the Local Group and numerous other clusters lying within about 20–30 Mpc of Virgo. Figure 25.20 shows a three-dimensional rendering of our extended cosmic neighborhood, illustrating the Virgo supercluster (near the center) relative to other "nearby" galaxy superclusters within a vast imaginary rectangle roughly 100 Mpc on its short side.

All told, the Local Supercluster is about 40–50 Mpc across, contains some 10^{15} solar masses of material (several tens of thousands of galaxies), and is very irregular in shape. The Local Supercluster is significantly elongated perpendicular to the line joining the Milky Way to Virgo, with its center lying near the Virgo Cluster. By now it should perhaps come as no surprise that the Local Group is *not* found at the heart of the Local Supercluster—we live far off in the periphery, about 18 Mpc from the center.

Redshift Surveys

The farther we peer into deep space, the more galaxies, clusters of galaxies, and superclusters we see. Is there structure on scales even larger than superclusters? To answer these

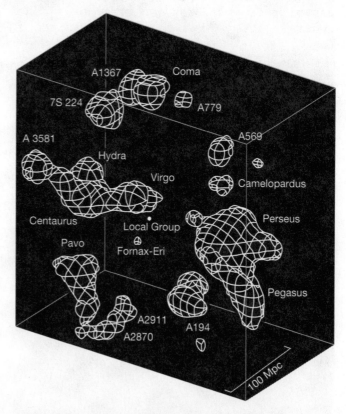

▲ **FIGURE 25.20 Virgo Supercluster in 3-D** The elongated structure of the Virgo supercluster (left center) is mapped relative to other neighboring galaxy superclusters within about 100 Mpc of the Milky Way. Individual galaxies are not shown; rather, smoothed contour plots outline galaxy clusters, each named or numbered by its most prominent member.

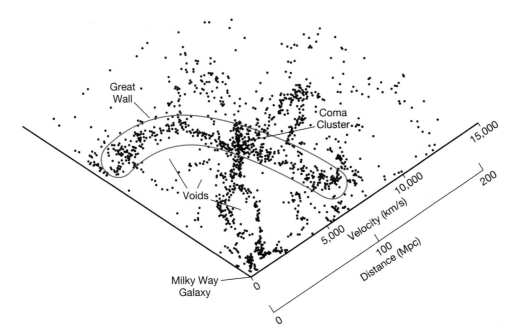

◀ FIGURE 25.21 Galaxy Survey
The first slice of a survey of the universe, covering 1732 galaxies out to an approximate distance of 200 Mpc, clearly shows that galaxies and clusters are not randomly distributed on large scales. Instead, they appear to have a filamentary structure, surrounding vast, nearly empty voids. We assume that $H_0 = 70$ km/s/Mpc. This slice covers 6° of the sky in the direction out of the plane of the paper. (*Harvard-Smithsonian Center for Astrophysics*)

questions, astronomers use Hubble's law to map out the distribution of galaxies in the universe.

Figure 25.21 shows part of an early survey of the universe performed by astronomers at Harvard University in the 1980s. Using Hubble's law as a distance indicator, the team systematically mapped out the locations of galaxies within about 200 Mpc of the Milky Way in a series of wedge-shaped "slices," each 6° thick, starting in the northern sky. The first slice (shown in the figure) covered a region of the sky containing the Coma Cluster (see Figure 24.1), which happens to lie in a direction almost perpendicular to our Galaxy's plane. Because redshift is used as the primary distance indicator, these studies are known as *redshift surveys.*

The most striking feature of maps such as that of Figure 25.21 is that the distribution of galaxies on very large scales is decidedly nonrandom. The galaxies appear to be arranged in a network of strings, or filaments, surrounding large, relatively unpopulated regions of space known as **voids.** The biggest voids measure some 100 Mpc across. The most likely explanation for the voids and the filamentary structure shown in the figure is that the galaxies and galaxy clusters are spread across the surfaces of vast "bubbles" in space. The voids are the interiors of these gigantic bubbles. The galaxies seem to be distributed like beads on strings only because of the way our slice of the universe cuts through the bubbles. Like suds on soapy water, these bubbles fill the entire universe. The densest clusters and superclusters lie in regions where several bubbles meet. The elongated shape of the Virgo Supercluster (Figure 25.20) is a local example of this same filamentary structure.

Most theorists think that this "frothy" distribution of galaxies, and in fact all structure on scales larger than a few megaparsecs, traces its origin directly to conditions in the very earliest stages of the universe (Chapter 27). Conse-

quently, studies of large-scale structure are vital to our efforts to understand the origin and nature of the cosmos itself.

The idea that the filaments are the intersection of a survey slice with much larger structures (the bubble surfaces) was confirmed when the next three slices of the survey, lying above and below the first, were completed. The region of Figure 25.21 indicated by the red outline was found to continue through both the other slices. This extended sheet of galaxies, which has come to be known as the *Great Wall,* measures at least 70 Mpc (out of the plane of the page) by 200 Mpc (across the page). It is one of the largest known structures in the universe.

Figure 25.22 shows a more recent redshift survey, considerably larger than the one presented in Figure 25.21. This survey includes nearly 24,000 galaxies within about 750 Mpc of the Milky Way. Numerous voids and "Great Wall-like" filaments can be seen (some are marked), but apart from the general falloff in numbers of galaxies at large distances—basically because the more distant galaxies are harder to see due to the inverse-square law—there is no obvious evidence for any structures on scales *larger* than about 200 Mpc. Careful statistical analysis confirms this impression. Apparently, voids and walls represent the largest structures in the universe. We will return to the far-reaching implications of this fact in Chapter 26.

Quasar Absorption Lines

How can we probe the structure of the universe on very large scales? As we have seen, much of the matter is dark, and even the "luminous" component is so faint that it is hard to detect at large distances. One way to study large-scale structure is to take advantage of the great distances, pointlike appearance, and large luminosities of quasars. Since quasars are so far

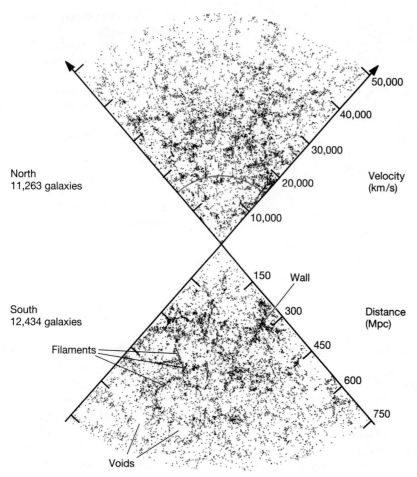

North
11,263 galaxies

Velocity
(km/s)

50,000
40,000
30,000
20,000
10,000

150
300
450
600
750

Wall

Distance
(Mpc)

South
12,434 galaxies

Filaments

Voids

◀ **FIGURE 25.22 The Universe on Larger Scales**
This large-scale galaxy survey, carried out at the Las Campanas Observatory in Chile, consists of 23,697 galaxies within about 1000 Mpc, in two 80° × 4.5° wedges of the sky. Many voids and "walls" on scales of up to 100–200 Mpc can be seen, but no larger structures are evident. For scale, the extent of the survey shown in Figure 25.21 is marked as a thin blue arc in the northern sky.

as ongoing and planned large-scale surveys scan the sky for fainter and fainter objects. Foremost among these surveys is the Sloan Digital Sky Survey (*Discovery 25-1*), which has constructed a map of much of the northern sky, including several million galaxies and more than 100,000 quasars.

In addition to exhibiting their own strongly redshifted spectra, many quasars show additional absorption features that are redshifted by substantially *less* than the lines from the quasar itself. For example, the quasar PHL 938 has an emission-line redshift of 1.954, placing the quasar at a distance of some 5700 Mpc, but it also shows absorption lines having redshifts of just 0.613. These lines with lesser redshifts are interpreted as arising from intervening gas that is much closer to us (only about 2400 Mpc away) than the quasar itself. Most probably, this gas is part of an otherwise invisible galaxy lying along the line of sight. Quasar spectra, then, afford astronomers a means of probing previously undetected parts of the universe.

The absorption lines of atomic hydrogen are of particular interest, since hydrogen makes up so much of all matter in the cosmos. Specifically, hydrogen's ultraviolet (122-nm) "Lyman-alpha" line, associated with transitions between the ground and first excited states, is often used in this context. ∞ (Sec. 4.3) As illustrated in Figure 25.23, when astronomers

away, light traveling from a quasar to Earth has a pretty good chance of passing through or near "something interesting" en route. By analyzing quasar images and spectra, it is possible to piece together a partial picture of the intervening space.

The quasar approach is reminiscent of the use of bright stars to probe the interstellar medium near the Sun, and it suffers from the same basic drawback: We can study only regions of the sky where quasars happen to be found. ∞ (Sec. 18.1) However, this problem will diminish in time

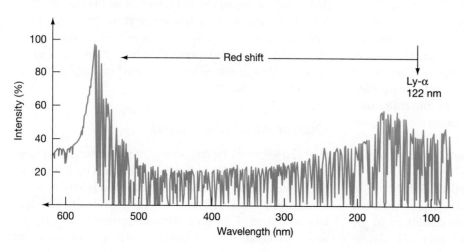

◀ **FIGURE 25.23 Lyman-Alpha "Forest"**
When light from a distant quasar passes through a foreground cloud of atomic hydrogen, the absorption lines produced bear the cloud's redshift. The huge number of absorption lines in the spectrum of this quasar (called QSO 1422 + 2309) are the ultraviolet Lyman-alpha lines from hundreds of clouds of foreground hydrogen gas, each redshifted by a slightly different amount (but less than the quasar itself). The peak at left marks the Lyman-alpha emission line from the quasar, emitted at 122 nm but redshifted here to a wavelength of 564 nm, in the visible range.

DISCOVERY 25-1

The Sloan Digital Sky Survey

Many of the photographs used in this book—not to mention most of the headline-grabbing imagery found in the popular media—come from large, high-profile, and usually very expensive instruments such as NASA's *Hubble Space Telescope* and the European Southern Observatory's *Very Large Telescope* in Chile. ∞ (Secs. 5.3, 5.4) Their spectacular views of deep space have revolutionized our view of the universe. Yet a less well-known, considerably cheaper, but no less ambitious, project currently underway may, in the long run, have every bit as great an impact on astronomy and our understanding of the cosmos.

The Sloan Digital Sky Survey (SDSS) is a 5-year project, now extended for a further 3 years, designed to systematically map out a quarter of the entire sky on a scale and at a level of precision never before attempted. It has catalogued more than 100 million celestial objects, recording their apparent brightnesses at five different colors (wavelength ranges) spread across the optical and near-infrared part of the spectrum. In addition, spectroscopic follow-up observations have determined redshifts and hence distances to 1 million galaxies and 100,000 quasars. These data have been used to construct detailed redshift surveys (see Section 26.1), and to probe the structure of the universe on very large scales. The sensitivity of the survey is such that it can detect bright galaxies like our own out to distances of more than 1 billion parsecs. Very bright objects, such as quasars and young starburst galaxies, are detectable almost throughout the entire observable universe.

The first figure shows the Sloan Survey telescope, a special-purpose 2.5-m instrument sited in Apache Point Observatory, near Sunspot, New Mexico. This reflecting telescope (whose boxlike structure protects it from the wind) is not space-based, does not employ active or adaptive optics, and cannot probe as deep (that is, far) into space as larger instruments. How can it possibly compete with these other systems? The answer is that, unlike most other large telescopes in current use, where hundreds or even thousands of observers share the instrument and compete for its time, the SDSS telescope was designed specifically for the purpose of the survey. It has a wide field of view and is dedicated to the task, carrying out observations of the sky on *every* clear night during the duration of the project.

The use of a single instrument night after night, combined with tight quality controls on which nights' data are actually incorporated into the survey (nights with poor seeing or other problematic conditions are discarded and the observations repeated) mean that the end product is a database of exceptionally high quality and uniformity spanning an enormous volume of space—a monumental achievement and an indispensible tool for the study of the universe. The survey field of view covers much of the sky away from the Galactic plane in the north, together with three broad "wedges" in the south.

Archiving images and spectra on millions of galaxies produces a lot of data. The full survey consists of roughly 15 *trillion* bytes of information—comparable to the entire Library of Congress! Data for some 350 million objects (more than 1,000,000 with measured spectra), has been released to the public. The second figure shows

the distant galaxy NGC 5792 and a bright red star much closer to us, in fact in our own Milky Way Galaxy, just one of hundreds of thousands of images that make up the full dataset. Among recent highlights, SDSS has detected the largest known structure in the universe, observed the most distant known galaxies and quasars, and has been instrumental in pinning down the key observational parameters describing our universe (see Chapter 26).

SDSS impacts astronomy in areas as diverse as the large-scale structure of the universe, the origin and evolution of galaxies, the nature of dark matter, the structure of the Milky Way, and the properties and distribution of interstellar matter. Its uniform, accurate, and detailed database is likely to be used by generations of scientists for decades to come.

R I V U X G

observe the spectrum of a high-redshift quasar, they typically see a "forest" of absorption lines, starting at the (redshifted) wavelength of the quasar's own Lyman-alpha emission line and extending to shorter wavelengths. These lines are interpreted as Lyman-alpha absorption features produced by gas clouds in foreground structures—galaxies, clusters, and so on—giving astronomers crucial information about the distribution of matter along the line of sight.

The quasar light thus explores an otherwise invisible component of cosmic gas. In principle, every intervening cloud of atomic hydrogen leaves its own characteristic imprint on the quasar's spectrum in a form that lets us explore the distribution of matter in the universe. By comparing these *Lyman-alpha forests* with the results of simulations, astronomers hope to refine many key elements of the theories of galaxy formation and the evolution of large-scale structure.

Quasar "Mirages"

In 1979, astronomers were surprised to discover what appeared to be a binary quasar—two quasars with exactly the same redshift and similar spectra, separated by only a few arc seconds on the sky. Remarkable as the discovery of such a binary would have been, the truth about this pair of quasars turned out to be even more amazing: Closer study of the quasars' radio emission revealed that they were *not* two distinct objects; instead, they were two separate images of the *same* quasar! Optical views of such a *twin quasar* are shown in Figure 25.24.

What could produce such a "doubling" of a quasar image? The answer is gravitational lensing—the deflection and focusing of light from a background object by the gravity of some foreground body (Figure 25.25). In Chapter 23, we saw how lensing by compact objects in the halo of the Milky Way Galaxy may amplify the light from a distant star,

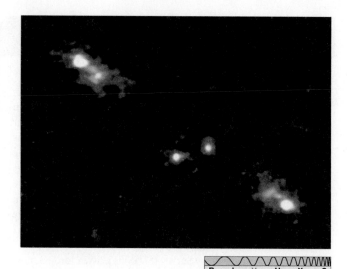

R I V U X G

▲ FIGURE 25.24 **Twin Quasar** This "double" quasar (designated AC114 and located about 2 billion parsecs away) is not two separate objects at all. Instead, the two large "blobs" (at upper left and lower right) are images of the same object, created by a gravitational lens. The lensing galaxy itself is probably not visible in this image—the two objects near the center of the frame are thought to be unrelated galaxies in a foreground cluster. *(NASA)*

allowing astronomers to detect otherwise invisible stellar dark matter. ∞ (Sec. 23.6) In the case of quasars, the idea is the same, except that the foreground lensing object is an entire galaxy or galaxy cluster, and the deflection of the light is so great (a few arc seconds) that several separate images of the quasar may be formed, as shown in Figure 25.26.* About two dozen such gravitational lenses are known. As telescopes

In fact, much of the theory of gravitational lensing was worked out after the first lensed quasar observations and subsequently applied to dark-matter searches in our Galaxy.

Image A

Lensing galaxy

Quasar

Image B

Observer on Earth

Interactive FIGURE 25.25 Gravitational Lens When light from a distant object passes close to a galaxy or a cluster of galaxies along the line of sight, the image of the background object (here, the quasar) can sometimes be split into two or more separate images (A and B). The foreground object is a gravitational lens.

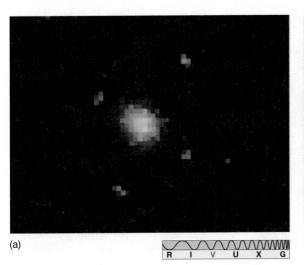

(a)

| R | I | V | U | X | G |

(b)

▲ FIGURE 25.26 **Einstein Cross** (a) The "Einstein Cross," a multiply-imaged quasar. In this *Hubble* view, spanning only a couple of arc seconds, four separate images of the same quasar have been produced by the galaxy at the center. (b) A simplified artist's conception of what might be occurring here, with Earth at right and the distant quasar at left. (*NASA; D. Berry*)

probe the universe with greater and greater sensitivity, astronomers are beginning to realize that gravitational lenses are relatively common features of the cosmos.

The existence of these multiple images provides astronomers with a number of useful observational tools. First, lensing by a foreground galaxy tends to amplify the light of the quasar, as just mentioned, making it easier to observe. At the same time, **microlensing** by individual stars within the galaxy may cause large fluctuations in the quasar's brightness, allowing astronomers to study both the quasar and the galaxy's stellar content. ∞ (Sec. 23.6) The amount of brightening due to microlensing depends on the size of the emitting region, and this in turn depends on the wavelength of the radiation observed—for example, the X rays are emitted from a smaller region closer to the central black hole than is the quasar's visible light (see Figure 24.34). By carefully comparing the amounts of brightening at different wavelengths, astronomers can probe the structure of the quasar's accretion disk on scales inaccessible by any other means.

Second, because the light rays forming the images usually follow paths of different lengths, there is often a *time delay,* ranging from days to years, between them. This delay provides advance notice of interesting events, such as sudden changes in the quasar's brightness. Thus, if one image flares up, in time the other(s) will, too, giving astronomers a second chance to study the event. The time delay also allows astronomers to determine the *distance* to the lensing galaxy. This method provides an alternative means of measuring Hubble's constant that is independent of any of the techniques discussed previously. The average value of H_0 reported by workers using this approach is 65 km/s/Mpc, a little less than the value we have assumed throughout the text.

Finally, by studying the lensing of background quasars and galaxies by foreground galaxy clusters, astronomers can obtain a better understanding of the distribution of dark matter in those clusters, an issue that has great bearing on the large-scale structure of the cosmos.

Mapping Dark Matter

Astronomers have extended the ideas first learned from studies of quasars to the lensing of any distant object in order to better probe the universe. Distant, faint irregular galaxies—the raw material of the universe if current theories are correct (see Section 25.3)—are of particular interest here; because they are far more common than quasars, they provide much better coverage of the sky. By studying the lensing of background quasars and galaxies by foreground galaxy clusters, astronomers can obtain a better understanding of the distribution of dark matter on large scales.

Figures 25.27(a) and (b) show how the images of faint background galaxies are bent into arcs by the gravity of a foreground galaxy cluster. The degree of bending allows the total mass of the cluster (*including* the mass of the dark matter) to be measured. The (mostly blue) loop- and arc-shaped features visible in Figure 25.27(b) are multiple images of a single distant (unseen) spiral or ring-shaped galaxy, lensed by the foreground galaxy cluster (the yellow-red blobs in the image).

It is even possible to reconstruct the foreground dark-matter distribution by carefully analyzing the distortions of the background objects, thereby providing a means of tracing out the distribution of mass on scales far larger than have previously been possible. Figure 25.28 is a reconstructed map showing the presence of dark matter many megaparsecs from the center of a small galaxy cluster (the

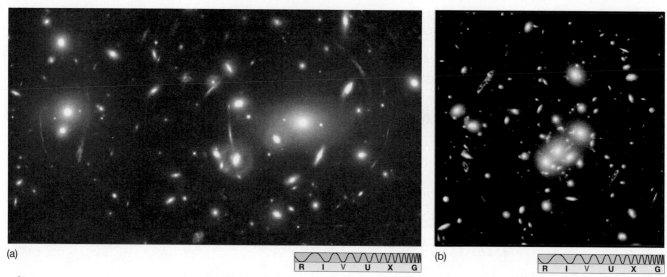

(a)

R I V U X G

(b)

R I V U X G

Interactive FIGURE 25.27 Galaxy Cluster Lensing (a) This spectacular example of gravitational lensing shows more than a hundred faint arcs from very distant galaxies. The wispy pattern spread across the foreground galaxy cluster (A 2218, about a billion parsecs distant) resembles a spider's web, but it is really an illusion caused by A 2218's gravitational field, which deflects the light from background galaxies and distorts their appearance. By measuring the extent of the distortion, astronomers can estimate the mass of the intervening cluster. (b) An approximately true-color image of the galaxy cluster known only by its catalog name, 0024 + 1654, residing some 1.5 billion pc away. The reddish-yellow blobs are mostly normal elliptical galaxy members of the cluster, concentrated toward the center of the image. The bluish looplike features are images of a single background galaxy. *(NASA)*

brightest blob near the center of the map). Notice the elongated structure of the dark-matter distribution, reminiscent of the Virgo Supercluster and filamentary structure seen in large-scale galaxy surveys.

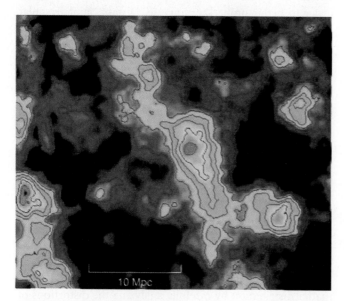

10 Mpc

▲ **FIGURE 25.28 Dark-Matter Map** By measuring distortions in the images of background objects, astronomers can make maps of dark matter in the universe. These contours show the distribution of dark matter in and near a small galaxy cluster, derived by analyzing the images of background galaxies. *(Data courtesy J. A. Tyson, Bell Labs)*

In 2006, astronomers used these techniques to obtain what may be the first direct observational evidence for dark matter. Figure 25.29 shows combined optical and X-ray images of a distant galaxy cluster called 1E 0657-56. The fuzzy red region shows the location of the hot X-ray-emitting gas in the system, the dominant *luminous* component of the mass. The blue regions indicate where most of the mass actually lies, as determined from lensing studies of background galaxies. Note that the bulk of the mass is *not* found in the form of hot gas, implying that the dark matter is distributed differently from the "normal" matter in the cluster.

The explanation for this odd state of affairs is that we are witnessing a collision between two clusters. Each initially contained hot gas and dark matter distributed throughout the cluster, but when the two collided the pressure of each gas cloud effectively stopped the other, leaving the gas behind in the middle as the galaxies and dark matter moved on. This separation between the gas and the dark matter directly contradicts some alternative theories of gravity that have been invoked to explain away the "dark matter problem" in galaxies and clusters and may prove to be a crucial piece of evidence in our understanding of large-scale structure in the universe.

CONCEPT CHECK

✔ How do observations of distant quasars tell us about the structure of the universe closer to home?

R I V U X G

ANIMATION/VIDEO Dark Matter Collision, Bullet Cluster Collision

Interactive FIGURE 25.29 Cluster Collision Clusters of galaxies must also occasionally collide, as is the case here. This cluster has the innocuous catalog name 1E 0657-56 and the nickname "bullet cluster." This is a composite image of a region about 1 billion parsecs away, showing optical light from the galaxies themselves in white and hot X-ray-emitting gas from the intracluster gas in red. By contrast, the blue color represents the inferred dark matter within the two large clusters that is distinctly displaced from the gas. The arrows indicate the approximate directions in which the two clusters are moving. By showing how dark matter behaves differently from intracluster gas, the bullet cluster provides critical evidence supporting the current theory of dark matter in the universe. Astronomers estimate that this might have been the most energetic collision in the universe since the Big Bang. (*NOAO/NASA*)

CHAPTER REVIEW

SUMMARY

1 The masses of nearby spiral galaxies can be determined by studying their rotation curves. Astronomers also use studies of binary galaxies and galaxy clusters to obtain statistical estimates of the masses of the galaxies involved.

2 Measurements of galaxy and cluster masses reveal the presence of large amounts of dark matter. The fraction of dark matter grows as the scale under consideration increases. More than 90 percent of the mass in the universe is dark. Large amounts of hot X-ray-emitting gas have been detected among the galaxies in many clusters, but not enough to account for the dark matter inferred from dynamical studies.

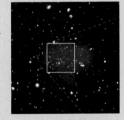

3 Researchers know of no simple evolutionary sequence that links spiral, elliptical, and irregular galaxies. Most astronomers think that large galaxies formed by the merger of smaller ones and that collisions and mergers among galaxies play very important roles in galactic evolution. A **starburst galaxy (p. 639)** may result when a galaxy has a close encounter or a collision with a neighbor. The strong tidal distortions caused by the encounter compress galactic gas, resulting in a widespread

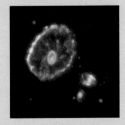

burst of star formation. Mergers between spirals most likely result in elliptical galaxies.

4 Quasars, active galaxies, and normal galaxies may represent an evolutionary sequence. When galaxies began to form and merge, conditions may have been suitable for the formation of large black holes at their centers, and a highly luminous quasar could have been the result. The brightest quasars consume so much fuel that their energy-emitting lifetimes must be quite short. As the fuel supply of such a quasar diminished, the quasar dimmed, and the galaxy in which it was embedded became intermittently visible as an active galaxy. At even later times, the nucleus became virtually inactive, and a normal galaxy was all that remained. Many normal galaxies have been found to contain massive central black holes, suggesting that most galaxies have the capacity for activity should they interact with a neighbor. **Quasar feedback (p. 648)** may provide a partial explanation of why the masses of black holes are correlated with the masses of their parent galaxies.

5 Galaxy clusters themselves tend to clump together into **superclusters (p. 650)**. The Virgo Cluster, the Local Group, and several other nearby clusters form the Local Supercluster. On even larger scales, galaxies and galaxy clusters are arranged on the surfaces of enormous "bubbles" of matter

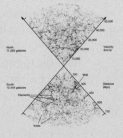

surrounding vast low-density regions called **voids** (p. 651). The origin of this structure is thought to be closely related to conditions in the very earliest epochs of the universe.

6 Quasar spectra can be used as probes of the universe along the observer's line of sight. Some quasars have been observed to have double or multiple images, caused by  gravitational lensing, in which the gravitational field of a foreground galaxy or galaxy cluster bends and focuses the light from the more distant quasar. Analysis of the images of distant galaxies, distorted by the gravitational effect of a foreground cluster, provides a means of determining the masses of galaxy clusters—including the dark matter within them—far beyond the information that the optical images of the galaxies themselves afford.

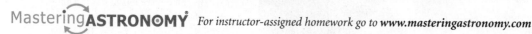

Mastering**ASTRONOMY**   *For instructor-assigned homework go to* **www.masteringastronomy.com**

Problems labeled **POS** explore the process of science | **VIS** problems focus on reading and interpreting visual information

REVIEW AND DISCUSSION

1. Describe two techniques for measuring the mass of a galaxy.

2. Why do astronomers think that galaxy clusters contain more mass than we can see?

3. Why are galaxies at great distances from us generally smaller and bluer than nearby galaxies?

4. **POS**   What evidence do we have that galaxies collide with one another?

5. Describe the role of collisions in the formation and evolution of galaxies.

6. Give an example of how mergers can transform one type of galaxy into another.

7. Do you think that collisions between galaxies constitute "evolution" in the same sense as the evolution of stars?

8. Do we have any evidence that our own Galaxy has collided with other galaxies in the past?

9. What are starburst galaxies, and what do they have to do with galaxy evolution?

10. What conditions result in a head–tail radio galaxy?

11. What evidence do astronomers have for supermassive black holes in galactic nuclei?

12. Why do astronomers think that quasars represent an early stage of galaxy evolution?

13. Why do astronomers think that quasars are relatively short-lived phenomena?

14. What happened to the energy source at the center of a quasar?

15. **POS**   Why does the theory of galaxy evolution suggest that there should be supermassive black holes at the centers of many normal galaxies?

16. What is a redshift survey?

17. What are voids?

18. Describe the distribution of galactic matter on very large (more than 100 Mpc) scales.

19. **POS**   How can observations of distant quasars be used to probe the space between them and Earth?

20. **POS**   How do astronomers "see" dark matter?

CONCEPTUAL SELF-TEST: MULTIPLE CHOICE

1. The more massive a galaxy is, **(a)** the more distant it is; **(b)** the faster star formation in it occurs; **(c)** the larger the proportion of old stars it contains; **(d)** the faster it rotates.

2. A galaxy containing substantial amounts of dark matter will **(a)** appear darker; **(b)** spin faster; **(c)** repel other galaxies; **(d)** have more tightly wound arms.

3. According to X-ray observations, the space between galaxies in a galactic cluster is **(a)** completely devoid of matter; **(b)** very cold; **(c)** very hot; **(d)** filled with faint stars.

4. Relative to luminous stellar matter, the fraction of dark matter in clusters is **(a)** greater than the fraction in galaxies; **(b)** less than the fraction in galaxies; **(c)** the same as the fraction in galaxies; **(d)** unknown.

5. **VIS**   The Hubble Deep Field (Figure 25.10) shows a patch of sky that has the same angular size as **(a)** the thickness of a piece of string; **(b)** a dime; **(c)** a clenched fist; **(d)** a basketball held at arm's length.

6. Galaxies evolve by **(a)** fragmenting into smaller galaxies; **(b)** merging to form larger galaxies; **(c)** ejecting their gas and dust into intergalactic space; **(d)** using up all their gas and eventually becoming ellipticals.

7. According to current theories of galactic evolution, quasars occur **(a)** early in the evolutionary sequence; **(b)** near the Milky Way; **(c)** when elliptical galaxies merge; **(d)** late in the evolutionary sequence.

8. Many nearby galaxies **(a)** will become black holes; **(b)** contain quasars; **(c)** have radio lobes; **(d)** were more active in the past.

9. **VIS**   If light from a distant quasar did not pass through any intervening atomic hydrogen clouds, then Figure 25.23 ("Lyman-Alpha 'Forest'") would have to be redrawn to show

(a) more absorption features; (b) few absorption features; (c) a single large absorption feature; (d) more features at short wavelengths, but fewer at long wavelengths.

10. **VIS** If Figure 25.25 ("Gravitational Lens") showed a more massive lensing galaxy, the quasar images would be (a) farther apart; (b) closer together; (c) fainter; (d) redder.

PROBLEMS

The number of dots preceding each Problem indicates its approximate level of difficulty.

1. •• The Andromeda Galaxy is approaching our Galaxy with a radial velocity of 120 km/s. Given the galaxies' present separation of 800 kpc, and neglecting *both* the transverse component of the velocity *and* the effect of gravity in accelerating the motion, estimate when the two galaxies will collide.

2. •• Based on the data in Figure 25.1, estimate the mass of the galaxy NGC 4984 inside 20 kpc.

3. •• Use Kepler's third law (Section 23.6) to estimate the mass required to keep a galaxy moving at 750 km/s in a circular orbit of radius 2 Mpc around the center of a galaxy cluster. Given the approximations involved in calculating this mass, do you think it is a good estimate of the cluster's true mass?

4. ••• Assuming that the average speed of the protons in the (ionized) X-ray-emitting gas in a galaxy cluster is the same as the mean orbital speed of the galaxies in the cluster, estimate the temperature of the X-ray-emitting intracluster gas in a galaxy cluster of mass 10^{15} solar masses and radius 3 Mpc. ∞ *(More Precisely 8-1)*

5. •• Calculate the average speed of hydrogen nuclei (protons) in a gas of temperature 20 million K. Compare your answer with the speed of a galaxy moving in a circular orbit of radius 1 Mpc around a galaxy cluster of mass 10^{14} solar masses.

6. • In a galaxy collision, two similar-sized galaxies pass through each other with a combined relative velocity of 1500 km/s. If each galaxy is 100 kpc across, how long does the event last?

7. ••• A small satellite galaxy is moving in a circular orbit around a much more massive parent and happens to be moving exactly parallel to the line of sight as seen from Earth. The recession velocities of the satellite and the parent galaxy are measured to be 6450 km/s and 6500 km/s, respectively, and the two galaxies are separated by an angle of 0.1° in the sky. Assuming that $H_0 = 70$ km/s/Mpc, calculate the mass of the parent galaxy.

8. ••• Galaxies in a distant galaxy cluster are observed to have recession velocities ranging from 12,500 km/s to 13,500 km/s. The cluster's angular diameter is 55′. Use these data to estimate the cluster's mass. What assumptions do you have to make in order to obtain this estimate? Take $H_0 = 70$ km/s/Mpc.

9. • Assuming an energy-generation efficiency (i.e., the ratio of energy released to total mass–energy available) of 10 percent, calculate how much mass a 10^{41}-W quasar would consume if it shone for 10 billion years.

10. • The spectrum of a quasar with a redshift of 0.20 contains two sets of absorption lines redshifted by 0.15 and 0.155, respectively. If $H_0 = 70$ km/s/Mpc, estimate the distance between the intervening galaxies responsible for the two sets of lines.

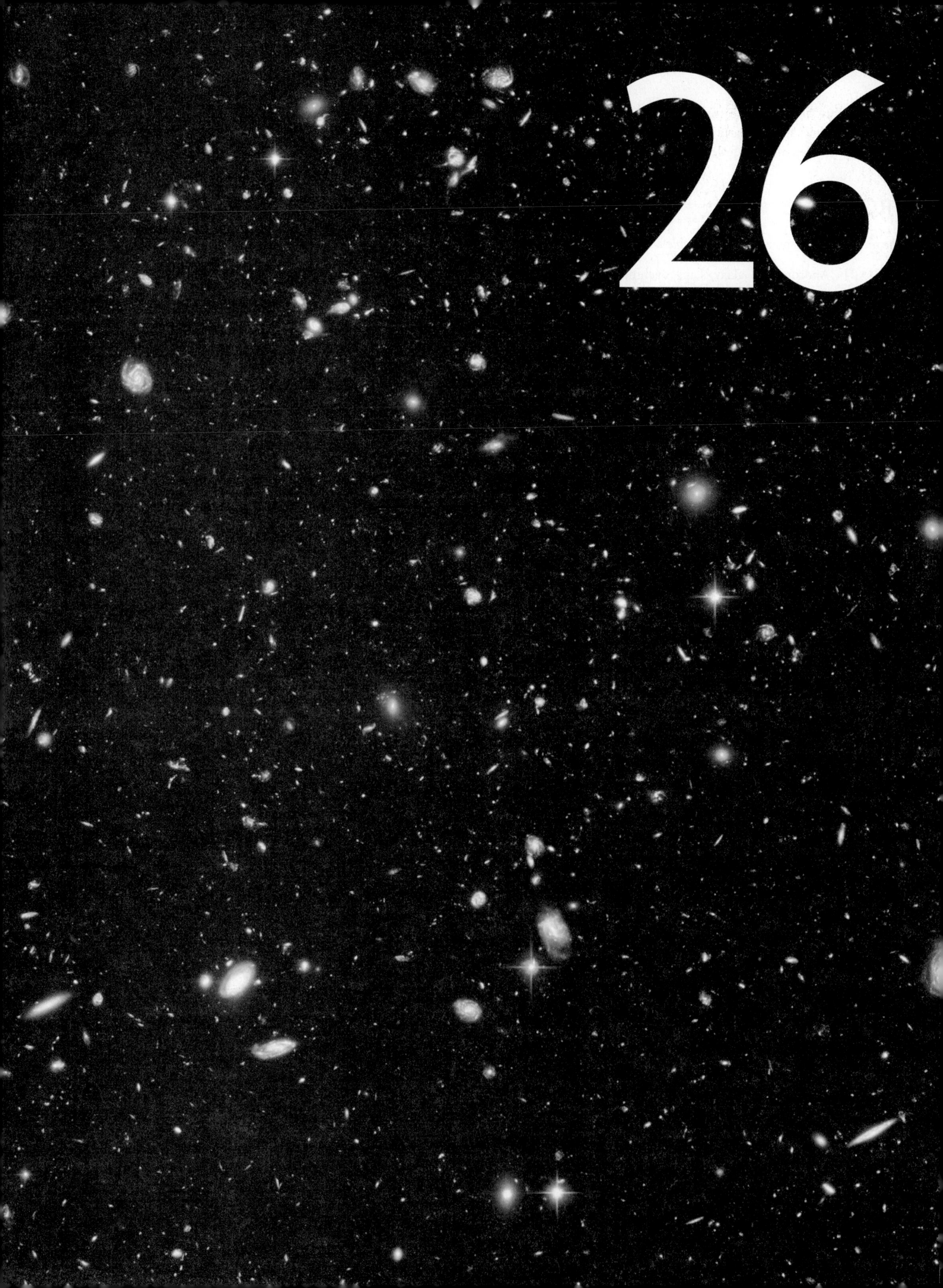
26

COSMOLOGY

THE BIG BANG AND THE FATE OF THE UNIVERSE

LEARNING GOALS

Studying this chapter will enable you to

1 State the cosmological principle, and explain its significance and observational underpinnings.

2 Explain what observations of the dark night sky tell us about the age of the universe.

3 Describe the Big Bang theory of the expanding universe.

4 Discuss the possible outcomes of the present cosmic expansion.

5 Describe the relationship between the density of the universe and the overall geometry of space.

6 Say why astronomers think the expansion of the universe is accelerating, and discuss the cause.

7 Explain what dark energy implies for the composition and age of the universe.

8 Describe the cosmic microwave background and explain its importance to the science of cosmology.

THE BIG PICTURE The universe began in a fiery expansion some 14 billion years ago, and that expansion continues today. Will it continue forever? Can we predict the future of the universe? This is the science of cosmology, the study of the origin, structure, and evolution of the cosmos on the largest scales.

Mastering ASTRONOMY

Visit the Study Area in www.masteringastronomy.com for quizzes, animations, videos, interactive figures, and self-guided tutorials.

Our field of view now extends for billions of parsecs into space and billions of years back in time. We have asked and answered many questions about the structure and evolution of planets, stars, and galaxies. At last we are in a position to address the central issues of the biggest puzzle of all: How big is the universe? How long has it been around, and how long will it last? What was its origin, and what will be its fate? Is the universe a one-time event, or does it recur and renew itself, in a grand cycle of birth, death, and rebirth? How and when did matter, atoms, and our Galaxy form? These are basic questions, but they are hard questions. Many cultures have asked them, in one form or another, and have developed their own cosmologies—theories about the nature, origin, and destiny of the universe—to answer them. In this and the next chapter, we will see how modern scientific cosmology addresses these important issues and what it has to tell us about the universe we inhabit. After more than 10,000 years of civilization, science may be ready to provide some insight regarding the origin of all things.

LEFT: *This image—called the Ultra Deep Field—was taken with the Advanced Camera for Surveys aboard the Hubble telescope. It is one of the finest photographs of deep space ever made. More than a thousand galaxies are crowded into this one image, displaying many different types, shapes, and colors. In all, astronomers estimate that the observable universe contains about 100 billion such galaxies. (NASA/ESA)*

26.1 The Universe on the Largest Scales

The universe shows structure on every scale we have examined so far. Subatomic particles form nuclei and atoms. Atoms form planets and stars. Stars form star clusters and galaxies. Galaxies form galaxy clusters, superclusters, and even larger structures—voids, filaments, and sheets that stretch across the sky. ∞ (Sec. 25.5) From the protons in a nucleus to the galaxies in the Great Wall, we can trace a hierarchy of "clustering" of matter from the very smallest to the very largest scales. It is natural to ask, "Does the clustering ever end? Is there some scale on which the universe can be regarded as more or less smooth and featureless?" Perhaps surprisingly, given the trend we have just described, most astronomers think the answer is yes. This turns out to be a crucial assumption in the science of **cosmology**—the study of the structure and evolution of the entire universe.

The End of Structure

We saw in Chapter 25 how astronomers use *redshift surveys* to explore the large-scale distribution of galaxies in space, combining position on the sky with distances inferred from Hubble's law to construct three-dimensional maps of the universe on truly "cosmic" scales. ∞ (Sec. 25.5) Figure 26.1 is a map similar in concept to those shown in Chapter 25, but based on data from the most extensive redshift survey to date: the Sloan Digital Sky Survey. ∞ *(Discovery 25-1)* The figure plots the positions of some 67,000 galaxies lying in a 120° wide "slice" of the sky within a few degrees of the celestial equator. The distances shown assume a Hubble constant of $H_0 = 70$ km/s/Mpc. ∞ (Sec. 24.3)

The region of the sky shown in Figure 26.1 extends out to a distance of approximately 1000 Mpc. This reach is comparable to the extent of Figure 25.22, but because the Sloan survey includes much fainter galaxies, the map in Figure 26.1 contains nearly three times more points than the earlier figure, making structure somewhat easier to discern, particularly at large distances. The extended "filament" of galaxies near the center of the wedge, some 300 Mpc from Earth, is called the *Sloan Great Wall*. Discovered in 2003 and measuring some 250 Mpc long by 50 Mpc thick, it is currently the largest known structure in the universe.

Plots such as this contain huge amounts of information about the structure and evolution of the universe. Yet, although they cover wide areas of the sky and enormous volumes of space, the studies on which they are based are

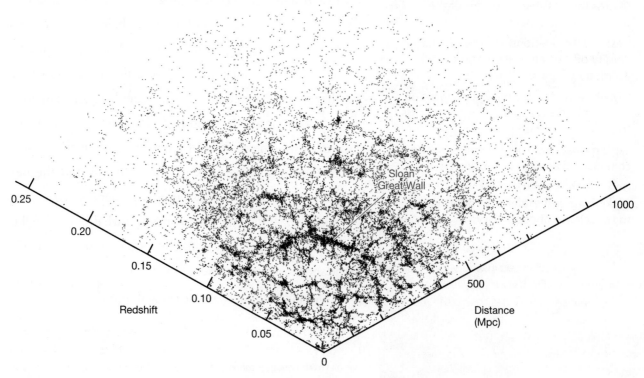

▲ **FIGURE 26.1 Galaxy Survey** This map of the universe is drawn using data from the Sloan Digital Sky Survey. ∞ *(Discovery 25-1)* It shows the locations of 66,976 galaxies lying within 12° of the celestial equator and extends out to a distance of almost 1000 Mpc (or a redshift of about 0.25). The largest known structure in the universe, the Sloan Great Wall, is marked, stretching nearly 300 Mpc across the center of the frame. There is no evidence for any structure on larger scales. *(Astrophysical Research Consortium/SDSS Collaboration)*

still relatively "local," in the sense that they span only about 10 percent of the distance to the farthest quasars (which lie over 9000 Mpc from Earth). ∞ (Sec. 24.4) The main obstacle to extending these wide-angle surveys to much greater distances is the sheer observational effort involved in measuring the redshifts of all the galaxies within larger and larger volumes of space.

An alternative approach is to narrow the field of view to only a few small patches of the sky, but to study extremely faint (and hence very distant) galaxies within those patches. The volume surveyed then becomes a long, thin "pencil beam" extending deep into space rather than a wide swath through the local universe. As illustrated in Figure 26.2, along the line of sight clusters and walls show up as "spikes" in the distribution—groups of galaxies with similar redshifts, separated by broad empty regions of space (the voids).

The data from both kinds of survey seem to agree that the largest known structures in the local universe are "only" 200–300 Mpc across. *No larger voids, superclusters, or walls of galaxies are seen.* Rich superclusters measure tens of megaparsecs across, whereas the largest voids are perhaps 100 Mpc in diameter. Most walls and filaments are less than 100 Mpc in length, and even the largest structures—the Great Walls mentioned previously—can be explained statistically as chance superpositions of smaller structures. Studies of Lyman-alpha forests in quasar spectra lead to generally similar conclusions. ∞ (Sec. 25.5) In short, there is *no* evidence for structure in the universe on scales greater than about 300 Mpc.

We will turn to the *origin* of large-scale cosmic structure in Chapter 27. In the current chapter, however, we focus on the *absence* of structure on the very largest scales to frame our discussion of the future of the universe.

The Cosmological Principle

The results of the large-scale studies just mentioned strongly suggest that the universe is **homogeneous** (the same everywhere) on scales greater than a few hundred megaparsecs. In other words, if we took a huge cube—300 Mpc on a side, say—and placed it anywhere in the universe, its overall contents would look much the same no matter where it was centered. Some of the galaxies it contained would be clustered and clumped into fairly large structures and some would not; we would see numerous walls and voids, but the total numbers of these objects would not vary much as the cube was moved from place to place. In this sense, the universe appears *smooth* on the largest scales.

The universe also appears to be **isotropic** (the same in all directions) on these large scales. Excluding directions that are obscured by our Galaxy, we count roughly the same number of galaxies per square degree in any patch of the sky we choose to observe, provided that we look deep (far) enough that local inhomogeneities don't distort our sample.

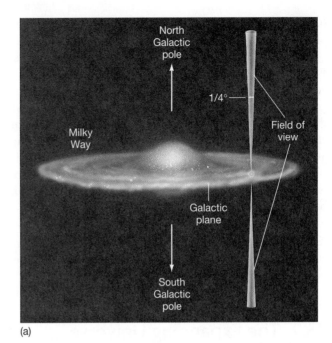

(a)

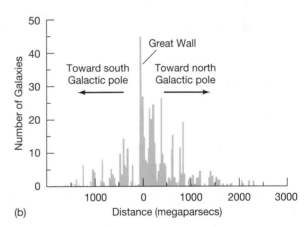

(b)

▲ **FIGURE 26.2 Pencil-Beam Survey** The results of a deep "pencil-beam" survey of two small portions of the sky in opposite directions from Earth, perpendicular to the Galactic plane (a), are plotted in (b). The graph shows the number of galaxies found at different distances from us, out to about 2000 Mpc. Wherever we look on the sky, the distinctive "picket fence" pattern highlights voids and sheets of galaxies on scales of 100–200 Mpc, but gives no indication of any larger structures.

In other words, any deep pencil-beam survey of the sky should count about the same number of galaxies, regardless of which patch of the sky is chosen.

Cosmologists generally assume that the universe is *homogeneous* and *isotropic* on sufficiently large scales. These twin assumptions are known as the **cosmological principle.** No one knows whether these assumptions are precisely correct, but we can at least say that they are consistent with current observations, and they provide helpful guidance to our studies of the cosmos. Note that the cosmological principle also includes the important assumption made throughout this book (and indeed throughout astronomy) that the laws of physics are the same everywhere. In this chapter, we simply assume that it holds.

The cosmological principle has far-reaching implications. For example, it implies that there can be no edge to the universe, because that would violate the assumption of homogeneity. Furthermore, it implies that there is no *center,* because that would mean that the universe would not be the same in all directions from any noncentral point, a violation of the assumption of isotropy. This is the familiar Copernican principle expanded to truly cosmic proportions—not only that are we not central to the universe, but that *no one* can be central, because *the universe has no center!* ∞ (Sec. 2.3)

CONCEPT CHECK

✔ In what sense, and on what scale, is the universe homogeneous and isotropic?

26.2 The Expanding Universe

Every time you go outside at night and notice that the sky is dark, you are making a profound cosmological observation. Here's why.

Olbers's Paradox

Let's assume that, in addition to being homogeneous and isotropic, the universe is infinite in spatial extent and unchanging in time—precisely the view of the universe that prevailed until the early part of the 20th century. On average, then, the universe is uniformly populated with galaxies filled with stars. In that case, when you look up at the night sky, your line of sight must *eventually* encounter a star, as illustrated in Figure 26.3. The star may lie at an enormous distance in some remote galaxy, but the laws of probability dictate that, in an infinite universe, sooner or later any line drawn outward from Earth will run into a bright stellar surface.

Of course, faraway stars appear fainter than those nearby because of the inverse-square law. ∞ (Sec. 17.2) However, they are also much more numerous, because the number of stars we see at any given distance in fact increases as the *square* of the distance. (Just consider the area of a sphere of increasing radius.) Thus, the diminishing brightnesses of distant stars are exactly balanced by their increasing numbers, and stars at all distances contribute equally to the total amount of light received on Earth. This fact has a dramatic implication: No matter where you look, the sky should be as bright as the surface of a star; in other words, the entire night sky should be as brilliant as the surface of the Sun! The obvious difference between this prediction and the actual appearance of the night sky is known as **Olbers's paradox,** after the 19th-century German astronomer Heinrich Olbers, who popularized the idea.

So why is it dark at night? Given that the universe appears to be homogeneous and isotropic, one (or both) of the

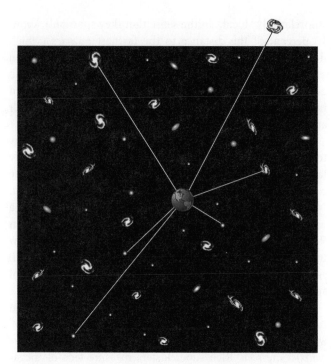

▲ FIGURE 26.3 **Olbers's Paradox** If the universe were homogeneous, isotropic, infinite in extent, and unchanging, then any line of sight from Earth should eventually run into a star and the entire night sky should be bright. This obvious contradiction of the facts is known as Olbers's paradox.

other two assumptions must be false: Either the universe is finite in extent, or it evolves over time. In fact, the answer involves aspects of each and is intimately tied to the behavior of the universe on the largest scales.

The Birth of the Universe

In Chapter 24, we saw that all the galaxies in the universe are rushing away from us in a manner described by Hubble's law,

$$\text{recession velocity} = H_0 \times \text{distance},$$

where we take Hubble's constant H_0 to be 70 km/s/Mpc. ∞ (Sec. 24.3) Up to now, we have used this relation as a convenient means of determining the distances to galaxies and quasars, but it is much more than that.

Assuming for the moment that all velocities have remained constant in time, we can ask a simple question: How long has it taken for any given galaxy to reach its present distance from us? The answer follows from Hubble's law. The time taken is simply the distance traveled divided by the velocity, so

$$
\begin{aligned}
\text{time} &= \frac{\text{distance}}{\text{velocity}} \\
&= \frac{\text{distance}}{H_0 \times \text{distance}} \quad \text{(using Hubble's law for the velocity)} \\
&= \frac{1}{H_0}.
\end{aligned}
$$

For $H_0 = 70$ km/s/Mpc, this time is about 14 billion years. Notice that it is *independent* of the distance: Galaxies twice as far away are moving twice as fast, so the time they took to cross the intervening distance is the same.

Hubble's law therefore implies that, at some time in the past—14 billion years ago, according to the foregoing simple calculation—*all* the galaxies in the universe lay right on top of one another. In fact, astronomers think that *everything* in the universe—matter and radiation alike—was confined at that instant to a single point of enormously high temperature and density, often referred to as the **primeval fireball.** Then the universe began to expand at a furious rate, its density and temperature falling rapidly as the volume increased. This stupendous, unimaginably violent event, involving literally everything in the cosmos, is known as the **Big Bang.** It marked the beginning of the universe.

Thus, by measuring Hubble's constant, we can estimate the age of the universe to be $1/H_0 \approx 14$ billion years. The range of possible error in this age is considerable, both because Hubble's constant is not known precisely and because the assumption that galaxies moved at constant speed in the past is not a good one. We will refine our estimate in a moment, but regardless of the details, the critical fact here is that the age of the universe is *finite.*

The Big Bang provides the resolution of Olbers's paradox. Whether the universe is actually finite or infinite in extent is irrelevant, at least as far as the appearance of the night sky is concerned. The finite *age* of the universe implied by Hubble's law is the key. We see only a finite part of the cosmos—the region lying within roughly14 billion light-years of us. What lies beyond is unknown—its light has not yet had time to reach us.

Note that, even though it appears to place us at the center of the expansion, Hubble's law does *not* violate the cosmological principle in any way. To see this, consider Figure 26.4, which shows how observers in five hypothetical galaxies perceive the motion of their neighbors. For simplicity, the galaxies are taken to be equally spaced, 100 Mpc apart, and they are separating in accordance with Hubble's law with $H_0 = 70$ km/s/Mpc, as seen by an observer in the middle galaxy, number 3. The first pair of numbers beneath each galaxy represents its distance and recessional velocity as measured by that observer. For definiteness, let's take galaxy number 3 to be the Milky Way and the observer to be an astronomer on Earth.

Now consider how the expansion looks from the point of view of the observer in galaxy 2. Galaxy 4, for example, is moving with velocity 7000 km/s to the right relative to galaxy 3, and galaxy 3 in turn is moving at 7000 km/s to the right as seen by observer 2. Therefore, galaxy 4 is moving at a velocity of 14,000 km/s to the right as seen by the observer in galaxy 2. But the distance between the two galaxies is 200 Mpc, so the Hubble constant measured by the observer on galaxy 2 is 14,000 km/s/200 Mpc = 70 km/s/Mpc, the same as the Hubble constant measured by the observer on galaxy 3. The distances and velocities that would be measured by observer 2 are noted in the second row. You can verify for yourself that the ratio of recession velocity to distance is the same for all galaxies.

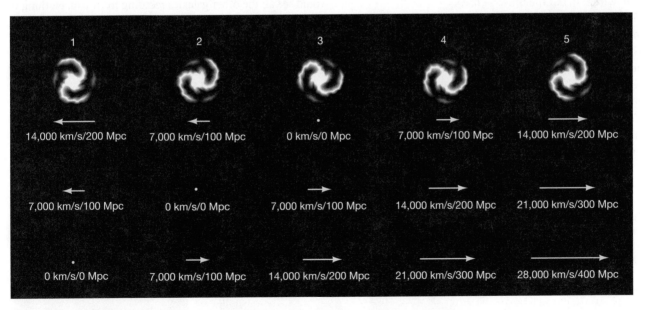

▲ **FIGURE 26.4 Hubble Expansion** Hubble's law is the same, regardless of who makes the measurements. The top numbers are the distances and recessional velocities as seen by an observer on the middle of five galaxies, galaxy 3. The bottom two sets of numbers are from the points of view of observers on galaxies 2 and 1, respectively. In all cases, Hubble's law holds: The ratio of the observed recession velocity to the distance is the same.

Similarly, the measurements made by an observer on galaxy 1 are noted in the third row. Again, the ratio of velocity to distance is the same. The conclusion is clear: *Each observer sees an overall expansion described by Hubble's law, and the constant of proportionality—Hubble's constant—is the same in all cases. Far from singling out any one observer as central, Hubble's law is in fact the *only* expansion law possible if the cosmological principle holds.

Where Was the Big Bang?

Now we know *when* the Big Bang occurred. Is there any way of telling *where*? We think that the universe is the same everywhere, yet we have just seen that the observed recession of the galaxies described by Hubble's law suggests that all the galaxies expanded from a point at some time in the past. Wasn't that point, then, different from the rest of the universe, violating the assumption of homogeneity expressed in the cosmological principle? The answer is a definite *no*!

To understand why there is no "center" to the expansion, we must make a great leap in our perception of the universe. If we were to imagine the Big Bang as simply an enormous explosion that spewed matter out into space, ultimately to form the galaxies we see, then the foregoing reasoning would be quite correct—there would be a center and an edge, and the cosmological principle would not apply. But the Big Bang was *not* an explosion in an otherwise featureless, empty universe. The only way that we can have Hubble's law hold *and* retain the cosmological principle is to realize that the Big Bang involved the entire universe—not just the matter and radiation within it, but the universe *itself*.

In other words, the galaxies are not flying apart into the rest of the universe. *The universe itself is expanding.* Like raisins in a loaf of raisin bread that move apart as the bread expands in an oven, the galaxies are just along for the ride.

Let's consider again some of our earlier statements in light of this new perspective. We now recognize that Hubble's law describes the expansion of the universe itself. Although galaxies have some small-scale, individual random motions, on average they are *not* moving with respect to the fabric of space—any such overall motion would pick out a "special" direction in space and violate the assumption of isotropy. On the contrary, the portion of the galaxies' motion that makes up the Hubble flow is really an expansion of space itself. The expanding universe remains homogeneous at all times. There is no "empty space" beyond the galaxies into which they rush. At the time of the Big Bang, the galaxies did not reside at a point located at some well-defined place within the universe. Rather, the *entire universe* was a point. That point was in no way different from the rest of the universe; that point *was* the universe. Therefore, there was no one point where the Big Bang "happened"—because the Big Bang involved the entire universe, it happened *everywhere* at once.

To illustrate these ideas, imagine an ordinary balloon with coins taped to its surface, as shown in Figure 26.5. (Better yet, do the experiment yourself!) The coins represent galaxies, and the two-dimensional surface of the balloon represents the "fabric" of our three-dimensional universe. The cosmological principle applies to the balloon because every point on the balloon looks pretty much the same as every other. Imagine yourself as a resident of one of the three dark-colored coin "galaxies" in the leftmost frame, and note your position relative to your neighbors. As the balloon inflates (i.e., as the universe expands), the other galaxies recede from you; more distant galaxies recede more rapidly. Notice, incidentally, that the coins themselves do *not* expand along with the balloon, any more than people, planets, stars, or galaxies—all of which are held together by their own internal forces—expand along with the universe. ∞ (Sec. 24.3)

Regardless of which galaxy you chose to consider, you would see all the other galaxies receding from you. Nothing is special or peculiar about the fact that all the galaxies are receding from you. Such is the cosmological principle: No observer anywhere in the universe has a privileged position. There is no center to the expansion and no position that can be identified as the location from which the universal expansion began. Everyone sees an overall expansion described by Hubble's law, with the same value of Hubble's constant in all cases.

Now imagine letting the balloon deflate. This corresponds to running the universe backward from the present time to the Big Bang. *All* the galaxies (coins) would arrive at the same

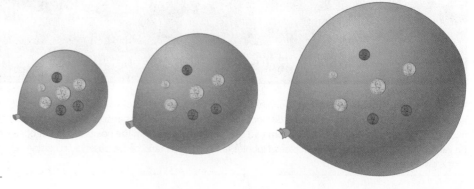

▶ **Interactive FIGURE 26.5**

Receding Galaxies Coins taped to the surface of a spherical balloon recede from one another as the balloon inflates (left to right). Similarly, galaxies recede from one another as the universe expands. As the coins recede, the distance between any two of them increases, and the rate of increase of this distance is proportional to the distance between them. Thus, the balloon expands according to Hubble's law.

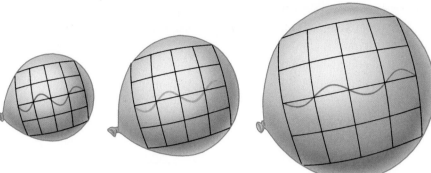

Interactive **FIGURE 26.6**
Cosmological Redshift As the universe expands, photons of radiation are stretched in wavelength, giving rise to the cosmological redshift. In this case, as the baseline in the diagram stretches, the radiation shifts from the short-wavelength blue region of the spectrum to the longer wavelength red region. ∞ (Sec. 3.1)

place at the same time—at the instant the balloon reached zero size. But there is no one point on the balloon that could be said to be *the* place where that occurred. The entire balloon expanded from a point, just as the Big Bang encompassed the entire universe and expanded from a point.

This analogy has its shortcomings. The main difficulty with it is that we see the balloon, which, in our illustration, we imagined as two dimensional, expanding into the third dimension of space. This might suggest that the three-dimensional universe is expanding "into" some fourth spatial dimension. It is not, so far as we know. At the very least, if higher spatial dimensions are involved, they are not relevant to our theory of the universe.

The Cosmological Redshift

This view of the expanding universe requires us to reinterpret the cosmological redshift. ∞ (Sec. 24.3) Previously, we discussed the redshift of galaxies as though it were a Doppler shift—a consequence of their motion relative to us. However, we have just argued that the galaxies are *not* in fact moving with respect to the universe, in which case the Doppler interpretation is incorrect. The true explanation is that, as a photon moves through space, its wavelength is influenced by the expansion of the universe. In a sense, we can think of the photon as being attached to the expanding fabric of space, so its wavelength expands along with the universe, as illustrated in Figure 26.6. Although it is common practice in astronomy to refer to the cosmological redshift in terms of recessional velocity, bear in mind that, strictly speaking, that is not the right thing to do. The cosmological redshift is a consequence of the changing size of the universe—it is *not* related to velocity at all.

The redshift of a photon measures the amount by which the universe has expanded since that photon was emitted. For example, when we measure the light from a quasar and find that it has a redshift of 5, it means that the observed wavelength is 6 times (1 plus the redshift) greater than the wavelength at the time of emission, and this in turn means that the light was emitted at a time when the universe was just one-sixth its present size (and we are observing the quasar as it was

at that time). ∞ *(More Precisely 24-1)* In general, the larger a photon's redshift, the smaller the universe was at the time the photon was emitted, so the longer ago that emission occurred. Because the universe expands with time and redshift is related to that expansion, cosmologists routinely use redshift as a convenient means of expressing time.

These concepts are difficult to grasp. The notion of the entire universe expanding from a hot, dense fireball—with *nothing*, not even space and time, outside—takes some getting used to. Nevertheless, this description of the universe lies at the heart of modern cosmology.

PROCESS OF SCIENCE CHECK
✔ Why does Hubble's law imply a Big Bang?

26.3 The Fate of the Cosmos

Will the universe expand forever? This fundamental question about the fate of the universe has been at the heart of cosmology since Hubble's law was first discovered. Until the late 1990s, the prevailing view among cosmologists was that the answer would most likely be found by determining the extent to which gravity would slow, and perhaps ultimately reverse, the current expansion. However, it now appears that the answer is more subtle—and perhaps a lot more profound in its implications—than was hitherto thought.

Critical Density

Let's begin with another analogy. Assume for the moment that gravity is the only force affecting large-scale motion in the universe, and consider a rocket ship launched from the surface of a planet. Until relatively recently, this scenario was much more than just an analogy—this basic picture and its implications represented the conventional wisdom among cosmologists. However, as we will see in Section 26.5, new observations have forced fundamental changes in astronomers' view of the universe. Nevertheless, the simplified view we now present is a convenient starting point, as it allows us to define some basic ideas and terminology.

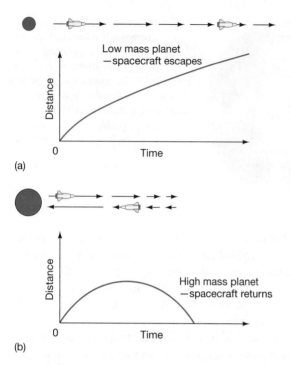

(a)

(b)

▲ FIGURE 26.7 **Escape Speed** (a) A spacecraft (arrows) leaving a planet (blue ball) with a speed greater than the planet's escape speed follows an unbound trajectory and escapes. The graph shows the distance between the ship and the planet as a function of time. (b) If the launch speed is less than the escape speed, the ship eventually drops back to the planet. Its distance from the planet first rises, then falls.

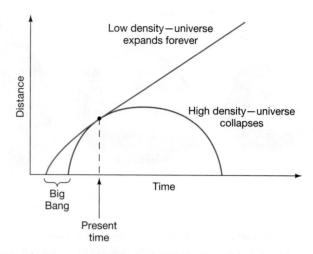

▲ FIGURE 26.8 **Model Universes** Distance between two galaxies as a function of time in each of the two basic universes discussed in the text: a low-density universe that expands forever and a high-density cosmos that collapses. The point where the two curves touch represents the present time.

What are the likely outcomes of the rocket ship's motion? According to Newtonian mechanics, there are just two basic possibilities, depending on the launch speed of the ship relative to the escape speed of the planet. ∞ (Sec. 2.8) If the launch speed is high enough, it will exceed the planet's escape speed, and the ship will never return to the surface. The speed will diminish because of the planet's gravitational pull, but it will never reach zero. The spacecraft leaves the planet on an unbound trajectory, as illustrated in Figure 26.7(a). Alternatively, if the launch speed is lower than the escape speed, the ship will reach a maximum distance from the planet and then fall back to the surface. Its bound trajectory is shown in Figure 26.7(b).

Similar reasoning applies to the expansion of the universe. Imagine two galaxies at some known distance from each other, moving apart with their current relative velocity given by Hubble's law. The same two basic possibilities exist for these galaxies as for our rocket ship: The distance between them can increase forever, or it can increase for a while and then start to decrease. What's more, the cosmological principle says that, whatever the outcome for galaxy A and galaxy B, it must be the same for *any* two galaxies—in other words, the same statement applies to the universe *as a whole*. Thus, as illustrated in Figure 26.8, the universe has only two options: It can continue to expand forever, or the

present expansion will someday stop and turn around into a contraction. The two curves in the figure are drawn so that they pass through the same point at the present time. Both are possible descriptions of the universe, given its current size and expansion rate.

What determines which of the two possibilities will actually occur? In the case of a rocket ship of fixed launch speed (analogous to a universe with a given expansion rate), the *mass* of the planet (for given radius) determines whether or not escape will occur—a more massive planet has a higher escape speed, making it less likely that the rocket can escape. For the universe, the corresponding factor is the *density* of the cosmos. A high-density universe contains enough mass to stop the expansion and eventually cause a collapse. A low-density universe, conversely, will expand forever.

The dividing line between these outcomes—the density corresponding to a universe in which *gravity acting alone* would be just sufficient to halt the present expansion—is called the universe's **critical density**. For $H_0 = 70$ km/s/Mpc the critical density is about 9×10^{-27} kg/m^3. That's an extraordinarily low density—just five hydrogen atoms per cubic meter, a volume the size of a small household closet. In more "cosmological" terms, it corresponds to about 0.1 Milky Way Galaxy (including the dark matter) per cubic megaparsec.

Two Futures

The two possibilities just presented represent radically different futures for our universe. If the cosmos emerged from the Big Bang with sufficiently high density, then it contains enough matter to halt its own expansion, and the recession of the galaxies will eventually stop. At some time in the

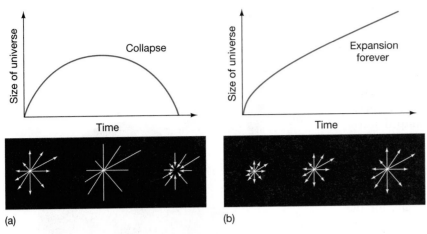

◀ **FIGURE 26.9 Two Futures** If gravity is the only force influencing the cosmic expansion, then the mass density of the universe determines its fate. (a) A high-density universe has a beginning, an end, and a finite lifetime. The lower frames illustrate its evolution, from initial expansion, to maximum size, to collapse. (b) A low-density universe expands forever, with galaxies getting farther and farther apart as time passes.

future, astronomers everywhere—on any planet within any galaxy—will announce that the radiation received from nearby galaxies is no longer redshifted. (The light from *distant* galaxies will still be redshifted, however, because we will see them as they were in the past, at a time when the universe was still expanding.) The bulk motion of the universe, and of the galaxies within, will be stilled—at least momentarily.

The expansion may stop, but the pull of gravity will not. The universe will begin to contract. Nearby galaxies will begin to show blueshifts, and both the density and the temperature of the universe will start to rise as matter collapses back onto itself. As illustrated in Figure 26.9(a), the universe will collapse to a point, requiring just as much time to fall back as it took to rise. First galaxies and then stars will collide with increasing frequency and violence as the available space diminishes and the entire universe shrinks toward a superdense, superhot singularity much like the one from which it originated. The cosmos will ultimately—billions of years from now—experience a "heat death," in which all matter and life are destined to be incinerated. Some astronomers call the final collapse of this high-density universe the "Big Crunch." Cosmologists do not know what will happen to the universe if it ever reaches the point of collapse. The laws of physics as we presently understand them are simply inadequate to describe those extreme conditions.

A quite different fate awaits a low-density universe whose gravity is too weak to halt the present expansion. As illustrated in Figure 26.9(b), such a universe will expand forever, the galaxies continually receding, their radiation steadily weakening with increasing distance. In time, an observer on Earth will see no galaxies in the sky beyond the Local Group (which is not itself expanding). Even with the most powerful telescope, the rest of the observable universe will appear dark, the distant galaxies too faint to be seen. Eventually, the Milky Way and the Local Group, too, will peter out as their fuel supply is consumed. This universe will ultimately experience a "cold death": All radiation, matter, and life are eventually destined to freeze.

How long might the "cold death" of the universe take? Astronomers estimate that our Galaxy probably contains enough gas to keep forming stars for several tens of billions of years, and the majority of stars (the low-mass red dwarfs) can shine for hundreds of billions of years or more. ∞ (Sec. 17.8) Thus, we can expect our Galaxy (and our neighbor Andromeda) to shine on—albeit feebly—for another trillion years or so.

We will see in a moment that the separation between never-ending expansion and cosmic collapse is not quite as straightforward as the foregoing simple reasoning would suggest. Several independent lines of evidence now indicate that gravity is *not* the only influence on the dynamics of the universe on large scales (Section 26.5). As a result, while the "futures" just described are still the only two possibilities for the long-term evolution of the universe, the distinction between them turns out to be more than just a matter of density alone. Nevertheless, the density of the universe—or, more precisely, the *ratio of the total density to the critical value*—is a vitally important quantity in cosmology.

CONCEPT CHECK

✔ What are the two basic possibilities for the future expansion of the universe?

26.4 The Geometry of Space

Our discussion in the previous section used the familiar notions of Newtonian mechanics and gravity because speaking in Newtonian terms made the evolution of the universe easier to understand. But in reality, the proper description of the universe as a dynamic, evolving object is far beyond the capabilities of Newtonian mechanics, which up to now has been our indispensable tool for understanding the cosmos. ∞ (Sec. 2.8) Instead, the more powerful techniques of Einstein's *general relativity,* with its built-in notions of warped space and dynamic spacetime, are needed. ∞ (Sec. 22.6)

Relativity and the Universe

We encountered general relativity in Chapter 22 when we discussed the strange properties of black holes. ∞ (Sec. 22.5) We can loosely summarize its description of the universe by saying that the presence of matter or energy causes a warping, or curvature, of *spacetime* and that the curved trajectories of freely falling particles within warped spacetime are what Newton thought of as orbits under the influence of gravity. The amount of warping depends on the amount of matter present.

When applied to the orbits of planets, stars, even of galaxies, the predictions of general relativity are, for the most part, in accord with those of Newtonian mechanics. But on the scale of the entire universe, relativity has some implications that simply have no counterpart within Newton's theory. Foremost among these non-Newtonian predictions is the fact that the space around us is *curved*, and that the degree of curvature is determined by the *total density* of the cosmos. ∞ (Sec. 22.6) This is the general relativistic restatement of the message in the previous section, but there is an important clarification to be made. In our earlier discussion, we did not specify exactly what we meant by "density." Our discussion implicitly focused on the density of matter, but according to the theory of relativity, mass (m) and energy (E) are equivalent, connected by Einstein's famous relation $E = mc^2$. ∞ (*More Precisely 22-1*)

What, then, actually contributes to the curvature of space? General relativity's answer is clear: Both matter *and* energy must be taken into account, with energy properly converted into matter units by division by the square of the speed of light. [That is, an energy of 1 joule is counted as its mass equivalent of $1 \, J/(3 \times 10^8 \, m/s)^2 = 1.1 \times 10^{-17} \, kg$—not much, but it adds up!] In this way, the density of the universe includes not just the atoms and molecules that make up the familiar "normal" matter around us, but also the invisible dark matter that dominates the masses of galaxies and galaxy clusters, as well as *everything* that carries energy—photons, relativistic neutrinos, gravity waves, and anything else we can think of.

Cosmic Curvature

In a homogeneous universe, the curvature (on sufficiently large scales) must be the same everywhere, so there are really only three distinct possibilities for the large-scale geometry of space. (For more information on the different types of geometry involved, see *More Precisely 26-1*.) General relativity tells us that the geometry of the universe depends only on the ratio of the density of the universe to the critical density (defined in the previous section). As just noted, for $H_0 = 70$ km/s/Mpc, the critical density is 9×10^{-27} kg/m^3. Cosmologists conventionally call the ratio of the universe's actual density to the critical value the *cosmic density parameter* and denote it by the

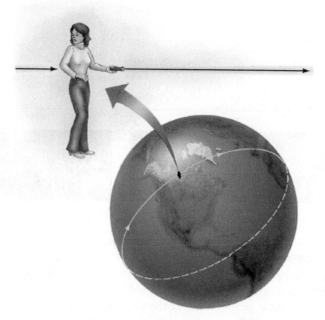

▲ **FIGURE 26.10 Einstein's Curve Ball** In a closed universe, a beam of light launched in one direction might return someday from the opposite direction after circling the universe, just as motion in a "straight line" on Earth's surface will eventually encircle the globe.

symbol Ω_0 ("omega nought"). In terms of this quantity, then, a universe with density equal to the critical value has $\Omega_0 = 1$, a "low-density" cosmos has Ω_0 less than 1, and a "high-density" universe has Ω_0 greater than 1.

In a high-density universe (Ω_0 greater than 1), space is curved so much that it bends back on itself and "closes off," making this universe *finite* in size. Such a universe is known as a **closed universe.** It is difficult to visualize a three-dimensional volume uniformly arching back on itself in this way, but the two-dimensional version is well known: It is just the surface of a sphere, like that of the balloon we discussed earlier. Figure 26.5, then, is the two-dimensional likeness of a three-dimensional closed universe. Like the surface of a sphere, a closed universe has no boundary, yet is finite in extent.* One remarkable property of a closed universe is illustrated in Figure 26.10: Just as a traveler on the surface of a sphere can keep moving forward in a straight line and eventually return to her starting point, a flashlight beam shone in some direction in space might eventually traverse the entire universe and return from the opposite direction!

The surface of a sphere curves, loosely speaking, "in the same direction," no matter which way we move from a given point. A sphere is said to have *positive curvature*. However, if the average density of the universe is below the critical value,

---

Notice that for the sphere analogy to work we must imagine ourselves as two-dimensional "flatlanders" who cannot visualize or experience in any way the third dimension perpendicular to the sphere's surface. Flatlanders and their light rays are confined to the sphere's surface, just as we are confined to the three-dimensional volume of our universe.

the surface curves like a saddle, in which case it has *negative curvature*. Most people have a good idea of what a saddle looks like—it curves "up" in one direction and "down" in another—but no one has ever seen a uniformly negatively curved surface, for the simple reason that it cannot be constructed in three-dimensional Euclidean space! It is just "too big" to fit. A low-density, saddle-curved universe is infinite in extent and is usually called an **open universe.**

The intermediate case, in which the density is precisely equal to the critical density (i.e., $\Omega_0 = 1$), is the easiest to

MORE PRECISELY 26-1

Curved Space

Euclidean geometry is the geometry of flat space—the geometry taught in high school. Set forth by one of the most famous of the ancient Greek mathematicians, Euclid, who lived around 300 B.C., it is the geometry of everyday experience. Houses are usually built with flat floors. Writing tablets and blackboards are also flat. We work easily with flat, straight objects, because the straight line is the shortest distance between any two points.

When we construct houses or any other straight-walled buildings on the surface of Earth, the other basic axioms of Euclid's geometry also apply: Parallel lines never meet, even when extended to infinity; the angles of any triangle always sum to 180°; the circumference of a circle equals π times the diameter of the circle. (See the accompanying figure.) If these axioms did not hold, walls and roof would never meet to form a house!

In reality, though, the geometry of Earth's surface is not really flat; it is curved. We live on the surface of a sphere, and on that *surface* Euclidean geometry breaks down. Instead, the rules for the surface of a sphere are those of *Riemannian geometry,* named after the 19th-century German mathematician Georg Friedrich Riemann. There are no parallel "straight" lines on a sphere. The analog of a straight line on a sphere's surface is a "great circle"—the arc formed when a plane passing through the center of the sphere intersects the surface. Any two such lines must eventually intersect. The sum of a triangle's angles, when drawn on the surface of a sphere, exceeds 180°—in the 90°–90°–90° triangle shown in the accompanying figure, the sum is actually 270°—and the circumference of a circle is less than π times the circle's diameter.

We see that the curved surface of a sphere, governed by the spherical geometry of Riemann, differs greatly from the flat-space geometry of Euclid. The two are approximately the same only if we confine ourselves to a small patch on the surface. If the patch is small enough compared with the sphere's radius, the surface looks "flat" nearby, and Euclidean geometry is approximately valid. This is why we can draw a usable map of our home, our city, and even our state, on a flat sheet of paper, but an accurate map of the entire Earth must be drawn on a globe.

When we work with larger parts of Earth, we must abandon Euclidean geometry. World navigators are fully aware of this. Aircraft do not fly along what might appear on most maps as a straight-line path from one point to another. Instead, they follow a great circle on Earth's surface. On the curved surface of a sphere, such a path is always the shortest distance between two points. For example, as illustrated in the figure, a flight from Los Angeles to London does not proceed directly across the United States and the Atlantic Ocean, as you might expect from looking at a flat map. Instead, it goes far to the north, over Canada and Greenland, above the Arctic Circle, finally coming in over Scotland for a landing at London. This is the great-circle route—the shortest path between the two cities, as you can easily see if you inspect a globe.

The "positively curved" space of Riemann is not the only possible departure from flat space. Another is the "negatively curved" space first studied by Nikolai Ivanovich Lobachevsky, a 19th-century Russian mathematician. In this geometry, there are an *infinite* number of lines through any given point that are parallel to another line, the sum of the angles of a triangle is *less* than 180° (see the first figure), and the circumference of a circle is *greater* than π times its diameter. This type of space is described by the surface of a curved saddle, rather than a flat plane or a curved sphere. It is a hard geometry to visualize!

Most of the local realm of the *three*-dimensional universe (including the solar system, the neighboring stars, and even our Milky Way Galaxy) is correctly described by Euclidean geometry. If the currently favored cosmology described in the text turns out to be correct, then the whole universe is, too!

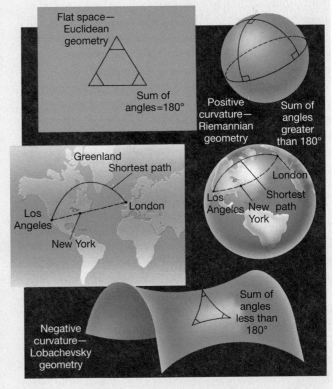

Flat space—Euclidean geometry
Sum of angles =180°

Positive curvature—Riemannian geometry
Sum of angles greater than 180°

Greenland
Shortest path
Los Angeles
New York
London

Los Angeles
New York
London
Shortest path

Negative curvature—Lobachevsky geometry
Sum of angles less than 180°

visualize. This universe, called a **critical universe,** has no curvature. It is said to be "flat" and is infinite in extent. In this case, and *only* in this case, the geometry of space on large scales is precisely the familiar Euclidean geometry taught in high schools. Apart from its overall expansion, this is basically the universe that Newton knew.

Euclidean geometry—the geometry of flat space—is familiar to most of us because it is a good description of space in the vicinity of Earth. It is the geometry of everyday experience. Does this mean that the universe is flat, which would in turn mean that it has exactly the critical density? Not necessarily: Just as a flat street map is a good representation of a city, even though we know Earth is really a sphere, Euclidean geometry is a good description of space within the solar system, or even the Galaxy, because the curvature of the universe is negligible on scales smaller than about 1000 Mpc. Only on the very largest scales would the geometric effects we have just discussed become evident.

CONCEPT CHECK
✔ How is the curvature of space related to the density of the universe?

26.5 Will the Universe Expand Forever?

Is there any way for us to determine which of the futures we have described actually applies to our universe (that is, apart from simply waiting to find out)? Will the universe end as a dense fireball much like that from which it began? Or will it expand forever? And can we hope to measure the geometry of the vast cosmos we inhabit? Finding answers to these questions has been the dream of astronomers for decades. We are fortunate to live at a time when astronomers can subject these questions to intensive observational tests and come up with definite answers—even though they aren't what most cosmologists expected! Let's begin by looking at the density of the universe (or, equivalently, the cosmic density parameter Ω_0).

The Density of the Universe

How might we determine the density of the universe? On the face of it, it would seem simple: Just measure the total mass of the galaxies residing within some large parcel of space, calculate the volume of that space, and then divide the mass by the volume to compute the average density. When astronomers do this, they usually find a little less than 10^{-28} kg/m^3 in the form of luminous matter. Largely independent of whether the chosen region contains many scattered galaxies or only a few rich galaxy clusters, the resulting density is about the same, within a factor of two or three.

Galaxy counts thus yield a value of Ω_0 of only a few percent. If that measure were correct, and galaxies were all that existed, then we would live in a low-density open universe destined to expand forever.

But there is a catch. We have noted (Chapters 23 and 25) that most of the matter in the universe is *dark*—it exists in the form of invisible material that has been detected only through its gravitational effect in galaxies and galaxy clusters. ∞ (Secs. 23.6, 25.1) Currently, we do not know what the dark matter is, but we *do* know that it is there. Galaxies may contain as much as 10 times more dark matter than luminous material, and the figure for galaxy clusters is even higher—perhaps as much as 95 percent of the total mass in clusters is invisible. Even though we cannot see it, dark matter contributes to the density of the universe and plays its part in opposing the cosmic expansion. Including all the dark matter that is known to exist in galaxies and galaxy clusters increases the value of Ω_0 to about 0.25.

Unfortunately, the distribution of dark matter on larger scales is not very well known. We can infer its presence in galaxies and galaxy clusters, but we are largely ignorant of its extent in superclusters, voids, or other larger structures. There are indications that dark matter may account for a greater fraction of the mass on large scales than it does in galaxy clusters. Observations of gravitational lensing by galaxy clusters suggest that dark matter may be considerably more extensive than is indicated by the motions of galaxies within the clusters. ∞ (Sec. 25.5) One of the few mass estimates for an object much larger than a supercluster comes from optical and infrared observations of the overall motion of galaxies (including the Local Group) within the Local Supercluster. The measured velocities suggest the presence of a nearby huge accumulation of mass known as the *Great Attractor,* with a total mass of about 10^{17} solar masses and a size of 100–150 Mpc. The average density of this structure may be quite close to the critical value.

However, it seems that, when averaged over the entire universe, even the densities of objects like the Great Attractor don't raise the overall cosmic density by much. In short, there doesn't seem to be much additional dark matter "tucked away" on very large scales. Most cosmologists agree that the density of matter (luminous plus dark) in the universe is not much greater than about 30 percent of the critical value—*not* enough to halt the universe's current expansion.

Cosmic Acceleration

Determining the mass density of the universe is an example of a *local* measurement that provides an estimate of Ω_0. But the result we obtain depends on just how local our measurement is—as we have just seen, there are many uncertainties in the result, especially on large scales. In an attempt to get around this problem, astronomers have devised alternative methods that rely instead on *global* measurements, covering much larger regions of the observable universe. In principle, such global tests should indicate the universe's overall density, not just its value in our cosmic neighborhood.

One such global method is based on observations of Type I (carbon-detonation) supernovae. ∞ (Sec. 21.3) Recall that these objects are very bright and have a remarkably narrow spread in luminosities, making them particularly useful as standard candles. ∞ (Sec. 24.2) They can be used as probes of the universe because, by measuring their distances (*without* using Hubble's law) and their redshifts, we can determine the rate of cosmic expansion in the distant past. Here's how the method works:

Suppose the universe is decelerating, as we would expect if gravity were slowing its expansion. Then, because the expansion rate is decreasing, objects at great distances—that is, objects that emitted their radiation long ago—should appear to be receding *faster* than Hubble's law predicts. Figure 26.11(a) illustrates this concept. If the universal expansion were constant in time, recessional velocity and distance would be related by the black line. (The line is not quite straight, because it takes the expansion of the universe properly into account in computing the distance.) ∞ (*More Precisely 24-1*) In a decelerating universe, the velocities of distant objects should lie *above* the black curve, and the deviation from that curve should be greater for a denser universe, in which gravity has been more effective at slowing the expansion.

How does theory compare with reality? In the late 1990s, two groups of astronomers announced the results of independent, systematic surveys of distant supernovae. Some of these supernovae are shown in Figure 26.11(b); the data are marked on Figure 26.11(a). Far from clarifying the picture of cosmic deceleration, these findings indicate that the expansion of the universe is not slowing, but actually accelerating! According to the supernova data, galaxies at large distance are receding *less* rapidly than Hubble's law would predict. The deviations from the decelerating curves appear small in the figure, but they are statistically very significant, and both groups report similar findings. Subsequent supernova observations, most recently from *HST* in 2006, are generally in agreement with the initial findings.

These observations are *inconsistent* with the "gravity only" Big Bang model just described and have sparked a major revision of our view of the cosmos. The measurements are difficult, and the results depend quite sensitively on just how "standard" the supernovae luminosities really are; some astronomers initially questioned the accuracy of the method. In particular, if supernovae at great distances (i.e., long ago) were for some reason slightly less luminous than those nearby, then we would think that these distant supernovae were farther away than they actually are, and the

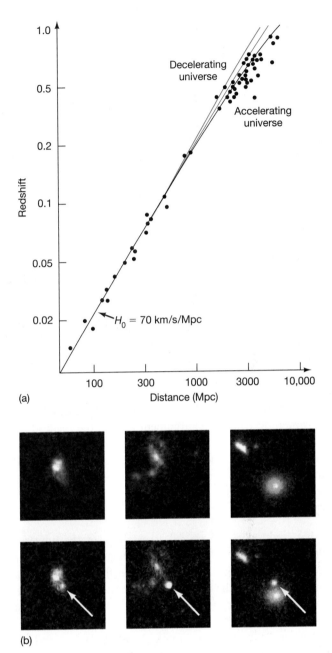

(a)

(b)

▲ FIGURE 26.11 **Accelerating Universe** (a) Observations of distant supernovae allow astronomers to measure changes in the rate of expansion of the universe. In a decelerating universe (purple and red curves), redshifts of distant objects are greater than would be predicted from Hubble's law (black curve). The reverse is true for an accelerating universe. The points showing observations of some 50 supernovae strongly suggest that the cosmic expansion is accelerating. The vertical scale shows redshift; for small velocities, redshift is just velocity divided by the speed of light. ∞ (*More Precisely 24-1*) (b) The bottom frames show three of the above-graphed supernovae (marked by arrows) that exploded in distant galaxies when the universe was nearly half its current age. The top frames show the same areas prior to the detonation of those supernovae, which were originally discovered in 1997 during a ground-based survey with the Canada–France–Hawaii Telescope on Mauna Kea. (*CfA/NASA*)

▶ FIGURE 26.12 **Dark Energy** The (THE BIG PICTURE) expansion of the universe is opposed by the attractive force of gravity and sped up by the repulsion due to dark energy. As the universe expands, the gravitational force weakens, whereas the force due to dark energy increases. Several billion years ago, dark energy came to dominate, and the expansion of the cosmos has been accelerating ever since. A universe dominated by dark energy is destined to expand forever.

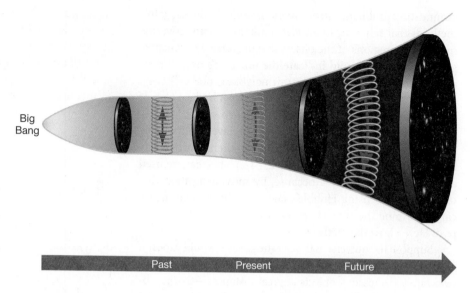

Big Bang

Past   Present   Future

error would appear as a deviation to the right of the black curve in Figure 26.11(a)—in other words, as an acceleration in the cosmic expansion rate.

Not surprisingly, since so much hangs on this measurement, the reliability of the supernova measurement technique has been the subject of intense scrutiny by cosmologists. However, no convincing argument against the method has yet been put forward, so there is no reason to think that we are somehow being "fooled" by nature. As far as we can tell, the measurements are good, and the acceleration is real.

Dark Energy

What could cause an overall acceleration of the universe? Frankly, cosmologists don't know, although several possibilities have been suggested. Whatever it is, the mysterious cosmic field causing the universe to accelerate is neither matter nor radiation. Although it carries energy, it exerts an overall *repulsive* effect on the universe, speeding up the expansion of empty space. It has come to be called **dark energy,** and it is perhaps *the* leading puzzle in astronomy today.

One leading dark-energy candidate is an additional "vacuum energy" force associated with empty space and effective only on very large scales. Known simply as the **cosmological constant,** it has a long and checkered history. It was first proposed by Einstein as a way to force his new theory of general relativity into "predicting" a static universe, but was subsequently dropped from Einstein's equations following Hubble's discovery that the universe is not static, but instead is expanding (see *Discovery 26-1*). Since the 1990s, the cosmological constant has arisen again, to become a staple of astronomers' models of the universe. Note, however, that although models that take this force into account can fit the observational data, as described in the next section, astronomers have *no* clear physical interpretation of what the force actually is.

It is neither required nor explained by any known law of physics.

As illustrated in Figure 26.12, the repulsive effect of dark energy is proportional to the size of the universe, so it increases as the universe expands. Thus, it was negligible at early times, but today, given the magnitude of the observed acceleration, it is the major factor controlling the cosmic expansion. Furthermore, since the effect of gravity *weakens* as the expansion proceeds, it follows that once dark energy begins to dominate, gravity can never catch up, and the universe will continue to accelerate at an ever-increasing pace. Thus, despite the considerable uncertainty as to the nature of dark energy, we can at least say that, by opposing the attractive force of gravity, dark energy's repulsion strengthens our earlier conclusion that the universe will expand forever.

One additional problem for cosmologists is the fact that the present value of the repulsive force is *comparable* to the attractive force of gravity opposing further expansion. Why is that a problem? Because, when we calculate the evolution of a universe containing a cosmological constant consistent with current observations (see Figure 26.14), we find that this state of affairs was not true in the early universe (when galaxies were forming, say), nor will it be true in 10 or 20 billion years' time. In other words, the observations seem to suggest that we live at a special time in the history of the universe—a conclusion viewed with great suspicion by astronomers who grew up with the Copernican principle as their guide.

One promising alternative dark-energy candidate, called *quintessence,** might offer a means of avoiding this problem. Whereas the cosmological constant is a property of empty space and is independent of any matter or "normal" energy

---

In ancient alchemy, quintessence was the "fifth element," after earth, air, fire, and water. It was believed to be the perfect substance composing the heavens and all heavenly bodies.

that space contains, quintessence evolves in time in a way that depends on the matter and radiation in the universe. By coupling the behavior of dark energy to the other contents of the cosmos, quintessence may provide a natural mechanism for dark energy to emerge as the dominant force as the universe expands and cools, and galaxies begin to form and grow.

With little hard data to constrain them, theorists currently have considerable freedom in constructing models of the dark-matter content of the universe. Cosmologists are searching for experimental and observational tests to refine their models and to distinguish between competing theories.

CONCEPT CHECK

✔ Why do astronomers think the universe will expand forever?

26.6 Dark Energy and Cosmology

As we proceed through the remainder of this chapter and the next, it is worth bearing in mind that the Big Bang *is* a scientific theory and, like any other, must continually be challenged and scrutinized. ∞ (Sec. 1.2) The Big Bang theory makes detailed, testable predictions about the state and history of the cosmos and must change—or be replaced—if these predictions are found to be at odds with observations. The supernova observations just described are a case in point.

Even though the supernova observations and their interpretation have so far withstood intense scrutiny, the idea of an accelerating universe driven by some completely unknown field called dark energy probably would not have gained such rapid and widespread acceptance among cosmologists were it not strongly supported by several other pieces of evidence. In this section, we discuss how the existence of dark matter fits in with observations of the universe and even helps resolve some long-standing riddles. Every independent piece of evidence, and every old puzzle solved, provides further support not just for the idea of dark energy, but also for the entire Big Bang theory of the universe.

The Composition of the Universe

In addition to measuring density and acceleration, astronomers have several other means of estimating the "cosmological parameters" that describe the large-scale properties of our universe.

Theoretical studies of the early universe (to be discussed in more detail in Chapter 27) strongly suggest that the geometry of the universe should be precisely flat—that is, that the total density of the cosmos should exactly equal the critical value. This idea first became widespread in the 1980s, and for many years there seemed to be a major discrepancy between it and observations that clearly showed a cosmic matter density of less than 30 percent of the critical

value, even taking the dark matter into account. Dark energy resolves that conflict by providing another form in which the "extra" density can exist, although not all astronomers are happy at the price of this resolution, which has introduced yet another unknown component into the cosmic mix!

Recent detailed measurements of the radiation field known to fill the entire cosmos (see Section 26.7 and Chapter 27) strongly support the theoretical prediction that $\Omega_0 = 1$ and are also consistent with the dark energy inferred from the supernova studies. Further independent corroboration comes from careful analyses of galaxy surveys such as those discussed in Section 26.1, which allow astronomers to measure the growth of large-scale structure in the universe. Simply put, the more mass there is in the universe, the easier it is for clusters, superclusters, walls, and voids to grow as gravity gathers matter into larger and larger clumps. Higher density implies more rapid formation of structure—or, equivalently, *less* structure in the past, given the structure we see around us today. Thus, structure measurements constrain the value of Ω_0.

Remarkably, all the approaches just described yield consistent results! As of early 2007, the consensus among cosmologists is that the universe is of precisely critical density, $\Omega_0 = 1$, but that this density is made up of both matter (mostly dark) *and* dark energy (converted into mass units as discussed earlier in Section 26.4). Radiation contributes negligibly to the total (see Section 27.1). Thus, the current best estimate, based on all the available data, is that normal "luminous" matter accounts for 4 percent of the total, dark matter for 23 percent, and dark energy for the remaining 73 percent. This is the assumption underlying Table 24.2 and used consistently throughout this book.

Note that such a universe will expand forever and, the heavy machinery of general relativity and curved spacetime notwithstanding, is perfectly flat (Figure 26.13)—an irony that would no doubt have amused Newton!

Cosmic Age Estimates

We have at least one other independent, noncosmological way of testing the preceding important conclusion. In Section 26.2, when we estimated the age of the universe

▲ FIGURE 26.13 **Geometry of the Universe** As best we can tell, the universe on the largest scales is geometrically flat—governed by the same familiar Euclidean geometry taught in high schools.

DISCOVERY 26-1

Einstein and the Cosmological Constant

Even the greatest minds are fallible. The first scientist to apply general relativity to the universe was, not surprisingly, the theory's inventor, Albert Einstein. When he derived and solved the equations describing the behavior of the universe, Einstein discovered that they predicted a universe that evolved in time. But in 1917, neither he nor anyone else knew about the expansion of the universe, as described by Hubble's law, which would not be discovered for another 10 years. ∞ (Sec. 24.3) At the time, Einstein, like most scientists, believed that the universe was static—that is, unchanging and everlasting. The discovery that there was no static solution of his equations seemed to Einstein to be a near-fatal flaw in his new theory.

To bring the theory into line with his beliefs, Einstein tinkered with the equations, introducing a "fudge factor" describing a hypothetical repulsive force operating on large scales in the universe. This factor is now known as the cosmological constant. One possible solution to Einstein's modified equations described a universe in which the repulsive cosmological constant just balanced the attractive force of gravity, allowing the size of the cosmos to remain constant for an indefinite period of time. Einstein took this to be the static universe he expected.

Instead of predicting an evolving cosmos, which would have been one of general relativity's greatest triumphs, Einstein yielded to a preconceived notion of the way the universe "should be," unsupported by observational evidence. Later, when the expansion of the universe was discovered and Einstein's equations—without the fudge factor—were found to describe it perfectly, he declared that the cosmological constant was the biggest blunder of his scientific career.

Scientists are reluctant to introduce unknown quantities into their equations purely to make the results "come out right."

Einstein introduced the cosmological constant to fix what he thought was a serious problem with his equations, but he discarded it immediately once he realized that no problem actually existed. As a result, the cosmological constant fell out of favor among astronomers for many years.

In the 1980s, the concept made something of a comeback with the realization by physicists that the very early universe may have gone through a phase when its evolution was determined by a "cosmological constant" of sorts (see Section 27.4), and this idea is now firmly entrenched in many cosmologists' models of the universe. Today, as discussed in the text, the cosmological constant has apparently been completely rehabilitated and identified as a leading candidate for the "dark energy" whose existence is inferred from studies of the universe on very large scales. As shown in Figure 26.14, inclusion of a suitable dark-energy term in Einstein's equations can cause the expansion of the universe to accelerate, instead of slowing down as it would if only gravity were involved.

For many researchers—Einstein included—the main problem with the cosmological constant was (and still is) the fact that we have no clear explanation for either its existence or its present value. The leading theories of the structure of matter do in fact predict repulsive forces of this sort, but these forces generally operate only under extreme conditions, and, in any case, their "natural" energy scale is vastly greater (by something like a factor of 10^{120}!) than anything consistent with cosmological observations.

As discussed in the text, theoretical efforts are underway to combine aspects of the cosmological constant with a more reasonable scale on which such a cosmic repulsive force might act. But before we make too many sweeping statements about the role of the cosmological constant in cosmology, we should probably remember the experience of its inventor and bear in mind that—at least for now—its physical meaning remains completely unknown.

---

from the accepted value of Hubble's constant, we made the assumption that the expansion speed of the cosmos was constant in the past. However, as we have just seen, this is a considerable oversimplification. Gravity tends to slow the universe's expansion, whereas dark energy acts to accelerate it, and the actual expansion of the universe is the result of the competition between the two. In the absence of a cosmological constant, the universe would have expanded faster in the past than it does today, so the assumption of a constant expansion rate leads to an overestimate of the universe's age—such a universe is younger than the 14 billion years calculated earlier. Conversely, the repulsive effect of dark energy tends to increase the age of the cosmos.

Figure 26.14 illustrates these points. It is similar to Figure 26.8, except that we have added two extra lines, one corresponding to a constant expansion rate at the present value—a completely *empty* 14-billion-year-old universe—the other to the

best-fit accelerating universe with the parameters just described. The age of a critical-density universe with no cosmological constant is about 9 billion years. A low-density open universe (again with no cosmological constant) is older than 9 billion years, but still less than 14 billion years old. The age corresponding to the accelerating universe is 13.7 billion years, coincidentally very close to the value for constant expansion.

How does this kind of calculation compare with an age estimated by other means? On the basis of the theory of stellar evolution, the oldest globular clusters formed about 12 billion years ago, and most are estimated to be between 10 and 12 billion years old. ∞ (Secs. 19.6, 20.5) This range is indicated in Figure 26.14. These ancient star clusters are thought to have formed at around the same time as our Galaxy, so they date the epoch of galaxy formation. More important, they can't be older than the universe! The figure shows that globular cluster ages are consistent with a

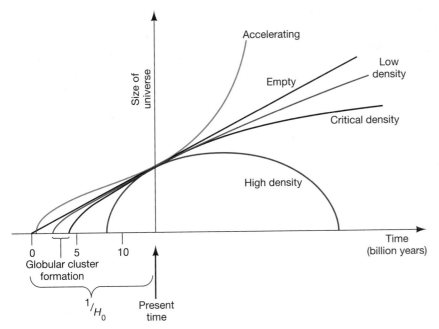

▲ FIGURE 26.14 **Cosmic Age** Both the expansion and the age of the cosmos are governed by the competition between gravity and dark energy. According to current observations, dark energy seems destined to win. The age of a universe without dark energy (represented by all three lower curves) is always less than $1/H_0$ and decreases for larger values of the present-day density. The existence of a repulsive cosmological constant (green curve) increases the age of the cosmos. The green curve is drawn using the best available cosmic parameters, as described in the text. Coincidentally, the age of the universe happens to be almost exactly $1/H_0$.

14-billion-year-old cosmos and even allow a couple of billion years for galaxies to form and grow, as discussed in Chapter 25. ∞ (Sec. 25.3) Note also that the cluster ages are *not* consistent with a critical-density universe without dark energy. This independent check of a key prediction is an important piece of evidence supporting the modern version of the Big Bang theory.

Thus, for $H_0 = 70$ km/s/Mpc, our current best guess of the history of the universe places the Big Bang at 14 billion years ago. The first quasars appeared about 13 billion years ago (at a redshift of 6), the peak quasar epoch (redshifts 2–3) occurred during the next 1 billion years, and the oldest known stars in our Galaxy formed during the 2 billion years after that. Even though astronomers do not currently understand the nature of dark energy, the good agreement between so many separate lines of reasoning has convinced many that the dark-matter, dark-energy Big Bang theory just described is the correct description of the universe. But astronomers aren't ready to relax just yet: The history of this subject suggests that there may be a few more unexpected twists and turns in the road before the details are finally resolved.

CONCEPT CHECK

✔ Why have astronomers concluded that dark energy is the major constituent of the universe?

26.7 The Cosmic Microwave Background

Looking out into space is equivalent to looking back into time. ∞ *(More Precisely 24-1)* But how far back in time can we probe? Is there any way to study the universe beyond the most distant quasar? How close can we come to perceiving the edge of time—the very origin of the universe—directly?

A partial answer to these questions was discovered by accident in 1964, during an experiment designed to improve the U.S. telephone system. As part of a project to identify and eliminate interference in planned satellite communications, Arno Penzias and Robert Wilson, two scientists at Bell Telephone Laboratories in New Jersey, were studying the Milky Way's emission at microwave (radio) wavelengths, using the horn-shaped antenna shown in Figure 26.15. In their data, they noticed a bothersome background "hiss" that just would not go away—a little like the background static on an AM radio station. Regardless of where and when they pointed their antenna, the hiss persisted. Never diminishing or intensifying, the weak signal was detectable at any time of the day, any day of the year, apparently filling all space.

What was the source of this radio noise? And why did it appear to come uniformly from all directions, unchanging

▲ FIGURE 26.15 **Microwave Background Discoverers** This "sugar-scoop" antenna was built to communicate with Earth-orbiting satellites, but was used by Robert Wilson (left) and Arno Penzias to discover the 2.7-K cosmic background radiation. *(Bell Labs)*

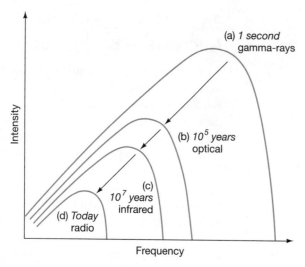

▲ FIGURE 26.16 **Cosmic Blackbody Curves** Theoretically derived blackbody curves for the entire universe (a) 1 second after the Big Bang, (b) 100,000 years after the Big Bang, (c) 10 million years after the Big Bang, and (d) today, approximately 14 billion years after the Big Bang.

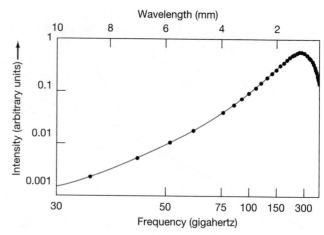

▲ FIGURE 26.17 **Microwave Background Spectrum** The intensity of the cosmic background radiation, as measured by the *COBE* satellite, agrees very well with theory. The curve is the best fit to the data, corresponding to a temperature of 2.725 K. The experimental errors in this remarkably accurate observation are smaller than the dots representing the data points.

in time? Unaware that they had detected a signal of great cosmological significance, Penzias and Wilson sought many different origins for the excess emission, including atmospheric storms, interference from the ground, short circuits of equipment—even pigeon droppings inside the antenna! Eventually, after conversations with colleagues at Bell Labs and theorists at nearby Princeton University, the two experimentalists realized that the origin of the mysterious static was nothing less than the fiery creation of the universe itself. The radio hiss that Penzias and Wilson detected is now known as the **cosmic microwave background.** Their discovery won them the 1978 Nobel Prize in physics.

In fact, researchers had predicted the existence and general properties of the microwave background well before its discovery. As early as the 1940s, physicists had realized that, in addition to being extremely dense, the early universe must also have been very hot, and shortly after the Big Bang the universe must have been filled with extremely high-energy thermal radiation—gamma rays of very short wavelength. Researchers at Princeton had extended these ideas, reasoning that the frequency of this primordial radiation would have been redshifted (simply by cosmic expansion) from gamma ray, to X ray, to ultraviolet, and eventually all the way into the radio range of the electromagnetic spectrum as the universe expanded and cooled (Figure 26.16). ∞ (Sec. 3.4) By the present time, they argued, this redshifted "fossil remnant" of the primeval fireball should have a temperature of no more than a few tens of kelvins, peaking in the microwave part of the spectrum. The Princeton group was in the process of constructing a microwave antenna to search for this radiation when Penzias and Wilson announced their discovery.

The Princeton researchers confirmed the existence of the microwave background and estimated its temperature at about 3 K. However, because of atmospheric absorption, this

part of the electromagnetic spectrum happens to be difficult to observe from the ground, and it was 25 years until astronomers could demonstrate conclusively that the radiation was described by a blackbody curve. In 1989, the *Cosmic Background Explorer* (*COBE*) satellite measured the intensity of the microwave background at wavelengths straddling the peak of the curve, from a half millimeter up to about 10 cm. The results are shown in Figure 26.17. The solid line is the blackbody curve that best fits the *COBE* data. The near-perfect fit corresponds to a universal temperature of about 2.7 K.

Figure 26.18 shows a *COBE* map of the microwave background temperature over the entire sky. The blue regions are hotter than average, by about 0.0034 K, the red regions

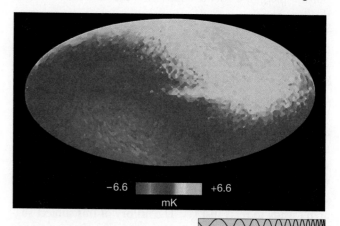

▲ FIGURE 26.18 **Microwave Sky** A *COBE* map of the microwave sky reveals that the microwave background appears a little hotter in the direction of the constellation Leo and a little cooler in the opposite direction. The maximum temperature deviation from the average is about 0.0034 K, corresponding to a velocity of 400 km/s in the direction of Leo. *(NASA)*

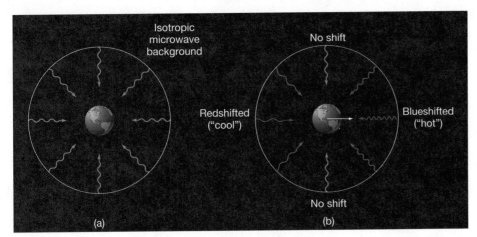

◀ FIGURE 26.19 **Earth's Motion through the Cosmos** (a) To an observer at rest with respect to the expanding universe, the microwave background appears isotropic. (b) A moving observer measures "hot" blueshifted radiation in one direction (the direction of motion) and "cool" redshifted radiation in the opposite direction.

cooler by the same amount. This temperature range is not an inherent property of the microwave background, however. Rather, it is a consequence of Earth's *motion* through space. If we were precisely at rest with respect to the universal expansion (like the coin taped to the surface of the expanding balloon in Figure 26.5), then we would see the microwave background as almost perfectly isotropic, as illustrated in Figure 26.19(a). However, if we are moving with respect to that frame of reference, as in Figure 26.19(b), then the radiation from in front of us should be slightly blueshifted by our motion, whereas that from behind should be redshifted. Thus, to a moving observer, the microwave background should appear a little hotter than average in front and slightly cooler behind.

The data indicate that Earth's velocity is about 380 km/s in the approximate direction of the constellation Leo. Once the effects of this motion are corrected for, the cosmic microwave background is found to be strikingly isotropic. Its intensity is virtually constant (in fact, to about one part in 10^5) from one direction on the sky to another, lending strong support to one of the key assumptions underlying the cosmological principle.

When we observe the microwave background, we are looking almost all the way to the very beginning of the universe. The photons that we receive as these radio waves today have not interacted with matter since the universe was a mere 400,000 years old, when, according to our models, it was less than a thousandth of its present size. To probe further, back to the Big Bang itself, requires that we enter the world of nuclear and particle physics. The Big Bang was the biggest and the most powerful particle accelerator of all! In the next chapter, we will see how studies of conditions in the primeval fireball aid us in understanding the present-day structure and future evolution of the universe in which we live.

CONCEPT CHECK

✔ When was the cosmic microwave background formed?

CHAPTER REVIEW

SUMMARY

1 Redshift surveys reveal that, on scales larger than a few hundred megaparsecs, the universe appears roughly **homogeneous** (the same everywhere, **p. 663**) and **isotropic** (the same in all directions, **p. 663**). In **cosmology (p. 662)**—the study of the universe

as a whole—researchers usually assume that the universe is homogeneous and isotropic. This assumption is known as the **cosmological principle (p. 663)** and implies that the universe cannot have a center or an edge.

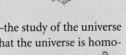

2 If the universe were homogeneous, isotropic, infinite, and unchanging, the night sky would be bright because any line of sight would eventually intercept a star.

The fact that the night sky is dark is called **Olbers's paradox (p. 664).** Its resolution lies in the fact that, regardless of whether the universe is finite or infinite, we see only a finite part of it from Earth—the region from which light has had time to reach us since the universe began.

3 Tracing the observed motions of galaxies back in time implies that some 14 billion years ago the universe consisted of a hot, dense **primeval fireball (p. 665)**

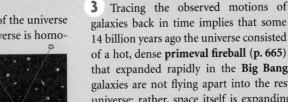

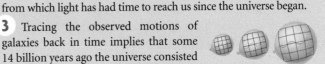

that expanded rapidly in the **Big Bang (p. 665).** However, the galaxies are not flying apart into the rest of an otherwise empty universe; rather, space itself is expanding. The Big Bang did not happen at any particular location in space, because space itself was compressed to a point at that instant—the Big Bang happened everywhere at once. The cosmological redshift occurs as a photon's wavelength is "stretched" by cosmic expansion. The extent of the observed redshift is a direct measure of the expansion of the universe since the photon was emitted.

4 There are only two possible outcomes to the current expansion: Either the universe will expand forever, or it will eventually recollapse. The **critical density (p. 668)** is the density of matter needed for gravity alone to overcome the present expansion and cause the universe to collapse. Most astronomers think that the total mass density of the universe today is no more than about 30 percent of the critical value.

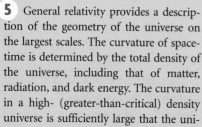

5 General relativity provides a description of the geometry of the universe on the largest scales. The curvature of spacetime is determined by the total density of the universe, including that of matter, radiation, and dark energy. The curvature in a high- (greater-than-critical) density universe is sufficiently large that the universe "bends back" on itself and is finite in extent, somewhat like the surface of a sphere. Such a universe is said to be a **closed universe (p. 670)**. A low-density **open universe (p. 671)** is infinite in extent and has a "saddle-shaped" geometry. The **critical universe (p. 672)** has a density precisely equal to the critical value and is spatially flat.

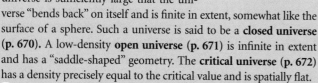

6 Observations of distant supernovae indicate that the expansion of the universe is accelerating, apparently driven by the effects of **dark energy (p. 674)**, a mysterious repulsive force that exists throughout all space. The physical nature of dark energy is unknown. Possible candidates include the **cosmological constant (p. 674)** and *quintessence*.

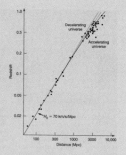

7 The best available observational data are consistent with the idea that the universe is flat—that is, of exactly critical density—with matter (mostly dark) making up 27 percent of the total and dark energy making up the rest. Such a universe is spatially flat and will expand forever. For $H_0 = 70$ km/s/Mpc, the age of a critical-density universe without dark energy would be about 9 billion years. This age estimate conflicts with the 10- to 12-billion-year ages of globular clusters derived from studies of stellar evolution. The inclusion of dark energy increases the age of the universe to 14 billion years, consistent with the cluster ages.

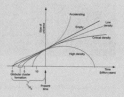
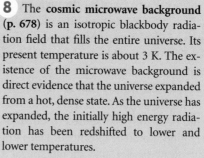

8 The **cosmic microwave background (p. 678)** is an isotropic blackbody radiation field that fills the entire universe. Its present temperature is about 3 K. The existence of the microwave background is direct evidence that the universe expanded from a hot, dense state. As the universe has expanded, the initially high energy radiation has been redshifted to lower and lower temperatures.

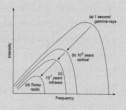

Mastering ASTRONOMY® *For instructor-assigned homework go to www.masteringastronomy.com*

Problems labeled **POS** explore the process of science | **VIS** problems focus on reading and interpreting visual information

REVIEW AND DISCUSSION

1. **POS** What evidence do we have that there is no structure in the universe on very large scales? How large is "very large"?

2. What is the cosmological principle?

3. What is Olbers's paradox? How is it resolved?

4. Explain how an accurate measure of Hubble's constant can lead to an estimate of the age of the universe.

5. We appear to be at the center of the Hubble flow. Why doesn't this observation violate the cosmological principle?

6. Why isn't it correct to say that the expansion of the universe involves galaxies flying outward into empty space?

7. Where did the Big Bang occur?

8. How does the cosmological redshift relate to the expansion of the universe?

9. What properties of the universe determine whether it will or will not expand forever?

10. What will be the ultimate fate of the universe if it does expand forever?

11. Is there enough luminous matter to halt the current cosmic expansion?

12. Is there enough dark matter to halt the current cosmic expansion?

13. What do observations of distant supernovae tell us about the expansion of the universe?

14. What is the cosmological constant, and what does it have to do with the future of the universe?

15. **POS** Why are measurements of globular cluster ages important to cosmology?

16. What is the significance of the cosmic microwave background?

17. Why does the temperature of the microwave background fall as the universe expands?

18. How can we measure Earth's motion with respect to the universe?

19. **POS** Many cultures throughout history have developed their own cosmologies. Do you think the modern scientific cosmology is more likely to endure than any other? Why or why not?

20. **POS** Do you think it constitutes good science to explain the universe mainly in terms of dark matter and dark energy, neither of which is known or understood?

CONCEPTUAL SELF-TEST: MULTIPLE CHOICE

1. If observations made from the middle of a large city are isotropic, then (a) there are tall buildings in every direction; (b) all buildings are exactly the same height; (c) all buildings are the same color; (d) some buildings are taller than others.

2. The cosmological principle would be invalidated if we found that (a) the universe is not expanding; (b) galaxies are older than currently estimated; (c) the number of galaxies per square degree is the same in every direction; (d) the observed structure of the universe depends on the direction in which we look.

3. When we use Hubble's law to estimate the age of the universe, the answer we get (a) depends on which galaxies we choose; (b) is the same for all galaxies; (c) depends on the direction in the sky toward which we are looking; (d) proves that we are at the center of the universe.

4. Olbers's paradox is resolved by (a) the finite size of the universe; (b) the finite age of the universe; (c) light from distant galaxies being redshifted so we can't see it; (d) the fact that there is an edge to the universe.

5. **VIS** The data points in Figure 26.11 ("Accelerating Universe") (a) prove that the universe is not expanding; (b) imply that the expansion is decelerating faster than expected; (c) allow a measurement of Hubble's constant; (d) indicate that the redshifts of distant galaxies are greater than would be expected if gravity alone were acting.

6. The galactic distances used to measure the acceleration of the universe are determined by observations of (a) trigonometric parallax; (b) line broadening; (c) Cepheid variable stars; (d) exploding white dwarfs.

7. The observed acceleration of the universe means that (a) we understand the nature of dark energy; (b) the amount of dark energy is small compared with the luminous mass in galaxies; (c) the amount of dark energy exceeds the total mass–energy of matter in the universe; (d) dark energy has a higher temperature than expected.

8. On the basis of our current best estimate of the present mass density of the universe, astronomers think that (a) the universe is finite in extent and will expand forever; (b) the universe is finite in extent and will eventually collapse; (c) the universe is infinite in extent and will expand forever; (d) the universe is infinite in extent and will eventually collapse.

9. The age of the universe is estimated to be (a) less than Earth's age; (b) the same as the age of the Sun; (c) the same as the age of the Milky Way Galaxy; (d) greater than the age of the Milky Way Galaxy.

10. The cosmic background radiation is observed to come from (a) the center of our Galaxy; (b) the center of the universe; (c) radio antennae in New Jersey; (d) all directions equally.

PROBLEMS

The number of dots preceding each Problem indicates its approximate level of difficulty.

1. • What is the greatest distance at which a galaxy survey sensitive to objects as faint as 20th magnitude could detect a galaxy as bright as the Milky Way (absolute magnitude –20)?

2. •• From Table 24.2, estimate the redshift of the Milky Way at the distance calculated in the previous question.

3. • If the entire universe were filled with Milky Way-like galaxies, with an average density of 0.1 galaxy per cubic megaparsec, calculate the total number of galaxies observable by the survey in Problem 1 if it covered the entire sky.

4. ••• Assuming that the entire universe is uniformly filled with Sun-like stars with a density of 5 billion stars per cubic megaparsec (corresponding to 50 billion stars per galaxy and critical mass density), calculate how far out into space one would have to look, on average, before the line of sight intersects a star.

5. •• Eight galaxies are located at the corners of a cube. The present distance from each galaxy to its nearest neighbor is 10 Mpc, and the entire cube is expanding according to Hubble's law, with $H_0 = 70$ km/s/Mpc. Calculate the recession velocity of one corner of the cube relative to the opposite corner.

6. • According to the Big Bang theory described in this chapter, *without a cosmological constant,* what is the maximum possible age of the universe if $H_0 = 60$ km/s/Mpc? 70 km/s/Mpc? 80 km/s/Mpc?

7. • For a Hubble constant of 70 km/s/Mpc, the critical density is 9×10^{-23} kg/m^3. (a) How much mass does this correspond to within a volume of 1 cubic astronomical unit? (b) How large a cube would be required to enclose 1 Earth mass of material?

8. •• The Virgo Cluster is observed to have a recession velocity of 1200 km/s. For $H_0 = 70$ km/s/Mpc and a critical-density universe, calculate the total mass contained within a sphere centered on Virgo and just enclosing the Milky Way. What is the escape speed from the surface of this sphere?

9. •• (a) What is the present peak wavelength of the cosmic microwave background? Calculate the size of the universe relative to its present size when the radiation background peaked in (b) the infrared, at 10 μm, (c) in the ultraviolet, at 100 nm, and (d) in the gamma-ray region of the spectrum, at 1 nm.

10. ••• Using the information given in Problem 8, calculate the distance from our location to the point that would one day become the center of the Virgo Cluster at the time when the temperature of the microwave background was equal to the present surface temperature of the Sun.

27

THE EARLY UNIVERSE

TOWARD THE BEGINNING OF TIME

THE BIG PICTURE Modern cosmology makes the mind-boggling prediction that the entire observable universe can be traced to microscopic "quantum" fluctuations in the very early universe a fraction of a second after the Big Bang. The large-scale structure of the cosmos is inextricably tied to the smallest scales known to physics.

What were the conditions during the first few seconds of the universe, and how did those conditions change to give rise to the universe we see today? In studying the earliest moments of our universe, we enter a truly alien domain. As we move backward in time toward the Big Bang, our customary landmarks slip away one by one. Atoms vanish, then nuclei, and then even the elementary particles themselves.

In the beginning, the universe consisted of pure energy at unimaginably high temperatures. As it expanded and cooled, the ancient energy gave rise to the particles that make up everything we see around us today. Modern physics has now arrived at the point where it can reach back almost to the instant of the Big Bang itself, allowing scientists to unravel some of the mysteries of our beginnings in time.

LEFT: *Underground on the Swiss-French border near Geneva, technicians are using the world's biggest physics experiment to explore matter more deeply than ever before. The Large Hadron Collider simulates events that occurred within the first second after the origin of the universe by violently smashing together subatomic protons. Scientists are now testing their best theoretical ideas about how the universe began against the data created in this machine. (CERN)*

27.1 Back to the Big Bang

On the very largest scales, the universe is a roughly homogeneous mixture of matter (mostly dark), radiation, and dark energy. ∞ (Sec. 26.5) As we have seen, "matter" includes both *normal matter,* made up of protons, neutrons, and electrons, and *dark matter,* whose composition is still being debated by astronomers. *Dark energy* is the mysterious repulsive force that permeates even the apparent vacuum of intergalactic space. As best we can tell, we live in a geometrically "flat" universe in which the total mass—energy density of *all* the constituents of the cosmos exactly equals the critical value. ∞ (Secs. 26.3, 26.4, 26.6) According to theoretical models, there is not enough matter in the cosmos for the attractive force of gravity to overcome the repulsion of dark energy and reverse the current expansion.

Thus, the future of the cosmos seems clear: The universe is destined to expand forever. In this chapter, we turn our attention to the past. To understand the early universe, just after the Big Bang, we must look more closely at the roles played by matter, radiation, and dark energy in the cosmos. We begin by taking stock of their contributions to the total energy density of the universe.

Cosmic Composition

On the basis of the best available observational data, cosmologists have concluded that, today, just over 70 percent of the total mass–energy of the universe exists in the form of dark energy. ∞ (Sec. 26.6) Virtually all of the remaining 30 percent is accounted for by matter. Thus, at the present moment, *dark energy dominates the density of the universe,* with matter a rather distant second. We can quantify this statement using the results of Chapter 26. For a Hubble constant $H_0 = 70$ km/s/Mpc, the critical density is 9×10^{-27} kg/m³. ∞ (Sec. 26.3) Thus, in round numbers, the density of dark energy in the universe today is just over 6×10^{-27} kg/m³; the current density of matter is slightly less than 3×10^{-27} kg/m³.

Most of the *radiation* in the universe is in the form of the cosmic microwave background—the low-temperature (3 K) radiation field that fills all space. ∞ (Sec. 26.7) Surprisingly, although the microwave background radiation is very weak, it still contains more energy than has been emitted by all the stars and galaxies that have ever existed! The reason is that stars and galaxies, though very intense sources of radiation, occupy only a tiny fraction of space. Averaged out over the volume of the entire universe, their energy falls short of the energy of the microwave background by at least a factor of 10. For our current purposes, then, we can ignore most of the first 26 chapters of this book and regard the cosmic microwave background as the only significant form of radiation in the universe!

Does radiation play an important role in the evolution of the universe on large scales? In order to compare matter and radiation, we must first as usual convert them to a "common currency"—either mass or energy. We will compare their

masses. We can express the energy in the microwave background as an equivalent density by first calculating the number of photons in any meter of space and then converting the total energy of these photons into a mass using the relation $E = mc^2$. ∞ (Sec. 16.6) When we do this, we arrive at an equivalent density for the microwave background of about 5×10^{-31} kg/m³. Thus, *at the present moment,* the densities of both dark energy and matter in the universe far exceed the density of radiation.

Radiation in the Universe

Was the universe always dominated by dark energy? To answer this question, we must ask how the densities of dark energy, matter, and radiation changed as the universe expanded. To this end, cosmologists construct *theoretical models* of the universe, taking into account the effects of Einstein's general relativity and incorporating both the known properties of matter and radiation and the assumed properties of dark energy. ∞ (Sec. 22.6) These models describe how cosmic quantities (such as the densities of the various components) change as the universe evolves. They also make detailed *predictions,* which can be compared directly with observations. ∞ (Sec. 1.2) The outstanding agreement between models and reality (see Section 27.5) is the main reason that astronomers attach so much weight to the measurements of cosmic density, composition, and evolution described in the previous chapter.

As illustrated in Figure 27.1, the models indicate that, as the scale of the universe increases, the densities of matter and radiation both decrease, with the expansion diluting the numbers of atoms and photons alike. But the radiation is also diminished in energy by the cosmological redshift, so its density falls *faster* than that of matter as the universe grows. The dark energy behaves in a very different way. According to theory, it is a large-scale phenomenon, increasing in importance as the universe expands (see Figure 26.12). In fact (at least, if it behaves like Einstein's cosmological constant), the density associated with the dark energy remains *constant* as the universe expands. ∞ *(Discovery 26-1)*

Hence, as we look *back* in time, closer and closer to the Big Bang, the density of the radiation increases faster than that of matter, and both increase faster than that of dark energy. These facts allow us to draw two important conclusions about the composition of the universe in the past.

1. Even though dark energy dominates the density of the universe today, it was *unimportant* at early times, and we can neglect it in our discussion of conditions in the very early universe. Astronomers estimate that the densities of matter and dark energy were equal about 4 billion years ago. Before then, in cosmological parlance, the universe was **matter dominated.**

2. Although the radiation density is currently much less than that of matter, there must have been a time even

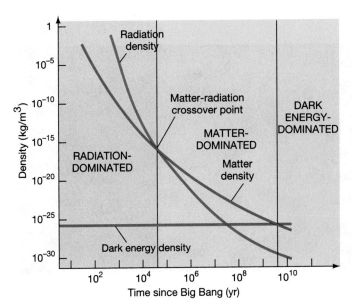

▲ FIGURE 27.1 **Radiation–Matter Dominance** As the universe expanded, not only did the number of both matter particles and photons per unit volume decrease, but the photons were also reduced in energy by the cosmological redshift, reducing their equivalent mass, and hence their density, still further. As a result, the density of radiation fell faster than the density of matter as the universe grew. Thus radiation dominated matter at early times, before the crossover point.

farther in the past when they, too, were equal. Before that time, radiation was the main constituent of the cosmos, which is said to have been **radiation dominated.** The crossover point—the time at which the densities of matter and radiation were equal—occurred about 50,000 years after the Big Bang, when the universe was about 6000 times smaller than it is today. The temperature of the background radiation at that time was about 16,000 K, so it peaked in the near-ultraviolet portion of the spectrum. ⚭ (Sec. 26.7)

Throughout this book, we have been concerned with the history of the universe long after it became dominated by matter and/or dark energy—the formation and evolution of galaxies, stars, and planets as the universe thinned and cooled toward the state we see today. In this chapter, we consider some important events in the early, hot, radiation-dominated universe, long before any star or galaxy existed, that played no less a role in determining the present condition of the cosmos.

Particle Production

The existence of the microwave background implies that the early universe was dominated by an intense radiation field whose temperature fell steadily as the cosmos expanded. The temperatures and densities prevailing at these times were far greater than anything we have encountered thus far, even in the hearts of supernovae. To understand conditions in the

universe shortly after the Big Bang, we must delve more deeply into the behavior of matter and radiation at very high temperatures.

The key to understanding events at very early times lies in a process called **pair production,** in which two photons give rise to a *particle–antiparticle* pair, as shown in Figure 27.2(a) for the particular case of electrons and positrons. Through pair production, matter is created directly from energy in the form of electromagnetic radiation. The reverse process can also occur: A particle and its antiparticle can *annihilate* each other to produce radiation, as depicted in Figure 27.2(b). Energy in the form of radiation can be converted into matter in the form of particles and antiparticles, and particles and antiparticles can be converted back into radiation, subject only to the law of conservation of mass and energy.

The higher the temperature of a radiation field, the greater the energy of the typical constituent photons, and the greater the masses of the particles that can be created by pair production. ⚭ (Secs. 3.4, 4.2, 16.6) For any given particle, the critical temperature above which pair production is possible and below which it is not is called the particle's *threshold temperature.* The threshold temperature increases as the mass of the particle increases. For electrons, it is about 6×10^9 K. For protons, which are nearly 2000 times more massive, it is just over 10^{13} K.

As an example of how pair production affected the composition of the early universe, consider the production of electrons and positrons as the universe expanded and cooled. At high temperatures—above about 10^{10} K—most photons had enough energy to form an electron or a positron, and pair production was commonplace. Space seethed with electrons and positrons, constantly created from the radiation field and annihilating one another to form photons again. Particles and radiation are said to have been in *thermal equilibrium:* New particle–antiparticle pairs were created by pair production at the same rate as they annihilated one another. As the universe expanded and the temperature decreased, so did the average photon energy. By the time the temperature had fallen below 1 billion kelvins, photons no longer had enough energy for pair production to occur, and only radiation remained. Figure 27.3 illustrates how this change took place.

Pair production in the very early universe was directly responsible for all the matter that exists in the universe today. *Everything we see around us was created out of radiation as the cosmos expanded and cooled.* Because we are here to ponder the subject, we know that some matter must have survived those early violent moments. For some reason, there was a slight excess of matter over antimatter at early times—about one extra proton for every billion proton–antiproton pairs. That small residue of particles that outnumbered their antiparticles was left behind as the temperature dropped below the threshold for creating them. With no antiparticles left to annihilate them, the number of particles has remained constant ever since. These survivors are said to have *frozen*

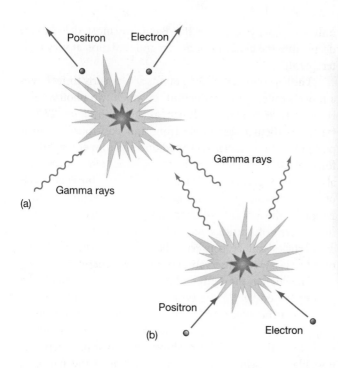

(a)

(b)

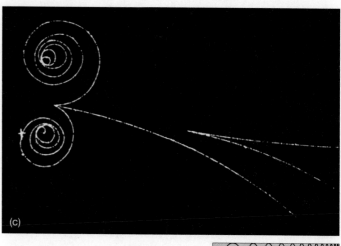

(c)

R I V U X G

◀ **FIGURE 27.2 Pair Production** (a) Two photons can produce a particle–antiparticle pair—in this case an electron and a positron—if their total energy exceeds the mass energy of the particles produced. (b) The reverse process is particle–antiparticle annihilation, in which an electron and a positron destroy each other, vanishing in a flash of gamma rays. (c) A particle detector (an instrument designed to show the tracks of otherwise invisible subatomic particles) allows us to visualize pair production. Here, a gamma ray, whose path is invisible because it is electrically neutral, arrives from the left, dislodging an atomic electron and sending it flying (the longest path). At the same time, the gamma ray provides the energy to produce an electron–positron pair (the spiral paths, which curve in opposite directions in the detector's magnetic field because of their opposite electric charges). The V-shaped tracks at the right are the result of a separate particle interaction. *(Fermi National Laboratory)*

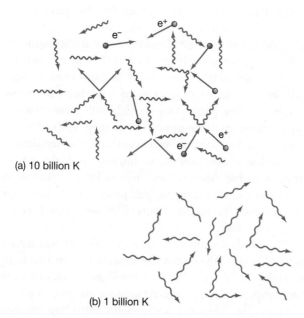

(a) 10 billion K

(b) 1 billion K

▲ **FIGURE 27.3 Thermal Equilibrium** (a) At 10 billion K, most photons have enough energy to create particle–antiparticle (electron–positron) pairs, so these particles exist in great numbers in equilibrium with the radiation. The label e⁻ refers to the electrons, e⁺ to positrons. (b) Below about 1 billion K, photons have too little energy for pair production to occur, so electrons and positrons are no longer in thermal equilibrium with the background radiation field.

out of the radiation field as the universe expanded and cooled.

According to the models, the first hundred or so seconds of the universe's existence saw the creation of all of the basic "building blocks" of matter we know today. Protons and neutrons froze out when the temperature dropped below 10^{13} K, when the universe was only 0.0001 s old; the lighter electrons froze out somewhat later, about a minute or so after the Big Bang, when the temperature fell below 10^9 K. This "matter-creation" phase of the universe's evolution ended when the electrons—the lightest known elementary particles—appeared out of the cooling primordial fireball. From then on, matter has continued to evolve, clumping into more and more complex structures, eventually forming the atoms, planets, stars, galaxies, and large-scale structure we see today, but for all practical purposes no new matter has been created since that early time.

CONCEPT CHECK

✔ What does it mean to say that the early universe was radiation dominated?

27.2 The Evolution of the Universe

For the first few thousand years after the Big Bang, the universe was small, dense, and dominated by radiation. We will refer to this period as the *radiation era.* Some matter existed during this time, but it was a mere contaminant in the blinding gamma-ray light of the primeval Big Bang fireball. Afterward, in the *matter era,* matter came to dominate. Atoms, molecules, and galaxies formed as the universe cooled and thinned toward the state we see today. Today we live in the *dark-energy era,* in which dark energy is becoming an increasingly important component of the cosmos.

Let's begin our study of the early universe by summarizing in broad terms the history of the cosmos, starting at the Big Bang. Figure 27.4 illustrates how the cosmic temperature and density dropped rapidly during the Radiation and Matter Eras and identifies eight significant epochs in the development of the universe. Notice how the time scale on the horizontal axis increases from tiny fractions of a second to thousands of years as we move from left to right—the rate of change of the cosmos slowed dramatically as the universe expanded. We will focus on these epochs in greater detail in the next few sections, but let's not lose sight of the big picture and the place of each epoch in it.

Before the Big Bang?

The Big Bang was a *singularity* in space and time—an instant when the present laws of physics imply that the universe had zero size and infinite temperature and density. As we saw in Chapter 22, where we discussed the singularities at the center of black holes, these predictions should not be taken too literally. ∞ (Sec. 22.7) The presence of singularities signals that, under extreme conditions, the theory making the predictions—in this case, general relativity—has broken down.

At present, no theory exists to let us penetrate the singularity at the start of the universe. We have no means of describing these earliest of times, so we have no way of answering the question "What came *before* the Big Bang?" Indeed, given the laws of physics as we currently know them, the question itself may be meaningless. The Big Bang represented the beginning of the entire universe—mass, energy, space, *and* time came into being at that instant. Without time, the notion of "before" does not exist. Consequently, some cosmologists maintain that asking what happened before the Big Bang is a little like asking what lies north of the North Pole! Others disagree, however, arguing that when the correct theory of quantum gravity—the "Theory of Everything" that unifies gravity and quantum mechanics—is constructed it will remove the singularity and allow us to address the question of what came before.

The Birth of the Cosmos

Although ignorant of the moment of creation itself, theorists nevertheless think that the physical conditions in the universe can be understood in terms of present-day physics back to an extraordinarily short time—a mere 10^{-43} s, in fact—after the Big Bang.

Why can't theorists push our knowledge back to the Big Bang itself? The answer is that we presently have no theory capable of describing the universe at these earliest of times. Under the extreme conditions of density and temperature within 10^{-43} s of the Big Bang, gravity and the other fundamental forces (electromagnetism, the strong force, and the weak force, as described in *More Precisely 27-1*) were indistinguishable from one another—a far cry from the radically different characteristics we see today, listed in Table 27.1. The four forces are said to have been *unified* at that early time—there was, in effect, only one force of nature.

The theory that combines quantum mechanics (the proper description of microscopic phenomena) with general relativity

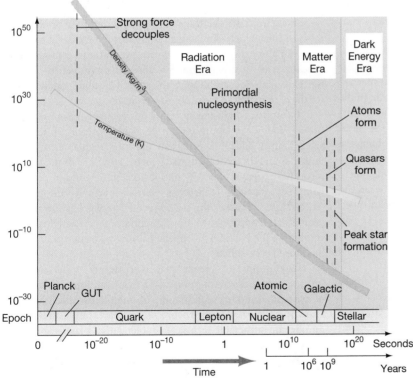

▲ **FIGURE 27.4 Epochs in Cosmic History** The average temperature and average density throughout the history of the universe. At the earliest times, the universe was a sea of radiation with a single unified force of nature. Some subsequent key events in the history of the cosmos as it expanded and cooled are marked; they are discussed later in the chapter.

MORE PRECISELY 27-1

More on Fundamental Forces

In *More Precisely 16-1*, we noted that the behavior of all matter in the universe is ruled by just three *fundamental forces*: gravity, the electroweak force (the unification of the electromagnetic and weak forces), and the strong (nuclear) force. In terrestrial laboratories, these forces display properties that are very different from one another (see Table 27.1). Gravity and electromagnetism are long-range, inverse-square forces, whereas the strong and weak forces have very short ranges—10^{-15} and 10^{-17} m, respectively. Furthermore, the forces do not all affect the same particles. Gravity affects everything. The electromagnetic force affects only charged particles. The strong force operates between nuclear particles, such as protons and neutrons, but it does not affect electrons and neutrinos. The weak force shows up in certain nuclear reactions and radioactive decays. The strong force is 137 times stronger than the electromagnetic force, 100,000 times stronger than the weak force, and 10^{39} times stronger than gravity.

In fact, there is more structure below the level of the nucleus. Protons and neutrons are not truly "elementary" in nature, but are actually made of subparticles called **quarks.** (The name derives from a meaningless word coined by novelist James Joyce in his book *Finnegans Wake.*) According to current theory, there are six distinct types of quark in the universe (with the obscure names *up, down, charm, strange, top,* and *bottom*). What we call the strong nuclear force is actually a manifestation of the interactions that bind quarks to one another.

On the face of it, one might not imagine that there could be any deep underlying connection between forces as dissimilar as those just described, yet there is strong evidence that they are really just different aspects of a single basic phenomenon. In the 1960s, theoretical physicists succeeded in explaining the electromagnetic and weak forces in terms of the electroweak force. Shortly thereafter, the first attempts were made at combining the strong and electroweak forces into a single all-encompassing "superforce." A central idea in the modern version of this superforce is that there is a one-to-one correspondence between the quarks, which interact via the strong force, and particles called **leptons,** which are affected only by the electroweak force. The six known types of quark are paired with six distinct types of lepton: the electron, two related "electronlike" particles (called muons and taus), and three types of neutrino.

Theories that combine the strong and electroweak forces into one are generically known as **Grand Unified Theories,** or GUTs for short. (Note that the term is plural—no one GUT has yet been proven to be "the" correct description of nature.) One general prediction of GUTs is that the three nongravitational forces are indistinguishable from one another only at enormously high energies, corresponding to temperatures in excess of 10^{28} K. Below that temperature, the superforce splits into two, displaying its separate strong and electroweak aspects. In particle-physics parlance, we say that there is a *symmetry* between the strong and the electroweak forces that is broken at temperatures below 10^{28} K, allowing the separate characters of the two forces to become apparent. At "low" temperatures—less than about 10^{15} K, a range that includes almost everything we know on Earth and in the stars—there is a second symmetry breaking, and the electroweak force splits to reveal its more familiar electromagnetic and weak natures.

The key predictions of the electroweak theory were experimentally verified in the 1970s, winning the theory's originators (Sheldon Glashow, Steven Weinberg, and Abdus Salam) the 1979 Nobel Prize in physics. The GUTs have not yet been experimentally verified (or refuted), in large part because of the extremely high energies that must be reached in order to observe their predictions.

An important idea that has arisen from the realization that the strong and the electroweak forces can be unified is the notion of **supersymmetry,** which extends the idea of symmetry between fundamental forces to place all particles—those that are acted on by forces (such as protons and electrons) and those that transmit those forces (such as photons and gluons; see Section 27.4)—on an equal footing. One particularly important prediction of supersymmetry is that all particles should have so-called *supersymmetric partners*—extra particles that must exist in order for the theory to remain self-consistent. None of these new particles has yet been detected, yet many physicists are convinced of the theory's essential correctness.

These new particles, if they exist, would have been produced in abundance in the Big Bang and should still be around today. They are also expected to be very massive—at least a thousand times heavier than a proton. So-called supersymmetric relics, the new particles, are among the current leading candidates for the dark matter in the universe (see Section 27.5).

Efforts to include gravity within this picture have so far been unsuccessful. Gravitation has not yet been incorporated into a single "Super GUT," in which all the fundamental forces are united. Some theoretical efforts to merge gravity with the other forces have tried to fit gravity into the quantum world by postulating extra particles—called *gravitons*—that transmit the gravitational force. However, this is a different view of gravity from the geometric picture embodied in Einstein's general relativity, and combining the two into a consistent theory of quantum gravity has proved very difficult.

One promising theory that is currently under active investigation seeks to interpret all particles and forces in terms of particular modes of vibration of submicroscopic objects known as *strings*. **String theory** is complex, but it solves a number of intractable technical problems that plagued previous efforts, and many theorists feel that it currently offers the greatest promise of unifying the forces of nature. Promise aside, though, realize that at present, *no* theory has yet succeeded in making any definite statement about conditions in the very early universe. A complete theory of quantum gravity continues to elude researchers.

TABLE 27.1 Fundamental Forces and Particles

Force	Range (m)	Particles Affected	Unification (temperature)		
strong	10^{-15}	matter composed of quarks (protons, neutrons, etc.)		GUT/superforce (10^{28} K)	quantum gravity (10^{32} K)
electromagnetic	infinite	charged particles (protons, electrons, etc.)	electroweak (10^{15} K)		
weak	10^{-17}	leptons (electrons, muons, taus, neutrinos)			
gravity	infinite	everything			

(which describes the universe on the largest scales) is generically known as quantum gravity. ∞ (Sec. 22.7) The period from the beginning to 10^{-43} s is often referred to as the *Planck epoch*, after Max Planck, one of the creators of quantum mechanics. Unfortunately, for now at least, there is no working theory of quantum gravity, so we simply cannot talk meaningfully about the universe during the Planck epoch.

By the end of the Planck epoch, the temperature was around 10^{32} K, and the universe was filled with radiation and a vast array of subatomic particles created by the mechanism of pair production. At around that time, gravity parted company with the other forces of nature—it became distinguishable from them and has remained so ever since. The strong, weak, and electromagnetic forces were still unified. The present-day theories that describe this epoch are collectively known as **Grand Unified Theories,** or GUTs for short (see *More Precisely 27-1*). Accordingly, we refer to this period as the *GUT epoch.*

Freeze-Out

Grand Unified Theories predict that three of the four basic forces of nature—electromagnetism and the strong and weak nuclear forces—are in reality aspects of a single, all-encompassing "superforce." However, this unification is evident only at enormously high energies, corresponding to temperatures in excess of 10^{28} K. At lower temperatures, the superforce reveals its separate electromagnetic, strong, and weak characters.

A fundamental concept in quantum physics is the idea that forces between elementary particles are exerted, or *mediated*, by the exchange of another type of particle, generically called a **boson.** We might imagine the two particles as playing a rapid game of catch, using a boson as a ball, as illustrated in Figure 27.5. As the ball is thrown back and forth, the force is transmitted. For example, in ordinary electromagnetism, the boson involved is the *photon*—a bundle of electromagnetic energy that always travels at the speed of light. The strong force is mediated by particles known as *gluons*. The electroweak theory includes a total of four bosons—the massless photon and three other massive particles, called (for historical reasons) W^{+}, W^{-}, and

Z^{0}, all of which have been observed in laboratory experiments. Gravity is (theoretically) mediated by *gravitons,* and so on. All of the particles we have encountered so far in this book—electrons, protons, neutrons, neutrinos—play "catch" with at least some of these "balls."

In Section 27.1 we saw how particles "froze out" of the universe as its temperature dropped below the threshold temperature for their creation by pair production. Now that we know that the basic forces of nature are mediated by particles, we can understand—in general terms, at least—how the fundamental forces froze out, too, as the universe cooled. According to the GUTs, the particle that unifies the strong and electroweak forces is extremely massive—at least 10^{15} times the mass of the proton (and possibly much more). It is because this particle is so massive that the unification of the strong and electroweak forces becomes evident only at extremely high temperatures.

At temperatures below 10^{28} K, the strong nuclear force becomes distinguishable from the electroweak force (the unified weak and electromagnetic forces). Once the universe had cooled to that temperature, about 10^{-35} s after the Big Bang, the GUT epoch ended. According to many GUTs, one important legacy of that epoch may have been the appearance and subsequent freeze-out of a veritable "zoo" of very massive (and as yet unobserved) elementary particles that interact only

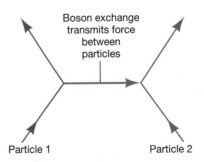

▲ **FIGURE 27.5 Fundamental Forces** Forces between elementary particles are transmitted through the exchange of other particles called bosons. As two particles interact, they exchange bosons, a little like playing catch with a submicroscopic ball.

very weakly with normal matter. These "exotic" particles are prime candidates for the dark matter of unknown composition thought to exist in abundance both within galaxies and in the unseen depths of intergalactic space. ∞ (Secs. 23.6, 25.1)

Quarks and Leptons

Our next major subdivision of the Radiation Era covers the period when all "heavy" elementary particles—that is, all the way down in mass to protons, neutrons, and their constituent quarks—were in thermal equilibrium with the radiation. We refer to this period as the *quark epoch*, since quarks are the fundamental components of all particles that interact via the strong force.

The universe continued to expand and cool. At a temperature of about 10^{15} K (10^{-10} s after the Big Bang), the weak and the electromagnetic components of the electroweak force began to display their separate characters. The W and Z particles responsible for the electroweak force have masses about 100 times the mass of a proton. The threshold temperature for their production—roughly 10^{15} K—marks the point at which the weak and electromagnetic forces parted company.

By about 0.1 millisecond (10^{-4} s) after the Big Bang, the temperature had dropped well below the 10^{13} K threshold for the creation of protons and neutrons (the lightest stable particles composed of quarks), and the quark epoch ended. The main constituents of the universe were now lightweight particles—muons (see *More Precisely 27-1*), electrons, neutrinos, and their antiparticles—all still in thermal equilibrium with the radiation. Compared with the numbers of these lighter particles, only very few protons and neutrons remained at this stage, because most had been annihilated.

Electrons, muons, and neutrinos are collectively known as *leptons,* after the Greek word meaning "light" (i.e., not heavy). Accordingly, we refer to this period in the history of the universe as the *lepton epoch.* During that epoch, at a temperature of about 3×10^{10} K—approximately 1 second after the Big Bang—the rapidly thinning universe became transparent to neutrinos, and these ghostly particles have been streaming freely through space ever since. (Most neutrinos have not interacted with any other particle since the universe was a few seconds old!) The lepton epoch ended when the universe was about 100 s old and the temperature fell to about 1 billion K—too low for electron–positron pair production to occur. The density of the universe by this time was about 10 times the density of water.

The final significant event in the Radiation Era occurred when protons and neutrons began to fuse into heavier nuclei. At the start of this period, which we will call the *nuclear epoch,* the temperature was a few hundred million kelvins, and fusion occurred very rapidly, forming deuterium ("heavy" hydrogen—see Section 16.6) and helium in quick succession before conditions became too cool for further reactions to occur. By the time the universe was about 15 minutes old, much of the helium we observe today had been formed.

The Matter and Dark-Energy Eras

Time passed, the universe continued to expand and cool, and radiation gave way to matter as the dominant constituent of the universe. Our next major epoch extends in time from 50,000 years (the end of the Radiation Era) to about 100 million years after the Big Bang. As the primeval fireball diminished in intensity, a crucial change occurred—perhaps the most important change in the history of the universe. At the end of the nuclear epoch, radiation still overwhelmed matter. As fast as protons and electrons combined, radiation broke them apart again, preventing the formation of even simple atoms or molecules. However, as the universe expanded and cooled, the early dominance of radiation eventually ended. Once formed, atoms remained intact. We will call this period the *atomic epoch*. It ended about 200 million years after the Big Bang, when the first stars formed and their intense radiation *reionized* the universe.

The last two epochs together bring us to the current age of the universe. During these late stages, change happened at a much more sedate pace. By the time the universe was about 3 billion years old, large-scale structure and most galaxies had formed. For the first time, the universe departed from homogeneity on macroscopic scales. The largely uniform universe of the Radiation Era became a universe containing large agglomerations of matter. We call the period from 200 million to 3 billion years after the Big Bang the *galactic epoch,* given that the main events at the time concerned galaxy construction. At its end, large-scale structure and the bulk of most galaxies had formed, quasars were shining brightly, and early generations of stars were burning and exploding, helping to determine the future shape of their parent galaxies.

Since then, galaxies have continued to merge and evolve, stars peaked their formation rate, and planets and life appeared in the universe. These last two epochs, including the current *stellar epoch*—so named for the myriad stars that are still forming within galaxies—have been the subject of the first 25 chapters of this book.

CONCEPT CHECK
✔ Why did lighter and lighter particles "freeze out" of the universe as the cosmos expanded?

27.3 The Formation of Nuclei and Atoms

We now have all the ingredients needed to complete our story of the creation of the elements, begun in Chapter 21, but never quite finished. ∞ (Sec. 21.4) The theory of stellar nucleosynthesis accounts very well for the observed abundances of heavy elements in the universe, but there are discrepancies between theory and observation when it comes to the abundances of the light elements, especially helium. Simply put, the total amount of helium in the universe today—about 25 percent by mass—is far too large to be explained by nuclear fusion in stars. The accepted explanation is that this base level of helium is *primordial*—that is, it was created during the early, hot epochs of the universe, before any stars had formed. The production of elements heavier than hydrogen by nuclear fusion shortly after the Big Bang is called **primordial nucleosynthesis**.

Helium Formation in the Early Universe

By about 100 s after the Big Bang, the temperature had fallen to about 1 billion K, and apart from "exotic" dark-matter particles, matter in the universe consisted of electrons, protons, and neutrons, with the protons outnumbering the neutrons by about five to one. The stage was set for nuclear fusion to occur. Protons and neutrons combined to produce deuterium nuclei (also called *deuterons*), containing 1 proton and 1 neutron:

$$^1\text{H (proton)} + \text{neutron} \rightarrow {}^2\text{H (deuteron)} + \text{energy}.$$

Although this reaction must have occurred frequently during the lepton epoch, the temperature then was still so high that the deuterium nuclei were broken apart by high-energy gamma rays as soon as they formed. The universe had to wait until it became cool enough for the deuterium to survive. This waiting period is sometimes called the *deuterium bottleneck*.

Only when the temperature of the universe fell below about 900 million K, roughly 2 minutes after the Big Bang, was deuterium at last able to form and endure. Once that

occurred, the deuterium was quickly converted into heavier elements by numerous reactions, including:

$$^2\text{H} + {}^1\text{H} \rightarrow {}^3\text{He} + \text{energy},$$

$$^2\text{H} + {}^2\text{H} \rightarrow {}^3\text{He} + \text{neutron} + \text{energy},$$

$$^3\text{He} + \text{neutron} \rightarrow {}^4\text{He} + \text{energy}.$$

The result was that, once the universe passed the deuterium bottleneck, fusion proceeded rapidly and large amounts of helium were formed. In just a few minutes most of the free neutrons were consumed, leaving a universe whose matter content was primarily hydrogen and helium. Figure 27.6 illustrates some of the reactions responsible for helium formation. Contrast it with Figure 16.27, which depicts how helium is formed today in the cores of main-sequence stars such as the Sun.* ∞ (Sec. 16.6)

We might imagine that fusion could have continued to create heavier and heavier elements, just as it does in the cores of stars, but that did not occur. In stars, the density and the temperature both *increase* slowly with time, allowing

The proton–proton chain that powers the Sun played no significant role in primordial helium formation. The proton–proton reaction that starts the chain is very slow compared with the proton–neutron discussed here and is important in the Sun only because the solar interior contains no free neutrons to make the latter reaction possible.

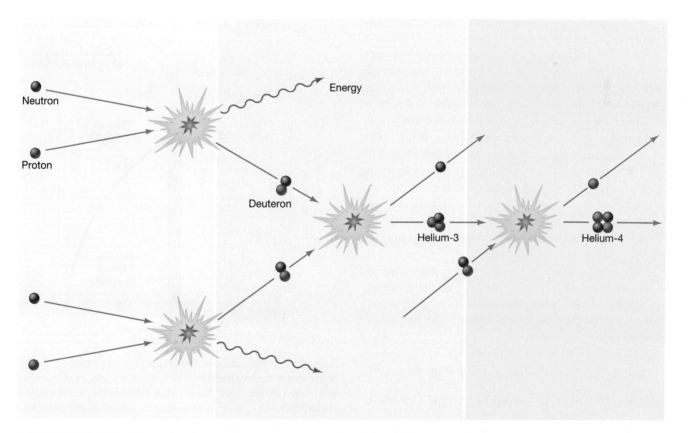

▲ **FIGURE 27.6 Helium Formation** Some of the reaction sequences that led to the formation of helium in the early universe. (A deuteron is a nucleus of deuterium, the heavy form of hydrogen.) Compare this figure with Figure 16.27, which depicts the proton–proton chain in the Sun.

more and more massive nuclei to form, but in the early universe exactly the opposite was true. The temperature and density were both *decreasing* rapidly, making conditions less and less favorable for fusion as time went on. Even before the supply of neutrons was completely used up, the nuclear reactions had effectively ceased. Reactions between helium nuclei and protons may also have formed trace amounts of lithium (the next element beyond helium) by this time, but for all practical purposes, the expansion of the universe caused fusion to stop at helium. The brief epoch of primordial nucleosynthesis was over about 15 minutes after it began.

By the end of the period of nucleosynthesis, some 1000 s after the Big Bang, the temperature of the universe was about 300 million K and the cosmic elemental abundances were set. Careful calculations indicate that about one helium nucleus had formed for every 12 protons remaining. Because a helium nucleus is four times more massive than a proton, helium accounted for about one-quarter of the total mass of matter in the universe:

$$\frac{1 \text{ helium nucleus}}{12 \text{ protons} + 1 \text{ helium nucleus}} = \frac{4 \text{ mass units}}{12 \text{ mass units} + 4 \text{ mass units}}$$
$$= \frac{4}{16} = \frac{1}{4}.$$

The remaining 75 percent of the matter in the universe was hydrogen. It would be almost a billion years before nucleosynthesis in stars would change these numbers. ∞ (Sec. 21.4)

The foregoing calculation implies that all stars and galaxies should contain *at least 25 percent* helium by mass. The figure for the Sun, for example, is about 28 percent. However, it is difficult to disentangle the contributions to the present-day helium abundance from primordial nucleosynthesis and later hydrogen burning in stars. Our best hope of determining the amount of primordial helium is to study the oldest stars known, since they formed early on, before stellar nucleosynthesis had had time to change the helium content of the universe significantly. Unfortunately, stars surviving from that early time are of low mass and hence quite cool, making the helium lines in their spectra very weak and hard to measure accurately. ∞ (Secs. 17.5, 17.8) Nevertheless, despite this uncertainty, the observations are generally consistent with the theory just described.

Bear in mind that while all this was going on, matter was just an insignificant "contaminant" in the radiation-dominated universe. Radiation outmassed matter by about a factor of 5000 at the time helium formed. The existence of helium is very important in determining the structure and appearance of stars today, but its creation was completely irrelevant to the evolution of the universe at the time.

Deuterium and the Density of the Cosmos

During the nuclear epoch, although most deuterium was quickly fused into helium as soon as it formed, a small amount was left over when the primordial nuclear reactions

ceased. Observations of deuterium—especially those made by orbiting satellites able to capture deuterium's strongest spectral feature, which happens to be emitted in the ultraviolet part of the spectrum—indicate a present-day abundance of about two deuterium nuclei for every 100,000 protons. However, unlike helium, deuterium is not produced to any significant degree in stars (in fact, deuterium tends to be destroyed in stars), so any deuterium we see today *must* be primordial.

This observation is of great importance to astronomers because it provides them with a sensitive method—and one that is completely independent of the techniques discussed in previous chapters—of probing the present-day density of matter in the universe. According to theory, as illustrated in Figure 27.7, the denser the universe is today, the more particles there were at early times to react with deuterium as it formed and the less deuterium was left over when nucleosynthesis ended. A comparison of the observed deuterium abundance (marked on the figure) with the theoretical results implies a present-day density of *at most* 3×10^{-28} kg/m^3—only a few percent of the critical density.

But before we jump to any far-reaching cosmic conclusions based on this number, we must make a very important qualification. As just described, primordial nucleosynthesis depends *only* on the presence of protons and neutrons in the early universe. Thus, measurements of the abundance of helium and deuterium tell us only about the density of "normal" matter—matter made up of protons and neutrons—in the

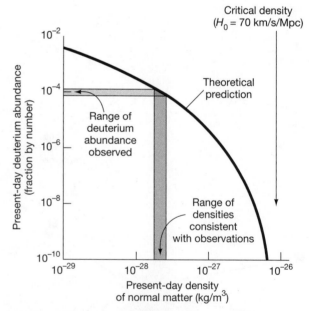

▲ **FIGURE 27.7 Deuterium Abundance** The present-day abundance of deuterium depends strongly on the amount of matter present at early times, and this, in turn, determines the present-day density of the universe. Thus, measuring the amount of deuterium in the universe gives us an estimate of the overall density of matter. The best deuterium measurements are marked; they imply that the density of matter in the universe is at most a few percent of the critical value.

cosmos. This finding has a momentous implication for the overall composition of the universe. As we saw earlier, astronomers have concluded, for a variety of reasons, that the total density of matter is about the critical value. ∞ (Sec. 26.5) In that case, if the density of normal matter is only a few percent of the critical value, then we are forced to admit that not only is most of the matter in the universe dark, but most of the dark matter is *not* composed of protons and neutrons.

We will see in Section 27.5 that, because normal matter and dark matter interact differently with the background radiation field, studies of the cosmic microwave background allow us to distinguish between these two types of matter. Observations made by the *WMAP* satellite have found the density of normal matter to be just 4 percent of the critical density, in excellent agreement with estimates based on the abundance of deuterium.

Thus, the bulk (about 90 percent) of the matter in the universe apparently exists in the form of elusive subatomic particles (for example, the WIMPs discussed as dark-matter candidates in Chapter 23) whose nature we do not fully understand and whose very existence has yet to be conclusively demonstrated in laboratory experiments. ∞ (Sec. 23.6) For the sake of brevity, from here on we will adopt the convention that the term "dark matter" refers only to these unknown particles and not to "stellar" dark matter, such as black holes and brown and white dwarfs (also discussed in Chapter 23), which are made of relatively well-understood normal matter.

The First Atoms

A few tens of thousands of years after the Big Bang, radiation ceased to be the dominant component of the universe. The Matter Era had begun. At the start of the atomic epoch, matter consisted of electrons, protons, helium nuclei (formed by primordial nucleosynthesis), and dark matter. The temperature was several tens of thousands of kelvins—far too hot for atoms of hydrogen to exist (although some helium ions may already have formed). During the next few hundred thousand years, a major change occurred: The universe expanded by another factor of 10, the temperature dropped to a few thousand kelvins, and electrons and nuclei combined to form neutral atoms. By the time the temperature had fallen to about 3000 K, the universe consisted of atoms, photons, and dark matter.

The period during which nuclei and electrons combined to form atoms is called the epoch of **decoupling,** for it was during this period that the radiation background parted company with normal matter. Many astronomers also refer to this period as **recombination** (although, technically speaking, protons and electrons had never previously been combined in the form of atoms).

At early times, when matter was ionized, the universe was filled with large numbers of free electrons that interacted frequently with electromagnetic radiation of all wavelengths.

As a result, a photon could not travel far before encountering an electron and scattering off it. In effect, the universe was opaque to radiation. Matter and radiation were strongly "tied," or *coupled,* to one another by these interactions. After the electrons combined with nuclei to form atoms of hydrogen and helium, only certain wavelengths of radiation—the ones corresponding to the spectral lines of those atoms—could interact with matter. ∞ (Sec. 4.2) Radiation of other wavelengths could travel virtually forever without being absorbed. Thus, the universe became nearly transparent. From that time on, most photons passed generally unhindered through space. As the universe expanded, the radiation simply cooled, eventually becoming the microwave background we see today.

The microwave photons now detected on Earth have been traveling through the universe ever since they decoupled. According to the models that best fit the observational data, the last interaction these photons had with matter (at the epoch of decoupling) occurred when the universe was about 400,000 years old and roughly 1100 times smaller (and hotter) than it is today—that is, at a redshift of 1100. As illustrated in Figure 27.8, the epoch of atom formation created a kind of "photosphere" in the universe, completely surrounding Earth at a distance of approximately 14,000 Mpc,

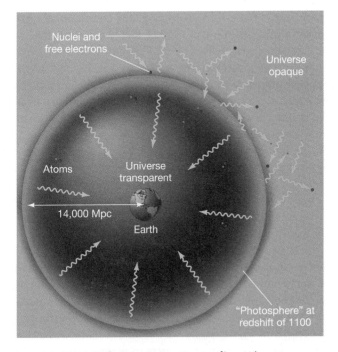

▲ FIGURE 27.8 **Radiation–Matter Decoupling** When atoms formed, the universe became virtually transparent to radiation. Thus, observations of the cosmic background radiation allow us to study conditions in the universe around a time when the redshift was 1100 and the temperature had dropped below about 3000 K. For an explanation of how we can see a region of space 14,000 Mpc (46 billion light-years) away when the universe is just 14 billion years old, see *More Precisely 24-1.*

the distance at which photons last interacted before they decoupled. ∞ *(More Precisely 24-1)*

On our side of the photosphere—that is, since decoupling—the universe is transparent. On the far side—before decoupling—it was opaque. Thus, by observing the microwave background, we are probing conditions in the universe almost all the way back in time to the Big Bang, in much the same way as studying sunlight tells us about the surface layers of the Sun.

PROCESS OF SCIENCE CHECK

✔ How do we know that most of the dark matter in the universe is not of "normal" composition?

27.4 The Inflationary Universe

In the late 1970s, cosmologists trying to piece together the evolution of the universe were confronted with two nagging problems that had no easy explanation within the standard Big Bang model. The resolution of these problems has caused cosmologists to completely rethink their views of the very early universe.

The Horizon and Flatness Problems

The first problem is known as the **horizon problem,** and it concerns the remarkable *isotropy* of the cosmic microwave background. ∞ (Sec. 26.7) Recall that the temperature of this radiation is virtually constant, at about 2.7 K, in all directions. Imagine observing the microwave background in two opposite directions of the sky, as illustrated in Figure 27.9. As we have just seen, that radiation last interacted with matter in the universe at around a redshift of 1100. Thus, in observing these two distant regions of the universe, marked A and B on the figure, we are studying regions that were separated by several million parsecs at the time they emitted this radiation. The fact that the background radiation is isotropic to high accuracy means that regions A and B had similar densities and temperatures at the time the radiation we see left them.

The problem is, according to the Big Bang theory as just described, there is *no* good reason why these regions should in fact be similar to each other. To take an everyday example, we all know that heat flows from regions of high temperature to regions of low temperature, but it takes time for this to occur. If we light a fire in one corner of a room, we have to wait a while for the other corners to warm up. Eventually, the room reaches a more or less uniform temperature, but only after the heat from the fire—or, more generally, the *information* that the fire is there—has had time to spread.

Similar reasoning applies to regions A and B in Figure 27.9. These regions are separated by many megaparsecs, and there has not been enough time for information, which can go no faster than the speed of light, to travel from one to the

other. In cosmological parlance, the two regions are said to be outside each other's *horizon.* But if that is so, then how do they "know" that they are supposed to look the same? With no possibility of communication between them, the only alternative is that regions A and B simply started off looking alike— an assumption that cosmologists are reluctant to make.

The second problem with the standard Big Bang model is called the **flatness problem.** Whatever the exact value of Ω_0, it appears to be very close to unity—the total density of the universe is fairly near the critical value. In terms of spacetime curvature, we can assert that the universe is remarkably close to being flat. ∞ (Sec. 26.4) We say "remarkably" here because, again, there is no particular reason that the universe should have formed with a density very close to the critical value. Why not a millionth of, or a million times, that value? Furthermore, as illustrated in Figure 27.10, a universe that starts off close to, but not exactly on, the critical curve soon deviates greatly from it, so if the universe is close to critical now, it must have been *extremely* close to critical in the past. (The acceleration due to dark energy *does* in fact tend to push the universe toward critical density, but dark energy has not dominated the expansion for long enough for this fact to change our basic conclusion.) For example, if $\Omega_0 = 0.3$ today (approximately the density of "known" dark matter), then the departure from critical density at the time of nucleosynthesis would have been only 1 part in 10^{15} (a thousand trillion)!

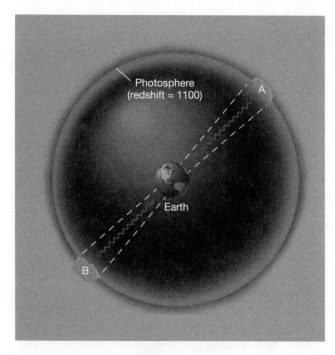

▲ **FIGURE 27.9 Horizon Problem** The isotropy of the microwave background indicates that regions A and B in the universe were very similar to each other when the radiation we observe left them, but there has not been enough time since the Big Bang for them ever to have interacted physically with one another. Why then should they look the same?

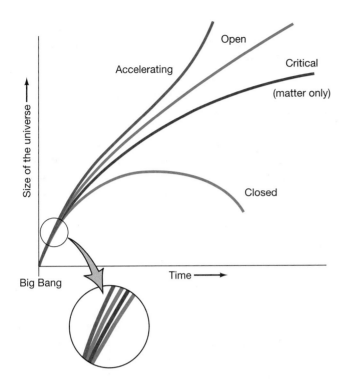

▲ **FIGURE 27.10 Flatness Problem** If the universe deviates even slightly from critical density, that deviation grows rapidly in time. For the universe to be as close to critical as it is today, it must have differed from the critical density in the past by only a tiny amount.

These observations constitute "problems" because cosmologists want to be able to *explain* the present condition of the universe, not just accept it "as is." They would prefer to resolve the horizon and flatness problems in terms of physical processes that could have taken a universe with no special properties and caused it to evolve into the cosmos we now see. The resolution of both problems takes us back in time even earlier than nucleosynthesis or the formation of any of the elementary particles we know today—back, in fact, almost to the instant of the Big Bang itself.

Cosmic Inflation

As we saw in Section 27.2, at very early times, during the GUT and Planck epochs, most or all of the fundamental forces of nature were unified—that is, indistinguishable from one another. The various theories that describe this unification (for example, those outlined in *More Precisely 27-1*) predict—and in fact, rely upon—the existence of certain quantum-mechanical fields, generically called **scalar fields** in particle physics jargon, whose interactions with particles in the theory determine those particles' properties. For our purposes, we can think of these fields as cosmic forces permeating all space, separate from, but closely related to, physical particles in the universe. These fields define the differences between the various forces of nature, and ultimately set the scale on which unification occurs.

What does all this have to do with cosmology? In the early 1980s, physicists realized that it was possible for these scalar fields to become temporarily *increased* in energy above their normal equilibrium states. Due to random fluctuations at the quantum level, regions of the universe could find themselves in this "elevated" state for some period of time. Under these circumstances, theory indicates that these parts of the universe would find themselves in a very odd and unstable condition—empty space would have acquired **vacuum energy.** Abstract as this may seem, these regions are of direct interest to us—if theorists are correct, we live in one!

The temporary appearance of vacuum energy within such a region—ours, say—had dramatic consequences. For a short while, as illustrated in Figure 27.11, the extra energy caused the region to expand at an enormously accelerated rate. The vacuum energy density remained almost constant as the region grew, and the expansion *accelerated* with time while this condition persisted. In fact, the size of the region doubled many times over. For definiteness, Figure 27.11 shows the expansion occurring near the end of the GUT epoch, with a doubling time of roughly 10^{-34} s. This period of unchecked cosmic expansion is known as the **epoch of inflation.**

Actually, bizarre though this may seem, we have already seen a universal expansion along these lines, albeit at a much more leisurely pace. The leading models for dark energy (the cosmological constant and quintessence) are both scalar fields. Their nonzero vacuum energy is responsible for the cosmic acceleration discussed in Chapter 26. ⚬ (Sec. 26.5)

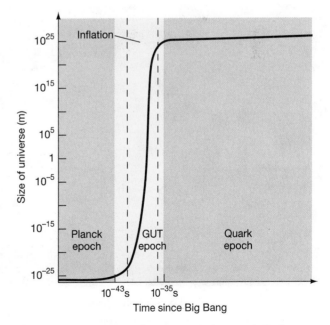

▲ **FIGURE 27.11 Cosmic Inflation** During the period of inflation at the end of the GUT epoch, the universe expanded enormously in a very short time. Afterward, it resumed its earlier "normal" expansion rate, except that the size of the cosmos had become about 10^{50} times bigger than it was before inflation, as submicroscopic scales had expanded to "cosmic" sizes.

Eventually, the scalar field returned to its equilibrium state, the region recovered its normal vacuum, and inflation stopped. For the example shown in Figure 27.11, the whole episode lasted a mere 10^{-32} s, but during that time the patch of the universe that had become unstable swelled in size by the incredible factor of about 10^{50}. After the inflationary phase, the universe once again resumed its (relatively) leisurely expansion. However, a number of important changes had occurred that would have far-reaching ramifications for the evolution of the cosmos.

The original theory of inflation was developed in the early 1980s and associated the inflationary period (as in Figure 27.11) with the end of the GUT epoch. The scalar field in that case was the one responsible for distinguishing between the strong and electroweak forces. However, since that time, researchers have realized that conditions suitable for inflation could have occurred under many different circumstances—and possibly *many times*—during the evolution of the early universe. This generalization actually strengthens inflation as a theory by loosening the restrictions on when it might have happened, although it blurs the question of exactly when the inflationary epoch(s) leading to "our" universe actually occurred.

Nevertheless, the basic idea of a *quantum fluctuation* expanding to become the universe we know is now quite well established. Some theorists have gone so far as to suggest that a quantum fluctuation during the Planck epoch may have been the trigger that caused the Big Bang. Others even speculate that we might be living in a sort of "self-creating universe" that erupted into existence spontaneously from inflation in one such random fluctuation! This sort of "statistical" creation of the primal cosmic energy from absolutely nothing has been dubbed "the ultimate free lunch."

Note that it is possible—even likely, according to many theorists—that not all of the universe underwent inflation. Only some regions became unstable, causing huge inflated "bubbles" to appear in the cosmos. We apparently live in one such bubble; the universe outside is probably unknowable to us. Henceforth, we will use the term "universe" to refer to just this bubble and its contents.

Implications for the Universe

The inflationary epoch provides a natural solution to the horizon and flatness problems. The horizon problem is solved because inflation took regions of the universe that had already had time to communicate with one another—and so had established similar physical properties—and then dragged them far apart, well out of communication range of one another. For example (again using the assumptions in Figure 27.11), regions A and B in Figure 27.9 have been out of contact since 10^{-32} s after creation, but they were in contact before then. As illustrated in Figure 27.12, their properties are the same today

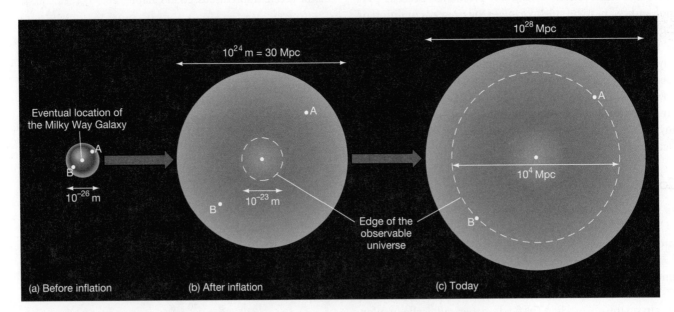

▲ FIGURE 27.12 **Inflation and the Horizon Problem** Inflation solves the horizon problem by taking a small region of the very early universe—whose parts had already had time to interact with one another and that had thus already become homogeneous—and expanding it to enormous size. In (a), points A and B are well within the (shaded) homogeneous region of the universe centered on the eventual site of the Milky Way Galaxy. In (b), after inflation, A and B are far outside the horizon (indicated by the dashed line), so they are no longer visible from our location. Subsequently, the horizon expands faster than the universe as a whole does, so that today (c) A and B are just reentering our field of view. They have similar properties now because they had similar properties before the inflationary epoch.

because they were the same long ago, before inflation separated them.

Figure 27.12(a) shows a small piece of the universe just before the onset of inflation. The point that will one day become the site of the Milky Way Galaxy is at the center of the shaded region, which represents the portion of space "visible" to that point at that time—that is, there has been enough time since the Big Bang for light to have traveled from the edge of this region to its center. That entire region is more or less homogeneous, because different parts of it have been able to interact with one another, so any initial differences between the parts have largely been smoothed out. The points A and B of Figure 27.9 are also marked. They lie within the homogeneous patch, so they have very similar properties. The actual size of the shaded region is about 10^{-26} m—only a trillionth the size of a proton.

Immediately after inflation, as shown in Figure 27.12(b), the homogeneous region has expanded by 50 orders of magnitude, to a diameter of about 10^{24} m, or 30 Mpc—larger than the largest supercluster. By contrast, the visible portion of the universe, indicated by the dashed line, has grown only by a factor of a thousand and is still microscopic in size. In effect, the universe expanded much faster than the speed of light during the inflationary epoch, so what was once well within the horizon of the point that is to become the site of our Galaxy now lies far beyond it. In particular, points A and B are no longer visible, either to us or to each other, at this time. (Note that, while the theory of relativity restricts matter and energy to speeds less than the speed of light, it imposes no such limit on the universe as a whole.)

Since the end of inflation, the universe has expanded by a further factor of 10^{27}, so the size of the homogeneous region of space surrounding us is now about 10^{51} m (10^{28} Mpc)—10 trillion trillion times greater than the distance to the most distant quasar. As shown in Figure 27.12(c), the horizon has expanded faster than the universe, so points A and B are just now becoming visible again. As the portion of the universe that is now observable from Earth grows in time, it remains homogeneous because our cosmic field of view is simply reexpanding into a region of the universe that was within our horizon long ago. We will have to wait a very long time—at least 10^{35} years—before the edge of the homogeneous patch surrounding us comes back into view.

To see how inflation solves the flatness problem, let's return to our earlier balloon analogy. ∞ (Sec. 26.2) Imagine that you are a 1-mm-long ant sitting on the surface of the balloon as it expands, as illustrated in Figure 27.13. When the balloon is just a few centimeters across, you can easily perceive the surface to be curved—its circumference is only a few times your own size. When the balloon expands to, say, a few meters in diameter, the curvature of the surface is less pronounced, but perhaps still perceptible. However, by the time the balloon has expanded to a few kilometers across, an "ant-sized" patch of the surface will look quite flat, just as the surface of Earth looks flat to us.

Now imagine that the balloon expands 100 trillion trillion trillion trillion trillion times, as the universe did during the period of inflation. Your local patch of the surface is now completely indistinguishable from a perfectly flat plane, deviating from flatness by no more than one part in 10^{50}. Exactly the same argument applies to the universe: Because it has expanded so much, for all practical purposes the universe is perfectly flat on all scales we can ever hope to observe.

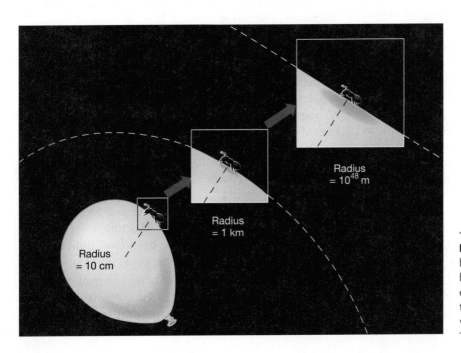

◀ FIGURE 27.13 **Inflation and the Flatness Problem** Inflation solves the flatness problem by taking a curved surface, here represented by the surface of the expanding balloon, and expanding it enormously in size. To an ant on the surface, the balloon looks virtually flat when the expansion is complete.

Notice that this resolution of the flatness problem—the universe appears close to being flat because the universe *is* in fact precisely flat, to very high accuracy—has a very important consequence: Because the universe is geometrically flat, relativity tells us that the total density must be exactly equal to the critical value of $\Omega_0 = 1$. ∞ (Sec. 26.4) This is the key result that led us to conclude in Chapter 26 that dark energy—whatever it is—must dominate the density of the universe. ∞ (Sec. 26.5)

Inflation as a Theory

Even though inflation solves the horizon and flatness problems in a quite convincing way, for nearly two decades after it was first proposed the theory was resisted by many astronomers. The main reason was that its prediction of $\Omega_0 = 1$ was clearly at odds with the growing evidence that the density of matter in the universe was no more than 30 percent or so of the critical value. Actually, many cosmologists *had* considered the possibility that a cosmological constant offered a way to account for the remaining 70 percent of the cosmic density, but without independent corroboration a conclusive case could not be made. That is why the supernova observations were so important: By providing empirical evidence for acceleration in the cosmic expansion rate, they established independent evidence for the effects of dark energy and, in doing so, reconciled inflation with the otherwise discrepant observations. ∞ (Sec. 26.5)

Thus, the combined weight of theory and observation force us to the conclusion that not only is most matter dark (Section 27.3), but *most of the cosmic density isn't made up of matter at all*. In a sense, this is the ultimate statement of the Copernican principle: Not only is Earth in no way central to the cosmos, but the stuff of which we are made is unrepresentative of matter in general, and matter itself is the minority constituent of the universe!

Physicists will probably never create in terrestrial laboratories conditions even remotely similar to those that existed in the universe during the inflationary epoch. The creation of our own vacuum energy is (safely) beyond our reach. Nevertheless, cosmic inflation seems to be a natural consequence of many Grand Unified Theories. It explains two otherwise intractable problems within the Big Bang theory—and, following the empirical observations of cosmic acceleration, it is now reconciled with measurements of the matter density of the universe with the inclusion of dark energy into the cosmic mix.

For all these reasons, despite the absence of direct evidence for the process, inflation theory has become an integral part of modern cosmology. Inflation makes definite, testable predictions about the large-scale geometry and structure of the universe that are critically important to current theories of galaxy formation. As we will see in the next section, astronomers are now subjecting these predictions to rigorous scrutiny.

CONCEPT CHECK

✔ Why does the theory of inflation imply that much of the energy density of the universe may be neither matter nor radiation?

27.5 The Formation of Structure in the Universe

Just as stars formed from *inhomogeneities*—deviations from perfectly uniform density—in interstellar clouds, so too are galaxies, galaxy clusters, and larger structures thought to have grown from small density fluctuations in the matter of the expanding universe. ∞ (Sec. 19.1)

Where did these density fluctuations come from? According to current theories of structure formation, they were the result of microscopic "quantum" fluctuations in the very early universe, expanded to macroscopic scales by the effects of inflation! In a very real sense, the quantum universe just after the Big Bang is the progenitor of all the cosmic structure we see around us today.

Given the conditions in the universe during the atomic and galactic epochs (Table 27.1), cosmologists calculate that regions of higher-than-average density that contained more than about a million times the mass of the Sun would have begun to contract. There was thus a natural tendency for million-solar-mass "pregalactic" objects to form. In Chapter 25, we learned how these pregalactic fragments might have interacted and merged to form galaxies. ∞ (Sec. 25.3) In the rest of this chapter, we concern ourselves mostly with the formation of structure on much larger scales.

The Growth of Inhomogeneities

By the early 1980s, cosmologists had come to realize that galaxies could not have formed from the contraction of inhomogeneities involving only *normal* matter. The following lines of reasoning led to this conclusion:

1. Calculations show that, before decoupling (which occurred at a redshift of 1100), the intense background radiation would have prevented clumps of normal matter from contracting. Matter and radiation were just too strongly coupled for structure to form. Thus, any such clumps would have had to wait until *after* decoupling before their densities could start to increase.

2. Because radiation was "tied" to normal matter up until decoupling, any variations in the matter density at that time would have led to temperature variations in the

cosmic background radiation—denser regions would have been a little hotter than less dense ones. The high degree of isotropy observed in the microwave background indicates that any density variations from one region of space to another at the time of decoupling must have been *small*—at most a few parts in 10^5. ∞ (Sec. 26.7)

3. Galaxies—or, at least, quasars—are known to have formed by a redshift of 6. Furthermore, some theorists think that, in order to produce the densest galactic nuclei we see today, the formation process must have already been well established as long ago as a redshift of 20. ∞ (Sec. 25.3) Thus, the initial fluctuations, which, as we have just seen, must have been very small at a redshift of 1100, had to grow to form the first stars and galaxies by a redshift of 20.

4. The contracting matter had to "fight" the general expansion of the universe. As a result, theory shows, these contracting pregalactic clumps could have increased in density by a factor of at most 50 to 100 in the time available. As a result, the small inhomogeneities permitted by observations of the microwave background *could not have grown into galaxies in the time available*—the universe would still have been almost perfectly homogeneous at a time when we know galaxies had already formed.

Put another way, if galaxies had grown from density fluctuations in the normal-matter component of the early universe, then the fluctuations would have had to be so large as to leave a clearly observable imprint on the cosmic microwave background. That imprint is not observed.

Dark Matter

Normal matter, then, cannot account for the large-scale structure we see today. Fortunately for cosmology (and for life on Earth), much of the universe is made of *dark* matter, which has properties quite different from those of normal matter and which provides a natural explanation for the large-scale structure we see today. Whatever the nature of dark matter, its defining property is that it interacts only very weakly with normal matter and radiation, so its natural tendency to clump and contract under gravity was not hindered by the radiation background. Dark matter started clumping well before decoupling (redshift 1100)—in fact, density inhomogeneities in the dark-matter component of

the universe probably began to grow as soon as matter first began to dominate the universe at a redshift of about 6000.

Because the dark matter was not directly tied to the radiation, these inhomogeneities could have been quite large at the time of decoupling, without having a correspondingly large effect on the microwave background. In short, dark matter could clump to form large-scale structure in the universe without running into any of the problems just described for normal matter.

Thus, as illustrated in Figure 27.14, dark matter determined the overall distribution of mass in the universe and clumped to form the observed large-scale structure without violating any observational constraints on the microwave background. Then, at later times, normal matter was drawn by gravity into the regions of highest density, eventually forming

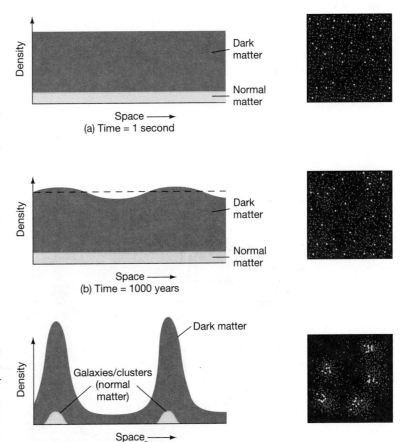

▲ **FIGURE 27.14 Structure Formation** The formation of structure in the cosmos depended crucially on the existence of dark matter. (a) The universe started out as a mixture of (mostly) dark and normal matter. (b) A few thousand years after the Big Bang, the dark matter began to clump. (c) Eventually, the dark matter formed large structures (represented here by the two high-density peaks) into which normal matter flowed, ultimately to form the galaxies we see today. The three frames at the right represent the densities of dark matter (red) and normal matter (yellow) in the respective graphs at left.

ANIMATION/VIDEO Cosmic Evolution, Cosmic Structure

galaxies and galaxy clusters. This picture explains why so much dark matter is found outside the visible galaxies. The luminous material is strongly concentrated near the density peaks and dominates the dark matter there, but the rest of the universe is largely devoid of normal matter. Like foam on the crest of an ocean wave, the universe we can see is only a tiny fraction of the total.

Given that the nature of the dark matter is still unknown, theorists have considerable freedom in choosing its properties when they attempt to simulate the formation of structure in the universe. Cosmologists conventionally classify dark matter as either "hot" or "cold," on the basis of its temperature at the time when galaxies began to form. The two types predict quite different kinds of structure in the present-day universe.

Hot dark matter consists of lightweight particles—much less massive than the electron. Neutrinos, which appear to have small, but nonzero, masses, are leading candidates for hot dark-matter particles. ∞ (Sec. 16.6) However, simulations of a universe filled with hot dark matter indicate that, whereas large structures, such as superclusters and voids, form fairly naturally, structure on smaller scales does not. Small amounts of hot material tend to disperse, not clump together. As a result, most cosmologists have concluded that models based on hot dark matter are unable to explain the observed structure of the universe.

Cold dark matter consists of very massive particles, possibly formed during the GUT epoch or even before. Computer simulations modeling the universe with these

particles as the dark matter easily produce small-scale structure. With the understanding that galaxies form referentially in the densest regions—as is predicted particularly by models that include a cosmological constant—these models also predict large-scale structure that is in excellent agreement with what is actually observed.

Figure 27.15 shows the results of a recent supercomputer simulation of a "best-bet" universe consisting of 4 percent normal matter, 23 percent cold dark matter, and 73 percent dark energy (in the form of a cosmological constant). ∞ (Sec. 26.6) Yellow dots represent regions in each frame where significant star formation is occurring—quasars at redshift 6 and bright interacting galaxies today. The similarities with real observations of cosmic structure, shown in Figures 25.21 and 26.1, are striking, and more detailed statistical analysis confirms that these models agree extremely well with reality. Notice the large-scale extended filamentary structure evident in the last two frames, which may be compared to the observed structure presented earlier in the text. The visible galaxies are also surrounded by extensive dark-matter halos.

Although calculations like this cannot prove that these models are the correct description of the universe, the agreement in detail between models and reality strongly favor the dark-energy/cold-dark-matter model of the cosmos.

CONCEPT CHECK

✔ Why was dark matter necessary for structure formation in the universe?

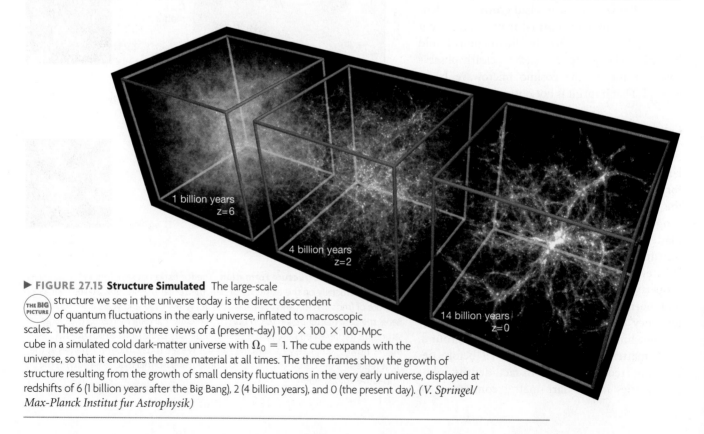

▶ **FIGURE 27.15 Structure Simulated** The large-scale structure we see in the universe today is the direct descendent of quantum fluctuations in the early universe, inflated to macroscopic scales. These frames show three views of a (present-day) 100 × 100 × 100-Mpc cube in a simulated cold dark-matter universe with $\Omega_0 = 1$. The cube expands with the universe, so that it encloses the same material at all times. The three frames show the growth of structure resulting from the growth of small density fluctuations in the very early universe, displayed at redshifts of 6 (1 billion years after the Big Bang), 2 (4 billion years), and 0 (the present day). *(V. Springel/ Max-Planck Institut fur Astrophysik)*

27.6 Cosmic Structure and the Microwave Background

Because dark matter does not interact directly with photons, its density variations do not cause large (and easily observable) temperature variations in the microwave background. However, that radiation is influenced slightly by the *gravity* of the growing dark clumps, which produces a slight gravitational redshift that varies from place to place, depending on the dark-matter density. As a result, cosmological models predict that there should be tiny "ripples" in the microwave background—temperature variations of only a few parts per million from place to place on the sky.

Until the late 1980s these ripples were too small to be measured accurately, although cosmologists were confident that they would be found. In 1992, after almost 2 years of careful observation, the *COBE* team announced that the expected ripples had indeed been detected. ∞ (Sec. 26.7) The temperature variations are tiny—only 30–40 millionths of a kelvin from place to place in the sky—but they are there. The *COBE* results are displayed as a temperature map of the microwave sky in Figure 27.16. The temperature variation due to Earth's motion (see Figure 26.18) has been subtracted out, as has the radio emission from the Milky Way, and temperature deviations from the average are displayed.

The ripples seen by *COBE,* combined with computer simulations such as that shown in Figure 27.15, predict present-day structure that is consistent with the superclusters, voids, filaments, and Great Walls we see around us. Although the *COBE* data were limited by relatively low (roughly 7°) resolution, detailed analysis of the ripples also supports the key prediction of inflation theory—that the universe is of exactly critical density and hence spatially flat. For these reasons, the *COBE* observations rank alongside the discovery of the microwave background itself in terms of their importance to the field of cosmology. The lead investigators of the *COBE* program won the 2006 Nobel Prize in physics for their groundbreaking work.

The decade following the end of the *COBE* mission in late 1993 has seen dramatic improvements in both the resolution and sensitivity of microwave background measurements. Figure 27.17 shows results returned by two more recent experiments that have radically improved our view of the microwave background, confirming and refining the *COBE* results.

The main image in Figure 27.17 shows an all-sky map made by the *Wilkinson Microwave Anisotropy Probe (WMAP),* launched by NASA in 2001. Permanently stationed some 1.5 million km outside Earth's orbit along the Sun–Earth line and always pointing away from the Sun to keep its delicate heat-sensitive detectors in shadow, *WMAP* completes a scan of the entire sky every 6 months. The instrument's angular resolution is roughly 20–30′, some 20 times finer than that of *COBE* (cf. Figure 27.16), allowing extraordinarily detailed measurements of many cosmological parameters to be made. The inset shows a smaller scale (just 2° wide), but even higher resolution (7′) image returned by *Cosmic Background Imager,* a ground-based microwave telescope located high in the Chilean Andes. (Recall from Chapter 3 that the microwave part of the spectrum is only partly transparent to electromagnetic radiation, so microwave detectors must be placed above as much of Earth's absorbing atmosphere as possible.) ∞ (Sec. 3.3)

Both maps show temperature fluctuations of a few hundred microkelvins, with a characteristic angular scale of about 1°. This temperature range is larger than the fluctuations seen by *COBE* because *COBE*'s low resolution effectively averaged the data over a large area of the sky, smearing out the peaks and troughs seen in the higher resolution observations.

The amount of structure seen in the data on different angular scales clearly shows a peak at about 1 degree—pretty

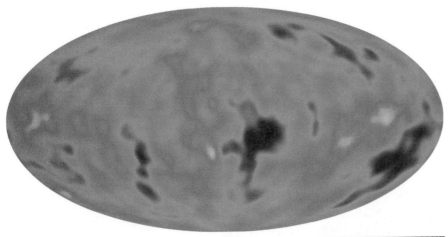

◄ **FIGURE 27.16 Cosmic Microwave Background Map** *COBE* map of temperature fluctuations in the cosmic microwave background over the entire sky. Hotter-than-average regions are in yellow, cooler-than-average regions in blue. The total temperature range shown is ±200 millionths of a kelvin. The temperature variation due to Earth's motion has been subtracted, as has the radio emission from the Milky Way Galaxy, and temperature deviations from the average are displayed. (*NASA*)

R I V U X G

ANIMATION/VIDEO Ripples in the Early Universe

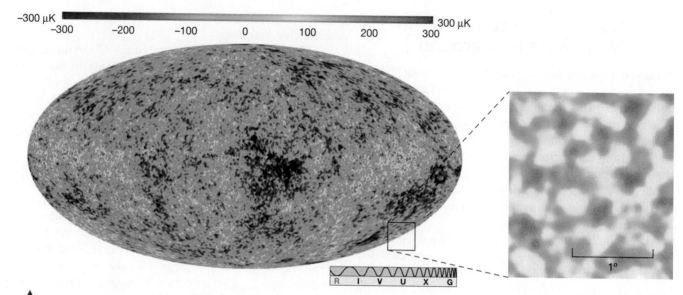

Interactive FIGURE 27.17 Early Structure The entire microwave sky, as seen by the *WMAP* spacecraft at frequencies between 22 and 90 GHz (wavelengths ranging from 14 to 3 mm). This new map may be compared directly with the *COBE* map shown in Figure 27.16. The effective resolution of this *WMAP* multiwavelength survey is about 25′ (compared with 7° for *COBE*). The inset at the right shows a 7′-resolution map of a roughly 2° patch of sky, obtained by the ground-based *Cosmic Background Imager* instrument at 30 GHz (1 cm wavelength). The bright blobs are slightly denser-than-average regions of the universe at an age of roughly 400,000 years—long after the epoch of inflation, but well before the formation of significant large-scale structure. They are destined to collapse to form clusters of galaxies. *(NASA)*

much what your eyes see in Figure 27.17. This is in excellent agreement with the theoretical prediction for a universe with $\Omega_0 = 1$ (actually, with 27 percent matter and 73 percent dark energy, as quoted in Chapter 26 and earlier in this chapter), strongly supporting the prediction from inflation that Ω_0 is very close to unity, with a small margin of error.

More detailed analysis provides a wealth of information on the history and composition of the universe. Indeed, the *WMAP* data are our primary source for the cosmological parameters used in *More Precisely 24-1* and throughout this text. The European Space Agency's *Planck* mission, launched in 2009, is further refining these observations, improving their accuracy and making still more precise measurements of the microwave sky. The first decade of the 21st century has seen the basic parameters of the universe measured (even if not yet fully understood) to an accuracy only dreamed of just a few years ago.

PROCESS OF SCIENCE CHECK

✔ What do observations of fluctuations in the microwave background tell us about the structure of the universe?

CHAPTER REVIEW

SUMMARY

1 At present, the universe is dominated by dark energy, which exceeds the
matter density by more than a factor of
2. The densities of dark energy and matter both greatly exceed the equivalent
mass density of radiation: The density
of matter was much greater in the past,

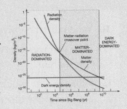

when the universe was smaller and the universe was **matter
dominated (p. 684)**, a few billion years ago. However, because
radiation is redshifted as the universe expands, the density of radiation at early times was greater still. Thus, the early universe
was **radiation dominated (p. 685)**.

2 During the first few minutes after the
Big Bang, matter was formed out of the
primordial fireball by the process of **pair
production (p. 685)**. In the early universe, matter and radiation were linked
by this process. Particles and forces
"froze out" of the radiation background
as the temperature fell below the threshold for creating them. The existence of

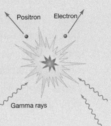

matter today means that there must have been unequal amounts
of matter and antimatter early on.

3 The physical state of the universe can
be understood in terms of present-day
physics back to about 10^{-43} s after the Big
Bang. Before that, the four fundamental
forces of nature—gravity, electromagnet-
ism, the strong force, and the weak force—were all indistinguish-

able. There is presently no theory that can describe these extreme
conditions. As the universe expanded and its temperature dropped,
the forces became distinct from one another. First gravity, then the
strong force, and then the weak and electromagnetic forces separated out. Subsequently, nuclei, then atoms, and eventually large
clumps of matter destined to become galaxies and stars formed as
the universe continued to cool.

4 All of the hydrogen in the universe is
primordial, formed from radiation as the
universe expanded and cooled. Most of
the helium observed in the universe
today is also primordial, created by
primordial nucleosynthesis (p. 691) in
the early universe a few minutes after the
Big Bang. Some deuterium was also

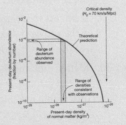

formed at these early times, and it provides a sensitive indicator of
the present density of the universe in the form of "normal" (as opposed to dark) matter. Studies of deuterium indicate that normal

matter can account for at most 3 or 4 percent of the critical density. The remaining mass inferred from studies of clusters must
then be made of dark matter, in the form of unknown particles
formed at some very early epoch.

5 When the universe was about 1100
times smaller than it is today, the temperature became low enough for atoms to
form. At that time, the (then-optical)
background radiation **decoupled (p. 693)**
from the matter. The universe became
transparent. The photons that now make
up the microwave background have been
traveling freely through space ever since.

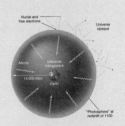

6 Very early on, the universe underwent
a brief period of rapid expansion called
the **epoch of inflation (p. 695)**, during
which the size of the cosmos increased by
a huge factor—10^{50} or more. The **horizon
problem (p. 694)** is the fact that, accord-

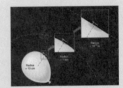

ing to the standard (that is, noninflationary) Big Bang model, there
is no good reason for widely separated parts of the universe to be as
similar as they are. Inflation solves the horizon problem by taking a
small homogeneous patch of the early universe and expanding it
enormously. The patch is still homogeneous, but it is now much
larger than the portion of the universe we can see today. Inflation
also solves the **flatness problem (p. 694)**, which is the fact that there
is no obvious reason why the present density of the universe is so
close to the critical value. Inflation implies that the cosmic density
is, for all practical purposes, exactly critical.

7 Large-scale structure in the universe
formed when density fluctuations in the
dark matter clumped and grew to form

the "skeleton" of the structure we observe. Normal matter then
flowed into the densest regions of space, eventually forming the
galaxies we now see. Cosmologists distinguish between hot dark
matter and **cold dark matter (p. 700)**, depending on the temperature of dark matter at the end of the Radiation Era. In order to
explain the observed large-scale structure in the universe, much of
the dark matter must be cold.

8 "Ripples" in the microwave background are the imprint of early density
inhomogeneities on the radiation field.
These ripples were observed by the *COBE*

satellite. Subsequent observations made by the *WMAP* spacecraft
provide accurate measurements of many cosmological parameters
and lend strong support to the inflationary prediction that we live
in a flat, critical-density universe.

Mastering ASTRONOMY® *For instructor-assigned homework go to www.masteringastronomy.com*

Problems labeled **POS** explore the process of science | **VIS** problems focus on reading and interpreting visual information

REVIEW AND DISCUSSION

1. For how long was the universe dominated by radiation? How hot was the universe when the dominance of radiation ended?

2. What was the role of dark energy in the very early universe?

3. Why is our knowledge of the Planck epoch so limited?

4. When and how did the first atoms form?

5. Describe the universe at the end of the galactic epoch.

6. Why do all stars, regardless of their abundance of heavy elements, seem to contain at least one-quarter helium by mass?

7. Why didn't heavier and heavier elements form in the early universe, as they do in stars?

8. If large amounts of deuterium formed in the early universe, why do we see so little deuterium today?

9. How do measurements of the cosmic abundance of deuterium provide a reliable estimate of the density of normal matter?

10. **POS** How do we know that most matter in the universe is not "normal"?

11. When did the universe become transparent to radiation?

12. How can we observe the epoch at which the universe became transparent?

13. What is the epoch of inflation, and what happened to the early universe during that time?

14. **POS** How does inflation solve the horizon problem?

15. **POS** How does inflation solve the flatness problem?

16. What does inflation tell us about the total density of the universe?

17. What is the difference between hot and cold dark matter?

18. What is the connection between dark matter and the formation of large- and small-scale structures?

19. **POS** Why were the observations made by the *COBE* satellite so important to cosmology?

20. **POS** What key measurement was made by the *WMAP* experiment?

CONCEPTUAL SELF-TEST: MULTIPLE CHOICE

1. Immediately after its birth, the universe (a) was dominated by photons; (b) was made mostly of protons; (c) had equal amounts of matter and antimatter; (d) formed stars and galaxies.

2. Present-day Grand Unified Theories unite all of the fundamental forces except (a) the strong force; (b) the weak force; (c) the electromagnetic force; (d) the gravitational force.

3. About half a million years after the Big Bang, the universe had cooled to the point that (a) protons and electrons could combine to form atoms; (b) particle–antiparticle annihilation ceased; (c) gas could condense to form stars; (d) carbon condensed to make dust.

4. One of the problems with the standard Big Bang model is that (a) galaxies are redshifted; (b) the temperature is almost exactly the same everywhere; (c) the universe is hottest in the center; (d) the galaxy will expand forever.

5. **VIS** According to our best estimates, the line that best describes the universe in Figure 27.10 ("Flatness Problem") is (a) accelerating; (b) open; (c) critical; (d) closed.

6. It is likely that the density of the universe is made up mostly of (a) hydrogen; (b) electromagnetic radiation; (c) dark energy; (d) cold dark matter.

7. The horizon problem in the standard Big Bang model is solved by having the universe (a) accelerate; (b) inflate rapidly early in its existence, (c) have tiny, but significant fluctuations in temperature; (d) be geometrically flat.

8. The structure we observe in the universe is the result of (a) dark matter clumping long ago; (b) galaxies colliding; (c) the freezing out of electrons; (d) radiation dominance in the early universe.

9. Elements more massive than lithium were not formed in the early universe because the temperature was (a) too high; (b) too low; (c) not related to density; (d) unstable.

10. Matter and energy clumping in the early universe result in (a) the formation of atoms; (b) rapid inflation; (c) small but observable red shifts; (d) lower temperatures.

PROBLEMS

The number of dots preceding each Problem indicates its approximate level of difficulty.

1. • What was the distance between the points that would some-day become, respectively, the center of the Milky Way Galaxy and the center of the Virgo Cluster at the time of decoupling? (The present separation is 18 Mpc.)

2. •• What was the equivalent mass density of the cosmic radiation field when the universe was one-thousandth its present size? (*Hint:* Don't forget the cosmological redshift!)

3. • Assuming critical density today, what were the temperature and density of the universe when the first quasars formed?

4. •• Of matter and radiation, which dominated the universe, and by what factor in density (assuming critical density today), at the start of (a) decoupling, (b) nucleosynthesis?

5. ••• Estimate the temperature needed for electron–positron pair production. The mass of an electron is 9.1×10^{-31} kg. Use $E = mc^2$ to find the energy involved in the reaction (*Discovery 22-1*). Then use $E = hf$ and $\lambda f = c$ to find the wavelength λ of the two photons contributing to that energy (see Section 4.2 and Section 27.1). Finally, use Wien's law to find the temperature at which a blackbody spectrum peaks at that wavelength (Section 3.4). How does your answer compare with the threshold temperature given in the text?

6. •• Given that the threshold temperature for the production of electron–positron pairs is about 6×10^9 K and that a proton is 1800 times more massive than an electron, calculate the threshold temperature for proton–antiproton pair production.

7. • At what wavelength did the background radiation peak at the start of the epoch of nucleosynthesis? In what part of the electromagnetic spectrum does this wavelength lie?

8. •• By what factor did the volume of the universe increase during the epoch of primordial nucleosynthesis, from the time when deuterium could first survive until the time at which all nuclear reactions ceased? By what factor did the matter density of the universe decrease during that period?

9. • From Table 24.1, the "photosphere" of the universe corresponding to the epoch of decoupling presently lies some 14,000 Mpc from us. (See Figure 27.8.) How far away was a point on the photosphere when the background radiation we see today was emitted?

10. •• The blobs evident in the inset to Figure 27.17 are about 20′ across. If those blobs represent clumps of matter around the time of decoupling (redshift = 1100), estimate the diameter of the clumps at the time of decoupling, assuming Euclidean geometry.

28

LIFE IN THE UNIVERSE

ARE WE ALONE?

LEARNING GOALS

Studying this chapter will enable you to

1 Summarize the process of cosmic evolution as it is currently understood.

2 Evaluate the chances of finding life elsewhere in the solar system.

3 Summarize the various probabilities used to estimate the number of advanced civilizations that might exist in the Galaxy.

4 Discuss some of the techniques we might use to search for extraterrestrials and to communicate with them.

THE BIG PICTURE Is there intelligent life elsewhere in the universe? The simple answer is that we don't know. None of the hundreds of extrasolar planets discovered in recent years have (yet) shown any sign of life, intelligent or otherwise. But that hasn't stopped astronomers from searching.

Are we unique? Is life on our planet the only example of life in the universe? These are difficult questions: Earth is the only place in the universe where we know for certain that life exists, and we have no direct evidence for extraterrestrial life of any kind. Yet these are important questions, with profound implications for our species.

In this chapter we take a look at how humans evolved on Earth and then consider whether those evolutionary steps might have happened elsewhere. We then assess the likelihood of our having galactic neighbors and consider how we might learn about them if they do exist.

LEFT: *This fanciful painting, entitled* Galaxyrise Over Alien Planet, *suggests a plurality of life—some perhaps extinct, some perhaps exotic—on alien worlds well beyond Earth. Despite blockbuster movies, science-fiction novels, and a host of claims for extraterrestrial contact, astronomers have so far found no unambiguous evidence for life of any kind anywhere else in the universe. (© Dana Berry)*

Mastering**ASTRONOMY**®

Visit the Study Area in www.masteringastronomy.com for quizzes, animations, videos, interactive figures, and self-guided tutorials.

28.1 Cosmic Evolution

In our study of the universe, we have been very careful to avoid any inference or conclusion that places Earth in a special place in the cosmos. This Copernican principle, or principle of mediocrity, has been our invaluable guide in helping us define our place in the "big picture." However, when discussing life in the universe, we face a problem: ours is the only planet we know of on which life and intelligence have evolved, making it hard for any discussion of intelligent life *not* to treat humans as special cases.

Accordingly, in this final chapter we adopt a decidedly different approach. We first describe the chain of events leading to the only technologically proficient, intelligent civilization we know—us. Then we try to assess the likelihood of finding and communicating with intelligent life elsewhere in the cosmos.

Life in the Universe

With this human-centered view clearly evident, Figure 28.1 identifies seven major evolutionary phases that have contributed to development of life on our planet: *particulate, galactic, stellar, planetary, chemical, biological,* and *cultural* evolution. Matter formed from energy in the early universe, then cooled and clumped to form galaxies and stars. Within galaxies, generation after generation of stars formed and died, seeding the interstellar medium with heavy elements so that, when our Sun formed billions of years after the first star blazed, the rocky planet Earth formed along with it. Eventually, on Earth, life appeared and slowly evolved into the diverse environment we see today.

Together, these evolutionary phases represent the grand sweep of **cosmic evolution**—the continuous transformation of matter and energy that has led to the appearance of life and civilization on our planet. The first four represent, in reverse order, the contents of this book. We now expand our field of view beyond astronomy to include the other three.

From the Big Bang, to the formation of galaxies, to the birth of the solar system, to the emergence of life, to the evolution of intelligence and culture, the universe has evolved from simplicity to complexity. We are the result of an incredibly complex chain of events that spanned billions of years. Were those events random, making us unique, or are they in some sense *natural*, so that *technological civilization*—which, as a practical matter, we will take to mean "civilization capable of off-planet communication by electromagnetic or other means"—is inevitable? Put another way, are we alone in the universe, or are we just one among countless other intelligent life-forms in our Galaxy?

Before embarking on our study, we need a working definition of *life*. Defining life, however, is not an easy task: The distinction between the living and the nonliving is not as obvious as we might at first think. Although most physicists would agree on the definitions of matter and energy, biologists have not arrived at a clear-cut definition of life. Generally speaking, scientists regard the following as characteristics of living organisms: (1) They can *react* to their environment and can often heal themselves when damaged; (2) they can *grow* by taking in nourishment from their surroundings and processing it into energy; (3) they can *reproduce,* passing along some of their own characteristics to their offspring; and (4) they have the capacity for genetic change and can therefore *evolve* from generation to generation and adapt to a changing environment.

These rules are not strict, and there is great leeway in interpreting them. Stars, for example, react to the gravity of their neighbors, grow by accretion, generate energy, and

▼ **FIGURE 28.1 Arrow of Time** Some highlights of cosmic history, as it relates to the emergence of life on Earth, are noted along this arrow of time, from the beginning of the universe to the present. Along the bottom of the arrow are seven "windows" outlining the major phases of cosmic evolution: evolution of primal energy into elementary particles; of atoms into galaxies and stars; of stars into heavy elements; of elements into solid, rocky planets; of those same elements into the molecular building blocks of life; of those molecules into life itself; and of advanced life forms into intelligence, culture, and technological civilization. *(D. Berry)*

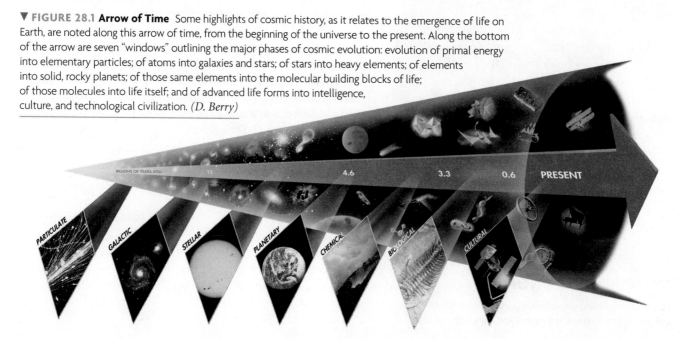

DISCOVERY 28-1

The Virus

The central idea of chemical evolution is that life evolved from nonliving molecules. But aside from insight based on biochemical knowledge and laboratory simulations of some key events on primordial Earth, do we have any direct evidence that life could have developed from nonliving molecules? The answer is yes.

The smallest and simplest entity that sometimes appears to be alive is a virus. We say "sometimes" because viruses seem to have the attributes of both nonliving molecules and living cells. *Virus* is the Latin word for "poison," an appropriate name, since viruses often cause disease. Although they come in many sizes and shapes—a representative example is the polio virus, shown here magnified 300,000 times—all viruses are smaller than the size of a typical modern cell. Some are made of only a few thousand atoms. In terms of size, then, viruses seem to bridge the gap between cells that are living and molecules that are not.

Viruses contain some proteins and genetic information (in the form of DNA or the closely related molecule RNA, the two molecules responsible for transmitting genetic characteristics from one generation to the next), but not much else—none of the material by which living organisms normally grow and reproduce. How, then, can a virus be considered alive? Indeed, alone, it cannot; a virus is absolutely lifeless when it is isolated from living organisms. But when it is inside a living system, a virus has all the properties of life.

Viruses come alive by transferring their genetic material into living cells. The genes of a virus seize control of a cell and establish themselves as the new master of chemical activity.

Viruses grow and reproduce copies of themselves by using the genetic machinery of the invaded cell, often robbing the cell of its usual function. Some viruses multiply rapidly and wildly, spreading disease and—if unchecked—eventually killing the invaded organism. In a sense, then, viruses exist within the gray area between the living and the nonliving.

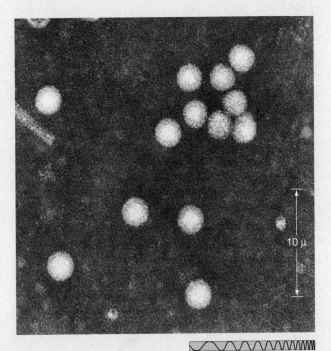

10 μ

(R. Williams)

"reproduce" by triggering the formation of new stars, but no one would suggest that they are alive. By contrast, a virus (see *Discovery 28-1*) is inert when isolated from living organisms, but once inside a living system, it exhibits all the properties of life, seizing control of a living cell and using the cell's own genetic machinery to grow and reproduce. Most researchers now think that the distinction between living and nonliving matter is more one of structure and complexity than a simple checklist of rules.

The general case in favor of extraterrestrial life is summed up in what are sometimes called the *assumptions of mediocrity:* (1) Because life on Earth depends on just a few basic molecules, and (2) because the elements that make up these molecules are (to a greater or lesser extent) common to all stars, and (3) if the laws of science we know apply to the entire universe, as we have supposed throughout this book, then—given sufficient time—life must have originated elsewhere in the cosmos. The opposing view maintains that intelligent life on Earth is the product of a series of extremely fortunate accidents—astronomical, geological, chemical, and

biological events unlikely to have occurred anywhere else in the universe. The purpose of this chapter is to examine some of the arguments for each of these viewpoints.

Chemical Evolution

What information do we have about the earliest stages of planet Earth? Unfortunately, not very much. Geological hints about the first billion years or so were largely erased by violent surface activity, as volcanoes erupted and meteorites bombarded our planet; subsequent erosion by wind and water has seen to it that little evidence of that era has survived to the present. Scientists think that the early Earth was barren, with shallow, lifeless seas washing upon grassless, treeless continents. Gases emanating from our planet's interior through volcanoes, fissures, and geysers produced an atmosphere rich in hydrogen, nitrogen, and carbon compounds and poor in free oxygen. As Earth cooled, ammonia, methane, carbon dioxide, and water formed. The stage was set for the appearance of life.

The surface of the young Earth was a very violent place. Natural radioactivity, lightning, volcanism, solar ultraviolet radiation, and meteoritic impacts all provided large amounts of energy that eventually shaped the ammonia, methane, carbon dioxide, and water on our planet into more complex molecules known as **amino acids** and **nucleotide bases**—organic (carbon-based) molecules that are the building blocks of life as we know it. Amino acids build *proteins,* and proteins control metabolism, the daily utilization of food and energy by means of which organisms stay alive and carry out their vital activities. Sequences of nucleotide bases form *genes*—parts of the DNA molecule—which direct the synthesis of proteins and thus determine the characteristics of the organism (Figure 28.2). These same genes, via the DNA contained within every cell in the organism, transfer hereditary characteristics from one generation to the next through reproduction. In all living creatures on Earth—from bacteria to amoebas to humans—genes mastermind life and proteins maintain it.

The idea that complex molecules could have evolved naturally from simpler ingredients found on the primitive Earth has been around since the 1920s. The first experimental verification was provided in 1953, when scientists Harold Urey and Stanley Miller, using laboratory equipment somewhat similar to that shown in Figure 28.3. The **Urey-Miller**

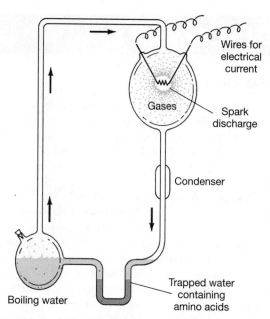

▲ FIGURE 28.3 **Urey-Miller Experiment** This chemical apparatus is designed to synthesize complex biochemical molecules by energizing a mixture of simple chemicals. A mixture of gases (ammonia, methane, carbon dioxide, and water vapor) is placed in the upper bulb to simulate Earth's primordial atmosphere and then energized by spark-discharge electrodes to simulate lightning. After about a week, amino acids and other complex molecules are found in the trap at the bottom, which simulates the primordial oceans into which heavy molecules produced in the overlying atmosphere would have fallen.

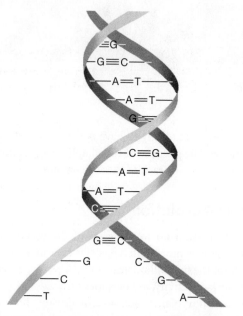

▲ FIGURE 28.2 **DNA Molecule** DNA (deoxyribonucleic acid) is the molecule containing all the genetic information needed for a living organism to reproduce and survive. Often consisting of literally tens of billions of individual atoms, its double-helix structure allows it to "unzip," exposing its internal structure to control the creation of proteins needed for a cell to function. The ordering of its constituent parts is unique to each individual organism. The letters represent the four types of nucleotide bases that make up DNA: adenine, cytosine, guanine, and thymine.

experiment took a mixture of the materials thought to be present on Earth long ago—a "primordial soup" of water, methane, carbon dioxide, and ammonia—and energized it by passing an electrical discharge ("lightning") through the gas. After a few days, they analyzed their mixture and found that it contained many of the same amino acids found in all living things on Earth. About a decade later, scientists succeeded in constructing nucleotide bases in a similar manner. These experiments have been repeated in many different forms, with more realistic mixtures of gases and a variety of energy sources, but always with the same basic outcomes.

Although none of these experiments has ever produced a living organism, or even a single strand of DNA, they do demonstrate conclusively that "biological" molecules—the molecules involved in the functioning of living organisms—can be synthesized by strictly *non*biological means, using raw materials available on the early Earth. More advanced experiments, in which amino acids are united under the influence of heat, have fashioned proteinlike blobs that behave to some extent like true biological cells. Such near-protein material resists dissolution in water (so it would remain intact when it fell from the primitive atmosphere into the ocean) and tends to cluster into small droplets called microspheres—a little like oil globules floating on the surface of water. Figure 28.4 shows some of these laboratory-made

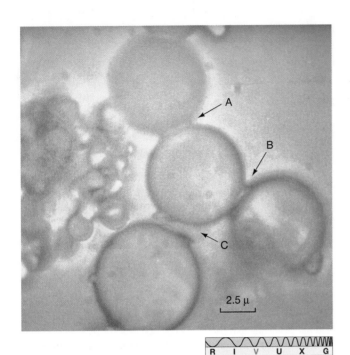

▲ **FIGURE 28.4 Chemical Evolution** These carbon-rich, proteinlike droplets display the clustering of as many as a billion amino acid molecules in a liquid. The droplets can "grow," and parts of them can separate from their "parent" droplet to become new individual droplets (as at A, B, and C). The scale of 2.5 microns noted here is 1/4000 of a centimeter. *(S. Fox)*

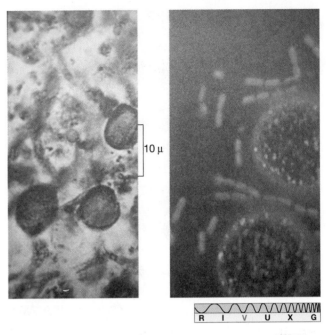

▲ **FIGURE 28.5 Primitive Cells** The photograph on the left, taken through a microscope, shows a fossilized organism found in sediments radioactively dated as 2 billion years old. This primitive system possesses concentric spheres or walls connected by smaller spheroids. The roundish fossils here measure about a thousandth of a centimeter in diameter. The photograph on the right, on approximately the same scale, displays modern blue-green algae. *(E. Barghoorn)*

proteinlike microspheres, whose walls permit the inward passage of small molecules, which then combine within the droplet to construct more complex molecules that are too large to pass back out through the walls. As the droplets "grow," they tend to "reproduce," forming smaller droplets.

Can we consider these proteinlike microspheres to be alive? Almost certainly not. Most biochemists would say that the microspheres are not life itself, but they contain many of the basic ingredients needed to form life. The microspheres lack the hereditary DNA molecule. However, as illustrated in Figure 28.5, they do have similarities to ancient cells found in the fossil record, which in turn have many similarities to modern organisms (such as blue-green algae). Thus, while no actual living cells have yet been created "from scratch" in any laboratory, many biochemists feel that the chain of events leading from simple nonbiological molecules almost to the point of life itself has been amply demonstrated.

An Interstellar Origin?

Recently, a dissenting view has emerged. Some scientists have argued that Earth's primitive atmosphere might *not* in fact have been a particularly suitable environment for the production of complex molecules. These scientists say that there may not have been sufficient energy available to power the

necessary chemical reactions and the early atmosphere may not have contained enough raw material for the reactions to have become important in any case. They suggest instead that much, if not all, of the organic (carbon-based) material that combined to form the first living cells was produced in *interstellar space* and subsequently arrived on Earth in the form of comets, interplanetary dust, and meteors that did not burn up during their descent through the atmosphere.

Several pieces of evidence support this idea. Interstellar molecular clouds are known to contain complex molecules—indeed, there have even been reports (still unconfirmed) of at least one amino acid (glycine) in interstellar space. ∞ (Sec. 18.5)

To test the interstellar space hypothesis, NASA researchers have carried out their own version of the Urey-Miller experiment in which they exposed an icy mixture of water, methanol, ammonia, and carbon monoxide—representative of many interstellar grains—to ultraviolet radiation to simulate the energy from a nearby newborn star. As shown in Figure 28.6, when they later placed the irradiated ice in water and examined the results, they found that the ice had formed droplets surrounded by membranes and containing complex organic molecules. As with the droplets found in earlier experiments, no amino acids, proteins, or DNA were observed in the mix, but the results, repeated numerous

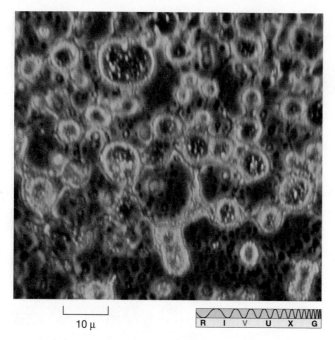

▲ FIGURE 28.6 **Interstellar Globules** These oily, hollow droplets rich in organic molecules were made by exposing a freezing mixture of primordial matter to harsh ultraviolet radiation. The larger ones span about 10 microns across and, when immersed in water, show cell-like membrane structure. Although they are not alive, they bolster the idea that life on Earth could have come from space. *(NASA)*

times, clearly show that even the harsh, cold vacuum of interstellar space can be a suitable medium in which complex molecules and primitive cellular structures can form.

As described in Chapter 15, these icy interstellar grains are thought to have formed the comets in our own solar system. ∞ *(Sec. 15.3)* Large amounts of organic material were detected on comet Halley by space probes when Halley last visited the inner solar system, and similarly complex molecules have been observed on many other well-studied comets, such as Hale-Bopp. ∞ *(Sec. 14.2)* Also as discussed in Chapter 15, there is reason to suspect that cometary impacts were responsible for most of Earth's water, and it is perhaps a small step to imagining that this water already contained the building blocks for life. ∞ *(Sec. 15.3)*

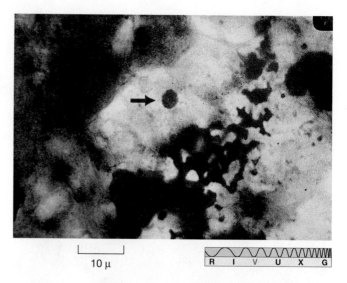

▲ FIGURE 28.7 **Murchison Meteorite** The Murchison meteorite contains relatively large amounts of amino acids and other organic material, indicating that chemical evolution of some sort has occurred beyond our own planet. In this magnified view of a fragment from the meteorite, the arrow points to a microscopic sphere of organic matter. *(CfA)*

In addition, a small fraction of the meteorites that survive the plunge to Earth's surface—including perhaps the controversial "Martian meteorite" discussed in Chapter 10—contain organic compounds. ∞ *(Discovery 10-1)* The Murchison meteorite (Figure 28.7), which fell near Murchison, Australia, in 1969, is a particularly well-studied example. Located soon after crashing to the ground, this meteorite has been shown to contain 12 of the amino acids normally found in living cells, although the detailed structures of these molecules indicate potentially important differences between those found in space and those found on Earth. At the very least, though, these discoveries argue that complex molecules can form in an interplanetary or interstellar environment and that they could have reached Earth's surface unscathed after their fiery descent.

Thus, the hypothesis that organic matter is constantly raining down on Earth from space in the form of interplanetary debris is quite plausible. However, whether this was the *primary* means by which complex molecules first appeared in Earth's oceans remains unclear.

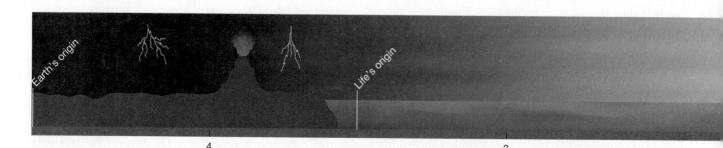

Diversity and Culture

However the basic materials appeared on Earth, we know that life *did* appear. The fossil record chronicles how life on Earth became widespread and diversified over the course of time. The study of fossil remains shows the initial appearance of simple one-celled organisms such as blue-green algae more than 3.5 billion years ago. These were followed by more complex one-celled creatures, such as the amoeba, about 2 billion years ago. Multicellular organisms such as sponges did not appear until about 1 billion years ago, after which there flourished a wide variety of increasingly complex organisms—insects, reptiles, and mammals. Figure 28.8 illustrates some of the key developments in the evolution of life on our planet.

The fossil record leaves no doubt that biological organisms have changed over time—all scientists accept the reality of *biological evolution.* As conditions on Earth shifted and Earth's surface evolved, those organisms that could best take advantage of their new surroundings succeeded and thrived—often at the expense of organisms that could not make the necessary adjustments and consequently became extinct.

What led to these changes? Chance. An organism that happened to have a certain useful genetically determined trait—for example, the ability to run faster, climb higher, or even hide more easily—would find itself with the upper hand in a particular environment. That organism was therefore more likely to reproduce successfully, and its advantageous characteristic would then be more likely to be passed on to the next generation. The evolution of the rich variety of life on our planet, including human beings, occurred as chance mutations—changes in genetic structure—led to changes in organisms over millions of years.

What about the development of intelligence? Many anthropologists think that, like any other highly advantageous trait, intelligence *is* strongly favored by natural selection. As humans learned about fire, tools, and agriculture, the brain became more and more elaborate. The social cooperation that went with coordinated hunting efforts was another important competitive advantage that developed as brain size increased.

Perhaps most important of all was the development of language. Indeed, some anthropologists have gone so far as to suggest that human intelligence *is* human language. Through language, individuals could signal one another while hunting for food or seeking protection. Even more importantly, now our ancestors could share ideas as well as food and shelter. Experience, stored in the brain as memory, could be passed down from generation to generation. A new kind of evolution had begun, namely, *cultural evolution,* the changes in the ideas and behavior of society. Within only the past 10,000 years or so, our more recent ancestors have created the entirety of human civilization.

To put all this into historical perspective, let's imagine the entire lifetime of Earth to be 46 years rather than 4.6 billion years. On this scale, we have no reliable record of the first decade of our planet's existence. Life originated at least 35 years ago, when Earth was about 10 years old. Our planet's middle age is largely a mystery, although we can be sure that life continued to evolve and that generations of mountain chains and oceanic trenches came and went. Not until about 6 years ago did abundant life flourish throughout Earth's oceans. Life came ashore about 4 years ago, and plants and animals mastered the land only about 2 years ago. Dinosaurs reached their peak about 1 year ago, only to die suddenly about 4 months later. ∞ *(Discovery 11-1)* Humanlike apes changed into apelike humans only last week, and the latest ice ages occurred only a few days ago. *Homo sapiens*—our species—did not emerge until about 4 hours ago. Agriculture was invented within the last hour, and the Renaissance—along with all of modern science—is just 3 minutes old!

▼ **FIGURE 28.8 Life on Earth** A simplified timeline of the origin and evolution of life on our planet. The scale begins at left with the origin of Earth about 4.6 billion years ago and extends linearly to the present day at right. Notice how life forms most familiar to us emerged relatively recently in the history of our planet. Technological civilization has existed on Earth for just a few millionths of 1 percent of our planet's lifetime.

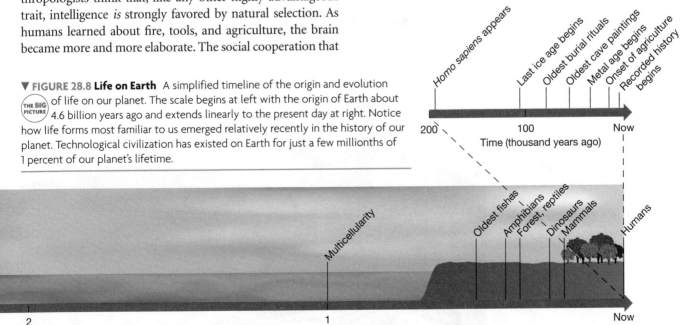

CONCEPT CHECK
✔ Has chemical evolution been verified in the laboratory?

28.2 Life in the Solar System

Simple one-celled life-forms reigned supreme on Earth for most of our planet's history. It took time—a great deal of time—for life to emerge from the oceans, to evolve into simple plants, to continue to evolve into complex animals, and to develop intelligence, culture, and technology. Have those (or similar) events occurred elsewhere in the universe? Let's try to assess what little evidence we have on the subject.

Life as We Know It

"Life as we know it" is generally taken to mean carbon-based life that originated in a liquid water environment—in other words, life on Earth. Might such life exist elsewhere in our solar system?

The Moon and Mercury lack liquid water, protective atmospheres, and magnetic fields, so these two bodies are subjected to fierce bombardment by solar ultraviolet radiation, the solar wind, meteoroids, and cosmic rays. Simple molecules could not survive in such hostile environments. Venus, by contrast, has *far too much* protective atmosphere! Its dense, dry, scorchingly hot atmospheric blanket effectively rules it out as an abode for life, at least like us.

The jovian planets have no solid surfaces (although some researchers have suggested that life might have evolved in their atmospheres), and Pluto and most of the moons of the outer planets are too cold. However, the likelihood of liquid water below the icy surfaces of Europa and other jovian moons has refueled speculation about the development of life there, making these bodies prime candidates for future exploration. ∞ (Sec. 11.5) For now at least, Europa is high on the priority list for missions by both NASA and the European Space Agency. Saturn's moon Titan, with its atmosphere of methane, ammonia, and nitrogen; liquid methane on its surface; and liquid water beneath it, is another site where life might conceivably have arisen. ∞ (Sec. 12.5) Titan's frigid surface conditions are inhospitable for anything familiar to us, but (as we discuss below) scientists are finding more and more examples of terrestrial organisms able to survive and thrive in environments once regarded as uninhabitable.

The planet most likely to harbor life (or to have harbored it in the past) seems to be Mars. The Red Planet is harsh by Earth standards: Liquid water is scarce, the atmosphere is thin, and the absence of magnetism and an ozone layer allows solar high-energy particles and ultraviolet radiation to reach the surface unabated. But the Martian atmosphere was thicker, and the surface probably warmer and much wetter, in the past. ∞ (Sec. 10.5) Indeed, there is strong photographic evidence from orbiters such as *Viking* and *Mars Global Surveyor* for flowing and standing water on Mars in the distant (and perhaps even relatively recent) past. In 2004, the European *Mars Express* orbiter confirmed the long-hypothesized presence of water ice at the Martian poles, and NASA's *Opportunity* rover reported strong geological evidence that the region around its landing site was once "drenched" with water for an extended period.

All of these lines of reasoning strongly suggest that Mars—at least at some time in its past—harbored large amounts of liquid water. However, none of the Mars landers has detected anything that might be interpreted as the remains (fossilized or otherwise) of large plants or animals, and only the *Viking* landers carried equipment capable of performing the detailed biological analysis needed to detect bacterial life (or its fossil remnants). The *Viking* robots scooped up Martian soil (Figure 28.9) and tested it for the

▶ FIGURE 28.9 **Search for Martian Life** Several trenches were dug by the *Viking* robots, such as *Viking 2* seen here on Mars's Utopia Planitia. Soil samples were scooped up and taken inside the robot, where instruments tested them for chemical composition and any signs of life. *(NASA)*

presence of life by conducting chemical experiments designed to detect the waste gases and other products of metabolic activity, but no unambiguous evidence of Martian life has emerged. ∞ *(Discovery 10-1)* Of course, we might argue that the landers touched down on the safest Martian terrain, not in the most interesting regions, such as near the moist polar caps.

Some scientists have suggested that a different type of biology may be operating on the Martian surface. They propose that Martian microbes capable of eating and digesting oxygen-rich compounds in the Martian soil could also explain the *Viking* results. This speculation would be greatly strengthened if recent announcements of fossilized bacteria in meteorites originating on Mars were confirmed (although it seems that the weight of scientific opinion is currently running against that interpretation of the data). The consensus among biologists and chemists today is that Mars does not house any life similar to that on Earth, but a solid verdict regarding past life on Mars will not be reached until we have thoroughly explored our intriguing neighbor.

In considering the emergence of life under adversity, we should perhaps not be too quick to rule out an environment based solely on its extreme properties. Figure 28.10 shows a very hostile environment on a deep-ocean floor, where hydrothermal vents spew forth boiling hot water from vertical tubes a few meters tall. The conditions are quite unlike anything on our planet's surface, yet life thrives in an environment rich in sulfur, poor in oxygen, and completely dark. Such underground hot springs might conceivably exist on alien worlds, raising the possibility of life-forms with much greater diversity over a much wider range of conditions than those known to us on Earth.

In recent years scientists have discovered many instances of so-called **extremophiles**—life-forms that have adapted to live in extreme environments. The superheated hydrothermal vents in Figure 28.10 are one example, but extremophiles have also been found in frigid lakes buried deep under the Antarctic glaciers, in the dark, oxygen-poor and salt-rich floor of the Mediterranean Sea, in the mineral-rich superalkaline environment of California's Mono lake, and even in the hydrogen-rich volcanic darkness far below Earth's crust. In many cases these organisms have evolved to create the energy they need by purely chemical means, using *chemosynthesis* instead of *photosynthesis*, the process whereby plants turn sunlight into energy. These environments may present conditions not so different from those found on Mars, Europa, or Titan, suggesting that even "life as we know it" might well be able to thrive in these hostile, alien worlds.

Alternative Biochemistries

Conceivably, some types of biology might be so different from life on Earth that we would not recognize them and would not know how to test for them. What might these other biologies be?

Some scientists have pointed out that the abundant element silicon has chemical properties somewhat similar to those of carbon and have suggested silicon as a possible alternative to carbon as the basis for living organisms.

◀ **FIGURE 28.10 Hydrothermal Vents** A small two-person submarine (the *Alvin*, partly seen at bottom) took this picture of a hot spring, or "black smoker"—one of many along the midocean ridge in the eastern Pacific Ocean. As hot water rich in sulfur pours out of the top of the vent's tube (near center), black clouds billow forth, providing a strange environment for many life-forms thriving near the vent. The inset shows a close-up of the vent base, where extremophilic life thrives, including, as seen here, giant red tube worms and huge crabs. *(WHOI)*

Ammonia (made of the common elements hydrogen and nitrogen) is sometimes put forward as a possible liquid medium in which life might develop, at least on a planet cold enough for ammonia to exist in the liquid state. Together or separately, these alternatives would surely give rise to organisms with biochemistries (the basic biological and chemical processes responsible for life) radically different from those we know on Earth. Conceivably, we might have difficulty even identifying these organisms as alive.

Although the possibility of such alien life-forms is a fascinating scientific problem, most biologists would argue that chemistry based on carbon and water is the one most likely to give rise to life. Carbon's flexible chemistry and water's wide liquid temperature range are just what are needed for life to develop and thrive. Silicon and ammonia seem unlikely to fare as well as bases for advanced life-forms. Silicon's chemical bonds are weaker than those of carbon and may not be able to form complex molecules—an apparently essential aspect of carbon-based life. Also, the colder the environment, the less energy there is to drive biological processes. The low temperatures necessary for ammonia to remain liquid might inhibit or even completely prevent the chemical reactions leading to the equivalent of amino acids and nucleotide bases.

Still, we must admit that we know next to nothing about noncarbon, nonwater biochemistries, for the very good reason that there are no examples of them to study experimentally. We can speculate about alien life-forms and try to make general statements about their characteristics, but we can say little of substance about them.

CONCEPT CHECK

✔ Which solar system bodies (other than Earth) are the leading candidates in the search for extraterrestrial life?

28.3 Intelligent Life in the Galaxy

With humans apparently the only intelligent life in the solar system, we must broaden our search for extraterrestrial intelligence to other stars and perhaps even other galaxies. At such distances, though, we have little hope of actually detecting life with current equipment. Instead, we must ask, "How likely is it that life in any form—carbon based, silicon based, water based, ammonia based, or something we cannot even dream of—exists?" Let's look at some numbers to develop estimates of the probability of life elsewhere in the universe.

The Drake Equation

An early approach to this problem is known as the **Drake equation** below, after the U.S. astronomer who pioneered the analysis (below). It attempts to express the probability of life in our Galaxy in terms of specific factors with roots in astronomy, biology, and anthropology.

Of course, several of the factors in this formula are largely a matter of opinion. We do not have nearly enough information to determine—even approximately—every factor in the equation, so the Drake equation cannot give us a hard-and-fast answer. Its real value is that it subdivides a large and difficult question into smaller pieces that we can attempt to answer separately. The equation provides the framework within which the problem can be addressed and parcels out the responsibility for the final solution among many different scientific disciplines. Figure 28.11 illustrates how, as our requirements become more and more stringent, only a small fraction of star systems in the Milky Way is likely to generate the advanced qualities specified by the combination of factors on the right-hand side of the equation.

Let's examine the factors in the equation one by one and make some educated guesses about their values. Bear in mind, though, that if you ask two scientists for their best estimates of any given factor, you will likely get two very different answers!

Rate of Star Formation

We can estimate the average number of stars forming each year in the Galaxy simply by noting that at least 100 billion stars now shine in the Milky Way. Dividing this number by the 10-billion-year lifetime of the Galaxy, we obtain a formation rate of 10 stars per year. This rate may be an overestimate, because we think that fewer stars are forming now than formed at earlier epochs of the Galaxy, when more interstellar gas was available. However, we do know that stars are forming today, and our estimate does not include stars that formed in the past and have since died, so our value of 10 stars per year is probably reasonable when averaged over the lifetime of the Milky Way.

| number of technological, intelligent civilizations now present in the Galaxy | = | rate of star formation, averaged over the lifetime of the Galaxy | × | fraction of stars having planetary systems | × | average number of habitable planets within those planetary systems | × | fraction of those habitable planets on which life arises | × | fraction of those life-bearing planets on which intelligence evolves | × | fraction of those intelligent-life planets that develop technological society | × | average lifetime of a technologically competent civilization. |

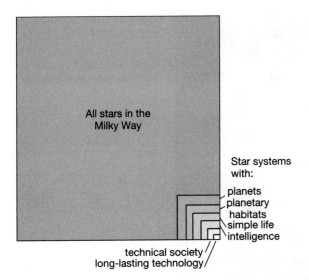

▲ FIGURE 28.11 Drake Equation Of all the star systems in our Milky Way Galaxy (represented by the largest box), progressively fewer and fewer have each of the qualities typical of a long-lasting technological society (represented by the smallest box at the lower right corner). The Drake Equation (presented in detail in this section) is not a solution to the problem of finding extraterrestrial intelligence, but it does provide a useful breakdown of the basic scientific issues involved.

Fraction of Stars Having Planetary Systems

Many astronomers regard planet formation as a natural result of the star-formation process. If the condensation theory (Chapter 16) or some variant of it is correct, and if there is nothing special about our Sun, as we have argued throughout this book, then we would expect many stars to have at least one planet. ∞ (Secs. 6.7, 15.5) Indeed, as we have seen, increasingly sophisticated observations indicate the presence of disks around young stars. Could these disks be protosolar systems? The condensation theory suggests that they are, and the short (theoretical) lifetimes of disks imply the existence of many planet-forming systems in the neighborhood of the Sun.

No planets like our own have yet been *seen* orbiting any other star. The light reflected by an Earth-like planet circling even the closest star would be too faint to detect even with the best equipment. The light would be lost in the glare of the parent star. Large orbiting telescopes may soon be able to detect Jupiter-sized planets orbiting the nearest stars, but even those huge planets will be barely visible. However, as described in Chapter 15, there is now overwhelming evidence for planets orbiting other stars. ∞ (Sec. 15.6) The planets found so far are Jupiter sized rather than Earth sized and generally have rather eccentric orbits, but astronomers are confident that an Earth-like planet (in an Earth-like orbit) will one day be detected, and plans to build the necessary equipment are well underway.

Accepting the condensation theory and its consequences, and without being either too conservative or naïvely optimistic,

we assign a value near unity to this factor—that is, we think that nearly all stars form with planetary systems of some sort.

Number of Habitable Planets per Planetary System

What determines the feasibility of life on a given planet? *Temperature* is perhaps the single most important factor, although the possibility of catastrophic external events, such as cometary impacts or even distant supernovae, must also be considered. ∞ (*Discovery 11-1*, Sec. 21.3)

The surface temperature of a planet depends on two things: the planet's distance from its parent star and the thickness of the planet's atmosphere. Planets with a nearby parent star (but not too close) and some atmosphere (though not too thick) should be reasonably warm, like Earth or Mars. Planets far from the star and with no atmosphere, like Pluto, will surely be cold by our standards. And planets too close to the star and with a thick atmosphere, like Venus, will be very hot indeed.

As discussed in Chapter 15, a three-dimensional *stellar habitable zone* of "comfortable" temperatures surrounds every star. ∞ (Sec. 15.7) It represents the range of distances within which a planet of mass and composition similar to Earth's would have a surface temperature between the freezing and boiling points of water. (Our Earth-based bias is again clear here!) The hotter the star, the larger is this zone (Figure 28.12). A- and F-type stars have rather large habitable zones, but the size of the zone diminishes rapidly as we proceed through G-, K-, and M-type stars (although we note that some of the "super-Earth" planets mentioned in Chapter 15 do in fact lie within the habitable zones of their low-mass parent stars). ∞ (Sec. 15.7)

In addition to their small habitable zones, lightweight M-type stars are thought to be prone to such violent surface activity that they are generally not considered likely hosts of life-bearing planets, despite their large numbers. ∞ (Sec. 17.8) At the other extreme, massive O- and B-type stars are also considered unlikely candidates, both because they are rare and because they are not expected to last long enough for life to develop, even if they do have planets.

Three planets—Venus, Earth, and Mars—reside in or near the habitable zone surrounding our Sun. Venus is too hot because of its thick atmosphere and proximity to the Sun. Mars is a little too cold because its atmosphere is too thin and it is too far from the Sun. But if the orbits of Venus and Mars were swapped—not inconceivable, since chance played such a large role in the formation of the terrestrial planets—then both of these nearby planets might conceivably have evolved surface conditions resembling those on Earth. ∞ (Secs. 10.5, 15.4) In that case our solar system would have had three habitable planets instead of one. Proximity to a giant planet may also render a moon (such as

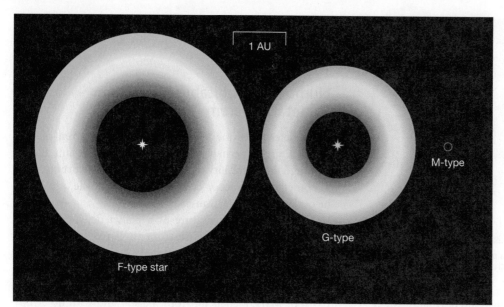

1 AU

M-type

G-type

F-type star

◀ FIGURE 28.12 **Stellar Habitable Zones** The extent of the habitable zone is much larger around a hot star than around a cool one. For a star like the Sun (a G-type star), the zone extends from about 0.85 to 2.0 AU. For a hotter F-type star, the range is 1.2 to 2.8 AU. For a cool M-type star, only Earth-like planets orbiting between about 0.02 and 0.06 AU would be habitable.

Europa) habitable, the planet's tidal heating making up for the lack of sunlight. ∞ (Sec. 11.5)

A planet moving on a "habitable" orbit may still be rendered uninhabitable by external events. Many scientists think that the outer planets in our own solar system are critical to the habitability of the inner worlds, both by stabilizing their orbits and by protecting them from cometary impacts, deflecting would-be impactors away from the inner part of the solar system. The theory presented in Chapter 15 suggests that a star with inner terrestrial planets on stable orbits would probably also have the jovian worlds needed to safeguard their survival. ∞ (Sec. 15.2) However, observations of extrasolar planets are not yet sufficiently refined to determine the fraction of stars having "outer planet" systems like our own. ∞ (Sec. 15.7)

Other external forces may also influence a planet's survival. Some researchers have suggested that there is a *galactic habitable zone* for stars in general, outside of which conditions are unfavorable for life (see Figure 28.13). Far from the Galactic center, the star formation rate is low and few cycles of star formation have occurred, so there are insufficient heavy elements to form terrestrial

planets or populate them with technological civilizations if any should form. ∞ (Sec. 21.5) Too close, and the radiation from bright stars and supernovae in the crowded inner part of the Galaxy might be detrimental to life. More importantly, the gravitational effects of nearby stars may send frequent showers of comets from the counterpart of the Oort cloud into the inner regions of a planetary system, striking the terrestrial planets and terminating any chain of evolution that might lead to intelligent life.

Thus, to estimate the number of habitable planets per planetary system, we must first take inventory of how many

▶ FIGURE 28.13 **Galactic Habitable Zone** Some regions of the Galaxy may be more conducive to life than others. Too far from the Galactic center, there may not be enough heavy elements for terrestrial planets to form or technological society to evolve. Too close, the radiative or gravitational effects of nearby stars may render life impossible. The result is a ring-shaped habitable zone, colored here in green. This zone presumably encloses the orbit of the Sun, but its radial extent in both directions is uncertain.

High density and intense radiation in here

Sun's orbit

Galactic habitable zone

Too few heavy elements out here

stars of each type shine in the Galactic habitable zone, then calculate the sizes of their stellar habitable zones, and, finally, estimate the number of planets likely to be found there. We must eliminate almost all of the 10 percent of surveyed stars around which planets have so far been observed, and, presumably, a similar fraction of stars in general, because the large jovian planets seen in most cases have eccentric orbits that would destabilize the motion of any inner terrestrial world, either ejecting it completely from the system or making conditions so extreme that the chances for the development of life are severely reduced. ∞ (Sec. 15.7)

Finally, we exclude the majority of binary-star systems, because a planet's orbit within the habitable zone of a binary would be unstable in many cases, as illustrated in Figure 28.14. Given the observed properties of binaries in our Galaxy, we expect that most habitable planetary orbits would be unstable, so there would not be time for life to develop.

The inner and outer radii of the Galactic habitable zone are not known with any certainty, and we still have insufficient data about most stars to make any definitive statement

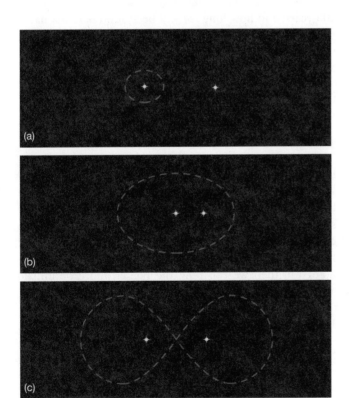

▲ **FIGURE 28.14 Binary-Star Planets** In binary-star systems, planets are restricted to only a few kinds of orbits that are gravitationally stable. (a) The orbit is stable only if the planet lies very close to its parent star, so that the gravity of the other star is negligible. (b) A planet circulating at a great distance about both stars in an elliptical orbit, in which case the orbit is stable only if it lies far from both stars. (c) Another possible but unstable path interweaves between the two stars in a figure-eight pattern.

about their planetary systems. Taking these uncertainties into account as best we can, we assign a value of 1/10 to this factor in our equation. In other words, we think that, on average, there is one potentially habitable planet for every 10 planetary systems that might exist in our Galaxy. Single F-, G-, and K-type stars are the best candidates.

Fraction of Habitable Planets on Which Life Actually Arises

The number of possible combinations of atoms is incredibly large. If the chemical reactions that led to the complex molecules that make up living organisms occurred completely at random, then it is extremely unlikely that those molecules could have formed at all. In that case, life is extraordinarily rare, this factor is close to zero, and we are probably alone in the Galaxy, perhaps even in the entire universe.

However, laboratory experiments (such as the Urey-Miller experiment described earlier) seem to suggest that certain chemical combinations are strongly favored over others—that is, the reactions are *not* random. Of the billions upon billions of basic organic groupings that could occur on Earth from the random combination of all sorts of simple atoms and molecules, only about 1500 actually do occur. Furthermore, these 1500 organic groups of terrestrial biology are made from only about 50 simple "building blocks" (including the amino acids and nucleotide bases mentioned earlier). This suggests that molecules critical to life are not assembled by chance alone; apparently, additional factors are at work on the microscopic level. If a relatively small number of chemical "evolutionary tracks" are likely to exist, then the formation of complex molecules—and hence, we assume, life—becomes much more likely, given sufficient time.

To assign a very low value to this factor in the equation is to think that life arises randomly and rarely. To assign a value close to unity is to think that life is inevitable, given the proper ingredients, a suitable environment, and a long enough time. No simple experiment can distinguish between these extreme alternatives, and there is little or no middle ground. To many researchers, the discovery of life (past or present) on Mars, Europa, Titan, or some other body in our solar system would convert the appearance of life from an unlikely miracle to a virtual certainty throughout the Galaxy. We will take the optimistic view and adopt a value of unity.

Fraction of Life-Bearing Planets on Which Intelligence Arises

As with the evolution of life, the appearance of a well-developed brain is a highly unlikely event if only chance is involved. However, biological evolution through natural selection is a mechanism that generates apparently highly improbable results by singling out and refining useful characteristics. Organisms that profitably use adaptations can develop more complex behavior, and complex behavior provides organisms with the *variety* of choices needed for more advanced development.

One school of thought maintains that, given enough time, intelligence is inevitable. On this view, assuming that natural selection is a universal phenomenon, at least one organism on a planet will always rise to the level of "intelligent life." If this is correct, then the fifth factor in the Drake equation equals or nearly equals unity.

Others argue that there is only one known case of intelligence: human beings on Earth. For 2.5 billion years—from the start of life about 3.5 billion years ago to the first appearance of multicellular organisms about 1 billion years ago—life did not advance beyond the one-celled stage. Life remained simple and dumb, but it survived. If this latter view is correct, then the fifth factor in our equation is very small, and we are faced with the depressing prospect that humans may be the smartest form of life anywhere in the Galaxy. As with the previous factor, we will be optimistic and simply adopt a value of unity here.

Fraction of Planets on Which Intelligent Life Develops and Uses Technology

To evaluate the sixth factor of our equation, we need to estimate the probability that intelligent life eventually develops technological competence. Should the rise of technology be inevitable, this factor is close to unity, given a long enough time. If it is not inevitable—if intelligent life can somehow "avoid" developing technology—then this factor could be much less than unity. The latter scenario envisions a universe possibly teeming with intelligent civilizations, but very few among them ever becoming technologically competent. Perhaps only one managed it—ours.

Again, it is difficult to decide conclusively between these two views. We don't know how many prehistoric Earth cultures failed to develop simple technology or rejected its use. We do know that the roots of our present civilization arose independently at several different places on Earth, including Mesopotamia, India, China, Egypt, Mexico, and Peru. Because so many of these ancient cultures originated at about the same time, it is tempting to conclude that the chances are good that some sort of technological society will inevitably develop, given some basic intelligence and enough time.

If technology is inevitable, then why haven't other life-forms on Earth also found it useful? Possibly the competitive edge given by intellectual and technological skills to humans, the first species to develop them, allowed us to dominate so rapidly that other species—gorillas and chimpanzees, for example—simply haven't had time to catch up.

The fact that only one technological society exists on Earth does not imply that the sixth factor in our Drake equation must be very much less than unity. On the contrary, it is precisely because *some* species will probably always fill the niche of technological intelligence that we will take this factor to be close to unity.

Average Lifetime of a Technological Civilization

The reliability of the estimate of each factor in the Drake equation declines markedly from left to right. For example, our knowledge of astronomy enables us to make a reasonably good stab at the first factor, namely, the rate of star formation in our Galaxy, but it is much harder to evaluate some of the later factors, such as the fraction of life-bearing planets that eventually develop intelligence. The last factor on the right-hand side of the equation, the longevity of technological civilizations, is totally unknown. There is only one known example of such a civilization: humans on planet Earth. Our own civilization has survived in its "technological" state for only about 100 years, and how long we will be around before a natural or human-made catastrophe ends it all is impossible to tell. ∞ (*Discovery 11-1*)

Number of Technological Civilizations in the Galaxy

One thing is certain: If the correct value for *any one factor* in the equation is very small, then few technological civilizations now exist in the Galaxy. In other words, if the pessimistic view of the development of life or of intelligence is correct, then we are unique, and that is the end of our story. However, if both life and intelligence are inevitable consequences of chemical and biological evolution, as many scientists think, and if intelligent life always becomes technological, then we can plug the higher, more optimistic values into the Drake equation. In that case, combining our estimates for the other six factors (and noting that $10 \times 1 \times 1/10 \times 1 \times 1 \times 1 = 1$), we have

number of technological intelligent civilizations now present in the Milky Way Galaxy	=	average lifetime of a technologically competent civilization in years.

Thus, if civilizations typically survive for 1000 years, there should be 1000 of them currently in existence scattered throughout the Galaxy. If they live for a million years, on average, we would expect there to be a million advanced civilizations in the Milky Way, and so on.

Note that, even setting aside language and cultural issues, the sheer size of the Galaxy presents a significant hurdle to communication between technological civilizations. The minimum requirement for a two-way conversation is that we can send a signal and receive a reply in a time shorter than our own lifetime. If the lifetime is short, then civilizations are literally few and far between—small in number, according to the Drake equation, and scattered over the vastness of the Milky Way—and the distances between them (in light-years) are much greater than their lifetimes (in years). In that case, two-way communication, even at the speed of light, will be impossible. However, as the lifetime increases, the distances get smaller as the Galaxy becomes more crowded, and the prospects improve.

Taking into account the size, shape, and distribution of stars in the Galactic disk (why do we exclude the halo?), and under the optimistic assumptions just made, we find that, unless the life expectancy of a civilization is *at least* a few thousand years, it is unlikely to have time to communicate with even its nearest neighbor.

PROCESS OF SCIENCE CHECK

✔ If most of the factors are largely a matter of opinion, how does the Drake equation assist astronomers in refining their search for extraterrestrial life?

28.4 The Search for Extraterrestrial Intelligence

Let us continue our optimistic assessment of the prospects for life and assume that civilizations enjoy a long stay on their parent planet once their initial technological "teething problems" are past. In that case, intelligent, technological, and perhaps also communicative cultures are likely to be plentiful in the Galaxy. How might we become aware of their existence? The ongoing search for extraterrestrial intelligence (known to many by its acronym, **SETI**) is the topic of this final section.

Meeting Our Neighbors

For definiteness, let's assume that the average lifetime of a technological civilization is 1 million years—only 1 percent of the reign of the dinosaurs, but 100 times longer than human civilization has survived thus far. Given the size and shape of our Galaxy and the known distribution of stars in the Galactic disk, we can then estimate the average *distance* between these civilizations to be some 30 pc, or about 100 light-years. Thus, any two-way communication with our neighbors—using signals traveling at or below the speed of light—will take at least 200 years (100 years for the message

to reach the planet and another 100 years for the reply to travel back to us).

One obvious way to search for extraterrestrial life would be to develop the capability to travel far outside our solar system. However, that may never be a practical possibility. At a speed of 50 km/s, the speed of the fastest space probes operating today, the round-trip to even the nearest Sun-like star, Alpha Centauri, would take about 50,000 years. The journey to the nearest technological neighbor (assuming a distance of 30 pc) and back would take 600,000 years—almost the entire lifetime of our species! Interstellar travel at these speeds is clearly not feasible. Speeding up our ships to near the speed of light would reduce the travel time, but doing that is far beyond our present technology.

Actually, our civilization has already launched some interstellar probes, although they have no specific stellar destination. Figure 28.15 is a reproduction of a plaque mounted on board the *Pioneer 10* spacecraft launched in the mid-1970s and now well beyond the orbit of Pluto, on its way out of the solar system. Similar information was included aboard the *Voyager* probes launched in 1978. Although these spacecraft would be incapable of reporting back to Earth the news that they had encountered an alien culture, scientists hope that the civilization on the other end would be able to unravel most of its contents using the universal language of mathematics. The caption to Figure 28.15 notes how the

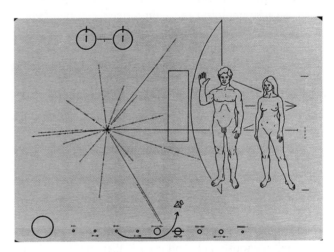

▲ **FIGURE 28.15** *Pioneer 10* **Plaque** A replica of a plaque mounted on board the *Pioneer 10* spacecraft. Included are scale drawings of the spacecraft, a man, and a woman; a diagram of the hydrogen atom undergoing a change in energy (top left); a starburst pattern representing various pulsars and the frequencies of their radio waves that can be used to estimate when the craft was launched (middle left); and a depiction of the solar system, showing that the spacecraft departed the third planet from the Sun and passed the fifth planet on its way to outer space (bottom). All the drawings have computer-coded (binary) markings from which actual sizes, distances, and times can be derived. *(NASA)*

aliens might discover from where and when the *Pioneer* and *Voyager* probes were launched.

Setting aside the many practical problems that arise in trying to establish direct contact with extraterrestrials, some scientists have argued that it might not even be a particularly good idea. Our recent emergence as a technological civilization implies that we must be one of the least advanced technological intelligences in the entire Galaxy. Any other civilization that discovers us will almost surely be more advanced than us. Consequently, a healthy degree of caution may be warranted. If extraterrestrials behave even remotely like human civilizations on Earth, then the most advanced aliens may naturally try to dominate all others. The behavior of the "advanced" European cultures toward the "primitive" races they encountered on their voyages of discovery in the seventeenth, eighteenth, and nineteenth centuries should serve as a clear warning of the possible undesirable consequences of contact. Of course, the aggressiveness of Earthlings may not apply to extraterrestrials, but given the history of the one intelligent species we know, the cautious approach may be in order.

Radio Communication

A cheaper and much more practical alternative to direct contact is to try to communicate with extraterrestrials by using only electromagnetic radiation, the fastest known means of transferring information from one place to another. Because light and other high-frequency radiation is heavily scattered while moving through dusty interstellar space, long-wavelength radio radiation seems to be the natural choice. We would not attempt to broadcast to *all* nearby candidate stars, however—that would be far too expensive and inefficient. Instead, radio telescopes on Earth would listen *passively* for radio signals emitted by other civilizations. Indeed, some preliminary searches of selected nearby stars are now underway, thus far without success.

In what direction should we aim our radio telescopes? The answer to this question, at least, is fairly easy: On the basis of our earlier reasoning, we should target all F-, G-, and K-type stars in our vicinity. But are extraterrestrials broadcasting radio signals? If they are not, this search technique will obviously fail. And even if they are, how do we distinguish their artificially generated radio signals from signals naturally emitted by interstellar gas clouds? To what frequency should we tune our receivers? The answer to this question depends on whether the signals are produced deliberately or are simply "waste radiation" escaping from a planet.

Consider how Earth would look at radio wavelengths to extraterrestrials. Figure 28.16 shows the pattern of radio signals we emit into space. From the viewpoint of a distant observer, the spinning Earth emits a bright flash of radio radiation every few hours. In fact, Earth is now a more intense

radio emitter than the Sun. The flashes result from the periodic rising and setting of hundreds of FM radio stations and television transmitters. Each station broadcasts mostly parallel to Earth's surface, sending a great "sheet" of electromagnetic radiation into interstellar space, as illustrated in Figure 28.16(a). (The more common AM broadcasts are trapped below our ionosphere, so those signals never leave Earth.)

Because the great majority of these transmitters are clustered in the eastern United States and western Europe, a distant observer would detect periodic blasts of radiation from Earth as our planet rotates each day (Figure 28.16b). This radiation races out into space and has been doing so since the invention of these technologies more than seven decades ago. Another civilization at least as advanced as ours might have constructed devices capable of detecting these blasts of radiation. If any sufficiently advanced (and sufficiently interested) civilization resides on a planet orbiting any of the thousand or so stars within about 70 light-years (20 pc) of Earth, then we have already broadcast our presence to them.

Of course, it may very well be that, having discovered cable and fiber-optics technology, most civilizations' indiscriminate transmissions cease after a few decades. In that case, radio silence becomes the hallmark of intelligence, and we must find an alternative means of locating our neighbors.

The Water Hole

Now let us suppose that a civilization has decided to assist searchers by actively broadcasting its presence to the rest of the Galaxy. At what frequency should we listen for such an extraterrestrial beacon? The electromagnetic spectrum is enormous; the radio domain alone is vast. To hope to detect a signal at some unknown radio frequency is like searching for a needle in a haystack. Are some frequencies more likely than others to carry alien transmissions?

Some basic arguments suggest that civilizations might communicate at a wavelength near 20 cm. As we saw in Chapter 18, the basic building blocks of the universe, namely, hydrogen atoms, radiate naturally at a wavelength of 21 cm. ∞ (Sec. 18.4) Also, one of the simplest molecules, hydroxyl (OH), radiates near 18 cm. Together, these two substances form water (H_2O). Arguing that water is likely to be the interaction medium for life anywhere and that radio radiation travels through the disk of our Galaxy with the least absorption by interstellar gas and dust, some researchers have proposed that the interval between 18 and 21 cm is the best range of wavelengths for civilizations to transmit or monitor. Called the **water hole,** this radio interval might serve as an "oasis" where all advanced galactic civilizations would gather to conduct their electromagnetic business.

The water-hole frequency interval is only a guess, of course, but it is supported by other arguments as well.

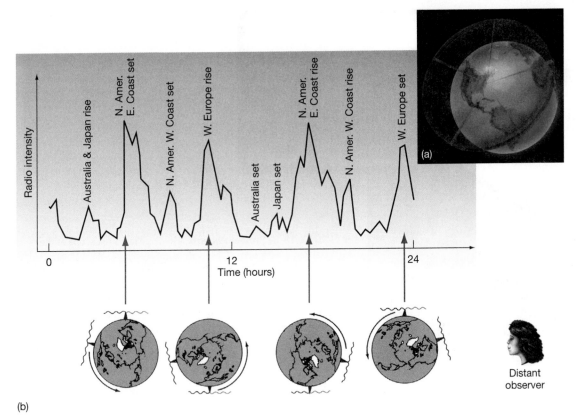

(b)

▲ FIGURE 28.16 Leakage Radio radiation leaks from Earth into space because of the daily activities of our technological civilization. (a) FM radio and television transmitters broadcast their energy parallel to Earth's surface, producing the strongest signal in any given direction when they lie on Earth's horizon as seen from afar. (The more common AM signals are trapped below our ionosphere and never leave Earth.) (b) Because most transmitters are clustered in the eastern United States and western Europe, a distant observer would detect blasts of radio radiation from Earth as our planet rotates.

Figure 28.17 shows the water hole's location in the electromagnetic spectrum and plots the amount of natural emission from our Galaxy and from Earth's atmosphere. The 18- to 21-cm range lies within the quietest part of the spectrum, where the galactic "static" from stars and interstellar clouds happens to be minimized. Furthermore,

the atmospheres of typical planets—or, at least, planets comparable to Earth—are also expected to interfere least at these wavelengths. Thus, the water hole seems like a good choice for the frequency of an interstellar beacon, although we cannot be sure of this reasoning until contact is actually achieved.

A few radio searches are now in progress at frequencies in and around the water hole. The latest generation currently under development will use arrays of hundreds of relatively

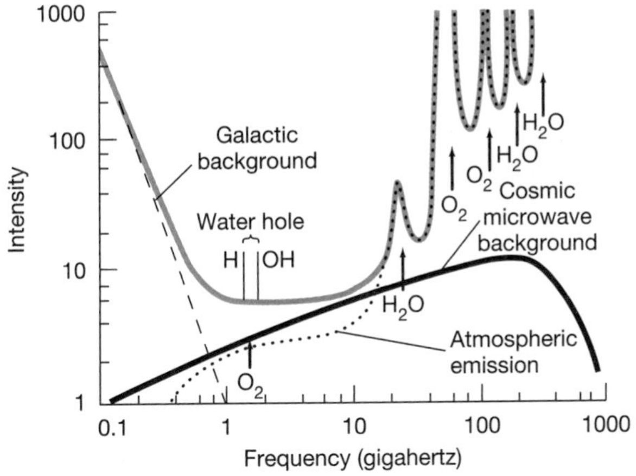

◀ FIGURE 28.17 Water Hole The "water hole" is bounded by the natural emission frequencies of the hydrogen (H) atom (21-cm wavelength) and the hydroxyl (OH) molecule (18-cm wavelength). ∞ (Secs. 18.4, 18.5) The topmost solid (blue) curve sums the natural emissions of our Galaxy (dashed line on left side of diagram, labeled "Galactic background") and Earth's atmosphere (dotted line on right side of diagram, denoted by various chemical symbols). ∞ (Sec. 26.7) This sum is minimized near the water-hole frequencies. Perhaps all intelligent civilizations conduct their interstellar communications within this quiet "electromagnetic oasis." If extraterrestrials want to be found, the reasoning goes, this is where we should look!

◀ **FIGURE 28.18 Project Phoenix** This large radio telescope at the National Radio Astronomy Observatory in Green Bank, West Virginia, was used by Project Phoenix during the 1990s to search for extraterrestrial intelligent signals. The inset shows a typical recording of an alien signal—here, as a test, the Doppler-shifted broadcast from the *Pioneer 10* spacecraft, now well beyond the orbit of Neptune. The diagonal line across the computer monitor, in contrast to the random noise in the background, betrays the presence of an intelligent signal in the incoming data stream. *(NRAO; SETI Institute)*

low-cost small antennas to mimic a much larger collecting area. One of the most sensitive and comprehensive SETI projects already completed was Project Phoenix, carried out during the late 1990s. Large radio antennas, such as that in Figure 28.18, were used to search millions of channels simultaneously in the 1- to 3-GHz spectrum. Actually, in these searches, computers do most of the "listening;" humans get involved only if the signals look intriguing. The inset in Figure 28.18 shows what a typical narrowband, 1-Hz signal— a potential "signature" of an intelligent transmission—would look like on a computer monitor. However, this observation was merely a test to detect the weak, redshifted radio signal emitted by the *Pioneer 10* robot, now receding into the outer realm of our solar system—a sign of intelligence, but one that

we put there. Nothing resembling an extraterrestrial signal has yet been detected.

Right now, the space surrounding all of us could be flooded with radio signals from extraterrestrial civilizations. If only we knew the proper direction and frequency, we might be able to make one of the most startling discoveries of all time. The result would likely provide whole new opportunities to study the cosmic evolution of energy, matter, and life throughout the universe.

PROCESS OF SCIENCE CHECK

✔ Why do many researchers regard the "water hole" as a likely place to search for extraterrestrial signals?

CHAPTER REVIEW

SUMMARY

1 **Cosmic evolution (p. 708)** is the continuous process that has led to the appearance of galaxies, stars, planets, and life on Earth. Living organisms may be  characterized by their ability to react to their environment, to grow by taking in nutrition from their surroundings, and to reproduce, passing along some of their own characteristics to their offspring. Organisms that can best take advantage of their new surroundings succeed at the expense of those organisms that cannot make the necessary adjustments. Intelligence is strongly favored by natural selection.

2 Powered by natural energy sources, reactions between simple molecules in the oceans of the primitive Earth are thought to have led to the formation of **amino acids (p. 710)** and **nucleotide bases (p. 710)**, the basic molecules of life.  Alternatively, some complex molecules may have been formed in interstellar space and then delivered to Earth by meteors or comets. The best hope for life beyond Earth in the solar system is the planet Mars, although no evidence for living organisms has been found there. Jupiter's moon Europa and Saturn's Titan may also be possibilities, but conditions on those bodies are harsh by terrestrial standards.

3 The **Drake equation (p. 716)** provides a means of estimating the probability of intelligent life in the Galaxy. The astronomical factors in the equation are the galactic star-formation rate, the likelihood of planets, and the number of habitable planets. Chemical and biological factors are the probability that life appears and the probability that it subsequently develops intelligence. Cultural and political factors are the probability that intelligence leads to technology and the lifetime of a civilization in the technological state. Taking an optimistic view of the development of life and intelligence  leads to the conclusion that the total number of technologically competent civilizations in the Galaxy is approximately equal to the lifetime of a typical civilization, expressed in years. Even with optimistic assumptions, the distance to our nearest intelligent neighbor is likely to be many hundreds of parsecs.

4 Currently, space travel is not a feasible means of searching for intelligent life. Existing programs to discover extraterrestrial intelligence involve scanning the electromagnetic spectrum for signals. So 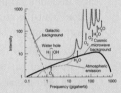 far, no intelligible broadcasts have been received. A technological civilization would probably "announce" itself to the universe by the radio and television signals it emits into space. Observed from afar, our planet would appear as a radio source with a 24-hour period, as different regions of the planet rise and set. The "**water hole**" **(p. 722)** is a region in the radio range of the electromagnetic spectrum, near the 21-cm line of hydrogen and the 18-cm line of hydroxyl, where natural emissions from the Galaxy happen to be minimized. Many researchers regard this region as the best part of the spectrum for communication purposes.

Mastering ASTRONOMY   *For instructor-assigned homework go to* **www.masteringastronomy.com**

Problems labeled **POS** explore the process of science | **VIS** problems focus on reading and interpreting visual information

REVIEW AND DISCUSSION

1. Why is life difficult to define?

2. What is chemical evolution?

3. What is the Urey-Miller experiment? What important organic molecules were produced in that experiment?

4. What other experiments have attempted to produce organic molecules by inorganic means?

5. What are the basic ingredients from which biological molecules formed on Earth?

6. Why do some scientists think life might have originated in space?

7. **POS**   How do we know anything at all about the early episodes of life on Earth?

8. What is the role of language in cultural evolution?

9. Where else, besides Earth, have organic molecules been found?

10. Where—besides Earth and the planet Mars—might we hope to find signs of life in our solar system?

11. **POS**   Do we know whether Mars ever had life at any time during its past? What argues in favor of the position that it may once have harbored life?

12. **POS**   What is generally meant by "life as we know it"? What other forms of life might be possible?

13. How many of the factors in the Drake equation are known with any degree of certainty? Which factor is least well known?

14. What factors determine the suitability of a star as the parent of a planet on which life might arise?

15. What is the relationship between the average lifetime of galactic civilizations and the possibility of our someday communicating with them?

16. How would Earth appear at radio wavelengths to extraterrestrial astronomers?

17. Do you think that advanced civilizations would continue to emit large amounts of radio energy as they evolved?

18. What are the advantages in using radio waves for communication over interstellar distances?

19. **POS** What is the "water hole"? What advantages does it offer for interstellar communication?

20. If you were designing a SETI experiment, what parts of the sky would you monitor?

CONCEPTUAL SELF-TEST: MULTIPLE CHOICE

1. The "assumptions of mediocrity" suggest that (a) life should be common throughout the cosmos; (b) lower forms of life must evolve to higher forms; (c) lower forms of life have lower intelligence; (d) viruses are actually life-forms.

2. The chemical elements that form the basic molecules needed for life are found (a) in the cores of Sun-like stars; (b) commonly throughout the cosmos; (c) only on planets that have liquid water; (d) only on Earth.

3. Fossil records of early life-forms on Earth suggest that life began about (a) 6000 years ago; (b) 65 million years ago; (c) 3.5 billion years ago; (d) 14 billion years ago.

4. The discovery of bacteria on another planet would be an important discovery because bacteria (a) can easily survive in high temperatures; (b) are the only life-form to exist on Earth for most of the planet's history; (c) are the lowest form of life known to exist; (d) eventually evolve into intelligent beings.

5. The least-well-known factor in the Drake equation is (a) the rate of star formation; (b) the average number of habitable planets within planetary systems; (c) the average lifetime of a technologically competent civilization; (d) the diameter of the Milky Way Galaxy.

6. Although the habitable zone around a large B-class star is large, we don't often look for life on planets there because the star (a) has too much gravity; (b) is too short lived for life to evolve; (c) is at too low a temperature to sustain life; (d) would have only gas giant planets.

7. **VIS** From the data shown in Figure 28.12 ("Stellar Habitable Zones") and your knowledge of stellar properties (Chapter 18), the habitable zone surrounding a main sequence K-type star (a) cannot be determined; (b) extends more than 3 AU from the star; (c) is larger than that of a G-type star; (d) is larger than that of an M-type star.

8. **VIS** If Figure 28.16 ("Leakage") were to be redrawn for a planet spinning twice as fast, the new jagged line would be (a) unchanged; (b) taller; (c) stretched out horizontally; (d) compressed horizontally.

9. Radio telescopes cannot simply scan the skies looking for signals, because (a) astronomers don't know what frequencies alien civilizations might use; (b) many nonliving objects emit radio signals naturally; (c) Earth's radio communications drown out extraterrestrial signals; (d) inclement weather in the winter prevents the use of radio telescopes.

10. The strongest radio-wavelength emitter in the solar system is (a) human-made signals from Earth; (b) the Sun; (c) the Moon; (d) Jupiter.

PROBLEMS

The number of dots preceding each Problem indicates its approximate level of difficulty.

1. • If Earth's 4.6-billion-year age were compressed to 46 years, as described in the text, what would be your age, in seconds? On that scale, how long ago was the end of World War II? The Declaration of Independence? Columbus's discovery of the New World? The extinction of the dinosaurs?

2. ••• According to the inverse-square law, a planet receives energy from its parent star at a rate proportional to the star's luminosity and inversely proportional to the square of the planet's distance from the star. ∞ (Sec. 17.2) According to Stefan's law, the rate at which the planet radiates energy into space is proportional to the fourth power of its surface temperature. ∞ (Sec. 3.4) In equilibrium, the two rates are equal. Based on this information, and given the fact that (taking into account the greenhouse effect) the Sun's habitable zone extends from 0.6 AU to 1.5 AU, estimate the extent of the habitable zone surrounding a K-type main sequence star of the luminosity of the Sun.

3. •• Using the data in the previous problem, how would the inner and outer radii of the Sun's habitable zone change if the solar luminosity increased by a factor of four?

4. • Based on the numbers presented in the text, and assuming an average lifetime of 5 billion years for suitable stars, estimate the total number of habitable planets in the Galaxy.

5. •• A planet orbits one component of a binary-star system at a distance of 1 AU. (See Figure 28.14a.) If both stars have the same mass and their orbit is circular, estimate the minimum distance between the stars for the tidal force due to the

companion not to exceed a "safe" 0.01 percent of the gravitational force between the planet and its parent star.

6. • Suppose that each of the fractional factors in the Drake equation turns out to have a value of 1/10, that stars form at an average rate of 20 per year, and that each star has exactly one habitable planet orbiting it. Estimate the present number of technological civilizations in the Milky Way Galaxy if the average lifetime of a civilization is (a) 100 years; (b) 10,000 years; (c) 1 million years.

7. ••• If we adopt the estimate from the text that the number of technological civilizations in the Milky Way Galaxy is equal to the average lifetime of a civilization, it follows that the distance to our nearest neighbor decreases as the average lifetime increases. Assuming that civilizations are uniformly spread over a two-dimensional Galactic disk of radius 15 kpc and that all have the same lifetime, calculate the *minimum* lifetime for which two-way radio communication with our nearest neighbor would be possible before our civilization ends. Is the assumption that civilizations are uniformly spread across the disk a reasonable one?

8. • Assuming that there are 10,000 FM radio stations on Earth, each transmitting at a power level of 50 kW, calculate the total radio luminosity of Earth in the FM band. Compare this value with the roughly 10^6 W radiated by the Sun in the same frequency range.

9. • Convert the water hole's wavelengths to frequencies. For practical reasons, any search of the water hole must be broken up into channels, much like those you find on a television, except that the water hole's channels are very narrow in radio frequency, about 100 Hz wide. How many channels must astronomers search in the water hole?

10. • There are 20,000 stars within 100 light-years that are to be searched for radio communications. How long will the search take if 1 hour is spent looking at each star? What if 1 day is spent per star?

APPENDIX 1 SCIENTIFIC NOTATION

The objects studied by astronomers range in size from the smallest particles to the largest expanse of matter we know—the entire universe. Subatomic particles have sizes of about 0.000000000000001 meter, while galaxies (like that shown in Figure 1.3) typically measure some 1,000,000,000,000,000,000,000 meters across. The most distant known objects in the universe lie on the order of 100,000,000,000,000,000,000,000,000 meters from Earth.

Obviously, writing all those zeros is both cumbersome and inconvenient. More important, it is also very easy to make an error—write down one zero too many or too few and your calculations become hopelessly wrong! To avoid this, scientists always write large numbers using a short-hand notation in which the number of zeros following or preceding the decimal point is denoted by a superscript power, or *exponent*, of 10. The exponent is simply the number of places between the first significant (nonzero) digit in the number (reading from left to right) and the decimal point. Thus, 1 is 10^0, 10 is 10^1, 100 is 10^2, 1000 is 10^3, and so on. For numbers less than 1, with zeros between the decimal point and the first significant digit, the exponent is negative: 0.1 is 10^{-1}, 0.01 is 10^{-2}, 0.001 is 10^{-3}, and so on. Using this notation we can shorten the number describing subatomic particles to 10^{-15} meter, and write the number describing the size of a galaxy as 10^{21} meters.

More complicated numbers are expressed as a combination of a power of 10 and a multiplying factor. This factor is conventionally chosen to be a number between 1 and 10, starting with the first significant digit in the original number. For example, 150,000,000,000 meters (the distance from Earth to the Sun, in round numbers) can be more concisely written at 1.5×10^{11} meters, 0.000000025 meters as 2.5×10^{-8} meter, and so on. The exponent is simply the number of places the decimal point must be moved *to the left* to obtain the multiplying factor.

Some other examples of scientific notation are:

- the approximate distance to the Andromeda Galaxy
  = 2,500,000 light-years = 2.5×10^6 light-years

- the size of a hydrogen atom
  = 0.00000000005 meter = 5×10^{-11} meter

- the diameter of the Sun
  = 1,392,000 kilometers = 1.392×10^6 kilometers

- the U.S. national debt (as of May 1, 2010)
  = \$12,948,739,000,000.00 = \$12.948739 trillion
  = 1.294874×10^{13} dollars.

In addition to providing a simpler way of expressing very large or very small numbers, this notation also makes it easier to do basic arithmetic. The rule for multiplication of numbers expressed in this way is simple: Just multiply the factors and add the exponents. Similarly for division: Divide the factors and subtract the exponents. Thus, 3.5×10^{-2} multiplied by 2.0×10^3 is simply $(3.5 \times 2.0) \times 10^{-2+3} = 7.0 \times 10^1$ —that is, 70. Again, 5×10^6 divided by 2×10^4 is just $(5/2) \times 10^{6-4}$, or $2.5 \times 10^2 (= 250)$. Applying these rules to unit conversions, we find, for example, that 200,000 nanometers is $200,000 \times 10^{2-9}$ meter (since 1 nanometer = 10^{-9} meter; see Appendix 2), or $2 \times 10^5 \times 10^{-9}$ meter, or $2 \times 10^{5-9} = 2 \times 10^{-4}$ meter = 0.2 mm. Verify these rules for yourself with a few examples of your own. The advantages of this notation when considering astronomical objects will soon become obvious.

Scientists often use "rounded-off" versions of numbers, both for simplicity and for ease of calculation. For example, we will usually write the diameter of the Sun as 1.4×10^6 kilometers, instead of the more precise number given earlier. Similarly, Earth's diameter is 12,756 kilometers, or 1.2756×10^4 kilometers, but for "ballpark" estimates we really don't need so many digits and the more approximate number 1.3×10^4 kilometers will suffice. Very often, we perform rough calculations using only the first one or two significant digits in a number, and that may be all that is necessary to make a particular point. For example, to support the statement, "The Sun is much larger than Earth," we need only say that the ratio of the two diameters is roughly 1.4×10^6 divided by 1.3×10^4. Since 1.4/1.3 is close to 1, the ratio is approximately $10^6/10^4 = 10^2$, or 100. The essential fact here is that the ratio is much larger than 1; calculating it to greater accuracy (to get 109.13) would give us no additional *useful* information. This technique of stripping away the arithmetic details to get to the essence of a calculation is very common in astronomy, and we use it frequently throughout this text.

APPENDIX 2 ASTRONOMICAL MEASUREMENT

Astronomers use many different kinds of units in their work, simply because no single system of units will do. Rather than the *Système Internationale* (SI), or meter-kilogram-second (MKS), metric system used in most high school and college science classes, many professional astronomers still prefer the older centimeter-gram-second (CGS) system. However, astronomers also commonly introduce new units when convenient. For example, when discussing stars, the mass and radius of the Sun are often used as reference points. The solar mass, written as $M_\odot$, is equal to 2.0×10^{33} g, or 2.0×10^{30} kg (since 1 kg = 1000 g). The solar radius, $R_\odot$, is equal to 700,000 km, or 7.0×10^8 m (1 km = 1000 m). The subscript $\odot$ always stands for the Sun. Similarly, the subscript $\oplus$ always stands for Earth. In this book, we try to use the units that astronomers commonly use in any given context, but we also give the "standard" SI equivalents where appropriate.

Of particular importance are the units of length astronomers use. On small scales, the *angstrom* (1 Å = 10^{-10} m = 10^{-8} cm), the *nanometer* (1 nm = 10^{-9} m = 10^{-7} cm), and the *micron* (1 μm = 10^{-6} m = 10^{-4} cm) are used. Distances within the solar system are usually expressed in terms of the *astronomical unit* (AU), the mean distance between Earth and the Sun. One AU is approximately equal to 150,000,000 km, or 1.5×10^{11} m. On larger scales, the *light-year* (1 ly = 9.5×10^{15} m = 9.5×10^{12} km) and the *parsec* (1 pc = 3.1×10^{16} m = 3.1×10^{13} km = 3.3 ly) are commonly used. Still larger distances use the regular prefixes of the metric system: *kilo* for one thousand and *mega* for one million. Thus 1 kiloparsec (kpc) = 10^3 pc = 3.1×10^{19} m, 10 megaparsecs (Mpc) = 10^7 pc = 3.1×10^{23} m, and so on.

Astronomers use units that make sense within a context, and as contexts change, so do the units. For example, we might measure densities in grams per cubic centimeter (g/cm^3), in atoms per cubic meter (atoms/m^3), or even in solar masses per cubic megaparsec ($M_\odot$/Mpc^3), depending on the circumstances. The important thing to know is that once you understand the units, you can convert freely from one set to another. For example, the radius of the Sun could equally well be written as $R_\odot = 6.96 \times 10^3$ m, or 6.96×10^{10} cm, or 109 $R_\oplus$, or 4.65×10^{-3} AU, or even 7.363×10^{-5} ly—whichever happens to be most useful. Some of the more common units used in astronomy, and the contexts in which they are most likely to be encountered, are listed below.

Length:		
1 angstrom (Å)	= 10^{-10} m	
1 nanometer (nm)	= 10^{-9} m	atomic physics, spectroscopy
1 micron (μm)	= 10^{-6} m	interstellar dust and gas
1 centimeter (cm)	= 0.01 m	
1 meter (m)	= 100 cm	in widespread use throughout all astronomy
1 kilometer (km)	= 1000 m = 10^5 cm	
Earth radius ($R_\oplus$)	= 6378 km	planetary astronomy
Solar radius ($R_\odot$)	= 6.96×10^8 m	
1 astronomical unit (AU)	= 1.496×10^{11} m	solar system, stellar evolution
1 light-year (ly)	= 9.46×10^{15} m = 63,200 AU	
1 parsec (pc)	= 3.09×10^{16} m = 206,000 AU	galactic astronomy, stars and star clusters
	= 3.26 ly	
1 kiloparsec (kpc)	= 1000 pc	
1 megaparsec (Mpc)	= 1000 kpc	galaxies, galaxy clusters, cosmology
Mass:		
1 gram (g)		
1 kilogram (kg)	= 1000 g	in widespread use in many different areas
Earth mass ($M_\oplus$)	= 5.98×10^{24} kg	planetary astronomy
Solar mass ($M_\odot$)	= 1.99×10^{30} kg	"standard" unit for all mass scales larger than Earth
Time:		
1 second (s)		in widespread use throughout astronomy
1 hour (h)	= 3600 s	
1 day (d)	= 86,400 s	planetary and stellar scales
1 year (yr)	= 3.16×10^7 s	virtually all processes occurring on scales larger than a star

APPENDIX 3 TABLES

TABLE 1 Some Useful Constants and Physical Measurements*	
astronomical unit	$1\ \text{AU} = 1.496 \times 10^8\ \text{km}\ (1.5 \times 10^8\ \text{km})$
light-year	$1\ \text{ly} = 9.46 \times 10^{12}\ \text{km}\ (10^{13}\ \text{km, about 6 trillion miles})$
parsec	$1\ \text{pc} = 3.09 \times 10^{13}\ \text{km} = 206{,}000\ \text{AU} = 3.3\ \text{ly}$
speed of light	$c = 299{,}792.458\ \text{km/s}\ (3 \times 10^5\ \text{km/s})$
Stefan-Boltzmann constant	$a = 5.67 \times 10^{-8}\ \text{W/m}^2 \cdot \text{K}^4$
Planck's constant	$h = 6.63 \times 10^{-34}\ \text{J s}$
gravitational constant	$G = 6.67 \times 10^{-11}\ \text{Nm}^2/\text{kg}^2$
mass of Earth	$M_\oplus = 5.98 \times 10^{24}\ \text{kg}\ (6 \times 10^{24}\ \text{kg, about 6000 billion billion tons})$
radius of Earth	$R_\oplus = 6378\ \text{km}\ (6500\ \text{km})$
mass of the Sun	$M_\odot = 1.99 \times 10^{30}\ \text{kg}\ (2 \times 10^{30}\ \text{kg})$
radius of the Sun	$R_\odot = 6.96 \times 10^5\ \text{km}\ (7 \times 10^5\ \text{km})$
luminosity of the Sun	$L_\odot = 3.90 \times 10^{26}\ \text{W}\ (4 \times 10^{26}\ \text{W})$
effective temperature of the Sun	$T_\odot = 5778\ \text{K}\ (5800\ \text{K})$
Hubble's constant	$H_0 = 70\ \text{km/s/Mpc}$
mass of an electron	$m_e = 9.11 \times 10^{-31}\ \text{kg}$
mass of a proton	$m_p = 1.67 \times 10^{-27}\ \text{kg}$

The rounded-off values used in the text are shown above in parentheses.

Conversions Between Common English and Metric Units

English	Metric
1 inch	= 2.54 centimeters (cm)
1 foot (ft)	= 0.3048 meters (m)
1 mile	= 1.609 kilometers (km)
1 pound (lb)	= 453.6 grams (g) or 0.4536 kilograms (kg) [on Earth]

TABLE 2 Periodic Table of Elements

Group	1	2	3	4	5	6	7	8	9	10	11	12	13	14	15	16	17	18
Period 1	1 **H** 1.0080 Hydrogen																	2 **He** 4.003 Helium
2	3 **Li** 6.939 Lithium	4 **Be** 9.012 Beryllium											5 **B** 10.81 Boron	6 **C** 12.011 Carbon	7 **N** 14.007 Nitrogen	8 **O** 15.9994 Oxygen	9 **F** 18.998 Fluorine	10 **Ne** 20.183 Neon
3	11 **Na** 22.990 Sodium	12 **Mg** 24.31 Magnesium											13 **Al** 26.98 Aluminum	14 **Si** 28.09 Silicon	15 **P** 30.974 Phosphorus	16 **S** 32.064 Sulfur	17 **Cl** 35.453 Chlorine	18 **Ar** 39.948 Argon
4	19 **K** 39.10 Potassium	20 **Ca** 40.08 Calcium	21 **Sc** 44.96 Scandium	22 **Ti** 47.87 Titanium	23 **V** 50.94 Vanadium	24 **Cr** 52.00 Chromium	25 **Mn** 53.94 Manganese	26 **Fe** 55.85 Iron	27 **Co** 58.93 Cobalt	28 **Ni** 58.69 Nickel	29 **Cu** 63.55 Copper	30 **Zn** 65.39 Zinc	31 **Ga** 69.72 Gallium	32 **Ge** 72.61 Germanium	33 **As** 74.92 Arsenic	34 **Se** 78.96 Selenium	35 **Br** 79.904 Bromine	36 **Kr** 83.80 Krypton
5	37 **Rb** 85.47 Rubidium	38 **Sr** 87.62 Strontium	39 **Y** 88.91 Yttrium	40 **Zr** 91.22 Zirconium	41 **Nb** 92.91 Niobium	42 **Mo** 95.94 Molybdenum	43 **Tc** (99) Technetium	44 **Ru** 101.07 Ruthenium	45 **Rh** 102.91 Rhodium	46 **Pd** 106.42 Palladium	47 **Ag** 107.87 Silver	48 **Cd** 112.41 Cadmium	49 **In** 114.82 Indium	50 **Sn** 118.71 Tin	51 **Sb** 121.76 Antimony	52 **Te** 127.60 Tellurium	53 **I** 126.904 Iodine	54 **Xe** 131.29 Xenon
6	55 **Cs** 132.91 Cesium	56 **Ba** 137.33 Barium	71 **Lu** 174.97 Lutetium	72 **Hf** 178.49 Hafnium	73 **Ta** 180.95 Tantalum	74 **W** 183.84 Tungsten	75 **Re** 186.21 Rhenium	76 **Os** 190.23 Osmium	77 **Ir** 192.22 Iridium	78 **Pt** 195.09 Platinum	79 **Au** 196.97 Gold	80 **Hg** 200.59 Mercury	81 **Tl** 204.38 Thallium	82 **Pb** 207.20 Lead	83 **Bi** 208.98 Bismuth	84 **Po** (209) Polonium	85 **At** (210) Astantine	86 **Rn** (222) Radon
7	87 **Fr** (223) Francium	88 **Ra** (226) Radium	103 **Lw** (262) Lawrencium	104 **Rf** (263) Rutherfordium	105 **Db** (262) Dubnium	106 **Sg** (266) Seaborgium	107 **Bh** (264) Bohrium	108 **Hs** (269) Hassium	109 **Mt** (268) Meitnerium	110 **Ds** (272) Darmstadtium	111 **Rg** (272) Roentgenium	112 **Cn** (277) Copernicium	113 **Uut** (284) Ununtrium	114 **Uuq** (289) Ununquadium	115 **Uup** (288) Ununpentium	116 **Uuh** (292) Ununhexium	117 **Uus** (294) Ununseptium	118 **Uuo** (294) Ununoctium

Key: 2 Atomic number / **He** Symbol of element / 4.003 Atomic weight / Helium Name of element

* Lanthanum series:

| 57 **La** 138.91 Lanthanum | 58 **Ce** 140.12 Cerium | 59 **Pr** 140.91 Praseodymium | 60 **Nd** 144.24 Neodymium | 61 **Pm** (145) Promethium | 62 **Sm** 150.36 Samarium | 63 **Eu** 151.96 Europium | 64 **Gd** 157.25 Gadolinium | 65 **Tb** 158.93 Terbium | 66 **Dy** 162.50 Dysprosium | 67 **Ho** 164.93 Holmium | 68 **Er** 167.26 Erbium | 69 **Tm** 168.93 Thulium | 70 **Yb** 173.04 Ytterbium |

** Actinium series:

| 89 **AC** (227) Actinium | 90 **Th** 232.04 Thorium | 91 **Pa** 231.03 Protactinium | 92 **U** 238.03 Uranium | 93 **Np** (237) Neptunium | 94 **Pu** (242) Plutonium | 95 **Am** (243) Americium | 96 **Cm** (247) Curium | 97 **Bk** (247) Berkelium | 98 **Cf** (249) Californium | 99 **Es** (252) Einsteinium | 100 **Fm** (257) Fermium | 101 **Md** (258) Mendelevium | 102 **No** (259) Nobelium |

Element 117 was discovered in 2010. Element 118 was "discovered" in 1999, retracted in 2002, and reported again in 2006.

TABLE 3A   Planetary Orbital Data

Planet	Semi-Major Axis (AU)	Semi-Major Axis (10⁶ km)	Eccentricity (e)	Perihelion (AU)	Perihelion (10⁶ km)	Aphelion (AU)	Aphelion (10⁶ km)
Mercury	0.39	57.9	0.206	0.31	46.0	0.47	69.8
Venus	0.72	108.2	0.007	0.72	107.5	0.73	108.9
Earth	1.00	149.6	0.017	0.98	147.1	1.02	152.1
Mars	1.52	227.9	0.093	1.38	206.6	1.67	249.2
Jupiter	5.20	778.4	0.048	4.95	740.7	5.46	816
Saturn	9.54	1427	0.054	9.02	1349	10.1	1504
Uranus	19.19	2871	0.047	18.3	2736	20.1	3006
Neptune	30.07	4498	0.009	29.8	4460	30.3	4537

Planet	Mean Orbital Speed (km/s)	Sidereal Period (tropical years)	Synodic Period (days)	Inclination to the Ecliptic (degrees)	Greatest Angular Diameter as Seen from Earth (arc seconds)
Mercury	47.87	0.24	115.88	7.00	13
Venus	35.02	0.62	583.92	3.39	64
Earth	29.79	1.00	—	0.01	—
Mars	24.13	1.88	779.94	1.85	25
Jupiter	13.06	11.86	398.88	1.31	50
Saturn	9.65	29.42	378.09	2.49	21
Uranus	6.80	83.75	369.66	0.77	4.1
Neptune	5.43	163.7	367.49	1.77	2.4

TABLE 3B   Planetary Physical Data

Planet	Equatorial Radius (km)	Equatorial Radius (Earth = 1)	Mass (kg)	Mass (Earth = 1)	Mean Density (kg/m³)	Surface Gravity (Earth = 1)	Escape Speed (km/s)
Mercury	2440	0.38	3.30×10^{23}	0.055	5430	0.38	4.2
Venus	6052	0.95	4.87×10^{24}	0.82	5240	0.91	10.4
Earth	6378	1.00	5.97×10^{24}	1.00	5520	1.00	11.2
Mars	3394	0.53	6.42×10^{23}	0.11	3930	0.38	5.0
Jupiter	71,492	11.21	1.90×10^{27}	317.8	1330	2.53	60
Saturn	60,268	9.45	5.68×10^{26}	95.16	690	1.07	36
Uranus	25,559	4.01	8.68×10^{25}	14.54	1270	0.91	21
Neptune	24,766	3.88	1.02×10^{26}	17.15	1640	1.14	24

Planet	Sidereal Rotation Period (solar days)*	Axial Tilt (degrees)	Surface Magnetic Field (Earth = 1)	Magnetic Axis Tilt (degrees relative to rotation axis)	Albedo†	Surface Temperature‡ (K)	Number of Moons**
Mercury	58.6	0.0	0.011	<10	0.11	100–700	0
Venus	−243.0	177.4	<0.001		0.65	730	0
Earth	0.9973	23.45	1.0	11.5	0.37	290	1
Mars	1.026	23.98	0.001		0.15	180–270	2
Jupiter	0.41	3.08	13.89	9.6	0.52	124	16
Saturn	0.44	26.73	0.67	0.8	0.47	97	18
Uranus	−0.72	97.92	0.74	58.6	0.50	58	27
Neptune	0.67	29.6	0.43	46.0	0.5	59	13

*A negative sign indicates retrograde rotation; †Fraction of sunlight reflected from surface; ‡Temperature is effective temperature for jovian planets; **Moons more than 10 km in diameter.

TABLE 4 Some Missions to Planets in Our Solar Sytem

Target Planet	Year(s) at Planet	Project	Launched by	Type of Mission	Scientific Achievements or Goals, and Other Comments
Mercury	1974–1975	*Mariner 10*	U.S.	flyby	Photographed 45 percent of the planet's surface
	2008–2011	*Messenger*	U.S.	flyby/orbit	Mapped 90% of the planet; studied the magnetosphere
Venus	1967	*Venera 4*	U.S.S.R.	atmospheric probe	First probe to enter the planet's atmosphere
	1970	*Venera 7*	U.S.S.R.	lander	First landing on the planet's surface
	1978–1992	*Pioneer Venus*	U.S.	orbiter	Radar mapping of the entire planet
	1983	*Venera 15, 16*	U.S.S.R.	orbiter	Radar mapping of northern hemisphere
	1990–1994	*Magellan*	U.S.	atmospheric probe, orbiter	High-resolution radar mapping of the entire planet
	2006–	*Venus Express*	European Space Agency	orbiter	Studies of the dynamics and chemistry of the atmosphere of Venus; volcanism and other surface studies
Mars	1965	*Mariner 4*	U.S.	flyby	Photographs showed cratered surface
	1969	*Mariner 6, 7*	U.S.	flyby	More photographs of the planet's surface
	1971	*Mariner 9*	U.S.	orbiter	First complete survey of the surface; revealed complex terrain and evidence of past geological activity
	1976–1982	*Viking 1, 2*	U.S.	orbiter, lander	Detailed surface maps, implying geological and climatic change; first surface landings, first atmospheric and soil measurements; search for life
	1997	*Mars Pathfinder*	U.S.	lander	First Mars rover, local geological survey
	1997–2006	*Mars Global Surveyor*	U.S.	orbiter	High-resolution surface mapping, remote analysis of surface composition
	2001–	*Mars Odyssey*	U.S.	orbiter	Remote surface chemical analysis, search for sub-surface water, measurement of radiation levels
	2003–	*Mars Express*	European Space Agency	atmospheric probe, orbiter	Studies of Martian atmosphere and geology; search for water and evidence of life
	2004–	*Mars Exploration Rover*	U.S.	two landers	Assessment of likelihood that life arose on Mars, measurement of Martian climate and detailed geological surveys near the landing sites
	2006–	*Mars Reconnaissance Orbiter*	U.S.	orbiter	History of water on Mars, detailed imaging of small-scale surface features
	2008	*Mars Phoenix*	U.S.	lander	Studied the Martian north polar region
Jupiter	1973	*Pioneer 10*	U.S.	flyby	First mission to the outer planets
	1974	*Pioneer 11*	U.S.	flyby	Detailed close-up images; gravity assist from Jupiter to reach Saturn
	1979	*Voyager 1*	U.S.	flyby	Detailed observations of planet and moons
	1979	*Voyager 2*	U.S.	flyby	Continued reconnaissance of the Jovian system
	1995–2003	*Galileo*	U.S.	atmospheric probe, orbiter	Atmospheric studies; long-term precision measurements of the planet's moon system
Saturn	1979	*Pioneer 11*	U.S.	flyby	First close-up observations of Saturn; now leaving the solar system
	1981	*Voyager 1*	U.S.	flyby	Observations of planet, rings and moons; close-up measurements of the moon Titan
	1982	*Voyager 2*	U.S.	flyby	Observations of planet and moons
	2004–	*Cassini-Huygens*	U.S., European Space Agency	atmospheric probe, orbiter	Atmospheric studies; repeated and detailed measurements of the planet's moons
Uranus	1986	*Voyager 2*	U.S.	flyby	Observations of planet and moons; "Grand Tour" of the outer planets
Neptune	1989	*Voyager 2*	U.S.	flyby	Observations of planet and moons; now leaving the solar system

TABLE 5 The Twenty Brightest Stars in Earth's Night Sky

Name	Star	Spectral Type* A	Spectral Type* B	Parallax (arc seconds)	Distance (pc)	Apparent Visual Magnitude* A	Apparent Visual Magnitude* B
Sirius	α CMa	A1V	wd†	0.379	2.6	−1.44	+8.4
Canopus	α Car	F0Ib–II		0.010	96	−0.62	
Arcturus	α Boo	K2III		0.089	11	−0.05	
Rigel Kentaurus (Alpha Centauri)	α Gen	G2V	K0V	0.742	1.3	−0.01	+1.4
Vega	α Lyr	A0V		0.129	7.8	+0.03	
Capella	α Aur	GIII	M1V	0.077	13	+0.08	+10.2
Rigel	β Ori	B8Ia	B9	0.0042	240	+0.18	+6.6
Procyon	α CMi	F5IV–V	wd†	0.286	3.5	+0.40	+10.7
Betelgeuse	α Ori	M2Iab		0.0076	130	+0.45	
Achernar	α Eri	B5V		0.023	44	+0.45	
Hadar	β Cen	B1III	?	0.0062	160	+0.61	+4
Altair	α Aql	A7IV–V		0.194	5.1	+0.76	
Acrux	α Cru	B1IV	B3	0.010	98	+0.77	+1.9
Aldebaran	α Tau	K5III	M2V	0.050	20	+0.87	+13
Spica	α Vir	B1V	B2V	0.012	80	+0.98	2.1
Antares	α Sco	M1Ib	B4V	0.005	190	+1.06	+5.1
Pollux	β Gem	K0III		0.097	10	+1.16	
Formalhaut	α PsA	A3V	?	0.130	7.7	+1.17	+6.5
Deneb	α Cyg	A2Ia		0.0010	990	+1.25	
Mimosa	β Cru	B1IV		0.0093	110	+1.25	

Name	Visual Luminosity* (Sun = 1) A	Visual Luminosity* (Sun = 1) B	Absolute Visual Magnitude A	Absolute Visual Magnitude B	Proper Motion (arc seconds/yr)	Transverse Velocity (km/s)	Radial Velocity (km/s)
Sirius	22	0.0025	+1.5	+11.3	1.33	16.7	−7.6‡
Canopus	1.4×10^4		−5.5		0.02	9.1	20.5
Arcturus	110		−0.3		2.28	119	−5.2
Rigel Kentaurus	1.6	0.45	+4.3	+5.7	3.68	22.7	−24.6
Vega	50		+0.6		0.34	12.6	−13.9
Capella	130	0.01	−0.5	+9.6	0.44	27.1	30.2‡
Rigel	4.1×10^4	110	−6.7	−0.3	0.00	1.2	20.7‡
Procyon	7.2	0.0006	+2.7	+13.0	1.25	20.7	−3.2‡
Betelgeuse	9700		−5.1		0.03	18.5	21.0‡
Achernar	1100		−2.8		0.10	20.9	19
Hadar	1.3×10^4	560	−5.4	−2.0	0.04	30.3	−12‡
Altair	11		+2.2		0.66	16.3	−26.3
Acrux	4100	2200	−4.2	−3.5	0.04	22.8	−11.2
Aldebaran	150	0.002	−0.6	+11.5	0.20	19.0	54.1
Spica	2200	780	−3.5	−2.4	0.05	19.0	1.0‡
Antares	1.1×10^4	290	−5.3	−1.3	0.03	27.0	−3.2
Pollux	31		+1.1		0.62	29.4	3.3
Formalhaut	17	0.13	+1.7	+7.1	0.37	13.5	6.5
Deneb	2.6×10^5		−8.7		0.003	14.1	−4.6‡
Mimosa	3200		−3.9		0.05	26.1	—

*Energy output in the visible part of the spectrum; A and B columns identify individual components of binary-star systems.
†"wd" stands for "white dwarf."
‡Average value of variable velocity.

TABLE 6 The Twenty Nearest Stars

Name	Spectral Type A	B	Parallax (arc seconds)	Distance (pc)	Apparent Visual Magnitude* A	B
Sun	G2V				−26.74	
Proxima Centauri	M5		0.772	1.30	+11.01	
Alpha Centauri	G2V	K1V	0.742	1.35	−0.01	+1.35
Barnard's Star	M5V		0.549	1.82	+9.54	
Wolf 359	M8V		0.421	2.38	+13.53	
Lalande 21185	M2V		0.397	2.52	+7.50	
UV Ceti	M6V	M6V	0.387	2.58	+12.52	+13.02
Sirius	A1V	wd†	0.379	2.64	−1.44	+8.4
Ross 154	M5V		0.345	2.90	+10.45	
Ross 248	M6V		0.314	3.18	+12.29	
ϵ Eridani	K2V		0.311	3.22	+3.72	
Ross 128	M5V		0.298	3.36	+11.10	
61 Cygni	K5V	K7V	0.294	3.40	+5.22	+6.03
ϵ Indi	K5V		0.291	3.44	+4.68	
Grm 34	M1V	M6V	0.290	3.45	+8.08	+11.06
Luyten 789-6	M6V		0.290	3.45	+12.18	
Procyon	F5IV–V	wd†	0.286	3.50	+0.40	+10.7
Σ 2398	M4V	M5V	0.285	3.55	+8.90	+9.69
Lacaille 9352	M2V		0.279	3.58	+7.35	
G51-15	MV		0.278	3.60	+14.81	

Name	Visual Luminosity* (Sun = 1) A	B	Absolute Visual Magnitude* A	B	Proper Motion (arc seconds/yr)	Transverse Velocity (km/s)	Radial Velocity (km/s)
Sun	1.0		+4.83				
Proxima Centauri	5.6×10^{-5}		+15.4		3.86	23.8	−16
Alpha Centauri	1.6	0.45	+4.3	+5.7	3.68	23.2	−22
Barnard's Star	4.3×10^{-4}		+13.2		10.34	89.7	−108
Wolf 359	1.8×10^{-5}		+16.7		4.70	53.0	+13
Lalande 21185	0.0055		+10.5		4.78	57.1	−84
UV Ceti	5.4×10^{-5}	0.00004	+15.5	+16.0	3.36	41.1	+30
Sirius	22	0.0025	+1.5	+11.3	1.33	16.7	−8
Ross 154	4.8×10^{-4}		+13.3		0.72	9.9	−4
Ross 248	1.1×10^{-4}		+14.8		1.58	23.8	−81
ϵ Eridani	0.29		+6.2		0.98	15.3	+16
Ross 128	3.6×10^{-4}		+13.5		1.37	21.8	−13
61 Cygni	0.082	0.039	+7.6	+8.4	5.22	84.1	−64
ϵ Indi	0.14		+7.0		4.69	76.5	−40
Grm 34	0.0061	0.00039	+10.4	+13.4	2.89	47.3	+17
Luyten 789-6	1.4×10^{-4}		+14.6		3.26	53.3	−60
Procyon	7.2	0.00055	+2.7	+13.0	1.25	2.8	−3
Σ 2398	0.0030	0.0015	+11.2	+11.9	2.28	38.4	+5
Lacaille 9352	0.013		+9.6		6.90	117	+10
G51-15	1.1×10^{-5}		+17.0		1.26	21.5	—

*A and B columns identify individual components of binary-star systems.
†"wd" stands for "white dwarf."

GLOSSARY

Key terms that are boldface in the text are followed by a page reference in the Glossary.

A

A ring One of three Saturnian rings visible from Earth. The A ring is farthest from the planet and is separated from the B ring by the Cassini division. (p. 292)

aberration of starlight Small shift in the observed direction to a star, caused by Earth's motion perpendicular to the line of sight. (p. 41)

absolute brightness The apparent brightness a star would have if it were placed at a standard distance of 10 parsecs from Earth.

absolute magnitude The apparent magnitude a star would have if it were placed at a standard distance of 10 parsecs from Earth. (p. 423)

Absolute Zero The lowest possible temperature that can be obtained; all thermal motion ceases at this temperature.

absorption line Dark line in an otherwise continuous bright spectrum, where light within one narrow frequency range has been removed. (p. 80)

abundance Relative amount of different elements in a gas.

acceleration The rate of change of velocity of a moving object. (p. 48)

accretion Gradual growth of bodies, such as planets, by the accumulation of other smaller bodies. (p. 149)

accretion disk Flat disk of matter spiraling down onto the surface of a neutron star or black hole. Often, the matter originated on the surface of a companion star in a binary-star system. (p. 518)

active galactic nucleus Region of intense emission at the center of an active galaxy, responsible for virtually all of the galaxy's nonstellar luminosity. (p. 620)

active galaxies The most energetic galaxies, which can emit hundreds or thousands of times more energy per second than the Milky Way, mostly in the form of long-wavelength nonthermal radiation. (p. 619)

active optics Collection of techniques used to increase the resolution of ground-based telescopes. Minute modifications are made to the overall configuration of an instrument as its temperature and orientation change; used to maintain the best possible focus at all times. (p. 112)

active region Region of the photosphere of the Sun surrounding a sunspot group, which can erupt violently and unpredictably. During sunspot maximum, the number of active regions is also a maximum. (p. 401)

active Sun The unpredictable aspects of the Sun's behavior, such as sudden explosive outbursts of radiation in the form of prominences and flares.

adaptive optics Technique used to increase the resolution of a telescope by deforming the shape of the mirror's surface under computer control while a measurement is being taken; used to undo the effects of atmospheric turbulence. (p. 112)

aerosol Suspension of liquid or solid particles in air.

alpha particle A helium-4 nucleus.

alpha process Process occurring at high temperatures, in which high-energy photons split heavy nuclei to form helium nuclei.

ALSEP Acronym for Apollo Lunar Surface Experiments Package.

altimeter Instrument used to determine altitude.

amino acids Organic molecules that form the basis for building the proteins that direct metabolism in living creatures. (p. 712)

Amor asteroid Asteroid that crosses only the orbit of Mars.

amplitude The maximum deviation of a wave above or below zero point. (p. 59)

angstrom Distance unit equal to 0.1 nanometer, or one ten-billionth of a meter.

angular diameter Angle made between the top (or one edge) of an object, the observer and the bottom (or opposite edge) of the object.

angular distance Angular separation between two objects as seen by some observer.

angular momentum Tendency of an object to keep rotating; proportional to the mass, radius, and rotation speed of the body. (p. 148)

angular resolution The ability of a telescope to distinguish between adjacent objects in the sky. (p. 105)

annular eclipse Solar eclipse occurring at a time when the Moon is far enough away from Earth that it fails to cover the disk of the Sun completely, leaving a ring of sunlight visible around its edge. (p. 19)

antiparallel Configuration of the electron and proton in a hydrogen (or other) atom when their spin axes are parallel but the two rotate in opposite directions.

antiparticle A particle of the same mass but opposite in all other respects (e.g., charge) to a given particle; when a particle and its antiparticle come into contact, they annihilate and release energy in the form of gamma rays.

aphelion The point on the elliptical path of an object in orbit about the Sun that is most distant from the Sun.

Apollo asteroid *See* Earth-crossing asteroid.

apparent brightness The brightness that a star appears to have, as measured by an observer on Earth. (p. 421)

apparent magnitude The apparent brightness of a star, expressed using the magnitude scale. (p. 422)

association Small grouping of (typically 100 or less) bright stars, spanning up to a few tens of parsecs across, usually rich in very young stars. (p. 482)

Assumption of Mediocrity Statements suggesting that the development of life on Earth did not require any unusual circumstances, suggesting that extraterrestrial life may be common.

asteroid One of thousands of very small members of the solar system orbiting the Sun between the orbits of Mars and Jupiter. Often referred to as "minor planets." (p. 334)

asteroid belt Region of the solar system, between the orbits of Mars and Jupiter, in which most asteroids are found. (p. 334)

asthenosphere Layer of Earth's interior, just below the lithosphere, over which the surface plates slide. (p. 168)

astrology Pseudoscience that purports to use the positions of the planets, Sun, and Moon to predict daily events and human destiny.

astronomical unit (AU) The average distance of Earth from the Sun. Precise radar measurements yield a value for the AU of 149,603,500 km. (p. 44)

astronomy Branch of science dedicated to the study of everything in the universe that lies above Earth's atmosphere. (p. 4)

asymptotic giant branch Path on the Hertzsprung–Russell diagram corresponding to the changes that a star undergoes after helium burning ceases in the core. At this stage, the carbon core shrinks and drives the expansion of the envelope, and the star becomes a swollen red giant for a second time. (p. 497)

Aten asteroid Earth-crossing asteroid with semimajor axis less than 1 AU.

atmosphere Layer of gas confined close to a planet's surface by the force of gravity. (p. 156)

atom Building block of matter, composed of positively charged protons and neutral neutrons in the nucleus surrounded by negatively charged electrons. (p. 82)

atomic epoch Period after decoupling when the first simple atoms and molecules formed.

aurora Event that occurs when atmospheric molecules are excited by incoming charged particles from the solar wind, then emit energy as they fall back to their ground states. Aurorae generally occur at high latitudes, near the north and south magnetic poles. (p. 175)

autumnal equinox Date on which the Sun crosses the celestial equator moving southward, occurring on or near September 21. (p. 15)

B

B ring One of three Saturnian rings visible from Earth. The B ring is the brightest of the three, and lies just past the Cassini division, closer to the planet than the A ring. (p. 292)

background noise Unwanted light in an image, from unresolved sources in the telescope's field of view, scattered light from the atmosphere, or instrumental "hiss" in the detector itself.

barred-spiral galaxy Spiral galaxy in which a bar of material passes through the center of the galaxy, with the spiral arms beginning near the ends of the bar. (p. 605)

basalt Solidified lava; an iron-magnesium-silicate mixture.

baseline The distance between two observing locations used for the purposes of triangulation measurements. The larger the baseline, the better the resolution attainable. (p. 23)

belt Dark, low-pressure region in the atmosphere of a jovian planet, where gas flows downward. (p. 262)

Big Bang Event that cosmologists consider the beginning of the universe, in which all matter and radiation in the entire universe came into being. (p. 665)

Big Crunch Point of final collapse of a bound universe.

binary asteroid Asteroid with a partner in orbit around it.

binary pulsar Binary system in which both components are pulsars.

binary-star system A system that consists of two stars in orbit about their common center of mass, held together by their mutual gravitational attraction. Most stars are found in binary-star systems. (p. 436)

biological evolution Change in a population of biological organisms over time.

bipolar flow Jets of material expelled from a protostar perpendicular to the surrounding protostellar disk. (p. 479)

blackbody curve The characteristic way in which the intensity of radiation emitted by a hot object depends on frequency. The frequency at which the emitted intensity is highest is an indication of the temperature of the radiating object. Also referred to as the Planck curve. (p. 66)

black dwarf The endpoint of the evolution of an isolated, low-mass star. After the white-dwarf stage, the star cools to the point where it is a dark "clinker" in interstellar space. (p. 504)

black hole A region of space where the pull of gravity is so great that nothing—not even light—can escape. A possible outcome of the evolution of a very massive star. (p. 551)

blazar Particularly intense active galactic nucleus in which the observer's line of sight happens to lie directly along the axis of a high-speed jet of particles emitted from the active region. (p. 624)

blue giant Large, hot, bright star at the upper-left end of the main sequence on the Hertzsprung–Russell diagram. Its name comes from its color and size. (p. 432)

blueshift Motion-induced changes in the observed wavelength from a source that is moving toward us. Relative approaching motion between the object and the observer causes the wavelength to appear shorter (and hence bluer) than if there were no motion at all.

blue straggler Star found on the main sequence of the Hertzsprung–Russell diagram, but which should already have evolved off the main sequence, given its location on the diagram; thought to have formed from mergers of lower mass stars. (p. 505)

blue supergiant The very largest of the large, hot, bright stars at the uppermost-left end of the main sequence on the Hertzsprung–Russell diagram. (p. 432)

Bohr model First theory of the hydrogen atom to explain the observed spectral lines. This model rests on three ideas: that there is a state of lowest energy for the electron, that there is a maximum energy beyond which the electron is no longer bound to the nucleus, and that within these two energies the electron can only exist in certain energy levels.

Bok globule Dense, compact cloud of interstellar dust and gas on its way to forming one or more stars.

boson Particle that exerts or mediates forces between elementary particles in quantum physics. (p. 689)

bound trajectory Path of an object with launch speed too low to escape the gravitational pull of a planet.

brown dwarf Fragments of collapsing gas and dust that did not contain enough mass to initiate core nuclear fusion. Such an object is then frozen somewhere along its pre-main-sequence contraction phase, continually cooling into a compact dark object. Because of their small size and low temperature, brown dwarfs are extremely difficult to detect observationally. (pp. 375, 474)

brown oval Feature of Jupiter's atmosphere that appears only at latitudes near 20° N, this structure is a long-lived hole in the clouds that allows us to look down into Jupiter's lower atmosphere. (p. 266)

C

C ring One of three Saturnian rings visible from Earth. The C ring lies closest to the planet and is relatively thin compared to the A and B rings. (p. 292)

caldera Crater that forms at the summit of a volcano.

capture theory (Moon) Theory suggesting that the Moon formed far from Earth but was later captured by it.

carbonaceous asteroid The darkest, or least reflective, type of asteroid, containing large amounts of carbon.

carbon-based molecule Molecule containing atoms of carbon.

carbon-detonation supernova *See* Type I supernova. (p. 523)

cascade Process of deexcitation in which an excited electron moves down through energy states one at a time.

Cassegrain telescope A type of reflecting telescope in which incoming light hits the primary mirror and is then reflected upward toward the prime focus, where a secondary mirror reflects the light back down through a small hole in the main mirror into a detector or eyepiece. (p. 101)

Cassini division A relatively empty gap in Saturn's ring system, discovered in 1675 by Giovanni Cassini. It is now known to contain a number of thin ringlets. (p. 292)

cataclysmic variable Collective name for novae and supernovae.

catalyst Something that causes or helps a reaction to occur, but is not itself consumed as part of the reaction.

catastrophic theory A theory that invokes statistically unlikely accidental events to account for observations.

celestial equator The projection of Earth's equator onto the celestial sphere.

celestial mechanics Study of the motions of bodies, such as planets and stars, that interact via gravity.

celestial pole Projection of Earth's North or South pole onto the celestial sphere. (p. 10)

celestial sphere Imaginary sphere surrounding Earth to which all objects in the sky were once considered to be attached. (p. 10)

Celsius Temperature scale in which the freezing point of water is 0 degrees and the boiling point of water is 100 degrees.

center of mass The "average" position in space of a collection of massive bodies, weighted by their masses. For an isolated system this point moves with constant velocity, according to Newtonian mechanics. (p. 51)

centigrade *See* Celsius.

centripetal force (literally "center seeking") Force directed toward the center of a body's orbit.

centroid Average position of the material in an object; in spectroscopy, the center of a spectral line.

Cepheid variable Star whose luminosity varies in a characteristic way, with a rapid rise in brightness followed by a slower decline. The period of a Cepheid variable star is related to its luminosity, so a determination of this period can be used to obtain an estimate of the star's distance. (p. 577)

Chandrasekhar mass Maximum possible mass of a white dwarf.

chaotic rotation Unpredictable tumbling motion that nonspherical bodies in eccentric orbits, such as Saturn's satellite Hyperion, can exhibit. No amount of observation of an object rotating chaotically will ever show a well-defined period.

charge-coupled device (CCD) An electronic device used for data acquisition; composed of many tiny pixels, each of which records a buildup of charge to measure the amount of light striking it. (p. 108)

chemical bond Force holding atoms together to form a molecule.

chemosynthesis Analog of photosynthesis that operates in total darkness.

chromatic aberration The tendency for a lens to focus red and blue light differently, causing images to become blurred.

chromosphere The Sun's lower atmosphere, lying just above the visible atmosphere. (p. 386)

circumnavigation Traveling all the way around an object.

cirrus High-level clouds composed of ice or methane crystals.

closed universe Geometry that the universe as a whole would have if the density of matter is above the critical value. A closed universe is finite in extent and has no edge, like the surface of a sphere. It has enough mass to stop the present expansion and will eventually collapse. (p. 670)

CNO cycle Chain of reactions that converts hydrogen into helium using carbon, nitrogen, and oxygen as catalysts.

cocoon nebula Bright infrared source in which a surrounding cloud of gas and dust absorb ultraviolet radiation from a hot star and reemits it in the infrared.

cold dark matter Class of dark-matter candidates made up of very heavy particles, possibly formed in the very early universe. (p. 700)

collecting area The total area of a telescope capable of capturing incoming radiation. The larger the telescope, the greater its collecting area, and the fainter the objects it can detect. (p. 103)

collisional broadening Broadening of spectral lines due to collisions between atoms, most often seen in dense gases.

color index A convenient method of quantifying a star's color by comparing its apparent brightness as measured through different filters. If the star's radiation is well described by a blackbody spectrum, the ratio of its blue intensity (B) to its visual intensity (V) is a measure of the object's surface temperature.

color-magnitude diagram A way of plotting stellar properties, in which absolute magnitude is plotted against color index. (p. 430)

coma An effect occurring during the formation of an off-axis image in a telescope. Stars whose light enters the telescope at a large angle acquire comet-like tails on their images. The brightest part of a comet, often referred to as the "head." (p. 341)

comet A small body, composed mainly of ice and dust, in an elliptical orbit about the Sun. As it comes close to the Sun, some of its material is vaporized to form a gaseous head and extended tail. (p. 340)

common envelope Outer layer of gas in a contact binary.

comparative planetology The systematic study of the similarities and differences among the planets, with the goal of obtaining deeper insight into how the solar system formed and has evolved in time. (p. 135)

composition The mixture of atoms and molecules that make up an object.

condensation nuclei Dust grains in the interstellar medium which act as seeds around which other material can cluster. The presence of dust was very important in causing matter to clump during the formation of the solar system. (p. 149)

condensation theory Currently favored model of solar system formation which combines features of the old nebular theory with new information about interstellar dust grains, which acted as condensation nuclei. (p. 149)

conjunction Orbital configuration in which a planet lies in the same direction as the Sun, as seen from Earth.

conservation of angular momentum *See* law of conservation of angular momentum. (p. 148)

conservation of mass and energy *See* law of conservation of mass and energy.

constellation A human grouping of stars in the night sky into a recognizable pattern. (p. 8)

constituents *See* composition.

continental drift The movement of the continents around Earth's surface.

continuous spectrum Spectrum in which the radiation is distributed over all frequencies, not just a few specific frequency ranges. A prime example is the blackbody radiation emitted by a hot, dense body. (p. 78)

convection Churning motion resulting from the constant upwelling of warm fluid and the concurrent downward flow of cooler material to take its place. (p. 157)

convection zone Region of the Sun's interior, lying just below the surface, where the material of the Sun is in constant convection motion. This region extends into the solar interior to a depth of about 20,000 km. (p. 386)

co-orbital satellites Satellites sharing the same orbit around a planet.

Copernican Principle The removal of Earth from any position of cosmic significance.

Copernican revolution The realization, toward the end of the sixteenth century, that Earth is not at the center of the universe. (p. 38)

core The central region of Earth, surrounded by the mantle. (p. 156); The central region of any planet or star. (p. 156)

core-accretion theory Theory that the jovian planets formed when icy protoplanetary cores became massive enough to capture gas directly from the solar nebula. *See* gravitational instability theory. (p. 364)

core-collapse supernova *See* Type II supernova. (p. 522)

core hydrogen burning The energy burning stage for main-sequence stars, in which the helium is produced by hydrogen fusion in the central region of the star. A typical star spends up to 90 percent of its lifetime in hydro-static equilibrium brought about by the balance between gravity and the energy generated by core hydrogen burning. (p. 492)

cornea (eye) The curved transparent layer covering the front part of the eye.

corona One of numerous large, roughly circular regions on the surface of Venus, thought to have been caused by upwelling mantle material causing the planet's crust to bulge outward (plural, *coronae*). (p. 219); The tenuous outer atmosphere of the Sun, which lies just above the chromosphere and, at great distances, turns into the solar wind. (p. 386)

coronal hole Vast region of the Sun's atmosphere where the density of matter is about 10 times lower than average. The gas there streams freely into space at high speeds, escaping the Sun completely. (p. 404)

coronal mass ejection Giant magnetic "bubble" of ionized gas that separates from the rest of the solar atmosphere and escapes into interplanetary space. (p. 402)

corpuscular theory Early particle theory of light.

cosmic density parameter Ratio of the universe's actual density to the critical value corresponding to zero curvature.

cosmic distance scale Collection of indirect distance-measurement techniques that astronomers use to measure distances in the universe. (p. 23)

cosmic evolution The collection of the seven major phases of the history of the universe—namely particulate, galactic, stellar, planetary, chemical, biological, and cultural evolution. (p. 710)

cosmic microwave background The almost perfectly isotropic radio signal that is the electromagnetic remnant of the Big Bang. (p. 678)

cosmic ray Very energetic subatomic particle arriving at Earth from elsewhere in the Galaxy.

cosmological constant Quantity originally introduced by Einstein into general relativity to make his equations describe a static universe. Now one of several candidates for the repulsive "dark energy" force responsible for the observed cosmic acceleration. (p. 674)

cosmological distance Distance comparable to the scale of the universe.

cosmological principle Two assumptions that make up the basis of cosmology, namely that the universe is homogeneous and isotropic on sufficiently large scales. (p. 663)

cosmological redshift The component of the redshift of an object that is due only to the Hubble flow of the universe. (p. 615)

cosmology The study of the structure and evolution of the entire universe. (p. 662)

cosmos The universe.

coudé focus Focus produced far from the telescope using a series of mirrors. Allows the use of heavy and/or finely tuned equipment to analyze the image.

crater Bowl-shaped depression on the surface of a planet or moon, resulting from a collision with interplanetary debris. (p. 186)

crescent Appearance of the Moon (or a planet) when less than half of the body's hemisphere is visible from Earth.

crest Maximum departure of a wave above its undisturbed state.

critical density The cosmic density corresponding to the dividing line between a universe that recollapses and one that expands forever. (p. 668)

critical universe Universe in which the density of matter is exactly equal to the critical density. The universe is infinite in extent and has zero curvature. The expansion will continue forever, but will approach an expansion speed of zero. (p. 672)

crust Layer of Earth which contains the solid continents and the seafloor. (p. 156)

C-type asteroid *See* carbonaceous asteroid.

cultural evolution Change in the ideas and behavior of a society over time.

current sheet Flat sheet on Jupiter's magnetic equator where most of the charged particles in the magnetosphere lie due to the planet's rapid rotation.

D

D ring Collection of very faint, thin rings, extending from the inner edge of the C ring down nearly to the cloud tops of Saturn. This region contains so few particles that it is completely invisible from Earth. (p. 296)

dark dust cloud A large cloud, often many parsecs across, which contains gas and dust in a ratio of about 1012 gas atoms for every dust particle. Typical densities are a few tens or hundreds of millions of particles per cubic meter. (p. 455)

dark energy Generic name given to the unknown cosmic force field thought to be responsible for the observed acceleration of the Hubble expansion. (p. 674)

dark halo Region of a galaxy beyond the visible halo where dark matter is thought to reside. (p. 592)

dark matter Term used to describe the mass in galaxies and clusters whose existence we infer from rotation curves and other techniques, but that has not been confirmed by observations at any electromagnetic wavelength. (p. 593)

dark matter particle Particle undetectable at any electromagnetic wavelength, but can be inferred from its gravitational influence.

Daughter/Fission theory Theory suggesting that the Moon originated out of Earth.

declination Celestial coordinate used to measure latitude above or below the celestial equator on the celestial sphere.

decoupling Event in the early universe when atoms first formed, after which photons could propagate freely through space. (p. 693)

deferent A construct of the geocentric model of the solar system which was needed to explain observed planetary motions. A deferent is a large circle encircling Earth, on which an epicycle moves. (p. 36)

degree Unit of angular measure. There are 360 degrees in one complete circle.

density A measure of the compactness of the matter within an object, computed by dividing the mass of the object by its volume. Units are kilograms per cubic meter (kg/m^3), or grams per cubic centimeter (g/cm^3). (p. 137)

detached binary Binary system where each star lies within its respective Roche lobe.

detector noise Readings produced by an instrument even when it is not observing anything; produced by the electronic components within the detector itself.

deuterium A form of hydrogen with an extra neutron in its nucleus.

deuterium bottleneck Period in the early universe between the start of deuterium production and the time when the universe was cool enough for deuterium to survive.

deuteron An isotope of hydrogen in which a neutron is bound to the proton in the nucleus. Often called "heavy hydrogen" because of the extra mass of the neutron. (p. 408)

differential rotation The tendency for a gaseous sphere, such as a jovian planet or the Sun, to rotate at a different rate at the equator than at the poles. More generally, a condition where the angular speed varies with location within an object. (p. 261)

differentiation Variation in the density and composition of a body, such as Earth, with low-density material on the surface and higher density material in the core. (p. 164)

diffraction The ability of waves to bend around corners. The diffraction of light establishes its wave nature. (p. 65)

diffraction grating Sheet of transparent material with many closely spaced parallel lines ruled on it, designed to separate white light into a spectrum.

diffraction-limited resolution Theoretical resolution that a telescope can have due to diffraction of light at the telescope's aperture. Depends on the wavelength of radiation and the diameter of the telescope's mirror. (p. 107)

direct motion *See* prograde motion.

distance modulus Difference between the apparent and absolute magnitude of an object; equivalent to distance, by the inverse-square law.

diurnal motion Apparent daily motion of the stars, caused by Earth's rotation.

DNA Deoxyribonucleic acid, the molecule that carries genetic information and determines the characteristics of a living organism.

Doppler effect Any motion-induced change in the observed wavelength (or frequency) of a wave. (p. 71)

double-line spectroscopic binary Binary system in which spectral lines of both stars can be distinguished and seen to shift back and forth as the stars orbit one another.

double-star system System containing two stars in orbit around one another.

Drake equation Expression that gives an estimate of the probability that intelligence exists elsewhere in the Galaxy, based on a number of supposedly necessary conditions for intelligent life to develop. (p. 718)

dust grain An interstellar dust particle, roughly 10^{-7}m in size, comparable to the wavelength of visible light. (p. 446)

dust lane A lane of dark, obscuring interstellar dust in an emission nebula or galaxy. (p. 451)

dust tail The component of a comet's tail that is composed of dust particles. (p. 342)

dwarf Any star with radius comparable to, or smaller than, that of the Sun (including the Sun itself). (p. 430)

dwarf elliptical Elliptical galaxy as small as 1 kiloparsec across, containing only a few million stars.

dwarf galaxy Small galaxy containing a few million stars.

dwarf irregular Small irregular galaxy containing only a few million stars.

dwarf planet A body that orbits the Sun and is massive enough that its own gravity has caused its shape to be approximately spherical, but which is insufficiently massive to have cleared other bodies from "the neighborhood" of its orbit. (p. 352)

dynamo theory Theory that explains planetary and stellar magnetic fields in terms of rotating, conducting material flowing in an object's interior. (p. 176)

E

E ring A faint ring, well outside the main ring system of Saturn, which was discovered by *Voyager* and is thought to be associated with volcanism on the moon Enceladus. (p. 296)

Earth-crossing asteroid An asteroid whose orbit crosses that of Earth. Earth-crossing asteroids are also called Apollo asteroids, after the first asteroid of this type discovered. (p. 337)

earthquake A sudden dislocation of rocky material near Earth's surface.

eccentricity A measure of the flatness of an ellipse, equal to the distance between the two foci divided by the length of the major axis. (p. 43)

eclipse Event during which one body passes in front of another, so that the light from the occulted body is blocked. (p. 18)

eclipse season Time of the year when the Moon lies in the same plane as Earth and Sun, so that eclipses are possible. (p. 20)

eclipse year Time interval between successive orbital configurations in which the line of nodes of the Moon's orbit points toward the Sun.

eclipsing binary Rare binary-star system that is aligned in such a way that from Earth we observe one star pass in front of the other, eclipsing the other star. (p. 437)

ecliptic The apparent path of the Sun, relative to the stars on the celestial sphere, over the course of a year. (p. 14)

effective temperature Temperature of a blackbody of the same radius and luminosity as a given star or planet.

ejecta (planetary) Material thrown outward by a meteoroid impact.

ejecta (stellar) Material thrown into space by a nova or supernova.

electric field A field extending outward in all directions from a charged particle, such as a proton or an electron. The electric field determines the electric force exerted by the particle on all other charged particles in the universe; the strength of the electric field decreases with increasing distance from the charge according to an inverse-square law. (p. 61)

electromagnetic energy Energy carried in the form of rapidly fluctuating electric and magnetic fields.

electromagnetic force Force (electric or magnetic) exerted between any two charged particles. (p. 410)

electromagnetic radiation Another term for light, electromagnetic radiation transfers energy and information from one place to another. (p. 58)

electromagnetic spectrum The complete range of electromagnetic radiation, from radio waves to gamma rays, including the visible spectrum. All types of electromagnetic radiation are basically the same phenomenon, differing only by wavelength, and all move at the speed of light. (p. 63)

electromagnetism The union of electricity and magnetism, which do not exist as independent quantities but are in reality two aspects of a single physical phenomenon. (p. 62)

electron An elementary particle with a negative electric charge; one of the components of the atom. (p. 61)

electron degeneracy pressure The pressure produced by the resistance of electrons to further compression once they are squeezed to the point of contact. (p. 496)

electrostatic force Force between electrically charged objects.

electroweak force Unification of the weak electromagnetic forces. (p. 410)

element Matter made up of one particular atom. The number of protons in the nucleus of the atom determines which element it represents. (p. 87)

elementary particle Technically, a particle that cannot be subdivided into component parts; however, the term is also often used to refer to particles such as protons and neutrons, which are themselves made up of quarks.

ellipse Geometric figure resembling an elongated circle. An ellipse is characterized by its degree of flatness, or eccentricity, and the length of its long axis. In general, bound orbits of objects moving under gravity are elliptical. (p. 43)

elliptical galaxy Category of galaxy in which the stars are distributed in an elliptical shape on the sky, ranging from highly elongated to nearly circular in appearance. (p. 607)

elongation Angular distance between a planet and the Sun.

emission line Bright line in a specific location of the spectrum of radiating material, corresponding to emission of light at a certain frequency. A heated gas in a glass container produces emission lines in its spectrum. (p. 78)

emission nebula A glowing cloud of hot interstellar gas. The gas glows as a result of one or more nearby young stars which ionize the gas. Since the gas is mostly hydrogen, the emitted radiation falls predominantly in the red region of the spectrum, because of the hydrogen-alpha emission line. (p. 449)

emission spectrum The pattern of spectral emission lines produced by an element. Each element has its own unique emission spectrum. (p. 80)

empirical Discovery based on observational evidence (rather than from theory).

Encke gap A small gap in Saturn's A ring. (p. 292)

energy flux Energy per unit area per unit time radiated by a star (or recorded by a detector).

epicycle A construct of the geocentric model of the solar system which was necessary to explain observed planetary motions. Each planet rides on a small epicycle whose center in turn rides on a larger circle (the deferent). (p. 36)

epoch of inflation Short period of unchecked cosmic expansion early in the history of the universe. During inflation, the universe swelled in size by a factor of about 10^{50}. (p. 695)

equinox *See* vernal equinox, autumnal equinox.

equivalence principle There is no experimental way to distinguish between a gravitational field and an accelerated frame of reference. (p. 556)

escape speed The speed necessary for one object to escape the gravitational pull of another. Anything that moves away from a gravitating body with more than the escape speed will never return. (p. 52)

euclidean geometry Geometry of flat space.

event horizon Imaginary spherical surface surrounding a collapsing star, with radius equal to the Schwarzschild radius, within which no event can be seen, heard, or known about by an outside observer. (p. 553)

evolutionary theory A theory which explains observations in a series of gradual steps, explainable in terms of well-established physical principles.

evolutionary track A graphical representation of a star's life as a path on the Hertzsprung–Russell diagram. (p. 470)

excited state State of an atom when one of its electrons is in a higher energy orbital than the ground state. Atoms can become excited by absorbing a photon of a specific energy, or by colliding with a nearby atom. (p. 83)

extinction The dimming of starlight as it passes through the interstellar medium. (p. 446)

extrasolar planet Planet orbiting a star other than the Sun. (p. 362)

extremophilic Adjective describing organisms that can survive in very harsh environments. (p. 717)

eyepiece Secondary lens through which an observer views an image. This lens is often chosen to magnify the image.

F

F ring Faint narrow outer ring of Saturn, discovered by *Pioneer 11* in 1979. The F ring lies just inside the Roche limit of Saturn and was found by *Voyager 1* to be made up of several ring strands apparently braided together. (p. 298)

Fahrenheit Temperature scale in which the freezing point of water is 32 degrees and the boiling point of water is 212 degrees.

false vacuum Region of the universe that remained in the "unified" state after the strong and electroweak forces separated; one possible cause of cosmic inflation at very early times.

fault line Dislocation on a planet's surface, often indicating the boundary between two plates.

field line Imaginary line indicating the direction of an electric or magnetic field.

fireball Large meteor that burns up brightly and sometimes explosively in Earth's atmosphere.

firmament Old-fashioned term for the heavens (i.e., the sky).

fission *See* nuclear fission. (p. 521)

flare Explosive event occurring in or near an active region on the Sun. (p. 401)

flatness problem One of two conceptual problems with the standard Big Bang model, which is that there is no natural way to explain why the density of the universe is so close to the critical density. (p. 694)

fluidized ejecta The ejecta blankets around some Martian craters, which apparently indicate that the ejected material was liquid at the time the crater formed.

fluorescence Phenomenon where an atom absorbs energy, then radiates photons of lower energy as it cascades back to the ground state; in astronomy, often produced as ultraviolet photons from a hot young star are absorbed by a neutral gas, causing some of the gas atoms to become excited and give off an optical (red) glow.

flyby Unbound trajectory of a spacecraft around a planet or other body.

focal length Distance from a mirror or the center of a lens to the focus.

focus One of two special points within an ellipse, whose separation from each other indicates the eccentricity. In a bound orbit, planets orbit in ellipses with the Sun at one focus. (p. 43)

forbidden line A spectral line seen in emission nebulae, but not seen in laboratory experiments because, under laboratory conditions, collisions kick the electron in question into some other state before emission can occur.

force Action on an object that causes its momentum to change. The rate at which the momentum changes is numerically equal to the force. (p. 48)

fragmentation The breaking up of a large object into many smaller pieces (for example, as the result of high-speed collisions between planetesimals and protoplanets in the early solar system). (p. 149)

Fraunhofer lines The collection of over 600 absorption lines in the spectrum of the Sun, first categorized by Joseph Fraunhofer in 1812.

frequency The number of wave crests passing any given point in a unit time. (p. 59)

full When the full hemisphere of the Moon or a planet can be seen from Earth.

full Moon Phase of the Moon in which it appears as a complete circular disk in the sky.

fusion *See* nuclear fusion.

G

G ring Faint, narrow ring of Saturn, discovered by *Pioneer 11* and lying just outside the F ring. (p. 296)

galactic bulge Thick distribution of warm gas and stars around the center of a galaxy. (p. 574)

galactic cannibalism A galaxy merger in which a larger galaxy consumes a smaller one.

galactic center The center of the Milky Way, or any other galaxy. The point about which the disk of a spiral galaxy rotates. (p. 581)

galactic disk Flattened region of gas and dust that bisects the galactic halo in a spiral galaxy. This is the region of active star formation. (p. 574)

galactic epoch Period from 100 million to 3 billion years after the Big Bang when large agglomerations of matter (galaxies and galaxy clusters) formed and grew.

galactic habitable zone Region of a galaxy in which conditions are conducive to the development of life.

galactic halo Region of a galaxy extending far above and below the galactic disk, where globular clusters and other old stars reside. (p. 574)

galactic nucleus Small, central, high-density region of a galaxy. Almost all the radiation from active galaxies is generated within the nucleus. (p. 597)

galactic rotation curve Plot of rotation speed versus distance from the center of a galaxy.

Galactic year Time taken for objects at the distance of the Sun (about 8 kpc) to orbit the center of the Galaxy, roughly 225 million years.

galaxy Gravitationally bound collection of a large number of stars. The Sun is a star in the Milky Way Galaxy. (p. 574)

galaxy cluster A collection of galaxies held together by their mutual gravitational attraction. (p. 614)

Galilean moons The four brightest and largest moons of Jupiter (Io, Europa, Ganymede, Callisto), named after Galileo Galilei, the seventeenth-century astronomer who first observed them. (p. 260)

Galilean satellites *See* Galilean moons.

gamma ray Region of the electromagnetic spectrum, far beyond the visible spectrum, corresponding to radiation of very high frequency and very short wavelength. (p. 58)

gamma-ray burst Object that radiates tremendous amounts of energy in the form of gamma rays, possibly due to the collision and merger of two neutron stars initially in orbit around one another. (p. 548)

gamma-ray spectrograph Spectrograph designed to work at gamma-ray wavelengths. Used to map the abundances of certain elements on the Moon and Mars.

gaseous Composed of gas.

gas-exchange experiment Experiment to look for life on Mars. A nutrient broth was offered to Martian soil specimens. If there were life in the soil, gases would be created as the broth was digested.

gene Sequence of nucleotide bases in the DNA molecule that determines the characteristics of a living organism.

general relativity Theory proposed by Einstein to incorporate gravity into the framework of special relativity. (p. 556)

general theory of relativity Theory proposed by Einstein to incorporate gravity into the framework of special relativity.

geocentric model A model of the solar system that holds that Earth is at the center of the universe and all other bodies are in orbit around it. The earliest theories of the solar system were geocentric. (p. 36)

giant A star with a radius between 10 and 100 times that of the Sun. (p. 430)

giant elliptical Elliptical galaxy up to a few megaparsecs across, containing trillions of stars.

gibbous Appearance of the Moon (or a planet) when more than half (but not all) of the body's hemisphere is visible from Earth.

globular cluster Tightly bound, roughly spherical collection of hundreds of thousands, and sometimes millions, of stars spanning about 50 parsecs. Globular clusters are distributed in the halos around the Milky Way and other galaxies. (p. 483)

gluon Particle that exerts or mediates the strong force in quantum physics.

gradient Rate of change of some quantity (e.g., temperature or composition) with respect to location in space.

Grand Unified Theories (GUTs) Class of theories describing the behavior of the single force that results from unification of the strong, weak, and electromagnetic forces in the early universe. (pp. 688, 689)

granite Igneous rock, containing silicon and aluminum, that makes up most of Earth's crust.

granulation Mottled appearance of the solar surface, caused by rising (hot) and falling (cool) material in convective cells just below the photosphere. (p. 392)

gravitational force Force exerted on one body by another due to the effect of gravity. The force is directly proportional to the masses of both bodies involved and inversely proportional to the square of the distance between them. (pp. 49, 410)

gravitational instability theory Theory that the jovian planets formed directly from the solar nebula via instabilities in the gas leading to gravitational contraction. *See* core-accretion theory. (p. 364)

gravitational lensing The effect induced on the image of a distant object by a massive foreground object. Light from the distant object is bent into two or more separate images. (p. 594)

gravitational radiation Radiation resulting from rapid changes in a body's gravitational field.

gravitational redshift A prediction of Einstein's general theory of relativity. Photons lose energy as they escape the gravitational field of a massive object. Because a photon's energy is proportional to its frequency, a photon that loses energy suffers a decrease in frequency, which corresponds to an increase, or redshift, in wavelength. (p. 558)

graviton Particle carrying the gravitational field in theories attempting to unify gravity and quantum mechanics.

gravity The attractive effect that any massive object has on all other massive objects. The greater the mass of the object, the stronger its gravitational pull. (p. 49)

gravity assist Using gravity to change the flight path of a satellite or spacecraft. (p. 142)

gravity wave Gravitational counterpart of an electromagnetic wave.

great attractor A huge accumulation of mass in the relatively nearby universe (within about 200 Mpc of the Milky Way).

Great Dark Spot Prominent storm system in the atmosphere of Neptune observed by *Voyager 2*, near the equator of the planet. The system was comparable in size to Earth. (p. 319)

Great Red Spot A large, high-pressure, long-lived storm system visible in the atmosphere of Jupiter. The Red Spot is roughly twice the size of Earth. (p. 262)

great wall Extended sheet of galaxies measuring at least 200 megaparsecs across; one of the largest known structures in the universe.

greenhouse effect The partial trapping of solar radiation by a planetary atmosphere, similar to the trapping of heat in a greenhouse. (p. 160)

greenhouse gas Gas (such as carbon dioxide or water vapor) that efficiently absorbs infrared radiation. (p. 160)

ground state The lowest energy state that an electron can have within an atom. (p. 82)

GUT epoch Period when gravity separated from the other three forces of nature.

gyroscope System of rotating wheels that allows a spacecraft to maintain a fixed orientation in space.

H

habitable zone Three-dimensional zone of comfortable temperature (corresponding to liquid water) that surrounds every star. (p. 378)

half-life The amount of time it takes for half of the initial amount of a radioactive substance to decay into something else. (p. 166)

hayashi track Evolutionary track followed by a protostar during the final pre-main-sequence phase before nuclear fusion begins.

heat Thermal energy, the energy of an object due to the random motion of its component atoms or molecules.

heat death End point of a bound universe, in which all matter and life are destined to be incinerated.

heavy element In astronomical terms, any element heavier than hydrogen and helium.

heliocentric model A model of the solar system that is centered on the Sun, with Earth in motion about the Sun. (p. 37)

helioseismology The study of conditions far below the Sun's surface through the analysis of internal "sound" waves that repeatedly cross the solar interior. (p. 389)

helium-burning shell Shell of burning helium gas surrounding a nonburning stellar core of carbon ash.

helium capture The formation of heavy elements by the capture of a helium nucleus. For example, carbon can form heavier elements by fusion with other carbon nuclei, but it is much more likely to occur by helium capture, which requires less energy. (p. 531)

helium flash An explosive event in the post-main-sequence evolution of a low-mass star. When helium fusion begins in a dense stellar core, the burning is explosive in nature. It continues until the energy released is enough to expand the core, at which point the star achieves stable equilibrium again. (p. 496)

helium precipitation Mechanism responsible for the low abundance of helium of Saturn's atmosphere. Helium condenses in the upper layers to form a mist, which rains down toward Saturn's interior, just as water vapor forms into rain in the atmosphere of Earth. (p. 291)

helium shell flash Condition in which the helium-burning shell in the core of a star cannot respond to rapidly changing conditions within it, leading to a sudden temperature rise and a dramatic increase in nuclear reaction rates.

Hertzsprung–Russell (H–R) diagram A plot of luminosity against temperature (or spectral class) for a group of stars. (p. 430)

HI region Region of space containing primarily neutral hydrogen. (p. 453)

hierarchical merging Widely accepted galaxy-formation scenario in which galaxies formed as relatively small objects in the early universe and subsequently collided and merged to form the large galaxies observed today. The present Hubble type of a galaxy depends on the sequence of merger events in the galaxy's past. (p. 641)

high-energy astronomy Astronomy using X- or gamma-ray radiation rather than optical radiation.

high-energy telescope Telescope designed to detect X- and gamma radiation. (p. 123)

highlands Relatively light-colored regions on the surface of the Moon that are elevated several kilometers above the maria. Also called terrae. (p. 186)

high-mass star Star with a mass more than 8 times that of the Sun; progenitor of a neutron star or black hole.

HII region Region of space containing primarily ionized hydrogen. (p. 453)

homogeneity Assumed property of the universe such that the number of galaxies in an imaginary large cube of the universe is the same no matter where in the universe the cube is placed. More generally, "the same everywhere." (p. 663)

horizon problem One of two conceptual problems with the standard Big Bang model, which is that some regions of the universe that have very similar properties are too far apart to have exchanged information within the age of the universe. (p. 694)

horizontal branch Region of the Hertzsprung–Russell diagram where post-main-sequence stars again reach hydrostatic equilib-

rium. At this point, the star is burning helium in its core and fusing hydrogen in a shell surrounding the core. (p. 496)

hot dark matter A class of candidates for the dark matter in the universe, composed of lightweight particles such as neutrinos, much less massive than the electron. (p. 700)

hot Jupiter A massive, gaseous planet orbiting very close to its parent star. (p. 373)

hot longitudes (Mercury) Two opposite points on Mercury's equator where the Sun is directly overhead at perihelion.

Hubble classification scheme Method of classifying galaxies according to their appearance, developed by Edwin Hubble. (p. 604)

Hubble diagram Plot of galactic recession velocity versus distance; evidence for an expanding universe.

Hubble flow Universal recession described by the Hubble diagram and quantified by Hubble's Law.

Hubble's constant The constant of proportionality that gives the relation between recessional velocity and distance in Hubble's law. (p. 616)

Hubble's law Law that relates the observed velocity of recession of a galaxy to its distance from us. The velocity of recession of a galaxy is directly proportional to its distance away. (p. 615)

hydrocarbon Molecule consisting solely of hydrogen and carbon.

hydrogen envelope An invisible sheath of gas engulfing the coma of a comet, usually distorted by the solar wind and extending across millions of kilometers of space. (p. 342)

hydrogen shell burning Fusion of hydrogen in a shell that is driven by contraction and heating of the helium core. Once hydrogen is depleted in the core of a star, hydrogen burning stops and the core contracts due to gravity, causing the temperature to rise, heating the surrounding layers of hydrogen in the star, and increasing the burning rate there. (p. 494)

hydrosphere Layer of Earth that contains the liquid oceans and accounts for roughly 70 percent of Earth's total surface area. (p. 156)

hydrostatic equilibrium Condition in a star or other fluid body in which gravity's inward pull is exactly balanced by internal forces due to pressure. (p. 388)

hyperbola Curve formed when a plane intersects a cone at a small angle to the axis of the cone.

hypernova Explosion where a massive star undergoes core collapse and forms a black hole and a gamma-ray burst. *See* supernova.

I

igneous Type of rock formed from molten material.

image The optical representation of an object produced when the object is reflected or refracted by a mirror or lens. (p. 98)

impact theory (Moon) Combination of the Capture and Daughter theories, suggesting that the Moon formed after an impact which dislodged some of Earth's mantle and placed it in orbit. (p. 206)

inertia The tendency of an object to continue moving at the same speed and in the same direction, unless acted upon by a force. (p. 48)

inferior conjunction Orbital configuration in which an inferior planet (Mercury or Venus) lies closest to Earth.

inflation *See* epoch of inflation.

infrared Region of the electromagnetic spectrum just outside the visible range, corresponding to light of a slightly longer wavelength than red light. (p. 58)

infrared telescope Telescope designed to detect infrared radiation. Many such telescopes are designed to be lightweight so that they can be carried above (most of) Earth's atmosphere by balloons, airplanes, or satellites. (p. 120)

infrared waves Electromagnetic radiation with wavelength in the infrared part of the spectrum. (p. 58)

inhomogeneity Deviation from perfectly uniform density; in cosmology, inhomogeneities in the universe are ultimately due to quantum fluctuations before inflation.

inner core The central part of Earth's core, thought to be solid, and composed mainly of nickel and iron. (p. 163)

instability strip Region of the Hertzsprung–Russell diagram where pulsating post-main-sequence stars are found.

intensity A basic property of electromagnetic radiation that specifies the amount or strength of the radiation.

intercloud medium Superheated bubbles of hot gas extending far into interstellar space.

intercrater plains Regions on the surface of Mercury that do not show extensive cratering but are relatively smooth. (p. 200)

interference The ability of two or more waves to interact in such a way that they either reinforce or cancel each other. (p. 65)

interferometer Collection of two or more telescopes working together as a team, observing the same object at the same time and at the same wavelength. The effective diameter of an interferometer is equal to the distance between its outermost telescopes. (p. 118)

interferometry Technique in widespread use to dramatically improve the resolution of radio and infrared maps. Several telescopes observe the object simultaneously, and a computer analyzes how the signals interfere with each other. (p. 118)

intermediate-mass black hole Black hole having a mass 100–1000 times greater than the mass of the Sun.

interplanetary matter Matter in the solar system that is not part of a planet or moon—cosmic "debris." (p. 140)

interstellar dust Microscopic dust grains that populate space between the stars, having their origins in the ejected matter of long-dead stars.

interstellar gas cloud A large cloud of gas found in the space among the stars.

interstellar medium The matter between stars, composed of two components, gas and dust, intermixed throughout all of space. (p. 446)

intrinsic variable Star that varies in appearance due to internal processes (rather than, say, interaction with another star).

inverse-square law The law that a field follows if its strength decreases with the square of the distance. Fields that follow the inverse-square law decrease rapidly in strength as the distance increases, but never quite reach zero. (p. 49)

Io plasma torus Doughnut-shaped region of energetic ionized particles, emitted by the volcanoes on Jupiter's moon Io and swept up by Jupiter's magnetic field.

ion An atom that has lost one or more of its electrons.

ionization state Term describing the number of electrons missing from an atom: I refers to a neutral atom, II refers to an atom missing one electron, and so on.

ionized State of an atom or molecule that has lost one or more of its electrons. (p. 82)

ionosphere Layer in Earth's atmosphere above about 80 km where the atmosphere is significantly ionized and conducts electricity. (p. 157)

ion tail Thin stream of ionized gas that is pushed away from the head of a comet by the solar wind. It extends directly away from the Sun. Often referred to as a plasma tail. (p. 342)

irregular galaxy A galaxy that does not fit into any of the other major categories in the Hubble classification scheme. (p. 608)

isotopes Nuclei containing the same number of protons but different numbers of neutrons. Most elements can exist in several isotopic forms. A common example of an isotope is deuterium, which differs from normal hydrogen by the presence of an extra neutron in the nucleus. (p. 408)

isotropy Looking the same in every direction. Often applied to the universe as part of the cosmological principle. (p. 663)

J

jet stream Relatively strong winds in the upper atmosphere channeled into a narrow stream by a planet's rotation. Normally refers to a horizontal, high-altitude wind.

joule The SI unit of energy.

jovian planet One of the four giant outer planets of the solar system, resembling Jupiter in physical and chemical composition. (p. 138)

K

Kelvin Scale Temperature scale in which absolute zero is at 0 K; a change of 1 kelvin is the same as a change of 1 degree Celsius.

Kelvin-Helmholtz contraction phase Evolutionary track followed by a star during the protostar phase.

Kepler's laws of planetary motion Three laws, based on precise observations of the motions of the planets by Tycho Brahe, that summarize the motions of the planets about the Sun.

kinetic energy Energy of an object due to its motion.

Kirchhoff's laws Three rules governing the formation of different types of spectra. (p. 81)

Kirkwood gaps Gaps in the spacings of orbital semimajor axes of asteroids in the asteroid belt, produced by dynamical resonances with nearby planets, especially Jupiter. (p. 340)

Kuiper belt A region in the plane of the solar system outside the orbit of Neptune where most short-period comets are thought to originate. (p. 347)

Kuiper-belt object Small icy body orbiting in the Kuiper belt. (p. 348)

L

labeled-release experiment Experiment to look for life on Mars. Radioactive carbon compounds were added to Martian soil specimens. Scientists looked for indications that the carbon had been eaten or inhaled.

Lagrangian point One of five special points in the plane of two massive bodies orbiting one another, where a third body of negligible mass can remain in equilibrium. (p. 309)

lander Spacecraft that lands on the object it is studying.

laser ranging Method of determining the distance to an object by firing a laser beam at it and measuring the time taken for the light to return.

lava dome Volcanic formation formed when lava oozes out of fissures in a planet's surface, creating the dome, and then withdraws, causing the crust to crack and subside. (p. 219)

law of conservation of angular momentum A fundamental law of physics that states that the total angular momentum of a system always remains constant in any physical process. This law ensures that a spinning gas cloud must spin faster as it contracts.

law of conservation of mass and energy A fundamental law of modern physics that states that the sum of mass and energy must always remain constant in any physical process. In fusion reactions, the lost mass is converted into energy, primarily in the form of electromagnetic radiation. (p. 407)

laws of planetary motion Three laws derived by Kepler describing the motion of the planets around the Sun. (p. 43)

lens (eye) The part of the eye that refracts light onto the retina.

lens Optical instrument made of glass or some other transparent material, shaped so that, as a parallel beam of light passes through it, the rays are refracted so as to pass through a single focal point.

lepton From the Greek word for *light*, referring in particle physics to low-mass particles such as electrons, muons, and neutrinos that interact via the weak force. (p. 688)

lepton epoch Period when the light elementary particles (leptons) were in thermal equilibrium with the cosmic radiation field.

lidar Light Detection and Ranging—a device that uses laser-ranging to measure distance.

light *See* electromagnetic radiation.

light curve The variation in brightness of a star with time. (p. 437)

light element In astronomical terms, hydrogen and helium.

light-gathering power Amount of light a telescope can view and focus, proportional to the area of the primary mirror.

lighthouse model The leading explanation for pulsars. A small region of the neutron star, near one of the magnetic poles, emits a steady stream of radiation that sweeps past Earth each time the star rotates. The period of the pulses is the star's rotation period. (p. 541)

light pollution Unwanted, upward-directed light from streets, homes and businesses, that scatters back to Earth from dust in the air, obscuring the faint objects that astronomers want to observe. (p. 111)

light-year The distance that light, moving at a constant speed of 300,000 km/s, travels in one year. One light-year is about 10 trillion kilometers. (p. 4)

limb Edge of a lunar, planetary, or solar disk.

line of nodes Intersection of the plane of the Moon's orbit with Earth's orbital plane.

line of sight technique Method of probing interstellar clouds by observing their effects on the spectra of background stars.

linear momentum Tendency of an object to keep moving in a straight line with constant velocity; the product of the object's mass and velocity.

lithosphere Earth's crust and a small portion of the upper mantle that make up Earth's plates. This layer of Earth undergoes tectonic activity. (p. 168)

Local Bubble The particular low-density intercloud region surrounding the Sun.

Local Group The small galaxy cluster that includes the Milky Way Galaxy. (p. 614)

local supercluster Collection of galaxies and clusters centered on the Virgo cluster. *See also* supercluster.

logarithm The power to which 10 must be raised to produce a given number.

logarithmic scale Scale using the logarithm of a number rather than the number itself; commonly used to compress a large range of data into more manageable form.

look-back time Time in the past when an object emitted the radiation we see today.

Lorentz contraction Apparent contraction of an object in the direction in which it is moving. (p. 557)

low-mass star Star with a mass less than 8 times that of the Sun; progenitor of a white dwarf.

luminosity One of the basic properties used to characterize stars, luminosity is defined as the total energy radiated by star each second, at all wavelengths. (p. 387)

luminosity class A classification scheme that groups stars according to the width of their spectral lines. For a group of stars with the same temperature, luminosity class differentiates between supergiants, giants, main-sequence stars, and subdwarfs. (p. 434)

lunar dust *See* regolith.

lunar eclipse Celestial event during which the moon passes through the shadow of Earth, temporarily darkening its surface. (p. 18)

lunar phase The appearance of the Moon at different points along its orbit. (p. 16)

Lyman-alpha forest Collection of lines in an object's spectra starting at the redshifted wavelength of the object's own Lyman-alpha emission line and extending to shorter wavelengths; produced by gas in galaxies along the line of sight.

M

macroscopic Large enough to be visible by the unaided eye.

Magellanic Clouds Two small irregular galaxies that are gravitationally bound to the Milky Way Galaxy. (p. 608)

magnetic field Field that accompanies any changing electric field and governs the influence of magnetized objects on one another. (p. 62)

magnetic poles Points on a planet where the planetary magnetic field lines intersect the planet's surface vertically.

magnetism The presence of a magnetic field.

magnetometer Instrument that measures magnetic field strength.

magnetopause Boundary between a planet's magnetosphere and the solar wind.

magnetosphere A zone of charged particles trapped by a planet's magnetic field, lying above the atmosphere. (p. 156)

magnitude scale A system of ranking stars by apparent brightness, developed by the Greek astronomer Hipparchus. Originally, the brightest stars in the sky were categorized as being of first magnitude, while the faintest stars visible to the naked eye were classified as sixth magnitude. The scheme has since been extended to cover stars and galaxies too faint to be seen by the unaided eye. Increasing magnitude means fainter stars, and a difference of five magnitudes corresponds to a factor of 100 in apparent brightness. (p. 422)

main sequence Well-defined band on the Hertzsprung–Russell diagram on which most stars are found, running from the top left of the diagram to the bottom right. (p. 431)

main-sequence turnoff Special point on the Hertzsprung–Russell diagram for a cluster, indicative of the cluster's age. If all the stars in the cluster are plotted, the lower mass stars will trace out the main sequence up to the point where stars begin to evolve off the main sequence toward the red giant branch. The point where stars are just beginning to evolve off is the main-sequence turnoff. (p. 508)

major axis The long axis of an ellipse.

mantle Layer of Earth just interior to the crust. (p. 156)

mare Relatively dark-colored and smooth region on the surface of the Moon (plural: *maria*). (p. 186)

marginally bound universe Universe that will expand forever but at an increasingly slow rate.

mass A measure of the total amount of matter contained within an object. (p. 48)

mass-energy equivalence Principle that mass and energy are not independent, but can be converted from one to the other according to Einstein's formula $E = mc^2$. (p. 407)

mass function Relation between the component masses of a single-line spectroscopic binary.

massive compact halo object (MACHO) Collective name for "stellar" candidates for dark matter, including brown dwarfs, white dwarfs, and low-mass red dwarfs.

mass-luminosity relation The dependence of the luminosity of a main-sequence star on its mass. The luminosity increases roughly as the mass raised to the third power.

mass–radius relation The dependence of the radius of a main-sequence star on its mass. The radius rises roughly in proportion to the mass.

mass transfer Process by which one star in a binary system transfers matter onto the other.

matter Anything having mass.

matter era (Current) era following the Radiation era when the universe is larger and cooler, and matter is the dominant constituent of the universe.

matter-antimatter annihilation Reaction in which matter and antimatter annihilate to produce high energy gamma rays. *See also* antiparticle.

matter-dominated universe A universe in which the density of matter exceeds the density of radiation. The present-day universe is matter dominated. (p. 684)

Maunder minimum Lengthy period of solar inactivity that extended from 1645 to 1715. (p. 401)

mean solar day Average length of time from one noon to the next, taken over the course of a year—24 hours.

medium Material through which a wave, such as sound, travels.

mesosphere Region of Earth's atmosphere lying between the stratosphere and the ionosphere, 50–80 km above Earth's surface. (p. 157)

Messier object Member of a list of "fuzzy" objects compiled by astronomer Charles Messier in the eighteenth century.

metabolism The daily utilization of food and energy by which organisms stay alive.

metallic Composed of metal or metal compounds.

metamorphic Rocks created from existing rocks exposed to extremes of temperature or pressure.

meteor Bright streak in the sky, often referred to as a "shooting star," resulting from a small piece of interplanetary debris entering Earth's atmosphere and heating air molecules, which emit light as they return to their ground states. (p. 353)

meteor shower Event during which many meteors can be seen each hour, caused by the yearly passage of Earth through the debris spread along the orbit of a comet. (p. 353)

meteorite Any part of a meteoroid that survives passage through the atmosphere and lands on the surface of Earth. (p. 353)

meteoroid Chunk of interplanetary debris prior to encountering Earth's atmosphere. (p. 353)

meteoroid swarm Pebble-sized cometary fragments dislodged from the main body, moving in nearly the same orbit as the parent comet. (p. 353)

microlensing Gravitational lensing by individual stars in a galaxy. (p. 655)

micrometeoroid Relatively small chunk of interplanetary debris ranging in size from a dust particle to a pebble. (p. 353)

microquasar Stellar sized source of energetic X and gamma radiation, powered by accretion onto a neutron star or black hole, somewhat like a quasar but on a much smaller scale. (p. 546)

microsphere Small droplet of protein-like material that resists dissolution in water.

midocean ridge Place where two plates are moving apart, allowing fresh magma to well up. (p. 170)

Milky Way Galaxy The spiral galaxy in which the Sun resides. The disk of our Galaxy is visible in the night sky as the faint band of light known as the Milky Way. (p. 574)

millisecond pulsar A pulsar whose period indicates that the neutron star is rotating nearly 1000 times each second. The most likely explanation for these rapid rotators is that the neutron star has been spun up by drawing in matter from a companion star. (p. 546)

molecular cloud A cold, dense interstellar cloud which contains a high fraction of molecules. It is thought that the relatively high density of dust particles in these clouds plays an important role in the formation and preservation of the molecules. (p. 460)

molecular cloud complex Collection of molecular clouds that spans as much as 50 parsecs and may contain enough material to make millions of Sun-size stars. (p. 461)

molecule A tightly bound collection of atoms held together by the atoms' electromagnetic fields. Molecules, like atoms, emit and absorb photons at specific wavelengths. (p. 89)

molten In liquid form due to high temperatures.

moon A small body in orbit around a planet.

mosaic (photograph) Composite photograph made up of many smaller images.

M-type asteroid Asteroid containing large fractions of nickel and iron.

multiple star system Group of two or more stars in orbit around one another.

muon A type of lepton (along with the electron and tau).

N

naked singularity A singularity that is not hidden behind an event horizon.

nanobacterium Very small bacterium with diameter in the nanometer range.

nanometer One billionth of a meter.

neap tide The smallest tide, occurring when the Earth-Moon line is perpendicular to the Earth-Sun line at the first and third quarters.

nebula General term used for any "fuzzy" patch on the sky, either light or dark. (p. 451)

nebular theory One of the earliest models of solar system formation, dating back to Descartes, in which a large cloud of gas began to collapse under its own gravity to form the Sun and planets. (p. 149)

nebulosity "Fuzziness," usually in the context of an extended or gaseous astronomical object.

neon-oxygen white dwarf White dwarf formed from a low-mass star with a mass close to the "high-mass" limit, in which neon and oxygen form in the core.

neutrino Virtually massless and chargeless particle that is one of the products of fusion reactions in the Sun. Neutrinos move at close to the speed of light, and interact with matter hardly at all. (p. 408)

neutrino oscillations Likely solution to the solar neutrino problem, in which the neutrino has a very tiny mass. In this case, the correct number of neutrinos can be produced in the solar core, but on their way to Earth some can "oscillate," or become transformed into other particles, and thus go undetected. (p. 411)

neutron An elementary particle with roughly the same mass as a proton, but which is electrically neutral. Along with protons, neutrons form the nuclei of atoms. (p. 87)

neutron capture The primary mechanism by which very massive nuclei are formed in the violent aftermath of a supernova. Instead of fusion of like nuclei, heavy elements are created by the addition of more and more neutrons to existing nuclei. (p. 532)

neutron degeneracy pressure Pressure due to the Pauli exclusion principle, arising when neutrons are forced to come into close contact. (p. 540)

neutron spectrometer Instrument designed to search for water ice by looking for hydrogen.

neutron star A dense ball of neutrons that remains at the core of a star after a supernova explosion has destroyed the rest of the star. Typical neutron stars are about 20 km across, and contain more mass than the Sun. (p. 540)

neutronization Process occurring at high densities, in which protons and electrons are crushed together to form neutrons and neutrinos.

new Moon Phase of the Moon during which none of the lunar disk is visible.

Newtonian mechanics The basic laws of motion, postulated by Newton, which are sufficient to explain and quantify virtually all of the complex dynamical behavior found on Earth and elsewhere in the universe. (p. 47)

Newtonian telescope A reflecting telescope in which incoming light is intercepted before it reaches the prime focus and is deflected into an eyepiece at the side of the instrument. (p. 101)

nodes Two points on the Moon's orbit when it crosses the ecliptic.

nonrelativistic Speed that is much less than the speed of light.

nonthermal spectrum Continuous spectrum not well described by a blackbody.

normal galaxy Galaxy whose overall energy emission is consistent with the summed light of many stars. (p. 618)

north celestial pole Point on the celestial sphere directly above Earth's North Pole.

Northern and Southern Lights Colorful displays produced when atmospheric molecules, excited by charged particles from the Van Allen belts, fall back to their ground state.

nova A star that suddenly increases in brightness, often by a factor of as much as 10,000, then slowly fades back to its original luminosity. A nova is the result of an explosion on the surface of a white-dwarf star, caused by matter falling onto its surface from the atmosphere of a binary companion. (p. 520)

nuclear binding energy Energy that must be supplied to split an atomic nucleus into neutrons and protons.

nuclear epoch Period when protons and neutrons fused to form heavier nuclei.

nuclear fission Mechanism of energy generation used in nuclear reactors on Earth, in which heavy nuclei are split into lighter ones, releasing energy in the process. (p. 521)

nuclear fusion Mechanism of energy generation in the core of the Sun, in which light nuclei are combined, or fused, into heavier ones, releasing energy in the process. (p. 407)

nuclear reaction Reaction in which two nuclei combine to form other nuclei, often releasing energy in the process. *See also* fusion.

nucleotide base An organic molecule, the building block of genes that pass on hereditary characteristics from one generation of living creatures to the next. (p. 712)

nucleus Dense, central region of an atom, containing both protons and neutrons, and orbited by one or more electrons. (p. 82); The solid region of ice and dust that composes the central region of the head of a comet. (p. 341); The dense central core of a galaxy. (p. 597)

O

obscuration Blockage of light by pockets of interstellar dust and gas.

Olbers's paradox A thought experiment suggesting that if the universe were homogeneous, infinite, and unchanging, the entire night sky would be as bright as the surface of the Sun. (p. 664)

Oort cloud Spherical halo of material surrounding the solar system out to a distance of about 50,000 AU; where most comets reside. (p. 347)

opacity A quantity that measures a material's ability to block electromagnetic radiation. Opacity is the opposite of transparency. (p. 64)

open cluster Loosely bound collection of tens to hundreds of stars, a few parsecs across, generally found in the plane of the Milky Way. (p. 482)

open universe Geometry that the universe would have if the density of matter were less than the critical value. In an open universe there is not enough matter to halt the expansion of the universe. An open universe is infinite in extent. (p. 671)

opposition Orbital configuration in which a planet lies in the opposite direction from the Sun, as seen from Earth.

optical double Chance superposition in which two stars appear to lie close together but are actually widely separated.

optical telescope Telescope designed to observe electromagnetic radiation at optical wavelengths.

orbital One of several energy states in which an electron can exist in an atom. (p. 83)

orbital period Time taken for a body to complete one full orbit around another.

orbiter Spacecraft that orbits an object to make observations.

organic compound Chemical compound (molecule) containing a significant fraction of carbon atoms; the basis of living organisms.

outer core The outermost part of Earth's core, thought to be liquid and composed mainly of nickel and iron. (p. 163)

outflow channel Surface feature on Mars, evidence that liquid water once existed there in great quantity; thought to be the relics of catastrophic flooding about 3 billion years ago. Found only in the equatorial regions of the planet. (p. 240)

outgassing Production of atmospheric gases (carbon dioxide, water vapor, methane, and sulphur dioxide) by volcanic activity.

ozone layer Layer of Earth's atmosphere at an altitude of 20–50 km where incoming ultraviolet solar radiation is absorbed by oxygen, ozone, and nitrogen in the atmosphere. (p. 158)

P

pair production Process in which two photons of electromagnetic radiation give rise to a particle–antiparticle pair. (p. 685)

parallax The apparent motion of a relatively close object with respect to a more distant background as the location of the observer changes. (p. 24)

parsec The distance at which a star must lie in order for its measured parallax to be exactly 1 arc second; 1 parsec equals 206,000 AU. (p. 419)

partial eclipse Celestial event during which only a part of the occulted body is blocked from view. (p. 18)

particle A body with mass but of negligible dimension.

particle accelerator Device used to accelerate subatomic particles to relativistic speeds.

particle-antiparticle pair Pair of particles (e.g., an electron and a positron) produced by two photons of sufficiently high energy.

particle detector Experimental equipment that allows particles and antiparticles to be detected and identified.

Pauli exclusion principle A rule of quantum mechanics that prohibits electrons in dense gas from being squeezed too close together.

penumbra Portion of the shadow cast by an eclipsing object in which the eclipse is seen as partial. (p. 19); The outer region of a sunspot, surrounding the umbra, which is not as dark and not as cool as the central region. (p. 396)

perihelion The closest approach to the Sun of any object in orbit about it.

period The time needed for an orbiting body to complete one revolution about another body. (pp. 44, 59)

period–luminosity relation A relation between the pulsation period of a Cepheid variable and its absolute brightness. Measurement of the pulsation period allows the distance of the star to be determined. (p. 579)

permafrost Layer of permanently frozen water ice thought to lie just under the surface of Mars. (p. 241)

phase Appearance of the sunlit face of the Moon at different points along its orbit, as seen from Earth. (p. 16)

photodisintegration Process occurring at high temperatures, in which individual photons have enough energy to split a heavy nucleus (e.g., iron) into lighter nuclei.

photoelectric effect Emission of an electron from a surface when a photon of electromagnetic radiation is absorbed. (p. 88)

photoevaporation Process in which a cloud in the vicinity of a newborn hot star is dispersed by the star's radiation.

photometer A device that measures the total amount of light received in all or part of the image. (p. 110)

photometry Branch of observational astronomy in which the brightness of a source is measured through each of a set of standard filters. (pp. 110, 425)

photomicrograph Photograph taken through a microscope.

photon Individual packet of electromagnetic energy that makes up electromagnetic radiation. (p. 83)

photosphere The visible surface of the Sun, lying just above the uppermost layer of the Sun's interior, and just below the chromo sphere. (p. 386)

photosynthesis Process by which plants manufacture carbohydrates and oxygen from carbon dioxide and water using chlorophyll and sunlight as the energy source.

pixel One of many tiny picture elements, organized into a two-dimensional array, making up a digital image. (p. 108)

Planck curve *See* blackbody curve.

Planck epoch Period from the beginning of the universe to roughly 10^{-43} second, when the laws of physics are not understood.

Planck's constant Fundamental physical constant relating the energy of a photon to its radiation frequency (color).

planet One of eight major bodies that orbit the Sun, visible to us by reflected sunlight.

planetary nebula The ejected envelope of a red-giant star, spread over a volume roughly the size of our solar system. (p. 499)

planetary ring system Material organized into thin, flat rings encircling a giant planet, such as Saturn.

planetary transit Orbital configuration in which an inferior planet (e.g. Mercury or Venus, as seen from Earth) is observed to pass directly in front of the Sun. (p. 371)

planetesimal Term given to objects in the early solar system that had reached the size of small moons, at which point their gravitational fields were strong enough to begin influencing their neighbors. (p. 149)

plasma A gas in which the constituent atoms are completely ionized.

plate tectonics The motions of regions of Earth's lithosphere, which drift with respect to one another. Also known as continental drift. (p. 168)

plutino Kuiper-belt object whose orbital period (like that of Pluto) is in a 3:2 resonance with the orbit of Neptune.

plutoid A dwarf planet orbiting beyond Neptune. (p. 352)

polarity A measure of the direction of the solar magnetic field in a sunspot. Conventionally, lines coming out of the surface are labeled "S" while those going into the surface are labeled "N." (p. 397)

polarization The alignment of the electric fields of emitted photons, which are generally emitted with random orientations. (p. 449)

polar vortex Long-lived pattern of circulating winds around the pole of a planet. (p. 223)

Population I and II stars Classification scheme for stars based on the abundance of heavy elements. Within the Milky Way, Population I refers to young disk stars and Population II refers to old halo stars.

positron Atomic particle with properties identical to those of a negatively charged electron, except for its positive charge. This positron is the antiparticle of the electron. Positrons and electrons annihilate one another when they meet, producing pure energy in the form of gamma rays. (p. 408)

prebiotic compound Molecule that can combine with others to form the building blocks of life.

precession The slow change in the direction of the rotation axis of a spinning object, caused by some external gravitational influence. (p. 15)

primary atmosphere The chemical components that would have formed Earth's atmosphere.

primary mirror Mirror placed at the prime focus of a telescope (*see* prime focus).

prime focus The point in a reflecting telescope where the mirror focuses incoming light to a point. (p. 98)

prime-focus image Image formed at the prime focus of a telescope.

primeval fireball Hot, dense state of the universe at very early times, just after the Big Bang. (p. 665)

primordial matter Matter created during the early, hot epochs of the universe.

primordial nucleosynthesis The production of elements heavier than hydrogen by nuclear fusion in the high temperatures and densities that existed in the early universe. (p. 691)

principle of cosmic censorship A proposition to separate the unexplained physics near a singularity from the rest of the well-behaved universe. The principle states that nature always hides any singularity, such as a black hole, inside an event horizon, which insulates the rest of the universe from seeing it.

progenitor "Ancestor" star of a given object; for example, the star that existed before a supernova explosion is the supernova's progenitor.

prograde motion Motion across the sky in the eastward direction.

prominence Loop or sheet of glowing gas ejected from an active region on the solar surface, which then moves through the inner parts of the corona under the influence of the Sun's magnetic field. (p. 401)

proper motion The angular movement of a star across the sky, as seen from Earth, measured in seconds of arc per year. This movement is a result of the star's actual motion through space. (p. 420)

protein Molecule made up of amino acids that controls metabolism.

proton An elementary particle carrying a positive electric charge, a component of all atomic nuclei. The number of protons in the nucleus of an atom dictates what type of atom it is. (p. 61)

proton–proton chain The chain of fusion reactions, leading from hydrogen to helium, that powers main-sequence stars. (p. 408)

protoplanet Clump of material, formed in the early stages of solar system formation, that was the forerunner of the planets we see today. (p. 149)

protostar Stage in star formation when the interior of a collapsing fragment of gas is sufficiently hot and dense that it becomes opaque to its own radiation. The protostar is the dense region at the center of the fragment. (p. 470)

protostellar disk Swirling disk of gas and dust within which a star (and possibly a planetary system) forms. The "solar nebula," in the case of our Sun. (p. 471)

protosun The central accumulation of material in the early stages of solar system formations, the forerunner of the present-day Sun. (p. 149)

Ptolemaic model Geocentric solar system model, developed by the second-century astronomer Claudius Ptolemy. It predicted with great accuracy the positions of the then known planets. (p. 36)

pulsar Object that emits radiation in the form of rapid pulses with a characteristic pulse period and duration. Charged particles, accelerated by the magnetic field of a rapidly rotating neutron star, flow along the magnetic field lines, producing radiation that beams outward as the star spins on its axis. (p. 541)

pulsating variable star A star whose luminosity varies in a predictable, periodic way. (p. 577)

P-waves Pressure waves from an earthquake that travel rapidly through liquids and solids.

pyrolific-release experiment Experiment to look for life on Mars. Radioactively tagged carbon dioxide was added to Martian soil specimens. Scientists looked for indications that the radioactive material had been absorbed.

Q

quantization The fact that light and matter on small scales behave in a discontinuous manner, and manifest themselves in the form of tiny "packets" of energy, called quanta. (p. 83)

quantum fluctuation Temporary random change in the amount of energy at a point in space.

quantum gravity Theory combining general relativity with quantum mechanics.

quantum mechanics The laws of physics as they apply on atomic scales. (p. 83)

quark A fundamental matter particle that interacts via the strong force; basic constituent of protons and neutrons. (p. 688)

quark epoch Period when all heavy elementary particles (composed of quarks) were in thermal equilibrium with the cosmic radiation field.

quarter Moon Lunar phase in which the Moon appears as a half disk.

quasar Starlike radio source with an observed redshift that indicates an extremely large distance from Earth. The brightest nucleus of a distant active galaxy. (p. 625)

quasar feedback The idea that some of the energy released by a quasar in an active galactic nucleus heats and ejects the gas in the surrounding galaxy, shutting off the quasar fuel supply and suppressing star formation in the galaxy, connecting the growth of the central black hole to the properties of the host galaxy. (p. 648)

quasi-stellar object (QSO) *See* quasar.

quiescent prominence Prominence that persists for days or weeks, hovering high above the solar photosphere.

quiet Sun The underlying predictable elements of the Sun's behavior, such as its average photospheric temperature, which do not significantly change in time.

quintessence Dark energy candidate that evolves in a way that depends on the density of matter and radiation in the universe, and which may provide a natural explanation for why dark matter seems to be just now emerging as the dominant force in the universe.

R

radar Acronym for radio detection and ranging. Radio waves are bounced off an object, and the time taken for the echo to return indicates its distance. (p. 46)

radial motion Motion along a particular line of sight, which induces apparent changes in the wavelength (or frequency) of radiation received.

radial velocity Component of a star's velocity along the line of sight.

radian Angular measure equivalent to $180/\pi = 57.3$ degrees.

radiant Constellation from which a meteor shower appears to come.

radiation A way in which energy is transferred from place to place in the form of a wave. Light is a form of electromagnetic radiation. (p. 58)

radiation darkening The effect of chemical reactions that result when high-energy particles strike the icy surfaces of objects in the outer solar system. The reactions lead to a buildup of a dark layer of material. (p. 322)

radiation era The first few thousand years after the Big Bang when the universe was small, dense, and dominated by radiation.

radiation zone Region of the Sun's interior where extremely high temperatures guarantee that the gas is completely ionized. Photons only occasionally interact with electrons, and travel through this region with relative ease. (p. 386)

radiation-dominated universe Early epoch in the universe, when the equivalent density of radiation in the cosmos exceeded the density of matter. (p. 685)

radio Region of the electromagnetic spectrum corresponding to the longest-wavelength radiation.

radioactivity The release of energy by rare, heavy elements when their nuclei decay into lighter nuclei. (pp. 165, 166)

radio galaxy Type of active galaxy that emits most of its energy in the form of long-wavelength radiation. (p. 621)

radiograph Image made from observations at radio wavelengths.

radio lobe Roundish extended region of radio-emitting gas, lying well beyond the center of a radio galaxy. (p. 621)

radio telescope Large instrument designed to detect radiation from space at radio wavelengths. (p. 114)

radio wave Electromagnetic radiation with wavelength in the radio part of the spectrum (p. 58)

radius–luminosity–temperature relationship A mathematical proportionality, arising from Stefan's Law, which allows astronomers to indirectly determine the radius of a star once its luminosity and temperature are known. (p. 428)

rapid mass transfer Mass transfer in a binary system that proceeds at a rapid and unstable rate, transferring most of the mass of one star onto the other.

ray The path taken by a beam of radiation.

Rayleigh scattering Scattering of light by particles in the atmosphere.

recession velocity Rate at which two objects are separating from one another.

recombination *See* decoupling.

recurrent nova Star that "goes nova" several times over the course of a few decades.

reddening Dimming of starlight by interstellar matter, which tends to scatter higher-frequency (blue) components of the radiation more efficiently than the lower-frequency (red) components. (p. 447)

red dwarf Small, cool faint star at the lower-right end of the main sequence on the Hertzsprung–Russell diagram. (p. 432)

red giant A giant star whose surface temperature is relatively low so that it glows red. (pp. 430)

red-giant branch The section of the evolutionary track of a star corresponding to intense hydrogen shell burning, which drives a steady expansion and cooling of the outer envelope of the star. As the star gets larger in radius and its surface temperature cools, it becomes a red giant. (p. 495)

red-giant region The upper-right corner of the Hertzsprung–Russell diagram, where red-giant stars are found. (p. 432)

redshift Motion-induced change in the wavelength of light emitted from a source moving away from us. The relative recessional motion causes the wave to have an observed wavelength longer (and hence redder) than it would if it were not moving.

redshift survey Three-dimensional survey of galaxies, using redshift to determine distance.

red supergiant An extremely luminous red star. Often found on the asymptotic-giant branch of the Hertzsprung–Russell diagram. (p. 430)

reflecting telescope A telescope which uses a mirror togather and focus light from a distant object. (p. 98)

reflection nebula Bluish nebula caused by starlight scattering from dust particles in an interstellar cloud located just off the line of sight between Earth and a bright star. (p. 451)

refracting telescope A telescope that uses a lens to gather and focus light from a distant object. (p. 98)

refraction The tendency of a wave to bend as it passes from one transparent medium to another. (p. 98)

regolith Surface dust on the moon, several tens of meters thick in places, caused by billions of years of meteoritic bombardment.

relativistic Speeds comparable to the speed of light.

relativistic fireball Leading explanation of a gamma-ray burst in which an expanding region of superhot gas radiates in the gamma-ray part of the spectrum.

remnant The object left behind after a supernova explosion. Can refer to (1) the expanding and cooling shell of glowing gas resulting from the event, or (2) the neutron star or black hole that remains at the center of the explosion. (p. 540)

residual cap Portion of Martian polar ice caps that remains permanently frozen, undergoing no seasonal variations. (p. 243)

resonance Circumstance in which two characteristic times are related in some simple way, for example, an asteroid with an orbital period exactly half that of Jupiter. (p. 193)

retina (eye) The back part of the eye onto which light is focused by the lens.

retrograde motion Backward, westward loop traced out by a planet with respect to the fixed stars. (p. 35)

revolution Orbital motion of one body about another, such as Earth about the Sun. (p. 13)

revolving *See* revolution.

Riemannian geometry Geometry of positively curved space (e.g., the surface of a sphere).

right ascension Celestial coordinate used to measure longitude on the celestial sphere. The zero point is the position of the Sun at the vernal equinox.

rille A ditch on the surface of the Moon where molten lava flowed in the past. (p. 200)

ring *See* planetary ring system.

ringlet Narrow region in Saturn's planetary ring system where the density of ring particles is high. *Voyager* discovered that the rings visible from Earth are actually composed of tens of thousands of ringlets. (p. 295)

Roche limit Often called the tidal stability limit, the Roche limit gives the distance from a planet at which the tidal force (due to the planet) between adjacent objects exceeds their mutual attraction. Objects within this limit are unlikely to accumulate into larger objects. The rings of Saturn occupy the region within Saturn's Roche limit. (p. 294)

Roche lobe An imaginary surface around a star. Each star in a binary system can be pictured as being surrounded by a teardrop-shaped zone of gravitational influence, the Roche lobe. Any material within the Roche lobe of a star can be considered to be part of that star. During evolution, one member of the binary system can expand so that it overflows its own Roche lobe and begins to transfer matter onto the other star. (p. 512)

rock Material made predominantly from compounds of silicon and oxygen.

rock cycle Process by which surface rock on Earth is continuously redistributed and transformed from one type into another. (p. 173)

rotation Spinning motion of a body about an axis. (p. 10)

rotation curve Plot of the orbital speed of disk material in a galaxy against its distance from the galactic center. Analysis of rotation curves of spiral galaxies indicates the existence of dark matter. (p. 592)

R-process "Rapid" process in which many neutrons are captured by a nucleus during a supernova explosion.

RR Lyrae variable Variable star whose luminosity changes in a characteristic way. All RR Lyrae stars have more or less the same average luminosity. (p. 577)

runaway greenhouse effect A process in which the heating of a planet leads to an increase in its atmosphere's ability to retain heat and thus to further heating, causing extreme changes in the temperature of the surface and the composition of the atmosphere. (p. 226)

runoff channel River-like surface feature on Mars, evidence that liquid water once existed there in great quantities. They are found in the southern highlands and are thought to have been formed by water that flowed nearly 4 billion years ago. (p. 239)

S

S0 galaxy Galaxy that shows evidence of a thin disk and a bulge, but which has no spiral arms and contains little or no gas. (p. 608)

Sagittarius A/Sgr A* Strong radio source corresponding to the supermassive black hole at the center of the Milky Way. (p. 597)

Saros cycle Time interval between successive occurrences of the "same" solar eclipse, equal to 18 years, 11.3 days.

satellite A small body orbiting another larger body.

SB0 galaxy An S0-type galaxy whose disk shows evidence of a bar. (p. 608)

scalar field Class of quantum-mechanical fields in Grand Unified Theories whose interactions with particles determine those particles' properties. Inflation in the early universe is thought to have occurred when one or more scalar fields temporarily increased in energy above their normal states. (p. 695)

scarp Surface feature on Mercury thought to be the result of cooling and shrinking of the crust forming a wrinkle on the face of the planet. (p. 201)

Schmidt telescope Type of telescope having a very wide field of view, allowing large areas of the sky to be observed at once.

Schwarzschild radius The distance from the center of an object such that, if all the mass were compressed within that region, the escape speed would equal the speed of light. Once a stellar remnant collapses within this radius, light cannot escape and the object is no longer visible. (p. 553)

scientific method The set of rules used to guide science, based on the idea that scientific "laws" be continually tested, and modified or replaced if found inadequate. (p. 7)

scientific notation Expressing large and small numbers using power-of-10 notation.

seasonal cap Portion of Martian polar ice caps that is subject to seasonal variations, growing and shrinking once each Martian year. (p. 243)

seasons Changes in average temperature and length of day that result from the tilt of Earth's (or any planet's) axis with respect to the plane of its orbit. (p. 15)

secondary atmosphere The chemicals that composed Earth's atmosphere after the planet's formation, once volcanic activity outgassed chemicals from the interior.

sedimentary Type of rock formed from the buildup of sediment.

seeing Term used to describe the ease with which good telescopic observations can be made from Earth's surface, given the blurring effects of atmospheric turbulence. (p. 110)

seeing disk Roughly circular region on a detector over which a star's point-like images is spread, due to atmospheric turbulence. (p. 110)

seismic wave A wave that travels outward from the site of an earthquake through Earth. (p. 162)

seismology The study of earthquakes and the waves they produce in Earth's interior.

seismometer Equipment designed to detect and measure the strength of earthquakes (or quakes on any other planet).

selection effect Observational bias in which a measured property of a collection of objects is due to the way in which the measurement was made, rather than being intrinsic to the objects themselves. (p. 373)

self-propagating star formation Mode of star formation in which shock waves produced by the formation and evolution of one generation of stars triggers the formation of the next. (p. 591)

semi-detached Binary system where one star lies within its Roche lobe but the other fills its Roche lobe and is transferring matter onto the first star.

semimajor axis One-half of the major axis of an ellipse. The semimajor axis is the way in which the size of an ellipse is usually quantified. (p. 43)

SETI Acronym for search for extraterrestrial intelligence. (p. 723)

Seyfert galaxy Type of active galaxy whose emission comes from a very small region within the nucleus of an otherwise normal-looking spiral system. (p. 620)

shepherd satellite Satellite whose gravitational effect on a ring helps preserve the ring's shape. Examples are two satellites of Saturn, Prometheus and Pandora, whose orbits lie on either side of the F ring. (p. 299)

shield volcano A volcano produced by repeated nonexplosive eruptions of lava, creating a gradually sloping, shield-shaped low dome. Often contains a caldera at its summit. (p. 219)

shock wave Wave of matter, which may be generated by a newborn star or supernova, that pushes material outward into the surrounding molecular cloud. The material tends to pile up, forming a rapidly moving shell of dense gas. (p. 480)

short-period comet Comet with an orbital period of less than 200 years.

SI Système International, the international system of metric units used to define mass, length, time, etc.

sidereal day The time needed between successive risings of a given star. (p. 12)

sidereal month Time required for the Moon to complete one trip around the celestial sphere. (p. 18)

sidereal year The time required for the constellations to complete one cycle around the sky and return to their starting points, as seen from a given point on Earth. Earth's orbital period around the Sun is one sidereal year. (p. 15)

single-line spectroscopic binary Binary system in which one star is too faint for its spectrum to be distinguished, so only the spectrum of the brighter star can be seen to shift back and forth as the stars orbit one another.

singularity A point in the universe where the density of matter and the gravitational field are infinite, such as at the center of a black hole. (p. 560)

sister/coformation theory (Moon) Theory suggesting that the Moon formed as a separate object close to Earth.

soft landing Use of rockets, parachutes, or packaging to break the fall of a space probe as it lands on a planet.

solar constant The amount of solar energy reaching Earth's atmosphere per unit area per unit time, approximately 1400 W/m^2. (p. 386)

solar core The region at the center of the Sun, with a radius of nearly 200,000 km, where powerful nuclear reactions generate the Sun's energy output.

solar cycle The 22-year period that is needed for both the average number of spots and the Sun's magnetic polarity to repeat themselves. The Sun's polarity reverses on each new 11-year sunspot cycle. (p. 400)

solar day The period of time between the instant when the Sun is directly overhead (i.e., noon) to the next time it is directly overhead. (p. 12)

solar eclipse Celestial event during which the new Moon passes directly between Earth and the Sun, temporarily blocking the Sun's light. (p. 19)

solar interior The region of the Sun between the solar core and the photosphere.

solar maximum Point of the sunspot cycle during which many spots are seen. They are generally confined to regions in each hemisphere, between about 15 and 20 degrees latitude.

solar minimum Point of the sunspot cycle during which only a few spots are seen. They are generally confined to narrow regions in each hemisphere at about 25–30 degrees latitude.

solar nebula The swirling gas surrounding the early Sun during the epoch of solar system formation, also referred to as the primitive solar system. (p. 147)

solar neutrino problem The discrepancy between the theoretically predicted flux of neutrinos streaming from the Sun as a result of fusion reactions in the core and the flux that is actually observed. The observed number of neutrinos is only about half the predicted number. (p. 411)

solar system The Sun and all the bodies that orbit it—Mercury, Venus, Earth, Mars, Jupiter, Saturn, Uranus, Neptune, their moons, the asteroids, the comets, and trans-Neptunian objects. (p. 134)

solar wind An outward flow of fast-moving charged particles from the Sun. (p. 386)

solstice *See* summer solstice, winter solstice.

sotropic Assumed property of the universe such that the universe looks the same in every direction.

south celestial pole Point on the celestial sphere directly above Earth's South Pole. (p. 10)

spacetime Single entity combining space and time in special and general relativity. (p. 554)

spatial resolution The dimension of the smallest detail that can be seen in an image.

special theory of relativity Theory proposed by Einstein to deal with the preferred status of the speed of light. (p. 553)

speckle interferometry Technique whereby many short-exposure images of a star are combined to make a high-resolution map of the star's surface.

spectral class Classification scheme, based on the strength of stellar spectral lines, which is an indication of the temperature of a star. (p. 427)

spectral window Wavelength range in which Earth's atmosphere is transparent.

spectrograph Instrument used to produce detailed spectra of stars. Usually, a spectrograph records a spectrum on a CCD detector, for computer analysis.

spectrometer Instrument used to produce detailed spectra of stars. Usually, a spectrograph records a spectrum on a photographic plate, or more recently, in electronic form on a computer. (p. 110)

spectroscope Instrument used to view a light source so that it is split into its component colors. (p. 78)

spectroscopic binary A binary-star system which appears as a single star from Earth, but whose spectral lines show back-and-forth Doppler shifts as two stars orbit one another. (p. 436)

spectroscopic parallax Method of determining the distance to a star by measuring its temperature and then determining its absolute brightness by comparing with a standard Hertzsprung–Russell diagram. The absolute and apparent brightness of the star give the star's distance from Earth. (p. 433)

spectroscopy The study of the way in which atoms absorb and emit electromagnetic radiation. Spectroscopy allows astronomers to determine the chemical composition of stars. (p. 81)

spectrum The separation of light into its component colors.

speed Distance moved per unit time, independent of direction. *See also* velocity.

speed of light The fastest possible speed, according to the currently known laws of physics. Electromagnetic radiation exists in the form of waves or photons moving at the speed of light. (p. 62)

spicule Small solar storm that expels jets of hot matter into the Sun's lower atmosphere.

spin–orbit resonance State that a body is said to be in if its rotation period and its orbital period are related in some simple way.

spiral arm Distribution of material in a galaxy forming a pinwheel-shaped design, beginning near the galactic center. (p. 588)

spiral density wave (1) A wave of matter formed in the plane of planetary rings, similar to ripples on the surface of a pond, which wrap around the rings forming spiral patterns similar to grooves in a record disk. Spiral density waves can lead to the appearance of ringlets. (p. 588) (2) Proposed explanation for the existence of galactic spiral arms, in which coiled waves of gas compression move through the galactic disk, triggering star formation.

spiral galaxy Galaxy composed of a flattened, star-forming disk component which may have spiral arms and a large central galactic bulge. (pp. 576, 604)

spiral nebula Historical name for spiral galaxies, describing their appearance.

spring tide The largest tides, occurring when the Sun, Moon, and Earth are aligned at new and full Moon.

S-process "Slow" process in which neutrons are captured by a nucleus; the rate is typically one neutron capture per year.

standard candle Any object with an easily recognizable appearance and known luminosity, which can be used in estimating distances. Supernovae, which all have the same peak luminosity (depending on type), are good examples of standard candles and are used to determine distances to other galaxies. (p. 611)

standard solar model A self-consistent picture of the Sun, developed by incorporating the important physical processes that are thought to be important in determining the Sun's internal structure into a computer program. The results of the program are then compared with observations of the Sun and modifications are made to the model. The standard solar model, which enjoys widespread acceptance, is the result of this process. (p. 388)

star A glowing ball of gas held together by its own gravity and powered by nuclear fusion in its core. (p. 386)

starburst galaxy Galaxy in which a violent event, such as near-collision, has caused an intense episode of star formation in the recent past. (p. 639)

star cluster A grouping of anywhere from a dozen to a million stars that formed at the same time from the same cloud of interstellar gas. Stars in clusters are useful to aid our understanding of stellar evolution because, within a given cluster, stars are all roughly the same age and chemical composition and lie at roughly the same distance from Earth. (p. 482)

Stefan's law Relation that gives the total energy emitted per square centimeter of its surface per second by an object of a given temperature. Stefan's law shows that the energy emitted increases rapidly with an increase in temperature, proportional to the temperature raised to the fourth power. (p. 68)

stellar epoch Most recent period when stars, planets, and life have appeared in the universe.

stellar nucleosynthesis The formation of heavy elements by the fusion of lighter nuclei in the hearts of stars. Except for hydrogen and helium, all other elements in our universe resulted from stellar nucleosynthesis. (p. 529)

stellar occultation The dimming of starlight produced when a solar system object such as a planet, moon, or ring passes directly in front of a star. (p. 326)

stratosphere The portion of Earth's atmosphere lying above the troposphere, extending up to an altitude of 40–50 km. (p. 157)

string theory Theory that interprets all particles and forces in terms of particular modes of vibration of submicroscopic strings. (p. 688)

strong nuclear force Short-range force responsible for binding atomic nuclei together. The strongest of the four fundamental forces of nature. (pp. 407, 410)

S-type asteroid Asteroid made up mainly of silicate or rocky material.

subatomic particle Particle smaller than the size of an atomic nucleus.

subduction zone Place where two plates meet and one slides under the other. (p. 170)

subgiant Star on the subgiant branch of the Hertzsprung–Russell diagram. (p. 495)

subgiant branch The section of the evolutionary track of a star corresponding to changes that occur just after hydrogen is depleted in the core, and core hydrogen burning ceases. Shell hydrogen burning heats the outer layers of the star, which causes a general expansion of the stellar envelope. (p. 495)

sublimation Process by which element changes from the solid to the gaseous state, without becoming liquid.

summer solstice Point on the ecliptic where the Sun is at its northernmost point above the celestial equator, occurring on or near June 21. (p. 14)

sunspot An Earth-sized dark blemish found on the surface of the Sun. The dark color of the sunspot indicates that it is a region of lower temperature than its surroundings. (p. 396)

sunspot cycle The fairly regular pattern that the number and distribution of sunspots follows, in which the average number of spots reaches a maximum every 11 or so years, then fall off to almost zero. (p. 400)

supercluster Grouping of several clusters of galaxies into a larger, but not necessarily gravitationally bound, unit. (p. 650)

super-Earth Extrasolar planet with mass between 2 and 10 Earth masses. Both rocky and ocean super-Earths have been detected. (p. 372)

superforce An attempt to combine the strong and electroweak forces into one single force.

supergiant A star with a radius between 100 and 1000 times that of the Sun. (p. 430)

supergranulation Large-scale flow pattern on the surface of the Sun, consisting of cells measuring up to 30,000 km across, thought to be the imprint of large convective cells deep in the solar interior. (p. 392)

superior conjunction Orbital configuration in which an inferior planet (Mercury or Venus) lies farthest from Earth (on the opposite side of the Sun).

supermassive black hole Black hole having a mass a million to a billion times greater than the mass of the Sun; usually found in the central nucleus of a galaxy. (p. 563)

supernova Explosive death of a star, caused by the sudden onset of nuclear burning (Type I), or an enormously energetic shock wave (Type II). One of the most energetic events of the universe, a supernova may temporarily outshine the rest of the galaxy in which it resides. (p. 522)

supernova remnant The scattered glowing remains from a supernova that occurred in the past. The Crab Nebula is one of the best-studied supernova remnants. (p. 526)

supersymmetric relic Massive particle that should have been created in the Big Bang if supersymmetry is correct.

supersymmetry Fundamental symmetry of the universe in which all particles that are acted on by forces (e.g., quarks and electrons) are paired with particles that transmit forces (e.g., gluons and photons). It predicts the existence of supersymmetric partners for all known elementary particles. (p. 688)

surface gravity The acceleration due to gravity at the surface of a star or planet.

S-waves Shear waves from an earthquake, which can travel only through solid material and move more slowly than P-waves.

synchronous orbit State of an object when its period of rotations is exactly equal to its average orbital period. The Moon is in a synchronous orbit, and so presents the same face toward Earth at all times. (p. 189)

synchrotron radiation Type of nonthermal radiation produced by high-speed charged particles, such as electrons, as they are accelerated in a strong magnetic field. (p. 630)

synodic month Time required for the Moon to complete a full cycle of phases. (p. 18)

T

T Tauri star Protostar in the late stages of formation, often exhibiting violent surface activity. T Tauri stars have been observed to brighten noticeably in a short period of time, consistent with the idea of rapid evolution during this final phase of stellar formation. (p. 472)

tail Component of a comet that consists of material streaming away from the main body, sometimes spanning hundreds of millions of kilometers. May be composed of dust or ionized gases. (p. 340)

tau A type of lepton (along with the electron and muon).

tectonic fracture Crack on a planet's surface, in particular on the surface of Mars, caused by internal geological activity.

telescope Instrument used to capture as many photons as possible from a given region of the sky and concentrate them into a focused beam for analysis. (p. 98)

temperature A measure of the amount of heat in an object, and an indication of the speed of the particles that comprise it. (p. 66)

tenuous Thin, having low density.

terminator The line separating night from day on the surface of the Moon or a planet.

terrae *See* highlands.

terrestrial planet One of the four innermost planets of the solar system, resembling Earth in general physical and chemical properties. (p. 138)

theoretical model An attempt to construct a mathematical explanation of a physical process or phenomenon, within the assumptions and confines of a given theory. In addition to providing an explanation of the observed facts, the model generally makes new predictions that can be tested by further observation or experimentation. (p. 6)

theories of relativity Einstein's theories, on which much of modern physics rests. Two essential facts of the theory are that nothing can travel faster than the speed of light and that everything, including light, is affected by gravity.

theory A framework of ideas and assumptions used to explain some set of observations and make predictions about the real world. (p. 6)

thermal equilibrium Condition in which new particle–antiparticle pairs are created from photons at the same rate as pairs annihilate one another to produce new photons.

thick disk Region of a spiral galaxy where an intermediate population of stars resides, younger than the halo stars but older than stars in the disk.

threshold temperature Critical temperature above which pair production is possible and below which pair production cannot occur.

tidal bulge Elongation of Earth caused by the difference between the gravitational force on the side nearest the Moon and the force on the side farthest from the Moon. The long axis of the tidal bulge points toward the Moon. More generally, the deformation of any body produced by the tidal effect of a nearby gravitating object. (p. 177)

tidal force The variation in one body's gravitational force from place to place across another body—for example, the variation of the Moon's gravity across Earth. (p. 177)

tidal locking Circumstance in which tidal forces have caused a moon to rotate at exactly the same rate at which it revolves around its parent planet, so that the moon always keeps the same face turned toward the planet.

tidal stability limit The minimum distance within which a moon can approach a planet before being torn apart by the planet's tidal force.

tides Rising and falling motion of terrestrial bodies of water, exhibiting daily, monthly, and yearly cycles. Ocean tides on Earth are caused by the competing gravitational pull of the Moon and Sun on different parts of Earth. (p. 176)

time dilation A prediction of the theory of relativity, closely related to the gravitational redshift. To an outside observer, a clock lowered into a strong gravitational field will appear to run slow. (p. 559)

total eclipse Celestial event during which one body is completely blocked from view by another. (p. 18)

transit *See* planetary transit.

transition zone The region of rapid temperature increases that separates the Sun's chromosphere from the corona. (p. 386)

trans-Neptunian object A small, icy body orbiting beyond the orbit of Neptune. Pluto and Eris are the largest currently known examples. (p. 348)

transverse motion Motion perpendicular to a particular line of sight, which does not result in Doppler shift in radiation received.

transverse velocity Component of star's velocity perpendicular to the line of sight.

triangulation Method of determining distance based on the principles of geometry. A distant object is sighted from two well-separated locations. The distance between the two locations and the angle between the line joining them and the line to the distant object are all that are necessary to ascertain the object's distance. (p. 23)

triple-alpha process The creation of carbon-12 by the fusion of three helium-4 nuclei (alpha particles). Helium-burning stars occupy a region of the Hertzsprung–Russell diagram known as the horizontal branch. (p. 496)

triple star system Three stars that orbit one another, bound together by gravity.

Trojan asteroid One of two groups of asteroids which orbit at the same distance from the Sun as Jupiter, 60 degrees ahead of and behind the planet. (p. 340)

tropical year The time interval between one vernal equinox and the next. (p. 15)

troposphere The portion of Earth's atmosphere from the surface to about 12 km. (p. 157)

trough Maximum departure of a wave below its undisturbed state.

true space motion True motion of a star, taking into account both its transverse and radial motion according to the Pythagorean theorem.

Tully-Fisher relation A relation used to determine the absolute luminosity of a spiral galaxy. The rotational velocity, measured from the broadening of spectral lines, is related to the total mass, and hence the total luminosity. (p. 612)

turnoff mass The mass of a star that is just now evolving off the main sequence in a star cluster.

21-centimeter line Spectral line in the radio region of the electromagnetic spectrum, associated with a change of spin of the electron in a hydrogen atom.

21-centimeter radiation Radio radiation emitted when an electron in the ground state of a hydrogen atom flips its spin to become parallel to the spin of the proton in the nucleus. (p. 459)

twin quasar Quasar that is seen twice at different locations in the sky due to gravitational lensing.

Type I supernova One possible explosive death of a star. A white dwarf in a binary-star system can accrete enough mass that it cannot support its own weight. The star collapses and temperatures become high enough for carbon fusion to occur. Fusion begins throughout the white dwarf almost simultaneously and an explosion results. (p. 523)

Type II supernova One possible explosive death of a star, in which the highly evolved stellar core rapidly implodes and then explodes, destroying the surrounding star. (p. 523)

U

ultraviolet Region of the electromagnetic spectrum, just beyond the visible range, corresponding to wavelengths slightly shorter than blue light. (p. 58)

ultraviolet telescope A telescope that is designed to collect radiation in the ultraviolet part of the spectrum. Earth's atmosphere is partially opaque to these wavelengths, so ultraviolet telescopes are put on rockets, balloons, and satellites to get high above most or all of the atmosphere. (p. 123)

umbra Central region of the shadow cast by an eclipsing body. (p. 19); The central region of a sunspot, which is its darkest and coolest part. (p. 396)

unbound An orbit which does not stay in a specific region of space, but where an object escapes the gravitational field of another. Typical unbound orbits are hyperbolic in shape. (p. 53)

unbound trajectory Path of an object with launch speed high enough that it can escape the gravitational pull of a planet.

uncompressed density The density a body would have in the absence of any compression due to its own gravity.

universe The totality of all space, time, matter, and energy. (p. 4)

unstable nucleus Nucleus that cannot exist indefinitely, but rather must eventually decay into other particles or nuclei.

upwelling Upward motion of material having temperature higher than the surrounding medium.

Urey-Miller experiment Groundbreaking experiment that simulated conditions on the early Earth and tested for the occurrence of chemical evolution—the production of complex molecules similar to those found in all living things on our planet. (p. 712)

V

vacuum energy Energy associated with empty space when a scalar field temporarily has higher than normal energy. (p. 695)

Van Allen belt One of at least two doughnut-shaped regions of magnetically trapped, charged particles high above Earth's atmosphere. (p. 174)

variable star A star whose luminosity changes with time. (p. 577)

velocity Displacement (distance plus direction) per unit time. *See* also speed. (p. 48)

vernal equinox Date on which the Sun crosses the celestial equator moving northward, occurring on or near March 21. (p. 15)

visible light The small range of the electromagnetic spectrum that human eyes perceive as light. The visible spectrum ranges from about 400–700 nm, corresponding to blue through red light. (p. 58)

visible spectrum The small range of the electromagnetic spectrum that human eyes perceive as light. The visible spectrum ranges from about 4000–7000 angstroms, corresponding to blue through red light.

visual binary A binary-star system in which both members are resolvable from Earth. (p. 436)

void Large, relatively empty region of the universe around which superclusters and "walls" of galaxies are organized. (p. 651)

volcano Upwelling of hot lava from below Earth's crust to the planet's surface.

W

wane (referring to the Moon or a planet) To shrink. The Moon appears to wane, or shrink in size, for two weeks after full Moon.

warm longitudes (Mercury) Two opposite points on Mercury's equator where the Sun is directly overhead at aphelion. Cooler than the hot longitudes by 150 degrees.

water hole The radio interval between 18 cm and 21 cm, the respective wavelengths at which hydroxyl (OH) and hydrogen (H) radiate, in which intelligent civilizations might conceivably send their communication signals. (p. 724)

water volcano Volcano that ejects water (molten ice) rather than lava (molten rock) under cold conditions.

watt/kilowatt Unit of power: 1 watt (W) is the emission of 1 joule (J) per second; 1 kilowatt (kW) is 1000 watts.

wave A pattern that repeats itself cyclically in both time and space. Waves are characterized by the speed at which they move, their frequency, and their wavelength. (p. 58)

wave period The amount of time required for a wave to repeat itself at a specific point in space.

wave theory of radiation Description of light as a continuous wave phenomenon, rather than as a stream of individual particles. (p. 63)

wavelength The distance from one wave crest (or trough) to the next, at a given instant in time. (p. 59)

wax (referring to the Moon or a planet) To grow. The Moon appears to wax, or grow in size, for two weeks after new Moon.

weak nuclear force Short-range force, weaker than both electromagnetism and the strong force, but much stronger than gravity; responsible for certain nuclear reactions and radioactive decays. (pp. 408, 410)

weakly interacting massive particle (WIMP) Class of subatomic particles that might have been produced early in the history of the universe; dark-matter candidates.

weight The gravitational force exerted on you by Earth (or the planet on which you happen to be standing). (p. 48)

weird terrain A region on the surface of Mercury with oddly rippled features. This feature is thought to be the result of a strong impact that occurred on the other side of the planet, and sent seismic waves traveling around the planet, converging in the weird region.

white dwarf A dwarf star with sufficiently high surface temperature that it glows white. (pp. 430, 501)

white-dwarf region The bottom-left corner of the Hertzsprung–Russell diagram, where white-dwarf stars are found. (p. 432)

white oval Light-colored region near the Great Red Spot in Jupiter's atmosphere. Like the red spot, such regions are apparently rotating storm systems. (p. 266)

Wien's law Relation between the wavelength at which a blackbody curve peaks and the temperature of the emitter. The peak wavelength is inversely proportional to the temperature, so the hotter the object, the bluer its radiation. (p. 68)

winter solstice Point on the ecliptic where the Sun is at its southernmost point below the celestial equator, occurring on or near December 21. (p. 15)

wispy terrain Prominent light-colored streaks on Rhea, one of Saturn's moons.

X

X ray Region of the electromagnetic spectrum corresponding to radiation of high frequency and short wavelength, far beyond the visible spectrum. (p. 58)

X-ray burster X-ray source that radiates thousands of times more energy than our Sun in short bursts lasting only a few seconds. A neutron star in a binary system accretes matter onto its surface until temperatures reach the level needed for hydrogen fusion to occur. The result is a sudden period of rapid nuclear burning and release of energy. (p. 544)

X-ray nova Nova explosion detected at X-ray wavelengths.

Z

Zeeman effect Broadening or splitting of spectral lines due to the presence of a magnetic field.

zero-age main sequence The region on the Hertzsprung–Russell diagram, as predicted by theoretical models, where stars are located at the onset of nuclear burning in their cores. (p. 473)

zodiac The 12 constellations on the celestial sphere through which the Sun appears to pass during the course of a year. (p. 14)

zonal flow Alternating regions of westward and eastward flow, roughly symmetrical about the equator of Jupiter, associated with the belts and zones in the planet's atmosphere. (p. 264)

zone Bright, high-pressure region in the atmosphere of a jovian planet, where gas flows upward. (p. 262)

ANSWERS TO CHECK QUESTIONS
CONCEPT CHECKS AND PROCESS OF SCIENCE CHECKS

Chapter 1

1.1 *(p. 8)* A theory can never become proven "fact," because it can always be invalidated, or forced to change, by a single contradictory observation. However, once a theory's predictions have been repeatedly verified by experiments over many years, it is often widely regarded as "true." **1.2** *(p. 12)* (1) Because the celestial sphere provides a natural means of specifying the locations of stars on the sky. Celestial coordinates are directly related to Earth's orientation in space, but are independent of Earth's rotation. (2) Distance information is lost. **1.3** *(p. 16)* (1) In the Northern Hemisphere, summer occurs when the Sun is near its highest (northernmost) point on the celestial sphere, or, equivalently, when Earth's North Pole is "tipped" toward the Sun, the days are longest and the Sun is highest in the sky. Winter occurs when the Sun is lowest in the sky (near its southernmost point on the celestial sphere) and the days are shortest. (2) We see different constellations because Earth has moved halfway around its orbit between these seasons and the darkened hemisphere faces an entirely different group of stars. **1.4** *(p. 22)* (1) The angular size of the Moon would remain the same, but that of the Sun would be halved, making it easier for the Moon to eclipse the Sun. We would expect to see total or partial eclipses, but no annular ones. (2) If the distance is halved, the angular size of the Sun would double and total eclipses would never be seen, only partial or annular ones. **1.5** *(p. 27)* Because astronomical objects are too distant for direct ("measuring tape") measurements, so we must rely on indirect means and mathematical reasoning.

Chapter 2

2.1 *(p. 38)* In the geocentric view, retrograde motion is the real backward motion of a planet as it moves on its epicycle. In the heliocentric view, the backward motion is only apparent, caused by Earth "overtaking" the planet in its orbit. **2.2** *(p. 42)* Mainly simplicity and elegance. Both theories made testable predictions, but until Newton developed his laws, neither could explain *why* the planets move as they do. However, Copernicus's model was much simpler than the Ptolemaic version, which became more and more convoluted as observations improved. **2.3** *(p. 45)* Because the laws were derived using only the orbits of the planets Mercury through Saturn, before the outermost planets were known. **2.4** *(p. 47)* Because Kepler determined the overall geometry of the solar system by triangulation using Earth's orbit as a baseline, so all distances were known only relative to the scale of Earth's orbit—the astronomical unit. **2.5** *(p. 52)* Kepler's laws were accurate descriptions of planetary motion based on observations, but they contained no insight into *why* the planets orbited the Sun or why the orbits are as they are. Newtonian mechanics explained the orbits

in terms of universal laws, and in addition made detailed predictions about the motion of other bodies in the cosmos—moons, comets, other stars—that were hitherto impossible to make. **2.6** *(p. 53)* In the absence of any force, a planet would move in a straight line with constant velocity (Newton's first law) and therefore tends to move along the tangent to its orbital path. The Sun's gravity causes the planet to accelerate toward the Sun (Newton's second law), bending its trajectory into the orbit we observe.

Chapter 3

3.1 *(p. 63)* Light displays properties that are characteristic of wave motion in other contexts. Light exhibits diffraction as it passes a corner or through a narrow opening, and two light rays can interfere (reinforce or cancel) one another. These effects are small in everyday experience, but are easily measured in the laboratory. They are consistent with the wave theory, and inconsistent with a particle description of radiation. **3.2** *(p. 66)* All are electromagnetic radiation and travel at the speed of light. In physical terms, they differ only in frequency (or wavelength), although their effects on our bodies (or our detectors) differ greatly. **3.3** *(p. 71)* As the switch is turned and the temperature of the filament rises, the bulb's brightness increases rapidly, by Stefan's law, and its color shifts, by Wien's law, from invisible infrared to red to yellow to white. **3.4** *(p. 73)* According to *More Precisely 2-2*, measuring masses in astronomy usually entails measuring the orbital speed of one object—a companion star, or a planet, perhaps—around another. In most cases, the Doppler effect is an astronomer's only way of making such measurements.

Chapter 4

4.1 *(p. 81)* They are characteristic frequencies (wavelengths) at which matter absorbs or emits photons of electromagnetic radiation. They are unique to each atom or molecule, and thus provide a means of identifying the gas producing them. **4.2** *(p. 84)* On macroscopic (everyday) scales, light exhibits wave properties, as discussed in Chapter 3. However, on microscopic scales, light displays particle characteristics, with the particles (photons) carrying specific amounts of energy—the photoelectric effect and atomic spectra rely on this fact. Thus scientists have to conclude that light behaves as *both* a wave and a particle, depending on circumstances. **4.3** *(p. 89)* Spectral lines correspond to transitions between specific orbitals within an atom. The structure of an atom determines the energies of these orbitals, hence the possible transitions, and hence the energies (colors) of the photons involved. **4.4** *(p. 90)* In addition to changes involving electron energies, changes involving a molecule's vibration or rotation can also result in emission or absorption of

radiation. **4.5** *(p. 93)* Because with few exceptions, spectral analysis is the only way we have of determining the physical conditions—composition, temperature, density, velocity, etc.—in a distant object. Without spectral analysis, astronomers would know next to nothing about the properties of stars and galaxies.

Chapter 5

5.1 *(p. 102)* Because reflecting telescopes are easier to design, build, and maintain than refracting instruments. **5.2** *(p. 108)* The need to gather as much light as possible, and the need to achieve the highest possible angular resolution. **5.3** *(p. 110)* First, photographic plates are inefficient; CCDs are almost always used. Second, because they want to make detailed measurements of brightness, variability, and spectra, which require the use of other, non-imaging detectors. **5.4** *(p. 114)* To reduce or overcome the effects of atmospheric absorption, instruments are placed on high mountains or in space. To compensate for atmospheric turbulence, adaptive optics techniques probe the air above the observing site and adjust the mirror surface accordingly to try to recover the undistorted image. **5.5** *(p. 117)* Radio observations allow us to see objects whose visible light is obscured by intervening matter, or which simply do not emit most of their energy in the visible portion of the spectrum. **5.6** *(p. 120)* The long wavelength of radio radiation. Astronomers use the largest radio telescopes possible and interferometry, which combines the signals from two or more separate telescopes to create the effect of a single instrument of much larger diameter. **5.7** *(p. 123)* Many astronomical objects have very different appearances when viewed at different wavelengths. Some emit only in certain bands, others are partly or wholly obscured by intervening material. By observing at many wavelengths, astronomers gain a much more complete picture of the object under study. **5.8** *(p. 126)* Benefits: they are above the atmosphere, so they are unaffected by seeing or absorption; they can also make round-the-clock observations. Drawbacks: cost, smaller size, inaccessibility, vulnerability to damage by radiation and cosmic rays.

Chapter 6

6.1 *(p. 135)* They provide new tests of our theories of solar system formation, and new examples of the sorts of planetary systems that are possible. **6.2** *(p. 137)* By applying the laws of geometry and Newtonian mechanics to observations made from Earth and (more recently) by visiting spacecraft. **6.3** *(p. 138)* The planets orbit in very nearly the same plane—the ecliptic. Viewed from outside the system, only the orbit of Mercury would deviate noticeably from this plane. **6.4** *(p. 140)* Because the two classes of planet differ in almost every physical property—orbit, mass, radius, composition, existence of rings, and number of moons. **6.5** *(p. 141)* Because it is thought to be much less evolved than material now found in planets, and hence a better indicator of conditions in the early solar system. **6.6** *(p. 146)* Spacecraft missions to other planets have returned detailed information on surfaces, atmospheres, and/or moon and ring systems that are simply unobtainable from Earth-based observations. These new data have revolutionized our view of the solar system. **6.7** *(p. 151)* The shapes and orientations of the orbits were set by the fact that the planets formed in a spinning disk; the composition differences result from the differing materials available to form planets at various distances from the heat of the Sun.

Chapter 7

7.1 *(p. 162)* It raises Earth's average surface temperature above the freezing point of water, which was critical for the development of life on our planet. Should the greenhouse effect continue to strengthen, however, it may conceivably cause catastrophic climate changes on Earth. **7.2** *(p. 166)* We could still construct models based on Earth's gravity and other surface properties, but we would have far less detailed direct (from volcanoes) or indirect (from seismic studies following earthquakes) information on our planet's interior. **7.3** *(p. 174)* Convection currents in the upper mantle cause portions of Earth's crust—plates—to slide around on the surface. As the plates move and interact, they are responsible for volcanism, earthquakes, the formation of mountain ranges and ocean trenches, and the creation and destruction of oceans and continents. **7.4** *(p. 176)* It tells us that the planet has a conducting, liquid core in which the magnetic field is continuously generated. **7.5** *(p. 179)* A tidal force is the *variation* in one body's gravitational force from place to place across another. Tidal forces tend to deform a body, rather than causing an overall acceleration, and they decrease proportional to the inverse cube, rather than the inverse square, of the distance.

Chapter 8

8.1 *(p. 186)* Both bodies have substantially lower escape speeds than Earth, and any atmosphere they may have once had has escaped into space long ago due to their high temperatures. **8.2** *(p. 188)* The maria are younger, denser, and much less heavily cratered than the highlands. **8.3** *(p. 195)* In the case of the Moon, Earth's tidal force has slowed the spin to the point where the rotation rate is now exactly synchronized with the Moon's orbital period around Earth. For Mercury, the Sun's tidal force has caused the rotation period to become exactly $\frac{2}{3}$ of the orbital period, and the rotation axis to be exactly perpendicular to the planet's orbit plane. A synchronous orbit is not possible because of Mercury's eccentric orbit around the Sun. **8.4** *(p. 200)* Heavy bombardment long ago created the basins which later filled with lava to form the maria. Subsequent impacts created virtually all the lunar craters, large and small, we see today. Meteoritic bombardment is the main agent of lunar erosion, although the rate is much

smaller than the erosion rate on Earth. **8.5** *(p. 202)* They are thought to have formed when the planet's core cooled and contracted, causing the crust to crumple. Faults on Earth are the result of tectonic activity. **8.6** *(p. 203)* Generation of a magnetic field is thought to require a rapidly rotating body with a conducting liquid core. Neither the Moon nor Mercury rotates rapidly and, while Mercury's core may still be partly liquid, the Moon's probably is not. **8.7** *(p. 205)* The impact theory holds that a collision created the Moon essentially from Earth's mantle, accounting for the composition similarities. If the collision occurred after Earth had already differentiated and the dense material had formed a core, relatively little of this material would have found its way into the newborn Moon.

Chapter 9

9.1 *(p. 214)* It is very slow and retrograde. The reason is unknown, but may simply be a matter of chance. **9.2** *(p. 215)* Because they were observing in the optical and could not see the surface. Their measurements pertained to the upper atmosphere, above the planet's reflective cloud layers. **9.3** *(p. 223)* No—they are mostly shield volcanoes, where lava upwells through a "hot spot" in the crust. There appears to be no plate tectonic activity on Venus. **9.4** *(p. 227)* Given Venus's other bulk similarities to Earth, the planet might well have had an Earth-like climate. **9.5** *(p. 227)* The dynamo model of planetary magnetism implies that both a conducting liquid core and rapid rotation are needed to generate a magnetic field. Venus's rotation is the slowest of any planet in the solar system.

Chapter 10

10.1 *(p. 233)* Because they occur when Martian opposition happens to coincide approximately with Martian perihelion. Such an alignment happens every 7 Martian synodic years. **10.2** *(p. 234)* Mars does have seasons, since its rotation axis is inclined to its orbit plane in much the same way as Earth's is. However, the Martian seasons are also affected by the planet's eccentric orbit. The appearance of the planet changes seasonally, although the changes have nothing to do with growing cycles, as was once thought. **10.3** *(p. 238)* The lowlands are much less heavily cratered, implying that they have been resurfaced by volcanism (or smoothed by erosion) since the highlands formed. **10.4** *(p. 242)* Before spacecraft visited Mars, the prevailing scientific view was that Mars was a dry, geologically dead world, unlikely ever to have been suitable for life. However, as more and more data have accumulated from spacecraft in orbit and on the planet's surface, scientists have realized that Mars has had a varied, and much wetter, past than was previously thought. Evidence exists for large quantities of flowing and standing water in the past, and water ice has been found at the poles and under the Martian surface. **10.5** *(p. 247)* Some of it may have escaped into space. Of the rest, some exists in the form of permafrost (or perhaps liquid water) below the surface. The rest is contained in the Martian polar caps. **10.6** *(p. 252)* Some of it was lost due to the planet's weak gravity, aided by violent meteoritic impacts. The rest became

part of the planet's surface rocks, the polar caps, or the subsurface permafrost. **10.7** *(p. 253)* The planet's small size, which allowed the planet's internal heat to escape, effectively shutting down the processes that drive mantle convection, volcanism, and plate tectonics. **10.8** *(p. 254)* They are much smaller, and orbit much closer to the parent planet. In addition, they seem to have a different formation history—their composition differences from Mars suggest that they were captured by the planet long after they (and Mars) formed; for Earth's Moon, this scenario does not seem to work.

Chapter 11

11.1 *(p. 262)* The magnetic field is generated by the motion of electrically conducting liquid in the deep interior, and therefore presumably shares the rotation of that region of the planet. **11.2** *(p. 268)* Like weather systems on Earth, the belts and zones are regions of high and low pressure and are associated with convective motion. However, unlike storms on Earth, they wrap all the way around the planet because of Jupiter's rapid rotation. In addition, the clouds are arranged in three distinct layers and the bright colors are the result of cloud chemistry unlike anything operating in Earth's atmosphere. Jupiter's spots are somewhat similar to hurricanes on Earth, but they are far larger and longer-lived. **11.3** *(p. 271)* By constructing theoretical models contrained by detailed spacecraft observations of the planet's bulk properties. **11.4** *(p. 273)* Because Jupiter rotates more rapidly than Earth and because the volume of conducting fluid responsible for the field is much greater. Both these factors result in a much stronger planetary magnetic field. **11.5** *(p. 279)* Because liquid water is thought to have played a critical role in the appearance of life on Earth, and is presumed to be similarly important elsewhere in the solar system. **11.6** *(p. 281)* Jupiter's gravitational field, via its tidal effect on the moons.

Chapter 12

12.1 *(p. 286)* Because Earth crossed Saturn's ring plane in 1995. The images from before then see the rings from above, those from later see the rings from below. **12.2** *(p. 290)* Saturn's cloud and haze layers are thicker than those on Jupiter because of Saturn's lower gravity, so we usually see only the upper level of the atmosphere even though the same basic features and cloud layers are there. **12.3** *(p. 292)* Roughly half of it has precipitated into the planet's interior, reducing the helium fraction in the outer layers (and releasing gravitational energy as it fell). **12.4** *(p. 299)* The Roche limit is the radius inside of which a moon will be torn apart by tidal forces. Planetary rings are found inside the Roche limit, (most) moons outside. Orbital resonances between ring particles and moons are responsible for much of the fine structure in the rings. They may also maintain the orbits of shepherd satellites that keep some ring features sharp. **12.5** *(p. 304)* It has a thick atmosphere (denser than Earth's), unlike any other moon in the solar system, and liquid methane on its surface. **12.6** *(p. 309)* Because they are tidally locked by Saturn's gravity into synchronous orbits, and therefore all have permanently leading and trailing faces that interact differently with the environment around the planet.

Chapter 13

13.1 *(p. 315)* Uranus's orbit was observed to deviate from a perfect ellipse, leading astronomers to try to compute the mass and location of the body responsible for those discrepancies. That body was Neptune. **13.2** *(p. 317)* For unknown reasons, Uranus rotates "on its side"—with its rotation axis almost in the plane of the ecliptic. **13.3** *(p. 320)* Neptune lies far from the Sun and is very cold, so the source of the energy for these atmospheric phenomena is not known. Uranus, closer to the Sun but with a similar atmospheric temperature, shows much less activity. **13.4** *(p. 320)* In each case, the field is significantly offset from the planet's center and inclined to the rotation axis. **13.5** *(p. 326)* Triton shows evidence for surface activity—nitrogen geysers and water volcanoes—that seems to have erased most of its impact craters. **13.6** *(p. 328)* Both are thin rings that require shepherd satellites to prevent them from spreading out and dispersing.

Chapter 14

14.1 *(p. 337)* Similarities: all orbit in the inner solar system, and are solid bodies of generally "terrestrial" composition. Differences: asteroids are much smaller than the terrestrial planets, and their orbits are much less regular. **14.2** *(p. 340)* Because most Earth-crossing asteroids will eventually come very close to or even collide with our planet, with potentially catastrophic results. **14.3** *(p. 346)* Asteroids are generally rocky. Comets are predominantly made of ice, with some dust and other debris mixed in. **14.4** *(p. 348)* Most comets never come close enough to the Sun for us to see them. **14.5** *(p. 352)* Because of its similarities to the known Kuiper belt objects, as well as to the icy moons of the outer planets, it is now regarded as the largest member of the Kuiper belt, rather than as the smallest planet. **14.6** *(p. 356)* They are fragments of comets or asteroids orbiting the Sun in interplanetary space. They are important to planetary scientists because they generally consist of ancient material, and contain vital information on conditions in the early solar system.

Chapter 15

15.1 *(p. 366)* Because the jovian planets could start to form only after the solar nebula had formed, but their formation was terminated when the Sun reached the T Tauri phase and dispersed the disk. **15.2** *(p. 368)* Yes—if a star formed with a disk of matter around it then the basic processes of condensation and accretion would probably have occurred there too, even if it doesn't have planets like Earth. **15.3** *(p. 369)* Because it must explain the general "orderly" features of solar system architecture, while still accommodating the fact that there are exceptions to many of them. **15.4** *(p. 372)* The two major techniques that have led to the discovery of almost all extrasolar planets are radial velocity measurements, where astronomers detect the back and forth movement of the parent star, and transits, where the planet temporarily

blocks the light of the star. We don't have many images because, in almost all cases, the planet is lost in the star's glare. **15.5** *(p. 376)* (1) Only planets significantly more massive than Earth have so far been detected; (2) The planets' orbits are often quite eccentric; (3) The "hot Jupiters" orbit very close to their parent star. **15.6** *(p. 378)* Hot Jupiters are explained as a consequence of giant planet migration in the nebular disk; eccentric Jupiters could be the result of gravitational interactions between planets or due to the tidal effects of other stars; the lack of Earth-sized planets is thought to be due to observational limitations.

Chapter 16

16.1 *(p. 388)* When we simply multiply the solar constant by the total area to obtain the solar luminosity, we are implicitly assuming that the same amount of energy reaches every square meter of the large sphere in Figure 16.3. **16.2** *(p. 392)* The energy may be carried in the form of (1) radiation, where energy travels in the form of light, and (2) convection, where energy is carried by physical motion of upwelling solar gas. **16.3** *(p. 396)* The spectrum shows (1) emission lines of (2) highly ionized elements, implying high temperature. **16.4** *(p. 401)* There is a strong field, with a well-defined east–west organization, just below the surface. The field direction in the southern hemisphere is opposite that in the north. However, the details of the fields are very complex. **16.5** *(p. 406)* Because high-energy particles from flares and coronal mass ejections can affect Earth's magnetosphere and atmosphere, disrupting communications and power transmission on our planet. On longer time scales, it appears that long-term changes in the solar cycle may also be correlated with climate variations on Earth. **16.6** *(p. 409)* Because the Sun shines by nuclear fusion, which converts mass into energy as hydrogen turns into helium. **16.7** *(p. 412)* When theory and observation come into conflict, observers must repeat and extend their experiments to check and improve the results, while theorists must look for errors or omissions in their calculations. In the case of the solar neutrino problem, new observations reinforced the disagreement with the theoretical calculations. It turned out that the calculations were also correct as far as they went, but the theory was found to be incomplete, as it omitted neutrino oscillations, which accounted for the discrepancy and resolved the problem.

Chapter 17

17.1 *(p. 421)* (1) Because stars are so far away that their parallaxes relative to any baseline on Earth are too small to measure accurately. (2) Because the transverse component must be determined by measuring the star's proper motion, which decreases as the star's distance increases, and is too small to measure for most distant stars. **17.2** *(p. 423)* Nothing—we need to know their distances before the luminosities (or absolute magnitudes) can be determined. **17.3** *(p. 428)* Because temperature controls which excited states the star's atoms and ions are in, and hence which atomic transitions are possible. **17.4** *(p. 430)* Yes, using the radius–luminosity–temperature

relationship, but only if we can find a method of determining the luminosity that doesn't depend on the inverse-square law (Sec. 17.3). **17.5** *(p. 430)* Because giants are intrinsically very luminous, and can be seen to much greater distances than the more common main-sequence stars or white dwarfs. **17.6** *(p. 435)* All stars would be further away, but their measured spectral types and apparent brightnesses would be unchanged, so their luminosities would be greater than previously thought. The main sequence would therefore move vertically upward in the H–R diagram. (We would then use larger luminosities in the method of spectroscopic parallax, so distances inferred by that method would also increase.) **17.7** *(p. 440)* We don't—we assume that their masses are the same as similar stars found in binaries.

Chapter 18

18.1 *(p. 449)* Because the scale of interstellar space is so large that even very low densities can add up to a large amount of obscuring matter along the line of sight to a distant star. **18.2** *(p. 455)* Because the UV light is absorbed by hydrogen gas in the surrounding cloud, ionizing it to form the emission nebulae. The red light is H radiation; part of the visible hydrogen spectrum emitted as electrons and protons recombine to form hydrogen atoms. **18.3** *(p. 457)* By studying absorption lines caused by atoms and molecules in the clouds, and the general extinction due to dust, it is possible to map out a cloud's properties—so long as enough stars are conveniently located behind it. **18.4** *(p. 459)* Because most of the interstellar matter in the galactic disk is made up of atomic hydrogen, and 21-centimeter radiation provides a probe of that gas largely unaffected by interstellar absorption. **18.5** *(p. 461)* Because the main constituent, hydrogen, is very hard to observe, so astronomers must use other molecules as tracers of the cloud's properties.

Chapter 19

19.1 *(p. 468)* The competing effects of gravity, which tends to make an interstellar cloud collapse, and heat (pressure), which opposes that collapse. **19.2** *(p. 473)* (1) The existence of a photosphere, meaning that the inner part of the cloud becomes opaque to its own radiation, signaling the slowing of the collapse phase. (2) Nuclear fusion in the core and equilibrium between pressure and gravity. **19.3** *(p. 474)* No—different parts of the main sequence correspond to stars of different masses. A typical star stays at roughly the same location on the main sequence of most of its lifetime. **19.4** *(p. 479)* We assume that we observe objects at many different evolutionary stages, and that the snapshot therefore provides a representative sample of the evolutionary stages that stars go through. **19.5** *(p. 482)* Star formation may be triggered by some external event, which might cause several interstellar clouds to start contracting at once. Alternatively, the shock wave produced when an emission nebula forms may be sufficient to send another nearby part of the same cloud into collapse. **19.6** *(p. 487)* Because stars form at different rates—high-mass stars reach the main sequence and start disrupting the parent cloud long before lower-mass stars have finished forming.

Chapter 20

20.1 *(p. 492)* We can't observe a single star evolve, but we can observe large numbers of stars at different stages of their lives, and hence build up an accurate statistical picture of stellar evolutionary tracks. **20.2** *(p. 498)* Because the nonburning inner core, unsupported by fusion, begins to shrink, releasing gravitational energy, heating the overlying layers and causing them to burn more vigorously, thus increasing the luminosity. **20.3** *(p. 505)* Because the core's contraction is halted by the pressure of degenerate (tightly packed) electrons before it reaches a temperature high enough for the next stage of fusion to begin. In fact, this statement is true whether the "next round" is hydrogen fusion (brown dwarf), helium fusion (helium white dwarf), or carbon fusion (carbon-oxygen white dwarf). **20.4** *(p. 508)* Fusion in high-mass stars is not halted by electron degeneracy pressure. Temperatures are always high enough that each new burning stage can start before degeneracy becomes important. Such stars continue to fuse more and more massive nuclei, faster and faster, eventually exploding in a supernova. **20.5** *(p. 511)* Because a star cluster gives us a "snapshot" of stars of many different masses, but of the same age and initial composition, allowing us to directly test the predictions of the theory. **20.6** *(p. 513)* Because many, if not most, stars are found in binaries, and stars in binaries can follow evolutionary paths quite different from those they would follow if single.

Chapter 21

21.1 *(p. 519)* No, because it is of low mass and not a member of a binary-star system. **21.2** *(p. 522)* Because iron cannot fuse to produce energy. As a result, no further nuclear reactions are possible, and the core's equilibrium cannot be restored. **21.3** *(p. 528)* Because the two types of supernova differ in their spectra and their light curves, making it impossible to explain them in terms of a single phenomenon. **21.4** *(p. 533)* Because they are readily formed by helium capture, a process common in evolved stars. Other elements (with masses not multiples of four) had to form via less common reactions involving proton and neutron capture. **21.5** *(p. 535)* Because it is responsible for creating and dispersing all the heavy elements out of which we are made. In addition, it may also have played an important role in triggering the collapse of the interstellar cloud from which our solar system formed.

Chapter 22

22.1 *(p. 541)* No—only Type II supernovae. According to theory, the rebounding central core of the original star in a Type II supernova becomes a neutron star. **22.2** *(p. 544)* Because (1) not all supernovae form neutron stars, (2) the pulses are beamed, so not all pulsing neutron stars are visible from Earth, and (3) pulsars spin down and become too faint to observe after a few tens of millions of years. **22.3** *(p. 548)* Some X-ray sources are binaries containing accreting neutron stars,

which may be in the process of being spun up to form millisecond pulsars. **22.4** *(p. 551)* They are energetic bursts of gamma rays, roughly isotropically distributed on the sky, occurring about once per day. They pose a challenge because they are very distant, and hence extremely luminous, but their energy originates in a region less than a few hundred kilometers across. There also appear to be two distinct types, with different energy generation mechanisms. **22.5** *(p. 557)* Newton's theory describes gravity as a force produced by a massive object that influences all other massive objects. Einstein's relativity describes gravity as a curvature of space-time produced by a massive object; that curvature then determines the trajectories of all particles—matter or radiation—in the universe. **22.6** *(p. 560)* Because the object would appear to take infinitely long to reach the event horizon, and its light would be infinitely redshifted by the time it got there. **22.7** *(p. 567)* By observing their gravitational effects on other objects, and from the X rays emitted as matter plunges toward the event horizon.

Chapter 23

23.1 *(p. 574)* The Milky Way is the thin plane of the Galactic disk, seen from within. When our line of sight lies in the plane of the Galaxy, we see many stars blurring into a continuous band. In other directions, we see darkness. **23.2** *(p. 582)* No, because even the brightest Cepheids are unobservable at distances of more than a kiloparsec or so through the obscuration of interstellar dust. **23.3** *(p. 585)* Because the compositions, ages, and orbits of the two classes of stars are quite different from one another. **23.4** *(p. 587)* The halo formed early in our Galaxy's history, before gas and dust had formed a spinning, flattened disk. Disk stars are still forming today. Consequently, halo stars are old and move in more or less random three-dimensional orbits, while the disk contains stars of all ages, all moving in roughly circular orbits around the Galactic center. **23.5** *(p. 591)* Because differential rotation would destroy the spiral structure within a few hundred million years. **23.6** *(p. 595)* Scientists are unwilling to suggest new physics to explain observations, and competing theories of galactic rotation curves and "missing mass" on large scales do exist. However, despite their reservations, most astronomers have concluded that dark matter, introduced as material with gravity but no interaction with electromagnetic radiation, best fits the observed facts. Several theories exist to explain how dark matter might have been formed in the early universe, and they lead to experimental tests that should some day detect dark matter, if it exists. **23.7** *(p. 599)* Observations of rapidly moving stars and gas, and the variability of the radiation emitted suggest the presence of a roughly 4 million-solar-mass black hole.

Chapter 24

24.1 *(p. 610)* Most galaxies are not large spirals—the most common galaxy types are dwarf ellipticals and dwarf irregulars. **24.2** *(p. 614)* Because distance-measurement techniques ultimately rely upon the existence of very bright objects whose luminosities can be

inferred by other means. Such objects become increasingly hard to find and calibrate the farther we look out into intergalactic space. **24.3** *(p. 617)* It doesn't use the inverse-square law. The other methods all provide a way of determining the luminosity of a distant object, which then is converted to a distance using the inverse-square law. Hubble's law gives a direct connection between redshift and distance. **24.4** *(p. 624)* It means that the energy source cannot simply be the summed energy of a huge number of stars—some other mechanism must be at work. **24.5** *(p. 626)* When quasars were discovered, they were thought to be faint, relative nearby stars, or starlike objects, although their unusual spectra posed problems for astronomers. Once astronomers realized that their odd spectra actually meant that they had very large redshifts, it became clear that quasars were actually among the most distant—and hence the most luminous—objects in the entire universe. **24.6** *(p. 630)* The energy is generated in an accretion disk in the central nucleus of the visible galaxy, then transported by jets out of the galaxy and into the lobes, where it is eventually emitted by the synchrotron process in the form of radio waves.

Chapter 25

25.1 *(p. 638)* First, that the galaxies are gravitationally bound to the cluster. Second, and more fundamentally, that the laws of physics as we know them in the solar system—gravity, atomic structure, the Doppler effect—all apply on very large scales, and to systems possibly containing a lot of dark matter. **25.2** *(p. 641)* Isolated galaxies do change in time as stars form and evolve, but collisions and mergers between galaxies are the main way in which galaxies grow. The large galaxies we see today are the result of repeated collisions between smaller galaxies in the past. **25.3** *(p. 645)* Stars form by collapse and fragmentation of a large interstellar cloud and subsequently evolve largely in isolation. Galaxies form by mergers of smaller objects, and interactions with other galaxies play a major role in their evolution. **25.4** *(p. 650)* Probably not. There may be galaxies (especially the smaller ones) that do not harbor supermassive central black holes, and in any case, only galaxies in clusters are likely to experience the encounters that trigger activity. **25.5** *(p. 656)* Light traveling from these objects to Earth is influenced by cosmic structure all along the line of sight. Light rays are deflected by the gravitational field of intervening concentrations of mass, and gas along the line of sight produces absorption lines whose redshifts tell us the distance at which each feature formed. The light received on Earth thus gives astronomers a "core sample" through the universe, from which detailed information can be extracted.

Chapter 26

26.1 *(p. 664)* On very large scales—more than 300 Mpc—the distribution of galaxies seems to be roughly the same everywhere and in all directions. **26.2** *(p. 667)* Because, tracing the motion backwards in time, it implies that all galaxies, and in fact everything in the entire universe, were located at a single point at the same instant in the past. **26.3** *(p. 669)* The universe can expand forever, in which case we die a cold death in which all activity gradually fades away, or the expansion can stop and the universe will recollapse to a fiery Big Crunch. **26.4** *(p. 672)* A low-density universe has negative curvature, a critical density universe is spatially flat (Euclidean), and a high-density universe has positive curvature (and is finite in extent). **26.5** *(p. 675)* There doesn't seem to be enough matter to halt the collapse and, in addition, the observed cosmic acceleration suggests the existence of a large-scale repulsive force in the cosmos that also opposes recollapse. **26.6** *(p. 677)* The observed acceleration of the expansion of the universe implies that some nongravitational force must be at work. Dark energy is our current best theory of the cause of that force. In addition, galactic and cosmological observations indicate that the universe is spatially flat, and hence that the density of matter (mostly dark) cannot account for the total. The amount of dark energy needed to account for the cosmic acceleration is in good agreement with the required extra density—about 70 percent of the critical value. **26.7** *(p. 679)* At the time of the Big Bang. It is the electromagnetic remnant of the primeval fireball.

Chapter 27

27.1 *(p. 686)* It means that the total mass-energy density of the universe, which today is made up almost entirely of matter and dark energy, was once comprised almost entirely of radiation. We know this because, going back in time toward the Big Bang, the dark energy density stays constant as the universe contracts, the matter density increases because the volume shrinks, but the radiation density increases even faster because of the cosmological redshift. Thus, at sufficiently early times, radiation was the dominant component of the cosmos. **27.2** *(p. 690)* Once the temperature of the expanding, cooling universe dropped below the point where particle–antiparticle pairs could no longer be created from the radiation background, the particles separated out of the radiation field. Particles and antiparticles annihilated one another, and any leftover "frozen out" matter has survived to the present day. **27.3** *(p. 694)* Because the amount of deuterium observed in the universe today implies that the present density of normal matter is at most a few percent of the critical value—much less than the density of dark matter inferred from dynamical studies. **27.4** *(p. 698)* Inflation implies that the universe is flat, and hence that the total cosmic density equals the critical value. However, the matter density seems to be only about one third of the critical value, and the density of electromagnetic radiation (the microwave background) is a tiny fraction of the critical density. The remaining density may be in the form of the "dark energy" thought to be powering the accelerating cosmic expansion (Section 27.3). **27.5** *(p. 700)* Because it could clump to form the large density fluctuations that would ultimately become galaxies and clusters, without leaving a correspondingly large imprint in the microwave background. After it decoupled from the radiation background, normal matter flowed into the dark matter structures, forming the galaxies we see. **27.6** *(p. 702)* They allow us to measure the value of Ω_0—and in fact imply that it is very close to 1.

Chapter 28

28.1 *(p. 716)* The formation of complex molecules from simple ingredients by nonbiological processes has been repeatedly demonstrated, but no living cell or self-replicating molecule has ever been created. **28.2** *(p. 718)* Mars remains the most likely site, although Europa and Titan also have properties that might have been conducive to the emergence of living organisms. **28.3** *(p. 723)* It breaks a complex problem up into simpler "astronomical," "biochemical," "anthropological," and "cultural" pieces, which can be analyzed separately. It also identifies the types of stars where a search might be most fruitful. **28.4** *(p. 726)* It is in the radio part of the spectrum, where Galactic absorption is least, at a region where natural Galactic background "static" is minimized, and in a portion of the spectrum characterized by lines of hydrogen and hydroxyl, both of which would likely have significance to a technological civilization.

ANSWERS TO SELF-TEST QUESTIONS

Chapter 1

Multiple Choice **1.1** b, **1.2** b, **1.3** d, **1.4** a, **1.5** c, **1.6** a, **1.7** c, **1.8** c, **1.9** a, **1.10** d
Odd-Numbered Problems **1.1** (c) the Moon (384,000 km) **1.3** it would decrease by roughly eight minutes **1.5** (a) 33 arc minutes, (b) 33 arc seconds, (c) 0.55 arc seconds; 57 minutes **1.7** (a) 57,300 km; (b) 3.44×10^6 km; (c) 2.06×10^8 km **1.9** 391.

Chapter 2

Multiple Choice **2.1** a, **2.2** d, **2.3** b, **2.4** c, **2.5** c, **2.6** a, **2.7** c, **2.8** b, **2.9** c, **2.20** a
Odd-Numbered Problems **2.1** (a) 110 km; (b) 44,000 km; (c) 370,000 km **2.3** 146 days **2.5** 8.1″, if Mercury is at aphelion and Earth is at perihelion at the point of closest approach **2.7** 9.57×10^{-4} times the mass of the Sun = 1.9×10^{27} km **2.9** 1.7 km/s; 2.4 km/s.

Chapter 3

Multiple Choice **3.1** a, **3.2** c, **3.3** b, **3.4** b, **3.5** d, **3.6** a, **3.7** b, **3.8** d, **3.9** a, **3.10** b
Odd-Numbered Problems **3.1** 1480 m/s **3.3** 3×10^{12} Hz, assuming a period diameter of 0.1 mm; infrared **3.5** 920 W, assuming my skin temperature is 300 K (27° C) and my surface area is roughly 2 m²; of course, I also absorb energy from my surroundings, so the net energy loss is less—about 2σ $(300^4 - 290^4) = 120$ W if my surroundings are 10 K cooler than me **3.7** 2.9 microns **3.9** 300 km/s away.

Chapter 4

Multiple Choice **4.1** c, **4.2** c, **4.3** d, **4.4** c, **4.5** b, **4.6** b, **4.7** b, **4.8** b, **4.9** b, **4.10** d
Odd-Numbered Problems **4.1** 2.8 eV, 6.2 eV **4.3** 620 nm, 12,400 nm, 0.25 nm **4.5** 51.8 microns, 5.7×10^{12} Hz, infrared; 4.5 cm, 6.5 GHz, radio; 46 m, 6.6 MHz, radio **4.7** the first six Balmer lines, ranging in wavelength from 656 nm (Hα) to 410 nm (Hζ) **4.9** 0.052 nm, to two significant figures.

Chapter 5

Multiple Choice **5.1** c, **5.2** d, **5.3** d, **5.4** b, **5.5** c, **5.6** c, **5.7** a, **5.8** d, **5.9** b, **5.10** c
Odd-Numbered Problems **5.1** 0.29 arc seconds; 6.8 pixels **5.3** 6.7 minutes; 1.7 minutes **5.5** (a) 0.022″; (b) 0.062″ **5.7** 155 light-years **5.9** 5.5 km, 92 m, 1.8 m.

Chapter 6

Multiple Choice **6.1** a, **6.2** d, **6.3** a, **6.4** b, **6.5** c, **6.6** a, **6.7** b, **6.8** b, **6.9** d, **6.10** c
Odd-Numbered Problems **6.1** 18 seconds of arc; 17 days **6.3** Mercury: 0.31 AU 0.47 AU, Mars: 1.38 AU, 1.67 AU **6.5** 7×10^{20}, 0.012 percent of Earth's mass **6.7** 0.21, 1.26 AU, 0.71 years **6.9** (a) 62.5 years; (b) 9 days.

Chapter 7

Multiple Choice **7.1** b, **7.2** d, **7.3** a, **7.4** c, **7.5** b, **7.6** b, **7.7** c, **7.8** b, **7.9** d, **7.10** b

Odd-Numbered Problems **7.1** 5.0×10^{18} kg; 8.4×10^{-7} times Earth's mass **7.3** 21 m **7.5** (a) 8.4×10^{-3}; (b) 0.16; (c) 0.82; (d) 0.014 **7.7** 1.9 billion years **7.9** 3.07×10^{-6} m/s², 3.07×10^{-7} of surface gravity.

Chapter 8

Multiple Choice **8.1** a, **8.2** b, **8.3** b, **8.4** c, **8.5** b, **8.6** a, **8.7** b, **8.8** d, **8.9** a, **8.10** a
Odd-Numbered Problems **8.1** 8.5 minutes **8.3** 1.71°, 1.13° **8.5** 1.7 hours **8.7** (a) 60.4 km/s, or 6.5°/day; (b) 39.7 km/s, or 2.8°/day **8.9** need at least 4.8×10^5 such craters, requiring at least 4.8 trillion years; rate would have to increase by a factor of about 1000.

Chapter 9

Multiple Choice **9.1** a, **9.2** c, **9.3** c, **9.4** b, **9.5** b, **9.6** c, **9.7** b, **9.8** c, **9.9** a, **9.10** a
Odd-Numbered Problems **9.1** (a) 35.1 arc seconds; (b) 22.9 arc seconds; (c) 9.7 arc seconds **9.3** 470 seconds, 280 seconds **9.5** yes—the smallest detectable feature would be about 20 km across, much smaller than the largest impact features **9.7** 400 km/h, 250 mph **9.9** 35 times.

Chapter 10

Multiple Choice **10.1** b, **10.2** c, **10.3** b, **10.4** c, **10.5** b, **10.6** d, **10.7** a, **10.8** c, **10.9** b, **10.10** a
Odd-Numbered Problems **10.1** 41° **10.3** 59 lbs, assuming you weigh 150 lbs on Earth **10.5** 3.2×10^{16} kg, taking Earth's atmospheric mass to be 5.0×10^{18} kg; 2.8 times the seasonal polar cap mass of 1.2×10^{16} kg **10.7** 1.3×10^{20} kg; 3900 times the current atmospheric mass **10.9** 2.9×10^{17} kg; 4.5×10^{-7} times the mass of Mars.

Chapter 11

Multiple Choice **11.1** c, **11.2** b, **11.3** a, **11.4** c, **11.5** a, **11.6** d, **11.7** b, **11.8** a, **11.9** a, **11.10** c
Odd-Numbered Problems **11.1** 2.5 times greater **11.3** 62 days **11.5** 1.2% of the actual mass **11.7** 4800 times greater; ratio for Earth–Moon is 81 **11.9** —.

Chapter 12

Multiple Choice **12.1** c, **12.2** c, **12.3** d, **12.4** c, **12.5** d, **12.6** d, **12.7** b, **12.8** c, **12.9** a, **12.10** c
Odd-Numbered Problems **12.1** 47″ **12.3** 7.3×10^{22} kg; 1.3×10^{-4} times the actual mass, 1.2 percent of Earth's mass **12.5** 74 K **12.7** 1.1×10^{18} particles **12.9** 1.33 m/s² about $\frac{1}{7}$ of Earth's; 2.6 km/s.

Chapter 13

Multiple Choice **13.1** b, **13.2** b, **13.3** c, **13.4** d, **13.5** c, **13.6** d, **13.7** d, **13.8** a, **13.9** a, **13.10** d
Odd-Numbered Problems **13.1** 1.7′, 13.2′; yes **13.3** 10,000 km/h—it moves halfway around the planet in just over eight hours; this is half the rotation period, it's mostly rotation **13.5** 0.42 Earth-Moon masses **13.7** yes, at 37 K, an escape speed of 1.08 km/s would be needed; Triton's escape speed is 1.46 km/s **13.9** 6 times farther out.

Chapter 14

Multiple Choice **14.1** c, **14.2** d, **14.3** a, **14.4** c, **14.5** a, **14.6** d, **14.7** b, **14.8** d, **14.9** d, **14.10** d
Odd-Numbered Problems **14.1** (a) a 150-lb person would weigh 4.8 lbs; (b) 0.41 km/s **14.3** 21 hours, taking the middle of the stated mass range and assuming a circular orbit **14.5** (a) 4.0 million years; (b) 49 AU **14.7** 3×10^{12} kg; 0.06 percent **14.9** 5 trillion.

Chapter 15

Multiple Choice **15.1** a, **15.2** a, **15.3** c, **15.4** c, **15.5** c, **15.6** c, **15.7** d, **15.8** c, **15.9** a, **15.10** b
Odd-Numbered Problems **15.1** 8 revolutions **15.3** 188 years **15.5** 0.3 AU **15.7** 2210 K **15.9** 1.7 g = 16.3 m/s²; 0.9 g = 8.9 m/s².

Chapter 16

Multiple Choice **16.1** c, **16.2** b, **16.3** c, **16.4** a, **16.5** a, **16.6** b, **16.7** b, **16.8** c, **16.9** c, **16.10** b
Odd-Numbered Problems **16.1** 14,600 W/m²; 52 W/m² **16.3** (a) 3000 km; (b) 1460; (c) 167 minutes **16.5** 36% as much light as the photosphere **16.7** 10,800,000 km, for a coronal temperature of 1,000,000 K **16.9** 310,000 years.

Chapter 17

Multiple Choice **17.1** a, **17.2** d, **17.3** d, **17.4** b, **17.5** c, **17.6** b, **17.7** c, **17.8** b, **17.9** c, **17.10** d
Odd-Numbered Problems **17.1** 83 pc; 0.36 arc seconds **17.3** 80 solar luminosities **17.5** B is three times farther away **17.7** (a) 100 pc; (b) 4200 pc; (c) 170,000 pc; (d) 1,000,000 pc **17.9** 3.5, 2.3 solar masses.

Chapter 18

Multiple Choice **18.1** a, **18.2** d, **18.3** c, **18.4** a, **18.5** d, **18.6** c, **18.7** b, **18.8** d, **18.9** a, **18.10** a
Odd-Numbered Problems **18.1** 1.9 grams **18.3** 7.1×10^{20} m³, or a cube of side 8900 km **18.5** 9.03 magnitudes **18.7** 100,000 pc, 1.3 magnitudes/kpc **18.9** escape speeds (km/s): 1.8, 1.1, 1.1, 0.80; average molecular speeds: 13.6, 14.0, 14.6, 14.2; no.

Chapter 19

Multiple Choice **19.1** c, **19.2** a, **19.3** d, **19.4** b, **19.5** b, **19.6** a, **19.7** a, **19.8** c, **19.9** a, **19.10** b
Odd-Numbered Problems **19.1** yes, barely: the escape speed is 0.93 km/s, molecular speed is 0.35 km/s **19.3** luminosity decreases by a factor of 7900; absolute magnitude increases by 9.7 **19.5** M 5 2.3 **19.7** 8.3 magnitudes **19.9** 10^{-6} solar luminosities.

Chapter 20

Multiple Choice **20.1** c, **20.2** b, **20.3** a,
20.4 b, **20.5** a, **20.6** c, **20.7** a, **20.8** b,
20.9 b, **20.10** d
Odd-Numbered Problems **20.1** 9.9×10^{26} kg
destroyed, 8.9×10^{43} J; emitted **20.3** (a) 3370;
(b) 84,300 solar luminosities **20.5** (a) 2.9
solar masses; (b) 1.7 solar masses **20.7** 5×10^{-5}; not noticeable to the naked eye, but (just)
measurable with a telescope and photometer
20.9 increases by a factor of 25.

Chapter 21

Multiple Choice **21.1** a, **21.2** d, **21.3** d,
21.4 b, **21.5** d, **21.6** c, **21.7** c, **21.8** b,
21.9 b, **21.10** c
Odd-Numbered Problems **21.1** 15 solar radii
21.3 m = 20, 3.2 Mpc **21.5** 0.44 pc; no—
there are no O or B stars (in fact no stars at all)
within that distance of us. For the Moon:
320 pc; yes, but rarely—see problem 14
21.7 1.2×10^{44} J; about 10 times greater than
a supernova's electromagnetic output, and
about $\frac{1}{10}$ the output in the form of neutrinos
21.9 roughly 1000 km/s; not a very good
assumption—the nebula is moving too fast
to be affected much by gravity, although it is
probably slowing down as it runs into the
interstellar medium.

Chapter 22

Multiple Choice **22.1** b, **22.2** c, **22.3** a,
22.4 b, **22.5** c, **22.6** d, **22.7** b, **22.8** b,
22.9 c, **22.10** b
Odd-Numbered Problems **22.1** 1 million
revolutions per day, or 11.6 revolutions
per second **22.3** 1.9×10^{12} m/s^2, or 200
billion Earth gravities; 190,000 km/s, or 64
percent of the speed of light; 306,000 km/s
22.5 2.4×10^{44} J, roughly twice the Sun's
total energy output; 2.4×10^{34} J; 1.4×10^{25} J
22.7 63,000 km/s or 21% of the speed
of light; 136,000 km/s or 45% of the speed
of light **22.9** (a) 26 microarcsec; (b) 0.016
arcsec; (c) 4.1 arcmin; (d) 8100 AU 5
0.039 pc.

Chapter 23

Multiple Choice **23.1** d, **23.2** b, **23.3** d,
23.4 b, **23.5** b, **23.6** c, **23.7** a, **23.8** c,
23.9 d, **23.10** c
Odd-Numbered Problems **23.1** 2.0″, much less
than the angular diameter of Andromeda
23.3 100 kpc **23.5** M = −6.4; 17 Mpc
23.7 0.014″/yr; proper motion has in fact been
measured for several globular clusters
23.9 (a) 3.9 kpc; (b) 19.7 kpc.

Chapter 24

Multiple Choice **24.1** c, **24.2** c, **24.3** a,
24.4 c, **24.5** c, **24.6** c, **24.7** c, **24.8** b,
24.9 a, **24.10** b
Odd-Numbered Problems **24.1** 320 Mpc
24.3 14,000 km/s, 57 Mpc; 12,000 km/s, 67
Mpc; 16,000 km/s, 50 Mpc **24.5** 58 arc
seconds **24.7** M = −22.5; 1.5×10^8 solar
luminosities **24.9** 1.5×10^8 solar masses.

Chapter 25

Multiple Choice **25.1** d, **25.2** b, **25.3** c,
25.4 a, **25.5** a, **25.6** b, **25.7** a, **25.8** d,
25.9 b, **25.10** a
Odd-Numbered Problems **25.1** 6.5 billion years
25.3 2.6×10^{14} solar masses; reasonable to
within a factor of perhaps 2–3 **25.5** 700 km/s;
650 km/s—comparable **25.7** 9.4×10^{10} solar
masses **25.9** 1.8×10^{12} solar masses.

Chapter 26

Multiple Choice **26.1** a, **26.2** d, **26.3** b,
26.4 b, **26.5** c, **26.6** d, **26.7** c, **26.8** c,
26.9 d, **26.10** d
Odd-Numbered Problems **26.1** 1000 Mpc
26.3 4×10^8 **26.5** 1200 km/s **26.7** (a)
30,000 tons; (b) 2.8 pc **26.9** (a) 1.1 mm;
(b) 0.01 mm; (c) 0.0001 mm; (d) 0.000001 mm.

Chapter 27

Multiple Choice **27.1** a, **27.2** d, **27.3** a,
27.4 b, **27.5** a, **27.6** c, **27.7** b, **27.8** a,
27.9 b, **27.10** c
Odd-Numbered Problems **27.1** 16 kpc
27.3 18.9 K; 2.7×10^{-24} kg/m^2
27.5 2.4×10^9 K, less than half the
temperature in the text **27.7** 3×10^{-12} m;
hard X ray/gamma ray **27.9** 12.7 Mpc.

Chapter 28

Multiple Choice **28.1** a, **28.2** b, **28.3** c,
28.4 c, **28.5** c, **28.6** b, **28.7** d, **28.8** d,
28.9 b, **28.10** a
Odd-Numbered Problems **28.1** 6.3 seconds
(for a 20-year old-reader); 20.9 seconds (in
2011); 73 seconds; 2.7 minutes; 240 days
28.3 They would both increase by a factor of
two **28.5** 27 AU **28.7** 3370 years; no,
civilizations are more likely to be found within
the galactic habitable zone **28.9** 1.43×10^9
to 1.67×10^9 Hz; 2.4×10^6 channels.

INDEX

Note: *f* indicates footnote

A 0620-00 (X-ray binary system), 562
A 2218 galaxy cluster, 656
Abell 39, 500
Abell 1689, 615
Abell 1835, 638
Aberration
    chromatic, 99–100
    spherical, 104
    of starlight, 41
Absolute brightness, 421. *See also* Luminosity
Absolute magnitude, 423, 425
Absolute temperature, 68
Absolute zero, 67
Absorption (Fraunhofer) lines, 80–81, 393–94
    atomic hydrogen, 652, 654
    quasar, 651–52, 654
Absorption spectra, 456–57, 458
AC114 quasar, 654
Acceleration, 48
    cosmic, 672–74
Accretion, 149, 362, 363–65, 366
Accretion disk, 518–19
Active galaxies/active galactic nuclei, 618–30, 648–50
    central engine of, 626–30
        energy emission, 629–30
        energy production, 627–29
    defined, 618–19, 620
    evolution of, 649
    properties of, 619–20
    quasars *vs.*, 625–26
    radiation emitted by, 618–20
    radio galaxies, 562–63, 620, 621–24, 638
    scientific method and, 650
    Seyfert galaxies, 620–21
    supermassive black holes at center of, 563, 597–99, 620
Active optics, 112, 113
Active prominence, 401
Active region, 401–3
Active Sun, 401–6
    active regions, 401–2
    changing solar corona, 403–6
    X-ray images of, 405
Adams, John, 315
Adams ring (Neptune), 328
Adaptive optics, 112, 113
AG Carinae, 507
Ahnighito (meteorite), 356

Aine (corona), 219, 221
Air, gravity and, 192–93
Aitken Basin, 188
Aldebaran, 427, 429–30, 432
Algol, 512, 513
ALH84001 (meteorite), 248–49
Almagest (Ptolemy), 36
Alpha Centauri, 385, 419, 420–21, 425, 440
Alpha particle, 496
Alpha process, 531–32
Alpha ring (Uranus), 327
ALSEP (*Apollo* Lunar Surface Experiments Package), 190–91
Alvin submarine, 715
AM 0644-741 galaxy, 609
Amalthea (moon), 273
Amino acid, 710, 712
Ammonia, 716
Amor (Mars-crossing) asteroid, 334, 337
Amplitude, 59
AM radio signals, 158, 722, 723
Andromeda galaxy, 58, 63, 103, 107, 570, 571, 574, 575, 577, 582, 584, 613
Andromeda Nebula, 570
Ångstrom, 61
Ångstrom, Anders, 61
Angular diameter, 26
Angular measurement, 11
Angular momentum, 147, 148
    conservation of, 147, 148, 362, 540
Angular resolution, 105–8, 110, 115, 116, 118, 119
Annular eclipse, 19, 20
Antarctic ozone hole, 158
Antenna, "sugar-scoop," 677
Antennae galaxies, 640
Antiparticle, 408
Apennine Mountains, 200
Aphelion, 194
Aphrodite Terra, 216, 218, 219, 221
Apollo (Earth-crossing) asteroid, 334, 337–40
Apollo Lunar Surface Experiments Package (ALSEP), 190–91
Apollo program, 183, 187, 188, 196, 197, 198, 200, 202, 206
    description of, 190–91

Apophis (MN4) (asteroid), 337–38
Apparent brightness, 421, 578–79, 582, 594
Apparent magnitude, 422, 423
Apparent motion, 37
Arago ring (Neptune), 328
Arc minute, 11
Arc (of ring), 299
Arc second, 11
Arcturus, 495, 584–85
Arecibo radio telescope, 115, 116, 191–92
Ariel (moon), 322, 323
A ring, 292, 293, 296, 297–98
Aristarchus of Samos, 37
Aristotle, 7–8, 36, 37, 39, 48
Armstrong, Neil, 190, 191
Asteroid, 134, 140, 141, 332, 333, 334–40
    Earth-crossing, 337–40
    observations from space, 335–37
    orbital resonances of, 340
    properties of, 136, 334–35
    stray, 354–56
    types of, 335
Asteroid belt, 134, 146, 334–35, 340, 366–67
Asthenosphere, 168, 169
Astrologer, 9
Astronomical unit (AU), 44–45, 46
Astronomy
    ancient, 32–34
    birth of modern, 39–41
    defined, 4
    ultraviolet, 122–26
Asymptote, 497*f*
Asymptotic-giant branch, 497, 498
Aten asteroid, 337
Atlantic Ocean, 170
Atlas (moon), 297–98, 299
Atmosphere, 156–62. *See also specific planets, stars, and other bodies*
    blue sky, reasons for, 159
    composition of, 193
    gravity and, 192–93
    heat and, 160–62
    opacity of, 64, 66
    origin of, 162
    ozone in, 158, 160

primary, 162
secondary, 162
solar, 393–96
structure of, 157–58
surface heating and, 160–62
Atmospheric blurring, 110–12
Atmospheric circulation on Earth, 404
Atmospheric opacity, 64, 66
Atmospheric turbulence, 110–11
Atomic epoch, 690, 698
Atomic motion, 468
Atomic spectra, 84
Atomic structure, 82–83
Atom(s), 82–84
    classical, 83
    excitation of, 83, 85
    first, 693–94
    formation of, 690–94
    ionization state of, 453
    ionized, 82, 453
    modern, 83
    nucleus of, 82, 166
    thermal motions of, 92
Aurora
    on Earth, 175–76
    on Jupiter, 272
    on Saturn, 292
Autolycus crater, 200
Autumnal equinox, 15

B ackground noise, 109
Ballistic Missile Defense Organization, 112–13
Balmer, Johann, 86
Balmer (hydrogen) series, 86
Bands, atmospheric, 262–64
Barnard, E.E., 273
Barnard 68 cloud, 69, 149, 447, 455, 535
Barnard's Star, 419, 420, 427, 435
Barred-spiral galaxy, 605–7, 610
Barringer Meteor Crater, 198, 199, 354
Basalt, 164
Baseline of interferometer, 119
Baseline triangle, 23
Becklin-Neugebauer object, 477–78
Bell, Jocelyn, 541
Belts, in atmospheric bands, 262–64
BeppoSAX satellite, 549

STAR CHARTS

Have you ever become lost in an unfamiliar city or state? Chances are you used two things to get around: a map and some signposts. In much the same way, these two items can help you find your way around the night sky in any season. Fortunately, in addition to the seasonal Star Charts on the following pages, the sky provides us with two major signposts. Each seasonal description will talk about the Big Dipper—a group of seven bright stars that dominates the constellation Ursa Major the Great Bear. Meanwhile, the constellation Orion the Hunter plays a key role in finding your way around the sky from late autumn until early spring.

Each chart portrays the sky as seen from near 35° north latitude at the times shown at the top of the page. Located just outside the chart are the four directions: north, south, east, and west. To find stars above your horizon, hold the map overhead and orient it so a direction label matches the direction you're facing. The stars above the map's horizon now match what's in the sky.

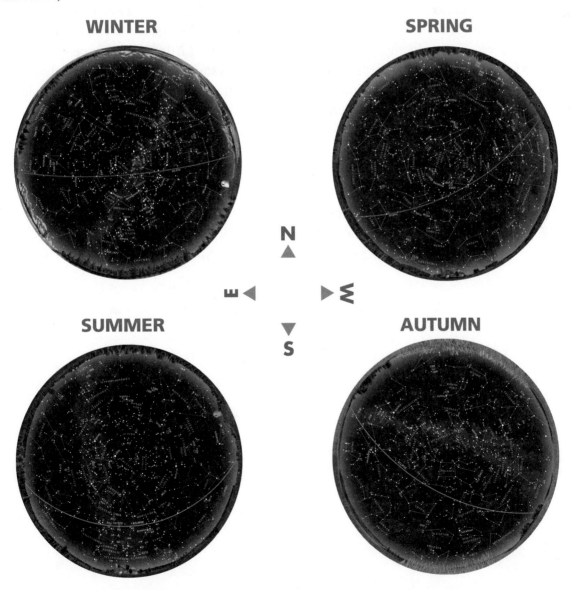

WINTER   SPRING

N

E   W

SUMMER   AUTUMN

S

EXPLORING THE WINTER SKY

Winter finds the Big Dipper climbing the northeastern sky, with the three stars of its handle pointing toward the horizon and the four stars of its bowl standing highest. The entire sky rotates around a point near Polaris, a 2nd-magnitude star found by extending a line from the uppermost pair of stars in the bowl across the sky to the left of the Dipper. Polaris also performs two other valuable functions: The altitude of the star above the horizon equals your latitude north of the equator, and a straight line dropped from the star to the horizon.

Turn around with your back to the Dipper and you'll be facing the diamond-studded winter sky. The second great signpost in the sky, Orion the Hunter, is central to the brilliant scene. Three closely spaced, 2nd-magnitude stars form a straight line that represents the unmistakable belt of Orion. Extending the imaginary line joining these stars to the upper right leads to Taurus the Bull and its orangish 1st-magnitude star, Aldebaran. Reverse the direction of your gaze to the belt's lower left and you cannot miss Sirius the Dog Star—brightest in all the heavens at magnitude −1.5.

Now move perpendicular to the belt from its westernmost star, Mintaka, and find at the upper left of Orion the red supergiant star Betelgeuse. Nearly a thousand times the Sun's diameter, Betelgeuse marks one shoulder of Orion. Continuing this line brings you to a pair of bright stars, Castor and Pollux. Two lines of fainter stars extend from this pair back toward Orion—these represent Gemini the Twins. At the northeastern corner of this constellation lies the beautiful open star cluster M35. Head south of the belt instead and your gaze will fall on the blue supergiant star Rigel, Orion's other luminary.

Above Orion and nearly overhead on winter evenings is brilliant Capella, in Auriga the Charioteer. Extending a line through the shoulders of Orion to the east leads you to Procyon in Canis Minor the Little Dog. Once you have these principal stars mastered, using the chart to discover the fainter constellations will be a whole lot easier. Take your time, and enjoy the journey. Before leaving Orion, however, aim your binoculars at the line of stars below the belt. The fuzzy "star" in the middle is actually the glorious Orion Nebula (M42), a stellar nursery illuminated by bright, newly formed stars.

WINTER

2 a.m. on December 1; midnight on January 1; 10 p.m. on February 1

Reproduced by permission. © 2007, Astronomy Magazine, Kalmbach Publishing Co.

EXPLORING THE SPRING SKY

The Big Dipper, our signpost in the sky, swings high overhead during the spring and lies just north of the center of the chart. This season of rejuvenation encourages us to move outdoors with the milder temperatures, and with the new season a new set of stars beckons us.

Follow the arc of stars outlining the handle of the Dipper away from the bowl and you will land on brilliant Arcturus. This orangish star dominates the spring sky in the kite-shaped constellation Boötes the Herdsman. Well to the west of Boötes lies Leo the Lion. You can find its brightest star, Regulus, by using the pointers of the Dipper in reverse. Regulus lies at the base of a group of stars shaped like a sickle or backward question mark, which represents the head of the lion.

Midway between Regulus and Pollux in Gemini, which is now sinking in the west, is the diminutive group Cancer the Crab. Centered in this group is a hazy patch of light that binoculars reveal as the Beehive star cluster (M44).

To the southeast of Leo lies the realm of the galaxies and the constellation Virgo the Maiden. Virgo's brightest star, Spica, shines at magnitude 1.0.

During springtime, the Milky Way lies level with the horizon, and it's easy to visualize that we are looking out of the plane of our Galaxy. In the direction of Virgo, Leo, Coma Berenices, and Ursa Major lie thousands of galaxies whose light is unhindered by intervening dust in our own Galaxy. However, all these galaxies are elusive to the untrained eye and require binoculars or a telescope to be seen.

Boötes lies on the eastern border of this galaxy haven. Midway between Arcturus and Vega, the bright "summer" star rising in the northeast, is a region where no star shines brighter than 2nd magnitude. A semicircle of stars represents Corona Borealis the Northern Crown, and adjacent to it is a large region that houses Hercules the Strongman, the fifth-biggest constellation in the sky. It is here we can find the northern sky's brightest globular star cluster, M13. A naked-eye object from a dark site, it looks spectacular when viewed through a telescope.

Returning to Ursa Major, check the second-to-last star in the Dipper's handle. Most people will see it as double, while binoculars show this easily. The pair is called Mizar and Alcor, and they lie just 0.2° apart. A telescope reveals Mizar itself to be a double. Its companion star shines at magnitude 4.0 and lies 14 arcseconds away.

SPRING

1 a.m. on March 1; 11 p.m. on April 1; 9 p.m. on May 1. Add one hour for daylight-saving time

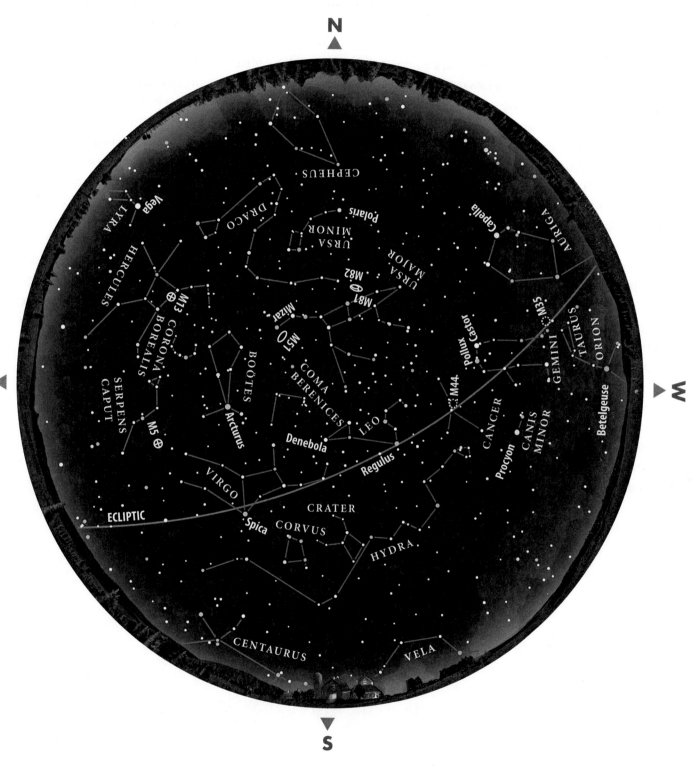

EXPLORING THE SUMMER SKY

The richness of the summer sky is exemplified by the splendor of the Milky Way. Stretching from the northern horizon in Perseus, through the cross-shaped constellation Cygnus overhead, and down to Sagittarius in the south, the Milky Way is packed with riches. These riches include star clusters, nebulae, double stars, and variable stars.

Let's start with the Big Dipper, our perennial signpost, which now lies in the northwest with its handle still pointing toward Arcturus. High overhead, and the first star to appear after sunset, is Vega in Lyra the Harp. Vega forms one corner of the summer triangle, a conspicuous asterism of three stars. Near Vega lies the famous double-double, Epsilon (ϵ) Lyrae. Two 5th-magnitude stars lie just over 3 arcminutes apart and can be split when viewed through binoculars. Each of these two stars is also double, but you need a telescope to split them.

To the east of Vega lies the triangle's second star: Deneb in Cygnus the Swan (some see a cross in this pattern). Deneb marks the tail of this graceful bird, the cross represents its outstretched wings, and the base of the cross denotes its head, which is marked by the incomparable double star Albireo. Albireo matches a 3rd-magnitude yellow star and a 5th-magnitude blue star and offers the finest color contrast anywhere in the sky. Deneb is a supergiant star that pumps out enough light to equal 60,000 Suns. Also notice that the Milky Way splits into two parts in Cygnus, a giant rift caused by interstellar dust blocking starlight from beyond.

Altair, the third star of the summer triangle and the one farthest south, is the second brightest of the three. Lying 17 light-years away, it's the brightest star in the constellation Aquila the Eagle.

Frequently overlooked to the north of Deneb lies the constellation Cepheus the King. Shaped rather like a bishop's hat, the southern corner of Cepheus is marked by a compact triangle of stars that includes Delta (Δ) Cephei. This famous star is the prototype of the Cepheid variable stars used to determine the distances to some of the nearer galaxies. It varies regularly from magnitude 3.6 to 4.3 and back again with a 5.37-day period.

Hugging the southern horizon, the constellations Sagittarius the Archer and Scorpius the Scorpion lie in the thickest part of the Milky Way. Scorpius's brightest star, Antares, is a red supergiant star whose name means "rival of Mars" and derives from its similarity to the planet in both color and brightness.

SUMMER

1 a.m. on June 1; 11 p.m. on July 1; 9 p.m. on August 1. Add one hour for daylight-saving time

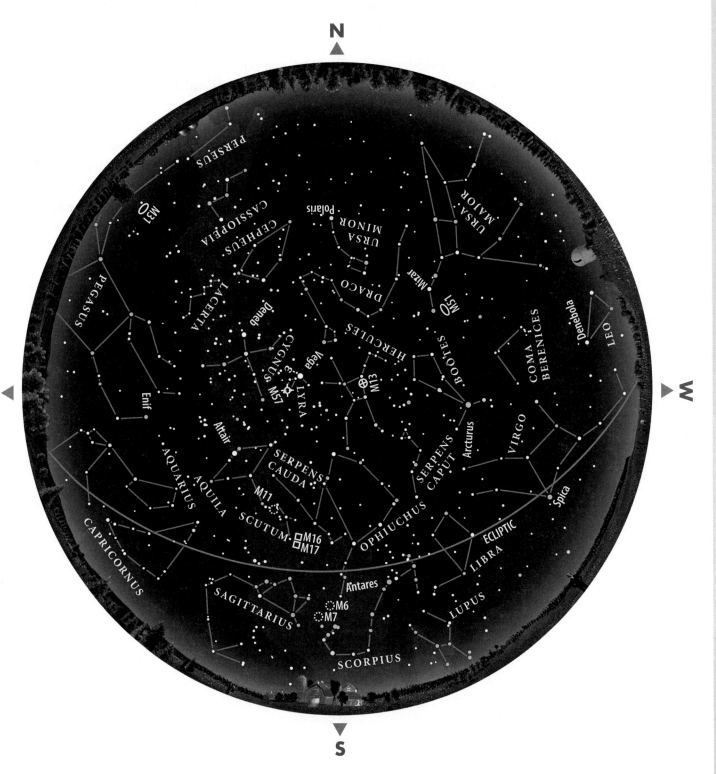

EXPLORING THE AUTUMN SKY

The cool nights of autumn are here to remind us the chill of winter is not far off. Along with the cool air, the brilliant stars of the summer triangle descend in the west to be replaced with a rather bland-looking region of sky. But don't let initial appearances deceive you. Hidden in the fall sky are gems equal to summertime.

The Big Dipper swings low this season, and for parts of the Southern United States it actually sets. Cassiopeia the Queen, a group of five bright stars in the shape of a "W" or "M," reaches its highest point overhead, the same spot the Big Dipper reached six months ago. To the east of Cassiopeia, Perseus the Hero rises high. Nestled between these two groups is the wondrous Double Cluster—NGC 869 and NGC 884—a fantastic sight in binoculars or a low-power telescope.

Our view to the south of the Milky Way is a window out of the plane of our Galaxy in the opposite direction to that visible in spring. This allows us to look at the Local Group of galaxies. Due south of Cassiopeia is the Andromeda Galaxy (M31), a 4th-magnitude smudge of light that passes directly overhead around 9 p.m. in mid-November. Farther south, between Andromeda and Triangulum, lies M33, a sprawling face-on spiral galaxy best seen in binoculars or a rich-field telescope.

The Great Square of Pegasus passes just south of the zenith. Four 2nd- and 3rd-magnitude stars form the square, but few stars can be seen inside of it. If you draw a line between the two stars on the west side of the square and extend it southward, you'll find 1st-magniude Fomalhaut in Piscis Austrinus the Southern Fish. Fomalhaut is the solitary bright star low in the south. Using the eastern side of the square as a pointer to the south brings you to Diphda in the large, faint constellation of Cetus the Whale.

To the east of the Square lies the Pleiades star cluster (M45) in Taurus, which reminds us of the forthcoming winter. By late evening in October and early evening in December, Taurus and Orion have both cleared the horizon and Gemini is rising in the northeast. In concert with the reappearance of winter constellations, the view to the northwest finds summertime's Cygnus and Lyra about to set. The autumn season is a great transition period, both on Earth and in the sky, and a fine time to experience the subtleties of these constellations.

AUTUMN

1 a.m. on September 1; 11 p.m. on October 1; 9 p.m. on November 1. Add one hour for daylight-saving time

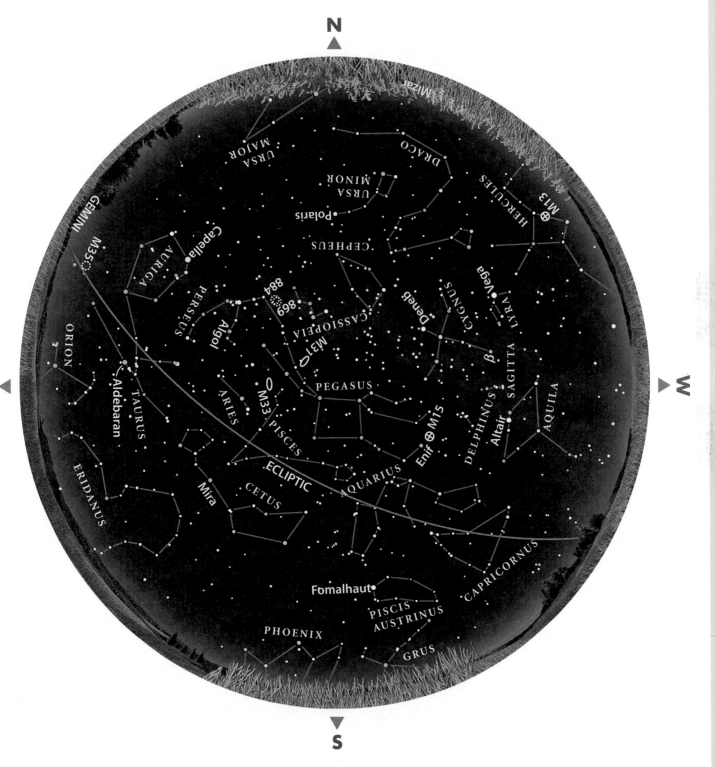

Reproduced by permission. © 2007, Astronomy Magazine, Kalmbach Publishing Co.